D0844694

Microwave Digital Radio

OTHER IEEE PRESS BOOKS

Oliver Heaviside: Sage in Solitude, *By P. J. Nahin*
Radar Applications, *Edited by M. I. Skolnik*
Principles of Computerized Tomographic Imaging, *By A. C. Kak and M. Slaney*
Selected Papers on Noise in Circuits and Systems, *Edited by M. S. Gupta*
Spaceborne Radar Remote Sensing: Applications and Techniques, *By C. Elachi*
Engineering Excellence, *Edited by D. Christiansen*
Selected Papers on Logic Synthesis for Integrated Circuit Design, *Edited by A. R. Newton*
Planar Transmission Line Structures, *Edited by T. Itoh*
Introduction to the Theory of Random Signals and Noise, *By W. B. Davenport, Jr. and W. L. Root*
Teaching Engineering, *Edited by M. S. Gupta*
Selected Papers on Computer-Aided Design of Very Large Scale Integrated Circuits, *Edited by A. L. Sangiovanni-Vincentelli*
Robust Control, *Edited by P. Dorato*
Writing Reports to Get Results: Guidelines for the Computer Age, *By R. S. Blicq*
Multi-Microprocessors, *Edited by A. Gupta*
Advanced Microprocessors, II, *Edited by A. Gupta*
Adaptive Signal Processing, *Edited by L. H. Sibul*
Selected Papers on Statistical Design of Integrated Circuits, *Edited by A. J. Strojwas*
System Design for Human Interaction, *Edited by A. P. Sage*
Microcomputer Control of Power Electronics and Drives, *Edited by B. K. Bose*
Selected Papers on Analog Fault Diagnosis, *Edited by R.-W. Liu*
Advances in Local Area Networks, *Edited by K. Kümmerle, J. O. Limb, and F. A. Tobagi*
Load Management, *Edited by S. Talukdar and C. W. Gellings*
Computers and Manufacturing Productivity, *Edited by R. K. Jurgen*
Selected Papers on Computer-Aided Design of Analog Networks, *Edited by J. Vlach and K. Singhal*
Being the Boss, *By L. K. Lineback*
Effective Meetings for Busy People, *By W. T. Carnes*
Selected Papers on Integrated Analog Filters, *Edited by G. C. Temes*
Electrical Engineering: The Second Century Begins, *Edited by H. Freitag*
VLSI Signal Processing, II, *Edited by S. Y. Kung, R. E. Owen, and J. G. Nash*
Modern Acoustical Imaging, *Edited by H. Lee and G. Wade*
Low-Temperature Electronics, *Edited by R. K. Kirschman*
Undersea Lightwave Communications, *Edited by P. K. Runge and P. R. Trischitta*
Multidimensional Digital Signal Processing, *Edited by the IEEE Multidimensional Signal Processing Committee*
Adaptive Methods for Control System Design, *Edited by M. M. Gupta*
Residue Number System Arithmetic, *Edited by M. A. Soderstrand, W. K. Jenkins, G. A. Jullien, and F. J. Taylor*
Singular Perturbations in Systems and Control, *Edited by P. V. Kokotovic and H. K. Khalil*
Getting the Picture, *By S. B. Weinstein*
Space Science and Applications, *Edited by J. H. McElroy*
Medical Applications of Microwave Imaging, *Edited by L. Larsen and J. H. Jacobi*
Modern Spectrum Analysis, *Edited by S. B. Kesler*
The Calculus Tutoring Book, *By C. Ash and R. Ash*
Imaging Technology, *Edited by H. Lee and G. Wade*
Phase-Locked Loops, *Edited by W. C. Lindsey and C. M. Chie*
VLSI Circuit Layout: Theory and Design, *Edited by T. C. Hu and E. S. Kuh*
Monolithic Microwave Integrated Circuits, *Edited by R. A. Pucel*
Next-Generation Computers, *Edited by E. A. Torrero*
Kalman Filtering: Theory and Application, *Edited by H. W. Sorenson*
Spectrum Management and Engineering, *Edited by F. Matos*
Digital VLSI Systems, *Edited by M. I. Elmasry*
Introduction to Magnetic Recording, *Edited by R. M. White*
Insights into Personal Computers, *Edited by A. Gupta and H. D. Toong*
Television Technology Today, *Edited by T. S. Rzeszewski*
The Space Station: An Idea Whose Time Has Come, *Edited by T. R. Simpson*
Marketing Technical Ideas and Products Successfully! *Edited by L. K. Moore and D. L. Plung*
The Making of a Profession: A Century of Electrical Engineering in America, *By A. M. McMahon*
VLSI: Technology and Design, *Edited by O. G. Folberth and W. D. Grobman*

Microwave Digital Radio

Edited by
Larry J. Greenstein
Department Head
AT&T Bell Laboratories

Mansoor Shafi
Research Engineer
Telecom Corporation of New Zealand

A volume in the IEEE PRESS Selected Reprint Series,
prepared under the sponsorship of the
IEEE Communications Society

The Institute of Electrical and Electronics Engineers, Inc., New York

PRINTED IN THE UNITED STATES OF AMERICA

IEEE Order Number: PC02345

Library of Congress Cataloging-in-Publication Data

Microwave digital radio.

(IEEE Press selected reprint series)
"IEEE order number: PC0234-5"—T.p. verso.
Includes bibliographies and indexes.
1. Digital communications. 2. Microwave communication systems. I. Greenstein, Larry J. II. Shafi, Mansoor, 1950– .
TK5103.7.M54 1988 621.38′0413 88-667

ISBN 0-87942-243-2

Contents

Preface

THE use of microwave line-of-sight radio for carrier telephony began in the early 1950's with the introduction of systems using analog modulation. Digital radio for carrier telephony did not appear until the early 1970's and was limited to modest spectral efficiencies (four-level modulations) and relatively short distances (routes under 400 km). The years since the mid-1970's have seen dramatic growth in both spectral efficiency and system length, spurred by the desire in many countries for high-capacity routes and nationwide digital connectivity. Today and into the future, efforts will be expanding to develop nationwide and worldwide networks for digital telecommunications. Microwave digital radio, along with optical fibers and communications satellites, will be an essential component in realizing this goal.

The prominence of this subject has been in evidence in numerous journals on communications and at communications conferences around the world. Since the mid-1970's, for example, every yearly IEEE International Conference on Communications (ICC) and IEEE Global Telecommunications Conference (GLOBECOM) has featured several technical sessions on digital radio, and much of the best work in the field has been reported at these conferences. Countless articles have also appeared over this period in prestigious journals, including the IEEE TRANSACTIONS ON COMMUNICATIONS, the IEEE JOURNAL ON SELECTED AREAS IN COMMUNICATIONS (JSAC), and IEEE COMMUNICATIONS MAGAZINE. Most recently, COMMUNICATIONS MAGAZINE presented a Special Series on Microwave Digital Radio that ran in several monthly issues from August 1986 to February 1987. In addition, the April 1987 JSAC was devoted to this subject.

The six articles that comprised the IEEE COMMUNICATIONS MAGAZINE'S Special Series on Microwave Digital Radio provide the structural framework for this book. Specifically, the tutorial that leads off each of the book's six parts is an updated version of one that was published in that series. It is then followed by reprints of several in-depth and informative papers dealing with the subject of the tutorial.

The set of reprints in each part was selected through consultations among the tutorial authors and the editors. Although selected with care, no such set should be regarded as exhaustive or singular but, rather, as a representative sampling of the many important papers on the subject. Many other outstanding contributions to the literature of this field are cited in the texts and reference lists of the leadoff tutorials.

This book should be of value to a wide range of audiences, from electrical engineering graduate students to novices in microwave digital radio to experts in the subject, the latter two groups encompassing designers, manufacturers, researchers, and field engineers. Students will require no sophisticated knowledge to understand the concepts presented in the tutorial papers, while practitioners will find here the in-depth material needed to assess the status of—and do the additional work in—this field.

Depending upon a given reader's background and interest, one might choose to first read the tutorial articles in sequence, thereby surveying the key issues in the field, and then focus on specific topics in more detail. Alternatively, one might deal with one topic at a time, first by means of the tutorial overview, and then through the set of specialized papers. The selection and organization of material are aimed at facilitating these and other possible approaches to the subject.

We now briefly review the six parts of the book.

Part I: Overview—There are two special conditions that drive the field of microwave digital radio. One has its roots in the regulatory environment common to all countries: The scarcity of available frequency spectrum invariably mandates that radio communications be efficient, as measured, say, by the number of telephone voice circuits per megahertz of bandwidth. When the resulting requirements on spectral efficiency are translated into bit rate per bandwidth for digital radio, we find there is a need for multilevel modulations, specifically, modulations with 64 levels or more.

The second special condition is the multipath propagation that often occurs on line-of-sight microwave links, which can play havoc with highly efficient modulations. The result is a field with special needs—not only circuit techniques for implementing multilevel modulations and robust reception, but analysis and measurement techniques for modeling multipath and estimating performance.

These conditions and requirements are reviewed in the leadoff tutorial by Taylor and Hartmann, "Telecommunications by Microwave Digital Radio." This is followed by reprints of two papers chosen to deepen the reader's understanding of the regulatory climate that affects this field. The first paper (DeWitt) emphasizes FCC requirements, while the second paper (Hayashi) emphasizes CCIR requirements. The more technical issues raised in the tutorial are dealt with in the remaining parts of the book.

Part II: Modulation Techniques—Driven by the need for more bit rate per bandwidth, digital radio engineers have, since the early 1970's, raised the number of modulation levels from 4 to as many as 256, or even more! The leadoff tutorial by Noguchi, Daido, and Nossek, "Modulation Techniques for Microwave Digital Radio," charts the growth of this area. It deals with modulation/demodulation techniques, spectral shaping, and synchronization schemes, as well as with the impact of the number of modulation levels on power efficiency, spectral efficiency, and vulnerability to impairments. Since no treatment of modulation can be complete without addressing methods to deal with multipath fading, this subject is dealt with briefly, with detailed treatment being deferred to Part IV.

The leadoff tutorial is followed by reprints of five papers that expand on the subject. The first three describe modern 64-level systems built in Japan (Noguchi, *et al*), the United States (Bates, *et al*) and Canada (McNicol, *et al*). The growth from

64 levels to 256 levels is highlighted by the fourth paper (Daido, *et al*), while the fifth paper (Saito, *et al*) describes the benefits and associated technical problems of transmitting multiple 256-level carriers within a single radio channel.

Part III: Multipath Channel Models—Anomalous meteorological conditions in the lower atmosphere, sometimes in concert with terrain reflections, can cause multipath propagation on radio paths that are designed to be line-of-sight. The result can be signal reduction (due to frequency-selective fading), intersymbol interference (due to pulse distortion), or both. Though generally infrequent, multipath occurs often enough on many paths throughout the world to cause serious degradations in transmission reliability. A key step in combating this phenomenon is to characterize both its effect on transmission and the statistical variability of that effect.

This leads us into the area of multipath channel modeling, which is the subject of Part III. The leadoff tutorial by Rummler, Coutts, and Liniger, "Multipath Fading Channel Models for Microwave Digital Radio," reviews a rich variety of statistical modeling approaches that have been developed and used. Virtually all of them amount to specifying a mathematical function for the propagation response within a radio channel bandwidth, and then describing the statistics of the function's numerical parameters. Also discussed are models for dual diversity links, wherein space diversity receiving antennas are used to combat fading; and models for dual-polarization links, wherein two signals are transmitted in the same radio channel in order to double link capacity. It is made clear that more modeling is needed for these two cases, particularly the latter.

The leadoff tutorial is followed by reprints of five papers on statistical modeling of multipath fading. The first two papers describe *ray models,* wherein the propagation frequency response is that of a medium with two or more discrete paths. One paper (Campbell and Coutts) describes and implements a two-path model, while the other (Rummler) presents a measurement-derived three-path model. The next two papers introduce the use of polynomial representation for the propagation frequency response. In one approach (Greenstein and Czekaj), polynomials are used to fit the complex signal response; in the other (Liniger), polynomials are used to fit the logarithm of that response. (The papers by Campbell and Coutts and by Liniger also underscore the close connection between multipath modeling and outage calculations, as detailed in Part V.) The fifth paper (Lavergnat and Sylvain) is a recent analytical treatment of several of these models, with the aim of contrasting and reconciling different published approaches.

Part IV: Receiver Techniques—Armed with an understanding of modulation methods and multipath fading channels, the reader is now prepared to study the receiver techniques used to combat fading. Among the important receiver techniques are adaptive equalization (both in the time and frequency domains); synchronization (both carrier phase and symbol timing); space diversity (both selection and continuous combining); and cross-polarization interference cancellation. These topics are reviewed in the leadoff tutorial by Chamberlain, Clayton, Sari, and Vandamme, "Receiver Techniques for Microwave Digital Radio."

Following this article are reprints of five papers on the subject. The first (Qureshi) is a general theoretical treatment of adaptive equalization, while the second (Takenaka, *et al*) describes an early practical application of equalization to microwave digital radio. The next two papers deal with space diversity combining. One of them (Komaki, *et al*) describes a novel approach that minimizes dispersion or maximizes received power, whichever need is greater, while the other (Yeh and Greenstein) describes an approach that continuously minimizes the composite effect of dispersion and noise. The fifth paper (Namiki and Takahara) was one of the first published treatments of adaptive cross-polarization interference cancellation in microwave digital radio.

Part V: Performance Calculations—The primary measure of link performance in microwave digital radio is *outage time,* which is essentially the number of seconds per year that the link suffers a short-term bit error rate in excess of 10^{-3}. Given the particulars of modulation, multipath fading and receiving equipment, one can estimate the outage time by a variety of methods. The leadoff tutorial by Greenstein and Shafi, "Outage Calculation Methods for Microwave Digital Radio," describes several methods that have been reported and used, primarily for nondiversity, single-polarization radio links. It also discusses methods for links that use diversity or that transmit dual polarizations, although a primary limit in such cases is the availability of adequate channel models.

Following the leadoff tutorial on outage time are reprints of five papers on the subject. The first paper (Barnett) deals with the all-important *multipath occurrence factor* used in calculations of outage time. It presents a simple, empirical formula for this factor that displays the influences of frequency band, path length, and terrain type. The remaining four papers describe specific methods for combining statistical fading models with equipment characteristics to compute outage time. The second paper (Emshwiller) invokes the two-path model and develops the *M-curves,* which are receiver signatures widely used to characterize vulnerability to multipath. The third paper (Lundgren and Rummler) invokes the three-path model and develops receiver signatures relevant to that case. It was one of the first papers, moreover, to report the value of in-band amplitude dispersion as a measure of outage. The fourth paper (Meyers) invokes the two-path model but extends earlier methods to include the case of equalized receivers, wherein noise can be more limiting than multipath distortion. The fifth paper (Foschini and Salz) describes a very different method—involving Monte Carlo simulations of the random channel responses—that is also useful for equalized receivers, and can be readily applied, as well, to links with space diversity and/or dual polarization.

Part VI: Future Trends—Having explored the history and current status of microwave digital radio, it is time to look forward, and we do that in this final part of the book. Three pairs of experts—from Asia (Kohiyama and Kurita), North America (Meyers and Prabhu), and Europe (Hart and Steinkamp)—provide the benefit of their foresight in three leadoff mini-papers. Originally published as a three-part article in the IEEE COMMUNICATIONS MAGAZINE to close our Special Series, these mini-papers were introduced with a foreword by Heiichi Yamamoto, and we repeat that introduction here.

We follow the leadoff article with five reprinted papers on recent advances that portend a bright future for microwave digital radio. The first paper (Ungerboeck) is an introduction to trellis-coded modulation, which many regard as a strong possibility for future digital radio systems. The second paper (Lee and Lin) and third paper (Lin, *et al*) describe important new applications of diversity to microwave digital radio. The former describes improved uses of frequency diversity, while the latter discusses *angle diversity,* one of a new class of approaches that is generating much interest. The fourth paper (Sebald, *et al*) and fifth paper (Matsue, *et al*) deal with advances in adaptive circuit techniques. The former concerns multipath equalization, while the latter concerns cross-polarization interference cancellation.

Many other trends in microwave digital radio, ranging from one-frequency repeaters to improved anti-interference techniques to advances in manufacturing methods, are also discussed in the leadoff mini-papers.

ACKNOWLEDGMENTS

We are grateful for the support this endeavor received from our respective organizations, AT&T Bell Laboratories and Telecom Corporation of New Zealand, and for the help and advice of many of our colleagues at those companies. We owe a particular debt to the experts who authored the tutorial articles that begin each part of this book. In producing the original Special Series for the IEEE COMMUNICATIONS MAGAZINE, they worked with us in an exemplary spirit of collegiality. More recently, they have cooperated in updating their articles and in recommending papers to be reprinted here. As we wrote in the editorial that began the Special Series, "In doing so, they have cheerfully coped with philosophical and technical differences among themselves and with the barriers created by national and corporate boundaries, not to mention mountains, oceans, and time zones." Once again, we extend to them our sincerest thanks for their efforts, talents, and good will.

LARRY J. GREENSTEIN
MANSOOR SHAFI
Editors

Part I
Overview

Telecommunications by Microwave Digital Radio

DESMOND P. TAYLOR AND PAUL R. HARTMANN

Editor's Note: This tutorial article was originally published in the August 1986 issue of the IEEE COMMUNICATIONS MAGAZINE, to initiate the Special Series on Microwave Digital Radio. It has been updated by the authors and edited for inclusion in this book.

INTRODUCTION

SINCE the early 1970's, microwave radio has been gaining in importance as a transmission medium for digital communications [1]–[4]. This growth has been driven to a great extent by the large-scale introduction of digital switching. The first digital microwave radio systems were somewhat rudimentary, utilizing quaternary phase-shift keying (QPSK) modulation and achieving a spectral efficiency of less than 2 bits/sec/Hz. However, since the mid-1970's major technical advances have been made, and today spectral efficiencies of more than 4 bits/sec/Hz are commonplace [5], with even higher efficiencies in the offing [6].

Digital radios are susceptible to propagation anomalies, such as dispersive fading and ducting, and are extremely sensitive to equipment imperfections. In order to develop high-capacity systems with the necessary performance levels, such problems have had to be overcome [7]. Through cooperative research efforts on a worldwide basis and the relatively free reporting of results in a timely manner, enormous gains in both efficiency and performance have been made.

This article provides an overview of the field of microwave digital radio, and will be expanded upon in the tutorial articles that begin the remaining five parts of this book. The other five articles will describe in some detail the problems of digital radio, the approaches that have been taken toward their solution, and the directions in which—in the judgment of the various authors—high-capacity digital radio systems are moving. In this introductory treatment, we will merely survey the various problem areas, touch lightly on some of the approaches that are being taken toward their solution, and then refer the reader to the other tutorials and the literature in general.

NETWORK AND REGULATORY ISSUES

The telecommunications administrations of most countries are now engaged in digitalizing their transmission and switching facilities, as a key step toward establishing an integrated digital network. The present transmission network consists of a mix of cable (fiber-optic and coaxial), satellite, and line-of-sight (analog and digital) radio systems. The introduction of digital radio systems can be successful only if they meet existing transmission performance standards, their spectral efficiencies are compatible with those of their analog counterparts, and their coexistence with other transmission systems does not lead to performance degradations.

To ensure that these criteria are adapted uniformly on a worldwide basis, regulatory bodies such as CCIR, CCITT, and FCC have approved a number of reports and recommendations [8], [9]. Three main issues discussed in such reports are (1) data rates, (2) frequency channelization, and (3) out-of-band emissions and interference. In this section, we will briefly examine these topics.

Data Rates

The standard bit rate for a digital voice circuit is 64 kb/s. In the United States, the digital hierarchy begins with the time-division multiplexing of 24 such circuits (plus framing bits) to form a 1.544-Mb/s signal called DS-1. The second-level signal, DS-2, is four multiplexed DS-1 signals, or 96 voice circuits at about 6.3 Mb/s; the DS-3 signal is 672 voice circuits at about 45 Mb/s; and so on. In most European nations, the first digital signal level is 30 multiplexed voice circuits at about 2 Mb/s. As a result, the third- and fourth-level signals contain 480 and 1920 voice circuits, at about 34 Mb/s and 140 Mb/s, respectively. The data rates 34 Mb/s, 140 Mb/s, and 45 Mb/s (or multiples thereof) figure prominently in digital radio transmission, as we shall see.

Frequency Channelization

Channel plans for carrier radio systems now exist for bands from 2 GHz to 40 GHz, those below 12 GHz being most commonly used for long-haul digital radio. Most countries have elected to use the same channelizations for analog and digital systems in order to facilitate coexistence on the same routes.

As a typical example, the 4-, 6-, and 11-GHz common-carrier bands in the United States are each 500 MHz wide, with channel spacings of 20, 29.65, and 40 MHz, respectively. In all cases, polarizations are assigned to channels in such a way that adjacent channels always operate on orthogonal polarizations.

The move to digital radio in the United States was aided by several important actions of the Federal Communications Commission (FCC) in the 1970's. In particular, Docket 19311 established the framework for manufacturers of digital microwave equipment [9]. Some of its major provisions included a requirement that radios operating in the 4-, 6-, and 11-GHz bands have minimum capacities of 1152 voice circuits per channel. To be consistent with the digital hierarchy, then, the nearest multiple of 672 voice circuits above 1152 (at a minimum) must be transmitted in each radio channel. This corresponds to 1344 voice circuits, or two DS-3 signals at 90 Mb/s.

To accommodate 90 Mb/s in the 20-MHz channels of a 4-GHz system requires a spectral efficiency of 4.5 bits/sec/Hz. The corresponding requirements for 6- and 11-GHz systems are 3.0 and 2.25 bits/sec/Hz, respectively. Earlier digital radio systems were designed for the latter two bands, and met the spectral efficiency requirements by using such modulations as 8-level phase shift keying (8-PSK), 9-level quadrature partial response signaling (9-QPRS), and 16-level quadrature amplitude modulation (16-QAM). More recent systems are using 64-QAM to achieve 4.5 bits/sec/Hz, and prototype systems using 256-QAM have been reported. Modulation techniques for digital radio are detailed in the tutorial article that beings Part II [10].

Out-of-band Emissions and Interference

Digital modulations are particularly notorious for producing high out-of-band emissions. To limit the resulting potential for adjacent channel interference, FCC Docket 19311 has defined a spectral emission "mask" which places limits on the transmitted spectral density-to-total power ratio as a function of frequency. Comparable rules are used by other administrations to control out-of-band emissions.

Other kinds of interference are important as well, for example, co-channel interference from other systems, and cross-polarization interference in systems with dually-polarized signals. An early argument in favor of digital radio was signal robustness, that is, high tolerance to interference. With the advent of high-level modulations, however, the consideration of interference has become increasingly important. System designers must take note of even relatively weak sources of interference, for example, satellite downlinks in shared bands, waveguide echos, etc. Also, antenna cross-polarization discriminations must be very good in order to control adjacent channel interference and permit the effective operation of dual-polarization systems.

Major Technical Issues

Digital radio must provide a transmission path which has essentially the same reliability and availability as its major competitors, namely, optical fiber and satellite systems. The major advantage of digital radio over optical fiber is its lower installation cost, and its primary advantage over satellite transmission is its much smaller propagation delay. The major technical issues facing the designer of high-capacity digital radio systems may be broadly classed as propagation problems and equipment requirements. In this section, we briefly examine both.

Propagation Problems

Most of the time, a line-of-sight microwave radio channel is a nondispersive transmission medium capable of highly reliable, high-speed digital transmission. However, because it is a natural medium, anomalous propagation conditions exist for some fraction of the time, and these can cause very severe degradations in radio system performance.

These conditions manifest themselves through what is referred to as multipath fading [11]. Fading almost always consists of a combination of flat and frequency-selective components. The flat component is a time-varying, frequency-independent attenuation of the channel response and, hence, of the signal. This attenuation can be thought of as the median, over a broad frequency range, of the radio path's power gain. It is referred to as the *median depression*, and is usually compensated for over a wide dynamic range by the automatic gain control of the receiver. The remaining (frequency-selective) component of the fade is the frequency variation of the power response about the median depression. It manifests itself either as a monotonic gain change (or *slope*) across the radio channel bandwidth, or as a dip (or *notch*) within that bandwidth. Fig. 1 illustrates a faded channel frequency response in which the various parameters that characterize fading are identified.

Multipath fading arises from the fact that the signal propagates along several paths, each of different electrical length. At the receiver, these relatively delayed signal components interfere with each other, and this leads to the frequency-selective effects described above. In order to design radio equipment to work in this environment, it is essential that the system designer have realistic and readily usable models of the multipath channel response. The development of statistical models for use in system design is the subject of the tutorial article that begins Part III [12].

The frequency-selective notches referred to above can vary in depth from 0 dB to more than 40 dB and, if left uncompensated, can cause severe degradation in the performance of a radio receiver. In fact, an uncompensated selective notch of as little as 5 dB in depth could effectively cause a radio link to be out of service, a condition that is usually referred to as an *outage*. (The 5 dB number is only representative; in reality, the notch depth that produces an outage would depend on such factors as notch frequency (measured with respect to the channel center frequency, Fig. 1), type of modulation, and reception technique.)

A number of convenient means of characterizing the sensitivity of radio equipment to propagation anomalies have been developed; the most widely used is the so-called signature curve [13] exemplified in Fig. 2. Each enclosed area represents, for a given modulation and type of reception, the notch depth/notch frequency region over which some specified bit error rate (usually 10^{-3}) is exceeded. The boundary curve for each such region is called the *equipment signature*. The shape of the signature for the unequalized receiver shown in Fig. 2 is fairly typical, and so, for obvious reasons, signature curves of notch depth vs. notch frequency are often called "M curves." Some digital radio engineers display the vertical scale with the reverse polarity, that is, with notch depth expressed as a *positive* quantity, in which case the signatures are inverted, and are hence called "W curves." Both M curves and W curves are in evidence throughout the literature.

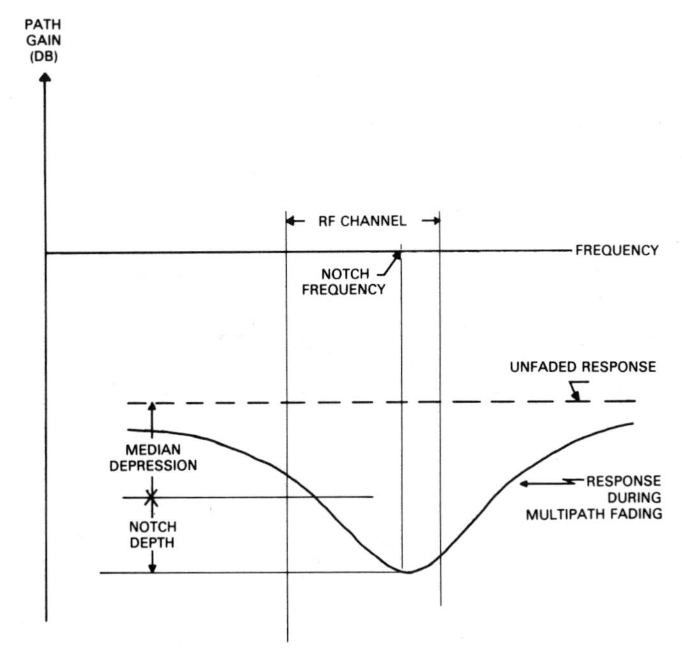

Fig. 1. Radio channel responses.

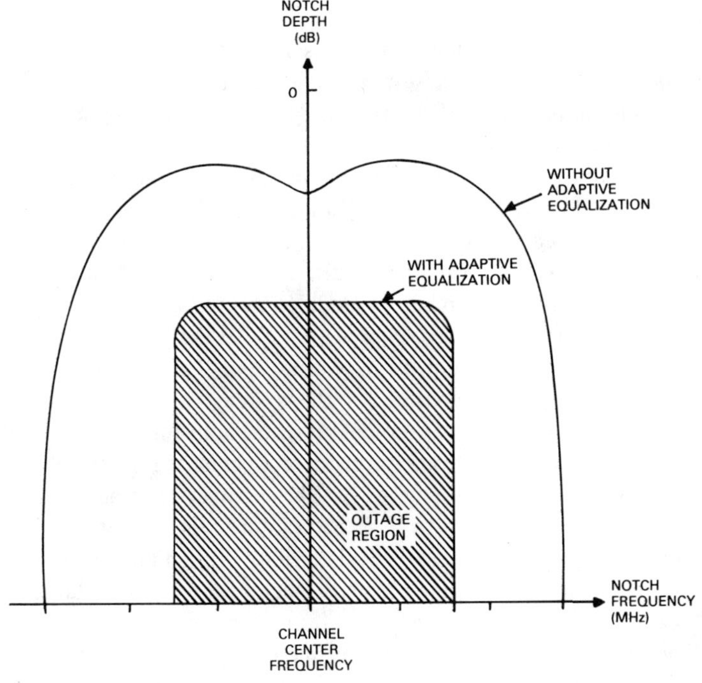

Fig. 2. Equipment signature curves.

During the past several years, one of the major achievements in radio design has been the development of sophisticated equalization techniques to combat the deleterious effect of multipath fading. Historically, the first equalizer strategy consisted of the combination of an amplitude slope equalizer, which compensates for inband gain slope, and a space diversity receiver, which compensates for notches or dips in the received frequency response (cf [14]). These techniques were sufficient for the relatively simple modulation techniques used in the early systems. With the continuing increase in bit rates and the concomitant increase in modulation complexity, however, such an approach soon proved to be inadequate, and more sophisticated equalizers were developed. These have mainly consisted of adaptive transversal equalizers (cf [15]) or decision feedback equalizers (cf [16]) used either alone or in combination with slope equalizers and (in most instances today) space diversity receivers. Many of these adaptive measures are based on the voiceband equalization techniques developed in the 1960's.

A block diagram of a receiver that includes space diversity combining, an IF slope equalizer, and a baseband transversal equalizer is shown in Fig. 3. Also shown are circuit blocks for carrier recovery and timing recovery, which could derive their inputs from the places shown or from other signal points in the receiver. These are critical receiver functions and, particularly in a complex fading environment, are interactive with each other and with the equalization strategy. These and the other adaptive techniques used in digital radio receivers are discussed in the tutorial article that begins Part IV [17].

A system designer armed with knowledge of possible modulations, statistical channel models, and adaptive receiver techniques is in a good position to estimate *outage time*, the expected number of seconds in a year for which outage will occur on a given link. The development of analytical/computational methods for outage time estimation (some of which are based on the use of signatures such as those shown in Fig. 2) has engaged many workers in the field (cf [18], [19]). Outage prediction techniques, and their relationship to the statistical behavior of the fading channel, are discussed in the tutorial article that begins Part V [20].

Equipment Requirements

Essentially, the equipment problems encountered in modern digital radio arise because of the need, amidst propagation anomalies, to transmit high-level modulations. We have seen that a digital radio system must achieve a spectral efficiency of 4.5 bits/sec/Hz in the 4-GHz band in order to provide a capacity of 1344 voice circuits. However, practical QPSK modulations can only achieve an efficiency somewhat below 2 bits/sec/Hz. It is thus clear that, to achieve the required efficiency, modulation formats that carry more than two information bits per channel symbol must be used. At the present stage of development, digital radios in the 4- and 6-GHz bands are, in fact, achieving spectral efficiencies of 4.5 bits/sec/Hz (cf [21]).

From the earliest days of digital radio, linear modulation formats have been employed almost exclusively. Thus, regardless of the number of modulation levels, the modulator and demodulator can always be modeled as a form of QAM. A general block diagram of a quadrature amplitude modulator is shown in Fig. 4. We see that a pair of quadrature-phased carrier signals are each modulated with a discrete number of information levels per symbol period and then summed to form the QAM signal. For example, in 16-QAM, each carrier is modulated to one of four discrete levels (corresponding to two information bits), so that, in each symbol period, the resulting modulated signal has 16 possible variations and carries four information bits. The demodulation of such a signal in the receiver essentially reverses the sequence of

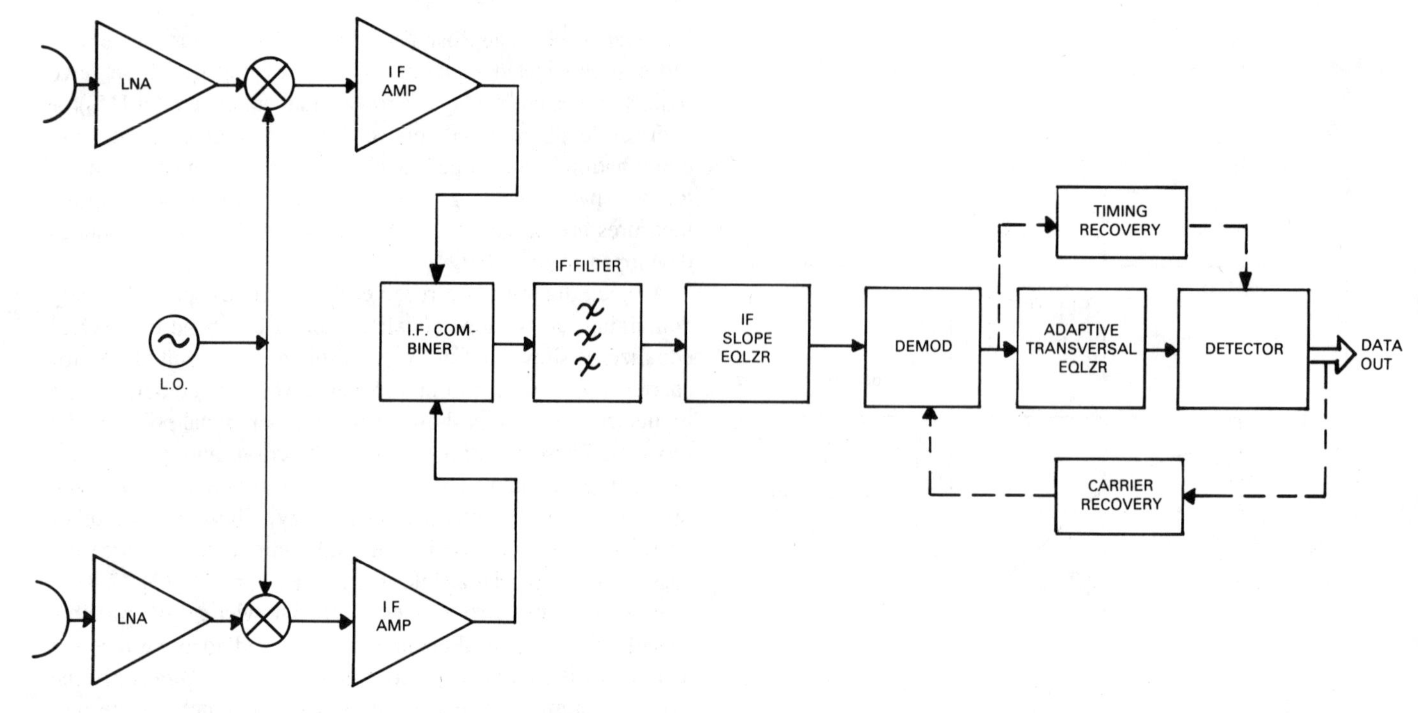

Fig. 3. Space diversity receiver.

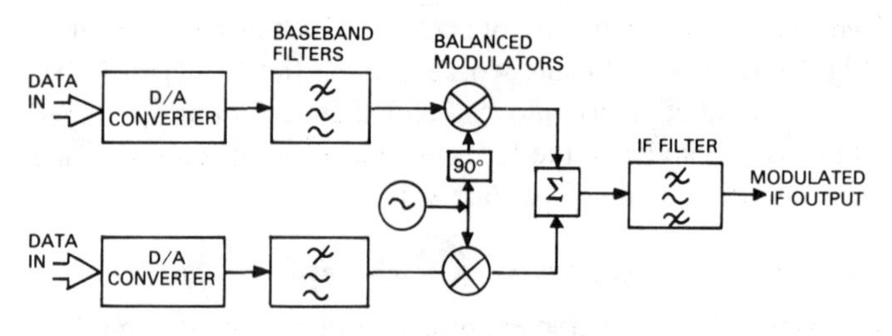

Fig. 4. Quadrature amplitude modulator.

operations: IF filtering, balanced demodulation using two quadrature-phased (IF) carriers, baseband filtering, and A/D conversion (detection).

As the number of modulation levels increases, so does the sensitivity of the resulting signal to equipment imperfections and impairments. For example, for a higher number of modulation levels, noise margins (and, hence, signal robustness) decrease. The result is that an equalizer must compensate for selective fading more accurately and to a greater degree.

In addition, high-level modulations require highly accurate carrier and timing recovery circuits, since the potential for intersymbol interference is greater for such signals. These synchronization schemes pose a complex design problem and, because of the interdependencies of carrier recovery, timing recovery, and data equalization, often must be decision-aided loop structures [15], [22], [23].

When high-level modulations are employed, system performance becomes increasingly sensitive to linear distortions in the transmission equipment. To a large extent, these distortions appear as asymmetries in the amplitude responses and variations in the group delay characteristics of the filters in the transmission path. In order to avoid overloading the adaptive equalizer in the receiver, it is therefore of importance that all filters in both the transmitter and receiver be equalized. In particular, they should have essentially symmetric amplitude resonses and flat group delay characteristics, at least within the passband of the signal. This necessity complicates both the design and the implementation of the required filters.

Finally, the high-level modulations used in modern digital radio systems have many amplitude levels and correspondingly high peak-to-average power ratios. Because of this, they are very sensitive to nonlinear distortion, and this sensitivity increases with the number of modulation levels. Nonlinearity may occur in any active component of either the transmitter or receiver. However, the primary source of degradation, and therefore the chief concern of the system designer, is the power amplifier (PA) at the output of the transmitter.

Most digital radio routes consist of a chain of regenerative repeaters, each containing a PA. These devices operate most efficiently when close to their saturation points, where they cause significant nonlinear distortion. To minimize this problem, it is necessary either to operate the PA at a large backoff from saturation, or to provide some means of nonlinearity compensation (for example, linearizing the amplifier or predistorting the signal so that it becomes insensitive to the nonlinear effects). From the standpoint of repeater power efficiency, the latter is the more attractive approach, and a number of compensation schemes have been investigated [24], [25]. These allow the PA to operate at or near saturation with a significant increase in power efficiency—an important consideration in remotely sited repeaters which rely on local power generation.

Current and New Systems

We now briefly cite some of the major digital radio systems in use today. In North America, there are common-carrier frequency allocations at 2, 4, 6, and 11 GHz, with an additional allocation at 8 GHz in Canada. In addition, there are allocations at 2, 6, and 12 GHz for industrial private networks. Common carrier designs have now been essentially standardized, so that the 4-GHz systems can carry two DS-3 streams per 20-MHz channel; the 6-GHz and 11-GHz systems carry three DS-3's in 29.65-MHz and 40-MHz channels, respectively; and the 8-GHz systems carry two DS-3's in a 40-MHz channel. In Canada, there is a long-haul system at 8 GHz which spans the entire country, and in the United States, there are well over 10,000 digital radio links in use in the telephone network. These include short- and medium-haul systems [26], as well as long-haul systems at both 4 and 6 GHz.

In Japan, there is currently an extensive network of digital radios in use in the 2-, 4/5-, 11/15-, and 20-GHz bands, at data rates ranging from 3 Mb/s to 400 Mb/s per radio channel. Background on the development and deployment of these is given in some detail in [27]. Most of the currently deployed radios utilize 4-PSK, 8-PSK, or 16-QAM. However, planning and development of 64- and 256-QAM radios is actively underway. In fact, by the year 2000, the Japanese expect to have an essentially completely digital telecommunications network, in which it is expected that high-capacity digital radio will play an important role.

In Australia and New Zealand, developments are also proceeding rapidly. In most instances, 140-Mb/s systems employing 16-QAM are being developed and deployed. However, some work is in progress on more efficient modulations, such as reduced bandwidth QPSK [28] and very high-level QAM.

In Europe, the 4-GHz band has been rechannelized to 40-MHz bandwidths, and 140-Mb/s systems are now being deployed in the 4-GHz and upper 6-GHz bands. Additional planning and deployment are underway at 11 GHz and 18 GHz.

Conclusions and Future Trends

There is now a great deal of activity worldwide on the development of high-capacity digital radio systems. While much of this activity is going on in research laboratories, we are also seeing the deployment in the field of a large number of systems of increasing capacity, sophistication, and reliability. The question naturally arises as to what the future holds, and this is discussed in some detail in the tutorial articles that begin Part VI [29]. We conclude this overview paper by briefly speculating on some of the possibilities.

It seems at this point that most of the developments will lie in the areas of (1) increasing the reliability of transmission, and (2) increasing the efficiency of use of the available radio spectrum. Increased transmission reliability can be achieved through more sophisticated equalization and diversity strategies. It can also be increased through the use of *forward error correction* (FEC) coding techniques and more robust modulation techniques. As VLSI implementations of error-correction decoders become available at the required signaling rates, it seems likely that the use of FEC coding will increase. Also, improved equipment reliability is anticipated from the increased use of monolithic microwave integrated circuits, VLSI, and other microwave techniques, such as dielectric resonators and solid-state amplifiers.

Increased efficiency of use of the available spectrum can be achieved through the use of still higher levels of modulation and through frequency reuse. From the above discussion, it is clear that higher-level modulations are being actively pursued. There is also significant research going on to increase capacity through the use of orthogonal polarizations (dual-polarization transmission) and through the development of technologies for higher frequency bands.

It is extremely difficult to forecast where these investigations will lead. However, it is safe to say that the next few years will bring continued rapid growth and development in the use of high-capacity digital radio systems.

References

[1] R. G. DeWitt, "Digital mcirowave radio—another building block for the integrated digital network," in *Conf. Record, ICC '75,* San Francisco, CA, pp. 21-5 to 21-9, June 16–18, 1975.

[2] I. Godier, "DRS-8 digital radio for long-haul transmission," in *Conf. Record, ICC '77,* Chicago, IL, pp. 102–105, June 12–15, 1977.

[3] Y. Hayashi, "CCIR work in radio relay systems," in *Conf. Record, ICC '79,* Boston, MA, pp. 9.5.1-5, June 10–14, 1979.

[4] N. Mokhoff, "Communications and microwave," *IEEE Spectrum,* vol. 17, no. 1, pp. 38–43, Jan. 1986.

[5] M. Borgne, "Comparison of high-level modulation schemes for high-capacity digital radio systems," *IEEE Trans. Comm.,* vol. COM-33, pp. 442–449, May 1985.

[6] Y. Takeda, N. Iizuka, Y. Daido, S. Takenaka, and H. Nakamura, "Performance of 256 QAM modem for digital radio system," in *Conf. Rec. GLOBECOM '85,* New Orleans, LA, pp. 1455–1459, Dec. 1-5, 1985.

[7] A. Walker, "Some recent advances in radio communications," *IEEE Commun. Mag.,* vol. 21, no. 6, pp. 51–53, Aug. 1983.

[8] Recommendations and Reports of the CCIR 1982 Study Group IX.

[9] FCC Docket 19311, FCC74-985, adopted Sept. 19, 1974, released Sept. 27, 1974, revised Jan. 25, 1975.

[10] T. Noguchi, Y. Daido, and J. A. Nossek, "Modulation techniques for microwave digital radio," Part II of this book. [An earlier version was published in *IEEE Commun. Mag.,* vol. 24, no. 11, Sept. 1986, pp. 21–30.]

[11] W. D. Rummler, "A new selective fading model: application to propagation data," *Bell Syst. Tech. J.,* vol. 58, pp. 1037–1071, May–June 1979.

[12] W. D. Rummler, R. P. Coutts, and M. Liniger, "Multipath fading channel models for microwave digital radio," Part III of this book. [An earlier version was published in *IEEE Commun. Mag.,* vol. 24, no. 11, Nov. 1986, pp. 30–42.]

[13] M. Emshwiller, "Characterization of the performance of PSK digital radio transmission in the presence of multipath fading," in *Conf. Rec. ICC '78,* Toronto, Canada, pp. 47.3.1-6, June 4–7, 1978.

[14] Y. Y. Wang, "Simulation and measured performance of a space diversity combiner for 6-GHz digital radio," *IEEE Trans. Commun.,* vol. COM-27, pp. 1896–1907, Dec. 1979.

[15] M. Shafi and D. J. Moore, "Further results on adaptive equalizer improvements for 16-QAM digital radio," *IEEE Trans. Commun.,* vol. COM-34, pp. 59–66, Jan. 1986.

[16] D. P. Taylor and M. Shafi, "Decision feedback equalization for multipath induced interference in digital microwave LOS links," *IEEE Trans. Commun.,* vol. COM-32, pp. 267–279, March 1984.

[17] J. K. Chamberlain, F. M. Clayton, H. Sari, and P. Vandamme, "Receiver techniques for microwave digital radio," Part IV of this book. [An earlier version was published in *IEEE Commun. Mag.,* vol. 24, no. 11, Nov. 1986, pp. 43–54.]

[18] C. W. Lundgren and W. D. Rummler, "Digital radio outage due to selective fading—observation versus prediction from laboratory simulation." *Bell Syst. Tech. J.,* vol. 58, pp. 1073–1100, May–June 1979.

[19] C. W. Anderson, S. G. Barber, and R. N. Patel, "The effect of selective fading on digital radio," *IEEE Trans. Commun.,* vol. COM-27, pp. 1870–18761, Dec. 1979.

[20] L. J. Greenstein and M. Shafi, "Outage calculation methods for microwave digital radio," Part V of this book. [An earlier version was published in *IEEE Commun. Mag.*, vol. 25, no. 2, Feb. 1987, pp. 30–39.]

[21] J. A. Crossett and P. R. Hartmann, "64-QAM digital radio transmission system integration and performance," in *Conf. Rec., ICC '84,* Amsterdam, The Netherlands, pp. 636–641, May 14–17, 1984.

[22] G. R. McMillen, M. Shafi, and D. P. Taylor, "Simultaneous adaptive estimation of carrier phase, symbol timing, and data for 49-QPRS DFE radio receiver," *IEEE Trans. Commun.,* vol. COM-32, pp. 429–443, April 1984.

[23] A. Leclert and P. Vandamme, "Universal carrier recovery loop for QASK and PSK signal sets," *IEEE Trans. Commun.,* vol. COM-31, pp. 130–136, Jan. 1983.

[24] A. A. M. Saleh and J. Salz, "Adaptive linearization of power amplifiers in digital radio systems," in *Conf. Rec., ICC '86,* Boston, MA, pp. 864–868, June 19–22, 1983.

[25] J. Grabowski and R. C. Davis, "An experimental M-QAM modem using amplifier linearization and baseband equalization techniques," in *Conf. Rec., NTC '82,* Galveston, TX, paper no. E3.2, Nov. 7–10, 1982.

[26] S. Barber and P. R. Hartmann, "An overview of high-capacity radio transmission systems in North America," in *Conf. Rec., ICC '84,* Amsterdam, The Netherlands, pp. 975–977, May 14–17, 1984.

[27] S. Katayama, M. Iwamoto, and J. Segawa, "Digital radio-relay systems in NTT," in *Conf. Rec., ICC '84,* Amsterdam, pp. 978–983, May 14–17, 1984.

[28] M. T. Dudek, J. M. Robinson, and J. K. Chamberlain, "Design and performance of a 4.5 bit/s/Hz digital radio using reduced bandwidth QPSK," in *Conf. Rec., GLOBECOM '82,* Miami, FL, paper no. D3.4, Nov. 29–Dec. 2, 1982.

[29] H. Yamamoto, K. Kohiyama, O. Kurita, M. H. Meyers, V. K. Prabhu, G. Hart, and J. A. Steinkamp, "Future trends in microwave digital radio," Part VI of this book. [Three mini-papers, introduced by Yamamoto, giving views from Asia, North America, and Europe. Earlier versions were published in a three-part article in *IEEE Commun. Mag.,* vol. 25, no. 2, Feb. 1987, pp. 40–52.]

[30] J. Steinkamp, "Radio relay systems for modern communication networks—an introduction, part 2: digital radio relay systems for different network levels," *Siemens AG Telecom Report,* vol. 7, no. 6, pp. 258–264, Nov.–Dec. 1984.

[31] J. C. Bellamy, *Digital Telephony.* New York: John Wiley and Sons, 1982.

[32] "Special issue on digital radio," *IEEE Trans. Commun.,* vol. COM-27, no. 12, Dec. 1979.

[33] K. Feher, *Digital Communications: Microwave Applications.* Englewood Cliffs, NJ: Prentice-Hall Inc., 1981.

DIGITAL MICROWAVE RADIO — ANOTHER BUILDING BLOCK FOR THE INTEGRATED DIGITAL NETWORK

R. G. DeWitt

Consultant to the Vidar Corp.
Potomac, Md.

ABSTRACT

The coming trend toward integrated digital networks provides the motivation and need for digital microwave radio as another building block. The scenerio calls for the next phase in the evolution of integrated digital networks to be the widespread use of digital switching. That phase starts in about 6 months when the first ESS #4 goes on line in Chicago. However, the digital microwave radio is needed to assist cable systems in providing the digital trunks to the tandem and toll digital switches and between these switches. It is also needed for local distribution where digital subscriber carrier (with or without concentration) makes economic sense. This latter use of digital radio will accelerate rapidly once the operating telephone companies realize that local digital switching is right around the corner and that by going digital in the subscriber plant right away they can be ready to reap additional benefits.

This article surveys the various modulation techniques that have already been used in commercially available digital microwave radios and digital transmission on analog radios to see how they compare with each other with regard to the all important parameters of bandwidth efficiency and S/N threshold (defined here as the S/N ratio at which the error rate is 10^{-7}).

The recently issued FCC Docket 19311 gives added emphasis, at least for the bands below 15 GHz, to the desire for higher bandwidth efficiencies obtainable without incurring significant increases in the threshold S/N ratio. This is so because it requires a minimum number of digital voice channels for each band. The 4 GHz band is the most difficult case because it can only be met using one polarization by designing equipment that achieves 4 b/s per Hz of RF bandwidth or using two polarizations and equipment that achieves 2 b/s per Hz of RF on each.

It was found in the survey that the Vicom/MA/Avantek digital radio approach of QAM-PR3 achieves the highest bandwidth efficiency with the lowest threshold S/N ratio (2 b/s per Hz of RF bandwidth and an average S/N ratio of 22 dB for a 10^{-7} error rate). AT&T's DUV (1A RDT) achieves the same high efficiency but has a threshold that is about 4 dB higher. It was observed that by going from 3 level partial response (QAM-PR3) to 7 level partial response (QAM-PR7) that a bandwidth efficiency of 4 b/s per Hz of RF could be achieved (if it could be implemented) and this would meet on one polarization the digital voice channel requirements of FCC Docket 19311 in all bands. Furthermore, at this efficiency the number of voice channels achieved in every band is close enough to the maximum number achieved by FDM that in conjunction with the 2 to 1 cost advantage of PCM-TDM terminal equipment would make PCM-TDM a serious candidate in all cases.

The equipment employing the QAM-PR3 modulation technique is available for the 2 GHz and 11 GHz bands. In the meantime, for the other bands (4,6, and 8 GHz) digital transmission on analog radios is available at 1 b/s per Hz of RF bandwidth. This equipment some of which has been in service for 3 years now uses 3 level partial response (PR3) on any analog radio currently available. Note that the threshold for this latter approach is only 2.5 dB higher than a currently available 4ϕ digital radio offering the same bandwidth efficiency.

INTRODUCTION

In order to explain the motivation for digital microwave radio, first, it is necessary to define an integrated digital network. There are two types of integration which can be involved in such a network. First, there can be physical integration of digital transmission and digital switching. Second, there can be service integration in which voice and data are handled by the same transmission and switching equipment. The first type of integration provides significant technical and economic benefits. The second type of integration may double the economic benefits because it eliminates the need for completely separate networks for both services.

The economic and technical advantages of an integrated digital network are numerous. Authors such as Duerdoth[5] of the BPO, and McDonald[6], Fleckenstein[9], Tuomenoksa and Vaughan[8] of BTL have done an excellent job of listing these advantages. However, highly efficient digital radio is the next major building block needed for the integrated digital network followed in the near future by the digital switch which has to be mentioned in advance because it provides the greatest motivation of all for the integrated digital network. There are savings enjoyed with a digital transmission network alone but when the digital switch is added a major payoff (both economical and technical) is obtained. Articles in the ISS 1974 Proceedings by McDonald[6], Ritchie and Smith[7], Tuomenoksa and Vaughan[8], and the articles by Duerdoth[5] all emphasize some of the same and some different technical and economic payoffs. That these payoffs are real can be attested to by the fact that at the present time more than 50 per cent of all telephone carrier equipment production in the U.S. is digital carrier (primarily T1). There is enough T1 lines installed in the U.S. that if they were placed end to end they would stretch

Reprinted from *IEEE 11th Int. Conf. Comm.*, pp. 21-5-21-9, June 1975.

around the circumference of the earth 40 times. It would be almost impossible nowadays to make a long distance call that did not go over at least one section of T1 carrier somewhere along the route. The introduction of digital switching by the Bell System starting with the first in service ESS #4 in Chicago in January 1976 (about 6 months from now and on site testing is in progress) will provide the impetus for the approach to 100 per cent digital carrier below the intertoll level because according to Ritchie and Smith[7] the ESS #4 moves the prove in for T1 from the present 6 miles down to 0 miles. Over 10 per cent of the toll connecting carrier is digital already according to Fleckenstein[9].

Duerdoth[5] says that there will be at least a 50 per cent savings in cost with digital transmission and digital switching over FDM and analog switching and at least a 3 to 1 savings of space. Many of the writers describe a fantastic savings in the amount of wiring required, ESS #4 will have 1/10 of the wiring of the 4A it replaces in an analog environment and 1/100 of the wiring in a digital environment[8]. McDonald[6] describes a huge savings in wiring in the local office.

In order to complete the readers understanding before discussing the bandwidth efficiency of digital transmission over microwave radio for terrestrial or satellite applications several concepts and definitions must be reviewed.

EXPLANATION OF ROLLOFF

A low pass transmission band that is rectangular corresponds to a SIN X/X time function whose zero crossings are at multiples of of 1/2f where f is the upper frequency limit of the transmission band. Thus pulses can be transmitted at a rate of twice the cutoff frequency of the low pass filter because the peak of each pulse falls where only zero crossings of the other pulses occur. However, a rectangular filter is not realizable because it requires infinite delay and no phase distortion. Also even if such a filter were possible, perfect timing would be needed because a slight difference in timing would result in the addition of all the pulse tails to large numbers that would prevent the detection of the desired pulses[1]. This impossible situation was solved when Nyquist showed that a gradual cutoff of the low pass channel symmetric about the original sharp cutoff would give practically realizable results. One form of rolloff characteristic that is convenient and useful is called the raised-cosine-characteristic[2]. It has a parameter α which is the amount of bandwidth used in excess of the Nyquist bandwidth divided by the Nyquist bandwidth. An α of unity implies that the total bandwidth used is twice the Nyquist bandwidth while an α equal to zero implies that the Nyquist bandwidth is used. Values of α between these limits can also be used but values closer to zero are harder to realize and values closer to one require more bandwidth.

One other type of low pass filtering that can achieve the objective of efficient digital transmission has to be described briefly. It is called partial response signaling. In the above discussion freedom from intersymbol interference was an objective. Partial response signaling is a method of using controlled intersymbol interference to achieve a higher efficiency. In particular the Nyquist rate of 2 bits per Hertz of low pass bandwidth can be achieved by allowing the intersymbol interference to create 3 levels that must be detected instead of the two original levels. The result is a system that is comparable in bandwidth efficiency to a binary system with α equal to zero as described above or a four level system with α equal to 1 as shown by Lucky et al[2]. The multilevel partial response signaling[11] such as four levels in and seven levels out and eight levels in and fifteen levels out have the same bandwidth efficiency as 4 and 8 levels with α equal to 0 respectively with S/N several dB higher required for the former to achieve the same error rate. Again it can be seen that partial response signaling is a method that can be used to achieve the Nyquist rate of 2f bits per second using realizable and pertubation tolerant filters at the cost of 2.1 dB in S/N ratio compared to binary PAM (Pulse Amplitude Modulation) with sharp rolloff (α=0) and it is 4.9 better than four level PAM with gradual rolloff (α=1).

EXPLANATION OF OTHER TERMS

QAM stands for Quadrature Amplitude Modulation. It is a technique in which some form of amplitude modulation (DSB-SC, PR7, or PR3) is impressed on the carrier and also on the same carrier after it is shifted 90 degrees. The modulation on each is an independent digital input so that the bit rate is doubled. The two independent modulations do not interfere with each other as long as the 90 degrees relationship between their carriers is precisely maintained.

The designation PR3 is shorthand for 3 level partial response (the input is two level NRZ). The designation PR7 is shorthand for 7 level partial response (the input is four level NRZ in this case). DSB-SC stands for double sideband suppressed carrier modulation.

DIGITAL MICROWAVE RADIO EFFICIENCY

Figure 1 is generated by starting with Figure 11-2 of Bennett and Davey. This is a plot of transmission efficiency in bits per second per Hertz of bandwidth versus S/N ratio with perfect rectangular filtering i.e. the rolloff factor α equals 0. Thus the Shannon Limit, QAM & VSB (α=0) and FM (α=0) curves in Figure 1 are the same as their Figure 11-2. Next note that with the rolloff factor α equal to 1 the bandwidth is doubled but the S/N ratio is unchanged. Therefore, the Figure 1 curves for QAM & VSB (α=1), PSK, (α=1) and FM (α=1) are the respective curves of Figure 11-2 with the ordinate values all reduced by a factor of 2 to correct for the doubled bandwidth. Next, the multilevel partial response points given by Lucky et al in Figure 4.27 are added for both the FM and the QAM cases (two level and four level for each gives four distinct points labeled QAM-PR7, QAM-PR3, PR7 and PR3). Everything plotted up to this point is theoretical. Next, points (designated by X's) are plotted based on experimental data for eight types of commercially available hardware for digital transmission on microwave radio.

The next order of business is some discussion and interpretation of Figure 1 which will help to compare the various approaches that have been used to date and to see some that could provide even higher efficiency in bits per second per Hertz of authorized RF bandwidth with the lowest possible S/N ratio in the future. Thus the design objective for digital transmission on microwave radio is to get as high on Figure 1 and simultaneously as far to the left as possible. Note that we are including two different categories of hardware for digital transmission on microwave radio. The first category is digital transmission on analog microwave radio and the second category is digital transmission using digital microwave radio. In the latter case the radio is designed to carry digital only and it is not possible to transmit analog signals. In the former case a modem is used to convert the digital signals into a form suitable for transmission over the analog radio but it is always possible to transmit analog signals if the modem is removed.

DISCUSSION OF FIGURE 1

A considerable amount of work has been done in the past to characterize the relative efficiencies, in the presence of white gaussian noise, of various modulation techniques that have been and are being used in voice bandwidth modems. These earlier results were found to be applicable to the much wider bandwidths of digital microwave radio. Bennett and Davey[1] gave a set of curves for most modulation techniques of interest with ideal rectangular filtering. Then Lucky et al[2] showed the effect of the rolloff factor (α=0 TO 1) and also showed where the multi-level partial response points fell. On Figure 1 the ordinate represents bandwidth efficiency. The numbers in circles signify the number of levels used.

The curve on the left labeled the Shannon Limit is the theoretical capacity of the channel with Nyquist bandwidth perturbed by white gaussian noise which can be achieved by sufficiently complex encoding with arbitrarily small error rate. Shannon suggested that the limit could be approached if the signals had the properties of white noise. Such an approach is not currently economically feasible; however, the best approach to the Shannon Limit is made by using QAM (Quadrature Amplitude Modulation) or VSB (Vestigal Sideband) with α equal to 0. This curve is about 10 dB from the Shannon Limit i.e. for a given ordinate the S/N values differ by 10dB. However, α equal to 0 specifies a rectangular filter which is not realizable. Thus the closest approach to the Shannon Limit at the highest bandwidth efficiency is made by using partial response signaling. Three level partial response (QAM-PR3) permits the efficiency of 2 b/s per Hz of RF bandwidth to be reached along with the closest approach to the Shannon Limit. This technique is currently used in equipment developed by Avantek[10] and Microwave Associates.[14] Note that by using QAM with 7 level partial response (QAM-PR7) it is possible in theory (implementation might be difficult) to move up to 4 b/s per Hz of RF bandwidth and achieve voice channel parity with FDM in all microwave radio bands even with 64 kb/s digital voice. Of course if lower bit rate digital voice (as provided by adaptive deltamodulation) is acceptable to a particular user then the parity line will drop down to a lower level. For example with 32 kb/s adaptive deltamodulation the present QAM-PR3 approach (Vicom/MA/Avantek) would provide 1152 voice channels in the most difficult 4 GHz band.

Figure 1 shows that the curve for multilevel QAM and VSB with α equal to 1 is further to the right indicating lower performance. The next best performer is PSK. Only the curve for a realizable filter (α=1) is shown on the graph. The Raytheon RDS 80 digital radio uses 4 phase PSK and thus achieves 1 b/s per Hz of RF bandwidth. Nippon Electric (NEC) with the radio being provided to DATRAN achieves 1.5 b/s per Hz of RF bandwidth by using 8 phase PSK. Next in performance comes FM. This is a very important category because all methods for digital transmission over existing analog microwave radio are contained within it. Canadian Marconi has achieved 1.5 b/s per Hz of RF bandwidth by using 8 level FM. Several years ago the Bell System achieved 1 b/s per Hz of RF bandwidth by using 4 level FM in their 2A RDT (20 mb/s over TD-2 radio)[3]. Going from standard multilevel to partial response signaling two more current systems are found. The Bell System DUV (1A RDT), used in their DDS network achieves the efficiency of 2 b/s per Hz of RF as shown by using 7 level partial response (PR7)[4]. The curve for FM with α equal to 0 although unrealizable itself is drawn to serve as a reference for the partial response points which are a couple of dB to the right of it. The Vicom system using 3 level partial response (PR3) achieves 1 b/s per Hz of RF bandwidth and has been used in many places for over three years now.

IMPLICATIONS OF FCC DOCKET NO. 19311

With reference to Figure 1 the FCC has put a floor at 1.0 bit per second per Hertz of RF authorized bandwidth (3.5 MHz in the 2 GHz band, 20 MHz at 4 GHz, 30 MHz at 6 GHz and 40 MHz at 11 GHz) for microwave transmitters operating below 15 GHz. Thus a horizontal line should be drawn at 1.0 bit per second per Hertz of RF bandwidth and the area below should be crosshatched to show that it is a forbidden region for bands below 15 GHz.

There is also a requirement for a given number of digital voice channels in each band below 15 GHz as shown in the table below:

Band (GHz)	No. Dig. Voice Ch.	Auth. BW	Bit Rate	B/S per Hz
2	96	3.5 MHz	6.3	2
4	1152	20 MHz	80 mb	4
6	1152	30 MHz	80 mb	2 2/3
11	1152	40 MHz	80 mb	2

Thus, to meet the requirements using a single polarization the efficiencies shown in the last column on the right are required. Avantek has accomplished this in the 2 GHz band while Microwave Associates has accomplished it in the 11 GHz band. No one has accomplished this feat in the 6 GHz band and in particular the very difficult 4 GHz band. As shown in Figure 1 if it is physically realizable QAM-PR7 would meet the requirement for an efficiency of 4 bits per Hertz of RF bandwidth. Of course the requirements can be met using two polarizations. Satisfactory transmission in the 2 GHz band is expected in almost all cases, but greater care must be used in the other three bands due to the problems with loss of polarization discrimination due to either rain or heavy multi-path fading.[12,13]

As implied earlier the FCC Docket No. 19311 digital voice channel requirements can be met in all bands (2,4,6,11 GHz) with one polarization using the Vicom/MA/Avantek QAM-PR3 digital radio approach if 32 kb/s adaptive delta-modulation is used in the aggregate bit stream instead of 64 kb/s PCM.

Another aspect of Docket 19311 which reqquires further discussion is those cases where mixtures of voice and data traffic are carried on the digital radio system rather than all voice. Hybrid radio systems (i.e. where a single RF channel carries FDM in part of its bandwidth and TDM in another part) for which the digital modulation contributes less than 50 per cent of the total peak frequency deviation of the transmitted RF carrier are excluded. In other words the rules of Docket 19311 do not apply to them. The Bell System DUV and the Western Union hybrid radio system between Atlanta and Cincinnati are examples of hybrid radio systems for which the rules of Docket 19311 do not apply. In all other cases up to and including 100 per cent digital modulation two rules apply. First, if the equipment meets the digital voice requirements you can use it for any combination of voice and data. Second, if the equipment is to be used only for data traffic and can't meet the digital voice requirements it can be type accepted <u>for data only</u> if it meets the minimum bandwidth efficiency requirements of 1 b/s per Hz of RF bandwidth. Thus, for the 4 GHz band the digital only equipment must provide 20 mb/s in the authorized 20 MHz of bandwidth and for the 6 GHz band the digital only equipment must provide 30 mb/s in the authorized 30 MHz of bandwidth. In order to use the digital only type of equipment, the carrier must convince the FCC that the system will carry only data. For example a carrier who only offers data services would have a sound argument. Also, a carrier who was splitting his traffic between two RF channels with one devoted entirely to voice using FDM and one devoted entirely to data using TDM would have a convincing case.

Finally, Figure 1 makes it clear why Docket 19311 states that seven level partial response is equivalent to a 16 state modulation format. It can be seen that the bandwidth efficiency of seven level partial response is equal to that of 16 level PM and 4 level QAM (which is really 16 levels because each quadrature carrier carries 4 levels and 4 times 4 equals 16). Also a realizable FM system ($\alpha=1$) would have 16 levels although that point is off the graph.

Summary and Conclusions

As shown in Figure 1 the Vicom/MA/Avantek digital radio approach of QAM-PR3 achieves the highest efficiency with the lowest S/N ratio (2 b/s per Hz of RF bandwidth and an average S/N of 22 dB for 10^{-7} error rate) that is currently available concurrent with the closest approach to the Shannon Limit. AT&T's DUV (1A RDT) achieves the same high efficiency but has an average S/N that is 4 dB higher for a 10^{-7} error rate. The QAM-PR3 technique is applicable to all microwave radio bands. The equipment for the 2 GHz and 11 GHz bands has already been shipped to the field for the first installations. In the meantime for the other bands (4,6, and 8 GHz) digital transmission at 1 b/s per Hz of RF bandwidth can be provided by using 3 level partial response (PR3) on any analog radio currently available. Note that this latter system is only 2.5 dB higher in average S/N for a 10^{-7} error rate than a currently available PM-4 level radio offering the same bandwidth efficiency (1 b/s per Hz of RF bandwidth).

REFERENCES

(1) Data Transmission, by W. R. Bennett and J. R. Davey, McGraw-Hill Book Company, 1965, p. 228, Figure 11-2.

(2) Principles of Data Communications, by R. W. Lucky, J. Salz, and E. J. Weldon, Jr. McGraw-Hill Book Company, 1968, p. 89, Figure 4.27.

(3) ICC 1969 Proceedings - Boulder, Colorado, "A 20 MB/S Digital Terminal for TD-2 Radio", by C. W. Broderick and R. W. Gutshall, p. 27-21 to p. 27-26.

(4) ICC 1972 Proceedings - Philadelphia, Pa., "Digital Transmission Over Analog Microwave Radio Systems", by K. L. Seastrand and L. L. Sheets, pp. 29-1 to 29-5.

(5) Proceedings of IEE, June 1974, "Development of an Integrated Digital Telecommunications Network", by W. T. Duerdoth, British Post Office.

(6) ISS 1974 Proceedings, Munich, Germany, September 9-13, 1974, "An Experimental Digital Local System", by H. S. McDonald, Bell Telephone Laboratories.

(7) ISS 1974 Proceedings, Munich, Germany, September 9-13, 1974, "System Planning for No. 4 ESS", by A. E. Ritchie and W. B. Smith, Bell Telephone Laboratories.

(8) ISS 1974 Proceedings, Munich, Germany, September 9-13, 1974, "Development of Features for No. 4 ESS", by L. S. Tuomenoksa and H. E. Vaughan, Bell Telephone Laboratories.

(9) ISS 1974 Proceedings, Munich, Germany, September 9-13, 1974, "Bell System ESS Family-Present and Future", by W. O. Fleckenstein, Bell Telephone Laboratories.

(10) Telephony, May 8, 1972, "Digital Microwave for T Carrier Transmission", by Walter J. Gill, Avantek.

(11) IEEE Transactions on Communication Techniques, Volume COM-14, February, 1966, "Generalization of a Technique for Binary Data Communication", by E. R. Kretzmer, Bell Telephone Laboratories.

(12) BSTJ, July-August, 1974, "Scattering of a Plane Electromagnetic Wave by Axisymmetric Raindrops", by J. A. Morrison and M. J. Cross, p. 955.

(13) BSTJ, October 1974, "Rain Induced Cross-Polarization at Centimeter and Millimeter Wavelengths", by T. S. Chu, p. 1557.

(14) NTC 1974 Proceedings, San Diego, California, December 2,3, and 4, 1974, "High Capacity Digital Microwave Radio Relay System", by R. H. Rearwin, Microwave Associates.

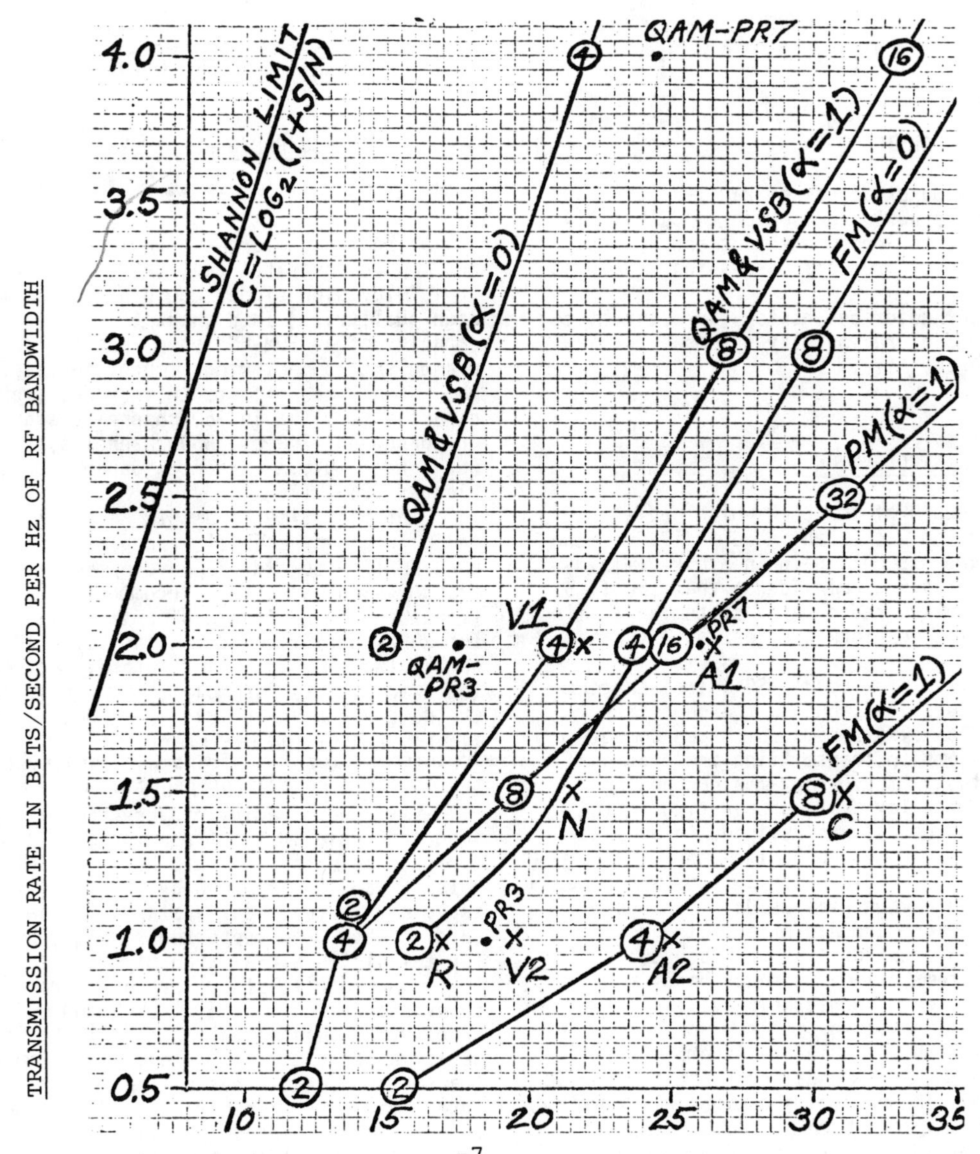

AVERAGE S/N FOR 10^{-7} ERROR RATE IN DB

FIGURE 1

LEGEND

QAM=Quadrature Amplitude Modulation
VSB= Vestigal Sideband
PR3= 3 Level Partial Response at Receiver
PR7= 7 Level Partial Response at Receiver
Ⓛ= No. in Circle is No. Levels Transmitted
α =Rolloff Factor(Equals 0 for Abrupt Cutoff and the BW Equals the Nyquist Band) (Equals 1 for Gradual Cutoff and the BW Equals Twice the Nyquist Band)

A1	= AT&T DUV (1A RDT)	PR7
A2	= AT&T (2A RDT)	FM - 4 Level
C	= Canadian Marconi	FM - 8 Level
N	= Nippon Electric	PM - 8 Level
R	= Raytheon RDS-80	PM - 4 Level
V1	= Vicom/MA/Avantek	QAM-PR3
V2	= Vicom	PR3
FM	= Frequency Modulation	
PM	= Phase Modulation	

CCIR WORK IN RADIO-RELAY SYSTEMS

YOSHIAKI HAYASHI

NIPPON TELEGRAPH & TELEPHONE PUBLIC CORP. TOKYO, JAPAN

ABSTRACT

Today's widespread use of radio-relay systems all over the world owes much to CCIR activities, especially to Recommendations and Reports which have come out from the discussion mainly at SG-9.

This paper describes history, present state and future trend of radio-relay sytems by reviewing the CCIR activities.

INTRODUCTION

Study Group 9 of the CCIR which deals with radio-relay system was established in 1948 at the Vth CCIR Plenary Assembly held in Stockholm. It was a very interesting coincidence that the first commercial microwave radio-relay system began its operation between New York and Boston in the same year.

For the first few years, SG-9 treated general technical questions, but as the necessity arised for radio-relay sytems which can compete with cable systems in transmitting international telephone calls, a Study Programme concerning "Wide-band radio-relay systems operating in the VHF, UHF and SHF bands" was assigned in 1951 at the VIth Plenary Assembly held in Geneva.

At that time, radio-relay systems were just introduced in such countries as United Kingdom, West Germany, France and Japan following U.S.A., and the main interest of SG-9 was focussed on radio-relay systems using frequency bands above 1 GHz where sufficient bandwidth is available for high capacity transmission.

At the VIIIth CCIR Plenary Assembly held in Warsaw in 1956, 22 Recommendations submitted by SG-9 were adopted. These Recommendations dealt with many important subjects such as radio frequency channel arrangement for 600 channel telephone system, interconnection at baseband and at intermediate frequencies, hypothetical reference circuit and allowable noise power, etc.. They formed the basis of the present Recommendations of SG-9 which made the largest number of Recommendations among the various CCIR Study groups. Thus, SG-9 contributed greatly in establishing international standard for radio-relay systems.

ANALOG RADIO-RELAY SYSTEMS

(1) History of development

The first microwave radio-relay system, TDX in AT&T, demonstrated that 480 telephone channels or 1 television signal could be transmitted in high quality.[1] In a short time after that, the capacity of an FDM-FM system was increased to 600 channels through various improvements.[2]

Ceaseless efforts to increase transmission capacity resulted in 960 channel systems using 4 GHz band and 1200 or 1800 channel systems using 6 GHz band in U.S.A., France, Japan and other countries by 1963. (Fig.1)

At that time, the CCIR Recommendations on radio frequency channel arrangement sometimes gave a guide line toward the developmental objectives of transmission capacity. Recommendation on radio frequency channel arrangement was first adopted for 2GHz and 4GHz bands at the VIIIth Plenary Assembly (Warsaw, 1956) and secondly for 6GHz band at the IXth Plenary Assembly (Los Angels, 1959). At the same Plenary Assembly, 2 Recommendations concerning limitation of interference and rf channel arrangement for T-H (Trans-Horizon) system using tropospheric-scatter propagation were adopted.

Microwave radio-relay systems are playing an important role in the transmission of not only multiplexed telephone signals but also television signals. Expansion of microwave transmission systems, especialy in their early stage of development, is said to have been supported strongly by the demand for television signal transmission.

In order to study long-haul transmission of television signal, a joint CCIR and CCITT Study Group, CMTT, was established in 1956. As for the performance objective of the television signal transmission, there was difficulty in establishing internationally agreed Recommendation because waveform of each broadcast television system was different. After a long and careful study, a voluminous Recommendation was adopted at the XIVth Plenary Assembly (Kyoto, 1978).[3]

As the exploitation of high frequency bands advanced, the Xth Plenary Assembly (Geneva, 1963) adopted Recommendations concerning the use of the upper 6GHz and 11GHz bands in which a 2700 channel system using the upper 6GHz band was already referred to.[4] It is worth noticing that this Recommendation was made 9 years before the first 2700 channel system was practically used.[5]

Reprinted from *IEEE 15th Int. Conf. Comm.*, pp. 9.5.1–9.5.5, June 1979.

Furthermore, at this Plenary Assembly, frequency sharing between satellite communication systems and terrestrial microwave systems became an important matter. There were many discussions on the range of frequency bands for common use, because the Recommendation stating that "in those frequency bands shared between both systems, the maximum value of the e.i.r.p. of any radio-relay system transmitter shall not exceed +55dBW" had strong effects on the realization of the above-mentioned 2700 channel system.

As the result of these discussions, the frequency band recommended for the 2700 channel system will not be used commonly with the satellite communication system. The 2700 channel systems using the upper 6GHz band (5GHz band in Japan) and the 11GHz band were developed in many countries and operated as high capacity trunk line systems.[6][7][8]

(2) Present state

Ratio of radio channel separation Δf to the highest baseband frequency f_h is often used as a measure of frequency utilization efficiency.(Fig.2) According to various investigations, interference noise from adjacent channels is negligible, provided the ratio $\Delta f/f_h$ is greater than 3. Most of the existing systems are designed to fulfil this condition. But some systems have overcome this limit of frequency spectrum utilization efficiency by using FDM/FM. Typical examples are introduced below.

In U.S.A., TD-2 system in the 4GHz band with 20MHz channel separation carrying 600 telephone channels was in operation for a long time. The new TD-3 system was planned to increase the capacity to 1500 channels. One of the critical problems was the interference due to direct coupling between the transmitting and the receiving antennas (in the same direction at the same station), because transmitting carriers and receiving carriers were separated by 20MHz in the frequency arrangement of this system.[9]

According to the studies at BTL, the coupling loss was mostly above 80dB, and the prospects were acquired for realizing the world's first system with $\Delta f/f_h$ smaller than 3, which was put into operation in 1973.[10]

In Japan, it was attempted to transmit 2700 channels using the lower 6GHz band with 29.65MHz channel separation which was allocated for systems carrying 1800 channels, and to transmit 3600 channels using the 5GHz band with 40MHz carrier separation. It was very important to clarify both propagation characteristics such as occurrence probability of fading and propagation distortion in order to design such high capacity transmission systems. In these fields, SG-5 has greatly contributed to the development of the radio-relay systems through its Report [11]. Nevertheless, problems are left for future investigations such as variation characteristics of XPD (Cross Polarization Discrimination) on propagation path, which directly influence noise power from adjacent channels.

The interference from adjacent channels can be reduced by improving the precision of antenna direction adjustment to achieve 38dB (average value evaluated in antilogarithm value) XPD in developing these systems. Moreover, field investigation of XPD variation on propagation path has been carried out, and it is expected to prove that system design can be achieved by allowing a certain variation margin on XPD objectives. As the results of these studies, lower 6 GHz band system carrying 2700 channels and 5GHz band system carrying 3600 channels were in operation since 1977 and 1979, respectively.[12]

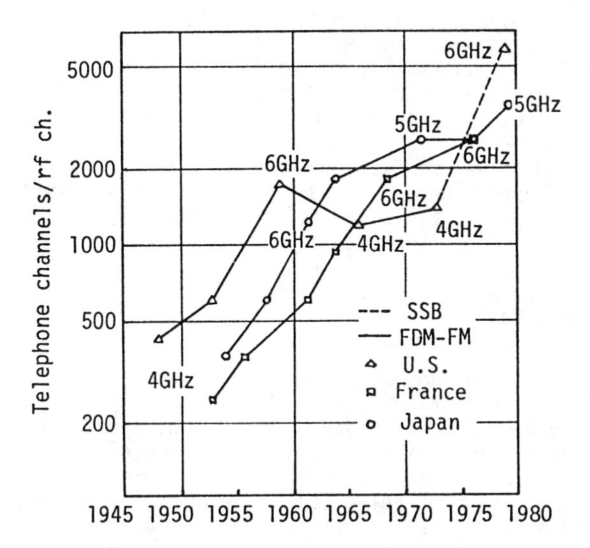

Figure 1 Trend of transmission capacity

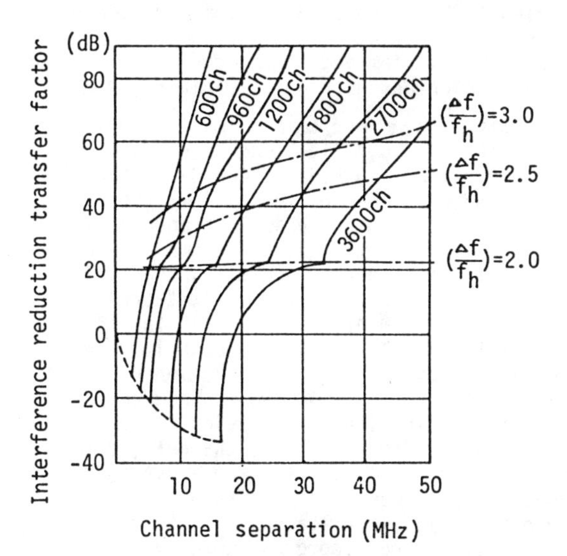

Figure 2 Adjacent channel interference characteristics at the worst baseband channel.

DIGITAL RADIO-RELAY SYSTEMS

(1) History of development

In the early day of development, digital modulation methods such as PPM (Pulse Position Modulation) were studied [13][14][15], but after the success of TD-2 system in Bell Telephone Laboratories [2], development of FDM-FM system was promoted energetically and digital system did not become operational in full scale.

However, later progress in semiconductor technology enabled economical PCM terminal equipment to bring the T1 carrier system [16] using pair cables in operation in AT&T, and studies were resumed to introduce PCM systems into microwave radio-relay systems.

It was at Extraordinary Administrative Radio Conference (Geneva, 1963) that the digital radio-relay system attracted general attention for the first time. Recommendation No.4A was adopted there which requested the CCIR to study modulation method such as pulse code modulation for line-of-sight radio-relay systems in relation to sharing problem with communication satellite systems.

Thereupon, at the interim meeting of the SG-9 in 1965, contributions on PCM radio-relay system were submitted from U.S.A. and Japan, and after lengthy discussions, a new Question on digital radio-relay system ("Radio-relay systems for the transmission of pulse-code modulation and other types of digital signals") was adopted.

The first microwave digital radio-relay system became operational in 1969 in Japan [17] which carried 240 channels (clock frequency at 7.876 MHz) using the 2GHz band. In order to achieve effective utilization of spectrum, this system adopted QPSK (Quadrature Phase Shift Keying) and coherent demodulation technic and used the same frequency in common by both vertically and horizontally polarized waves. This system became the guidepost for the later development of digital radio-relay systems.

Since then, medium and small capacity digital systems using 2,6,7,11,13 and 15GHz bands have become operational in U.S.A., Canada, Japan, Italy, France and U.K. Most of the systems use QPSK and the rest of them use 8 phase PSK and PRS (Partial Response Signalling). These systems are mainly applied to short-haul circuits, where maximum advantage can be taken of the merit that terminal equipment for PCM are cheaper than those for FDM.

In 1976, NTT put a high capacity digital radio-relay system [18] in operation using the 20 GHz band (17.7 to 21.2GHz) which had not been developed till then. The transmission capacity of this system is 400 Mbps per radio channel (corresponding to 5760 telephone channels) and about 46000 telephone channels can be transmitted per route by 9 radio channels (interleaved channel arrangement).

Besides, developmental work on digital radio-relay system using the 18GHz band (17.7 to 19.7GHz) are being promoted in U.S.A.[19], Italy and other European countries.

On the other hand, in order to transmit data signal economically using existing analog transmission lines, systems for simultaneous transmission of digital and analog signals are developed in many countries. DUV (Data Under Voice) system [20] and PDUV (Parallel DUV) system [21] are introduced in U.S.A. and Canada, respectively, which can carry 1.544Mbps digital signal by 7 level PRS using the lower frequency band of the FDM-FM baseband signal (below 500KHz). In Japan, STD-1 (Simultaneous Transmission of Data and Telephone signal) system [22] is in operation, which can transmit 1.544Mbps by 8 level roll-off shaping, using the lower frequency band of the baseband of the FDM-FM signal as recommended by CCITT. There are also systems under study which can transmit 1.5Mbps or 2.0Mbps digital signal using the frequency band above the FDM-FM baseband signal or television signal which are called DAV (Data Above Voice)/DAVID (Data Above Video) system in Canada, U.S.A., Italy[23], etc..

(2) Present state

(A) Radio frequency channel arrangement

The most important thing to be considered in designing a digital radio-relay system is the radio frequency channel arrangement. In order to determine the arrangement, it is necessary to consider various details such as transmission capacity, modulation method and its realizable performance, propagation characteristics in the frequency band, etc.. But, according to the results of studies at SG-9, it is reported that adjacent channel spacing Xs, spacing between adjacent transmitter and receiver Ys, and guard band at the edges of the band Zs will be within the following range in cases of medium and large capacity systems using binary or QPSK (Report 608-1).

$$1.5\ f_p \leq Xs \leq 2.5\ f_p$$
$$2.0\ f_p < Ys \leq 4.0\ f_p$$
$$Zs \approx 1.0\ f_p$$

,where f_p indicates pulse repetition frequency.

From the standpoint of effective spectrum utilization, it is desirable that Xs, Ys and Zs are small, but in view of degradation in performance due to distortion and interference from adjacent channels, these parameters are limited to following values in the existing systems.

$$Xs \approx 1.6\ f_p,\ Ys \approx 2.5\ f_p,\ Zs \approx 0.8\ f_p.$$

The above-mentioned discussions hold good when the entire frequency band can be used by digital systems, but as most frequency bands are already used for analog transmission systems, it is necessary to verify the compatibility with analog systems when introducing digital transmission systems (Report 610-1).

(B) Hypothetical reference circuit and performance objectives

The fundamentals in designing a radio-relay system are hypothetical reference circuit (HRC) and performance objectives. As for the former, new Recommendation 556 was adopted at the interim meeting of the SG-9 (Geneva, 1976), which recommends that HRC for a digital radio-relay system should be 2500 km long and include 9 multiplexer sections in accordance with CCITT Recommendation G721. At every 3 sections, connection is achieved at 64kbps.

As for the performance objectives on HRC, a draft Recommendation, as mentioned below, has been proposed and further studies are requested to examine the propriety of these figures and other objectives for burst errors, jitter, etc..

(i) The 10-minute mean value of the bit error rate should not exceed 10-7 for more than 5% of any month.
(ii) The 1-second mean value of the bit error rate should not exceed 10^{-3} for more than 0.05% of any month.

(C) Improvement of spectrum utilization efficiency

The efficiency of spectrum utilization of radio-relay systems has been discussed only in the frequency domain. However, it becomes important to raise the spatial spectrum utilization efficiency, that is, to increase the density of transmission circuits in a certain area. An example of the definition of spectrum utilization efficiency taking into account the spatial efficiency is expressed by SG-1 (Report 662).

$$\eta = \frac{M \cdot A}{B}$$

where, A : Transmission capacity (e.g. number of telephone channels) per radio channel.
B : Required bandwidth per radio channel.
M : Number of branching links (e.g. two-way radio routes) in a repeater station.

Fig. 3 shows the spectrum utilization efficiency of a point-to-point radio-relay system for various modulation methods. As shown in the figure, digital system, in general, is more profitable than analog system from the standpoint of the spectrum utilization efficiency. This tendency is remarkable, especially in short hop length.

FUTURE TREND OF STUDY

(1) Increase of transmission capacity

Instead of improving conventional FM systems, increase of transmission capacity of an analog system is being investigated from an entirely new point of view with the proposal of SSB-AM systems. [24][25] It aims at realizing the lower 6GHz band (5925MHz-6425MHz) with 29.65MHz channel separation. Studies are continued in many countries in order

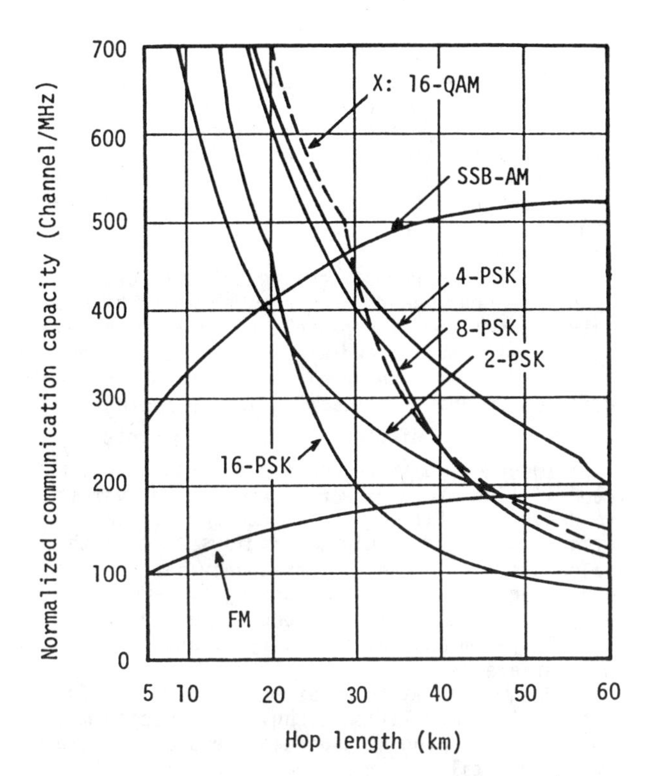

Figure 3 Spectrum use efficiency for radio-relay systems

Circuit length: 2500 km
Frequency band: 6 GHz
Antenna: Horn reflector type
Note: Dashed curve X for 16-QAM is newly calculated under the same assumptions used in Report 662

to solve the problems such as method to improve and compensate the non-linearity of the repeater, method to analyze and equalize propagation distortion, and the synchronization of carrier frequency for demodulator.

For digital radio systems to play an important role in the future integrated digital network, it is necssary to develop not only systems operating in higher frequency bands but also large capacity digital radio-relay systems using frequency bands ranging from 4GHz to 8GHz which are employed in large capacity analog systems. As the available bandwidth of each frequency band is only 500 or 600 MHz, further improvement of spectrum utilization efficiency is necessary, and, therefore, spectrally efficient digital systems are studied in many countries.

In Canada, development for an 8GHz band system conveying 90Mbps with QPRS is carried out to put it in operation by 1980.[26] In Japan, field researches on 200 Mbps systems using the 4 and 5 GHz bands with 16 QAM have been performed since 1978[27], which is scheduled to begin service in 1982.

In response to these circumstances, SG-9 decided in the XIVth Plenary Assembly (1978, Kyoto) to revise the present Study Programme and investigate the frequency arrangement for digital systems in frequency bands below 12GHz.

(2) Exploitation of higher frequency bands

Frequency bands below 20GHz are becoming stringent, more and more, because most of the radio -relay systems applied to current telecommunication network utilize these frequency bands.

On the other hand, recent development of satellite communication technology is requiring additional assignment of frequency bands, especially of those below 20GHz.

In these circumstances, studies to exploit higher frequency bands are required where high capacity transmission is possible by overcoming severe rainfall attenuation and atmospheric absorption. These studies will be accelerated through re-examination of frequency allocation in the coming WARC-G.

(3) Co-existence of digital and analog systems

In order to introduce digital radio systems into existing radio network where analog systems are densely installed, it is necessary to have compatible operation of digital radio systems with analog systems in the same frequency band. For the co-existence of digital and analog systems, it is necessary to cope with the spectrum expansion of digital system and vulnerability of analog system to interference.

One method is to adopt multi-phase and multi-level digital modulation technic. The second method is spectrum shaping to suppress the spectrum expansion of digital system as much as possible. This is realized by roll-off filtering in the baseband and/or limitting transmission spectrum by using QPRS technics which positively utilize the inter-symbol interference caused by band limitation. The third method is to reduce the interference under normal condition by controlling the digital system RF output. Furthermore, there is a method to reduce adjacent channel interference by improving the cross-polarization discrimination characteristics of the antenna. Since co-existence is an important subject, it is necessary to pursue further studies.

(4) Co-existence of terrestrial and space communication systems

With the advance of various space communication systems, frequency sharing between terrestrial and space systems are becoming an important subject. Study of digital radio-relay system was originally started from this point, and it still remains significant now. It will also be required to investigate conditions for compatible operation by estimating atmospheric absorption, rain scatter, etc., because propagation characteristics are different in each frequency band. The improvement of antenna directivity and attitude control technic of the satellite will also help to realize the co-existence in each frequency band.

REFERENCE

[1] Thayer G.N. et al.:"The New York-Boston microwave radio-relay system"-Proceedings of the IRE, Vol. 37(February 1949).

[2] Roetken A.A. et al.:"The TD-2 microwave radio-relay system"-Bell System Technical Journal, Vol. 30, No.4(October 1951).

[3] CCIR/CCITT: Recommendation 567 "Transmission performance of television circuits designed for use in international connections."

[4] Carassa F.:"Research on radio-relay systems having a very high transmission capacity (2700 telephone channels or the equivalent)"-Alta Frequenza, Vol. XXXI, No.2(1962).

[5] Matsuhashi S.:"2700-channel radio-relay system operating in the 5 GHz band."-Japan Telecommunication Review, Vol.14, No.3(July 1972).

[6] Magre P.:"Faisceaux hertziens à grande capacité pour 1800 et 2700 voies"-Câbles et transmission, 30th year, No.4 (special issue, October 1976).

[7] Myrseth E.:"An economical microwave link system for 2700 telephones"-Conference Proceedings of 6th European Microwave Conference,1976.

[8] Lupke G. et al.:"Radio-relay system FM 2700/6700"-Electrical Communication, Vol. 51 No.3 (1976).

[9] CCIR: Recommendation 382-2, Annex.

[10] Hathaway W.G. et al.:"TD-3 Microwave Radio-relay System."-Bell System Technical Journal, Vol. 47, No.7(1968).

[11] CCIR: Report 338-2 "Propagation data required for line-of-sight radio-relay systems.

[12] Ohi J. et al.:"High capacity FDM/FM microwave sytem with increased efficiency of fequency utilization"-ICC, 1977.

[13] Black H.S.:"AN/TRC-6,A microwave relay system." -Bell Laboratories Record, Vol.23, No.12 (December 1945).

[14] Feldman C.B.:"A 96-ch pulse code modulation system."- Bell Laboratories Record, Vol.26, No.9 (September 1948).

[15] Gilman G.W.:"Systems engineering in Bell Telephone Laboratories."-Bell Laboratories Record, Vol. 31, No.1 (January 1953).

[16] Hoth D.F.:"The T1 carrier system"-Bell Laboratories Record, Vol.40, No.11 (November 1962).

[17] Yoshida K. et al.:"2 GHz microwave PCM system." -Japan Telecommunications Review, Vol.11, No.1 (1969).

[18] Nishino K. et al.:"20 GHz PCM radio-relay system."-Japan Telecommunications Review, Vol.18, No.1 (1976).

[19] Longton A.C.:"DR-18A high speed QPSK system at 18 GHz."-ICC (1976).

[20] Prime R.C. et al.:"The 1A radio digital system makes 'data under voice' a reality."-Bell Laboratories Record, Vol.51, No.12 (December 1973).

[21] CCIR: Doc. 9/218, Canada (May 1977).

[22] Hosoda A. et al.:"The STD-1 system"-Japan telecommunications Review, Vol.17, No.3 (July 1975).

[23] CCIR: Doc. 9/280, Italy August 1977).

[24] CCIR: Report 781 "Radio-relay systems for telephone using single sideband amplitude modulation (SSB-AM)"

[25] Markle R.E.:"The ARGA single sideband long haul radio system."-ICC,1979.

[26] Roadhouse R.A. et al.:" The trans-Canada digital network."-ICC (1977).

[27] Okamoto Y. et al.:"Characteristics of a high capacity 16 QAM digital radio system on a multpath fading channel."-ICC (1979).

Part II
Modulation Techniques

Modulation Techniques for Microwave Digital Radio

TOSHITAKE NOGUCHI, YOSHIMASA DAIDO, AND JOSEF A. NOSSEK

Editor's Note: This tutorial article was originally published in the October 1986 issue of the IEEE COMMUNICATIONS MAGAZINE, as part of the Special Series on Microwave Digital Radio. It has been updated by the authors and edited for inclusion in this book.

INTRODUCTION

THIS tutorial article deals with digital radio modulations, a technology that has experienced remarkable growth in a very short time. The first-generation digital radio systems, introduced scarcely over a decade ago, used low-level modulations such as 2- and 4-level phase shift keying (2-PSK and 4-PSK). Shortly after, systems were introduced using 8-PSK [1], [2] and 9-level quadrature partial response signaling (9-QPRS) [3]. This was followed by systems using high-level quadrature amplitude modulation (QAM). Specifically, 16-QAM [4]–[10] now enjoys widespread use and 64-QAM [11]–[22] is becoming commonplace. Moreover, the feasibility of 256-QAM [23]–[26]—and even 1024-QAM—as candidate modulations are being actively investigated by various manufacturers.

What are the features of radio systems that use such high-level modulations? Why has multilevel QAM become the most popular type of modulation? What are the important associated technologies for realizing such high-level modulations?

MODULATION/DEMODULATION TECHNIQUES

Special Features of Microwave Digital Radio

Digital modulation techniques are used in several applications, notably: microwave line-of-sight radio, satellite communications, and data transmission over voiceband channels. The first of these applications is the one that concerns us here, and it is distinguished from the others by a combination of three special features. The first is the need for high digital speeds, which poses serious challenges to hardware realization. Bit rates in microwave digital radio vary from about 2 Mb/s to 400 Mb/s, with emphasis on rates between 34 Mb/s and 140 Mb/s. In voiceband data modems, by contrast, bit rates lie below 20 kb/s, and so hardware requirements are much easier to satisfy.

The second feature is a primary emphasis on bandwidth (or spectral) efficiency, with power efficiency being a secondary factor. This emphasis arises from the widespread demand for use of the limited radio spectrum, and leads to the design of high-level modulations that trade power efficiency for spectral efficiency. As a result, modern line-of-sight systems have spectral efficiencies of better than 4 bits/sec/Hz. In satellite systems, by contrast, on-board weight limitations place a premium on power efficiency, and so low-level modulations (for example, 4-PSK), with spectral efficiencies closer to 1 bit/sec/Hz, tend to be used instead.

The third feature is the susceptibility of microwave digital radio to frequency-selective (*dispersive*) fading, which is caused by multipath fading on line-of-sight links. The consequences are more severe for high-level modulations (second feature) and more difficult to combat at high speeds (first feature). By contrast, voiceband data transmission is subject to channel dispersion but not fading, while satellite links can encounter signal fading (such as, due to rain) but seldom of the dispersive kind.

The above three features in combination profoundly influence the choice of modulation techniques in microwave digital radio. We now turn to specific techniques and their properties.

Modulation

Nearly all microwave digital radio systems use *linear* modulations, those that are formed by translating baseband pulse streams to IF or RF using balanced amplitude modulators. These modulations can be used to achieve considerable spectral efficiency and, at the same time, good power efficiency. In addition, they are easy to analyze, which facilitates design and assessment, and they lend themselves to practical methods of equalization.

The principle of linear modulation is demonstrated in the left half of Fig. 1. The inputs I and Q represent sequences of data values in digital form, with data values in each stream separated by T seconds. The D/A converters change the digital streams into analog sequences, $\{a_n\}$ and $\{b_n\}$; and these are converted by the low pass filters (LPF_M) into pulse streams, $\Sigma a_n g(t - nT)$ and $\Sigma b_n g(t - nT)$, where $g(t)$ is the filter impulse response.

The local oscillator (LO) produces a sinusoidal carrier, cos $\omega_0 t$, where ω_0 is either at IF or RF, and applies it to two balanced modulators. The carrier input to the upper modulator is in phase with the LO output and the carrier input to the lower modulator is, because of the 90° phase shift, in quadrature phase. This is the origin of the designations I and Q in Fig. 2 and subsequent discussions.

The modulator outputs are added and bandpass-filtered (in

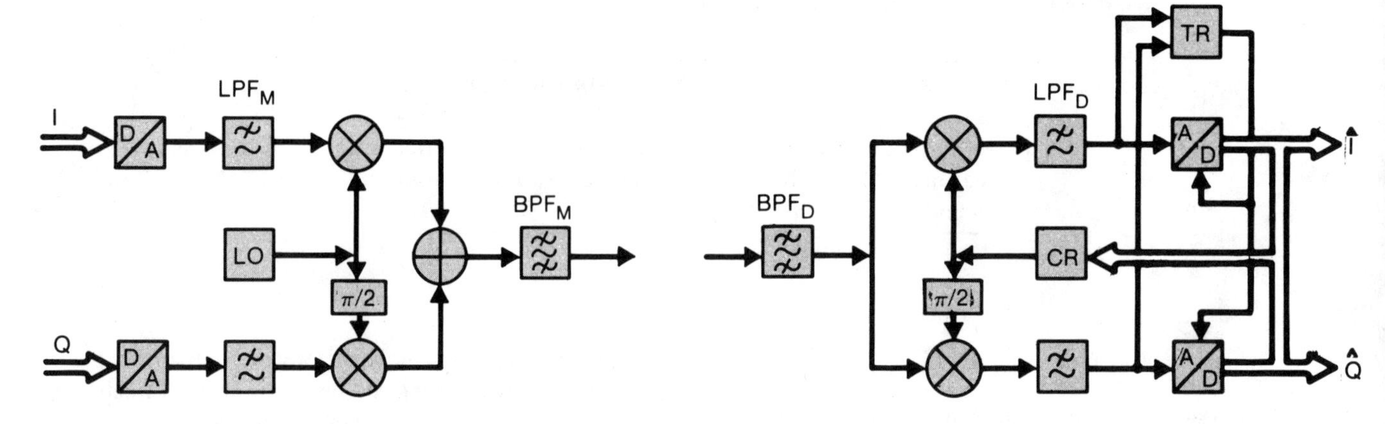

Fig. 1. Modulator and demodulator sections used in digital radio links.

BPF$_M$) to produce the IF or RF signal

$$S(t)=\left[\sum_n a_n h(t-nT)\right]\cos\omega_0 t - \left[\sum_n b_n h(t-nT)\right]\sin\omega_0 t \quad (1)$$

where $h(t)$ is the pulse shape produced by the combined filtering of LPF$_M$ and BPF$_M$. This modulated signal might be translated to another frequency (from IF to RF) and is, in any case, boosted in a high-power amplifier and sent to the transmitter antenna.

Modulation Categories

The data value corresponding to the nth signaling period can be designated by the complex number $a_n + jb_n$. The modulation used by a particular system is classified according to the discrete alphabet (or constellation) of complex numbers from which the data values are taken. For example, suppose that all a_n's are zero and all b_n's are taken from one of M evenly-spaced values, $\pm 1, \pm 3, \cdots, \pm(M-1)$. In this case, the modulation is called M-level amplitude shift keying (M-ASK), and the constellation consists of M points distributed uniformly on the vertical axis of the complex plane. Now suppose, instead, that $a_n = \cos\phi_n$, $b_n = \sin\phi_n$, where each ϕ_n is taken from M evenly-spaced values between $-180°$ and $180°$. In this case, the modulation is called M-level phase shift keying (M-PSK), and the constellation consists of M points distributed uniformly on the unit circle. Finally, suppose that each a_n is taken from the possibilities $\pm 1, \pm 3, \cdots, \pm(\sqrt{M}-1)$, and similarly for each b_n. This is called M-level quadrature amplitude modulation (M-QAM), and the constellation is a square lattice of M points.*

* Strictly speaking, any modulation derived by adding the outputs of two balanced modulators in phase quadrature can be called a quadrature amplitude modulation. In common usage, however, the QAM label is applied to those modulations for which M is a perfect square and the constellation is a square lattice. We will follow that convention here.

Examples of these modulations are shown in the I-Q planes of Fig. 2. Note that 2-PSK is the same as 2-ASK, and that 4-PSK is the same as 4-QAM. All performance data given in this paper will be for M-PSK or M-QAM systems. However, a variation on QAM called quadrature partial response signaling (QPRS) is also of interest in digital radio; we will discuss it later in connection with pulse shaping.

For both M-PSK and M-QAM, the transmitted bit rate is

$$R_b=\frac{1}{T}\log_2 M \text{ bits/sec.} \quad (2)$$

Thus, bit rate grows linearly with the symbol rate ($1/T$, measured in bauds) and logarithmically with M.

Demodulation

The right half of Fig. 1 shows the process of demodulation. The input signal (at carrier frequency ω_0) is bandpass-filtered in BPF$_D$ and applied to two balanced modulators. The other input to each is a local carrier, $\cos(\omega_0 t + \theta)$ for the top modulator and $-\sin(\omega_0 t + \theta)$ for the bottom one. Since each modulator is fed by two signals at the same frequency, it acts as a coherent demodulator. The two outputs are low-pass filtered in LPF$_D$ and applied to A/D converters to produce the output digital streams $\hat{I}$ and $\hat{Q}$.

The local carrier signal, $\cos(\omega_0 t + \theta)$, is delivered by the carrier recovery circuit (CR), which may be a voltage-controlled oscillator driven by inputs from the detected streams. If the carrier recovery is properly done, θ will be zero. To see the implication of this, assume that the input to BPF$_D$ is the modulator output, (1). [For present purposes, we ignore additive noise and other channel effects.] If $\theta = 0$, the outputs of the top and bottom low pass filters will be the baseband signals

$$S_I(t)=\sum_n a_n p(t-nT),\ S_Q(t)=\sum_n b_n p(t-nT) \quad (3)$$

where the pulse shape $p(t)$ is the result of the combined effects

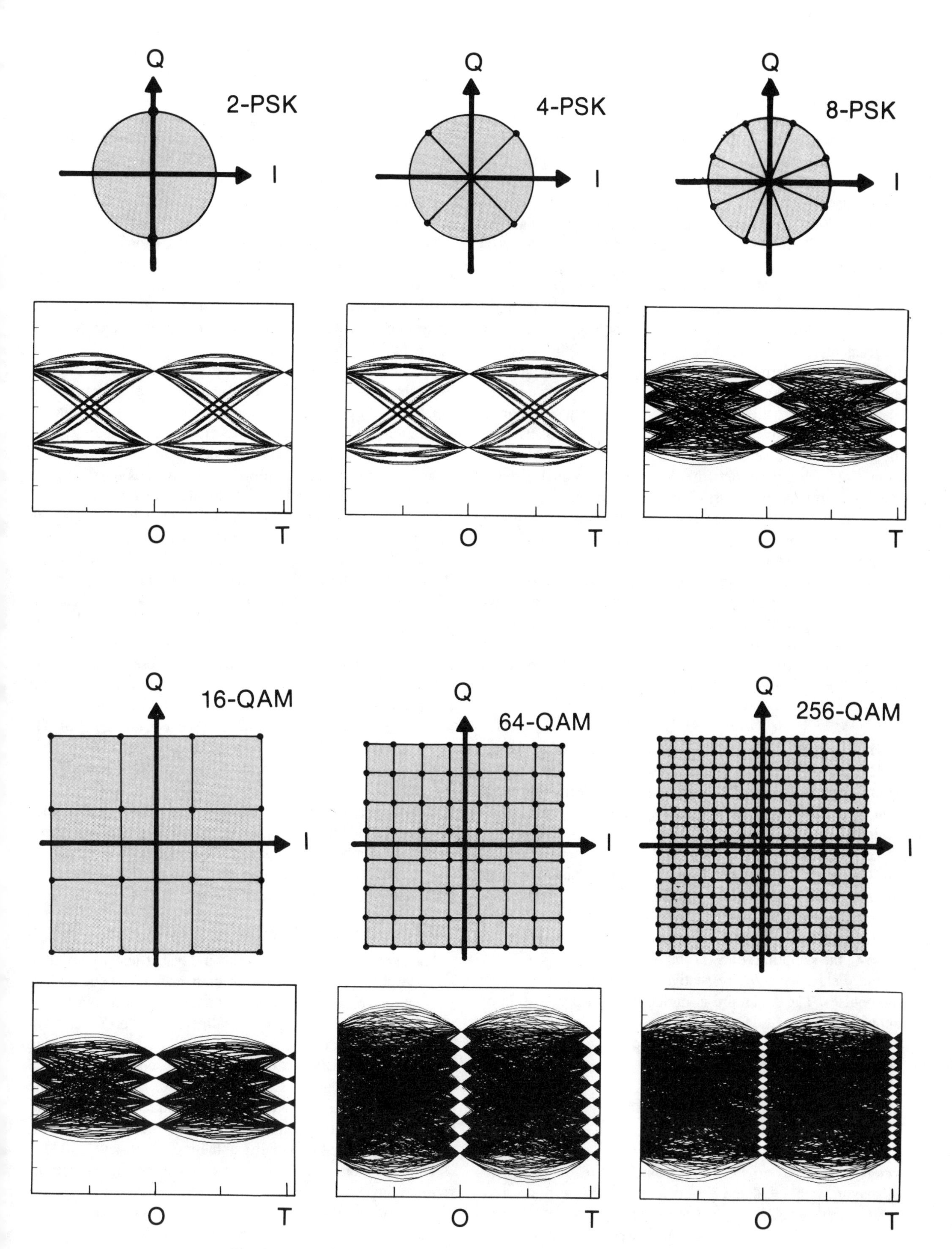

Fig. 2. Constellations and eye diagrams for PSK and QAM signals.

of BPF_D and LPF_D on $h(t)$ in (1). For nonzero θ, the output in the top branch will be $S_I(t)\cos\theta + S_Q(t)\sin\theta$ and the output in the bottom branch will be $S_Q(t)\cos\theta - S_I(t)\sin\theta$. In each case, the second term is cross-rail interference, which should be as small as possible for good detection.

The timing recovery circuit (TR) extracts a sinusoid at frequency $1/T$ from the baseband signals, and its phase determines the sampling epoch, t_s. Thus, to detect the data in the nth period, the low pass filter outputs are sampled at time $nT + t_s$ and the two samples are quantized in the A/D converters to obtain digital estimates of a_n and b_n.

To demonstrate the importance of carrier phase, pulse shape, and sampling epoch on detection, data "eye" diagrams are shown below each data constellation in Fig. 2. Each such diagram represents the output of either one of the low pass filters, with superimposed traces representing outputs for different sequences of data values. The "eyes" are the diamond-shaped white spaces that occur every T seconds. Signal thresholds located at the vertical midpoints of the eyes dictate the data value decided on for each sample. The time instant of each sample determines how much signal distance (that is, margin against noise) is available for ensuring correct decisions. In the diagrams of Fig. 2, the optimal sampling instant in every data period is located at the horizontal midpoint of the eyes.

For all of the eye diagrams in Fig. 2, carrier recovery is ideal ($\theta = 0$). In addition, $p(t)$ has the property that, at the ideal sampling instants, there are no contributions from data pulses other than the one being detected (there is no *intersymbol interference,* or ISI). If $p(t)$ lacks this property and/or $\theta \neq 0$, the eye openings will be smaller.

Pulse/Spectrum Shaping

The combined effects of the transmitter and receiver filters, and of the propagation medium, determine $p(t)$ in (3). We ignore for now the possibility of frequency-selective propagation, so that the shape of $p(t)$ is totally within the control of the designer.

Let us consider a pulse shape that has a maximum value of unity at some time instant t_0 and has zero values at all time instants $t_0 + kT$, $k = \pm 1, \pm 2$, and so forth. Any pulse shape having this property is called a Nyquist pulse. If such a pulse could be used for $p(t)$ and the sampling epoch t_s could be made to be t_0, the samplings of $S_I(t)$ and $S_Q(t)$ in (3) at $t = t_s + nT$ would yield a_n and b_n, respectively, with no ISI from other data pulses. The usual aim in digital radio design is to bring about such a condition.

The most popular of all possible Nyquist pulse shapes is the *cosine rolloff* pulse,

$$p(t) = \frac{\sin(\pi t/T)}{\pi t/T} \frac{\cos(\pi\alpha t/T)}{1-(2\alpha t/T)^2} \tag{4}$$

where $0 \leq \alpha \leq 1$, and α is the *rolloff factor*. This pulse has its maximum at $t = 0$ and is zero at all $t = kT$, as desired. Its Fourier transform, $P(f)$, is T for all $|f| < (1 - \alpha)/2T$; zero for all $|f| > (1 + \alpha)/2T$; and, in between, falls from T to 0 as the decreasing half of a cosine-squared pulse. The cosine rolloff pulse is strictly bandlimited; when translated to IF or RF, its bandpass spans a bandwidth of $(1 + \alpha)/T$.

Among all possible Nyquist pulses, the smallest RF bandwidth that can be attained is $1/T$, and this is often referred to as the Nyquist bandwidth. The pulse shape that achieves it is the cosine rolloff pulse with $\alpha = 0$. The amount by which the bandwidth of a given pulse exceeds the Nyquist bandwidth (for example, α/T for the cosine rolloff pulse) is called the excess bandwidth.

The choice of α involves many considerations. Choosing α near 1 compromises spectral efficiency, for example. On the other hand, choosing α near 0 makes manufacturing more difficult and costly; also, transmission is made more vulnerable to impairments, as we will see. The usual choice for α in digital radio is close to 0.5.

Another important choice is how to apportion the shaping of $P(f)$ between the transmitter and receiver. The popular approach is to provide an overall spectral shape $\sqrt{P(f)}$ in the transmitter, and the same in the receiver. This even division of the spectral shaping yields the best detection performance for a given average transmitter power and a white Gaussian noise channel, as well as for adjacent channel interference.

Finally, we can now explain QPRS modulation. Suppose that the transmitted signal has a QAM-type square constellation but that $p(t)$ is a *partial response* pulse. Specifically, let $P(f)$ be a real half-cosine function on the interval $|f| \leq 1/2T$. The inverse transform, $p(t)$, has the property that the samples of $S_I(t)$ and $S_Q(t)$ in the nth interval will yield $(a_n + a_{n-1})$ and $(b_n + b_{n-1})$, rather than a_n and b_n. By means of suitable data precoding in the transmitter, these detected levels can be made to represent the original data values. This is the essence of QPRS. Instead of $\sqrt{M}$ levels per baseband output, as in M-QAM, there are $(2\sqrt{M} - 1)$ levels, or $(2\sqrt{M} - 1)^2$ levels all together. Thus, transmissions of 4, 16, and 64 levels lead to output levels numbering 9, 49, and 225, respectively [3], [27], [28].

QPRS modulations are inherently spectrum-efficient since the radio bandwidth is $1/T$, not $(1 + \alpha)/T$. At the same time, the larger numbers of output levels lead to power penalties, that is, necessary increases in power to achieve a given bit error rate.

Carrier and Timing Recovery

The carrier recovery process should be sufficiently accurate that the phase error θ is low in both its static value and its fluctuations (phase jitter). One method of carrier recovery is to put the IF signal through a nonlinearity, chosen so as to produce a spectral line at the carrier frequency or some multiple thereof, and to extract that line component with a phase-locked loop (PLL). Alternatively, the carrier line component can be generated by using the detected data stream to "remodulate" the IF signal. A third approach, the decision-directed method, is especially suitable for high-level QAM.

The A/D converter outputs, $\hat{I}$ and $\hat{Q}$, are digital streams representing the transmitted data sequences. Because of additive noise, ISI and other distortions, the digital output values will lie *not* at M discrete points in the $\hat{I}$-$\hat{Q}$ plane but, rather, in small disc-like regions around these points. These

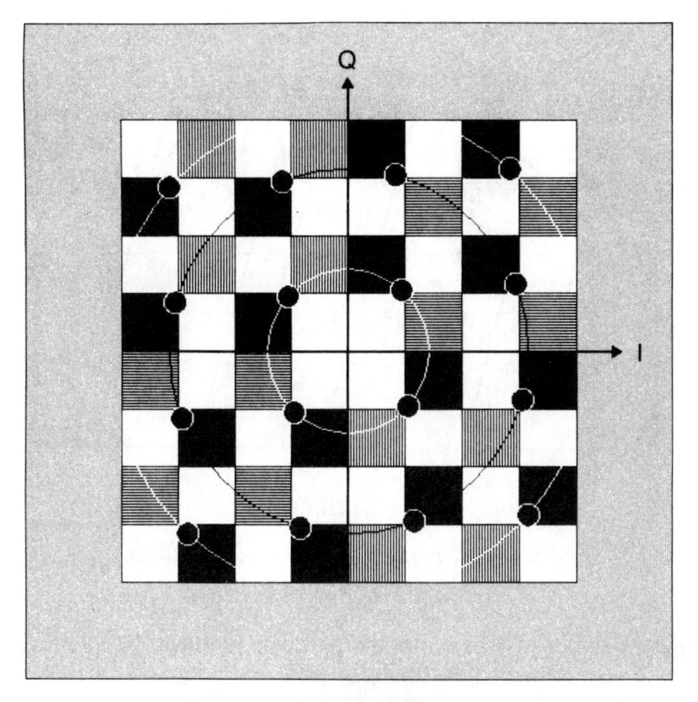

Fig. 3. Carrier recovery template for 16-QAM.

are the small, dark circular areas shown in Fig. 3 for the case of 16-QAM. They are superimposed on a *template* consisting of 64 small square regions, the coordinates of which are stored in digital memory in the carrier recovery circuit. Output samples falling into the black (gray) regions indicate need for a clockwise (counterclockwise) rotation of the output constellation, and they lead to positive (negative) control increments. The time sequence of control increments is low-pass-filtered and applied to a VCO to control the carrier frequency and phase. Thus, the carrier recovery process acts like a phase-locked loop, where the control increments obtained using the template replace the phase detector function in a conventional PLL. By appropriately choosing the circuit parameters, the desired acquisition and tracking of the received carrier frequency can be achieved. Frequency-sweeping can be used, in addition, to aid the acquisition process.

Timing recovery is less critical, in terms of performance sensitivity, than carrier recovery, but is important nonetheless. A popular approach is to square the outputs of the low pass filters (LPF_D in Fig. 2) and extract from their sum the spectral line component at frequency $1/T$. This line component could be generated, instead, via square-law envelope detection of the IF signal. It is also possible to apply decision-directed methods to timing recovery, which may be useful for detection in multipath fading.

Factors Influencing the Modulation Choice

Spectral Efficiency

Assume that a common carrier band is subdivided into channels of width W. (For example, the 4-, 6- and 11-GHz bands in the USA have channel bandwidths of roughly 20, 30, and 40 MHz, respectively.) The spectral efficiency, η, can be defined as the ratio of bit rate to bandwidth, R_b/W, expressed in bits/sec/Hz. Using (2),

$$\eta = \frac{1}{WT} \log_2 M. \tag{5}$$

In theory, WT can be as low as 1 without adjacent channel interference (ACI). Using cosine rolloff shaping, this could be achieved by using a rolloff factor $\alpha = 0$. As a practical matter, however, it is more appropriate to choose α close to 0.5 and to allow the total RF bandwidth, $(1 + \alpha)/T$, to exceed W somewhat. The resulting out-of-band emissions are subject to formal constraints imposed by spectrum regulation. These are based on considerations of interference into adjacent channels, which may contain either analog or digital modulations. The emission rules take into account the fact that adjacent channels generally operate on orthogonal polarizations.

In the USA, the FCC has codified the emission rules in terms of mask functions that digital spectra must lie below. The FCC mask for the 6-GHz band is shown in Fig. 4. The ordinate represents the transmitted power in a 4-kHz bandwidth centered on frequency f, in dB above the total transmitted power. A given transmission satisfies the emission rules if this quantity, computed at each f, lies below the mask.

Fig. 5 shows transmitted spectra for a system in which half of the cosine rolloff shaping is done in the transmitter. It can be shown that, with $1/T = 3/4\ W$ and α near 0.5, the FCC masks for 4, 6, and 11 GHz are all satisfied. These are the choices, therefore, for most digital radio systems in the USA. The resulting η, from (5), is $3/4 \log_2 M$, so that systems using 4, 16, 64, and 256 levels have spectral efficiencies of 1.5, 3.0, 4.5, and 6.0 bits/sec/Hz, respectively.

Power Efficiency

Once a choice is made of desired spectral efficiency, the power efficiency of the contending modulations becomes important. Power efficiency is a measure of how much received power is needed to achieve a specified bit error rate (BER). Let the received carrier-to-noise ratio, C/N, be defined as the average received signal power divided by the input thermal noise in the Nyquist bandwidth, $1/T$. Using well known analytical techniques for signals in Gaussian noise, it is quite simple to derive BER as a function of C/N for any modulation. In this type of derivation, it is generally assumed that 1) all modulation levels occur with equal probability, $1/M$; 2) Gray encoding is used to map the original data bits into modulation levels (this ensures that virtually all errors in detecting levels lead to one-bit errors); and 3) differential encoding is used as well (this safeguards, for example, against the effects of quadrant uncertainties in the recovered carrier). Assuming all of the above, Fig. 6 shows curves of BER vs. C/N for several modulations.

Of particular interest is the C/N required to achieve a performance threshold, say BER $= 10^{-4}$, above which the radio link is considered to be in outage. Under normal propagation conditions, the received C/N is usually well above that required for threshold performance, with correspondingly lower BER. The amount of further signal reduction allowed before BER exceeds its threshold is called the *flat fade*

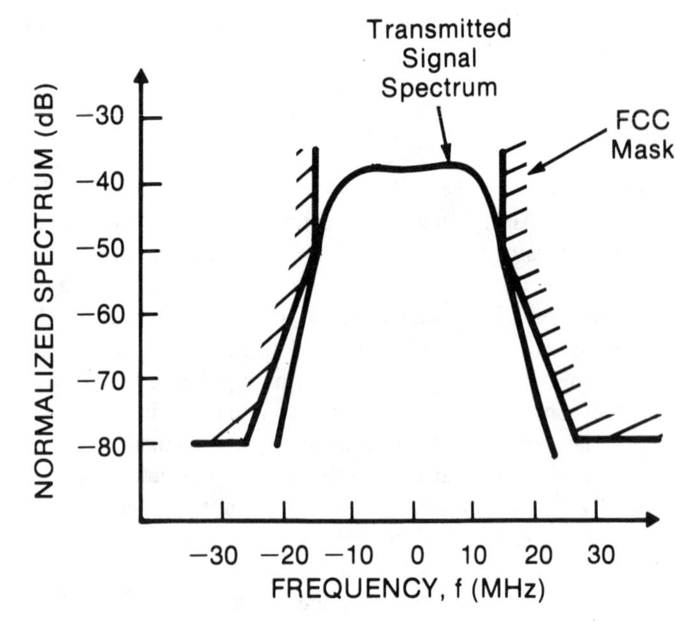

Fig. 4. Transmitted spectrum and FCC mask (6-GHz band).

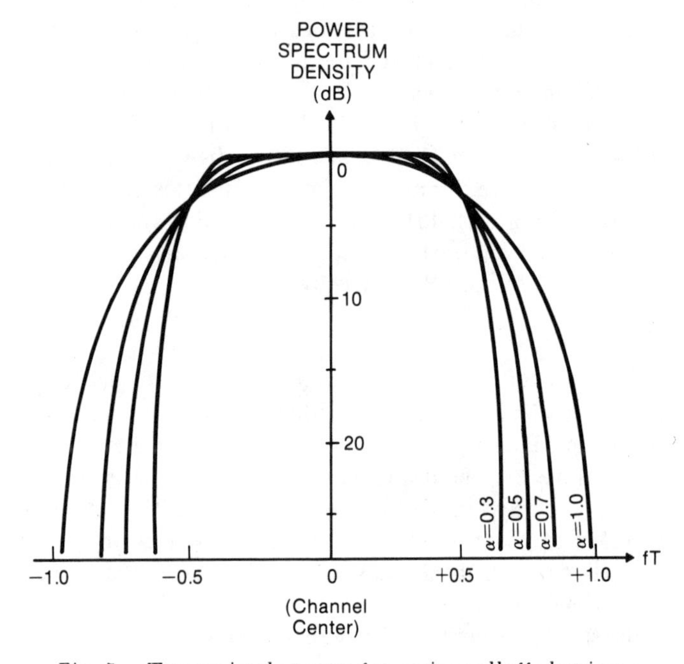

Fig. 5. Transmitted spectra for cosine rolloff shaping.

margin. Also, the *system gain* is the flat fade margin (in dB) plus the dB loss, during normal propagation, between transmitter and receiver. In typical digital radio links, the flat fade margin is in the vicinity of 40 dB and the system gain is in the vicinity of 100 dB. Fig. 6 reveals how, with all other factors the same, flat fade margin and system gain decrease with increasing M. It also shows that QAM is more power-efficient than PSK for $M \geq 16$. This is because, for the same average power, QAM achieves larger signal distance in the I-Q plane (Fig. 2) than does PSK.

Tolerance to Impairments

Digital radio signals are susceptible to impairments from both equipment and the channel. The former include linear distortion, nonlinear distortion, and synchronization errors, while the latter include interference and multipath fading.

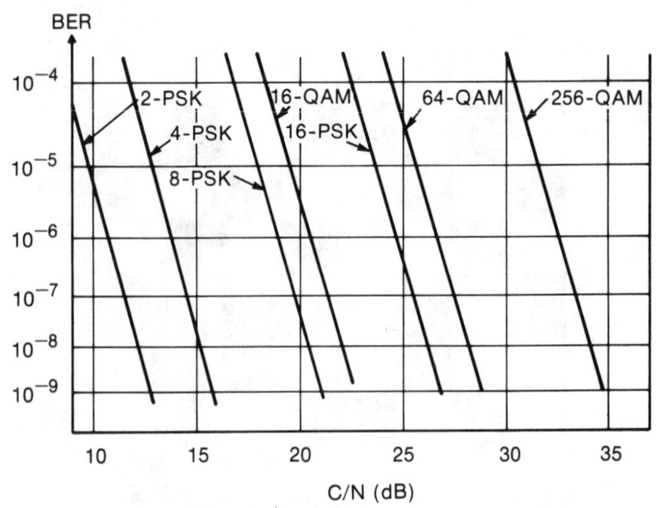

Fig. 6. BER curves for several modulations.

Linear Distortion—Linear distortion is the departure of the end-to-end spectral shaping in a link from the design goal, $P(f)$. To a first order, the departure from ideal of both the dB amplitude response and the group delay response can be described entirely in terms of linear and quadratic variations in f. The top of Fig. 7 displays the linear and quadratic coefficients (A_1, A_2, d_1, d_2) for the two responses. Each coefficient is defined to be the maximum change in the response (dB amplitude or group delay) over all frequencies within $\pm 1/2T$ of the channel center. The performance degradation for a given coefficient is defined to be the increase in C/N required to maintain a specified BER when the other three coefficients are zero.

The four graphs of Fig. 7 show the C/N increases needed (the power penalties), for each of the four distortions taken separately, to maintain a BER of 10^{-6}. The sensitivity to the number of modulation levels is evident.

Nonlinear Distortion—Nonlinear distortion can come from any amplifier or mixer stage in the transmitter and receiver. The primary source of such distortion, however, is the final amplifier of the transmitter. This is a peak-power-limited device that becomes increasingly nonlinear as the signal level approaches saturation. It is instructive, therefore, to compare the required peak input powers for different modulations and spectral shapings.

The starting point is the *average* input power needed to achieve a specified bit error rate in the receiver. Relative values of this quantity for different modulations can be obtained from curves like those in Fig. 6. For QAM signals, there is an added peak factor (peak-to-average power ratio) because different data pulses have different magnitudes, in contrast to PSK. For both modulations, the presence of overlapping pulses at the amplifier input produces additional peaking. Assuming cosine rolloff shaping, with half the shaping done in the transmitter, one can find the peak instantaneous power over all time and all data sequences as a function of rolloff factor, α.

All these considerations are accounted for in the curves of

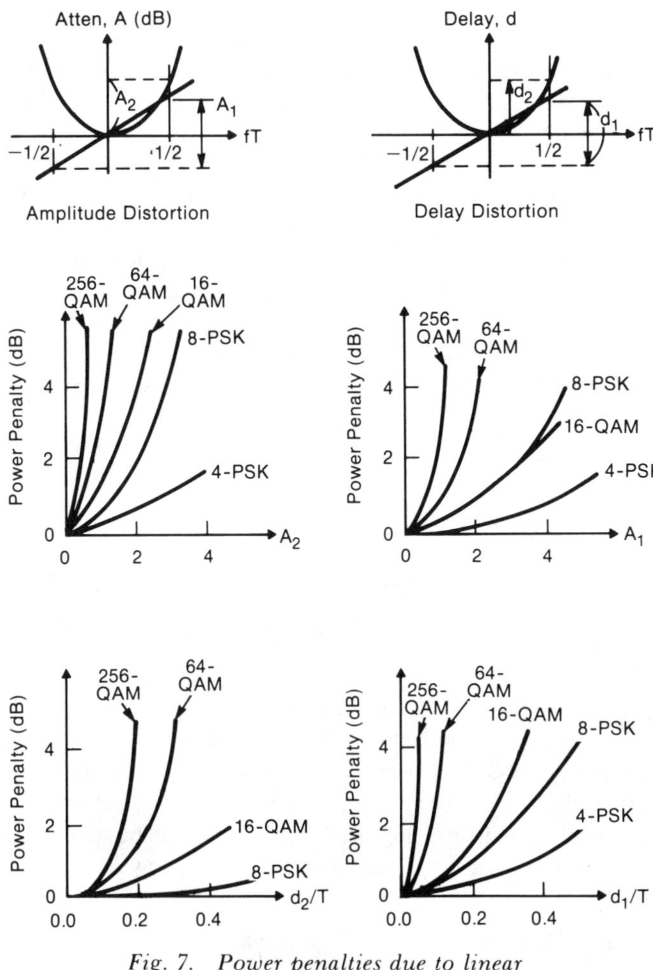

Fig. 7. Power penalties due to linear distortions (BER = 10^{-6}).

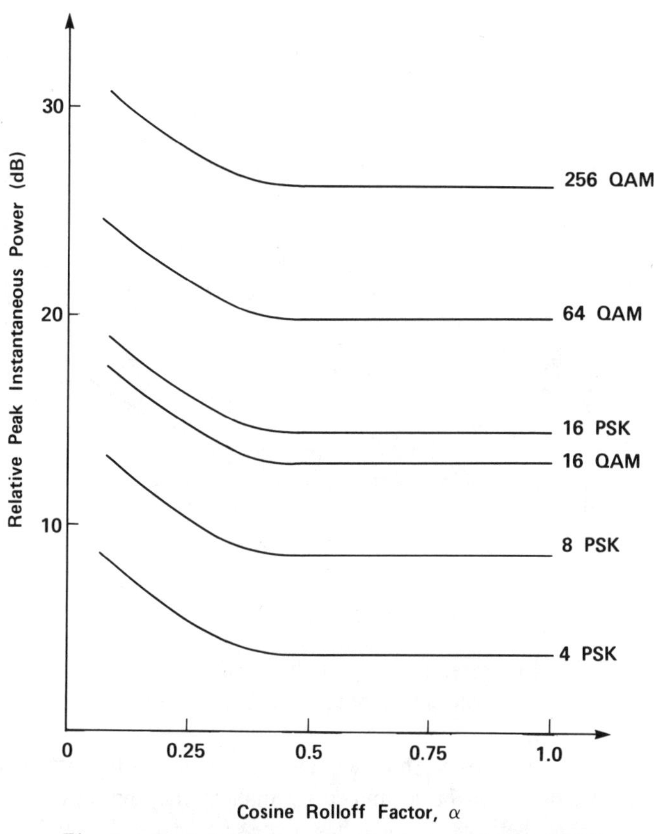

Fig. 8. Peak instantaneous power (relative to QPSK with square pulses) for several modulations.

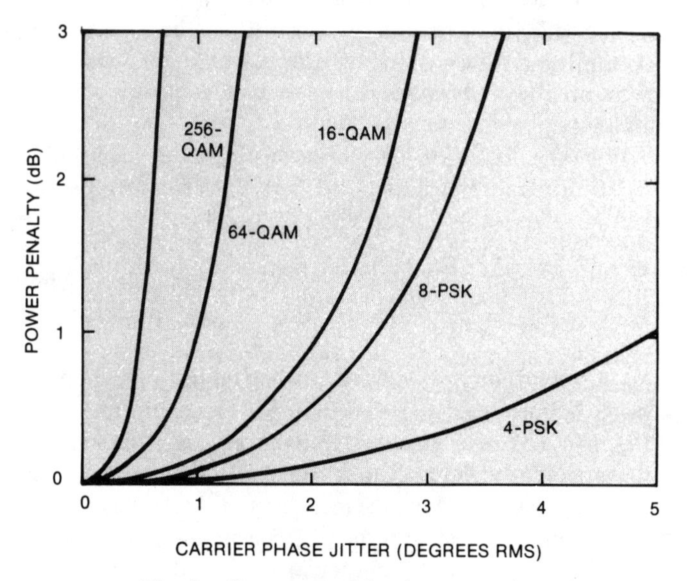

Fig. 9. Power penalties due to carrier phase jitter (BER = 10^{-6}).

Fig. 8. The ordinate is peak instantaneous power relative to the case of 4-PSK with square T-second pulses. These relative results are for a required BER of 10^{-6}, but they would hardly be different for other values.

To accommodate a given modulation and rolloff factor, the saturation power of the transmitter amplifier must be sufficiently large that the peak input power lies in the linear range. Alternatively, nonlinear predistortion can be used to improve the linearity of the overall response. Such measures become increasingly important as M increases, because higher-level modulations are more sensitive to nonlinear distortion.

Another reason to limit nonlinear distortion is that it causes spectral spreading of the transmitter output. This spreading can lead to violations of the emission rules (see Fig. 4). In addition to the measures cited above, post-amplifier filtering can be used to minimize this problem.

Synchronization Errors—Both carrier and timing recovery errors can cause performance degradations. Each kind of error can have a static component, but these are usually made quite small by the recovery circuits. Harder to minimize are the fluctuations (or jitter) of the carrier and timing phases, which are due to both randomness of the data patterns and additive noise.

Fig. 9 shows the increase in C/N required to achieve BER $= 10^{-6}$, plotted against rms carrier phase error for various modulations. Fig. 10 does the same for rms timing phase errors. In the latter case, the rolloff factor has a strong influence, as shown for 16-QAM. For carrier recovery, α does not influence degradation for a given rms error, but the rms errors themselves tend to increase as α decreases.

Interference—Interference on digital radio links can arise from various sources, notably, adjacent channel interference (ACI), as noted above, and co-channel interference (CCI). Co-channel interference can be particularly strong when dually polarized transmissions are used to double radio channel

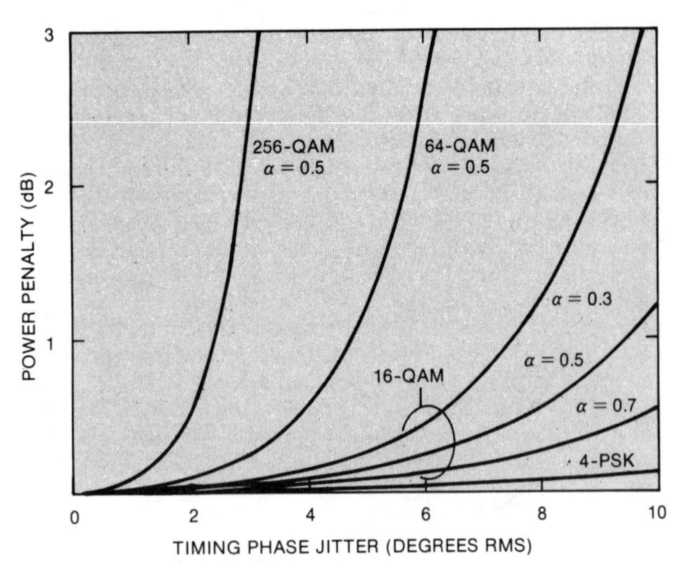

Fig. 10. Power penalties due to timing phase jitter (BER = 10^{-6}).

capacity. Interference arises, in this case, from the mechanism of cross-polarization coupling; cross-pol interference is labeled XPI.

The carrier-to-interference ratio (C/I) required for different modulations can be estimated by making the approximation that the net interference has Gaussian statistics, and by postulating that the interference is strong compared to thermal noise. Under these conditions, detection performance is dominated by the interference, and BER can be estimated by substituting C/I for C/N on the abscissa of Fig. 6. Doing so gives an idea of how much interference is permissible for different modulations.

As the number of modulation levels increases, controlling ACI requires careful design of filters and antenna cross-pol discrimination (XPD); controlling CCI requires care in route engineering and in design of antenna characteristics (sidelobe patterns and XPD); and controlling XPI in particular requires careful XPD design and the use of adaptive cross-pol cancelers.

Multipath Fading—Multipath fading on line-of-sight paths occurs frequently enough to be a major source of link outage [29], [30]. When fading occurs, the propagation response across the bandwidth of a channel can be very low in amplitude, which reduces the received signal relative to noise. The response can also be quite frequency-selective, which distorts the end-to-end frequency response and produces both cross-rail and intersymbol interferences. To maintain satisfactory link operation during multipath fading, measures such as space diversity and equalization must be used, as discussed in the next section.

Associated Techniques

To achieve high capacities in modern digital radio systems, high-level modulations must be used, as we have seen. To meet stringent performance standards despite the sensitivities of these modulations to impairments, various associated techniques must be deployed. Some are briefly noted here; for more detailed treatments, see [31].

Diversity Methods

Space diversity and frequency diversity are both widely used methods for enhancing digital radio availability. Frequency diversity is implemented by setting aside one or more backup channels in the same common carrier band and switching one of them into service when fading or equipment failures cause an outage in a working channel. Space diversity is usually implemented using two vertically-spaced antennas on a radio tower, with a separation large enough to ensure independent fadings. The outputs of the two antennas can be added together in some way, or the "best" of the two outputs can be selected for processing by the receiver. Either approach serves to improve the statistics of the signal strength into the receiver.

Equalization

Where space diversity improves the probability of a strong received signal, equalization improves the end-to-end frequency response in the presence of multipath fading, linear distortion, or both. In the early days of digital radio, IF equalizers were used to adjust the slope of the amplitude response across the band or to reduce the effects of in-band notches, or both. No direct action was taken to compensate for group delay distortion. Today, both transversal and decision feedback equalizers are used, and these adaptive circuits reshape the pulse so as to end up with ISI as low as possible.

Cross-Pol Interference Cancellation

Systems using dually polarized signals achieve two-fold use of the same radio channel. Dual-pol systems have been built with 4-PSK modulations, for which case special measures against XPI are not necessary. However, for $M \geq 16$, the isolation provided by typical antenna XPDs is not sufficient and adaptive cross-pol cancelers are needed. Moreover, canceler circuits must be capable of providing frequency-selective responses, since XPI tends to be most serious during multipath fading. In addition, they should be able to operate on plesiochronous symbol sequences in the two polarizations.

Forward Error Correction (FEC) Coding

Given the emphasis on spectral efficiency in digital radio, FEC coding, with its requirement of bit redundancy, has not been widely used. The use of high-level modulations, however, requires special measures to maintain very low bit error rates (for example, BER $= 10^{-10}$ or less during normal propagation). FEC coding would help to relax the stringent hardware design requirements that might otherwise be imposed, particularly for $M > 64$. High-efficiency codes must be used, however, to limit the erosion of spectral efficiency. One example is a rate 18/19 convolutional code, for a 64-QAM radio, that provides a *coding gain* (that is, reduction in required C/N) of about 3 dB at a BER of 10^{-6} [32], [33]. Another example is a rate 247/255 block code, for a 256-QAM radio, that provides a coding gain of about 3.4 dB at a BER of 10^{-6} [23].

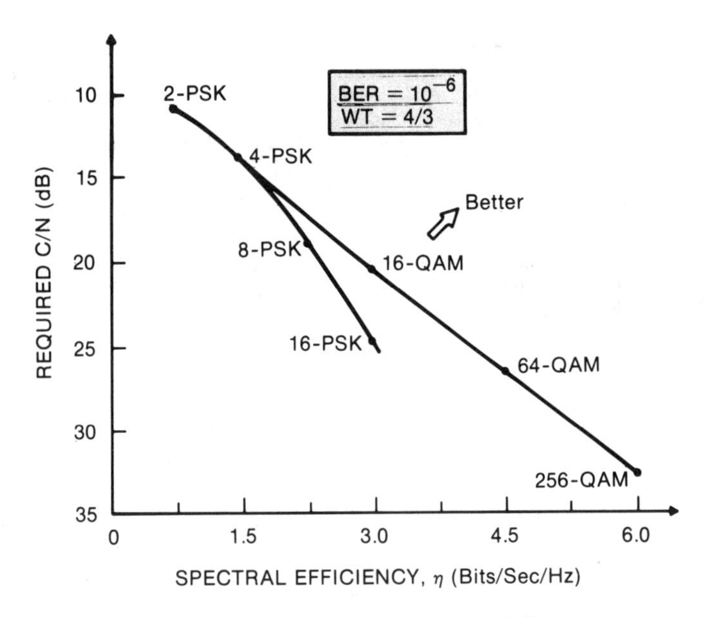

Fig. 11. Comparisons of several modulations.

TABLE I
CANDIDATE MODULATIONS FOR SYSTEMS IN EUROPE AND NORTH AMERICA

	Bit Rate \ Channel BW	20 MHz	30 MHz	40 MHz
EUROPE	34 Mb/s	4-PSK	4-PSK	4-PSK
	68 Mb/s	16-QAM	8-PSK	4-PSK
	140 Mb/s	256-QAM	64-QAM	16-QAM
	280 Mb/s	----	1024-QAM	256-QAM
NORTH AMERICA	45 Mb/s	**	**	**
	90 Mb/s	64-QAM	16-QAM	8-PSK
	135 Mb/s	256-QAM	64-QAM	16-QAM
	180 Mb/s	1024-QAM	256-QAM	64-QAM
	270 Mb/s	----	1024-QAM	256-QAM

**FCC requires a minimum of 78 Mb/s for common carriers.

CONCLUSION

The choice of a digital radio modulation for a given application involves many factors, including technical, economic, and regulatory issues. One consideration is the tradeoff between spectral efficiency and power efficiency, theoretical curves for which are shown in Fig. 11. Also important are tolerances to various impairments, as quantified by data such as those in Figs. 7, 9, and 10. Finally, the choice is influenced by the channelization plan, the digital hierarchy, and the costs of different alternatives, including the associated techniques required.

So far, the use of higher-level modulations has always led to lower costs per channel, and this fact has motivated the exploration of ever-higher numbers of levels. For large M, moreover, both technical factors and considerations of circuit complexity and cost dictate the choice of QAM over other modulations.

Table I shows candidate modulations for several channel bandwidths and bit rates, for systems in Europe and North America. The bit rates shown reflect the differences in digital hierarchies between the two continents.

In conclusion, the introduction of very high-capacity systems in microwave digital radio has been made possible by rapid advances in the design of modulation/demodulation circuitry and the associated techniques. The result is that 256-QAM is now a practical modulation candidate, and 1024-QAM may become practical before too long.

REFERENCES

[1] P. R. Hartmann and J. A. Crosset, "A 90 MBS digital transmission system at 11 GHz using 8 PSK modulation," in *ICC '76*, pp. 18.8–18.13, June 1976.
[2] Y. Tan, *et al.*, "An 8-Level PSK modem with cosine roll-off spectrum for microwave communications," in *ICC '76*, pp. 29.13–29.18, June 1976.
[3] S. Barber and C. W. Anderson, "Modulation considerations for the RD-3 91 Mb/s digital radio," in *ICC '77*, pp. 111–115, June 1977.
[4] S. Komaki, *et al.*, "Characteristics of a high capacity 16 QAM digital radio system in multipath fading," *IEEE Trans. Comm.*, vol. COM-27, no. 12, pp. 1854–1861, Dec. 1979.
[5] M. A. Byington, and C. J. R. Pallemaerts, "Design and performance of a 16-state digital modem," in *ICC '79*, pp. 5.4.1–5.4.6, June 1979.
[6] Y. Yoshida, *et al.*, "Six GHz-90 Mbps digital radio system with 16 QAM modulation," in *ICC '81*, pp. 52.4.1–52.4.5, June 1981.
[7] C. P. Bates, *et al.*, "DR6-30 system design and application," in *ICC '81*, pp. 3.1.3–3.1.8, June 1981.
[8] P. Dupuis, *et al.*, "16 QAM modem for a high capacity microwave system: design and performance," in *ICC '81*, pp. 3.2.1–3.2.6, June 1981.
[9] J. A. Nossek, *et al.*, "16-state QAM modem for a 140 Mbit/s digital radio system family," in *GLOBECOM '83*, pp. 9.1.1–9.1.6, Nov. 1983.
[10] Y. Saito, *et al.*, "Feasibility considerations of high-level QAM multi-carrier system," in *ICC '84*, pp. 665–671, June 1984.
[11] C. P. Bates, *et al.*, "Impact of technology on high-capacity digital radio systems," in *ICC '83*, pp. F2.3.1–2.3.5, June 1983.
[12] T. Noguchi, *et al.*, "6 GHz 135 MBPS digital radio system with 64 QAM modulation," in *ICC '83*, pp. F2.4.1–F2.4.5, June 1983.
[13] P. R. Hartmann, *et al.*, "135 MBS-6 GHz transmission system using 64 QAM modulation," in *ICC '83*, pp. F2.6.1–F2.6.7, June 1983.
[14] E. Fukuda, *et al.*, "Design of 64 QAM modem for high capacity digital radio systems," *GLOBECOM '83*, pp. 25.5.1–25.5.5, Nov. 1983.
[15] J. A. Crosset, *et al.*, "64-QAM digital radio transmission system integration and performance," in *ICC '84*, pp. 636–641, May 1984.
[16] J. D. McNicol, *et al.*, "Design and application in the RD-4A and RD-6A 64 QAM digital radio system," in *ICC '84*, pp. 646–652, May 1984.
[17] S. Takenaka, *et al.*, "A new 4 GHz 90 Mbps digital radio system using 64-QAM modulation," in *ICC '84*, pp. 642–645, May 1984.
[18] Y. Ito, *et al.*, "Design and performance of 6 GHz 135 Mb/s with 64 QAM," in *ICC '84*, pp. 632–635, May 1984.
[19] K. Aoki, *et al.*, "The adaptive transversal equalizer for 90 Mbps 64 QAM radio relay system," in *ICC '84*, pp. 1003–1006, May 1984.
[20] C. P. Bates, *et al.*, "DR6-30-135 system design and application," in *GLOBECOM '84*, pp. 16.7.1–16.7.8, Nov. 1984.
[21] W. R. Brouillette, *et al.*, "Microwave radio design for 64-QAM digital radio," in *GLOBECOM '84*, pp. 535–538, Nov. 1984.
[22] W. Grafinger, *et al.*, "Design and realization of a high speed multilevel QAM digital radio modem with time-domain equalizer," in *ICC '85*, pp. 31.5.1–31.5.6, June 1985.
[23] Y. Daido, *et al.*, "256 QAM modem for high capacity digital radio systems," in *GLOBECOM '84*, pp. 547–551, Nov. 1984.
[24] Y. Takeda, *et al.*, "Performance of 256 QAM modem for digital radio system," in *GLOBECOM '85*, pp. 47.2.1–47.2.5, Dec. 1985.
[25] T. Ryu, *et al.*, "A stepped square 256 QAM for digital radio system," in *ICC '86*, pp. 46.6.1–46.6.5, June 1986.

[26] Y. Yoshida, *et al.,* "6 GHz 140 Mbps digital radio system with 256 QAM modulation," in *ICC '86,* pp. 46.7.1–46.7.5, June 1986.
[27] D. P. Taylor and M. Shafi, "Fade margin and outage computation of 49-QPRS radio employing decision feedback equalization," in *ICC '83,* pp. 1453–1458, June, 1983.
[28] I. Sasase, *et al.,* "Comparison of improved efficiency 225-QPRS and 256-QAM in distorted channels," in *ICC '85,* pp. 448–452, June 1985.
[29] W. D. Rummler, R. P. Coutts, and M. Liniger, "Multipath fading channel models for microwave digital radio," Part III of this book. [An earlier version was published in *IEEE Commun. Mag.,* vol. 24, no. 11, Nov. 1986, pp. 30–42.]
[30] L. J. Greenstein and M. Shafi, "Outage calculation methods for microwave digital radio," Part V of this book. [An earlier version was published in *IEEE Commun. Mag.,* vol. 25, no. 2, Feb. 1987, pp. 30–39.]
[31] J. K.Chamberlain, F. M. Clayton, H. Sari, and P. Vandamme, "Receiver techniques for microwave digital radio," Part IV of this book. [An earlier version was published in *IEEE Commun. Mag.,* vol. 24, no. 11, Nov. 1986, pp. 43–54.]
[32] G. D. Martin, "Optimal convolutional self-orthogonal codes with an application to digital radio," in *ICC '85,* pp. 39.4.1–39.4.5, June 1985.
[33] M. Kavehrad, "Convolutional coding for high-speed microwave radio communications," *AT&T Tech. J.,* vol. 64, no. 7, pp. 1625–1637, Sept. 1985.

6GHz 135MBPS DIGITAL RADIO SYSTEM WITH 64 QAM MODULATION

T.NOGUCHI*, T.RYU*, Y.KOIZUMI*
S.MIZOGUCHI*, M.YOSHIMOTO*, K.NAKAMURA**

Microwave & Satellite Comm. Division*, C&C System Research Labs.**
NEC Corporation, Tokyo, Japan

ABSTRACT

The 64 QAM digital radio system for DMR application is discussed. Many new technologies have been developed to achieve this multi-level system, which is capable of transmitting three DS3 singals in the 6 GHz common carrier band at 135 MBPS. The 64 QAM modem, with a new joint control loop, IF band transversal equalizer, forward error correction technology and transmitter and receiver, is discussed with such things as BER performance, carrier recovery performance and signature.

1. INTRODUCTION

The development of digital modulation technology over the last few years has been aimed at the primary objective of achieving faster digital transmission rates. This objective of obtaining more bits per hertz has led to the development of 4 PSK, 8 PSK and 16 QAM systems which are now widely used for commercial digital radio service. Fig. 1 illustrates the Bell System's hierachy of digital modulation technologies and shows the bit rate, or channel capacity, possible in a given frequency band. This diagram indicates the next logical step from 16 QAM is the 64 QAM system, which more than 4.5 bits per hertz.
This paper discusses a newly developed 6 GHz 135 MBPS 64 QAM system.
To develop this system, the following problems must be overcome:

(1) complexity of the hardware,
(2) susceptability to nonlinear distortion,
(3) susceptability to frequency response distortion due to circuit imperfection and selective fading over the transmission path, and
(4) a high carrier to noise ratio.

The technologies developed to overcome these problems are:

(1) joint control of the AGC circuit, transversal equalizer and carrier recovery loop,
(2) an error correcting code with low redundancy,
(3) an IF band transversal equalizer, and
(4) a high power TWT amplifier with predistortor.

The application of these technologies makes it possible to attain 4.5 bits per hertz with a system gain of more than 100 dB at a BER of 10^{-6}

2. SYSTEM OVERVIEW

2-1 Basic Configuration

This section describes signal flow for each operation. Fig. 2 is a simple block diagram of the equipment. Three trains of asynchronous DS3 signals (44.736 MBPS) are converted at the B-U CONV into three unipolar trains, followed by synchronization in the TX DPU stuffing system. Frame synchronization bits, parity bits, digital SC bits and stuffing information bits are time-division-multiplexed at this point, and the six-train signal is speed converted. Two of these six tranis are differentially encoded at the DIFF ENCOD to solve phase ambiguity of the recovered carrier.

The two trains are then fed with the other four data streams to the CODE CONV where signal mapping is performed to minimize bit error rate. The output of the CODE CONV is split into three, and sent to two independent error correcting encoders (E.C. ENCOD). The two encoded signals, corresponding to in-phase and quadrature signals, are directed to the 64 QAM MOD, and converted into two 8-level signals. These signals are then separetely shaped to a roll-off spectrum characteristic by low pass filters and then converted into a 64 QAM IF signal.

On the receiving side, the modulated signal is fed to the adaptive and transversal equalizers where linear distortion is cancelled to minimize intersymbol interference. The modulated signal is then delivered to the 64 QAM demodulator where two baseband signals are coherently detected. These signals are converted into six data streams at the decision circuit by the recovered clock signal. Error signals are obtained in the same process for joint control of the AGC, carrier recovery and transversal equalizer loops.

The six signals are split into three, and sent to the error correction decoder. Code convertion is performed at the RX CODE CONV, followed by differential decoding. These signals are supervised by a frame synchronizer panel. The time-division multiplexed signal on the transmitting side is demultiplexed by the RX DPU; it undergoes destuffing and speed conversion from 6-train to 3-train, asynchronous, unipolar signals. The unipolar signals are then restored to their original bipolar signal form and fed to the output.

Reprinted from *IEEE Int. Conf. Comm.*, vol. 3, pp. 1472–1477, June 1983.

2-2 64 QAM Modulator/Demodulator

The modem is the key portion of the transmission system since the performance of the total system depends on it. Filtering, signal mapping and carrier recovery techniques are now discussed.

2-2-1 Filtering

The following conditions for suitable filtering must be considered:

1) conformance with the FCC rules and regulations,
2) immunity from adjacent channel interference,
3) minimum excess thermal noise, and
4) ability of circuits to minimize intersymbol inteference.

NEC has adopted a method involving the square root of Nyquist shaping at the transmitter and the receiver sides. Consequently, thermal noise and adjacent channel interference is reduced to a minimum. Fig. 3 and Fig. 4 respectively show the TX LPF characteristics and TX output power spectrum with the FCC mask.

2-2-2 Signal Mapping

Fig. 5 shows signal mapping in the phasor plane. Since the outermost signals are most susceptible to AM/AM and AM/PM distortion due to nonlinearity, the inter-code distance is designed to be as short as possible to optimize the bit error rate performance. The first two bits denote a quadrant signal which requires diffrential coding. The last four bits are allocated to be free from phase-ambiguity, and are represented as a decimal number corresponding to the Gray code in Fig. 5.

2-2-3 Carrier Recovery Technique

The quality of coherently detected signals depends on jitter components of the recovered carrier. Multi-level QAM systems theoretically require a high carrier to noise ratio; therefore, jitter must be suppressed sufficiently. In the newly developed method, two dimensional error signals, Ep and Eq, are generated by monitoring the eye aperture in the baseband. These signals jointly control the carrier recovery circuit, the transversal equalizer and the AGC circuit. Baseband signals can be represented by the equivalent baseband signal B.

When the phase error is θ, detected signals are expressed by $Be^{j\theta}$.

The error signal E between the input and output of the decision circuit is given by

$$E = Be^{j\theta} - D \quad \text{.......................... (1)}$$

Where D is the output of the decision circuit. Therefore,

$$\mathrm{Im}\{E \cdot D^*\} = \mathrm{Im}\ \{(Be^{j\theta}) \cdot D^*\} \quad \text{.......... (2)}$$

Where D^* is the complex conjugate of D.

If B = D, then:

$$\mathrm{Im}\{E \cdot D^*\} = |D|^2 \sin\theta \quad \text{.................. (3)}$$

Where, $|D|^2$ is the square of the absolute value of D and is constant in time average. This demonstrates that equation (2) is a model for the APC signal.

The right side of equation (2) corresponds to a convertional COSTAS loop. The left side is derived from the control algorithm for the transversal equalizer described in 2-4.
Fig. 6 shows the APC characteristics using this algorithm. There are fewer false lock points than in a conventional COSTAS loop.

To accomplish this, most associated circuits are digital. Furthermore, the developed circuitry is very small and easy to adjust. Photo 1 shows jitter distributed near the recovered carrier.

The D/U ratio is about 42 dB, and, the resulting signal quality degradation (at BER = 10^{-6}) caused by jitter is less than 0.5 dB in terms of C/N ratio.

2-2-4 Modulation and Demodulation

Fig. 7 is a block diagram of the modulator. Six data trains are converted by a PLS AMP into two eight-level baseband signals for the quadrature channels P and Q.
These signals are roll-off spectrum shaped by LPF's. The signals go to two linear mixers where a 70 MHz OSC supplies the quadrature local signals to produce an IF band QAM signal. This method is known as linear quadrature amplitude modulation.

The BPF suppresses harmful spurious components and delivers a signal to the IF AMP to obtain a normal output level.

Fig. 8 is a block diagram of the demodulator. After the IF signal is equalized by the ADP EQL and TRSV EQL, it is sent to the demodulator for quadrature coherent detection. The two baseband signals obtained are filtered in LPF's for whole roll-off spectrum shaping and excess thermal noise elimination. The receiver LPF is almost identical to that of the transmitter.

The LPF output is forwarded to a decision circuit to regenerate six data trains using a recovered clock signal (sampling). A two dimensional error signal is generated by taking the difference of the input and output of the decision circuit. This signal controls the TRSV EQL and a carrier recovery circuit.

Fig. 9 shows BER performance of the modem back-to-back; Table 1 and Talbe 2 show the characteristics of the modulator and demodulator.

Service channel (SC) signals can be simultaneously transmitted by frequency modulation of the QAM signal. The SC signal is obtained by filtering the APC signal of the carrier recovery circuit.

2-3 Forward Error Correction with a Low Redundancy Rate

Even an ideal 64 QAM system, requires a high power transmitter since a carrier-to-noise ratio greater than 26 dB is needed for a BER less than 10^{-6}. Nonlinearity and frequency response variation along the transmission path also affetct the system, possibly causing dribble error. Forward error correction (FEC) may be useful to eliminate these problems.

The application of FEC codes has been limited because the conventionally used codes have had high redundancy rates.

FEC codes have not been used since band limitation is strict and in DMR band efficiency is important.
To meet these requirements, a new FEC code with a low redundancy rate is used in the 64 QAM system. The FEC is a "transparent single lee-error correcting (SLEC) code".

The main features of the FEC code used in 64 QAM are:

(1) application of an independent octal SLEC code for each in-phase and quadra-phase data stream.
(2) use of FEC in a modem without differential encoding-decoding to avoid the effect of double errors from differential coding. FEC is transparent to recovered carrier phase ambiguity.
(3) an FEC coding rate of 81/84; three redundancy symbols are included with eighty-one information symbols for 96.43% efficiency.
(4) correction of all single Lee-errors and 50% of double Lee-errors.

Symbol error rate after correction is given by the formula:

$$Pc = 1.5N\,Ps^2$$

where Pc = Symbol error rate after error correction
Ps = Symbol error rate before error correction
N = Code length (=84)

Therefore, Pc is approximately 1×10^{-6} when Ps is 1×10^{-4}.
The measured FEC coding gain is 3.3 dB at a BER of 10^{-6}, which is good enough.

2-4 Transversal Equalizer

Multi-level modulation systems are susceptible to frequency response fluctuations along the transmission path. To maintain high transmission quality of digital signals, either frequency domain or time domain adaptive equalizers are used.

A five-tap transversal equalizer has been developed for the IF stage of this 64 QAM system. Fig. 10 is a block diagram of the filter.

In-phase distortion is equalized by R_{-2}, R_{-1}, R_o, R_1 and R_2 and quadrature phase distortion by I_{-2}, I_{-1}, I_o, I_1 and I_2. The center taps (R_o, I_o) generate the AGC and APC signals.

The control algorithm is given by the equation:

$$Cm^{i+1} = Cm^{i} - \alpha\ \mathrm{sgn}\ (Ei)\ \mathrm{sgn}\ (B^{*}_{i-m}\ \mathrm{EXP}\ (jm\omega\tau)) \quad \ldots\ldots\ldots\ldots\ldots (7)$$

Where Cm^i = m-th tap coeffient at time slot i
α = constant value
Ei = error signal at time slot i
B_{i-m} = complex conjugate of input data at time slot i-m
ω = carrier frequency offset
τ = delay time at delay CKT

and on the condition:

$$\mathrm{sgn}\ (D^{*}_{i-m}) = \mathrm{sgn}\ (B^{*}_{i-m}\ \mathrm{EXP}\ (jm\omega\tau))$$

Equation (7) can be reduced to:

$$Cm^{i+1} = Cm^{i} - \alpha\ \mathrm{sgn}\ (Ei)\ \mathrm{sgn}\ (D^{*}_{i-m})$$

where D^{*}_{i-m} = complex conjugate of the regenerated data at time slot i-m.

When m equals zero, the tap coefficient corresponds to the AGC CKT control signal and phase control signal of the carrier recovery loop. This algorithm is found by integrating the correlation between data signals and error signals, over the time domain.

The transversal filter is located at the IF stage for easy circuit design and isolation from digital baseband circuitry. Performance of the transversal equalizer is evaluated from its signature. The signature for a 2-ray model of the 64 QAM system with a path difference of 6.3 ns is shown in Fig. 11. Photo 2 and photo 3 respectively show the unequalized and equalized eye diagrams for a notch depth of 6.7 dB in the same model. The conclusion from Yoshida, Guiffrida and Fig. 11 is that performance of a fully equipped 64 QAM system with space diversity may be superior to a 16 QAM system with space diversity and ADP EQL.

2-5 Transmitter/Receiver

As the modulation level increases, the QAM syst becomes less immune to nonlinearity. However, excessively reducing the transmitter output to maintain linearity sacrifices system gain. To cope with this difficulty, NEC has adopted a predistorter for the TWT amplifier. Optimal system gain is achieved with the output reduced to approximately 10 dB.

The TWT used in the experimental system had a saturated power of 40 watts. A system gain of up to 103.4 dB was achieved, when FEC was used.

Fig. 12 shows relative system gain vs. output reduction. Fig. 13 shows BER performance of a TX/RX loop back, and photo 4 shows TX output power spectrum vs. output reduction. Note that the configurations of the TX filter and the entire receiver are the same as in NEC 6G-90MB 16 QAM equipment.

3. Performance

Performance degradation allocation and a break down of system gain for the 6 GHz 64 QAM system are shown in Table 3 and Table 4. Table 4 shows results using alternately on experimental 40-watt TWT and a 20-watt TWT now undergoing development.

4. Conclusion and Acknowledgement

This paper has described a new 6 GHz 64 QAM system. Many recent technologies have been applied to this system: joint control of the AGC CKT, transversal equalizer and carrier recovery loop, error correction code with low redundancy, IF band transversal equalizer, and high power TWT amplifier with predistortor. System performance is very satisfactory. These technologies can be applied to other 64 QAM systems such as, 4 GHz 90 MBPS and 11 GHz 180 MBPS.
The authors are especially grateful to Messrs. K. Kinoshita, Y. Tagashira and Y. Matsuo for their advice and encouragement.

REFERENCES

[1] Y. Tan, et al. "The 8-level PSK Modem with Cosine Roll-off Spectrum for Digital Microwave Communications," ICC'76, pp.29-13 to 29-18.

[2] I. Horikawa, et al. "Characteristics of a High Capacity Digital Radio System on a Multipath Condition," ICC'79, pp.48.4.1 to 48.4.6.

[3] Y. Yoshida, et al. "6G-90Mbps Digital Radio System with 16 QAM Modulation," ICC'80, pp.52.4.1 to 52.4.5.

[4] Y. Saito, et al. "5L-D1 Digital Radio System," ICC '82, pp.28.1.1 to 28.1.7.

[5] F. Akashi, et al. "A High Performance Digital QAM 9600 bit/s Modem," NEC R & D, April 1979, No.45, pp.38 to 49.

[6] D.D. Falconer. "Joint Adaptive Equalization and Carrier Communication Systems," BSTJ, March 1976, pp.317 to 334.

[7] K. Nakamura. "A Class of Error Correcting Codes for DPSK Channels," ICC'79, pp.45.4.1 to 45.4.5.

[8] T. Murase, et al. "200 MB/s 16 QAM Digital Radio System with New Contermeasure Techniques for Multipath Fading," ICC'81, pp.46.1.1 to 46.1.5.

[9] T.S. Giuffrida and W.W.Toy. "16 QAM and Adjacent Channel Interference," ICC'81, pp.13.1.1 to 13.1.5.

Table 1 MODULATOR CHARACTERISTICS

Modulator		
IF Frequency	(MHz)	70
Symbol Rate	(MB)	23.4
Phase Errors	(deg)p-p	1.5
Amplitude Error	(dB) p-p	0.2

Table 2 DEMODULATOR CHARACTERISTICS

Demodulator		
Carrier Jitter Recovered	(dB)	42
IF Frequency	(MHz)	F_o = 70
Pull-in Range of Carr. Recov. Loop	(MHz)	F_o±0.4
Lock Range of Carr. Recov. Loop	(MHz)	F_o±1.5

Table 3 DEGRADATION ALLOCATION

Degradaton (or Improvement) factor	C/N Degradation (dB) at BER=10^{-6}
1) Phase and Amplitude Errors of the MOD at the Sampling Point	0.8
2) Intersymbol Interference by Spectrum Shaping Error	0.8
3) Carrier Jitter Recovered	0.3
4) Other Impairment	0.6
* Total Degradation for MODEM	2.5
5) TWT Amplifier Non-Linearity	0.9
6) RF Filtering in Transmitter	0.5
Improvement by Error Correcting	-3.3
* Total Degradation of TR, MODEM (E/W E.C)	0.6
7) Temperature Variance Over The 0 to 50°C Range (Overall)	1.5

Table 4 64 QAM RADIO SYSTEM GAIN

Item	6-GHz 2016 CH	
1) Modulation	64 QAM	
2) Roll Off Factor (α)	0.4	
3) Output Power (at TX FIL OUT)	1.5 W* (TWT)	3 W (TWT)
o Saturated Output	20 W*	40 W
o Output Back Off	10 dB*	10 dB
4) Receiver Threshold (at BER = 10^{-6})	-68.4 dBm	
o Transmission Bandwidth	23.4 MHz	
o Noise Figure (Including RX FIL Loss = 1 dB)	3.5 dB (4.5 dB)	
o Normalized C/N (at BER = 10^{-6})	27.2 dB	
System Gain	100.4 dB*	103.4 dB

(* under development)

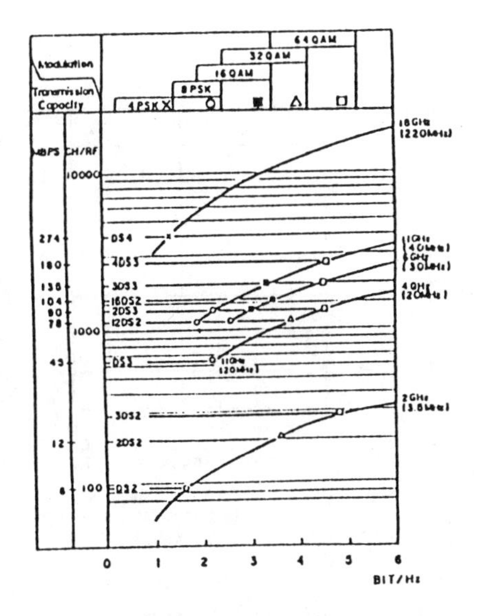

Fig. 1 THE TREND OF MULTI-LEVEL IN NORTH AMERICA

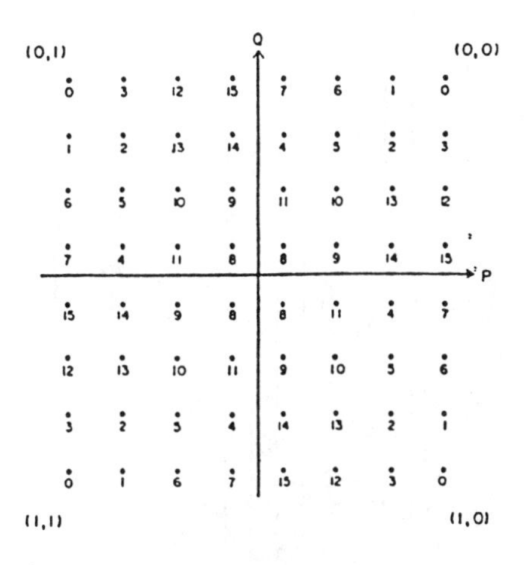

Fig. 5 SIGNAL MAPPING

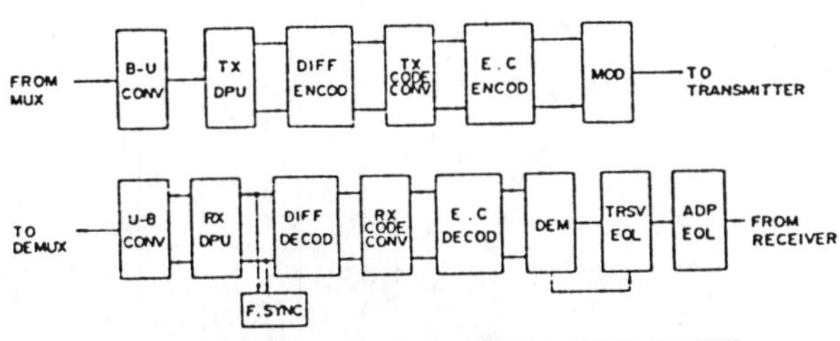

B-U/U-B CONV ; BIPOLAR to UNIPOLAR / UNIPOLAR to BIPOLAR CONVERTER.
TX/RX DPU ; TRANSMIT / RECEIVE DATA PROCESSOR UNIT.
DIFF ENCOD/DECOD ; DIFFERENTIAL ENCODER / DECODER.
TX/RX CODE CONV ; TRANSMIT/RECEIVE CODE CONVERTER.
E.C ENCOD/DECOD ; ERROR CORRECTING ENCODER / DECODER.
MOD/DEM ; 64QAM MODULATOR / DEMODULATOR.
F.SYNC ; FRAME SYNCHRONIZER.
TRSV EQL ; TRANSVERSAL EQUALIZER.
ADP EQL ; ADAPTIVE EQUALIZER.

Fig. 2 MODEM EQUIPMENT

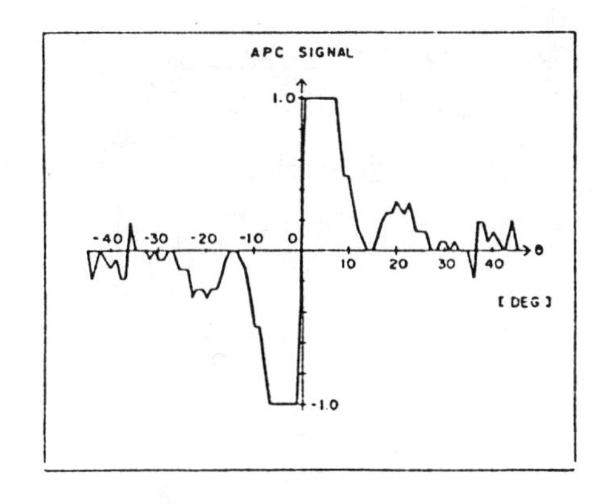

Fig. 6 APC CHARACTERISTICS

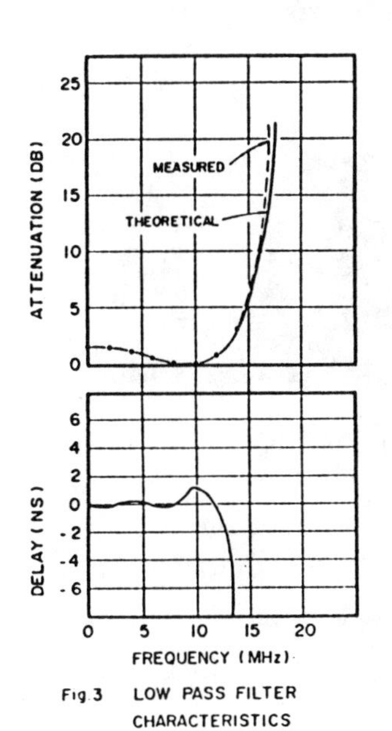

Fig.3 LOW PASS FILTER CHARACTERISTICS

Fig. 4 OUTPUT POWER SPECTRUM VS. FCC SPECIFICATION

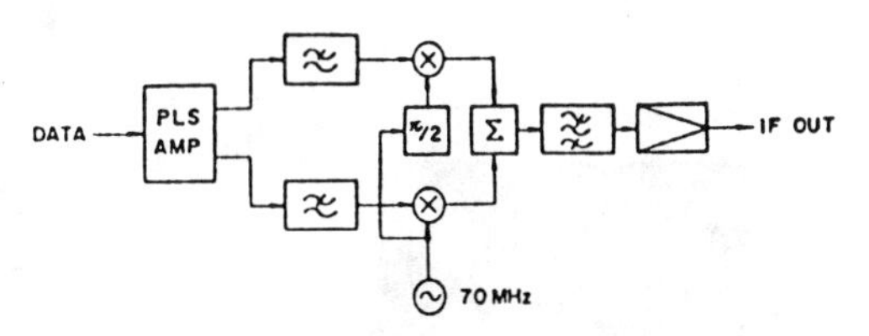

Fig. 7 MODULATOR

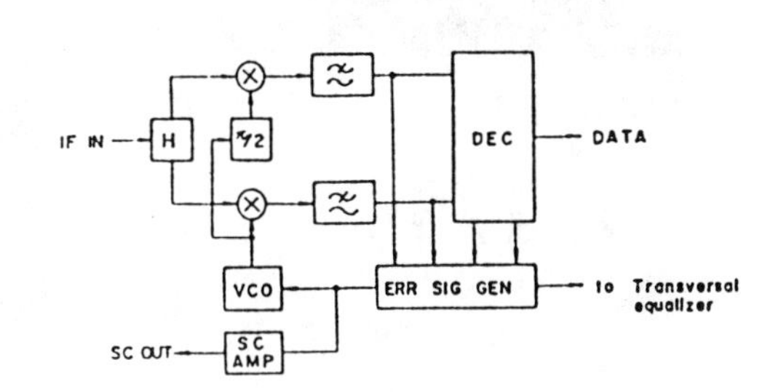

Fig. 8 DEMODULATOR

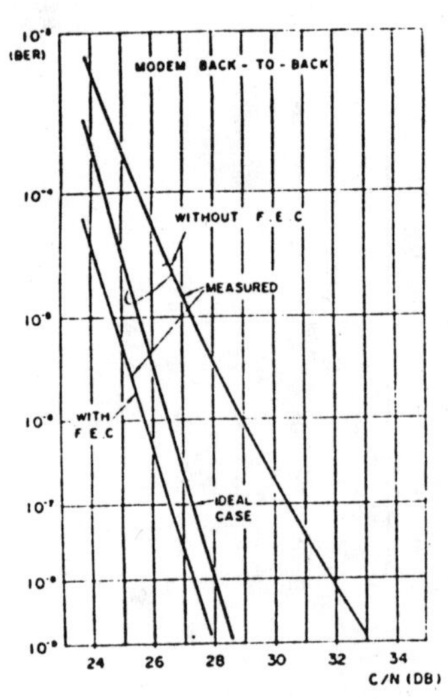

Fig. 9 BER PERFORMANCE

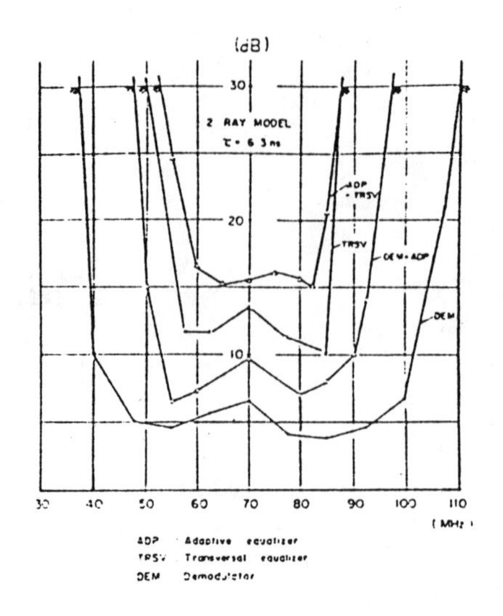

Fig. 11 SIGNATURE OF 64 QAM SYSTEM

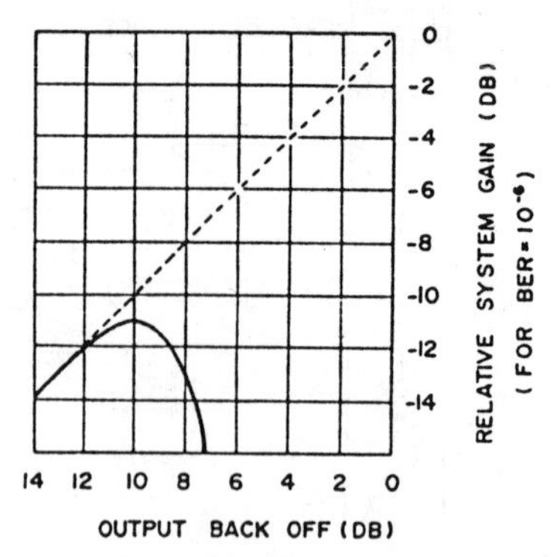

Fig. 12 SYSTEM GAIN REDUCTION DUE TO TRANSMITTER NONLINEARITY

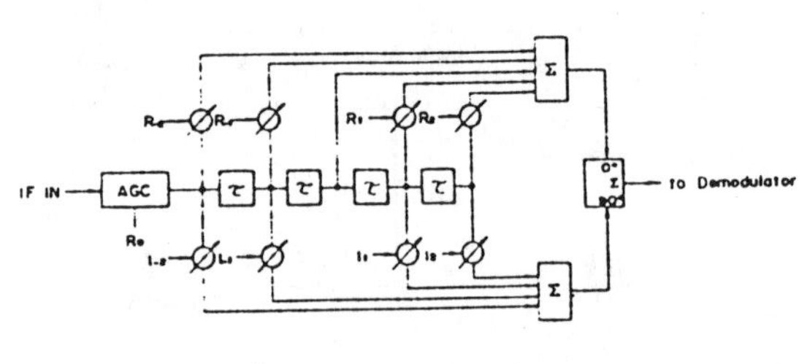

Fig. 10 TRANSVERSAL FILTER

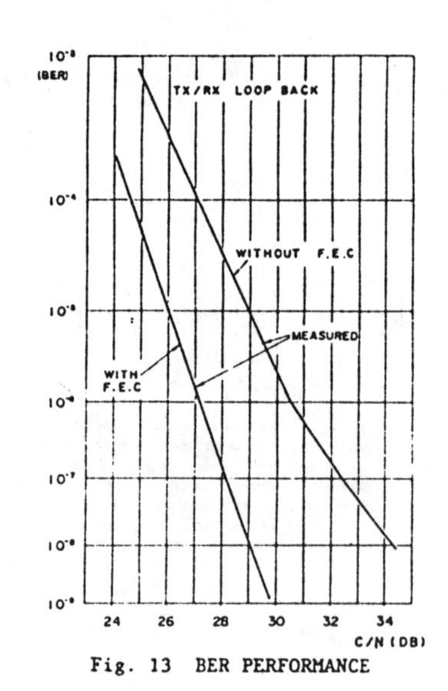

Fig. 13 BER PERFORMANCE

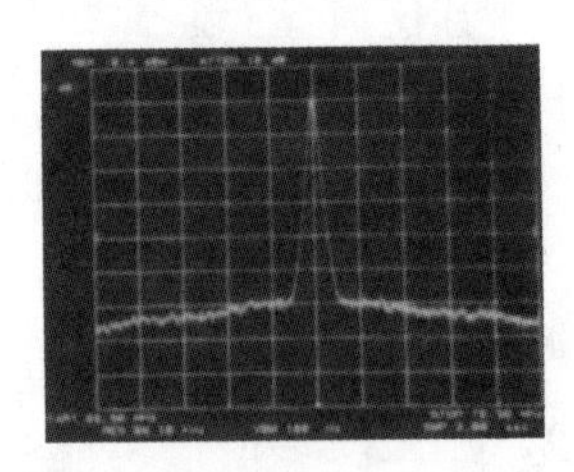

V: 10 dB/DIV
H: 200 kHz/DIV
RES.BW: 10 kHz

Photo 1 JITTER OF RECOVERED CARRIER

Photo 2 UNEQUALIZED EYE DIAGRAM

Photo 3 EQUALIZED EYE DIAGRAM

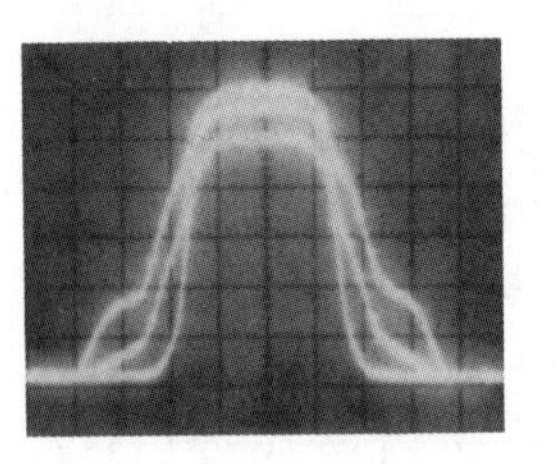

V: 10 dB/DIV
H: 10 MHz/DIV
FROM TOP TO BOTTOM
0, 3, 10 dB

Photo 4 TX POWER SPECTRUM vs. BACK OFF

DR 6-30-135 SYSTEM DESIGN AND APPLICATION

C. P. Bates, W. G. Robinson, III, and M. A. Skinner

AT&T Bell Laboratories
North Andover, Massachusetts 01845

ABSTRACT

A new digital radio system, DR 6-30-135, has been developed to provide 135-Mb/s capacity (three DS3, 2016 VF channels) in a 30-MHz channel within the 6-GHz common carrier band. This paper describes salient features and design considerations. The system configuration is also covered along with operational and maintenance features. In addition, impairments are examined in order to establish the quality of performance expected in real-world application.

INTRODUCTION

A new digital radio system, DR 6-30-135, has been developed to provide 135-Mb/s capacity (three DS3, 2016 VF channels) in a 30-MHz channel within the 6-GHz common carrier band (see Fig. 1 and 2). A very effective modulation format, 64 QAM, has been selected to obtain 4-1/2 bps per hertz, primarily because of the negligible impact on adjacent RF channels. Hence, full route development of eight RF channels is possible.

This system uses the latest available technology. Hybrid integrated circuits are used extensively in baseband and intermediate frequency amplifiers. GaAs FET devices are used in low-noise amplifiers, as well as in optional power amplifiers. Barium titanate material is used for RF filters and dielectric resonator local oscillators. In addition, microwave hybrid integrated circuit technology is employed and permits compact circuits with minimal adjustment and high reliability.

Some of the major features are:

a. Compact physical design
b. Complete front access
c. Adaptive transversal equalization
d. Per-hop and per-section cyclic redundancy check bits
e. Performance monitoring with error rate reporting from 10^{-3} to 10^{-9}
f. Digital service channels
g. Space diversity with IF combining and adaptive slope equalizer
h. Intelligent microprocessor-controlled terminals and regenerators
i. Errorless line switching

SYSTEM DESIGN CONSIDERATIONS

Modulation Technique

The choice of a modulation scheme for this new digital radio system had to take into account many important and sometimes co flicting constraints. Of particular significance is RF interference. It must be controlled to ensure the

FIGURE 1 - DR 6-30-135 DIGITAL TERMINAL

Reprinted from *IEEE 3rd Global Telecomm. Conf.*, vol. 1, pp. 539–546, Nov. 1984.

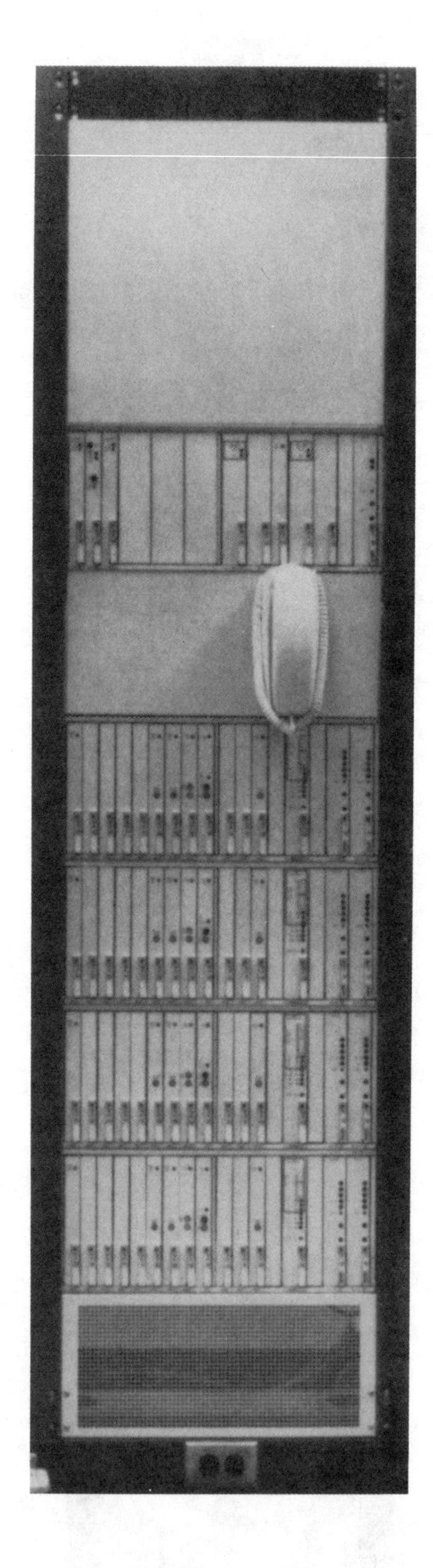

FIGURE 2 - DR 6-30-135 DIGITAL REGENERATOR

compatibility of the system with itself and with other radio systems and to ensure compliance with FCC regulations.

In general, higher order modulation schemes result in smaller bandwidths, thus reducing adjacent channel interference. On the other hand, in the most simplistic comparison, a multiple-level signal is more sensitive to thermal noise because of the closer spacing between states. This is mitigated to an extent by the narrower noise bandwidth used at the receiver. Such tradeoffs are listed in Table I for some of the more popular modulation schemes. The condition for this table is a specified bit rate. The "Relative State Separation" is the reduced margin incurred with the various modulation schemes relative to a simple binary 2-phase system. The "Relative Noise Bandwidth" is the reduction of the noise power by the narrower bandwidth permissible. The "Relative Signal" is the resulting increased average signal power needed, relative to a 2-phase system, for a comparable error performance.

TABLE I

COMPARISON OF KEYING SYSTEMS FOR FIXED BIT RATE

Modulation	Relative State Separation (dB)	Relative Noise Bandwidth (dB)	Relative Signal (dB)
2ϕ PSK	0 (REF)	0 (REF)	0 (REF)
4ϕ PSK	-3.0	-3.0	0
8ϕ PSK	-8.3	-4.8	+3.5
16ϕ PSK	-14.2	-6.0	+8.2
16 QAM	-10.0	-6.0	+4.0
32 PSK	-20.2	-7.0	+13.2
64 QAM	-16.2	-7.8	+8.4

For a design objective of 135 Mb/s in a 30-MHz channel, with appropriate filtering, a modulation scheme of higher complexity than 16 QAM is required. From comparisons such as those of Table I, it can be concluded that 64 QAM shows minimal noise degradation penalty, compared to other high-order modulation schemes. In addition, 64 QAM, because of its narrower bandwidth, should be less vulnerable to multipath fading and should have less impact on adjacent digital or analog channels.

The important parameters for the DR 6-30-135 Digital Radio System are presented in Table II.

TABLE II

IMPORTANT PARAMETERS FOR THE DR 6-30-135 DIGITAL RADIO SYSTEM

Overall bit rate	136.603 Mb/s
Modulation format	64 QAM
Symbol rate	22.767 Mbaud
Raised cosine spectrum roll-off	45 percent
Transmitter/Receiver filtering split	50/50
Transmitter power (antenna port)	3.5 watts
Noise figure	4.0 dB
IF frequency	70 MHz
Receiver threshold (typical at BER = 10^{-3})	-72 dBm
Receiver threshold (typical at BER = 10^{-6})	-67 dBm
Typical system gain at BER = 10^{-3}	107 dB
Typical system gain at BER = 10^{-6}	102 dB
Emission designation	30,000 A9Y

Interference Compatibility

With a signaling rate of 22.767 Mbaud, DR 6-30-135 yields a controlled transmitter spectrum to fit into 30 MHz. Baseband raised cosine Nyquist filtering (roll-off factor of 0.45) with half the filtering at the transmitter and half at the receiver is used. This highly selective receiver filter controls interference from adjacent channels. The baseband transmit filter, along with stringent control of the nonlinear performance of the transmitter (at the specified 3.5-watt output) and modest additional RF filtering, controls the transmitter spectrum (Fig. 3).

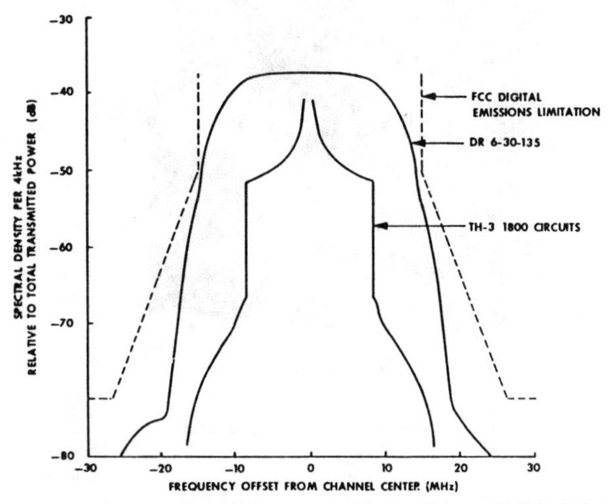

FIGURE 3 - DR 6-30-135 AND TH-3 SPECTRA

The isolation between a DR 6-30-135 transmitter and an adjacent channel (30-MHz spacing) DR 6-30-135 receiver, due to spectral shape and filter alone, is about 34 dB. Even without the aid of cross-polarization discrimination, the S/N degradation this would contribute at an error rate of 10^{-6} is less than 1 dB. If the interferer adjacent to DR 6-30-135 were a TH-3 transmitter, 1800 voice circuits (FCC emission 28,000 F9Y), the DR 6-30-135 receiver would provide approximately 65-dB rejection of the FM signal. Thus, DR 6-30-135 is quite resistant to adjacent channel interference from either DR 6-30-135 or TH-3.

The spectrum in Fig. 3 shows that DR 6-30-135 has more energy near the edge of the channel than TH-3, 1800 voice circuits, does. The C/I requirement for DR 6-30-135 adjacent channel interference into TH-3, 1800 voice circuits, at 30-MHz channel spacing is 42 dB for a 4-dBrnc0 exposure. The transmitter power for TH-3 is 10 watts. However, DR 6-30-135 with its low-noise receiver can achieve excellent system gain (Table II) with a transmit power of 1 watt. Accounting for this transmit power difference, the XPD required is only 32 dB. Therefore, if adjacent channel operation of TH 1800 is desired at a 4-dBrnc0 exposure, DR 6-30-135 operated at 1 watt will meet this goal.

SYSTEM ARCHITECTURE

DR 6-30-135 has three major equipment units: a 135A line terminal bay, an IF/RF radio bay, and a 135A regenerator bay. These arrangements are shown in Fig. 4. All signal interconnections between these units are at an intermediate frequency of 70 MHz.

Two protection arrangements are available with DR 6-30-135 which automatically protect against equipment failure:

1. 1 x 1 hot standby
2. 1 x N frequency diversity

These are both available with space diversity to protect against propagation anomalies.

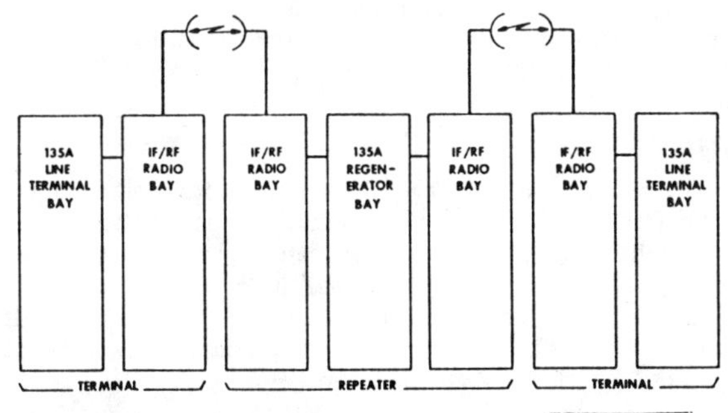

FIGURE 4 - TYPICAL BAY ARRANGEMENT

135A Line Terminal Bay

The 135A line terminal bay incorporates digital terminals and line protection switching. A digital terminal transmitter receives three asynchronous bipolar DS3-rate bitstreams and produces a modulated IF carrier at 70 MHz. A digital terminal receiver in turn performs the corresponding inverse functions with circuitry somewhat more complex because of carrier recovery, timing recovery, and the need for framing functions.

Radio line protection switching is also integrated into the 135A line terminal bay. Transfer of service to the protection channel occurs when frame is lost or when the error rate exceeds 10^{-6} as established by cyclic redundancy check bits. Line switch operation is errorless during switching whether initiated manually or automatically. It is available for both 1 x 1 hot standby and 1 x 1 frequency diversity (growable to a 1 x 7 frequency-diversity arrangement). A 135A line terminal bay is shown in Fig. 1.

Radio Transmitter/Receiver Bay

The IF/RF DR 6-30-135 radio bay interconnects at an IF frequency of 70 MHz (block diagram shown in Fig. 5). At a terminal station, 70-MHz inputs and outputs are connected to a 135A line terminal bay. At a repeater, the inputs and outputs are connected to a DR 6-30-135 regenerator bay. The radio bay design is described in detail in a companion paper at this conference (Ref. 1).

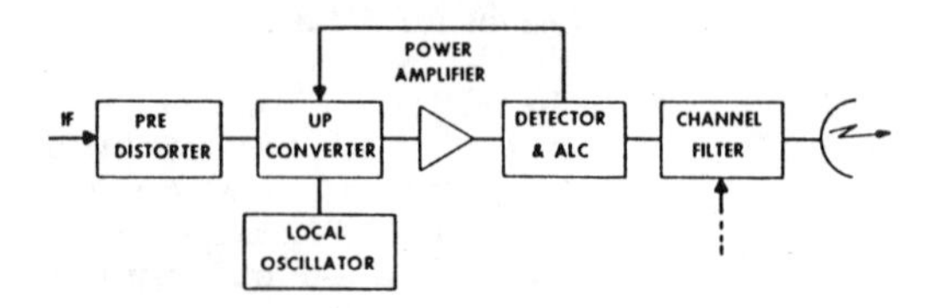

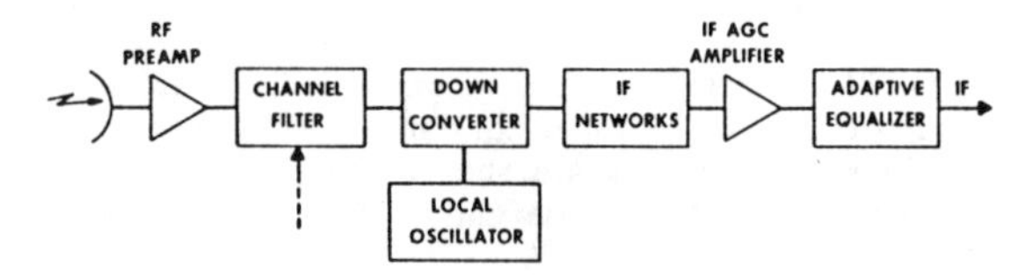

FIGURE 5 - DR 6-30-135 IF/RF RADIO BLOCK DIAGRAM

The radio bay, as shown in Fig. 6, contains four transmitters and four receivers in a 7-foot frame that is 23-5/8 inches wide by 10-1/4 inches deep. Very few adjustments are required, and maintenance access is from the front.

On hops which require improvement against multipath fading, space diversity can be added at IF to combine received signals from two antennas, using a specific algorithm based on their relative amplitude and phase.

Regenerator Bay

At a repeater station, regeneration of the transmitted signal occurs to preserve maximum immunity to accumulated noise and other distortions. A repeater site contains two regenerators, serving one channel in each direction with protection of transmission. This bay is shown in Fig. 2. It has been possible to design this configuration into a 7-foot bay, 23-5/8 inches wide by 10-1/4 inches deep.

DIGITAL TERMINALS

Digital terminals provide appropriate functions that allow for synchronizing, encoding, modulating input data information, and performing associated inverse functions to suitably reproduce the desired signal.

Digital Terminal Transmitter

The block diagram of the digital terminal transmitter is shown in Fig. 7. A radio line digital processing unit receives three bipolar signals. They are synchronized and added to overhead and control bits forming six synchronized "rails" each having a bit rate of 22.767 Mb/s. Overhead bits contain stuffing indicator, framing, cyclic redundancy check bits, and service channel information bits. DR 6 cyclic redundancy check bits are unique to the DR 6 signal and are used for error-rate measurements on each hop and are completely independent of the DS3 parity bits. The composite service channel signal is in the form of a 384-kb/s serial bitstream generated within a separate service channel multiplex unit.

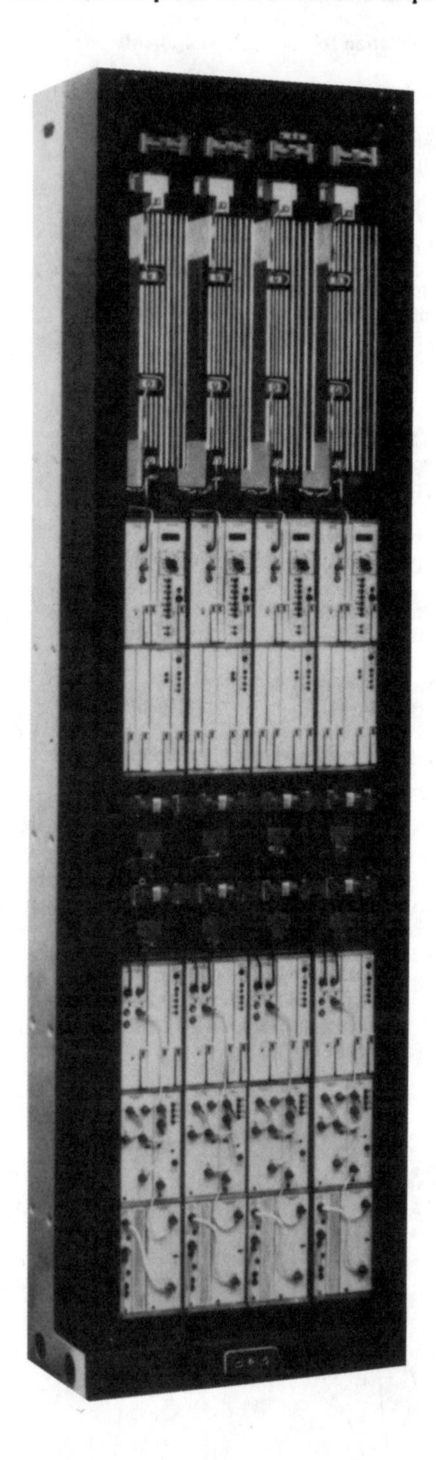

FIGURE 6 - IF/RF RADIO BAY

Within the transmitter these synchronized rails undergo a number of steps of digital processing, such as scrambling, serial-to-parallel conversion, and quadrant encoding. Finally, each rail is filtered and applied to balanced mixers in a modulator to form a 64-QAM modulated IF signal.

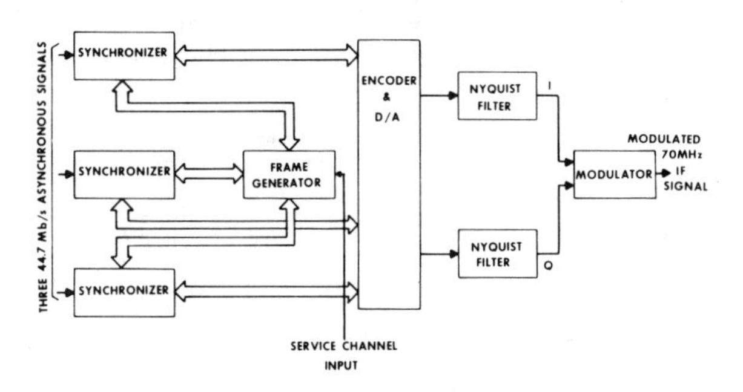

FIGURE 7 - 64QAM DIGITAL TERMINAL TRANSMITTER (135Mb/s, 3 DS3's)

Digital Terminal Receiver

The block diagram of the digital terminal receiver is shown in Fig. 8. The demodulator, using the recovered 70-MHz reference carrier, outputs the two quadrature baseband rails. These signals are low-pass filtered, modified by an adaptive baseband transversal equalizer, and fed to the decision circuitry for regeneration and timing recovery. The regenerated data rails are quadrant decoded, fed to the framer circuitry, and then desynchronized to provide the original DS3 signals. Digital service channel signals are recovered from the framers.

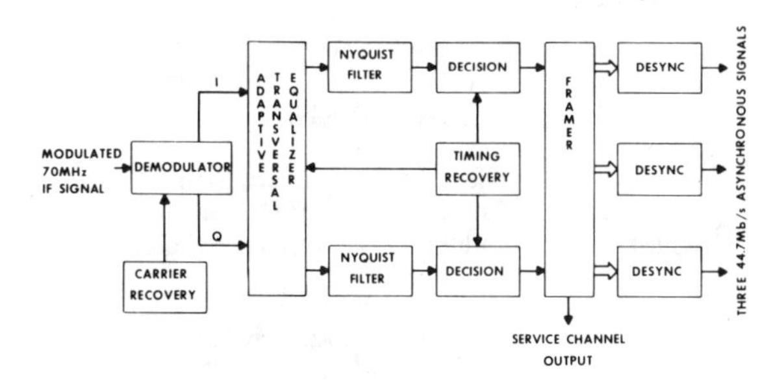

FIGURE 8 - 64QAM DIGITAL TERMINAL RECEIVER (135Mb/s, 3DS3's)

REGENERATOR

Figure 9 shows the block diagram of the 135-Mb/s regenerator. Since the DS3 signal is not required at a repeater, some simplification of the digital terminal equipment previously described is possible.

A regenerator signal from the demodulator is applied to a framer circuit which checks the DR 6 cyclic redundancy check bits, generates error information and misframe indications, and extracts and inserts service channel information bits. The resulting signal is applied to a 64-QAM modulator.

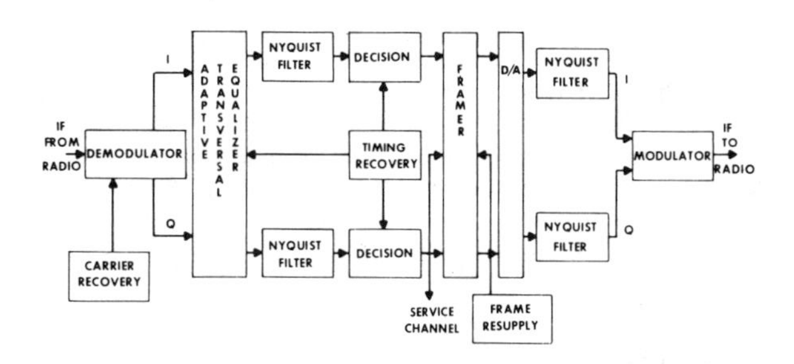

FIGURE 9 - DIGITAL REGENERATOR

CIRCUIT DESCRIPTION

The design of the 64-QAM modem has been greatly simplified by the use of integrated baseband and IF processing amplifiers (known as SAMs, for semiconductor amplifier modules). Figure 10 is a photograph of the 64-QAM transversal equalizer showing the compact design that results by the use of these highly reliable integrated amplifiers.

Self-adjusting circuitry used throughout the design results in a modem which requires no field adjustments. For example, the eye slicing circuits use the output statistics of the data in a novel circuit to provide highly accurate threshold voltages which are independent of operating temperatures, absolute signal variations, and manufacturing tolerances.

The carrier recovery uses a decision-directed feedback method which requires no manual adjustment of the recovered carrier reference phase, which can be a source of difficulties. Thus, the QAM modem has been designed to be both highly reliable and self-adjusting for trouble-free operation.

FIGURE 10 - TRANSVERSAL EQUALIZER BOARD

Figure 11 is a photograph of the frame generator and scrambler board used to align and provide overhead bit insertion into the DS3 signals. The use of multilayer printed wiring boards has resulted in both excellent digital waveform control with improved timing margin and provides for reduced spurious radiation and cross-coupling of the various clock and data signals into the transmitted IF spectrum. Additionally, the use of screened and "burned in" integrated circuits results in improved reliability.

FIGURE 11 - FRAME GENERATOR AND SCRAMBLER BOARD

OPERATIONS AND MAINTENANCE FEATURES

As with other transmission facilities, increased significance is being attributed to operation and maintenance features of digital radio systems. They are designed to reduce maintenance costs and to improve service by use of operations support systems located at maintenance centers. For example, all switch, test, and reset operations that can be done locally are executable remotely from a maintenance center. This capability allows corrective maintenance to proceed without the need for personnel at terminal stations.

To improve trouble isolation, most equipment plug-ins which are field replaceable have built-in test circuits to indicate whether or not the unit is functioning properly. Local alarm and status indications from these units are first conditioned and then transferred on a station-by-station basis to centralized telemetry equipment via a single data link. In addition, modern digital radio systems feature continuous, in-service monitoring of their transmission quality (Ref. 2). This assures that the maintenance center is able to determine at all times the performance of the radio systems. The arrangement of a system with performance monitoring is shown in Fig. 12. This monitoring is accomplished by microcomputer analysis and storage of cyclic redundancy check and frame data for the previous 24 hours. Thus, low bit error rates and brief intermittent failures that may not have been directly reported by built-in test equipment are made available to the maintenance center for scrutiny. Performance monitoring is performed at each station, allowing the precise determination of the faulty transmitter-receiver pair in a multihop system. A summary of a modern maintenance philosophy of a digital radio system, both with and without performance monitors, is given in Table III.

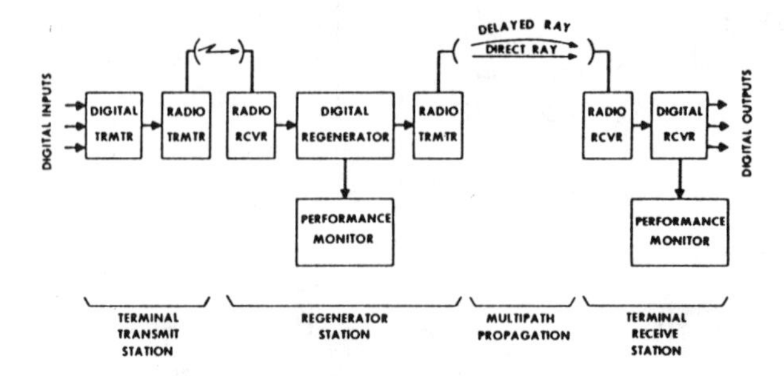

FIGURE 12 - DIGITAL RADIO SYSTEM

As part of maintenance support features, self-contained voice-frequency auxiliary channels may be made available. Such auxiliary channel facilities are, in essence, party-line communications connections. Multifrequency pushbutton signaling is possible, as well as connection to a central office line in order to provide both dial-in and dial-out capability to the direct distance dialing network. Four additional service channels can be added for a variety of customer data transmission needs.

TABLE III

TROUBLE CLEARING PROCEDURES

	Type of Trouble		
	Clear, Simple Failure	Intermittent or Degraded Performance Failure	
Task		Without Performance Monitors	With Performance Monitors
Detection	Clear, simple alarms	Customer complaints Switching machine reports Test set data	Remotely reported alarms
Isolation	Clear, simple telemetry	Local use of test sets	Remote telemetry procedures
Verification	Usually obvious	Local use of test sets	Remote telemetry procedures

SYSTEM SPECIFICATION

Stressed Bit Error Rate Performance

The typical performance of a DR 6-30-135 Digital Radio System has been presented in Table II. These data are in the form most appropriate for a user, including, for example, system gain parameters. It is of interest to examine error performance versus the signal-to-noise ratio (after the baseband filter). This is shown in Fig. 13 for a typical system. Also plotted on this figure is the theoretical curve. The difference is the degradation including modulator/demodulator and IF/RF radio equipment. The 23.5-dB S/N ratio at a BER of 10^{-3} translates to a system gain of 107 dB with a 4-dB noise figure and +35.5 dBm transmitter power.

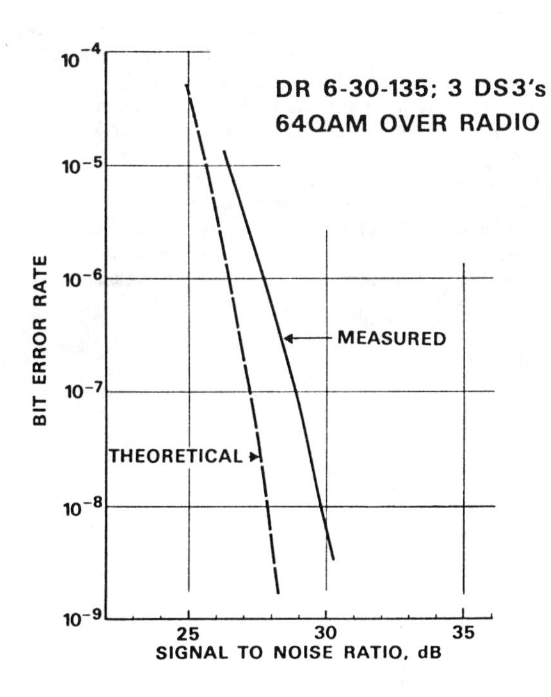

FIGURE 13 - DR 6-30-135 DIGITAL RADIO BER CHARACTERISTICS

Equipment Signatures For 64 QAM

A 64-QAM modulation scheme is more sensitive to dispersion caused by multipath fading than a 16-QAM modulation scheme. The adaptive transversal equalizer is the key technology for mitigating this effect.

The ability of a system to withstand multipath fading is determined by its equipment signature. The measured equipment signatures for 16- and 64-QAM systems equipped with an adaptive slope equalizer only are shown in Fig. 14. The outage, which is proportional to the linear area (Ref. 3) under the equipment signature in Fig. 14, is much greater for the 64-QAM system. For the examples given, the 64-QAM system would have seven times more outage.

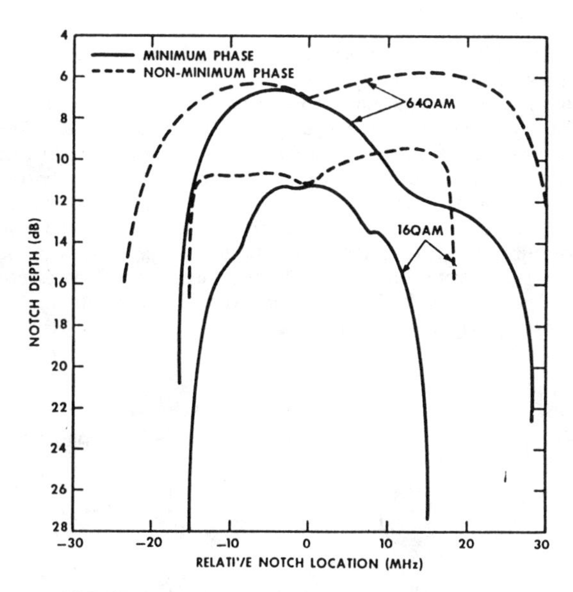

FIGURE 14 - EQUIPMENT SIGNATURES FOR 16QAM AND 64QAM WITH ASE

We have found that the application of a transversal equalizer will significantly improve the performance. The comparison of a 16-QAM system equipped with an adaptive slope equalizer only and a 64-QAM system equipped with an adaptive slope equalizer and a transversal equalizer is shown in Fig. 15. The adaptive transversal equalizer technology permits the higher modulation 64-QAM system to perform as well or better than the 16-QAM system equipped with slope equalization only.

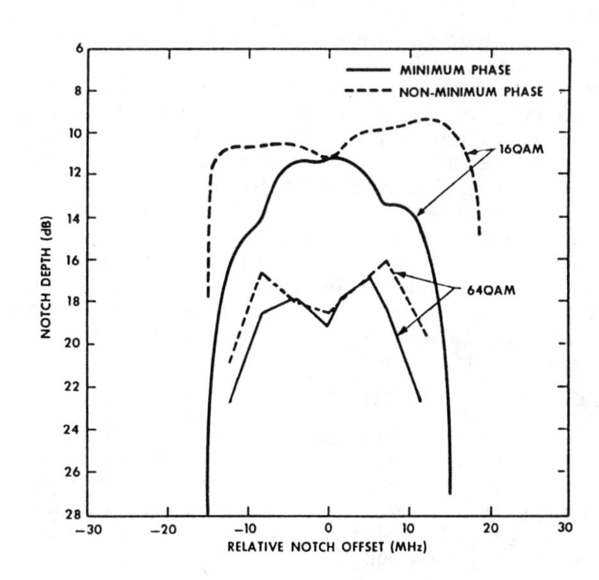

FIGURE 15 - EQUIPMENT SIGNATURES FOR 16QAM WITH ASE AND 64QAM WITH ASE + ATE

OTHER DESIGN CONSIDERATIONS

In addition to a system's ability to perform with dispersive inputs and have an appropriate flat fade margin, it must also have a low background error rate. This is controlled by having an adequate eye opening as defined in Fig. 16. The approach used is to determine the eye opening at an error rate with a reasonable measuring time interval (10^{-8}, for example) and then to use it to predict the expected long-term performance. Our experience has been that if the eye opening at a BER of 10^{-8} is greater than 20 percent, then one can guarantee a long-term background error rate (unstressed) of less than 10^{-10}. This performance will be suitable for the most demanding digital data customers.

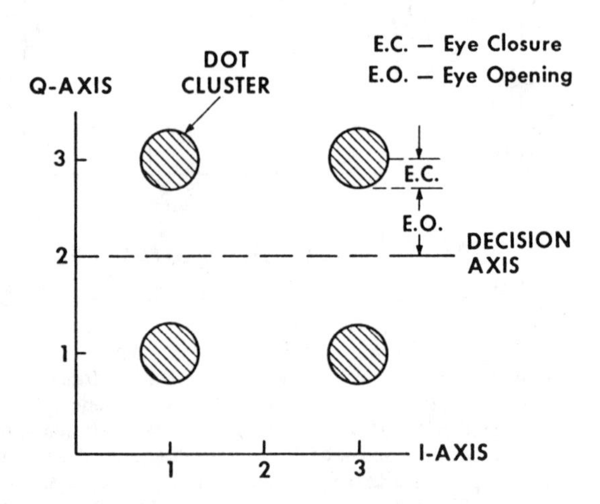

FIGURE 16 - EYE OPENING DEFINITION

Adaptive transversal equalizers have a significant impact on eye opening. The linear distortions are significantly reduced by the transversal equalizer. However, this alone is not adequate. First of all, advanced techniques for carrier recovery to control its peak jitter are employed. Second, radio nonlinearities cause three times as much eye closure for a 64-QAM signal as for a 16-QAM signal. Radio linearity, therefore, must be excellent to transmit this 64-QAM signal. Analog predistortion techniques provide this excellent linearity and are implemented at the intermediate frequency, as indicated in Fig. 5. In summary, the objectives as discussed are met with adequate margin using these techniques and result in a 64-QAM system capable of meeting digital data service performance criteria.

ADDITIONAL APPLICATIONS

A 64-QAM modulation system has applications in all the domestic common carrier bands. The bit rate capable of being transmitted in these bands is shown in Table IV. The resulting baud for the various systems can be transmitted within the authorized bandwidths using well-established Nyquist filtering techniques.

TABLE IV

APPLICATIONS FOR 64-QAM DIGITAL RADIO SYSTEMS

Frequency GHz	Authorized Bandwidth MHz	Bit Rate Mb/s	Number DS3s	Symbol Rate Mbaud
4	20	90	2	15
6	30	135	3	22.5
11	40	180	4	30

SUMMARY

Technological advances in equalization, predistortion, timing, and carrier recovery permit the realization of very high-capacity digital radio systems. Use of the 64-QAM modulation format results in a system with a spectral efficiency of 4-1/2 bps per hertz in the 6-GHz common carrier band. This paper has described a new digital radio system for such an application. Design features have been discussed. Requirements and a system architecture have been presented. Overall system performance based upon a specific realization has been presented. This system can be engineered to meet both stringent dispersive outage requirements and provide excellent (10^{-10}) long-term bit error rate performance.

REFERENCES

1. W. R. J. Brouillette, R. C. Heidt, W. E. Schroeder, and J. S. Bitler, "Microwave Radio Design for 64-QAM Digital Radio," Globecom '84, this conference.

2. W. G. Robinson, III, "Maintenance of Digital Radio using Remoted Performance Monitoring," ICC '84, Amsterdam, Netherlands, May 1984.

3. C. W. Lundgren and W. D. Rummler, "Digital Radio Outage Due to Selective Fading - Observation vs. Prediction from Laboratory Simulation," B.S.T.J., 58, No. 5, May-June 1979.

DESIGN AND APPLICATION OF THE RD-4A AND RD-6A 64 QAM DIGITAL RADIO SYSTEMS

J.D. McNicol, S.G. Barber

BNR
P.O. Box 3511, Station C
Ottawa, Canada K1Y 4H7

F. Rivest

Northern Telecom
9300 TransCanada Hwy
St. Laurent, Canada H4S 1K5

The Northern Telecom RD-4A and RD-6A long haul digital radio systems operate in the North American 4 GHz and 6 GHz common carrier bands. The 64 QAM modulation scheme used provides the spectral efficiency to transmit two DS-3 rate signals in the 4 GHz band and three DS-3 rate signals in the 6 GHz band while maintaining full compatibility with analog radio operating in the same frequency plans. This paper discusses the RD-4A and RD-6A system design considerations, key system performance parameters, details of system architecture and hardware design considerations. Measured data which verified that the system met all long-haul performance objectives is presented.

1. INTRODUCTION

Experience derived from the successful introduction of DRS-8 microwave digital radio in a long-haul (6560 km) network [1] coupled with recent technological advances has led to the development and manufacture of the RD-4A and RD-6A digital radio systems. The RD-4A and RD-6A are second generation systems using 64 state quadrature amplitude modulation (64 QAM). Operating in the 6 GHz band the RD-6A has a transmission capacity of 3 DS-3 rate (135 Mb/s) and in the 4 GHz band the RD-4A has a capacity of 2 DS-3 per RF channel.

This paper presents system design considerations, key system performance parameters, details of system architecture and hardware design considerations for a 64 QAM system. In addition, maintenance and operational considerations are discussed and measured system performance data is presented.

2. SYSTEM ARCHITECTURE

2.1 Design Objectives

The specific characteristics required in the digital radio system area capacity of 2 DS-3 streams per 20 MHz bandwidth and 3 DS-3 streams per 30 MHz bandwidth with the ability to overbuild existing analog radio systems and performance consistent with long-haul objectives. The long-haul objective specifies an availability of better than 99.98% on 6560 km 2-way circuit for a service threshold of 10^{-3} BER. Of the allowable .02% outage, .01% is allocated for equipment failure and .01% for multipath fading. This leads to a per hop allocation for propagation outage of 11 seconds per year or 3.7 seconds in the worst fading month.

During non-fading periods, end-to-end bit error rate better than 10^{-10} is desired which means that individual hops must operate essentially error-free ($<10^{-13}$ BER).

2.2 System Design

The RD-4A and RD-6A digital radio systems use 64 QAM. The key property of 64 QAM is that 6 bits of information are transmitted in each symbol time interval; this contrasts with 8 PSK and 16 QAM in which 3 and 4 bits, respectively, are sent in every symbol interval. The result of the increased word size is that a high bit rate may be achieved while maintaining a narrow spectrum.

The relatively low ratio of symbol rate to authorized bandwidth, 0.75, of the RD-4A and RD-6A systems means that compliance with FCC emissions regulations is achieved without compromise of basic system performance. This same low symbol rate also allows the use of highly selective filters in order that the radio system be compatible with existing analog radio systems and antenna structures.

It has been observed [2] that when 64 QAM is implemented using first generation technology, the results are increased sensitivity to the thermal and dispersive effects of multipath fading and poor background bit error rate. Through innovative design procedures and the application of new technology these potential weaknesses of a high level modulation technique have been overcome in the design of the RD-4A and RD-6A digital radio systems yielding performance which fully meets long-haul requirements. These techniques include: adaptive time domain equalization, a new signal control technology in the modem, r.f. preamplification in the radio receiver and pre-distortion type linearization in the transmitter. Sufficient system gain is provided in the RD-4A and RD-6A radio systems to overcome power fading but without compromising equipment power consumption.

A block diagram of the RD-4A and RD-6A systems is given in Figure 1. Protection switching is performed at the DS-3 level in the "multiline" arrangement where the 2 or 3 DS-3 signals associated with a particular radio are switched together as a group. The protection switch also provides parity correction and "blue signal insertion" features. The physical integration of the radio and muldem function arrangement allows the use of a standard DS-3 interface to the radio equipment racks for flexibility and ease of test.

Reprinted from *IEEE Int. Conf. Comm.*, pp. 646-652, May 1984.

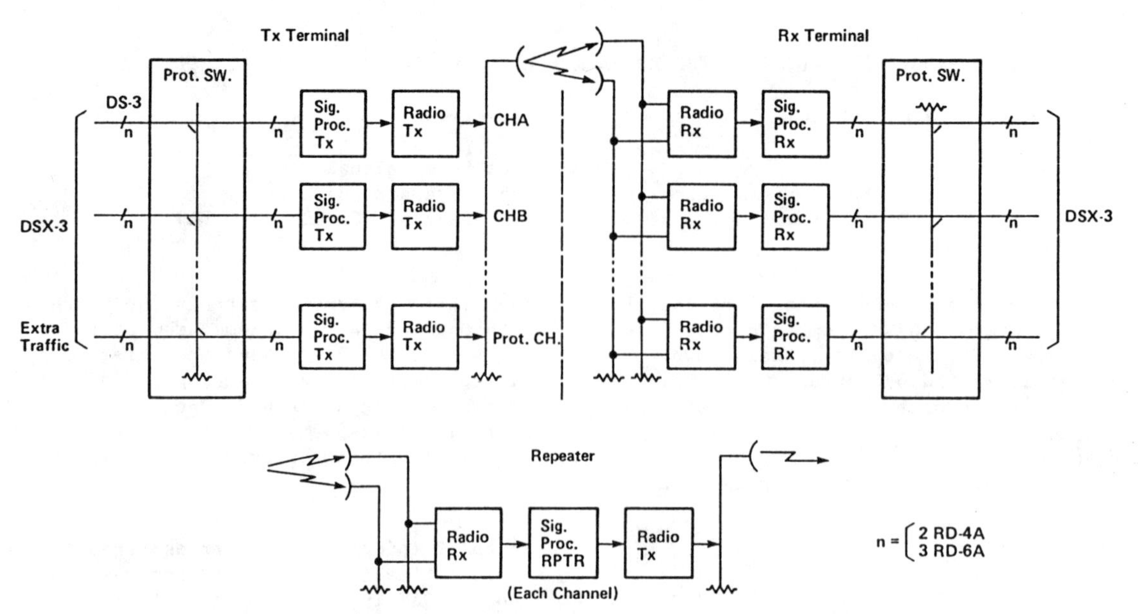

Figure 1: The Typical RD-4A and RD-6A System Configurations for One Switching Section

RD-4A and RD-6A provide a variety of operational and maintenance features including a protected service channel (to carry voice and data), order wire functions and data stream monitoring at each radio hop.

The key parameters of the RD-4A and RD-6A systems are shown in Table 1.

TABLE 1: RD-4A AND RD-6A EQUIPMENT PARAMETERS

	RD-4A	RD-6A
Frequency Range	3.7(3.54)[1] to 4.2 GHz	5.925 to 6.425 GHz
Max. Two-way RF Channels	11(15)[1] working + 1 standby	7 working + 1 standby
Channel spacing	20 MHz	29.65 MHz
Modulation	64 QAM	64 QAM
Data Rate	90.77 Mb/s	135.5 Mb/s
Digital Interface	2 DS-3	3 DS-3
Intermediate Frequency	70 MHz	70 MHz
Power Output[2]	25.4 dBm	29.7 dBm
Receiver Threshold[2] at 10^{-3} BER	-75.0 dBm	-73.0 dBm
System Gain[2] at 10^{-3} BER typical	100.4 dB	102.7 dB
BER (normal propagation)	$<10^{-13}$	$<10^{-13}$
C/I for 10^{-3} BER due to		
co-channel interference	23 dB	23 dB
adjacent ch. interference	-20 dB	-13 dB
Distortion Fade Margin		
with ATDE[3]	47.5 dB	46.5 dB
without ATDE[3]	38.0 dB	34.0 dB
Power Consumption -24 or -48 Vdc		
repeater T-R(R)	94 Watts	154 Watts
terminal T-R(R)	134 Watts	204 Watts

NOTES:
1. including the down-band expansion (Canada only)
2. measured at the antenna port(s) of branching circulator(s)
3. adaptive time domain equalizer.

3.0 HARDWARE DESIGN

The RD-4 and RD-6A digital radio systems consists of two main assemblies (Figure 1): a microwave radio transmitter/receiver and a signal processor. These units interface at a 70 MHz intermediate frequency. System filtering is arranged such that the primary filters, defining transmitter spectrum width and receiver selectivity, are located in the signal processing unit; thus the filtering in the radio portion is relatively benign.

3.1 Transmitter/Receiver

The transmitter and diversity receiver subsystems are shown in Figure 2.

A key feature of the transmitter is the use of pre-distortion within the up-converter assembly to compensate for the GaAs FET power amplifier nonlinearities. Predistortion allows output power requirements to be met with less powerful devices than would be possible without predistortion thereby conserving DC power (up to 50%). Low noise down converter assemblies incorporate up-fade AGC for signal overload protection.

3.2 Signal Processor

The signal processor shown in Figure 3 is provided in terminal transmitter, terminal receiver and repeater configurations. The baseband equipment, or muldem, used at terminals, implements the functions of signal (DS-3) synchronization, scrambling, differential coding, and generation of the overhead frame structure. The format adopted uses 6 bit parallel processing and a direct interface with the modem to maintain a low maximum logic rate (15 Mb/s for RD-4A, 22 Mb/s for RD-6A) which permits reduced power consumption by avoiding ECL logic. The frame structure carries four order wires, each equipped with 32 Kb/s CVSD codecs, and five data channels. The service channels are equipped with an integrated protection switching scheme.

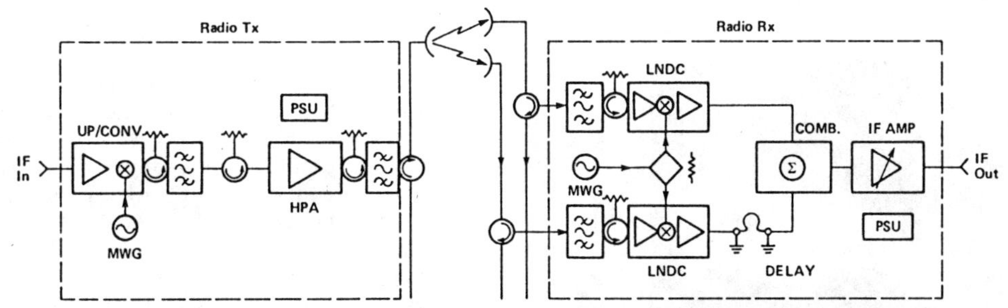

Figure 2: The RD-4A and RD-6A Microwave Radio Transmitter and Receiver

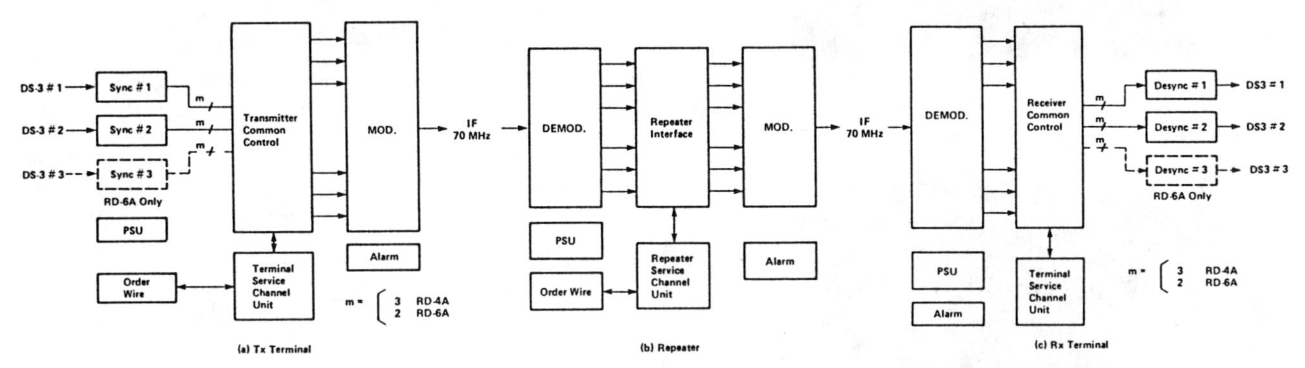

Figure 3: Signal Processor Configurations of RD-4A and RD-6A

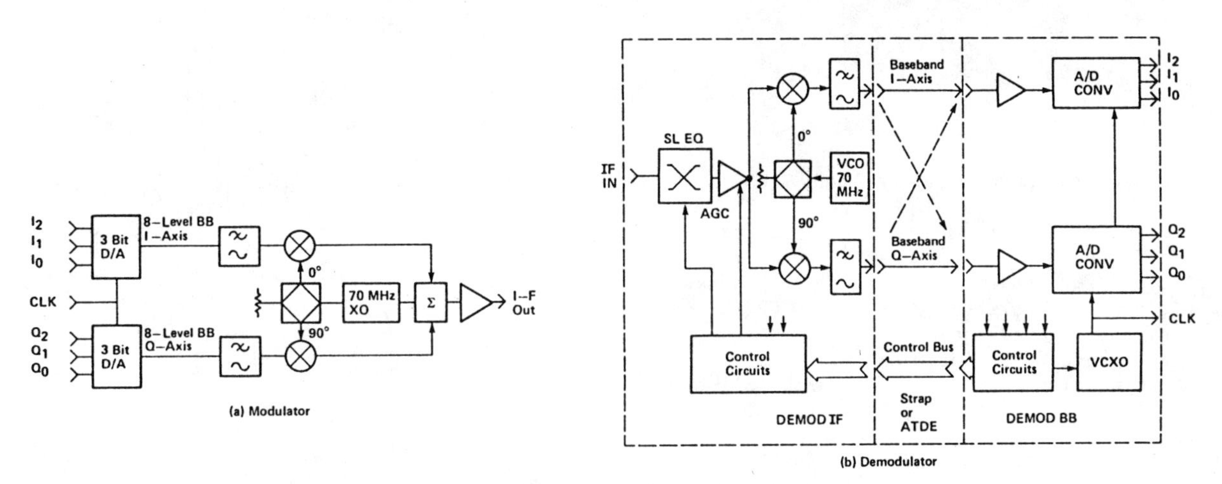

Figure 4: The RD-4A and RD-6A Modem Block Diagrams

The key portion of the RD-4A and RD-6A 64 QAM digital radio systems is the modem (Figure 4) which interfaces the data stream to the 70 MHz channel of the radio and which corrects for time varying signal distortion causes by multipath fading.

Responses of the transmitter and receiver baseband filters are shown in Figure 5. Filters were designed to jointly optimize for low intersymbol interference and high adjacent channel interference immunity. The eye pattern of a production RD-4A system is given in Figure 7.

The requirement for error free operation ($<10^{-13}$ BER) during non-fading periods means that with 64 QAM signalling the modem must maintain >30 dB equivalent signal to noise ratio at the decision circuit. As a result, degradations within the modem must be carefully controlled as indicated in Table 2. Extensive use of data-directed signal control allowed bit

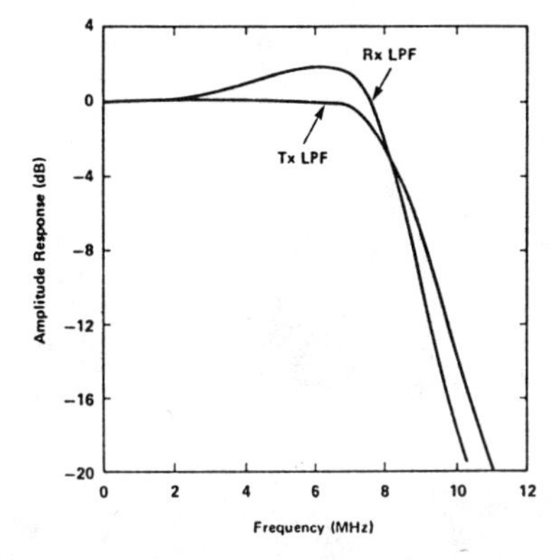

Figure 5: The Frequency Response of the RD-4A Baseband Filters

error rate objectives to be met over the 0 to 50°C temperature range; in addition, this has facilitated elimination of all customer adjustments in the modem and has made possible compact size as shown in Figure 7.

Figure 6: The 64-QAM Eye Pattern

Figure 7: The RD-4A Modem Cards

TABLE 2: MODEM CIRCUIT PARAMETERS

carrier phase jitter	≤0.3° rms
timing jitter	≤0.5° rms
static carrier phase error	≤0.1° peak
static amplitude errors	≤0.02 dB peak
intersymbol interference (eye closure with binary data)	≤1.0% rms
spectrum shaping	25% roll-off factor

An important feature of the RD-4A and RD-6A digital radio systems is the adaptive time domain equalizer; this optional unit significantly enhances the resistance of the demodulator to multipath channel distortion. The RD-6A adaptive time domain equalizer (Figure 8) is

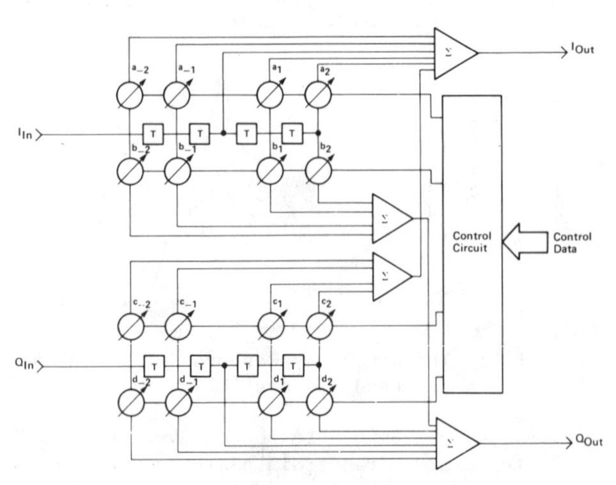

Figure 8: The RD-6A Adaptive Time Domain Equalizer

a five tap transversal type implemented at baseband. (The RD-4A unit requires only four taps to achieve the same performance due to the lower symbol rate). In designing the adaptive time domain equalizer extensive use was made of computer simulation and of hardware testing on an automated multipath simulator in order to optimize performance to the expected channel conditions. The equalizer exhibits no hysteresis in the outage characteristic and can track fast-moving notches (test speeds up to 20 MHz/second).

3.3 Measured Performance

Both production and prototype hardware have been extensively tested at the unit and system level.

Figure 9 shows the equipment signatures as measured with a two ray multipath simulator.

Figure 10 shows the transmitter spectra. The

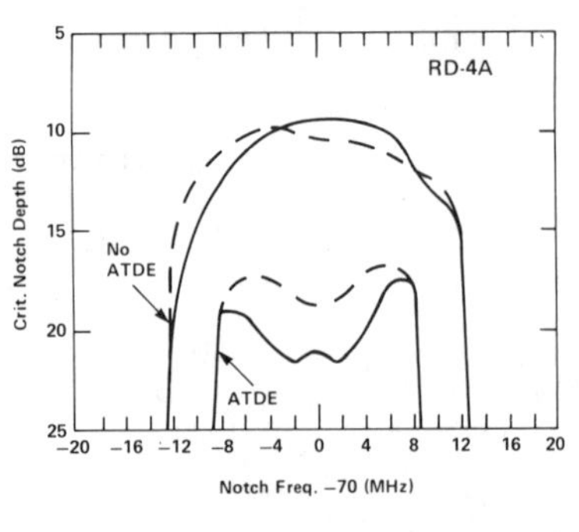

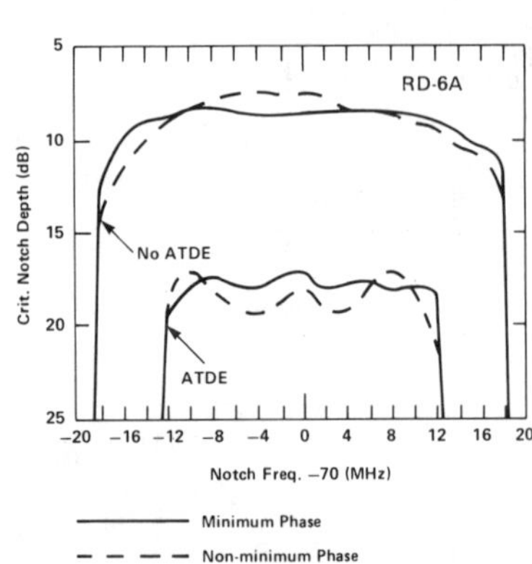

Figure 9: Measured Equipment Signatures (BER = 10^{-3}, τ = 6.3 ns)

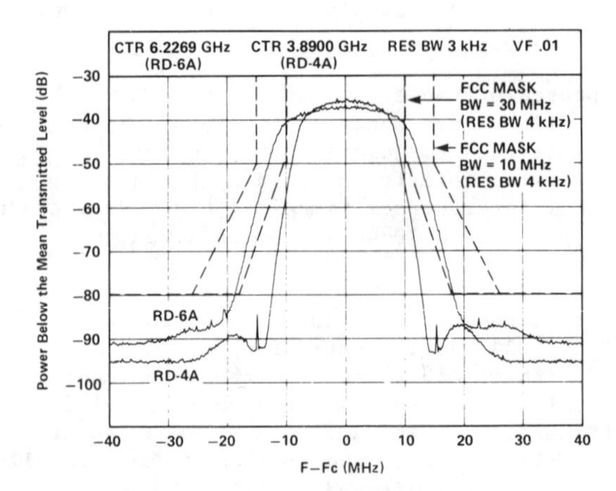

Figure 10: The Transmitted Spectra

FCC mask is met with margin providing the needed spectral confinement for coexistence with analog systems.

Figure 11 displays a measurement of DS-3 bit error rate versus received signal level for typical production units. Long term error rate measurements have shown the system to exhibit an unfaded error rate better than 10^{-13} per hop for ambient temperatures in the 0 to 50°C range.

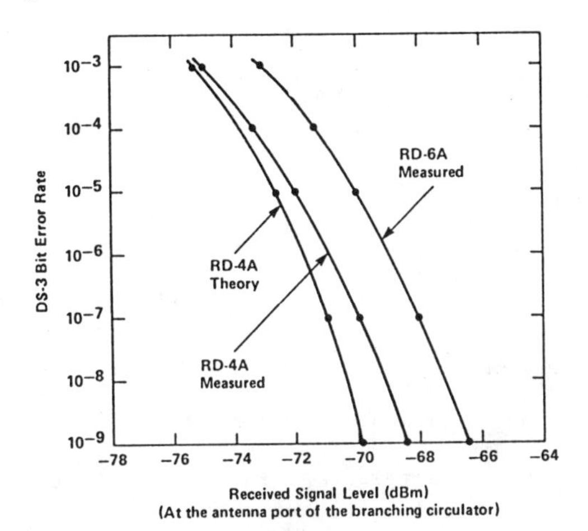

Figure 11: BER as a Function of Signal Level

3.4 Equipment Configurations

The RD-4A and RD-6A radio systems may be configured with either 1:N section switching (Figure 1) or hot-standby equipment protection. Distinct arrangements of equipment are provided for terminal and repeater applications and each configuration may be equipped with or without space diversity reception.

All subsystems (radio transmitter and receiver and signal processor transmitter, receiver and repeater) are packaged as shelves which are mounted in standard 19 inch (483 mm) frames as illustrated in Figure 12. Modular design and low power consumption leads to high density packaging with up to three repeaters or two terminals mounting in a 7 ft. (2.1 m) rack.

3.5 Maintenance Features

Most routine and diagnostic tests for the maintenance of the radio equipment can be performed with a handheld digital voltmeter.

Each equipment shelf is equipped with a separate metering and alarm unit. All alarms and telemetry points can be connected to remote alarm reporting systems. Diagnosis of fault conditions is done through interpretation of alarms and test point readings.

To simplify alarm interpretation, repeaters may be optionally equipped with frame resupply. This option operates in a manner similar to carrier re-insert in analog radio in order to mute downstream repeater alarms during equipment failure or propagation outage which exceeds one second duration. For remote location of transient faults, the data stream is monitored to record the total number of seconds

Figure 12: Photographs of the Terminal and Repeater Assemblies

with intermittent signal continuity and a remote alarm is raised if these exceed user set thresholds.

4.0 SYSTEM ENGINEERING CONSIDERATIONS

4.1 Propagation Unavailability

Design for long-haul availability requires consideration of the thermal and dispersive components of multipath fading. The engineering of RD-4A and RD-6A systems in an effective and economic manner is assisted by reference to an unavailability calculation formula:

$$U_m = 3.3 \times 10^{-4} \times C \times D^4 \times \left[\frac{10^{-F_C/10}}{S} + \frac{10^{-F_D/10}}{6.0}\right]^2 \quad [1]$$

where U_m is the annual outage probability of a single hop which is equipped with space diversity reception, C the terrain and climate factor (equals unity with typical terrain and climate), D the path length [km], S the diversity antenna spacing [m], F_C the combined channel flat fade margin [dB] including interference effects, F_D the distortion fade margin [dB] (Table 1) derived from the measured equipment signature using the fading statistics from [3]

and $$F_C = F + 2.6 \text{ dB} - \Delta, \quad [2]$$

where F is the nondiversity flat fade margin

(10^{-3} BER), Δ the degradation of fade margin due to the interference and 2.6 dB the threshold improvement due to the combiner.

The formula is based on the standard equation for thermal fading in a space diversity system with the addition of a term independent of antenna spacing to account for dispersive fading. The dispersion parameter is derived from the measured equipment signature of the radio equipment (Table 1) and includes an empirical factor to account for the extent of multipath distortion in the space diversity combined channel. The latter factor was determined for RD-4A on the basis of field experiments described in Section 5.

Using equation 1 for a typical hop (D=47 km, S=15 m) with typical values of combined flat fade margin (38 dB for RD-4A, 43 dB for RD6A), the calculated unavailabilities are 9 and 3 seconds per year respectively for RD-4A and RD-6A systems. This result compares with an objective of 11 seconds per year.

4.2 Intrasystem Interference Considerations

The main considerations are: the effects of antenna system performance on the digital radio fade margin and, where digital channels are added to an existing analog system, the additional interference noise in the analog channels due to digital radio traffic. Table 1 shows the C/I requirements for operation in an all-digital environment. Due to the low sensitivity to adjacent-channel interference, antenna cross polarization discrimination (XPD) performance has negligible impact on the performance of RD-4A and RD-6A. Figure 13 shows the impact of other antenna system parameters on the digital radio fade margin. In typical existing installations engineering for analog radio [4], only small degradations (typically less than ½ dB) will result.

Table 3 summarizes the C/I requirements necessary to limit noise to 4 dBrnC0 per exposure in a mixed analog and digital environment (calculated using the convolution technique [5]). Taking into account the power level differences between the digital and analog systems (approximately 10 dB) and assuming antennas of medium XPD performance (25 to 30 dB) it may be seen that RD-4A and RD-6A are compatible with adjacent analog channels of capacities up to 1320 voice channels and 1800 voice channels respectively.

TABLE 3: C/I LIMITS FOR INTERFERENCE INTO FM ANALOG RADIO

(CRITERIA: 4 dBrnC0 PER EXPOSURE)

DIGITAL CASE	ANALOG LOADING	TEST TONE [kHz RMS]	FREQ. SPACING [MHz] 0	20	29.65
RD-4A	960 CH	200	70 dB	26 dB	-
RD-4A	1200 CH	140	75 dB	40 dB	-
RD-4A	1320 CH	140	75 dB	40 dB	-
RD-4A	1500 CH	100	80 dB	50 dB	-
RD-6A	1800 CH	140	77 dB	-	36 dB
RD-6A	2100 CH	140	78 dB	-	47 dB
RD-6A	2400 CH	100	82 dB	-	49 dB
RD-6A	2700 CH	70	86 dB	-	51 dB

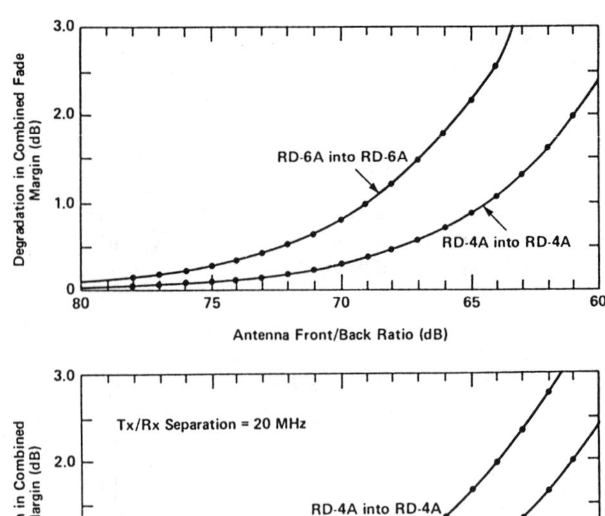

Figure 13: The Effect of the Antenna System Performance on the Fade Margin

5.0 FIELD PERFORMANCE OF THE RD-4A

The field verification of the RD-4A design had two phases: first a single hop technology trial at the engineering model stage, second an extensive multi-hop system trial using production equipment.

5.1 Technology Trial

In the summer of 1982 an engineering model of RD-4A was installed on an operational Bell Canada hop near Ottawa which has been previously shown to exhibit close to average fading [6]. The objective of the trial was to verify the performance predictions for a 90 Mb/s 64 QAM radio equipped with an adaptive time domain equalizer.

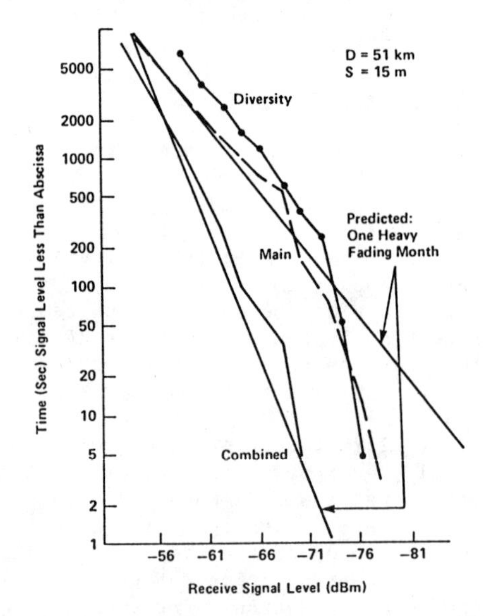

Figure 14: Fade Statistics for the 1982 Technology Trial

5.2 Technology Trial Results

Data was recorded from August 2, 1982 to October 9, 1983 and during this time multipath fading equivalent to approximately one heavy fading month occurred. Figure 15 shows the fade statistics for this period (nominal receive signal level was -39 dBm). There was no outage recorded during the test interval and this compares favourably with a predicted outage of 1.0 seconds.

One 30 minute ducting event was recorded during the test interval; following the practise of previous experiments [6] the data was removed from that present in Figure 14 because of its strongly atypical characteristics.

5.3 System Trial

In May 1983, 32 transmitter/receivers from the first production run were shipped to MCI Communications for installation on the first four-hop switch section of their New York to Chicago route (see Figure 15 for route map). The section was equipped with four channels in each direction: three working and one protection. All hops were equipped with adaptive time domain equalizers and space diversity. The primary aims of the trial were to verify that system performance objectives were met and to assist the customers in setting operational standards.

Instrumentation of the radio equipment was restricted to standard radio alarms and BER of the DS-3 streams; however, even with this seemingly modest level of instrumentation the total number of parameters monitored was nearly 900. As a measure of the level of multipath fading encountered, the time that the receive signal was below 15 dB was recorded at each site on each antenna. The trial ran from August 22 to October 2 and the expected time below 15 dB (prorated for distance and season) was 16500 seconds per month. The measured fading, shown in Table 4 for the eastbound channels was as predicted for one channel and somewhat less than predicted for the others. The westbound channels showed similar fading.

Examination of the raw data showed that system outage was significantly in excess of that expected; however, close scrutiny of the

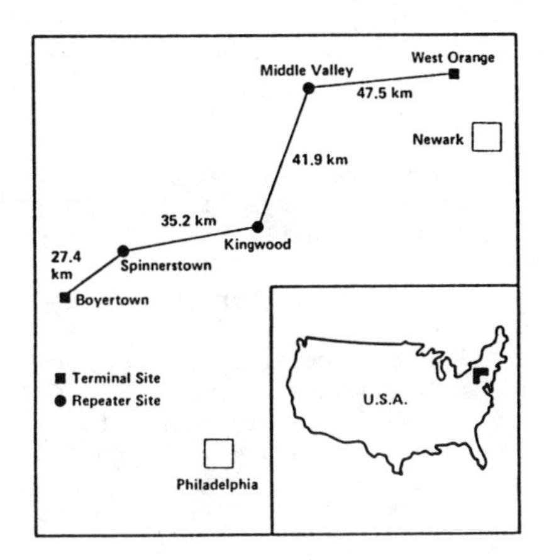

Figure 15: Route Map of the 1983 System Trial

TABLE 4: SYSTEM TRIAL FADE STATISTICS

predicted time[1] below 15 dB (one half heavy fading month)	16500 sec
predicted outage[1]	2 sec
measured time below 15 dB	
channel E1	4472 sec
channel E3	13334 sec
channel E4	6479 sec
measured outage due to fading	0 sec

NOTES: 1. each channel, 4 hops.

results indicated that the bulk of the outage was due either to faults in the monitoring system or to system line-up and testing activity that was done while the monitoring system was active. No outage was associated with multipath fading. Ten incidents of equipment-related transmission impairments were recorded and 80% of these were intermittent or so-called "soft failures". With the exception of three semi-rigid cable failures these faults were randomly distributed and have been attributed to "infant mortality". Cable failures were eliminated in subsequent shipments by a design change. The background BER of the system was well within the design objective of 10^{-13} per hop with periods of up to one week passing without a single error being recorded.

6.0 OPERATONAL STATUS

The RD-4A radio has been in production since March 1983 and to date 1100 T-R's have been shipped for use in the MCI long haul network. First production of the RD-6A is due for completion in July 1984.

REFERENCES

[1] C.W. Anderson and S.G. Barber, Modulation Considerations for a 91 Mbit/s Digital Radio, IEEE Transactions on Communications, Vol. COM-26, No. 5, May 1978.

[2] C.P. Bates and M.A. Skinner, Impact of Technology on High-Capacity Digital Radio Systems, IEEE International Conference on Communications, Boston, June 1983.

[3] C.W. Lundgren and W.D. Rummler, Digital Radio Outage Due to Selective Fading - Observation vs Prediction From Laboratory Simulation, Bell System Technical Journal Vol. 58, No. 5, May-June 1979.

[4] S.D. Hathway, W.A. Hensel, D.R. Jordon, and R.C. Prime, TD-3 Microwave Radio Relay System, Bell System Technical Journal, Vol. 47, No. 7, Sept. 1968.

[5] B.A. Pontano et al, Interference into Angle-Modulated Systems Carrying Multichannel Telephony Signals, IEEE Transactions on Communications, June 1973.

[6] C.W. Anderson, S.G. Barber and R.N. Patel, The Effect of Selective Fading on Digital Radio, IEEE Transactions on Communications, Vol. COM-27, No. 12, December 1979.

256 QAM MODEM FOR HIGH CAPACITY DIGITAL RADIO SYSTEMS

Y. Daido, Y. Takeda, E. Fukuda, S. Takenaka and H. Nakamura

Fujitsu Laboratories Ltd.
1015 Kamikodanaka Nakahara-ku Kawasaki, 211 Japan

ABSTRACT

Feasibility of 256 QAM modulation for high capacity digital radio systems is described both theoretically and experimentally. CNR impairment is estimated theoretically for typical impairment factors.

Considering results of estimation, two techniques are adopted in an experimentally fabricated modem. One is a carrier injection method to obtain very high SNR of recovered carrier. The other is a newly developed predistorter in the modulator which adjusts symbol points in constellation very accurately.

Clearly opened eyes and a reasonable value of CNR impairment of 3.7 dB at a bit error ratio of 10^{-6} are obtained by using the above two techniques. Good reproducibilities are obtained for the high SNR of recovered carrier and accurate adjustment of constellation.

1. INTRODUCTION

Improvement of spectrum efficiency has been one of the primary motives for developing digital radio systems for these several years. 16 QAM digital radio systems are already in practical use in the field (Refs. 1-4). For attaining higher spectrum efficiency, development of 64 QAM systems has been discussed, recently (Refs.5-7). Since systems with higher spectrum efficiency are more vulnerable to linear path distortion or multipath fading, more precise adjustment of each subsystem is required for a 64 QAM system than for the 16 QAM system.

Though very precise adjustment is required for 256 QAM systems, Y. Saito et. al. have reported its feasibility (Ref. 8). Since the possibility of 256 QAM is verified, key techniques have to be established to obtain good reproducibility and stable performance. Carrier to noise ratio (CNR) impairment is estimated theoretically to know what factors cause significant impairment to the system. The estimation suggests the importance of the following two factors; (a) signal to noise ratio (SNR) of recovered carrier, (b) arrangement of symbol points in the space diagram. These two factors are the same as pointed out in Ref. 8.

This paper recommends two methods as the key techniques. One is pilot carrier injection method (Ref. 7) which guarantees recovered carrier SNR higher than 46 dB. The other is predistorter in the modulator for adjusting the symbol points in constellation. The predistorter is adjusted very precisely and quickly by using a micro-computer. The 256 QAM modem is designed, taking estimated tolerances and above two techniques into account. Remarkable improvement of eye pattern and performance is obtained by using the predistorter.

2. THEORETICAL ESTIMATION OF TOLERANCES OF MODEM

As already mentioned, systems with higher spectrum efficiency are more vulnerable to various impairment factors. Tolerances for important factors have to be estimated, theoretically, before designing hardware. The following impairment factors are selected; (1) linear and quadratic amplitude distortions, (2) linear and quadratic delay distortions, (3) filter bandwidth error, (4) timing error, and (5) recovered carrier phase error. In items (4) and (5), jitter as well as fixed errors are taken into account.

2.1 CNR impairment

Our method for estimating tolerances has already been verified by experiments in our previous paper for 64 QAM modem (Ref. 7). The same method is used to estimate tolerance. Among the above mentioned impairment factors, items (1) and (2) are important for the design of IF amplifiers and filters.

Figure 1 shows carrier to noise ratio (CNR) impairment caused by items (1) and (2). There is a difference of CNR impairment between the quadratic amplitude distortions with a peak and a valley in the frequency response of the transmission line. Distortions are defined as variations within Nyquist bandwidth. Delay is normalized by time slot duration. As is seen from this figure, precise adjustment of IF circuits is required. For example, tolerance for CNR impairment of 0.5 dB is 0.45 dB for linear amplitude distortion. Roughly speaking, tolerance of the 256 QAM system for each type of distortion is half of that in 64 QAM systems.

Figure 2 shows CNR impairment caused by items (3), (4) and (5). For items (4) and (5), fixed errors are considered in this figure. Filter bandwidth and timing errors are normalized by Nyquist bandwidth and time slot duration, respectively. Small tolerance for filter bandwidth requires accurate design of the

Reprinted from *IEEE 3rd Global Telecomm. Conf.*, pp. 547-551, Nov. 1984.

spectrum shaping filter. Tolerances for timing and recovered carrier phase error are especially severe.

Despite severe tolerances, requirements for items from (1) to (3) can be attained by the present state of the art. Items (4) and (5) have to be discussed in more detail. For these impairment factors, reduction of jitter is much more difficult than elimination of fixed errors. CNR impairment by jitter is described in the next subsection.

2.2 Timing jitter and recovered carrier phase jitter

Figure 3 shows CNR impairment by timing jitter. Jitter is given in rms. Abrupt increases of CNR impairment at jitter of 2.8 and 3.7 deg. mean that bit error ratio can not be reduced lower than the specified value for each curve in the presence of timing jitter. It was confirmed by experiment that reducing filter bandwidth in clock recovery improves modem performance.

Figure 4 shows CNR impairment by recovered carrier phase jitter. The pilot carrier injection method (Ref. 7) gives a signal to noise ratio (SNR) of more than 46 dB for recovered carrier. Though requirement for carrier phase jitter is severe, CNR impairment less than 0.5 dB can be expected at a bit error ratio of 1×10^{-6}, by applying the pilot injection method.

3. DESIGN OF 256 QAM MODEM

3.1 Modulation and demodulation

The block diagrams of the modulator and the demodulator are shown in Figs.5 and 6, respectively. In the modulator two digital data streams of 60 Mbps, In-phase channel (I-CH), and Quadrature channel (Q-CH), are converted into 8 binary streams by serial to parallel converter circuits (S/P), each of which has a data rate of 15 Mbps. These 8 binary data streams are fed to the DC-control circuits (DC-CONT) to suppress the lower frequency component in the output spectrum of the D/A converters. The 8 binary parallel main data streams, inverted or not, are converted to a pair of 16-level signals by high speed D/A converters. Signals with a symbol rate of 15.6 Mbaud pass through the spectrum shaping filters with a 50 % cosine roll-off factor and are converted up to a 70 MHz IF band. The I-channel signal is biased by two apertures of 15 eyes to inject a pilot tone of carrier frequency which can be a reference carrier in the carrier recovery circuits.

In the demodulator, the received IF signal is divided into a pair of parallel channels and multiplied by the two orthogonal recovered carriers. The I- and Q-channel signals are filtered by the low pass filters to remove noise and the higher order harmonic components. The level of each signal is decided by an A/D converter using the clock recovered by bit timing recovery circuits (BTR). The 8 regenerated binary data streams are frame-synchronized. These 8 binary parallel data streams are fed to the parallel-to-serial converter circuits (P/S) and the regeneration of two 60 Mbps data streams is completed.

3.2 Predistorter

Extremely severe tolerance for carrier phase error shown in Fig.2 implies that precise adjustment is also required for the arrangement of symbol points in constellation. The predistorter is adopted in our modem to improve constellation. Figure 7 is a block diagram of the predistorter with the automatic adjustment method. In this figure, the predistorter memorizes deviation of the constellation from the ideal. The predistorter uses a conversion matrix to convert equally spaced input signal into one arranged arbitrarily. The feed-back loop including the computer minimizes constellation deviation caused by nonlinearity of double balanced mixers, iteratively. After iterative adjustment, the conversion matrix is written in the ROM of the predistorter and the computer is removed. The maximum error in the constellation is determined by digital level of D/A converters. In the present modem, the maximum error is 1/16 of symbol spacing. Adjustment by using computer is accurate, speedy and reproducible.

3.3 Carrier Recovery

Since pilot injection is used in the modulator, carrier recovery is simple, similar to that described in our previous paper (Ref. 7). The calculated spectrum shown in Fig.8 reflects suppression of signal power near the reference carrier by DC-control circuits in the modulator. This figure shows that SNR of recovered carrier is more than 46 dB, using a 5 kHz bandwidth filter in the carrier recovery phase locked loop. Calculated spectrum for 256 QAM differs very little from one for 64 QAM. Validity of the calculation method is confirmed by direct comparison of Fig.8 with the measured one shown in our previous paper (Ref. 7).

4. PERFORMANCE OF 256 QAM MODEM

4.1 Measurement

Table 1 shows the 256 QAM modem system parameters, determined by taking the preceding basic considerations into account. This 256 QAM modem is operating at a 70 MHz carrier frequency and a bit rate of 120 Mbps.

Figure 9 shows the 256 state space diagram observed at the demodulator output, with the clock frequency reduced to about 10 kHz for measurement. Accuracy of symbol arrangement is shown by one dimensional level diagram in Fig.9. This figure is the space diagram obtained without predistorter. Figure 10 shows the space diagram improved by predistorter. Comparison of one dimensional level diagrams in Figs.9 and 10 makes it clear that remarkable improvement of space diagram is obtained by using the predistorter.

The eye diagram observed at the demodulator

output is shown in Fig.11. Clearly opened eyes correspond to improvement of space diagram by the predistorter.

4.2 Bit error performance of 256 QAM modem

To evaluate the performance of the modulator and demodulator quantitatively, the bit error ratio (BER) was measured with a modem back-to-back connection. Though hard wired carrier was used in this measurement, it is believed that the high SNR of recovered carrier ensures almost the same performance. As pointed out in subsection 3.3, recovered carrier SNR of 46 dB is readily obtained by the pilot injection method (Ref. 7). Theoretical and measured BERs are shown in Fig. 12. The SNR in the figure is the ratio of average signal power to average noise power at the decision circuits. The 3.7 dB degradation from the theoretical value at BER of 1×10^{-6} suggests possibility of a 256 QAM system. The degradation is considered to be caused mainly by timing jitter, considering the severe requirement shown in Fig.3.

5. CONCLUSION

256 QAM modulation was studied theoretically to find out which impairment factors significantly affect the performance. To reduce impairment caused by two major factors, a carrier injection method and a predistorter in modulator are adopted in an experimentally fabricated 120 Mbps modem in which subsystems developed for 64 QAM are converted to 256 QAM.

The predistorter adjusted by micro-computer improves the eye diagram and BER performance remarkably. Theoretically estimated SNR of recovered carrier for 256 QAM system is almost the same as that for 64 QAM systems. Since recovered carrier SNR higher than 46 dB is reproducible for 64 QAM systems, adoption of pilot carrier injection method for 256 QAM system is expected to ensure the same SNR of recovered carrier.

ACKNOWLEDGMENT

The authors are especially grateful to Messrs. Y. Hayashi, T. Shoji and Dr. Komizo, Fujitsu Ltd. for their guidance and encouragement.

REFERENCES

1. H. Yamamoto ,"Advanced 16-QAM Techniques for Digital Microwave radio", IEEE, Trans. Comm., CM-19 No.3, pp36-45, 1981.
2. Ph. Dupuis, M.Joindot, A.Leclert, and P.Vandamme,"16 QAM Modem for High Capacity Microwave System: Design and Performance", ICC '81, 1981.
3. V.K.Prabhu,"The Detection Efficiency of 16-ary QAM", BSTJ, Vol.59 No.4, pp639-656, April 1980.
4. T.S.Giuffrida and W.W.Toy,"16 QAM and Adjacent Channel Interference", ICC '81, 1981.
5. T.Hill and K.Feher,"NLA 64-state QAM: A Power and Spectral Efficient Modulation Technique for Digital Radio", ICC '82, 1982.
6. P.R.Hartmann and J.A.Crossett ,"135 MBS - 6 GHz Transmission System using 64-QAM Modulation", ICC '83, 1983.
7. E. Fukuda et. al. "DESIGN OF 64 QAM MODEM FOR HIGH CAPACITY DIGITAL RADIO SYSTEMS", GLOBCOM '83, San Diego 1983.
8. Y. Saito et. al. "FEASIBILITY CONSIDERATIONS OF HIGH-LEVEL QAM MULTI-CARRIER SYSTEM", ICC '84, Amsterdam, 1984.

Table 1 256 QAM modem system parameters.

Item	Contents
Modulation technique	256 QAM
Carrier frequency	70 MHz
Symbol rate	15 MHz
Bit rate	120 Mbps
Spectrum shaping	50% Nyquist roll-off

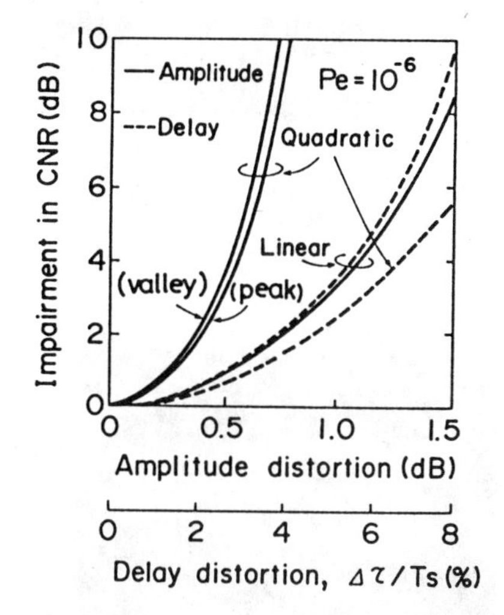

Fig.1 Carrier to noise ratio (CNR) impairment caused by amplitude and delay distortions.

Distortion due to the transmission line is defined as variation range within the Nyquist bandwidth. Delay is normalized by time slot duration.

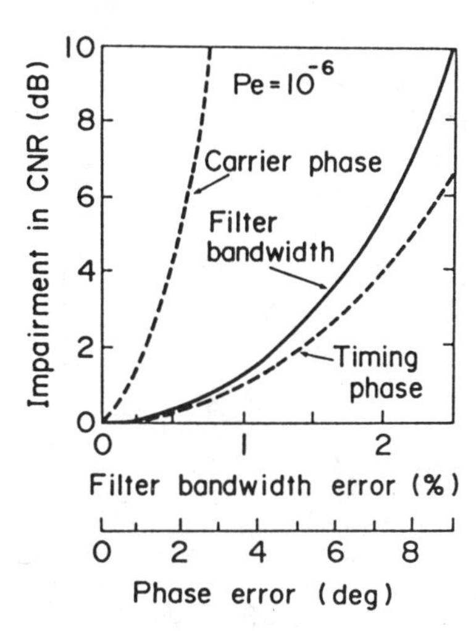

Fig.2 CNR impairment caused by filter bandwidth, timing and recovered carrier phase error.

Filter bandwidth error is normalized by the Nyquist bandwidth.

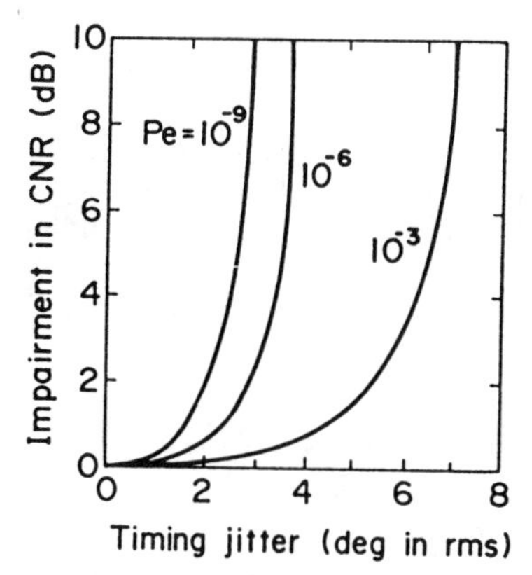

Fig.3 Degradation of performance by timing jitter.

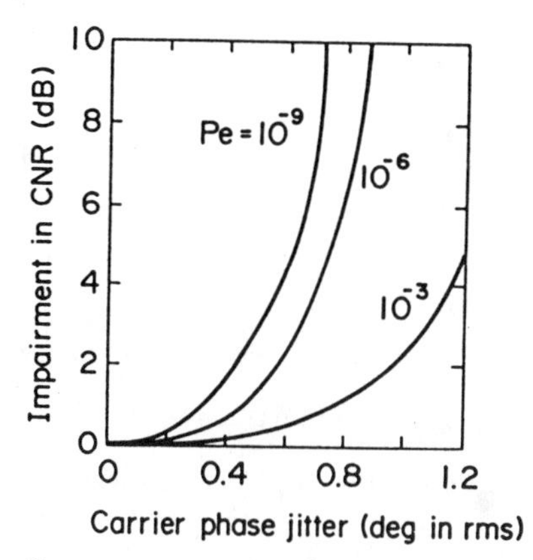

Fig.4 Degradation of performance by recovered carrier phase jitter.

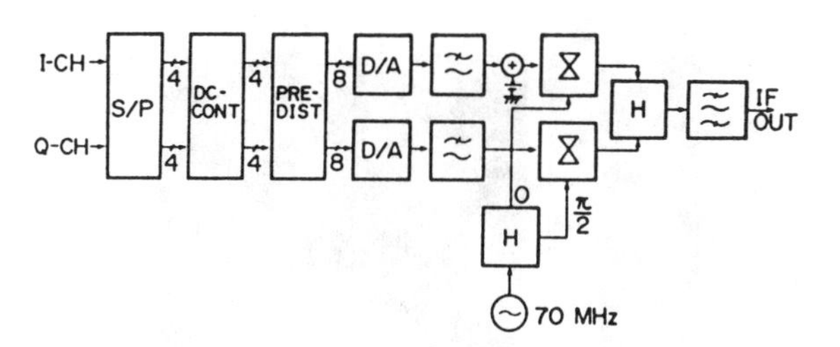

Fig.5 Block diagram of modulator.

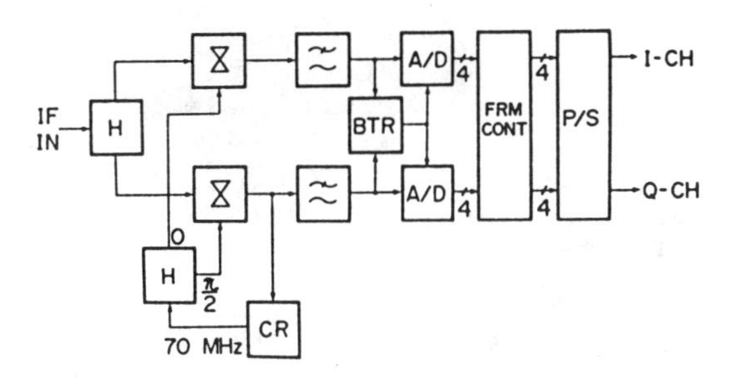

Fig.6 Block diagram of demodulator.

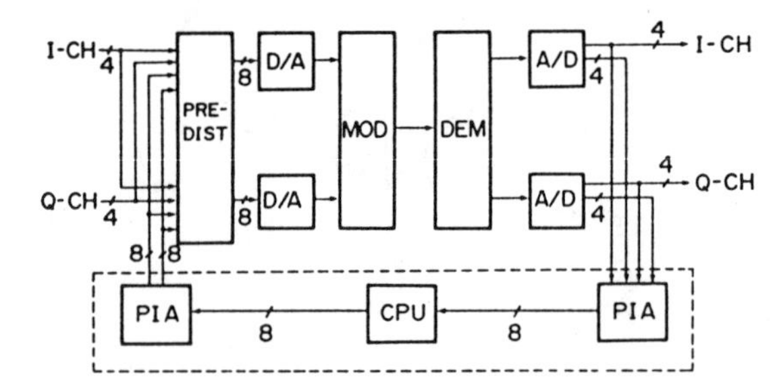

Fig.7 Block diagram of predistorter.

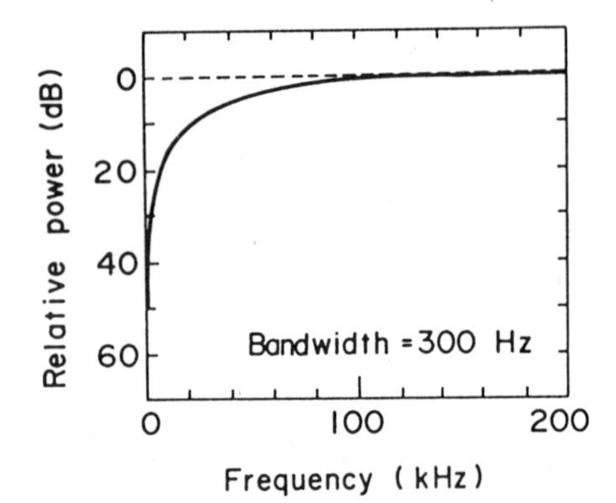

Fig.8 Modulator output spectrum near the carrier.

Spectrum is calculated analytically, taking polarity control by DC-control unit into account.

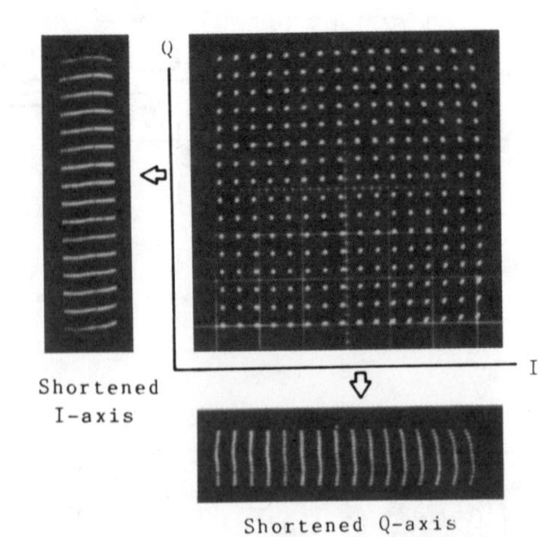

Fig.9 256-state space diagram obtained without predistorter. There are too many signal points to see the accuracy of their arrangement. The size of diagram is reduced to one-tenth along one axis, keeping the size along the other axis constant. The shortened-axis diagrams are shown along with the space diagram, and arrangement errors are clearly seen.

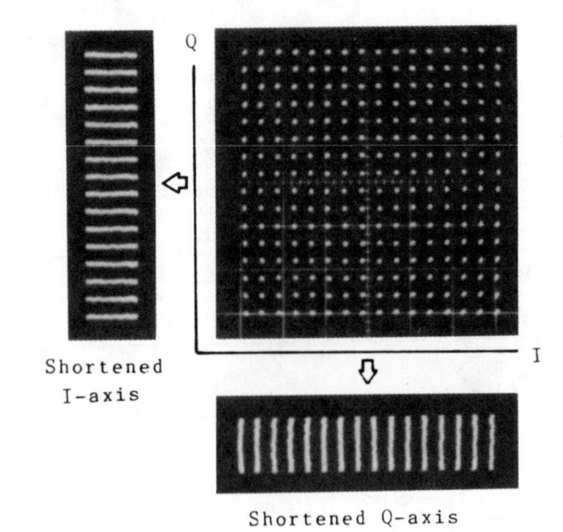

Fig.10 Space diagram improved by predistorter.

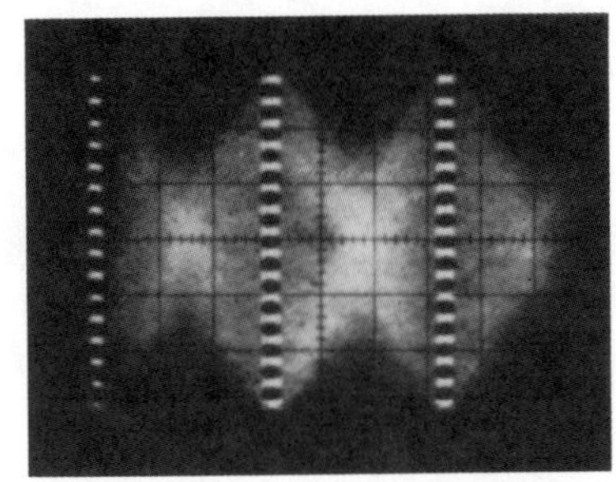

Fig.11 Eye diagram of demodulator.

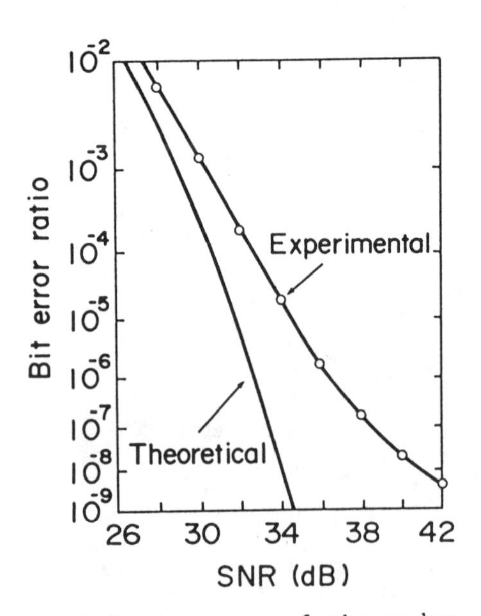

Fig.12 Performance of the modem.

FEASIBILITY CONSIDERATIONS OF HIGH-LEVEL QAM MULTI-CARRIER SYSTEM

Yoichi SAITO, Shozo KOMAKI, and Masayoshi MUROTANI

Yokosuka Electrical Communication Laboratory,
N.T.T. Public Corp.
Take, Yokosuka-shi, Kanagawa-ken, 238-03, Japan

High-level modulation schemes are effective ways to increase frequency utilization efficiency. This paper discusses the feasibility of such high-level modulation, i.e., 64/256 QAM systems, under a multipath environment. It is demonstrated that the multi-carrier transmission method is useful for high-level modulation schemes. The requirements for a 64/256 QAM modem and experimental result are also presented. These studies lead to the conclusion that 10 bits/s/Hz will become possible if a four- to eight-carrier 256 QAM system and dual polarization operation are achieved.

1. INTRODUCTION

More bits per Hertz has become a major concern of digital microwave radio system designers in recent years. In order to increase the number of bits assigned to one symbol, multi-level modulation systems such as 4 PSK, 8 PSK and 16 QAM have been developed [1,2,3]. Today, high-level modulation schemes, such as 32 QAM or 64 QAM, are being investigated [4,5,6]. Because such schemes are more sensitive to waveform distortion and interference noise, it is necessary to establish integrated multipath fading countermeasures, sophisticated modem technology and various interference cancelling techniques.

This paper discusses the feasibility of high-level modulation, i.e., 64/256 QAM, systems under a multipath environment. The basis is 5L-D1 digital radio system [3] which has a capacity of 200 Mb/s using 16 QAM with a frequency spacing of 40 MHz. This corresponds to a frequency utilization efficiency of 5 bits/s/Hz. The target of our research is to achieve 64 or 256 QAM with the same radio frequency arrangement. The first topic to be discussed is the upper baud limit for high-level QAM that meets the outage objective for multipath waveform distortion. The probability of dispersive outage for Nyquist spectral shaped high-level QAM of various clock rate can be estimated from the allowable Nyquist-band amplitude dispersion and its probability of occurrence.

The second topic is the required performance of a high-level QAM modem that meets the criteria for low CNR degradation. Degradation factors, such as modulation phase error, carrier recovery steady-state phase error, recovered carrier jitter, recoverd clock phase error and threshold level drift, cause drastic impairment as the number of modulation levels increases. Equivalent CNR degradation of high-level QAM is theoretically estimated.

As the last topic, characteristics of a 256 QAM prototype modem that was designed in our laboratory and operates at 12.5 MB are described.

From the above considerations and experimental results, frequency utilization efficiency of 10 bits/s/Hz is considered possible if a multi-carrier transmission method [7] is employed and dual polarization operation is achieved in 256 QAM modulation system.

2. UPPER BAUD LIMIT FOR HIGH-LEVEL QAM

2.1 Waveform Distortion Due to Multipath Fading

Most multipath fading events can be represented by a two-ray-path model, with ρ and τ defined as the rays' amplitude ratio and delay difference, respectively. Then, the direct ray, $v_d(t)$, and the interfering ray, $v_i(t)$, of the QAM signals can be expressed as:

$$v_d(t) = (a-jb)r(t)e^{j2\pi f_c t} \tag{1}$$

$$v_i(t) = \rho(a-jb)r(t-\tau)e^{j(2\pi f_c t+\theta)} \tag{2}$$

where

a, b = ±1, ±3, ... , ±(2N-1)
r(t) : NRZ pulse response,
fc : carrier frequency, and
θ : phase difference between the direct and interfering rays.

At the receiving point, $v_d(t)$ and $v_i(t)$ are superposed and then detected by the reference carrier, $\mathrm{Re}[\exp\{-j(2\pi f_c t + \theta_0)\}]$, where θ_0 is the reference carrier phase as reconstructed by a carrier recovery circuit. The optimum carrier phase is:

$$\theta_0 = \tan^{-1}\frac{\rho\sin\theta}{1+\rho\cos\theta} \tag{3}$$

The demodulated in-phase signal, d(t), can be represented as:

$$\begin{aligned} d(t) &= \mathrm{Re}[\{(a-jb)r(t)e^{j2\pi f_c t} \\ &\quad +\rho(a-jb)r(t-\tau)e^{j(2\pi f_c t+\theta)}\}e^{-j(2\pi f_c t+\theta_0)}] \\ &= [ar(t)\cos\theta_0 + a\rho r(t-\tau)\cos(\theta-\theta_0)] \\ &\quad -[br(t)\sin\theta_0 - b\rho r(t-\tau)\sin(\theta-\theta_0)] \end{aligned} \tag{4}$$

The first bracket represents the demodulated signal with in-phase distortion and the second, quadrature distortion. Because the frequency notch falls into the in-band of the transmitted spectrum during deep fading, phase difference nearly equals π radians and θ_0 is nearly zero. Therefore, Eq. (4) yields:

$$d(t) \fallingdotseq a[r(t) - \rho r(t-\tau)] \tag{5}$$

After some calculations [8] based on Eq.(5),

Reprinted from *IEEE Int. Conf. Comm.*, pp. 665–671, May 1984.

the eye opening, L, at the optimum decision time can be approximated as:

$$L \fallingdotseq 1 - \rho - W_f \rho \tau \quad (6)$$

where,

$$W_f = A \sum_{\substack{n=-\infty \\ n \neq 0}}^{\infty} \left| \left\{ \frac{dr(t)}{dt} \right\}_{t=nT} \right| \quad (7)$$

$$A = \begin{cases} q - 1 & \text{for } q \times q - \text{QAM} \\ \cot(\pi/p) & \text{for p-PSK} \end{cases}$$

r(t): Nyquist impulse response.
We call W_f the waveform factor. A system is resistant to waveform distortion if the waveform factor is small.

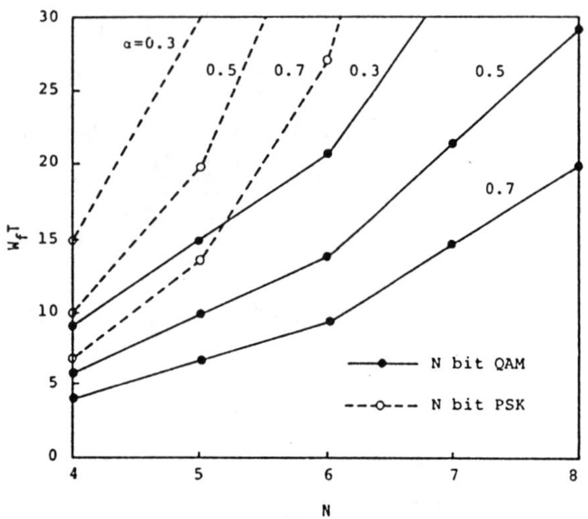

Fig.1 Waveform factors of high-level modulation systems (α: Roll-off factor)

Table 1. Waveform factors for binary system (Nyquist impulse response; α = 0.5)

	Basic system	SD	SD & IF EQL	SD & Tr. EQL	SD, IF EQL & Tr. EQL
W_fT	1.95	1.95	0.996	0.664	0.527

Fig. 1 shows the Nyquist channel waveform factor normalized by clock rate 1/T. From this result, it is clear that QAM with a large roll-off factor is superior to PSK or QAM with a small roll-off factor in multipath environments.

Table 1 shows the waveform factors for a binary system in the Nyquist channel with various countermeasures against multipath fading. These values were estimated from the data obtained in 5L-D1 field experiments [9].

2.2 Probability of Dispersive Outage

It is assumed that dispersive outages occur when the eye opening equals zero. The allowable Nyquist-band amplitude dispersion, Z_0(dB), can be obtained from the two-ray-path model as [9]:

$$Z_0 = \begin{cases} 10 \log_{10}\left(1 + \frac{4}{W_f T}\right)^2 & \tau/T \leq \frac{1}{2} \\ 10 \log_{10}\left(1 + \frac{2}{W_f \tau}\right)^2 & \tau/T > \frac{1}{2} \end{cases} \quad (8)$$

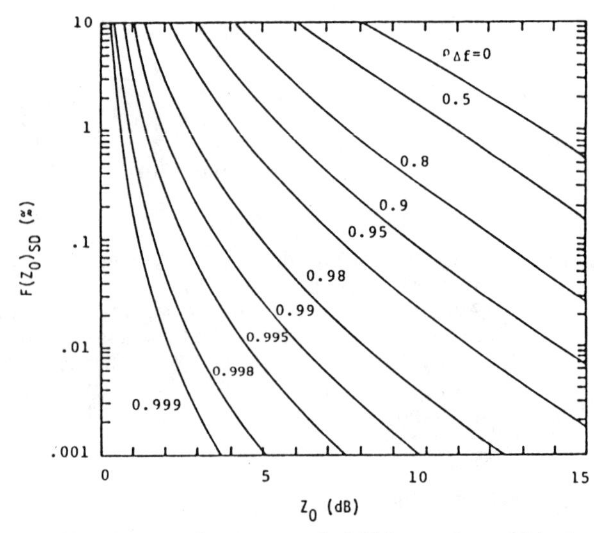

Fig.2 Cummulative probability of amplitude dispersion (SD reception)

The cummulative probabilities of Z_0 dB amplitude dispersion for single-antenna reception and dual space-diversity reception can be given as [10]:

$$F(Z_0) = 1 - \frac{1 - Z}{\sqrt{(1+Z)^2 - 4\rho_{\Delta f} Z}} \equiv 2m \quad (9)$$

$$F(Z_0)_{SD} = 6m^2 - 4m^3 \quad (10)$$

where,
$\rho_{\Delta f}$: frequency correlation factor in the Nyquist-band, and
$Z = 10^{Z_0/10}$.

Fig.2 shows the cummulative probability of amplitude dispersion for space diversity reception.

The probability of dispersive outage is easily determined as the same probability of allowable Nyquist-band amplitude dispersion, so it can be given as:

$$P_d = 2m \quad (11)$$

$$P_{dSD} = 6m^2 - 4m^3 \quad (12)$$

For minimum in-band dispersion combining space-diversity (MID SD) reception, the frequency correlation factor can be transformed as [11]:

$$\rho_{\Delta fMID} = 0.46277\rho_{\Delta f} + 0.53723 \quad (13)$$

Other countermeasures against multipath fading, such as use of an adaptive equalizer or transversal equalizer, reduce the system waveform factor, as shown in Table 1.

2.3 Upper Baud Limit

To demonstrate the relationship between the probability of outage and the clock rate, we now consider two representative paths. Fig.3 (a) and (b) show a typical mountain path and a severe over-the-sea path in Japan. The probabilities of dispersive outages for these paths have been estimated and are shown in Figs.4 (a) and (b). The outage objectives during fading are also presented in these figures. These objectives are per hop outage probabilities derived from the CCIR performance objective (0.05 %/2500 km) divided by Rayleigh fading occurrence probabilities over the paths.

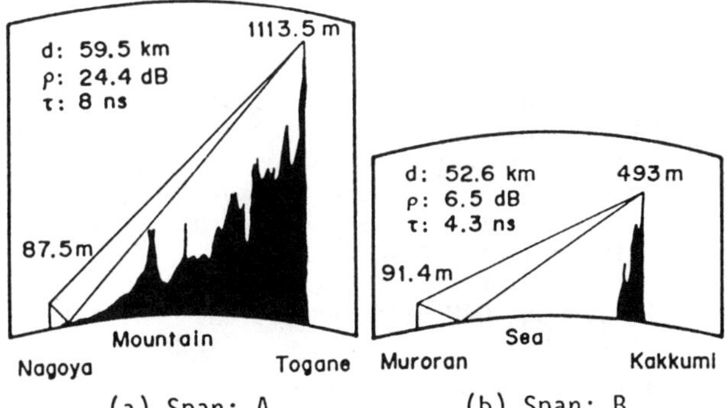

Fig.3 Profiles of transmission path models

5L-D1, using 16 QAM, has the transmission capacity of 200 Mb/s in 80 MHz same polarization channel spacing [3]. Under the same condition, the transmission capacity of 300 Mb/s and 400 Mb/s can be achieved when 64 QAM and 256 QAM are employed, respectively. IF dispersive outage is allowed to be half the outage objective, the following would be concluded:

i) In case of 64 QAM 300 Mb/s system, 25 MB two-carrier system with MID SD is required over the mountain path, and 12.5 MB four-carrier system with MID SD and Tr. EQL over the sea path.

ii) In case of 256 QAM 400 Mb/s system, 12.5 MB four-carrier system with MID SD is required over the mountain path, and 6.25 MB eight-carrier system with MID SD and Tr. EQL over the sea path.

3. PERFORMANCE REQUIRED FOR A 256 QAM MODEM

3.1 Bit Error Rate

The 256 QAM has 16 levels in the baseband signal. The baseband pulse is obtained from the D/A conversion rule as:

$$S_1 = 2^3 a_1 + 2^2 a_2 + 2^1 a_3 + 2^0 a_4 \qquad (14\text{-a})$$

$$S_2 = 2^3 b_1 + 2^2 b_2 + 2^1 b_3 + 2^0 b_4 \qquad (14\text{-b})$$

where, $(a_1, b_1), \ldots, (a_4, b_4)$ are binary and defined as "path 1",..,"path 4" signals.

The BER depends on differential encoding. Circular-symmetry differential encoding is employed to obtain a 256 QAM eight-bit signal set. The first two bits, "path 1" signals, are differentially encoded by a modulo-four adder and represent the change in quadrant. The remaining six bits within the quadrant are Gray coded. Code assignments in quadrants other than the first quadrant is rotationally symmetric to those in the first quadrant. In this encoding scheme, the BER of the "path 1" signal, P_{e1}, is increased by a factor of two, whereas the others are increased by the BER of the pre-decoded "path 1" signals, 1/2 P_{e1}, as:

$$P_{e1} = 2 \cdot \frac{1}{8} \left\{ \sum_{n=1}^{8} \frac{1}{2} \operatorname{erfc}\left[\frac{(2n-1)\delta}{\sqrt{2}\sigma} \right] \right\}$$

$$\fallingdotseq \frac{1}{8} \operatorname{erfc}\left(\frac{\delta}{\sqrt{2}\sigma} \right) \qquad (15)$$

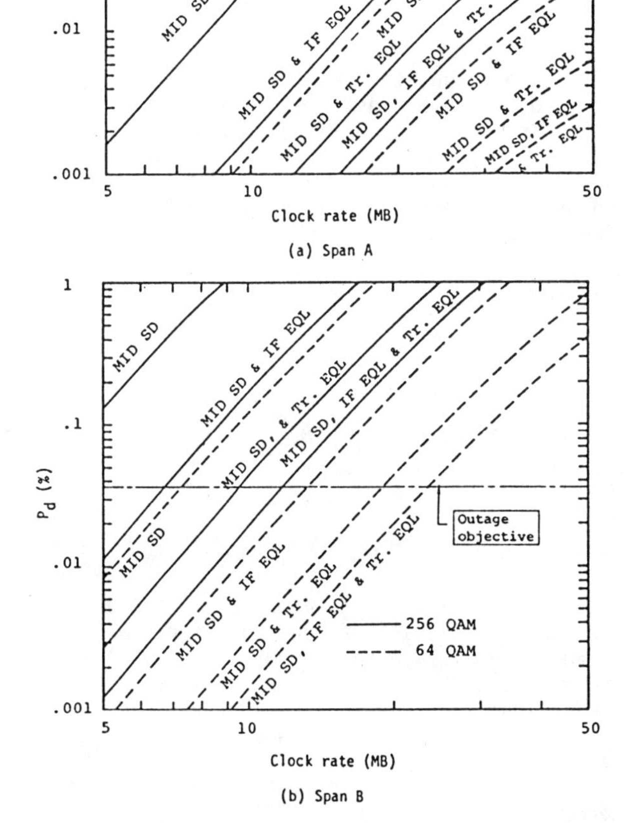

Fig.4 Probability of 64/256 QAM dispersive outage (during fading)

$$P_{e2} = \frac{1}{4} \left\{ \sum_{n=1}^{4} \frac{1}{2} \operatorname{erfc}\left[\frac{(2n-1)\delta}{\sqrt{2}\sigma} \right] \right\} + \frac{1}{2} P_{e1}$$

$$\fallingdotseq \frac{3}{16} \operatorname{erfc}\left(\frac{\delta}{\sqrt{2}\sigma} \right) \qquad (16)$$

$$P_{e3} = \frac{1}{2} \left\{ \sum_{n=1}^{2} \frac{1}{2} \operatorname{erfc}\left[\frac{(2n-1)\delta}{\sqrt{2}\sigma} \right] \right\} + \frac{1}{2} P_{e1}$$

$$\fallingdotseq \frac{5}{16} \operatorname{erfc}\left(\frac{\delta}{\sqrt{2}\sigma} \right) \qquad (17)$$

$$P_{e4} = \frac{1}{2} \operatorname{erfc}\left(\frac{\delta}{\sqrt{2}\sigma} \right) + \frac{1}{2} P_{e1}$$

$$\fallingdotseq \frac{9}{16} \operatorname{erfc}\left(\frac{\delta}{\sqrt{2}\sigma} \right) \qquad (18)$$

where

2δ: minimum distance between signal points,

σ^2: white Gaussian noise power, and

$$\operatorname{erfc}(x) = \frac{2}{\sqrt{\pi}} \int_x^{\infty} e^{-t^2}\, dt.$$

The average CNR of 256 QAM can be obtained as:

$$C/N = 85\delta^2/\sigma^2 \equiv K_0^2 \quad (19)$$

Then, the average BER for 256 QAM becomes

$$P_e = \frac{1}{4}\sum_{n=1}^{4} P_{en} = \frac{19}{64}\,\mathrm{erfc}\left(\frac{\delta}{\sqrt{2}\sigma}\right)$$

$$= \frac{19}{64}\,\mathrm{erfc}\left(\frac{K_o}{\sqrt{170}}\right) \quad (20)$$

3.2 Recovered Carrier Steady-State Phase Error/Modulation Phase Error

To locate a signal set in one quadrant, let us use r and θ, the polar coordinates. The amplitude of demodulated signal set increases in one quadrant but decreases in another as detected by a reference carrier with phase error θ_e. The 256 QAM probability of error is:

$$P_e(\theta_e) = \frac{1}{4}\sum_{n=1}^{4}\frac{1}{2}\{P_{en}(\theta_e) + P_{en}(-\theta_e)\}$$

$$\leq \frac{1}{2}\{P_{e4}(\theta_e) + P_{e4}(-\theta_e)\}$$

$$= \frac{1}{2}\cdot\frac{9}{16}\cdot\frac{1}{64}\sum_{i=1}^{8}\sum_{j=1}^{8}\mathrm{erfc}\left[\frac{K_o}{\sqrt{170}}\right.$$

$$\left.|||\frac{r_{ij}}{\delta}\cos(\theta_{ij} \pm \theta_e) - 8|-4|-2||\right] \quad (21)$$

The modulation phase error can be handled in the same way as reference carrier phase error; however, parameter θ_e in Eq. (21) must be replaced by $\theta_{mmax} - \overline{\theta}_m$, which is the maximum deviation from the mean value of the modulation phase error. This is because the reference carrier is phase-locked to $\overline{\theta}_m$.

Equivalent CNR degradation due to phase error is derived from Eq. (21) as the CNR increase necessary to obtain the required BER performance from the theoretical ideal value and is shown in Fig. 5. From this figure, we see that phase control within $\pm$ 0.5 deg. is necessary for 256 QAM to obtain less than 0.8 dB.

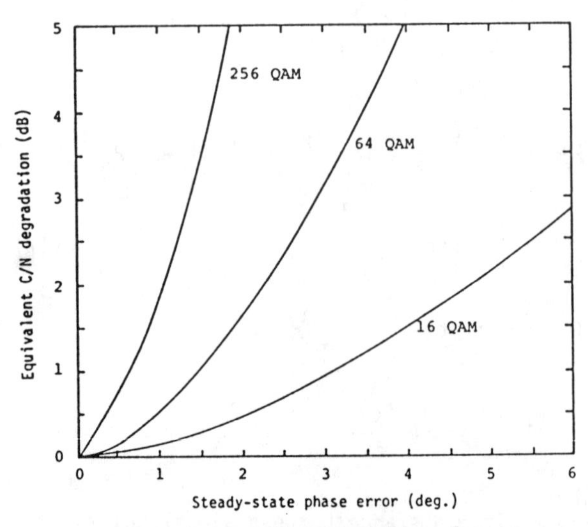

Fig.5 Equivalent C/N degradation due to reference carrier phase error (θ_e) or modulation phase error ($\theta_{mmax}-\overline{\theta}_m$) (BER: 10^{-6})

3.3 Carrier Jitter/Interference Noise

It can be presumed that recovered carrier jitter, local oscillator phase noise, echo noise and interference noise cause the similar performance degradation. These degradation factors are independent random variables that occur at various places in a repeater. Therefore, they can be treated as Gaussian noise from the view point of the central-limit theorem.

If the power sum of these factors is N_I, the BER for 256 QAM with interference noise is:

$$P_e(N_I) = \frac{19}{64}\,\mathrm{erfc}\left(\frac{\delta^2}{\sqrt{2}\sqrt{\sigma^2 + N_I}}\right)$$

$$= \frac{19}{64}\,\mathrm{erfc}\left(\frac{K_o}{\sqrt{170}}\,\frac{1}{\sqrt{1 + \frac{N_I}{\sigma^2}}}\right) \quad (22)$$

The equivalent CNR degradation is:

$$D_N = -10\log_{10}\{1 - \frac{(C/N)_{min}}{(C/N_I)}\} \text{ (dB)} \quad (23)$$

where

$(C/N)_{min}$: theoretical ideal value to obtain the required BER.

Fig. 6 shows the equivalent CNR degradation at a BER of 10^{-6}. From this figure, carrier jitter performance of more tan 42 dB is required for 256 QAM to obtain less than 0.5 dB CNR degradation.

3.4 Clock Phase Error

Phase offset from the best sampling time reduces eye opening and degrades the BER, particulary in a multi-level Nyquist system. If we define η and p(η) as the amplitude increase factor at the time of sampling and its probability of occurrence, the BER for 256 QAM with clock phase error becomes:

$$P_e(\eta) = \sum_{\eta} p(\eta)\,\frac{19}{64}\,\mathrm{erfc}\left(\frac{\eta K_o}{\sqrt{170}}\right) \quad (24)$$

The minimum value of η with clock phase offset, θ_{cL}, is obtained from the Nyquist system impulse response, g(t):

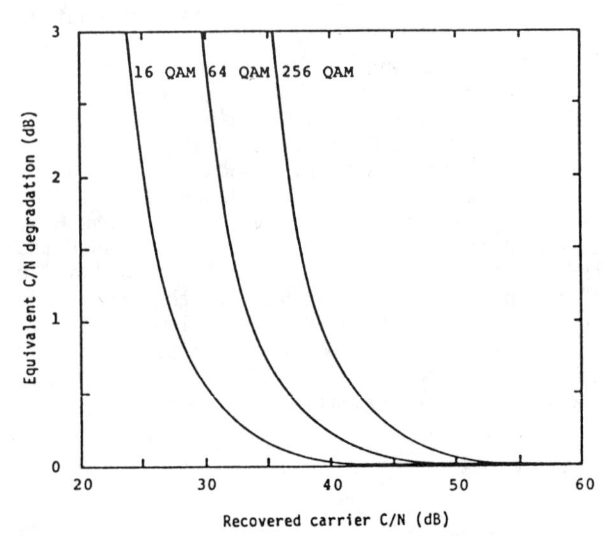

Fig.6 Equivalent C/N degradation due to recovered carrier jitter (BER: 10^{-6})

$$\eta_0(\theta_{cL}) = \mathrm{Min}(S)g(\theta_{cL}) - \mathrm{Max}(S)\sum_{n\neq 0} |g(nT+\theta_{cL})|$$

$$= g(\theta_{cL}) - 15 \sum_{n\neq 0} |g(nT+\theta_{cL})| \qquad (25)$$

The maximum probability of error occurs for the data sequence that gives the minimum eye opening. For this sequence, equivalent CNR degradation due to clock phase error yields:

$$D_{cL} = -20 \log_{10} \eta_0 \quad (\mathrm{dB}) \qquad (26)$$

It should be noted from Fig. 7 that an accuracy of ± 0.5 degrees is required for a clock synchronizer to obtain less than 0.3 dB equivalent CNR degradation in 256 QAM.

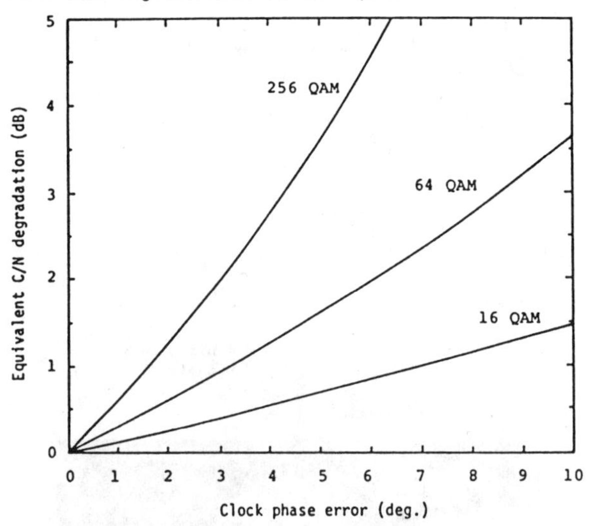

Fig.7 Equivalent C/N degradation due to clock phase error (BER: 10^{-6}, $\alpha = 0.5$)

3.5 Decision Threshold Drift/Decision Uncertain-Region

Decision threshold or demodulated signal level drifts according to temperature variation, power supply voltage variation and aging. Moreover, the A/D decision circuit for high speed operation has a decision uncertain-region in which the probability of error reaches 0.5.

The previous degradation factors are represented by v_d. The 256 QAM probability of error is easily derived, when the demodulated signal peak-to-peak amplitude is $2V_0$, as:

$$P_e(v_d) = \frac{1}{2}\cdot\frac{19}{64}\{\mathrm{erfc}(\frac{\delta - v_d}{\sqrt{2}\sigma}) + \mathrm{erfc}(\frac{\delta + v_d}{\sqrt{2}\sigma})\}$$

$$= \frac{19}{128}\,\mathrm{erfc}[\frac{K_o}{\sqrt{170}}(1 \pm \frac{15v_d}{V_o})] \qquad (27)$$

$$(\because V_o = 15\delta)$$

Due to the threshold uncertain region, v_w, the BER deteriorates to:

$$P_e(v_w) = \frac{19}{128}\,\mathrm{erfc}[\frac{K_o}{\sqrt{170}}(1 \pm \frac{15v_w}{2V_o})] \qquad (28)$$

From Eqs. (27) and (28), degradation factors v_d and v_w can be treated as one parameter:

$$v_T = \frac{|v_d| + 0.5v_w}{V_o} \qquad (29)$$

In this case, the BER for 256 QAM can be represented as:

$$P_e(v_T) = \frac{19}{128}\,\mathrm{erfc}[\frac{K_o}{\sqrt{170}}(1 \pm 15v_T)] \qquad (30)$$

The equivalent CNR degradation at a BER of 10^{-6}, caluculated using the above equation,is shown in Fig. 8. In that figure, we see that 0.25 % resolution and a 0.5 % uncertain-region are required for an A/D decision circuit to obtain 0.3 dB equivalent CNR degradation in 256 QAM.

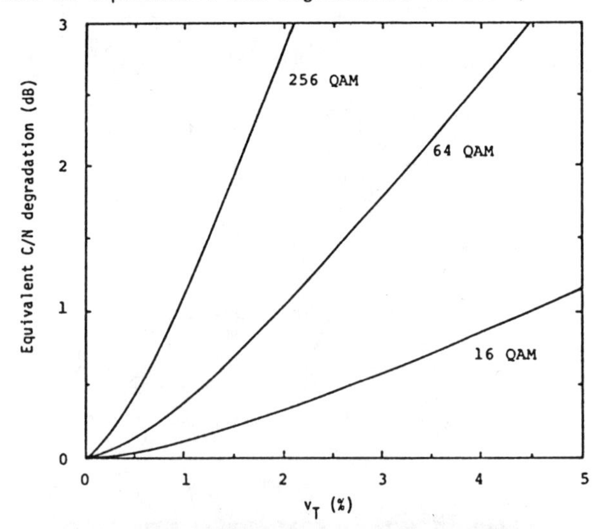

Fig.8 Equivalent C/N degradation due to decision threthold drift

4. EXPERIMENTAL CONSIDERATIONS

4.1 256 QAM Multi-carrier Modem

The multi-carrier transmission method is a powerful tool for overcoming severe propagation impairment. It has been demonstrated that a four- to eight-carrier 256 QAM modem would be required in order to double the transmission capacity of a 200 Mb/s 5L-D1 digital radio system [3] with the same bandwidth.

The configuration of a four-carrier 256 QAM modem is shown in Fig. 9. The transmitting terminal equipment converts 400 Mb/s into eight rails of 12.5 MB x 4 pulse streams. Four of these pulse streams are converted into 16-level signals by a D/A converter. Each 16-level signal is roll-off spectral shaped and modulates a local oscillator. The IF frequencies of the local oscillators are seperated by 20 MHz. The 256 QAM four-carriers are combined by a hybrid circuit and supplied to the transmitter.

At the receiver, each 256 QAM signal is coherently detected, and regenerated by the 8 bit A/D converters. The most significant bits, $(\hat{a}_1, \hat{b}_1)$, are the estimated "path 1" signals. The estimated fifth bits, $(\hat{a}_5, \hat{b}_5)$, contain the error signal which control the VCO of carrier recovery PLL. Phase control voltage is derived from the product of these signals and "path 1" signals as follows:

$$e_v = \hat{a}_1\hat{b}_5 - \hat{b}_1\hat{a}_5 \qquad (31)$$

One of these units has been constructed in our laboratory. This prototype 256 QAM modem operates at a 12.5 MB clock rate and a 140 MHz IF frequency.

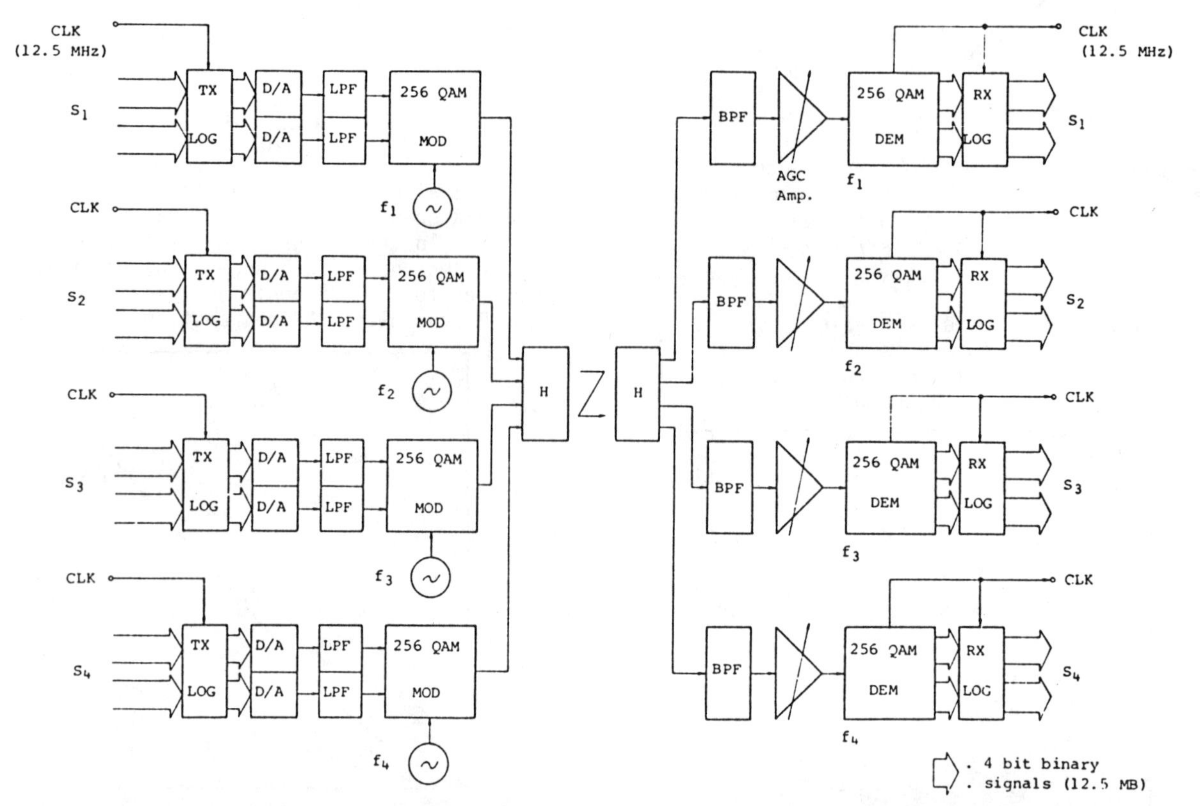

Fig.9 Block diagram of four-carrier 256 QAM modem

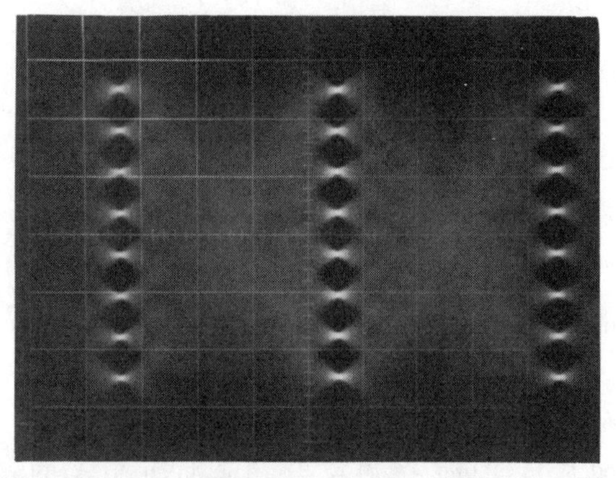

(a) 64 QAM demodulated eye pattern
(H:20 ns/div)

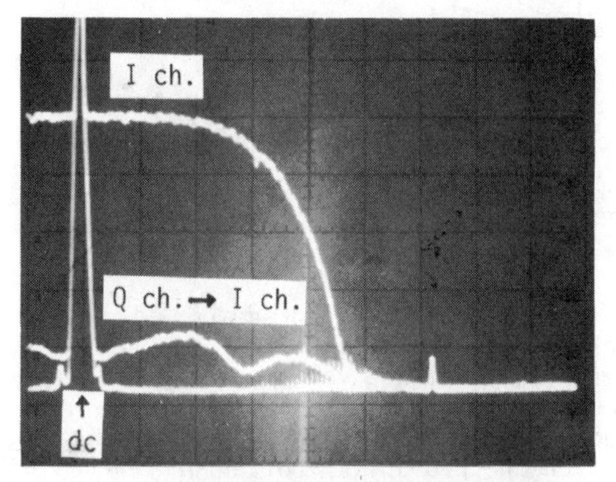

(c) 256 QAM demodulated spectrum
H:2 MHz/div, V:10 dB/div
100 kHz IF BW
100 Hz video filter

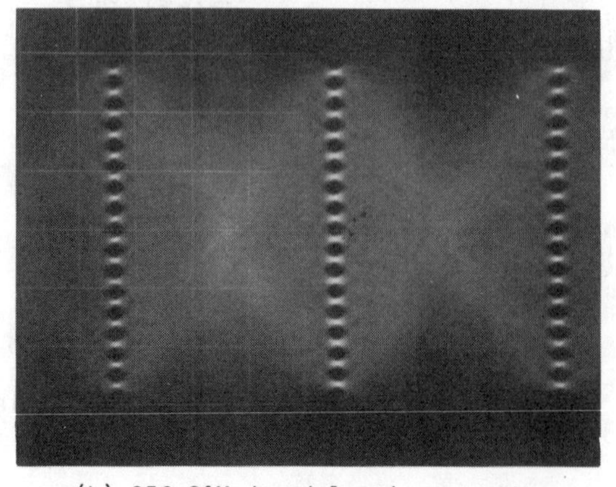

(b) 256 QAM demodulated eye pattern
(H:20 ns/div)

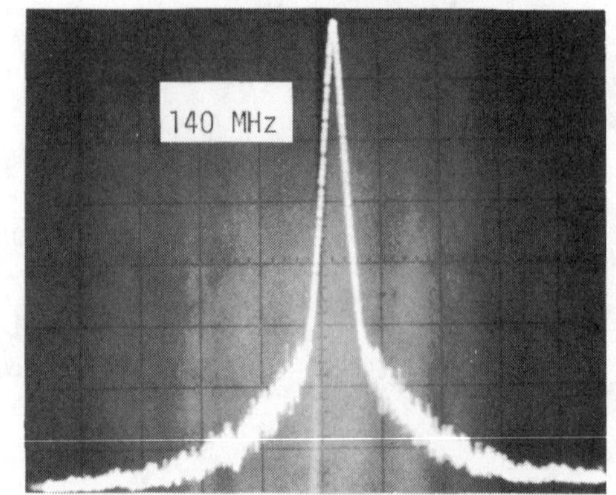

(d) 256 QAM recovered carrier spectrum
H:10 kHz/div, V:10 dB/div
1 kHz IF BW
100 Hz video filter

Fig.10 Demodulated eye pattern and spectrum

4.2 Measured Performance

The 64/256 QAM demodulated eye patterns, demodulated power spectrum, orthogonal interference power spectrum and recovered carrier spectrum are shown in Fig.10. Orthogonal interference, which is leakage from the quadrature phase to the in-phase channel, can be viewed by modulating only the quadrature channel and detecting the in-phase channel, taking advantage of the cunning carrier. This interference, a D/U ratio of about 40 dB, is produced by modulation phase error, phase detection nonlnearity, and amplitude and delay distortions in the filter system. Recovered carrier to noise power ratio of more than 40 dB is obtained. The equivalent CNR degradation due to carrier jitter is about 0.8 dB.

Fig. 11 shows the BER performance. Equivalent CNR degradation of less than 3 dB has been realized for 64 QAM. Long term bit error count measurement in a noise free state has shown that the 256 QAM modem produses no errors. Equivalent CNR degradation of about 4 dB has been obtained by precise adjustment of the phase orthogonality, decision level, decision timing, etc.

5. CONCLUSION

Two basic items that influence realization of high-level modulation schemes, baud boundary and required modem performance, have been discussed. Additionally, the performance of a 64/256 QAM prototype modem was described. However, this study is only a first step toward development of a 256 QAM multi-carrier system. Further work on multipath countermeasures and interference cancellers, in addition to that on a multi-carrier modem, are necessary.

Certain technical breakthroughs for development of a 256 QAM multi-carrier system should come from LSI, or other device, technology. By utilizing such advanced technology, a 256 QAM multi-carrier system with 10 bits/s/Hz frequency utilization efficiency can be realized in the near future.

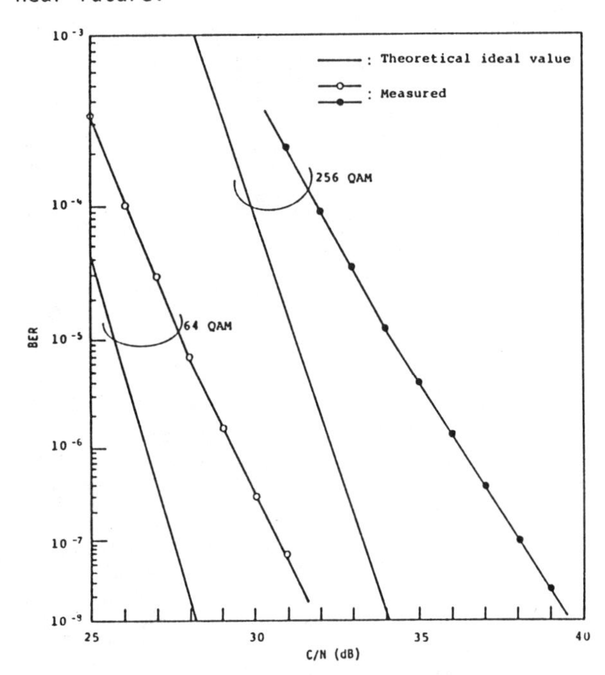

Fig.11 64/256 QAM BER performance (Clock rate: 12.5 MB)

ACKNOWLEDGEMENT

The authors wish to thank Dr. K. Kohiyama and Mr. T. Murase for their helpful comments and discussions.

REFERENCES

[1] H.D.HYAMSON, et.al., "An 11 GHz High Capacity Digital Radio System for Overlaying Existing Microwave Routes", IEEE Trans. Commun., vol. COM-27, No.12, PP. 1928-1937, Dec.1979.

[2] Y.TAN, et.al., "2 GHz Band Digital Radio Equipment Employing 8-Level PSK with Cosine Roll-off Spectrum Shaping", ICC'78, pp. 33.3.1-33.3.5.

[3] Y.SAITO, et.al., "5L-D1 Digital Radio System", ICC'82, pp. 2B.1.1-2B.1.7.

[4] M.ACX, et.al., "140 Mbit/s 32 QAM Modem for High Capacity Digital Radio Systems", ICC'83, pp. F2.7.1-F2.7.5.

[5] T.NOGUCHI, et.al., "6 GHz 135 MBPS Digital Radio System with 64 QAM Modulation", ICC'83, pp. F2.4.1-F2.4.6.

[6] P.R.HARTMANN, et.al., "135 MBS-6GHz Transmission System Using 64-QAM Modulation", ICC'83, pp. F2.6.1-F2.6.7.

[7] T.YOSHIDA, et.al., "System Design and New Techniques for an Over-Water 100 Km Span Digital Radio", ICC'83, pp. C.2.7.1-C.2.7.7.

[8] T.MURASE, "A Unified Analysis of Multi-path Degradation in Multi-level Modulation Digital Radio Systems", in Japanese, Trans. IECE JAPAN, to be published.

[9] T.MURASE, Yokosuka ECL Technical Document, in Japanese.

[10] S.SAKAGAMI, et.al., "Some Experimental Results on In-band Amplitude Dispersion and a Method for Estimating In-band Linear Amplitude Dispersion", IEEE Trans. Commun., vol. COM-30, No.8, pp. 1875-1888, Aug. 1982.

[11] K.TAJIMA, et.al., "Performance Evaluation of Minimum Dispersion Combiner", in Japanese, Trans. IECE JAPAN, vol.J66-B, No.3, pp. 367-374, March 1983.

Part III
Multipath Channel Models

Multipath Fading Channel Models for Microwave Digital Radio

WILLIAM D. RUMMLER, REGINALD P. COUTTS AND MARKUS LINIGER

Editor's Note: This tutorial article was originally published in the November 1986 issue of the IEEE COMMUNICATIONS MAGAZINE, as part of the Special Series on Microwave Digital Radio. It has been updated by the authors and edited for inclusion in this book.

INTRODUCTION

PREVIOUS tutorials in this book [1], [2] have presented an overview of microwave digital radios and their modulations. In the present paper, we consider the problem of modeling the effects that the atmosphere has on digital radio signals as they pass through. Such a characterization is necessary to evaluate the performance of digital radios. Each of the authors has contributed to various aspects of this problem, although from somewhat different viewpoints. The objective of this article is to give an overview and provide a framework within which the work in this area can be put into perspective. In so doing, we hope to provide a useful introduction for the general reader and to clarify the underlying concepts and differences for specialists in the area.

Modern digital radios operating on line-of-sight paths under normal propagation conditions will operate essentially error free, i.e., with bit error ratios less than 10^{-10}. Radio designs achieve this performance by taking advantage of the margins engineered into the equipment so that it will meet network outage objectives under fading conditions. The allocation of these objectives to a radio on a path of, say, 40 km requires it to maintain a bit error ratio of better than 10^{-3} for all but tens of seconds in a year. Thus, the essence of the modeling problem is to represent the effect of propagation defects, or anomalous propagation conditions, which may be present on the path during some hundreds of seconds in a year.

The study of propagation effects on line-of-sight paths began with the introduction of frequency modulated (FM) systems in the early 1950's. Although much work was done to develop propagation models for predicting the performance of FM systems, the introduction of digital radios in the 1970's fueled an interest in more detailed models, because digital radios are sensitive to aspects of the channel distortions that have little effect on FM systems.

An FM signal contains most of its energy at the carrier frequency and carries its information redundantly encoded in its sidebands. Hence, it is adversely affected by loss of power at the carrier frequency, but relatively unaffected by loss of signal in only one of the sidebands. For this reason radio channel modeling for FM systems focuses on single-frequency fading statistics.

Under many propagation conditions, the loss or gain of signal due to the atmosphere is uniform across the radio channel bandwidth; an example is the loss due to extreme rain rate. In these cases, the approaches developed for the FM systems are applicable to digital radios. However, unlike FM signals, a spectrally efficient digital radio signal does not have redundant information in its sidebands. Consequently, the selective loss of some of its frequency components can affect the detectability of the received signal. Thus, the modeling of frequency selective effects is crucial for evaluating the performance of digital radios.

When the signal loss imposed by the atmosphere varies across the frequency band occupied by a radio channel, the channel is said to be experiencing selective fading. The primary cause of selective fading on line-of-sight microwave radio paths is multipath propagation. This occurs when the index of refraction of the atmosphere allows energy leaving the transmitting antenna at different angles to travel to the receiving antenna along slightly different paths with different time delays. In the next section, we provide a brief description of the multipath fading phenomenon and the requirements for measuring and modeling it. The section following that one provides a view of models that are used to characterize multipath fading in a single channel. Models used for diversity and cross-polarized channels are described in the two subsequent sections, followed by a discussion of channel dynamics.

MULTIPATH FADING–OVERVIEW

Free Space Propagation

High capacity digital radios operate over networks of paths in assigned frequency bands from 2 to 18 GHz [3]. The frequency bands are subdivided into channels with bandwidths of about 0.5 percent (e.g., 20 MHz in the 4 GHz band). Typically, links in the bands below 10 GHz use antennas with beam widths of about 1 degree, have path lengths of about 40 km, and antenna tower heights of 50 to 100 meters. For higher frequencies, considerations of the attenuation due to rain lead to significantly shorter path lengths.

Under normal conditions, the decrease in the index of refraction with height imparts a downward curvature to the

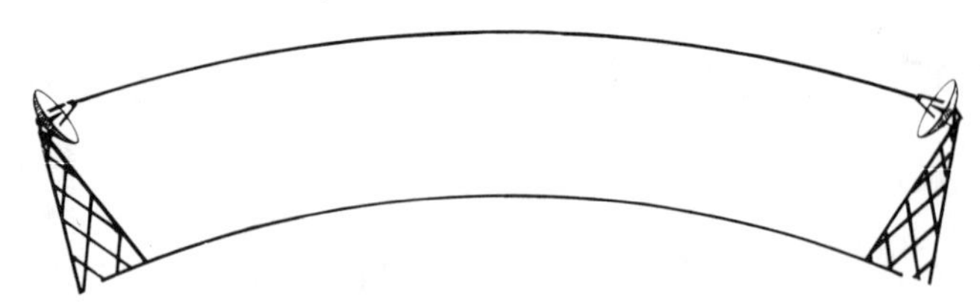

Fig. 1. Free Space Propagation.

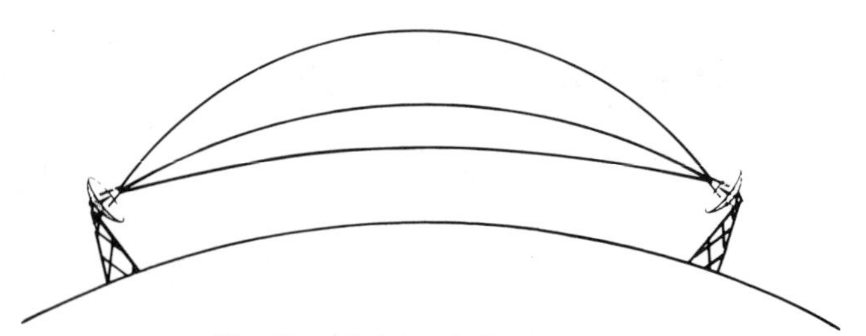

Fig. 2. Multipath Propagation.

direction of propagation over the path, as shown in Fig. 1. Indeed, variations in the index of refraction are the dominant cause of propagation problems on line-of-sight paths. Allowances for reduced rates of decrease with height (subrefractive conditions), which sometimes occur, are the controlling consideration in determining tower heights to clear the intervening terrain.

Multipath Phenomena

For a large fraction of the time in most locations, propagation conditions do not vary significantly from those shown in Fig. 1, and digital radios operate essentially error free. However, at other times, meteorological conditions lead to the formation of unusual temperature and humidity profiles; these, in turn, impose structure on the index of refraction profile. There are various atmospheric conditions that are conducive to the formation of such layered atmospheres; for example, in temperate climates, hot, humid summer evenings with no clouds and little wind frequently provide appropriate conditions.

In a layered atmosphere, energy that would normally be radiated into space can be refracted down to the receiving antennas by other paths, as shown in Fig. 2. The receiver sees a weighted sum of time shifted replicas of the transmitted signal from these multiple atmospheric layers, and possibly from ground reflections as well. The impulse response of such a channel can be represented by a weighted sum of delta functions, i.e.,

$$h(t)=\sum_{k=1}^{N}\alpha_k\delta(t-\tau_k). \tag{1}$$

The corresponding frequency response, or complex voltage transfer function at radian frequency ω, is given by

$$H(j\omega)=\sum_{k=1}^{N}\alpha_k e^{-j\omega\tau_k}. \tag{2}$$

The essence of the channel modeling problem is to characterize $H(j\omega)$ in the frequency interval (channel bandwidth) of interest in a way that accounts for its temporal variability. Fig. 2 provides an elementary physical description of the multipath fading problem. It is important to note that this description, in itself, is a model of atmospheric propagation conditions. Such atmospheric propagation models are developed using ray tracing techniques from assumptions or measurements of the variation of index of refraction with height [4]–[8]. It is necessary to distinguish between *atmospheric* models, which describe the physical propagation, and *channel* models, which describe the frequency response of the channel. The description of atmospheric propagation is the point of departure for the multipath fading channel models to be described in this paper. The interrelation between these two types of models is one of the most difficult aspects of characterizing multipath phenomena.

The Channel Characterization Problem

Since the representation of multipath transfer functions, $H(j\omega)$, is a common concern of all multipath channel models, it is worthwhile to provide a brief description of how this function is measured. The transfer function of a radio channel can be described in magnitude-phase form,

$$H(j\omega)=|H(j\omega)|\,e^{j\Phi(\omega)}. \tag{3}$$

Specifically, one might measure the dB attenuation $A(\omega)$ and the delay distortion (or group delay) $D(\omega)$, where

$$A(\omega)=-20\log|H(j\omega)|$$
$$D(\omega)=-\frac{\partial\Phi(\omega)}{\partial\omega}. \tag{4}$$

Attenuation is measured in decibels relative to the free space value for the path, while the reference for delay distortion is arbitrary.

Fig. 3 shows an example of wideband swept measurements of attenuation and delay during a multipath fading period [9]. Although there is considerable structure in the responses shown, they serve to illustrate the basic defect introduced by multipath fading. Frequency selective minima in power are accompanied by maxima or minima in the delay distortion. The delay distortion trace with a minimum corresponds to a minimum phase condition, in the circuit theory sense.* The trace with a maximum, measured three seconds later, corresponds to a nonminimum phase condition. This transformation is equivalent to replacing a given complex voltage transfer function, $H(j\omega)$, by its complex conjugate; this operation leaves the amplitude of the transfer function unchanged, but reverses the sign of the delay distortion.

A precise physical description of the channel during a transition from the minimum to the nonminimum phase state is difficult to determine. Although the phase transition of ray models can be described in several different ways with different consequences [10], the most convincing explanations of the event shown involve at least three distinct rays [11]. No statistical studies have described how often such transitions

* As we shall see, in the narrowband channels of interest, $H(s)$ may be characterized as a single zero function. If the zero is in the left (right) half plane, the corresponding notch is minimum (nonminimum) phase.

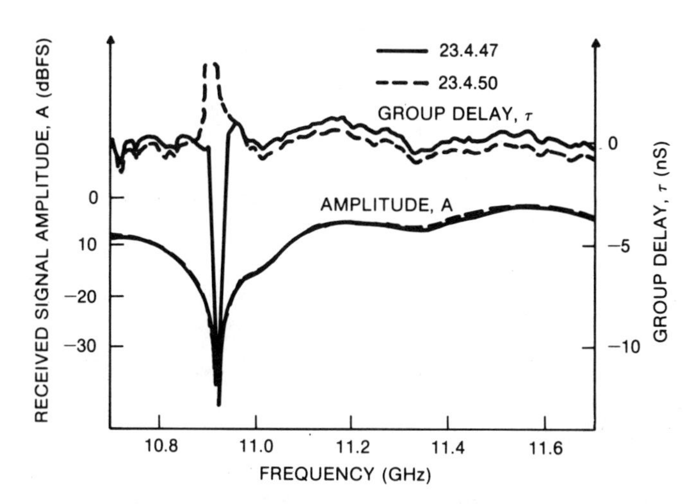

Fig. 3. Wide Band Measurement of a Multipath Propagation Event.

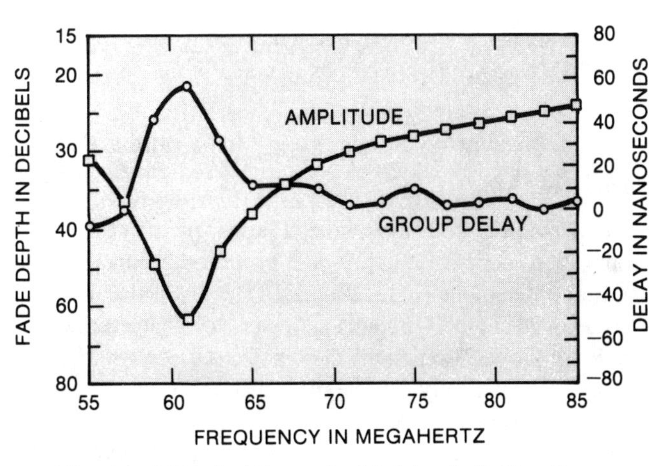

Fig. 4. A Typical Scan of a Multipath Fading Event in a 6-GHz Radio Channel.

occur; however, several studies have reported the relative probability of occurrence of the two states [12]–[15]. Generally, the minimum phase state is more probable for shallow fading (less selective events), and the states are more nearly equally probable for severe fading (more selective events).

Propagation conditions are usually stable enough that scanning the frequency axis several times a second produces a meaningful sequence of observations of the type shown in Fig. 3. However, conditions are unstable enough that the observed shape is constantly changing and the transmission defects move and change unpredictably. The wideband modeling of such conditions typically requires multiple rays of rapidly changing amplitudes and delays.

To fully characterize the medium, one would need to measure attenuation and delay distortion over a bandwidth of several GHz. Such wide bandwidths offer the prospect of identifying the physical parameters of the multipath. However, practical considerations of equipment complexity limit the observed bandwidths to about 1 GHz, and few experiments of this type have been reported [9], [12], [16]–[20].

The wideband measurements show that, within a radio channel, which typically has a bandwidth of 50 MHz or less, there is usually only a single amplitude minimum. Consequently, measurements are more frequently made over a narrow channel bandwidth, as shown in Fig. 4. Within such a narrow band one cannot hope to resolve and estimate the propagation path delays, but only to represent the response with a modeling function. Moreover, if one is interested in measuring propagation conditions while simultaneously observing the performance of a digital radio, the power spectral energy of the digital signal can be used to measure the channel amplitude response [$A(\omega)$ in (4)]; however, no delay distortion information can be obtained by this method.

What is a Multipath Fading Channel Model?

A multipath fading channel model provides the means for estimating the amount or fraction of time that the propagation conditions on a path will be too severe for a radio to meet an acceptable performance criterion. There are three basic components to such a model:

1. A *Channel Modeling Function* which approximates $H(j\omega)$ over a finite frequency interval by suitable choices of the parameters of the function;
2. The *Joint Probability Distributions* for the parameters of the function, conditioned on the presence of multipath fading on the path; and
3. A *Scale Factor* which represents the probability of multipath fading on the path, or, equivalently, the multipath fading time for the path. This multiplicative factor is applied to the conditional probabilities (2). Usually, results are scaled to an annual or to a "worst month" basis.

The path specifics, primarily terrain characteristics and climatology, determine the values of the scale factor and the parameters of the probability distributions.

To complete a calculation of outage time, one must characterize the radio in terms of the model function parameters. Such a characterization, called a radio *signature*, may be defined for any specified performance threshold, e.g., BER = 10^{-3}. A signature is the locus of points in parameter space that separates acceptable performance from unacceptable performance. We use the term signature in this general sense to accommodate the description of all the models considered.

Since some equalizers perform quite differently under minimum phase conditions than for nonminimum phase conditions, signatures are usually measured for both. In general, these signatures are noticeably different, and the total outage is taken as the weighted sum of the contributions from the two situations. Although the appropriate weighting of the two situations may depend on the path and propagation conditions as well as on the radio equipment, an equal weighting is often taken as a worst case condition. Whether this is appropriate for systems employing diversity is less clear.

Note that models fitting the preceding description are not the only methods of predicting digital radio outage. Other methods do not make explicit use of a modeling function. They are based on observed fading statistics [21], [22], and/or on the joint statistics of single frequency fading and dispersion [23]-[26]. One of these methods is noted in the next section.

NONDIVERSITY SINGLE-POLARIZATION MODELS

Three-Path Models

Many investigators have studied the properties of three-path propagation models [19], [27] and proposed channel models based on three paths [25], [28]-[30]. The distinction between *atmospheric* models, which attempt to characterize the physics of propagation on the radio path, and *channel* models, which mathematically fit the propagation response over a finite bandwidth, can be illustrated by discussing two published models. They are the *general three-path model* [25], which exemplifies the former, and the *simplified three-path model* [28], which exemplifies the latter.

The general three-path model postulates a radio hop in which a low-amplitude surface-reflected ray is always present at the receiver. The effect of this ray is negligible except when atmospheric anomalies produce an additional ray, which interferes with the direct ray. This three-ray (or three-path) picture is particularly representative of propagation over water. The associated frequency response is given by (2) with three terms ($N = 3$). This representation has been used to develop statistics of linear amplitude dispersion in the channel, and estimates of outage have been developed by relating this dispersion to equipment performance [25]. Since this model does not fully characterize the channel response, it lies outside the direct concerns of this paper; however, outage calculations based on this model are considered in the tutorial article that begins Part V [31].

A direct modeling approach is to choose a modeling function which can accurately represent all propagation responses over the bandwidth of interest by a suitable choice of its adjustable (or free) parameters. The modeling method in this case consists of measuring many responses, choosing parameter values for each, and empirically deriving parameter statistics.

One such function, used in the simplified three-path model [28], is

$$H(j\omega) = a[1 - be^{-j(\omega - \omega_0)\tau}]. \tag{5}$$

The reason for calling this a three-path model will be explained shortly. The amplitude response for this function is shown in Fig. 5. The parameter a is a flat loss term and the quantity in brackets suggests the interference between two rays with relative delay τ producing a minimum in the response at a frequency ω_0. The frequency of the minimum is called the notch frequency. Both this and the frequency variable ω are measured from a common reference, usually the channel midfrequency. The loss term (or fade level) is measured in decibels as $A = -20 \log a$, and the relative notch depth as $B = -20 \log (1 - b)$. Thus, $A + B$ gives the total fade depth at the response minimum. The relative amplitude b ranges from zero to one, providing a minimum phase function for τ positive. The response is nonminimum phase when the sign of the delay is reversed ($0 < b < 1, \tau < 0$). The nonminimum

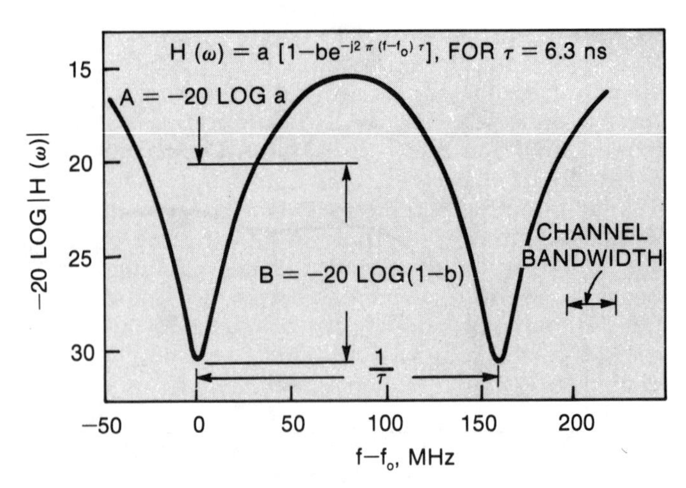

Fig. 5. Attenuation of the Modeling Function Used in the Simplified Three-Path Model.

phase state is also obtained when the relative amplitude of the delayed ray is greater than unity ($b > 1, \tau > 0$).

Note that (5) has the appearance of a two-path response, which is an occasional source of confusion. This response can be visualized as arising from three paths, wherein the direct path is unfaded; a second path is similar in strength and close enough in delay to the first that their composite response over the channel width is constant (the parameter a); and a third path at relative delay τ provides the frequency shaping of $H(j\omega)$. This picture of propagation is a commonly recurring one in models of atmospheric multipath propagation conditions, and it gives the simplified three-path model its name.

The channel modeling function (5) has been found to provide a good fit to almost all measured responses of narrowband radio channels. During multipath fading, these channels usually have only simple transmission defects that can be described as either attenuation slopes or single notches. For the purposes of representing the responses of such channels, this modeling function has too many parameters if all four are regarded as free. That is, within measurement errors, one cannot uniquely determine all four parameters from a given channel response measurement.

To avoid this difficulty, the delay parameter τ can be fixed at a convenient value that insures that the period of $H(j\omega)$ in frequency is large compared to the measurement bandwidth. In the original development of this model [28], τ was chosen to be the reciprocal of six times the measurement bandwidth, or 6.3 ns. This value has been accepted as standard by many workers, while others have followed the factor-of-six rule [13], [32]. The important point is that any fixed τ suffices if it permits the channel responses in almost all fading events to be fitted by (5), with some choice of (a, b, and ω_0). Note also that the joint statistics of these fitting parameters will depend on the choice of τ.

The methods used to estimate the parameters of the modeling function (5) from measurements of received power at a set of uniformly spaced frequencies have been thoroughly reported [28]. The joint statistics of the function parameters were carefully considered and their probability distributions were developed using simple representations. In particular, the distribution of the notch frequency is independent of the other parameters and is uniform over all frequencies of interest. The relative notch depth, B, and the fade level, A, parameters are partially correlated, with more shapely fades (larger B) occurring at more depressed levels (larger A). The B distribution is a one-sided exponential, and the A distribution is Gaussian (a is log-normal).

In addition to the function (5) and the statistics of A, B, and ω_0, there is the time scale factor, representing the number of fading seconds in a heavy fading month or year. Several papers have dealt with this issue [33]–[35], indicating how it can be related to the multipath occurrence factor [36] which was developed to characterize single-frequency fading in terms of path length, climate, and terrain.

Since its introduction, the simplified three-path model has been widely used in theoretical studies and simulations. Various investigators [13], [32], [33], [37] have reported the development of models based on the approach in [28]. The resulting parameter distributions have been surprisingly similar. The major differences are in the degree of correlation between fade levels (A) and notch depth (B), and in the mean of the fade level distribution. This latter parameter determines the relative occurrence of loss of power and occurrence of dispersion.

Finally, a word is in order about signatures for this model. The complete characterization of a radio in terms of this model requires a three-parameter signature, which interrelates the values of A, B, and ω_0 that cause a prescribed bit error ratio (BER), usually 10^{-3}. The characterization is presented as a family of A-B curves for a set of values of the notch frequency parameter. These curves can be developed from appropriate measurements using a channel simulator [34], or from a pair of dispersion signatures, one of which is measured at a lower BER [35].

The dispersion signatures, which are more commonly used, characterize the relative sensitivity of a given radio to fade shapes (neglecting thermal noise) in terms of the notch frequency, ω_0, and the relative notch depth, B. Fig. 6 shows the dispersion signatures for a 16-QAM radio for a BER of 10^{-3}. For all points below the curve, the radio will exhibit a higher BER. Since the notch frequency is uniformly distributed over the width of the signature, the probability of BER being greater than 10^{-3} is obtained by integrating the area under the curve weighted by the notch depth probability distribution. It will be noted that the constant delay of the modeling function allows the fabrication of a simple simulation circuit to implement these measurements [38].

Two-Path Models

Two-path models describe the multipath propagation of Fig. 2 in terms of a primary and a dominant interfering ray. Although many authors have used two-path or single echo models [39]–[47], they have used slightly different forms of the modeling functions. However, there is a consistency in viewpoint that allows the different forms to be considered together and to be put into perspective.

The earliest application of two-path models to digital radio [39], [40] used the simplest form of the modeling function,

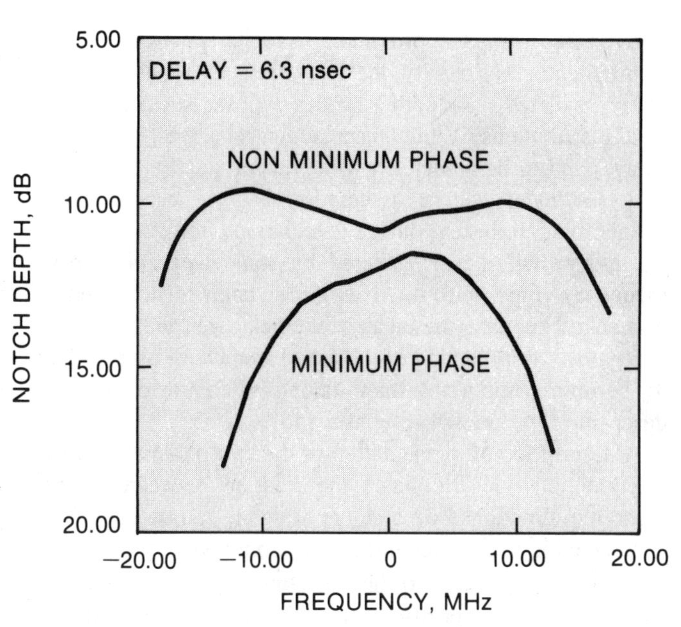

Fig. 6. Signature Curves for a 16 QAM Digital Radio at 10^{-3} Bit Error Ratio.

i.e.,

$$H(j\omega) = 1 + be^{-j\omega\tau}. \tag{6}$$

This function has strictly only two parameters, the relative echo delay τ, and the relative amplitude b. Furthermore, the frequency ω is measured at RF. In most subsequent applications of this model, a random phase component has been added to the delayed ray. This is achieved by introducing a notch frequency offset so that (6) can be written as

$$H(j\omega) = 1 - be^{-j(\omega-\omega_0)\tau}. \tag{7}$$

The addition of the notch frequency term allows ω to be measured from any convenient frequency (e.g., the RF or IF center frequency), since $H(j\omega)$ depends only on frequency differences. It will be shown in the discussion of signatures that, for identical probability distributions of the b and τ parameters, the modeling functions (6) and (7) lead to an equivalent fading channel model. Both channel responses are minimum phase for a positive delay ($\tau > 0$) if the relative amplitude (b) is less than 1, that is, for the same conditions on b and τ described in the preceding subsection.

The two-path model has been most often used to represent the dispersive component of fading. This approach is taken for simplicity and because dispersion is usually the dominant cause of system outage for high capacity systems. While one recent study suggests that the two-path model can also account for thermal noise [44], others have introduced a constant factor a to be applied to the modeling function of (7), to represent a median depression or flat fade component [41], [45]. In the two-path model, this factor is assumed to be independent of the dispersive component characterized by the other parameters. With its inclusion, the modeling function is identical to that used in the simplified three-path model; however, the models themselves remain distinct.

The statistics of the parameters of the two-path function are derived from simple approximations to the atmospheric model of propagation. Normally, the delay and relative amplitude are both considered as random variables with statistically independent distributions. While measurements suggest that larger delays tend to be associated with smaller relative amplitudes [46], the independence assumption is used because of its simplicity. For system outage calculations, the exact form of the b-distribution is considered by some workers to be of secondary importance to that of the delay distribution. In particular, knowing the mean of the delay distribution, which is regarded as a function of the path parameters, is considered to be more important than having a complete statistical description of the path responses [45].

A description of the signature of the two-parameter modeling function will illuminate the relation between the two modeling functions (6) and (7). Digital radios experience excessive bit error ratios under stress only when the response has a deep notch close to the operating frequency band. This happens in (6) for specific values of delay. Consequently, the signature consists of a series of disjoint outage regions [40] in b-τ space, as illustrated in Fig. 7, which shows the values of relative amplitude that cause excessive bit error ratios for the respective delays.

The centers of these regions are the delays for which the notch is at the radio channel center frequency, ω_c. Thus, the nth region is centered at delay $\tau_n = 2n\pi/\omega_c$. At 6 GHz, for example, adjacent regions are separated by 0.17 ns. The width of the outage regions in delay is exaggerated in Fig. 7 for clarity. For instance, if a radio was sensitive only to notches occurring within a 30 MHz channel, the width of the nth region would be $0.05\tau_n$. Within each region, small changes in τ have the effect of moving the frequency of the notch across the band. If τ is assumed to be constant in a region, the variation can be obtained, equivalently, by introducing a variable notch frequency [ω_0 in (7)] with an independent and uniform probability distribution. The two representations are equivalent for τ-distributions that are smooth on a scale of tenths of nanoseconds.

Several approaches to outage calculation have been developed for two-path models. The exact one for the two-parameter form of the model is to consider the sum of the contributions from the large number of disjoint regions in Fig. 7. However, it has been shown that the model parameters of delay and amplitude for the same notch offset frequency in two different outage regions can be interrelated [40], [48], [49]. This leads to the concept of "scaling" where the profile of one outage region (i.e., centered at one τ) can be used to derive an approximate profile for all the other regions (i.e., centered at other τ's). Scaling allows the measurement of equipment at one τ to be used to derive the performance over the total two-parameter plane. One such outage region, usually at the median delay value for a particular hop, is used to define a "system signature" which best characterizes equipment performance.

Another technique has been developed for the two-path model using the modeling function of (7) and an exponential delay distribution [50]. It involves measuring constant delay signatures at a set of delays specified by the Gauss-Laguerre

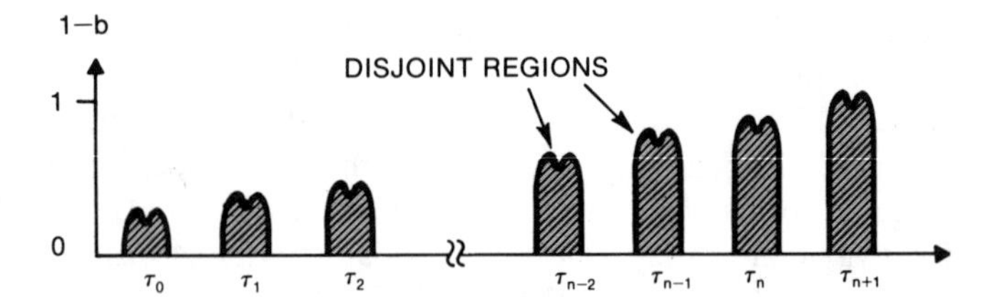

Fig. 7. An Example of the Disjoint Outage Regions of a Radio for the Two-Parameter, Two-Path Modeling Function, Equation (6).

quadrature formulas to evaluate the integral over the outage regions.

Polynomial Models

Experiments to investigate propagation effects on line-of-sight links are most easily instrumented using channels in the existing radio transmission network where the available bandwidth is limited to several tens of megahertz. For measurements within such narrow-band channels, it is generally not possible to resolve the individual rays in the propagation medium (i.e., the α's and τ's of (2)). The alternative, then, is to fit measured responses with suitable mathematical functions, but these need not evoke images of multiray propagation; any function that closely approximates the responses will do. An attractive possibility is to use polynomial functions of frequency, since low-order polynomials can be adequate for narrow-band channel responses.

One well-studied approach is to describe the attenuation and group delay responses, (4), by such polynomials [51]–[53]. (This is equivalent to describing the logarithm of $H(j\omega)$ by a complex polynomial.) We will exemplify this approach by concentrating on the attenuation response, $A(f)$.

Scan-by-scan records of the attenuation response of a channel can be fitted with a polynomial of order M, i.e.,

$$A(f) = a_0 + a_1 f + a_2 f^2 + \cdots + a_M f^M. \quad (8)$$

Coefficients have been obtained using least-squares fitting with polynomials of orders $M = 2$, 4, and 6. For highly selective fading events, the most suitable polynomial order is $M = 4$; however, for most fading periods, polynomials of order $M = 2$ can be used to provide acceptable accuracy. The representation with a second order polynomial has the advantages of familiarity and simplicity. The channel defects, amplitude slope and parabolic distortion, are well known in transmission theory, and their impact on equipment performance is easily measured and quantified. In contrast, the determination of whether a response is minimum or nonminimum phase can only be made by examining both the amplitude and phase responses at a given time.

The $M+1$ coefficients derived from fitting the channel response can be used to provide a statistical description of the channel. One set of coefficients $(a_0, a_1, \cdots, a_M)$ represents an element in an $(M+1)$-dimensional probability density function $p(a_0, a_1, \cdots, a_M)$. Fig. 8 shows an example of a two-dimensional density function $p(a_1, a_2|\text{all } a_0)$ obtained from an 11 GHz experiment [14]. Large positive and negative slope distortions combined with negative parabolic ones occur, (①

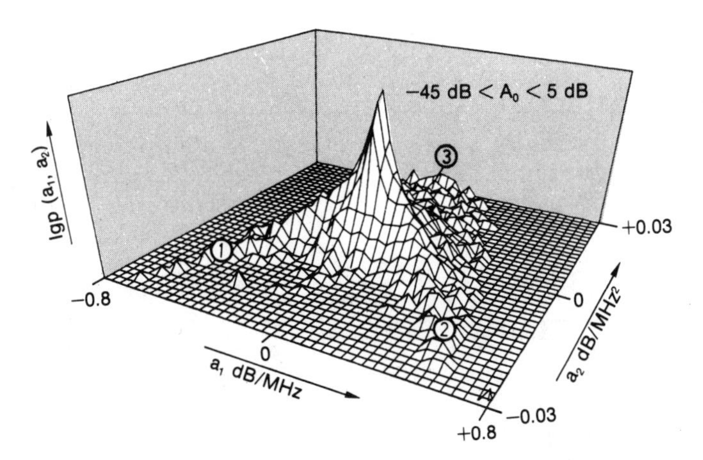

Fig. 8. A Probability Density Function of the Linear and Quadratic Coefficients of the Polynomial Model of Attenuation, Based on Observations on an 111 km Path at a Frequency of 11.285 GHz with a 40-MHz Bandwidth. Observations Spanned One Year of Activity (1,193,785 Scans).

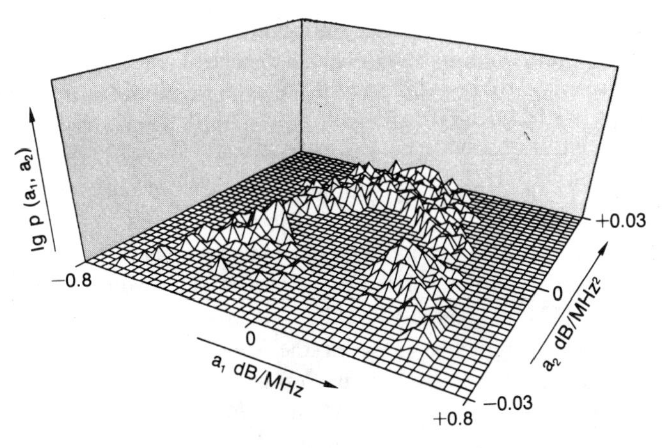

Fig. 9. Probability Density Function of the Coefficients of the Data of Figure 8 After Multiplication by a Signature.

and ②), if a notch is situated above or below the measured channel. If a notch falls within the channel, it produces positive parabolic distortion, ③.

While joint probability plots such as those of Fig. 8 are useful for qualitatively representing dispersive fading, more quantitative descriptions have been developed in terms of the conditional distributions of the coefficient values [54]. The results are presented as plots of the quantiles of the conditional distribution of slope, $p(a_1 | a_0$, all $a_2)$, and of the conditional distribution of the parabolic term, $p(a_2 | a_0$, all $a_1)$. This manner of presentation neglects the correlation between the slope and parabolic coefficients. A comparison of the behavior of four different hops under multipath conditions has been reported in this way for a 30 MHz bandwidth [55].

A possible method of outage prediction based on the above polynomial representation is easily described [14]. Because the description of the propagation uses the probability density function of polynomial coefficients, a signature is derived in terms of these coefficients; that is, the joint values of a_1 and a_2 for which the BER is at the threshold of outage are determined. This signature is described as a Boolean function, $S(a_1, a_2)$, which is zero for a_1, a_2 values within the contour and unity for values outside. The joint probability density function of a_1 and a_2 is multiplied by this function to give a result like the one shown in Fig. 9. The remaining samples are integrated, and the result gives the outage probability due to distortion. Results recently calculated for a 64-QAM radio and four different hops showed good agreement with bit error probabilities measured in an adjacent channel [55].

A quite different approach to polynomial modeling begins with a representation suggested by Bello a generation ago [56], in which $H(j\omega)$ itself (not its logarithm) is characterized by a complex polynomial. Thus,

$$H(j\omega) = A_0 + \sum_{n=1}^{N} (A_n + jB_n)(j\omega)^n \tag{9}$$

where ω is measured from the center of the channel. In one modeling study [57], this expression was used to fit essentially the same experimental data as was used to develop the simplified three-path model. It was found that, over the 26.4 MHz bandwidth of the measurements, the three coefficients A_0, A_1, and B_1 were sufficient to satisfactorily represent all scans. Closed form descriptions of the parameter statistics were developed, along with the time scale factor.

In this first-order complex polynomial model, the minimum phase (nonminimum phase) condition corresponds to $A_1/A_0 > 0$ $(A_1/A_0 < 0)$. The equipment signature corresponds to a closed contour in the complex $(A_1 + jB_1)/A_0$ plane, with the outage region being outside that contour. This model has proved useful for theoretical studies and simulations where the form of the modeling function is an asset [58], [59]. However, over wider bandwidths or more dispersive channels, its first-order nature would be inadequate and at least a quadratic term in $j\omega$ would be additionally needed [60], [61]. This would raise the number of function parameters to five and complicate the statistical modeling process.

Diversity Models

Diversity techniques are widely used as a countermeasure to the effects of multipath propagation. Both space and frequency diversity are commonly used to obtain a second replica of the transmitted signal at the receiver. In space diversity, the second signal is derived from a second antenna at the receiving site. The receiver may either adaptively combine the two signals or switch between them (selection diversity). Frequency diversity is implemented by switching operation, hitlessly, to an alternate radio channel. Much of the work on diversity channel models is based on propagation experiments using multiple receivers to observe simultaneously both the diversity and nondiversity performance. The results of such experiments are invaluable for this purpose, but they are confounded to some extent by the detailed effects of the combiner/switch implementation.

The most detailed model of the joint propagation conditions for diversity channels is based on a space diversity experiment and uses the simplified three-path model [37]. The spectrum of the received power on two antennas was simultaneously observed five times a second during multipath fading periods. At each observation, both received spectra were fitted by the three-parameter function in (5). The model provides a joint statistical description of the resulting six parameters. To summarize briefly, it was found that the fade shapes in the two channels, determined by the notch frequencies and relative notch depths, were independent. The fade level parameters of the two channels were partially correlated with each other and with their respective notch depths. The statistics were parameterized [62] to include the effects of antenna separation by requiring congruence between the single-frequency fading statistics predicted by the model and those predicted by classical formulas [36], [63], [64].

This space diversity model was extended to provide a description of frequency diversity on the premise that the fading on pairs of frequency diversity channels has a similar description in terms of fade shape and level correlations [62]. This conjecture has been supported by subsequently reported results [65].

Extensions of the two-path model to space diversity have also been made based on applying the two-path model separately to each diversity channel. This space diversity model is derived by invoking assumptions about the joint statistics of the resulting channel parameters, in particular, that the relative delays associated with each diversity channel are strongly correlated, while the notch depths and notch offsets are independent [49], [66], [67] or possess some correlation [43]. The outage estimation procedure in this case is analogous to that for the nondiversity case, but with two extra dimensions.

Another view of space diversity derived from detailed experimental measurements was obtained using the polynomial model of dB attenuation [68]. Selected detailed joint statistics of the modeling parameters of the two channels have been presented, but a complete statistical description has not been developed.

Dual-Polarization Models

Radio links carrying several channels use both vertical and horizontal polarizations, with adjacent channels traditionally being cross-polarized. This usage, among others, spurs a continuing interest in cross-polarization modeling and performance. Thus, although the modeling of a single channel does not depend on the transmitted polarization, the interactions between adjacent channels depend on the joint characterization of orthogonal polarizations.

For digital applications, no comprehensive dual-polarization channel model has been derived directly from measurements; however, single-frequency models have been developed [69], [70]. Also, detailed spectral data have been obtained in several experiments. One of these provides detailed spectral information and summary distributions [71]. Another, described in a series of papers [14], [54], [68], uses a polynomial expansion to fit the transfer functions (dB attenuation) of both polarizations in successive 0.2 second intervals. Results have been presented in the form of joint probability statistics of the coefficients, plots similar to those presented in Fig. 8. To clarify discussions of the application of this data,

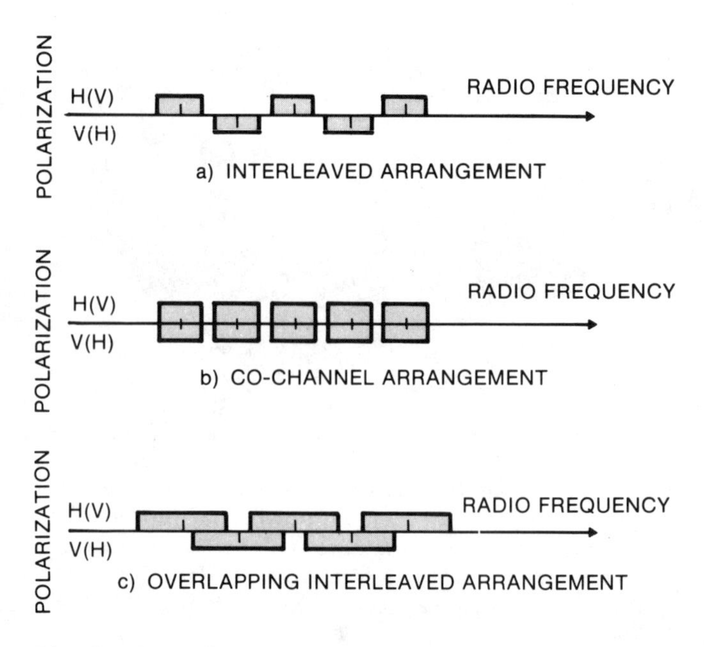

Fig. 10. Examples of Recommended Channel Arrangements, Showing the Spectral Content of the Signals Transmitted in One Direction on the Channels in a Frequency Band.

we distinguish between the three usages of a radio frequency band shown in Fig. 10. Case (a) is typical of the common carrier bands originally channelized for analog radio. Here, the cross-polarized signal is due to energy from the "tails" of the power spectrum of the signal in the adjacent channels. Usually the effect is minor [72], [73], and is treated in performance calculations as a noise or interference power [74].

The desire to increase the information carrying capacity of the radio frequency bands has led to considerations of channel arrangements of the type shown in Fig. 10(b) and (c) [75]. Here, signal energy from one or two adjacent cross-polarized channels ((b) or (c), respectively) occupies the same frequency interval as the desired signal. The coupling of cross-polarized signals into the desired channels is usually attributed to terrain, antenna, and waveguide characteristics [76], [77]. While some systems may perform satisfactorily in these configurations [78], higher level modulations may require improved antenna designs [79] and/or cross-polarization cancelers [80]–[82].

Advanced antenna designs are based on the understanding (atmospheric model) of multipath propagation. Most of the antennas in current use provide a cross-polarization discrimination (XPD) of close to 40 dB on boresight under free space conditions. That is, the antenna gain of the cross-polarized signal into the desired channel is 40 dB less than that for the desired signal. The variation of XPD with angle of arrival depends on the antenna design, and the decreases in XPD during severe multipath fading have been shown to be dependent on this variation [83]. The advanced antenna designs attempt to maintain a high XPD over a range of elevation angles, corresponding to the arrival angles of the propagation raypaths and, hence, to minimize the loss of XPD.

Cross-polarization cancelers, as recently reported [81]–[86], are complex devices resembling transversal equalizers.

Their performance can be evaluated using conventional techniques, given an appropriate channel model. Although the propagation data that would allow a comprehensive model to be developed is not available, some investigators [84]–[86] have constructed dual-polarization channel models by extending the simplified three-path model to have properties consistent with both single frequency models [69], [70] and wideband observations [14], [54], [68], [71]. While such models are speculative, their use permits the comparative assessment of the impact of the physical coupling parameters on the effectiveness of cancellation techniques.

If the theoretical predictions for antenna designs and cross-polarization cancelers are substantiated by tests of practical designs in the field, fundamental questions will be raised. For example: How should the channel modeler describe the severity of the cross-polarization problem in the channel? How should the antenna or cross-polarization canceler designer describe the effectiveness of his design? Such questions bring us to the limits of current knowledge in these areas.

Channel Dynamics

The models considered in this paper are static models that describe the fraction of time that channels have a prescribed response. Underlying their application are two assumptions. The first is that the time variation of the channel response is much slower than the dynamic response of the radio equipment to be used. The second, intimately associated with the first, is that the performance of a radio at any time is uniquely related to the current state of the channel, i.e., there is no hysteresis in the system behavior. These considerations become increasingly important as more complex equalizers are used in radios employing more complex modulations.

Circuits for carrier recovery, timing recovery, and equalization that employ differential correction can suffer from both dynamic and hysteresis problems. For instance, in a differentially corrected transversal equalizer, when the difference between the current settings and those required to decrease intersymbol interference becomes too large because of rapid changes, or requires settings beyond the range of compensation, the circuit will not function properly. In such cases, the system may lose lock and subsequently be unable to recover from this condition until the channel reaches a state that would normally cause only a low BER. Although some work has been done to describe the dynamics of fading [37], [38] and to account for hysteresis in outage calculations [87], much remains to be done in this area.

Channel dynamics may be handled by using simple measures of the speed of variation of model parameters to qualify (if not quantify) the dynamic performance of digital radio. The simplest measure of the dynamics of multipath fading, which has been used both for analog and digital systems, is the rate of change of the received signal power at a single frequency, expressed in dB/s. Reported rates of fading vary widely, up to over 100 dB/s in some climatic conditions. Note that, for digital radio systems on highly dispersive paths, the power of the received signal fades at a lower rate because the signal covers a wide bandwidth [42].

Another common measure of change is the speed at which a notch moves across the digital radio channel. Typical measurements of notch speed [88] are from 10–30 MHz/s, with some measurements indicating notch speeds ranging up to 100 MHz/s [89]. However, estimates of the impact of dynamic effects require not only a distribution function of notch speeds, but also a means of accounting for the simultaneous variations of the other model parameters. Thus, the threshold for outage in a signature measurement can depend on the trajectory as well as the rate of change of the model parameters [90]. For example, an equalizer could behave differently for a notch of constant depth passing through the band than for a fixed in-band notch varying in depth.

The combination of dynamic behavior limitations and hysteresis effects can significantly affect system performance but is not characterized by any of the channel models described in this paper. There is still much debate over the "correct" model to use for the quasistationary narrow-band channel without attempting to derive the more appropriate stochastic model of multipath fading. Even if such a complex model were available, specifying or measuring the values of the increased number of parameters and relating them to the path characteristics would be an even more difficult problem.

Conclusions

We have provided an overview of several models used to characterize multipath fading in radio channels on line-of-sight microwave radio paths. The approaches differ in emphasis and are not easily reconciled. The user's preference may be determined by available data and the application at hand. Moreover, all approaches suffer from the common need to relate the model parameters to the root physical causes: path parameters; local terrain conditions; and local climatology. Current propagation work continues to enhance our understanding of the multipath phenomena in terms of raypath characterization [91], and recent efforts to reconcile the channel models have produced encouraging results [92].

The application of these channel models to space diversity, frequency diversity, and cross-polarization characterizations was also described. The diversity configurations have been well represented; the treatment of cross-polarized channels is less complete. Analyses of co-channel cross-polarization operation, while encouraging, need to be tempered with more detailed propagation data, or by performance data derived from the field testing of actual cross-polarization cancelers.

The models described all claim success in predicting digital radio outages where the effects of channel dynamics and equipment hysteresis are of little importance. With the development of increasingly complicated transversal equalizers and the introduction of cross-polarization cancelers, dynamic effects, particularly the phase transitions of the channel, may become increasingly important considerations in performance prediction. Providing adequate improved methods for measuring, modeling, and quantifying these effects poses a continuing challenge for those concerned with microwave digital radio.

References

[1] D. P. Taylor and P. R. Hartmann, "Telecommunications by microwave digital radio," Part I of this book. [An earlier version was

published in *IEEE Commun. Mag.*, vol. 24, no. 8, Aug. 1986, pp. 11-16.]

[2] T. Noguchi, Y. Daido, and J. A. Nossek, "Modulation techniques for microwave digital radio," Part II of this book. [An earlier version was published in *IEEE Commun. Mag.*, vol. 24, no. 11, Sept. 1986, pp. 21-30.]

[3] CCIR, XVIth Plenary Assembly, Dubrovnik, Yugoslavia, 1986, Reports 934, 782, 936; Recommendations 635, 383, 384, 387, 595.

[4] F. Ikegami, M. Haga, T. Fukuda, and H. Yoshida, "Experimental studies on atmospheric ducts and microwave fading," *Rev. Elec. Commun. Lab.*, vol. 14, no. 7-8, pp. 505-533, July-Aug. 1966.

[5] D. C. Livingston, *The Physics of Microwave Propagation.* Englewood Cliffs, NJ: Prentice-Hall, 1970.

[6] C. L. Ruthroff, "Multiple-path fading on line-of-sight microwave radio systems as a function of path length and frequency," *Bell Syst. Tech. J.*, vol. 50, no. 7, pp. 2375-98, Sept. 1971.

[7] A. R. Webster, "Angles-of-arrival and delay times on terrestrial line-of-sight microwave links," *IEEE Trans. Antennas Propagat.*, vol. AP-31, no. 1, pp. 12-17, Jan. 1983.

[8] J. Claverie and C. Klapisz, "Meteorological features leading to multipath propagation observed in the PACEM 1 experiment," *Ann. Télécommun.*, vol. 40, no. 11-12, pp. 660-71, Nov.-Dec. 1985.

[9] A. J. Bundrock and J. V. Murphy, "A broadband 11 GHz radio propagation experiment," *IEEE Trans. Antennas Propagat.*, vol. AP-32, no. 5, May 1984.

[10] M. Kavehrad, "Cross-polarization interference cancellation and non-minimum phase fades," *AT&T Bell Lab. Tech. J.*, vol. 64, no. 10, pp. 2247-59, Dec. 1985.

[11] L. Martin, "Phase distortions of multipath transfer functions," in *Conf. Rec., 1984 Int. Conf. Commun.*, vol. 3, paper 46.2, pp. 1437-41.

[12] L. Martin, "Etude de la sélectivité des évanouissements dus aux trajets multiples," *Ann. Télécommun.*, vol. 35, no. 11-12, pp. 482-487, Nov.-Dec. 1980.

[13] P. Balaban, "Statistical model for amplitude and delay of selective fading," *AT&T Bell Lab. Tech. J.*, vol. 64, no. 10, pp. 2525-2550, Dec. 1985.

[14] M. Liniger, "One year results of sweep measurements of a radio link," in *Conf. Rec., 1983 Int. Conf. Commun.*, vol. 2, pp. C2.3.1-5.

[15] A. Leclert and P. Vandamme, "Non-minimum phase fadings effects on equalization techniques in digital radio systems," in *Conf. Rec., GLOBECOM'83,* vol. 1, paper 1.2, pp. 8-12.

[16] A. B. Crawford and W. C. Jakes, Jr., "Selective fading of microwaves," *Bell Syst. Tech. J.*, vol. 31, no. 1, pp. 68-90, Jan. 1952.

[17] R. L. Kaylor, "A statistical study of selective fading of super-high frequency radio signals," *Bell Syst. Tech. J.*, vol. 32, no. 5, pp. 1187-1202, Sept. 1953.

[18] J. Sandberg, "Extraction of multipath parameters from swept measurements on a line-of-sight path," *IEEE Trans. Antennas Propagat.*, vol. AP-28, no. 6, pp. 743-750, Nov. 1980.

[19] W. I. Lam and A. R. Webster, "Microwave propagation on two line-of-sight oversea paths," *IEEE Trans. Antennas Propagat.*, vol. AP-33, no. 5, pp. 510-16, May 1985.

[20] M. Sylvain, "Panorama des etudes sur les trajets multiples," *Ann. Télécommun.*, vol. 40, no. 11-12, pp. 547-564, Nov.-Dec. 1985.

[21] P. Dupuis, M. Joindot, A. Leclert, and M. Rooryck, "Fade margin of high capacity digital radio system," *Conf. Rec., 1979 Int. Conf. Commun.*, vol. 3, pp. 48.6.1-5.

[22] A. W. Muir and M. J. De Belin, "The field evaluation of a 140 Mbit/s digital radio relay system in the 11 GHz band," IEE Conf. Publ. No. 193, *Proc. IEE Conf. Telecommun. Transmission,* Mar. 1981.

[23] A. Ranade and P. E. Greenfield, "An improved method of digital radio characterization from field experiment," *Conf. Rec., 1983 Int. Conf. Commun.*, vol. 1, paper C2.6, pp. 659-63.

[24] S. Komaki, I. Horikawa, K. Morita, and Y. Okamoto, "Characteristics of a high capacity 16 QAM digital radio system in multipath fading," *IEEE Trans. Commun.*, vol. COM-27, no. 12, pp. 1854-1861, Dec. 1979.

[25] S. Sakagami and Y. Hosoya, "Some experimental results on in-band amplitude dispersion and a method for estimating in-band linear amplitude dispersion," *IEEE Trans. Commun.*, vol. COM-30, no. 8, pp. 1875-1888, Aug. 1982.

[26] Y. Serizawa and S. Takeshita, "A simplified method for prediction of multipath fading outage of digital radio," *IEEE Trans. Commun.*, vol. COM-31, no. 8, pp. 1017-1021, Aug. 1983.

[27] W. D. Rummler, "Time- and frequency domain representation of multipath fading on line-of-sight microwave paths," *Bell Syst. Tech. J.*, vol. 59, no. 5, pp. 763-96, May-June 1980.

[28] W. D. Rummler, "A new selective fading model: Application to propagation data," *Bell Syst. Tech. J.*, vol. 58, no. 5, pp. 1037-1071, May-June 1979.

[29] T. C. Lee and S. H. Lin, "More on frequency diversity for digital radio," *Conf. Rec., GLOBECOM'85,* vol. 3, paper 36.7, pp. 1108-12.

[30] M. Shafi and D. P. Taylor, "Influence of terrain induced reflections on the performance of high capacity digital radio systems," *Conf. Rec., 1986 Int. Conf. Commun.*, vol. 3, paper 51.2, pp. 1627-31.

[31] L. J. Greenstein and M. Shafi, "Outage calculation methods for microwave digital radio," Part V of this book. [An earlier version was published in *IEEE Commun. Mag.*, vol. 25, no. 2, Feb. 1987, pp. 30-39].

[32] M. Sylvain and J. Lavergnat, "Modelling the transfer function in medium bandwidth radio channels during multipath propagation," *Ann. Télécommun.*, vol. 40, no. 11-12, pp. 584-603, Nov.-Dec. 1985.

[33] W. D. Rummler, "More on the multipath fading channel model," *IEEE Trans. Commun.*, vol. COM-29, no. 3, pp. 346-352, Mar. 1981.

[34] C. W. Lundgren and W. D. Rummler, "Digital radio outage due to selective fading-observation vs prediction from laboratory simulation," *Bell Syst. Tech. J.*, vol. 58, no. 5, pp. 1073-1100, May-June 1979.

[35] W. D. Rummler, "A simplified method for the laboratory determination of multipath outage of digital radios in the presence of thermal noise," *IEEE Trans. Commun.*, vol. COM-30, no. 3, pp. 487-494, Mar. 1982.

[36] W. T. Barnett, "Multipath propagation at 4, 6, and 11 GHz," *Bell Syst. Tech. J.*, vol. 51, no. 2, pp. 321-61, Feb. 1972.

[37] W. D. Rummler, "A statistical model of fading on a space diversity radio channel," *Bell Syst. Tech. J.*, vol. 61, no. 9, pp. 2185-2219, Nov. 1982.

[38] A. J. Rustako, Jr., C. B. Woodworth, R. S. Roman, and H. H. Hoffman, "A laboratory simulation facility for multipath fading microwave radio channels," *AT&T Bell Lab. Tech. J.*, vol. 64, no. 10, pp. 2281-2317, Dec. 1985.

[39] W. C. Jakes, Jr., "An approximate method to estimate an upper bound on the effect of multipath delay distortion on digital transmission," in *Conf. Rec., 1978 Int. Conf. Commun.*, vol. 3, pp. 47.1.1-5.

[40] M. Emshwiller, "Characterization of the performance of PSK digital radio transmission in the presence of multipath fading," in *Conf. Rec., 1978 Int. Conf. Commun.*, vol. 3, pp. 47.3.1-6.

[41] J. C. Campbell and R. P. Coutts, "Outage prediction of digital radio systems," *Electron. Lett.*, vol. 18, no. 25/26, pp. 1071-1072, Dec. 1982.

[42] A. L. Martin, R. P. Coutts, and J. C. Campbell, "Results of a 16 QAM digital radio field experiment," in *Conf. Rec., 1983 Int. Conf. Commun.*, vol. 3, pp. F2.2.1-8.

[43] E. Damosso, "A new approach to outage prediction for radio digital systems," CSELT Technical Reports, vol. XIII, no. 3, June 1985.

[44] M. H. Meyers, "Multipath fading characteristics of broadband radio channels," in *Conf. Rec., GLOBECOM'84,* vol. 3, paper 45.1, pp. 1460-65.

[45] J. C. Campbell, A. L. Martin, and R. P. Coutts, "140 Mbit/s digital radio field experiment-further results," in *Conf. Rec., 1984 Int. Conf. Commun.*, vol. 3, pp. 46.8.

[46] E. H. Lin and A. J. Giger, "Radio channel characterization by three tones," *IEEE J. Selected Areas Commun.*, vol. SAC-5, no. 3, pp. 402-15, Apr. 1987.

[47] J. C. Campbell, R. P. Coutts, A. L. Martin, and R. L. Reid, "Observations and conclusions from a three year digital radio field experiment in Australia," *IEEE J. Selected Areas Commun.*, vol. SAC-5, no. 3, pp. 416-26, Apr. 1987.

[48] W. D. Rummler, "A multipath channel model for line-of-sight digital radio systems," in *Conf. Rec., 1978 Int. Conf. Commun.*, vol. 3, pp. 47.5.1-4.

[49] J. C. Campbell, "Outage prediction for the route design of digital radio systems," *Aust. Telecommun. Res. (ATR)*, vol. 18, no. 2, pp. 37-49, 1984.

[50] M. H. Meyers, "Multipath fading outage estimates incorporating path and equipment characteristics," in *Conf. Rec., GLOBECOM'84,* vol. 3, paper 45.2, pp. 1466-70.

[51] C. W. Anderson, S. Barber, and R. Patel, "The effect of selective fading on digital radio," in *Conf. Rec., 1978 Int. Conf. Commun.*, vol. 2, pp. 33.5.1-6.

[52] D. R. Smith and J. J. Cormack, ''Measurements and characterization of a multipath fading channel for application to digital radio links,'' in *Conf. Rec., 1982 Int. Conf. Commun.*, vol. 3, pp. 7B.4.1–6.

[53] M. Liniger, ''Sweep measurements of the transfer function of an RF-channel and their representation by polynomials,'' in *Conf. Rec., 1982 Int. Conf. Commun.*, vol. 3, pp. 7B.3.1–5.

[54] M. Liniger, ''More results on the transfer function of cross-polarized and diversity-protected RF-channels,'' *Int. Symp. Antennas and Propagat. (ISAP'85)*, Kyoto, Japan.

[55] M. Liniger and D. Vergeres, ''Field test results for a 16-QAM and a 64-QAM digital radio, compared with the prediction based on sweep measurements,'' in *Conf. Rec., IEEE Int. Conf. Commun. (ICC'86)*, Toronto, Canada.

[56] P. A. Bello, ''Characterization of randomly time-variant linear channels,'' *IEEE Trans. Commun. Sys.*, vol. CS-11, no. 4, pp. 360–393, Dec. 1963.

[57] L. J. Greenstein and B. A. Czekaj, ''A polynomial model for multipath fading channel responses,'' *Bell Syst. Tech. J.*, vol. 59, no. 7, pp. 1197–1226, Sept. 1980.

[58] Y. S. Yeh and L. J. Greenstein, ''A new approach to space diversity combining in microwave digital radio,'' *AT&T Bell Lab. Tech. J.*, vol. 64, no. 4, pp. 885–905, Apr. 1985.

[59] L. J. Greenstein and B. A. Czekaj-Agun, ''Performance comparisons among digital radio techniques subjected to multipath fading,'' *IEEE Trans. Commun.*, vol. COM-30, no. 5, pp. 1184–97, May 1982.

[60] P. A. Bello and K. Pahlavan, ''Performance of adaptive equalization for staggered QPSK and QPR over frequency-selective LOS microwave channels,'' in *Conf. Rec., 1982 Int. Conf. Commun.*, vol. 2, pp. 3H.1.1–6.

[61] L. J. Greenstein and B. A. Czekaj, ''Modeling multipath fading responses using multitone probing signals and polynomial approximation,'' *Bell Syst. Tech. J.*, vol. 60, no. 2, pp. 193–214, Feb. 1981.

[62] W. D. Rummler, ''A rationalized model for space and frequency diversity line-of-sight radio channels,'' in *Conf. Rec., 1983 Int. Conf. Commun.*, vol. 3, pp. E2.7.1–5.

[63] W. T. Barnett, ''Microwave line-of-sight propagation with and without frequency diversity,'' *Bell Syst. Tech. J.*, vol. 49, no. 8, pp. 1827–1871, Oct. 1970.

[64] A. Vigants, ''Space-diversity engineering,'' *Bell Syst. Tech. J.*, vol. 54, no. 1, pp. 103–142, Jan. 1975.

[65] P. L. Dirner and S. H. Lin, ''Measured frequency diversity improvement for digital radio,'' *IEEE Trans. Commun.*, vol. COM-33, no. 2, pp. 106–109, Jan. 1985.

[66] J. C. Campbell, ''Digital radio outage prediction with space diversity,'' *Electron. Lett.*, vol. 19, no. 23, pp. 1003–1004, Nov. 1983.

[67] J. Steel, J. C. Campbell, and L. Davey, ''Measurement and prediction of digital radio performance on long overwater path,'' *Electron. Lett.*, vol. 21, no. 25/26, pp. 1212–14, Dec. 1985.

[68] M. Liniger, ''Sweep measurements of multipath effects on cross-polarized RF-channels including space diversity,'' in *Conf. Rec., GLOBECOM'84*, vol. 3, no. 45.7, pp. 1492–6.

[69] S. H. Lin, ''Impact of microwave depolarization during multipath fading on digital radio performance,'' *Bell Syst. Tech. J.*, vol. 56, no. 5, pp. 645–674, May–June 1977.

[70] T. O. Mottl, ''Dual polarized channel outages during multipath fading,'' *Bell Syst. Tech. J.*, vol. 56, no. 5, pp. 675–702, May–June 1977.

[71] K. T. Wu, ''Measured statistics on multipath dispersion of cross polarization interference,'' *Conf. Rec., 1984 Int. Conf. Commun.*, vol. 3, paper 46.3.

[72] T. S. Giuffrida, ''The effects on multipath fading upon adjacent channel operation of an 8-PSK 6 GHz digital radio,'' in *Conf. Rec., 1980 Int. Conf. Commun.*, vol. 2, pp. 34.1.1–5.

[73] T. S. Giuffrida and W. W. Toy, ''16 QAM and adjacent channel interference,'' in *Conf. Rec., 1981 Int. Conf. Commun.*, vol. 1, pp. 13.1.1–5.

[74] W. D. Rummler, ''A comparison of calculated and observed performance of digital radio in the presence of interference,'' *IEEE Trans. Commun.*, vol. COM-30, no. 7, pp. 1693–1700, July 1982.

[75] CCIR, XVIth Plenary Assembly, Dubrovnik, Yugoslavia, 1986, Reports 934, 936; Recommendations 635, 595.

[76] R. L. Olsen, ''Cross polarization during clear-air conditions on terrestrial links: A review,'' *Radio Sci.*, vol. 16, no. 5, pp. 631–647, Sept.–Oct. 1981.

[77] M. L. Steinberger, ''Design of a terrestrial cross pol canceler,'' in *Conf. Rec., 1982 Int. Conf. Commun.*, vol. 1, pp. 2B.6.1–5.

[78] S. Barber, ''Cofrequency cross-polarized operation of a 91 Mb/s digital radio,'' in *Conf. Rec., 1981 Int. Conf. Commun.*, vol. 3, pp. 46.6.1–6.

[79] G. D. Richman, ''The variation in XPD during multipath fading and its effect on co-frequency cross-polarized RBQPSK digital radio,'' in *Conf. Rec., GLOBECOM'83*, vol. 1, paper 1.6, pp. 29–34.

[80] K. T. Wu and T. S. Giuffrida, ''Feasibility study of an interference canceler for co-channel cross polarization operation of digital radio,'' in *Conf. Rec., 1982 Int. Conf. Commun.*, vol. 1, pp. 2B.7.1–5.

[81] T. Ryu, M. Tahara, and T. Noguchi, ''IF band cross polarization canceller,'' in *Conf. Rec., 1984 Int. Conf. Commun.*, vol. 3, paper 46.4, pp. 1442–6.

[82] Y. Aono, Y. Daido, S. Takanaka, and H. Nakamura, ''Cross polarization interference canceler for high-capacity digital radio systems,'' in *Conf. Rec., 1985 Int. Conf. Commun.*, vol. 3, paper 39.5, pp. 1254–8.

[83] K. Morita, S. Sakagami, S. Murata, T. Mukai, and N. Ohtani, ''A method for estimating cross polarization discrimination ratio during multipath fading,'' *Trans. IECE Japan*, vol. E62, no. 11, pp. 810–1, Nov. 1979.

[84] N. Amitay and J. Salz, ''Linear equalization theory in digital data transmission over dually polarized fading radio channels,'' *AT&T Bell Lab. Tech. J.*, vol. 63, no. 10, part 1, pp. 2215–59, Dec. 1984.

[85] M. Kavehrad and J. Salz, ''Cross-polarization cancellation and equalization in digital transmission over dually polarized multipath fading channels,'' *AT&T Bell Lab. Tech. J.*, vol. 64, no. 10, pp. 2211–45, Dec. 1985.

[86] L. J. Greenstein, ''Analysis/simulation study of cross-polarization cancellation in dual polarization digital radio,'' *AT&T Bell Lab. Tech. J.*, vol. 64, no. 10, pp. 2261–80, Dec. 1985.

[87] A. Ranade, ''Statistics of the time dynamics of dispersive multipath fading and its effects on digital microwave radios,'' *Conf. Rec., 1985 Int. Conf. Commun.*, vol. 3, paper 47.7, pp. 1537–40.

[88] M. F. Gardina and A. Vigants, ''Measured multipath dispersion of amplitude and delay at 6 GHz in a 30 MHz bandwidth,'' in *Conf. Rec., 1984 Int. Conf. Commun.*, vol. 3, paper 46.1, pp. 1433–6.

[89] L. Martin, ''Rates of change of propagation medium transfer functions during selective fadings,'' in *Proc. URSI (Commission F) Int. Symp. Wave Propagat. and Remote Sensing*, Louvain, Belgium, June 1983, pp. 31–35.

[90] A. L. Martin, ''Digital microwave radio–A new system measurement technique,'' in *Conf. Rec., 1986 Int. Conf. Commun.*, vol. 1, paper 15.6, pp. 472–6.

[91] A. R. Webster and A. M. Scott, ''Angles-of-arrival and tropospheric multipath microwave propagation,'' *IEEE Trans. Antennas Propagat.*, vol. AP-35, no. 1, pp. 94–99, Jan. 1987.

[92] J. Lavergnat and M. Sylvain, ''Selective fading radio channels: modeling and prediction,'' *IEEE J. Selected Areas Commun.*, vol. SAC-5, no. 3, pp. 378–88, Apr. 1987.

OUTAGE PREDICTION OF DIGITAL RADIO SYSTEMS

Indexing terms: Telecommunication, Digital radio

It is well established that the performance of digital radio systems is often limited by the amplitude and group delay distortions which accompany frequency selective fading, and the concept of the 'system signature' has been developed for predicting the outage probability of such systems. The letter presents a new outage prediction technique and introduces the concept of a 'normalised system signature' which enables comparisons of modulation schemes and equaliser implementations.

Introduction: From the results of field experiments it is known that the performance of digital radio systems is often limited by the amplitude and group delay distortions which accompany frequency selective fading, and the concept of the system signature has been developed for predicting the outage probability of such systems.[1,2] This letter presents a new outage prediction technique and introduces the concept of a 'normalised system signature' which enables comparisons of modulation schemes and equaliser implementations.

Outage equation: The outage prediction technique expresses the probability of outage (i.e. that the BER $\geq 10^{-3}$) in the worst month as

$$\Pr[\text{outage}] = \eta P_c \tag{1}$$

where $\eta \triangleq$ probability of frequency selective fading in worst month, and $P_c \triangleq$ probability of outage given frequency selective fading.

It is noted that η (which is related to the activity factor r documented in CCIR[3]) is solely dependent on the propagation conditions while P_c is dependent on both the propagation conditions and the radio equipment. Further reasons for this separation are:

(i) From chart records of AGC and pilot levels one can determine η. All that remains is to determine P_c, and as will be shown this can be determined from the 'system signature' and the 'mean echo delay', the latter of which can also be determined from AGC and pilot level records.[4]

(ii) A relative comparison of two radio equipments is simply given by the ratio of the respective probabilities P_c.

Fading model: To describe the channel mathematically during periods of frequency selective fading the well known 'single echo model' is used.[1,2] Simply, it is assumed that the transmitted signal arrives at the receiver via two paths: a 'direct path' and a 'reflected path'. The impulse response shall be written as

$$c(t) = a[\delta(t) + b\delta(t - \tau)] \tag{2}$$

where:

$a \triangleq$ flat relative gain of the received signal with respect to the nonfaded signal. As the statistics of this parameter influence the probability η but not P_c, it can be neglected in this letter

$b \triangleq$ relative amplitude of 'echo signal' with respect to the 'direct signal'. This parameter is assumed to possess a uniform distribution over the range zero to unity (this corresponds to the case $\alpha = 0$ in Greenstein's model[5])

$\tau \triangleq$ relative delay of the echo signal with respect to the direct signal. This parameter is assumed to possess a negative exponential distribution with mean τ_0, which in general will be a function of the hop distance and other path characteristics.

Also define f_0 as the frequency difference between the carrier and the neighbouring notch, i.e. the notch frequency offset. For typical echo delays ($\tau < 6$ ns) a small increment in the echo delay is sufficient to move the notch frequency offset right across its range ($-1/2\tau$ to $1/2\tau$). Consequently, to a very good approximation, the notch frequency offset is assumed to possess a uniform distribution across this range and further, assumed independent of τ.

Finally, it is assumed that the channel parameters are statistically independent.

Signature scaling: The concept of the system signature is well known.[1,2] As will become apparent, a technique necessary for the evaluation of P_c is that of 'scaling' system signatures to arbitrary echo delays τ and baud periods T. A full description of the scaling procedure is given by the authors,[6] where the scaling procedure employed in this letter has been termed the 'approximate scaling technique'. Briefly, considering the echo delay, scaling is achieved by utilising the result that at a fixed notch frequency offset the relationship $\tau/(1 - b)$ remains approximately constant. Secondly, one can scale the baud period by simply 'time scaling' and involves holding the products τ/T and $f_0 T$ constant. Letting:

T_r, τ_r = baud period and echo delay, respectively, corresponding to the measured or 'reference signature'

$\lambda_c^r(f_0)$ = critical notch depth $(1 - b^r(f_0))$ as a function of notch frequency offset as given by the 'reference signature'

T, τ, F = baud period, echo delay and notch frequency offset, respectively, at which the critical notch depth is required

then it is shown that

$$\lambda_c(\tau, F, T) = k_c'(F) \frac{\tau}{T} \tag{3}$$

where

$$k_c'(F) = \lambda_c^r\left(F \frac{T}{T_r}\right) \frac{T_r}{\tau_r} \tag{4}$$

where λ_c is the required critical notch depth, i.e. $1 - b_c^r$.

Outage probability given frequency selective fading: The probability P_c is given by the probability that the state of the channel is 'under the signature'. Thus

$$P_c = \int_{\tau=0}^{\infty} \frac{1}{\tau_0} e^{-\tau/\tau_0} \int_{F=-1/2\tau}^{1/2\tau} \tau \int_{\lambda=0}^{\lambda_c(\tau,F,T)} d\lambda \, dF \, d\tau \tag{5}$$

Reprinted with permission from *Electronics Letters,* vol. 18, no. 25, pp. 1071–1072, Dec. 1982.

Integrating with respect to λ and then τ yields

$$P_c = \frac{2\tau_0^2}{T} \int_S k_c'(F)\, dF \tag{6}$$

where it is assumed for typical delays that the integration is over the full signature 'S'. Now,

$$\int_S k_c'(F)\, dF \approx \frac{1}{T} \int_S \lambda_c^r \left(F \frac{1}{T_r}\right) \frac{T_r}{\tau_r}\, dF$$

$$= \frac{1}{T} \int_S \lambda_c(F) \bigg|_{T=1\cdot 0,\, \tau=1\cdot 0} dF \tag{7}$$

Define

$$K = 2 \int_S \lambda_c(F) \bigg|_{T=1\cdot 0,\, \tau=1\cdot 0} dF = 2T \int_S k_c'(F)\, dF \tag{8}$$

It is apparent that K is a parameter which is proportional to the area under the signature under the normalisation of baud period and echo delay equal to unity. Consequently, K can be considered as the area under a 'normalised system signature', and one may expect K to be dependent on the modulation method and equaliser type employed. Substituting for K into eqn. 6 yields

$$P_c = \left(\frac{\tau_0}{T}\right)^2 K \tag{9}$$

From signatures presented in the literature, values of K have been evaluated for various modulation schemes and the results are presented in Table 1.

Table 1 VALUES OF K FOR VARIOUS MODULATION METHODS

Modulation method	K
16 QAM	11·7[1], 10·9[7]
8 PSK	14·8[1], 12·6[2], 14·0[7]
4 PSK	1·8[8]

It is evident that the values of K are strongly dependent on the modulation method—a convenient result for system design. Further, the effect of an equaliser is to reduce the value of K to an extent which is characteristic of the equaliser type and modulation method employed. Typical 'equaliser improvements' range from 2 to 10, the lower values being characteristic of simple adaptive amplitude equalisers, while the larger improvements are characteristic of time-domain equalisers. Finally, results from a field experiment being conducted by Telecom Australia have shown the prediction technique to be within 30% of the measured seconds where the BER exceeded 10^{-3}.

Conclusion: The concept of utilising a 'normalised system signature' leads to a very simple expression for the outage probability and is applicable to hops of arbitrary length. Further, this approach enables comparisons of modulation methods to be made, independent of the baud rate.

Acknowledgment: The permission of the Director, Research, Telecom Australia Research Laboratories, to publish this correspondence is acknowledged.

J. C. CAMPBELL *19th October 1982*
R. P. COUTTS

Telecom Australia Research Laboratories
762–772 Blackburn Road
Clayton North, Victoria 3168, Australia

References

1 GIGER, A., and BARNETT, W. T.: 'Effects of multipath propagation on digital radio', *IEEE Trans.*, 1981, **COM-29**, pp. 1345–1352
2 LUNDGREN, C. W., and RUMMLER, W. D.: 'Digital radio outage due to selective fading—observation vs. predictions from laboratory simulation', *Bell Syst. Tech. J.*, 1979, **58**, pp. 1073–1100
3 DAMOSSO, E., and DE PADOVA, S.: 'A statistical model for the evaluation of the impairments due to multipath fades on digital radio links'. ICC, 1981
4 ROORYCK, M.: 'Validity of two-path model for calculating quality of digital radio links: determination of model from measurements on analogue links', *Electron. Lett.*, 1979, **15**, pp. 783–784
5 GREENSTEIN, L. J., and PRABHU, V. K.: 'Analysis of multipath outage with application to 90 Mbit/s PSK systems at 6 and 11 GHz', *IEEE Trans.*, 1979, **COM-27**, pp. 68–75
6 COUTTS, R. P., and CAMPBELL, J. C.: 'Mean square error analysis of QAM digital radio systems subject to frequency selecting fading', *ATR*, 1982, **16**, pp. 23–38
7 YASHIDA, Y., KITAHORA, Y., and YOKOYAMA, S.: '6 GHz–90 Mbit/s digital radio systems with 16 QAM modulation'. ICC, 1980
8 HYAMSON, H. D., MUIR, A. W., and ROBINSON, J. M.: 'An 11 GHz high capacity digital radio system for overlaying existing microwave routes', *IEEE Trans.*, 1979, **COM-27**, pp. 1928–1937

A New Selective Fading Model: Application to Propagation Data

By W. D. RUMMLER

(Manuscript received September 21, 1978)

Channel transmission models for use in estimating the performance of radio systems on line-of-sight paths at 6 GHz are explored. The basis for this study is the simple three-ray multipath fade, which provides a channel transfer function of the form $H(\omega) = a[1 - b \exp - j(\omega - \omega_0)\tau]$, *where a is the scale parameter, b is a shape parameter,* τ *is the delay difference in the channel, and* ω_0 *is the (radian) frequency of the fade minimum. This model is indistinguishable from an ideal channel model, within the accuracy of existing measurements. The propagation data that confirm the model were obtained in summer 1977 from a 26.4-mile hop near Atlanta, Georgia. The received power at 24 sample frequencies spaced at 1.1 MHz and centered on 6034.2 MHz was continuously monitored and recorded during periods of anomalous behavior. The model is applied to estimating the statistics of the channel delay difference,* τ. *The average delay difference giving rise to significant selectivity in the channel is between 5 and 9 ns. The distribution of delay difference is obtained for delay differences greater than 10 ns. The channel is found to have more than 3 dB of selectivity (difference between maximum and minimum attenuation in band) due to delay differences greater than 20 ns for more than 70 seconds in a heavy fading month. (This is comparable to the time the channel attenuation of a single frequency exceeds 40 dB.) The three-path model requires further simplification for narrowband channel application. For a channel with 30 MHz bandwidth, a model with fixed delay of 6.3 ns provides a sufficiently accurate representation of all observed channel conditions. The resulting nonphysical model is used to statistically characterize the condition of the fading channel. The statistics of the parameters of the fixed delay model are almost independent and of relatively simple form. The distribution of the shape parameter b is of the form* $(1 - b)^{2.3}$. *The distribution of a is lognormal. For* $b > 0.5$, *the mean and standard deviation of* $-20 (\log a)$ *are 25 and 5 dB, respectively; the mean decreases to 15 dB for smaller values of b. The probability density function of* ω_0 *is uniform at two levels; measuring* ω_0 *from the center of the band, the magnitude of* $\omega_0\tau$ *is five times as likely to be less than* $\pi/2$ *than to be greater. A companion paper describes the use of this model for determining the bit error rate statistics of a digital radio system on the modeled path.*

I. INTRODUCTION

Performance prediction of a digital radio system on a line-of-sight microwave channel requires an accurate statistical model of the channel. Because different digital radio systems may have different sensitivities to the various channel impairments, the model must be complete to the extent that it must be capable of duplicating the amplitude and phase (at least approximately) of all observed channel conditions. To facilitate laboratory measurements and computer simulations for calculating outage, the model should be realizable as a practical test circuit and should have as few parameters as possible. Most important, the parameters should be statistically well behaved.

Two types of models have been generally considered for line-of-sight microwave radio channels: power series type models[1-3] and multipath models.[4-6] A power series model will require a few terms only if the channel is a multipath medium with a small spread of delays relative to the reciprocal bandwidth of the channel.[3] This implies that one must understand the channel as a multipath medium to understand the behavior of a power series model. Hence, we have limited our characterization efforts to multipath models.

The basis for this study is the simple three-ray multipath fade.[7] If the fading in a channel can be characterized by a simple three-path model, the channel will (as shown in Section II) have a voltage transfer function of the form

$$H(\omega) = a[1 - be^{\pm j(\omega - \omega_0)\tau}]. \tag{1}$$

where the real positive parameters a and b control the scale and shape of the fade, respectively, τ is the delay difference in the channel, and ω_0 is the radian frequency of the fade minimum. The plus and minus signs in the exponent correspond, respectively, to the channel being in a nonminimum phase or minimum phase state. Note that, with appropriate choices of parameters, this model can be reduced to a two-path model or a scaled two-path model, etc.

It has been shown previously,[7] and is illustrated in Section II, that the simple three-path fade overspecifies the channel transfer function if the delay is less than ⅙ B, where B is the observation bandwidth. The critical value of τ for a 30-MHz channel is about 5.5 ns, which is comparable to the mean delay in the channel. As a consequence, unless

Reprinted with permission from *Bell System Technical Journal,* vol. 58, no. 5, pp. 1037–1071, May/June 1979.

the channel response can be determined to an accuracy on the order of 0.001 dB, a unique set of parameters a, b, τ, and f_o cannot be determined for more than half the faded channel conditions encountered. To avoid this problem, one must suppress or fix one of the model parameters. Section II shows that the delay, τ, is the only parameter which, when fixed, produces a reasonable model.

While a model with a fixed delay may appear to be a strange choice, it has all the required characteristics for modeling the channel transfer function. Figure 1 shows the amplitude of the channel transfer function of eq. (1) on a power scale and on a decibel scale for $\tau = 6.31$ ns. With τ fixed, the response minimum is shifted with respect to frequency by varying f_o. Varying a changes the overall level and b changes the

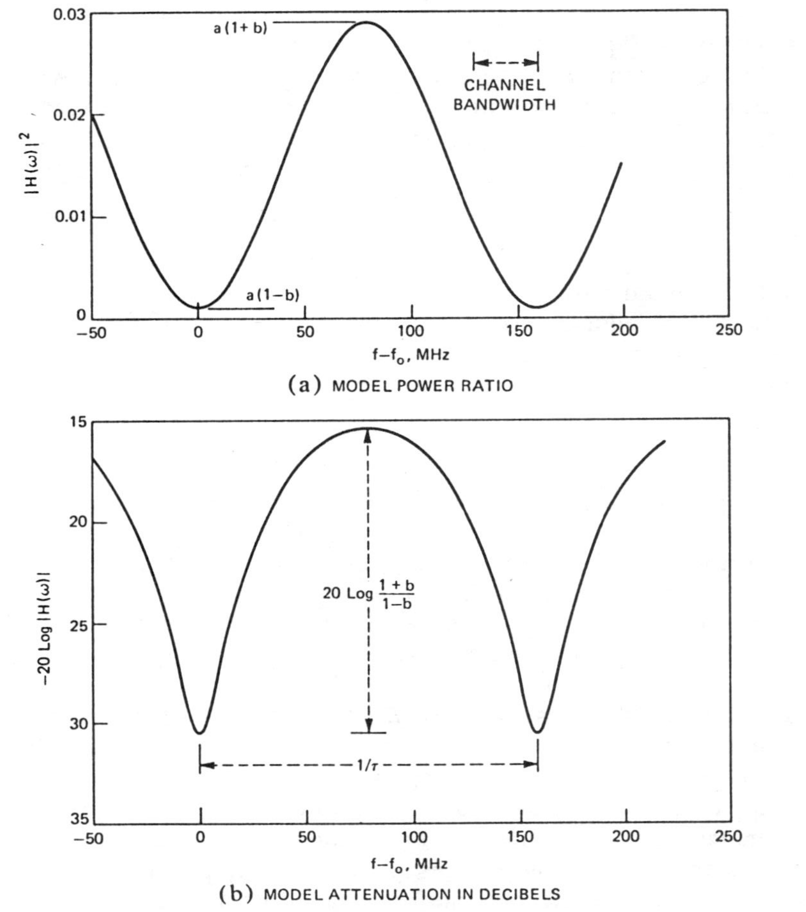

Fig. 1—Channel model function. $H(\omega) = a[1 - b\exp(-j2\pi(f - f_o)\tau)]$, for $\tau = 6.3$ ns, $a = 0.1$, $b = 0.7$.

"shapeliness." If the minimum is within the 30-MHz bandwidth of a channel, the fixed delay model can generate notches with a wide range of levels and notch widths. With the minimum out of band, it can generate a wide range of combinations of levels, slopes, and curvatures within the channel bandwidth. Section VI shows that the model versatility, with τ chosen to be 6.31 ns, is sufficient to characterize a 30-MHz channel in the 6-GHz common carrier band.

Section II provides a brief discussion of the simple three-path fade. A comparative discussion of the relative merits of the different possible simplifications of this model leads to the choice of the fixed delay model.

The data used for detailed evaluation of models were obtained from a 6-GHz experiment in Palmetto, Georgia, in June 1977. The radio channel was equipped with a general trade 78-Mbit/s, 8-PSK digital radio system, and the received spectrum was monitored with a set of 24 filters with bandwidths of 200 kHz spaced at a 1.1-MHz separation across this channel. During fading activity, the received power of each of these frequencies was measured five times each second, or once every 2 seconds, depending on how rapidly the channel was changing; sampled power, quantized in 1-dB steps, was recorded by the MIDAS system.* The data base used for this study consists of approximately 25,000 scans representing 8400 seconds of fading activity; about 8700 scans were recorded during periods when the equipment was indicating errors. These data represent about 60 percent of the fading activity of a heavy fading month; therefore, the derived statistics must be viewed as provisional and subject to some modification as additional data are processed. At the very least, the data base is sufficiently large to indicate what can happen on the channel and to form a basis for choosing and validating a model.

As described in Section III, the model parameters were estimated for each scan by fitting the magnitude squared of the transfer characteristic [eq. (1)] to the observed channel shape as characterized by the power received at the sampling frequencies. Phase is subsequently derived by assuming the channel is minimum phase. Problems are encountered in realizing a minimum-phase solution because of quantization noise and the presence of certain channel shapes caused by large delays. The procedure for handling these difficulties is described.

The statistics of the parameters of the fixed delay model are discussed in Section IV. Equations providing an idealized description of the statistics of the parameters of the model are also given here.

In Section V, the determination of the delay difference present in the channel is considered. In the first subsection, it is demonstrated

* Multiple Input Data Acquisition System, constructed by G. A. Zimmerman; see Ref. 1.

that, during the observed period of fading activity, the average delay is 9 ns. A lower bound on the distribution of delay difference for large delays is developed in the second subsection. A third subsection provides an example of a channel scan that can best be approximated by a three-path fade with a delay difference of 26 ns. Fades with at least this delay and with a more moderate amount of shape (2 dB or more) were encountered for about 60 seconds of the data base studied. Thus, one might expect 26-ns delays to be present during about 100 seconds of a heavy fading month.

The presence of such large apparent delays raises questions as to the accuracy with which the fixed delay model represents the channel. These questions are addressed in Section VI where the statistics of the errors in modeling scan fits are described. The errors are small and do not compromise the usefulness of the model.

Results and conclusions are briefly summarized in Section VII.

II. CHOICE OF MODEL

In this section, we provide a brief description of the simple three-path model and show why it cannot be used to estimate delays when the delay bandwidth product is less than ⅙. In a comparative discussion, we show why the fixed delay model is the only simplification of the model that is manageable.

2.1 *Simple three-path model*

Consider a channel characterized by three paths or rays. The amplitude of the signal on each of these three paths, as seen by the receiver, is 1, a_1, and a_2. The second and third paths are delayed with respect to the first by τ_1 and τ_2 seconds, respectively, where $\tau_2 > \tau_1$. We define the simple three-path model by requiring the delay between the first two paths to be sufficiently small, i.e.,

$$(\omega_2 - \omega_1)\tau_1 \ll 1, \tag{2}$$

where ω_2 and ω_1 are the highest and lowest (radian) frequencies in the band. The complex voltage transfer function of the channel at a frequency ω may be illustrated with a phasor diagram. Figure 2a shows the phasor diagrams for ω_1 and ω_2 superimposed. By designating the amplitude of the (vector) sum of the first two paths by a; the angle of the sum by $\phi = \omega_0\tau - \pi$, where τ is equal to τ_2, the delay difference in the channel; and the amplitude of the third ray by ab, we obtain the simplified diagram in Fig 2b.*

* Note that, if the third amplitude is greater than the sum of the first two, we interchange the assignments of amplitudes a and ab and obtain a nonminimum phase fade.

The simple three-path fade cannot be used for a channel model because the path parameters lack uniqueness. The basic difficulty is illustrated by the two superimposed fades in Fig. 3. Note that the amplitudes of the transfer functions of these two fades match, at

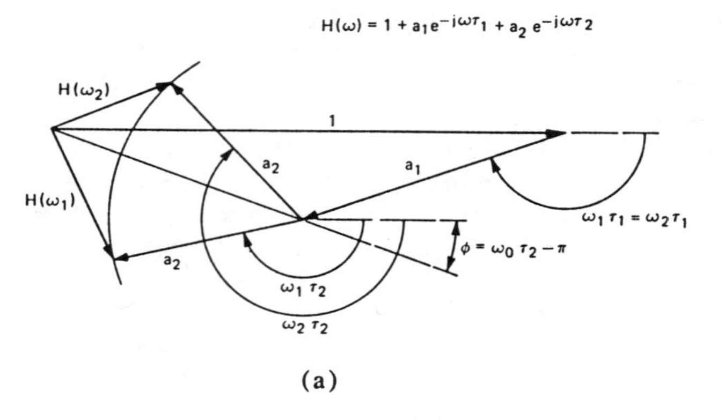

(a)

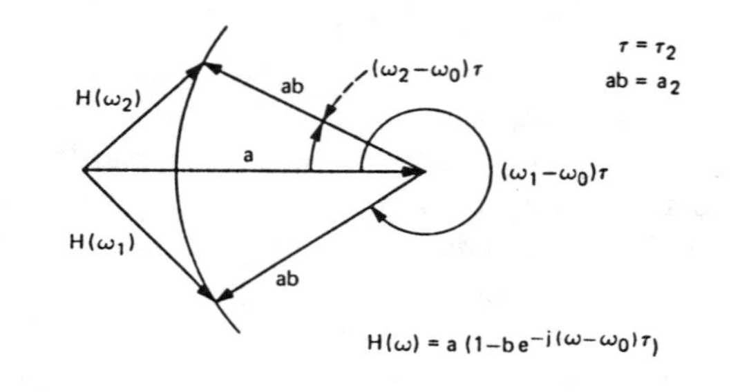

(b)

Fig. 2—Simple three-path fade. (a) Three rays shown. (b) Simplified.

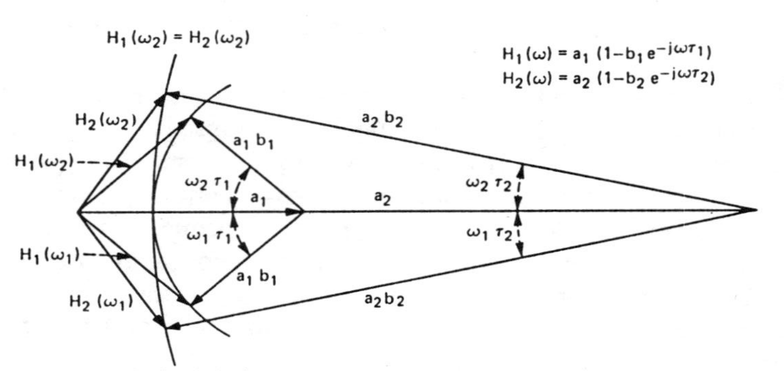

Fig. 3—Two degenerate simple three-path fades with $\omega_0\tau = 0$.

midband and at both edges. It has been shown elsewhere[7] that fades matched in this way will be identical in band to within a few tenths of a decibel at most, and will have almost identical envelope delay distortion. Given noisy quantized measurements of $|H(\omega)|$ over the band, it is impossible to distinguish between such fades unless we fix one of the four parameters. Let us consider each of the four possibilities.

2.2 Pseudo two-path fade

If one fixes the amplitude, a, at unity, the simple three-path fade reduces to a two-path fade with independent control of the frequency of the minimum of the response. The difficulty with this model, as may be seen by referring to Fig. 2b, is that it can provide in-band minima only for $|H(\omega)| < 1$ and maxima in-band only for $|H(\omega)| > 1$. In other words, the model cannot match an in-band maximum at an arbitrary fade level. In addition, it was found that during approximately half the periods when the radio equipment was indicating errors, the channel could not be well modeled with a pseudo two-path model.

2.3 Scaled two-path fade

If one fixes the phase, $\phi = \omega_0\tau - \pi$, in the simple three-path model at 0, the fade reduces to a scaled two-path fade. (For a two-path fade, we require the additional condition $a = 1$.) This is the most physically desirable of the reduced three-path models because it may be derived without recourse to the three-path formalism. Unfortunately, it is mathematically intractable, particularly when dealing with amplitude data only. In fitting the model to a given channel shape (in the manner described in Section III for the fixed delay model), one obtains a function of a, b, and τ that must be minimized to obtain the best fit. Because of the $\omega\tau$ term in the exponent of the model, this function has a local minimum in every interval of τ of length 0.17 ns, the reciprocal of 6 GHz. Since the possible range of τ extends to about 30 ns, one may have to perform hundreds of minimizations to find the best fit to a single channel scan. Even then this "best fit" may have no minimum phase realization, and there is no known procedure that leads to one.

2.4 Fixed b model

If one fixes the amplitude b in the simple three-path model, the resulting reduced model has all the mathematical difficulties of the scaled two-path model and no satisfactory physical interpretation.

2.5 Fixed delay model

It is demonstrated in the remainder of this paper that the fixed delay model described in Section I is useful and effective in characterizing the channel.

III. ESTIMATION PROCEDURES

This section describes how the model parameters are estimated from the channel scans and how realizability difficulties are surmounted.

3.1 Parameter estimation

The channel data consist of a set of 25,000 scans of the channel power spectrum. Each scan consists of a power measurement at each of 24 frequencies at 1.1-MHz spacing across the channel. (Actually, only 23 frequencies are used since the 19th was inoperative during this test period). The power measurements are recorded in decibels, and each must be referenced to the average power level of that frequency at mid-day. With proper conversion and calibration, the basic data characterizing a scan are a set of power ratios. We designate the power ratio at nth frequency by Y_n, where

$$Y_n = Y(\omega_n) \qquad n = 1, 2, \cdots, 24. \tag{3}$$

We wish to model the channel with a voltage transfer function of the form given in eq. (1), which we repeat here for convenience

$$H(\omega) = a[1 - be^{\pm j(\omega-\omega_0)\tau}], \tag{1}$$

Thus our estimate of Y_n will be

$$\hat{Y}_n = |H(\omega_n)|^2 = \alpha - \beta \operatorname{Cos}(\omega_n - \omega_0)\tau, \tag{4}$$

where

$$\alpha = a^2(1 + b^2)$$
$$\beta = 2a^2b. \tag{5}$$

For convenience, we measure frequency in the units of the frequency separation of the power measurements. Thus,

$$\omega_n = 2\pi f_n = 2\pi n(1.1 \times 10^6) \quad n = 1, 2, 3, \cdots, 24. \tag{6}$$

If we choose

$$\tau = \frac{1}{N(1.1 \times 10^6)}, \tag{7}$$

then

$$\omega_n\tau = 2\pi\frac{n}{N}. \tag{8}$$

For the fixed delay model, we choose $N = 144$ which gives a model τ of 6.31 ns. Thus, the in-band frequencies correspond to n values between 1 and 24, and the channel transfer function given by the model is periodic for n modulo 144, corresponding to a frequency shift of $144 \times 1.1 \times 10^6 = 158.4$ MHz.

The weighted mean-square error between the estimated and observed power is given by

$$E = \frac{\sum_{n=1}^{24} C_n (Y_n - \hat{Y}_n)^2}{\sum_{n=1}^{24} C_n}, \tag{9}$$

where the summation skips $n = 19$ as described above, and where C_n is a weighting applied to the measurement at frequency ω_n. Since the original data, from which the Y_n's were derived, were uniformly quantized on a logarithmic scale, it is appropriate to use a weighting that is approximately logarithmic. Hence, we use the weighting function

$$C_n = \frac{1}{Y_n^2}. \tag{10}$$

A number of different weighting functions were tested, but the one given by (10) is, generally, the most satisfactory.

Estimates of a, b, and f_o may be obtained by minimizing the weighted mean-square error, E. It is shown in the appendix that one may obtain closed form estimators for α, β, and f_o by substituting eq. (4) into (9) and minimizing E, first with respect to α, then with respect to β (or vice versa), and last with respect to f_o. In the resulting scheme, the estimator of f_o, the frequency of the model minimum, is a function of data only. The estimators of α and β are functions of the estimated f_o and the data.*

After estimates of α and β have been calculated, the parameters a and b of the model are obtained by inverting the relationships given by eq. (5).

$$b = \frac{\alpha}{\beta} - \left[\left(\frac{\alpha}{\beta}\right)^2 - 1\right]^{1/2} \tag{11}$$

$$a = \left[\frac{\beta}{2b}\right]^{1/2}. \tag{12}$$

It is clear from (11) and (12) that we can realize the channel shape with the model only if $\alpha \geq \beta$. This is to be expected. Since $|H(\omega)|^2$ is a power transfer function, it must be positive for all frequencies, which is possible only if $\alpha \geq \beta$ [see eq. (4)]. Thus, the condition $\alpha \geq \beta$ allows us to obtain a minimum (or nonminimum) phase transfer function whose magnitude squared is the minimum weighted mean-square error fit to the observed power transfer response of the channel.

* For mathematical simplicity, we actually use an estimator for β conditioned on f_o, α, and the data.

3.2 Application of estimators

If the procedure described above is strictly applied to the set of 25,000 scans in the data base, one finds that about 35 percent of the scans cannot be modeled with real values of a and b. A study of these problem scans revealed that the estimator for f_o, the frequency of the fade minimum, was biased for two types of scans. One type is a scan with little shape, dominated by quantization noise; the other is a selective channel shape having a steep slope across the band. Both types of scan are illustrated in Fig. 4. The scan in Fig. 4, which is almost flat, was fabricated to illustrate the severity of the quantization problem. The other scan is typical of the more shapely troublesome scans.

To obtain a good realizable fit to such channel shapes requires degrading the quality of the fit; that is, moving the parameters away from the values that minimize the fit error, eq. (9). Given the form of the estimation scheme, this is easily accomplished by moving the frequency of the fade minimum, f_o, away from its original "optimum" value and reoptimizing the remaining parameters to obtain values of a and b that are optimum for the new value of f_o. Figures 5 and 6 illustrate the results of such a quasi-optimization regarding f_o as a free parameter. They show the fit error E and the values of a and b as f_o is varied from its original optimum value. Figure 5 corresponds to the flat fade in Fig. 4 and Fig. 6 to the sloped fade.

The shapes of the curves in Figs. 5 and 6 are typical of those obtained when the channel has no minimum in band. The weighted error in the fit, E, is not very sensitive to the estimate of f_o, the

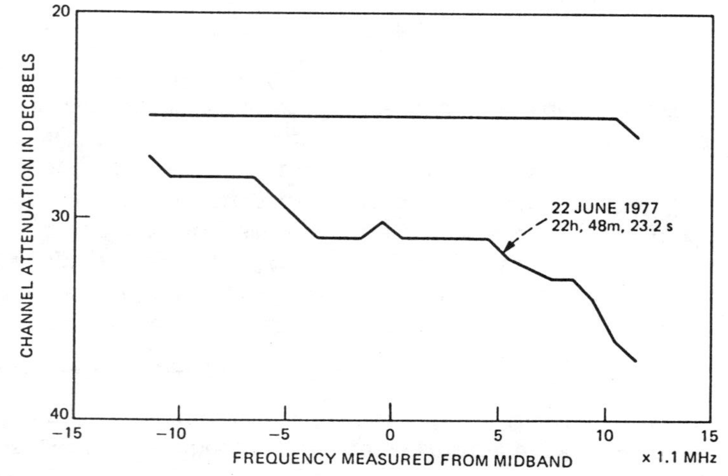

Fig. 4—Two channel scans that produce realization difficulties.

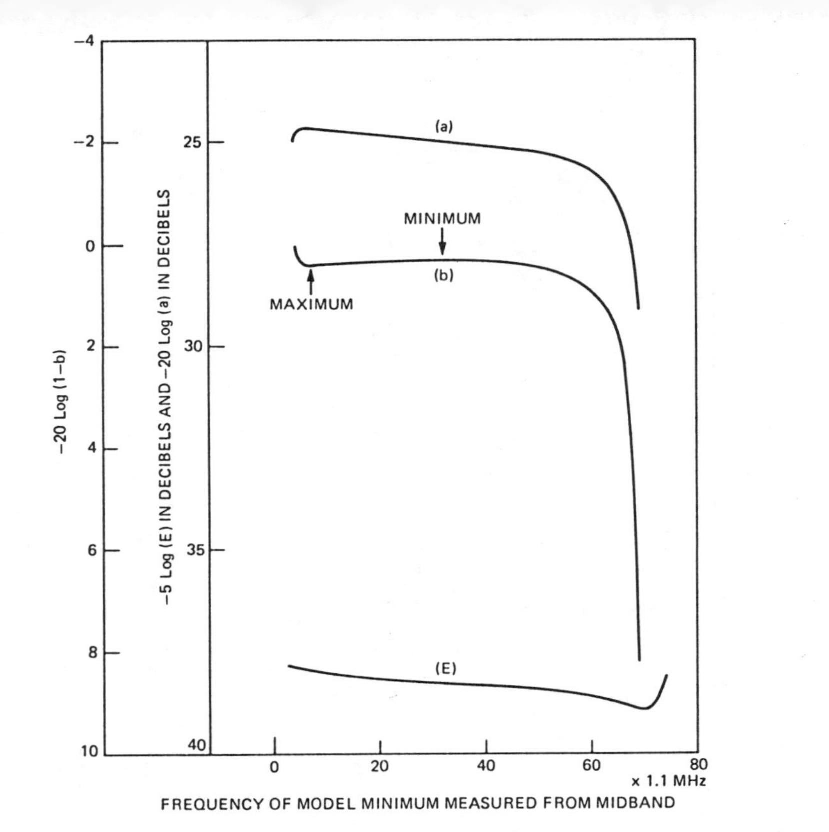

Fig. 5—Locus of weighted fit error and model parameters with f_o as a free variable for flat fade in Fig. 4.

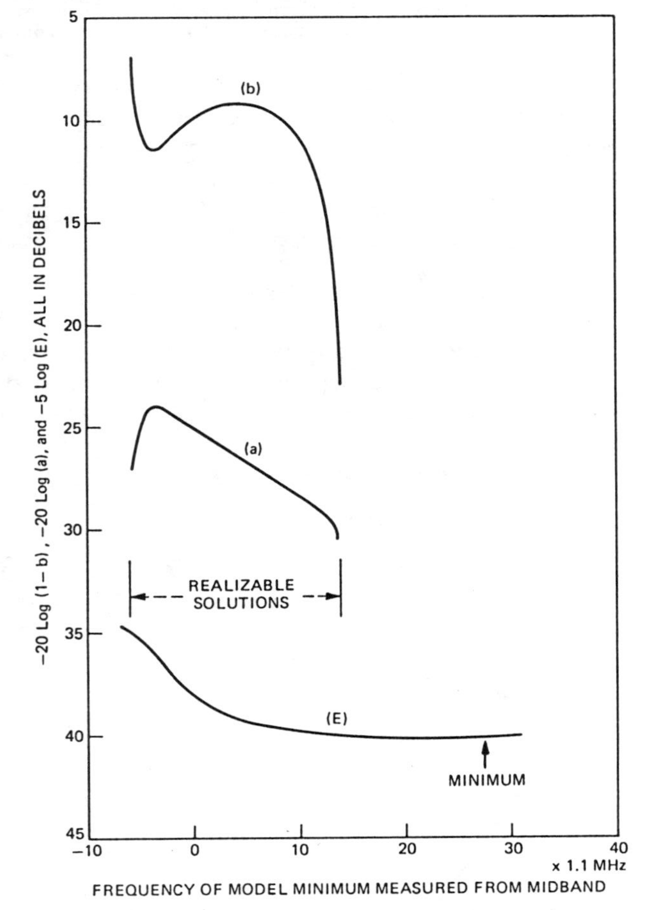

Fig. 6—Locus of weighted fit error and model parameters with f_o as a free variable for typical scan in Fig. 4.

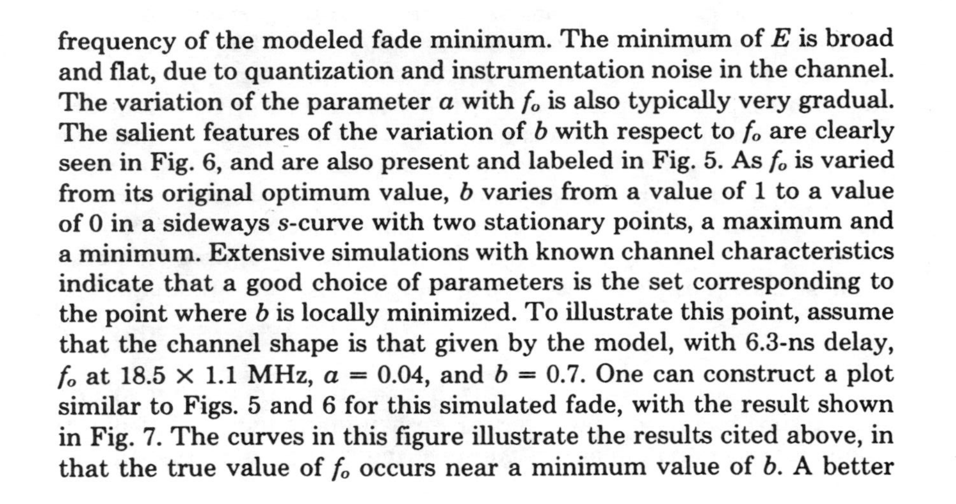

frequency of the modeled fade minimum. The minimum of E is broad and flat, due to quantization and instrumentation noise in the channel. The variation of the parameter a with f_o is also typically very gradual. The salient features of the variation of b with respect to f_o are clearly seen in Fig. 6, and are also present and labeled in Fig. 5. As f_o is varied from its original optimum value, b varies from a value of 1 to a value of 0 in a sideways s-curve with two stationary points, a maximum and a minimum. Extensive simulations with known channel characteristics indicate that a good choice of parameters is the set corresponding to the point where b is locally minimized. To illustrate this point, assume that the channel shape is that given by the model, with 6.3-ns delay, f_o at 18.5 × 1.1 MHz, a = 0.04, and b = 0.7. One can construct a plot similar to Figs. 5 and 6 for this simulated fade, with the result shown in Fig. 7. The curves in this figure illustrate the results cited above, in that the true value of f_o occurs near a minimum value of b. A better choice for the case shown and for others that have been simulated would be "on the shoulder" between the minimum and $b = 1$; however, such a criterion is difficult to quantify.

To summarize, if the standard routine does not provide a realizable fit to a scan, one merely varies f_o, the position of the minimum, until one obtains a realizable solution with a value of b that is stationary* with respect to variations in f_o. We recognize that this procedure introduces additional sources of error into the estimates of the model parameters. The errors in a and b are small because b is near a stationary value and a is slowly varying. The error in f_o is also small,

* Since b is a monotone function of α/β, it is only necessary to invert solutions with stationary values of the ratio, α/β.

usually less than 3 MHz, but is always in the direction corresponding to moving the minimum nearest to the band closer. We consider the effects of these errors in Section VI.

IV. MODEL STATISTICS

Applying the procedures described in Section III to the scans in the data base results in 25,000 sets of values of a, b, and f_o. The relative joint frequency of occurrence of these three parameters may be described by the set of distribution functions shown in Figs. 8 to 12. The distribution of the parameter b is described in Fig. 8 in terms of the distribution of $-20 \log (1 - b)$, which is approximately exponentially distributed with a mean of 3.8 dB. This distribution gives the time that b exceeds the value given by the abscissa as a fraction of the time in a heavy fading month that the rms level in the channel is depressed by more than 15 dB. For instance, we see that 40 percent of the time

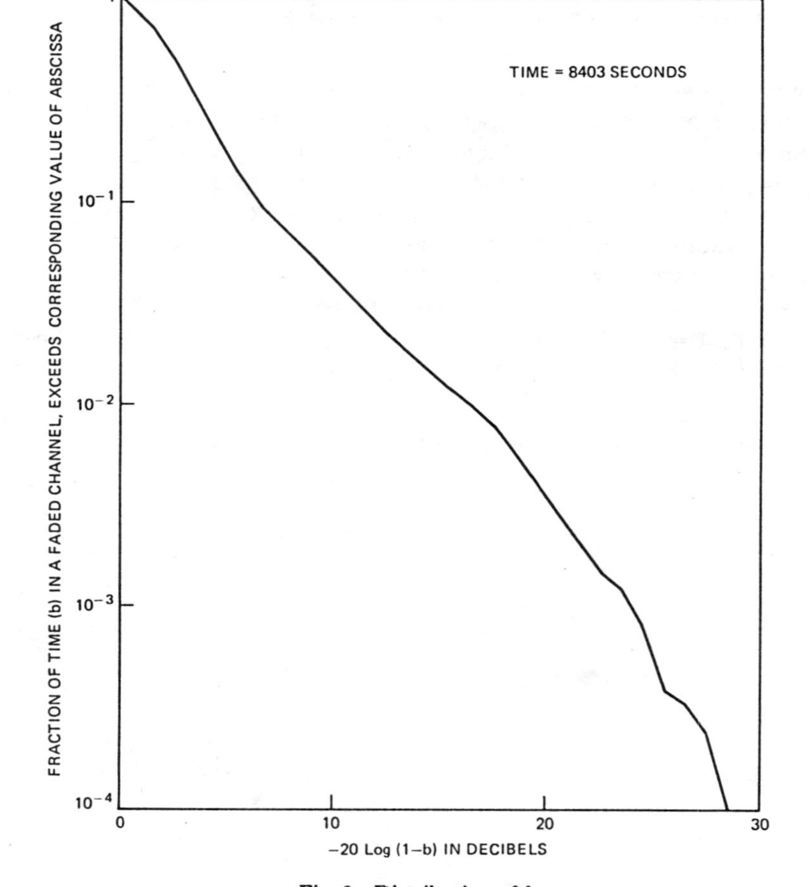

Fig. 7—Locus of weighted fit error and model parameters with f_o as a free variable. For channel given by model with $\tau = 6.31$ ns, $a = 0.04$, $b = 0.7$, $f_o = 18.5 \times 1.1$ MHz.

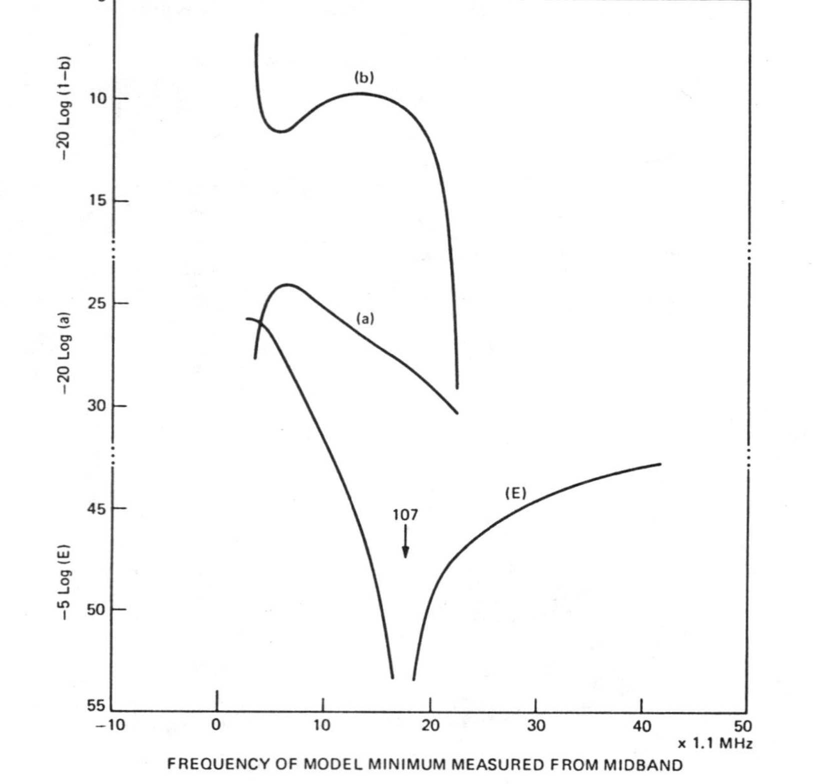

Fig. 8—Distribution of b.

when the channel is depressed the value of b exceeds 0.3. It exceeds 0.7 for 4 percent of that time, and 0.99 about 0.3 percent of that time. The distribution of b can be modeled in the form

$$P(1 - b < X) = X^{\frac{20}{3.8 \text{ Log } 10}} = X^{2.3}. \tag{13}$$

The distribution of a is conditioned on b and is approximately lognormal as shown in Figs. 9 and 10. The mean and standard deviation of the distributions in Figs. 9 and 10 are plotted in Fig. 11. From Figs. 9 to 11 it is apparent that a and b are almost independent; however, less shapely fades tend to occur at less depressed values. We note that shape occurs when the average depression is 20 to 25 dB,* that the

* The value of a corresponds to average power level over a large frequency span and not strictly to the average power in a narrowband channel.

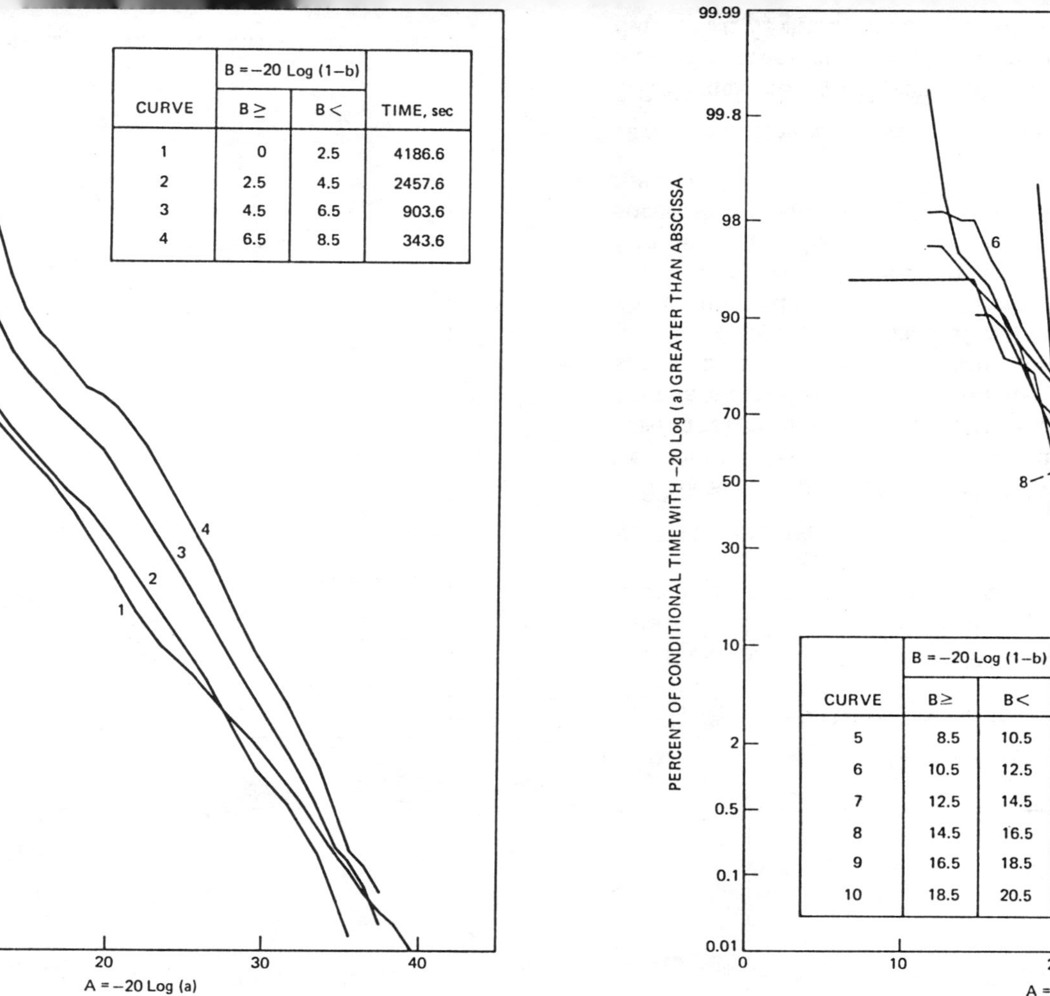

CURVE	B = −20 Log (1−b) B ≥	B <	TIME, sec
1	0	2.5	4186.6
2	2.5	4.5	2457.6
3	4.5	6.5	903.6
4	6.5	8.5	343.6

Fig. 9—Distribution of a conditioned on the value of b for $-20 \log (1 - b)$ less than 8.5 dB.

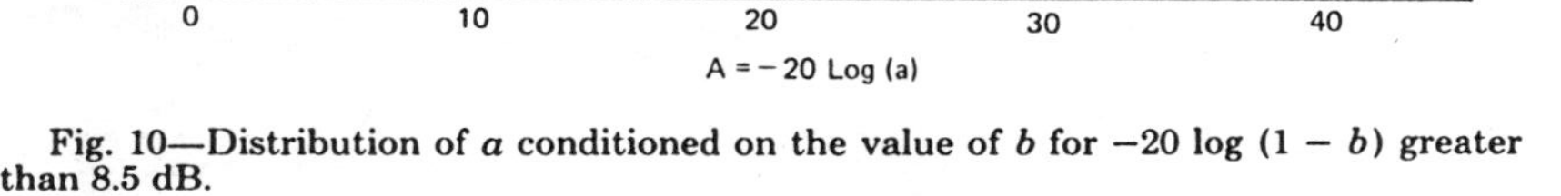

CURVE	B = −20 Log (1−b) B ≥	B <	TIME, sec
5	8.5	10.5	200.0
6	10.5	12.5	120.4
7	12.5	14.5	65.4
8	14.5	16.5	42.8
9	16.5	18.5	36.4
10	18.5	20.5	23.4

Fig. 10—Distribution of a conditioned on the value of b for $-20 \log (1 - b)$ greater than 8.5 dB.

average depression is near 25 dB for b greater than 0.7, and that it falls off gradually to 15 dB for small b. The distribution of $A = -20 \log a$ is conditioned on b and may be modeled as

$$P(A > Y) = 1 - P\left[\frac{Y - A_o(b)}{5}\right], \tag{14}$$

where P is the cumulative distribution function of a zero mean, unit variance, and Gaussian random variable, and $A_o(b)$ is the mean of A for a given value of b as given in Fig. 11. We see from Fig. 11 that the standard deviation of A may be taken as 5 dB regardless of the value of b; the variations near $-20 \log (1 - b) = 20$ are due to small sample problems.

Figure 12 shows the time during which scans had f_o in 4×1.1-MHz frequency intervals. It is, in effect, an estimate of the density function of the distribution of f_o and is, consequently, quite noisy. The maxima near $\pm 30 \times 1.1$ MHz from the center of the band are due in part to the movement of estimates of f_o to achieve realizability. While, on physical grounds, one would expect f_o to have a uniform distribution, the fixed delay model is decidedly not a physical model. Consider a simulated set of simple three-path fades having a uniform distribution of f_o, fixed values for a and b, and a delay τ, fixed at a value other than 6.31 ns. This set of fades will engender a nonflat probability density function for the f_o's obtained in fitting to the 6.31-ns model. The probability density function is flat within the band regardless of the fixed delay of the set of simulated fades; however, it will more nearly resemble that shown in Fig. 12 if the delay of the set is greater than 6.31 ns than if it is less than 6.31 ns. In short, Fig. 12 is characteristic of a channel with a considerable fraction of delay differences greater than 6 ns.

Based on Fig 12, we approximate the probability density function of f_o by a two-level function. Note that f_o is defined on an interval of length $1/\tau$, where τ is 6.3 ns the delay of the model. Thus, with f_o measured from the center of the band, the probability density function for f_o may be approximated by

$$p_{f_o}(f_o) = \begin{cases} \dfrac{5\tau}{3} & |f_o| \le \dfrac{1}{4\tau} \\ \dfrac{\tau}{3} & \dfrac{1}{4\tau} < |f_o| < \dfrac{1}{2\tau}. \end{cases} \tag{15}$$

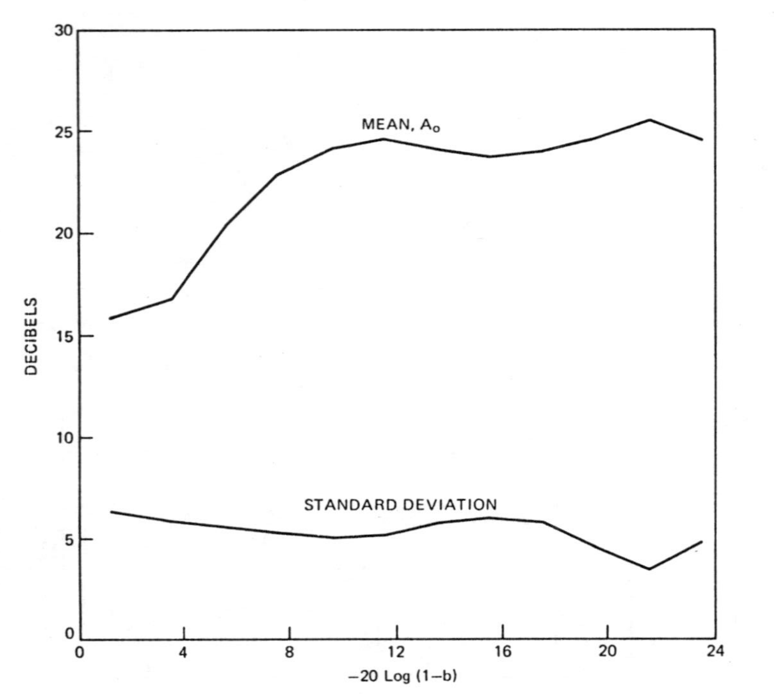

Fig. 11—Mean and standard deviation of the distribution of $-20 \log a$ as a function of $-20 \log (1 - b)$.

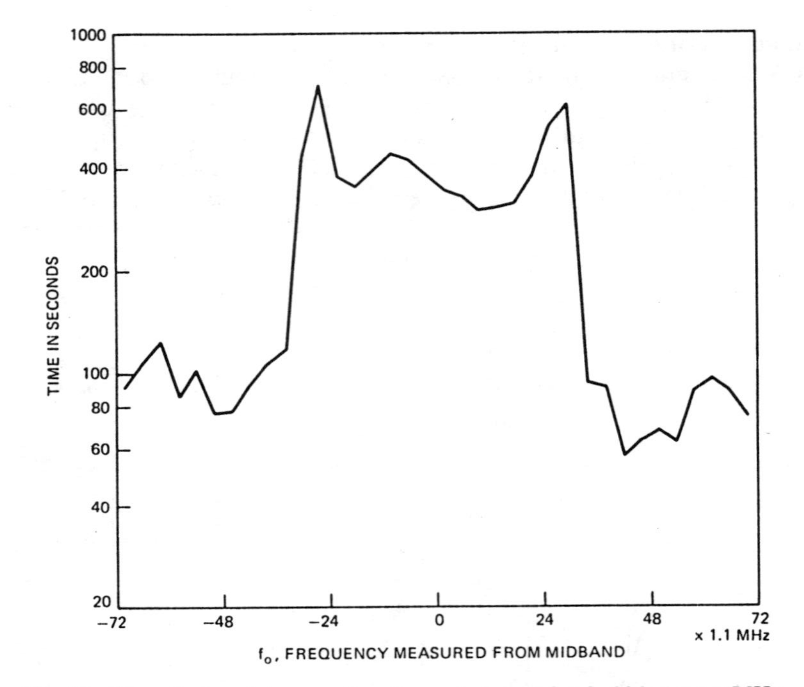

Fig. 12—Time that model parameter, f_o, was in intervals of width 4 × 1.1 MHz.

An extensive examination of various conditional distributions has established that there are no other obvious and pervasive dependencies among the statistics of the parameters.

V. CHANNEL DELAY DIFFERENCE

This section presents some results obtained in estimating the channel delay difference. Some techniques described here are used in the error analysis in Section VI. Three topics are considered in this section. First a simple method is presented of estimating the average delay spread in the channel. A second subsection shows that the distribution of large delays (larger than 10 ns) can be obtained for a simple three-path fade model. The delay distribution is shown to be consistent with the estimate of average delay. A third subsection illustrates the problem with an observed channel shape that can be matched most successfully using a simple three-path model with a delay of approximately 26 ns.

5.1 Mean delay difference in the channel

The mean delay difference of a channel that can be characterized by a simple three-path model is easily estimated. Consider a fade with a delay, τ. If f_o, the frequency of the minimum, is uniformly distributed, the probability that such a fade produces a minimum in a band B Hz wide is equal to the ratio of the bandwidth to the spacing of the minima, or

$$\frac{B}{1/\tau} = B\tau. \tag{16}$$

If $p(\tau_k)\Delta\tau$ is the fractional number of fades having delays between $(k-1)\Delta\tau$ and $k\Delta\tau$, then the fractional number of fades having a minimum in band will be $P_{\min}$, where

$$P_{\min} = \sum_k B\tau_k p(\tau_k)\Delta\tau = B\bar{\tau} \tag{17}$$

and

$$\bar{\tau} = \sum \tau_k p(\tau_k)\Delta\tau = \int \tau p(\tau)\, d\tau. \tag{18}$$

It follows from eq. (17) that one may estimate the mean delay, $\bar{\tau}$, from a knowledge of $P_{\min}$, the fractional number of scans having a minimum in a band of width B. Since any method of determining $P_{\min}$ is acceptable, consider estimates of $P_{\min}$ from the parameters estimated using the fixed delay model. The method of estimating the frequency parameter in the model involved moving null positions of some fades that had out-of-band minima. These fades can be excluded by using only the central two-thirds of the band in estimating $\bar{\tau}$. Of the 24,920 scans in the data base, 3974 had minima between the $4th$ and $20th$ frequencies. Hence,

$$\bar{\tau} = \left[\frac{3974}{24920}\right] \frac{1}{16 \times 1.1 \times 10^6} = 9.1 \text{ ns}. \tag{19}$$

One might argue that the mean delay should be estimated for a more carefully screened set of scans. Table I shows the mean delay estimates obtained from scan populations qualified by having the estimate of the model parameter a in a given 5-dB interval. Table II shows mean delay estimates qualified by the model parameter b, which specifies the shapeliness of the fade.

Table I—Mean delay for scans selected by value of parameter, a

$-20 \log a$, dB	Number of Scans	Scans with Min. in Band	Delay, $\bar{\tau}$, ns
0–5	101	31	17.4
5–10	725	235	18.4
10–15	4299	875	11.6
15–20	6891	1161	9.6
20–25	7644	906	6.7
25–30	4184	606	8.2
30–35	1019	159	8.9
All	24920	3974	9.1

Table II—Mean delay for scans selected by value of parameter, b

$-20 \log 1\text{-}b$, dB	Number of Scans	Scans With Min. in Band	Delay, $\bar{\tau}$, ns
0–2	10,442	1186	6.5
2–4	7040	1712	13.8
4–6	3721	538	8.2
6–8	1474	191	7.4
8–10	892	118	7.5
10–12	527	68	7.3
12–14	282	28	5.6
14–16	190	21	6.3
16–18	146	46	17.9
18–20	99	32	18.4
All	24920	3974	9.1

With several exceptions, the estimated delay spreads given in Tables I and II are reasonably constant. One exception is seen for large values of b ($-20 \log 1 - b$ greater than 16). This is consistent with a channel for which large differential attenuation across the channel is more likely to occur when long delays are present. The existence of such a correlation should not be surprising. The other exception is the large delays estimated for small values of b and for values of a between 0 and 10 dB. We provide strong evidence of the existence of such a class of fades in the next subsection. The existence of this subclass of fades suggests that they have a different physical source than the other fades in the population.

5.2 Distribution of delay difference

To further enhance our knowledge of the distribution of delay in the channel, the data base was processed to extract a delay estimate. Recall that, for the fixed delay model, parameter estimates are chosen to minimize the weighted fit error [E in eq. (9)] for a given fixed τ. The present calculation was performed for a set of different values of τ and the value which produced the smallest weighted fit error and corresponded to a realizable fade was designated as the delay for that scan if it met certain qualifications.

Because of the degeneracy in the simple three-path model, changing the delay in the fixed delay model will not appreciably improve the fit for any scan that can be well approximated by a fixed delay of 6 ns or less.[7] In performing the optimum delay calculation, the weighted fit error was minimized for a predetermined set of delays; the differences between adjacent delay values were chosen to be approximately 15 percent. A given scan was assigned a delay different from 6.3 ns only if the third best value of the weighted fit error was at least 0.1 dB worse than the best value. (We use the third best value because we must examine three values to detect a minimum.) This criterion sets a threshold on the acceptable sharpness of the minimum in the fit error with respect to changes in delay.

The selection criterion was chosen, after several iterations, to insure regularity in the estimates derived from successive scans. With the chosen criterion, the scans that were assigned a new delay occurred in groups of consecutive scans and may be said to constitute fading events. During any of these events, the delay was consistent in that indicated delays were within ±15 percent. If we assume that the physical channel does not change between scans, we can associate a time with each scan and plot the distribution of the time periods during which the characterizing delay was greater than a specified delay.

A series of such plots, conditioned on the concurrently estimated value of b, is given in Fig. 13. The uppermost curve contains the data derived from all scans which met the selection criterion; its shape is

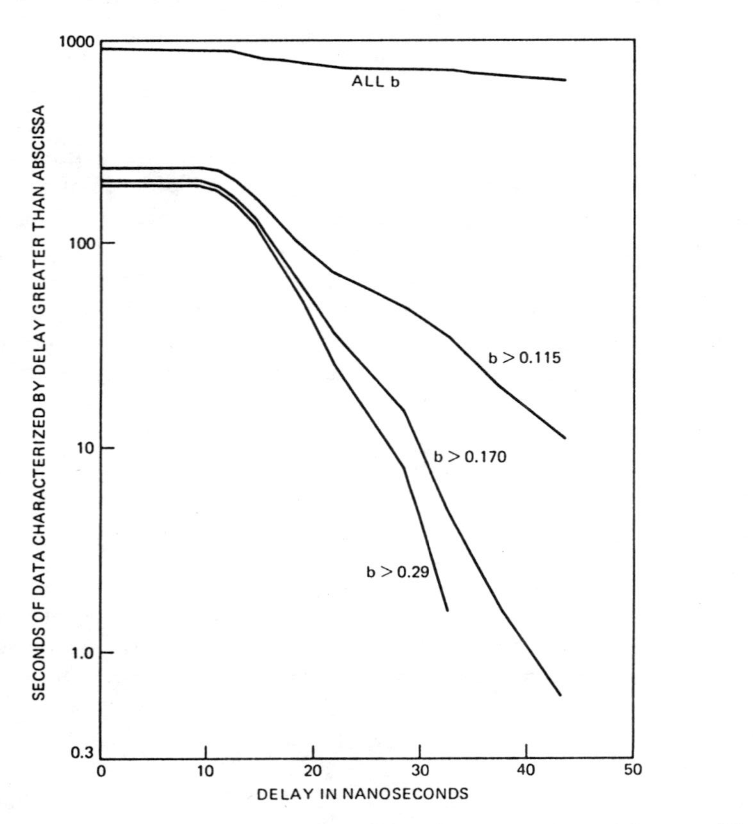

Fig. 13—Distribution of optimum delay for simple three-path model, as qualified by realizability, the sharpness of minimum, and by several values of the model parameter b.

dominated by the 627 seconds during which the channel was best modeled by a delay of 43 ns (the largest delay in the test set) or more, but had little shape ($b < 0.115$). These characteristics contribute to the large (18 ns) mean delays noted in the previous subsection for small values of b. They may be due to quantization but are apparently not artifacts of the estimation scheme. Although the origin of this type of channel defect is currently not understood, it should not trouble any existing radio system.

It is apparent from the distributions in Fig. 13 that very few scans qualified for a new delay with delays less than 10 ns. Consequently, the distribution should not be trusted for delays less than 12 or 15 ns; beyond 15 ns, it may be interpreted as a lower bound to the true distribution. The three curves qualified by the parameter b correspond to fades with peak-to-peak variability of 2, 3, and 6 dB. (Peak-to-peak variability is $20 \log (1 + b/1 - b)$, as may be seen in Fig. 1.) If the delay were exponentially distributed, the distribution of delay would be a straight line on Fig. 13 and would have the form

$$P(\tau > x) = e^{-x/\bar{\tau}}. \tag{20}$$

Fitting a straight line to the three distributions in Fig. 13 for which $b > 0.115$ shows that the average delay decreases with increasing b. The corresponding values are 5, 5.5, and 11 ns. Note that this implies that b and τ in a simple three-path model are not independent.

5.3 *An example of a long delay scan*

To confirm the existence of long delay scans, consider an event that covered approximately 10 seconds on 22 June 1977, from 23 h, 28 m, 54 s. A representative scan from the middle of this period is shown with the fit obtained with the fixed delay (6.3 ns) model in Fig. 14a. To emphasize the consistency of this channel condition, an average of the channel condition for the central 4.2 seconds (21 scans) of this event is compared to the selected scan in Fig. 14b.

It is apparent from Fig. 14*a* that the 6.3 ns delay does not have enough curvature (delay is too short) to precisely model the channel shape. Figure 15 shows the same scan modeled by three-path fades having delays of 22.7, 26, and 30.3 ns. The 26-ns fit is the best; it has a weighted fit error 0.4 dB better than the 22.7-ns fit and 0.8 dB better than the 30.3 ns fit. However, the closeness of all three fits illustrates the difficulties in estimating channel delay differences. Visually, one would choose the 26-ns model on the basis that the 30.3-ns fit has too much curvature and the 22.7-ns fit too little.

VI. ERROR ANALYSIS

To verify that the model adequately represents the transmission characteristics of the channel, we examine the errors between the

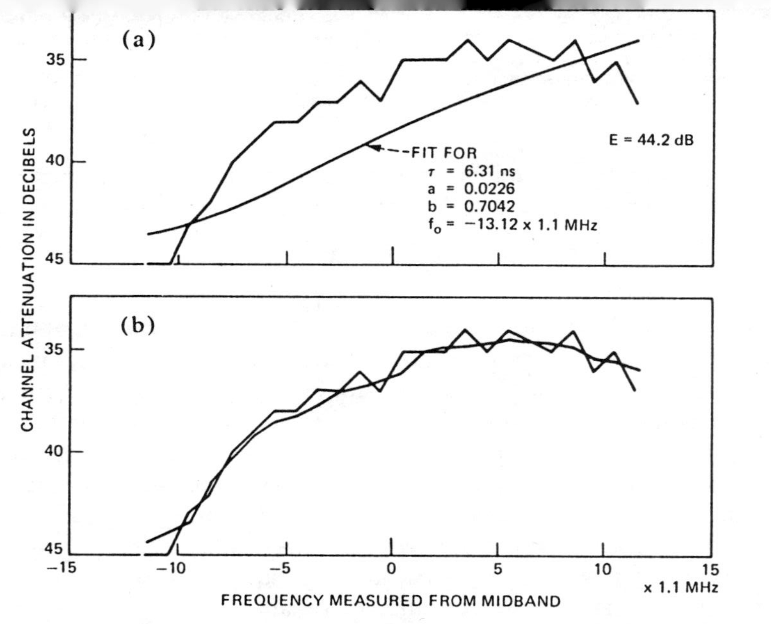

Fig. 14—Scan from 22 June 1977, 23 h, 28 m, 48.6 s. (a) Comparison with fixed delay model. (b) Comparison with average of scans from 23 h, 28 m, 46.4 s to 23 h, 28 m, 50.4 s.

channel as observed and as modeled. In this section we consider the statistics of the rms errors and the maximum errors.

6.1 RMS errors

A useful measure of the quality of the fit of the model to a given channel scan is the root-mean-square value of the decibel error at each of the sampled frequencies. Denoting this error as E_{rms}, we have

$$E_{rms} = \left[\frac{1}{23} \sum_{\substack{n=1 \\ n \neq 19}}^{24} (\text{dB error at } f_n)^2 \right]^{1/2}. \tag{21}$$

The model parameters were estimated, as described in Section III, to minimize the error, E, which is a weighted sum of the squares of the power differences at each frequency [see eq. (9)]. The weighting was chosen [eq. (10)] so that the error E would approximate the error E_{rms} as given by eq. (21).* Indeed, one may show directly that the two expressions are equivalent as long as

$$\left| 1 - \frac{\hat{Y}_n}{Y_n} \right| \ll 1 \text{ for all } n. \tag{22}$$

* Note that the parameter estimation problem cannot be solved in closed form by minimizing E_{rms}.

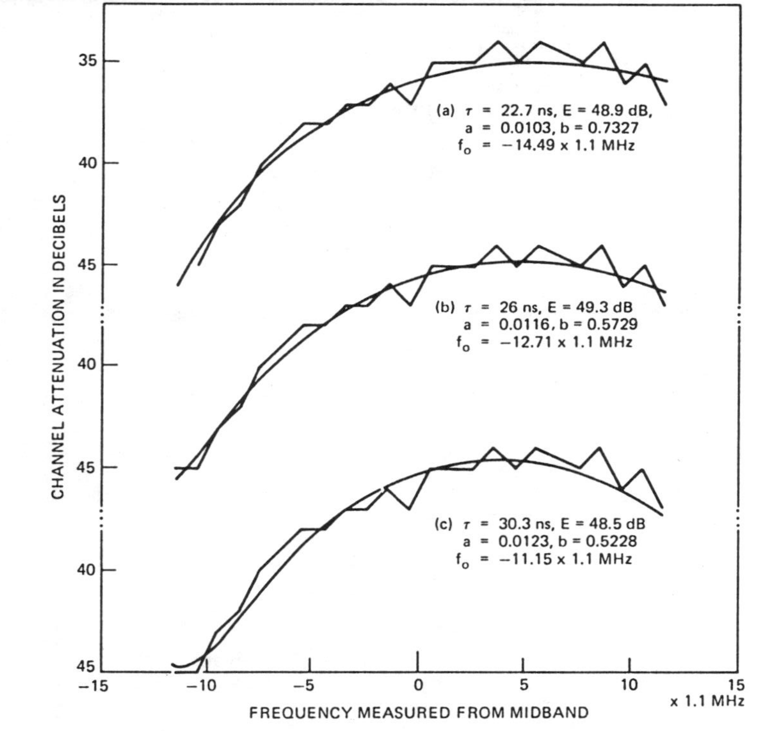

Fig. 15—Model fits to long delay scan for three different model delays.

As we have seen in Fig. 14, this inequality is not always satisfied. Consequently, in using E_{rms} as a standard of comparison, we are evaluating not only how well the model fits the observed channel, but also how well we have chosen the parameters to make the match.

The error E_{rms} is a desirable quantity to work with because we can estimate its distribution under the assumption of perfect matching. We observe that if the decibel error were Gaussian with unit variance and zero mean, $23\ E_{rms}^2$ would be a χ^2 variable with 20 degrees of freedom (to account for the three parameters estimated per scan). Observations of a simulated channel with the transmitter and receiver back-to-back indicate that the instrumentation errors are approximately Gaussian with a standard deviation, σ_i, of about 0.65 dB. Observations of the channel at mid-day with the channel nominally flat and unfaded indicate that the standard deviation of the errors is between 0.68 and 0.73, varying frequency to frequency and day to day by a few hundredths of a decibel. Hence, if we enter a table of the χ^2 distribution, $Q(\chi^2 | 20)$, with

$$\chi^2 = \frac{23\ E_{rms}^2}{\sigma_i^2}, \tag{23}$$

we can determine the distribution of E_{rms} under the assumption of perfect matching.* This distribution is shown as a reference on Figs. 16 and 17. It is indicated by a solid curve labeled "ideal" for $\sigma_i = 0.70$ and by o's for $\sigma_i = 0.75$.

Figure 16 presents the distribution of the rms error for two scan subpopulations using the fixed delay (6.3 ns) model. The subpopulation of the distribution labeled "standard" consists of all scans that could be modeled directly; the distribution labeled "modified" shows the rms error distribution for all scans which required an adjustment of the frequency of the modeled fade to achieve realizability. Figure 17 shows the distribution of the rms error for the composite of all samples using the fixed delay (6.3 ns) model. The distribution labeled simple three-path model indicates the error distribution that was obtained when the scan fitting allowed unqualified variation in model delay to achieve the best fit. That is, the calculation described in Section 5.2 was performed and the results were qualified only on the basis of realizability.†

In each case described above, the mean value of the rms error is close to the median value. For the two subpopulations shown in Fig. 16, the calculated mean fit errors correspond to σ_i values of 0.76 and 0.85 dB, or the errors are about 0.09 dB larger when a realizable fit is obtained by varying the frequency of the model minimum. Comparing the composite distributions in Fig. 17, we find that the mean error in the fixed delay (6.3 ns) model corresponds to σ_i = 0.78 dB or about 0.08 dB higher than that observed when the channel is quiescent. The simple three-path model has a distribution of rms error that very nearly matches the ideal distribution (with 19 degrees of freedom) for $\sigma_i = 0.75$. This is consistent with the instrumentation error imputed to the standard distribution in Fig. 16 and is indicative of the instrumentation error in the presence of multipath fading. It is exceptionally good considering that the data are obtained from time sequential measurements on a dynamically changing channel. One concludes that the modeling error is negligible for the simple three-path model. For the fixed delay model under the assumption that the instrumentation and modeling errors add in quadrature, the modeling error has a tolerable value on the order of 0.2 dB. That is,

$$[(0.75)^2 + (0.2)^2]^{1/2} = 0.776.$$

The tails of the distributions in Figs. 16 and 17 for large errors are of considerable interest. The tails near small values are of little

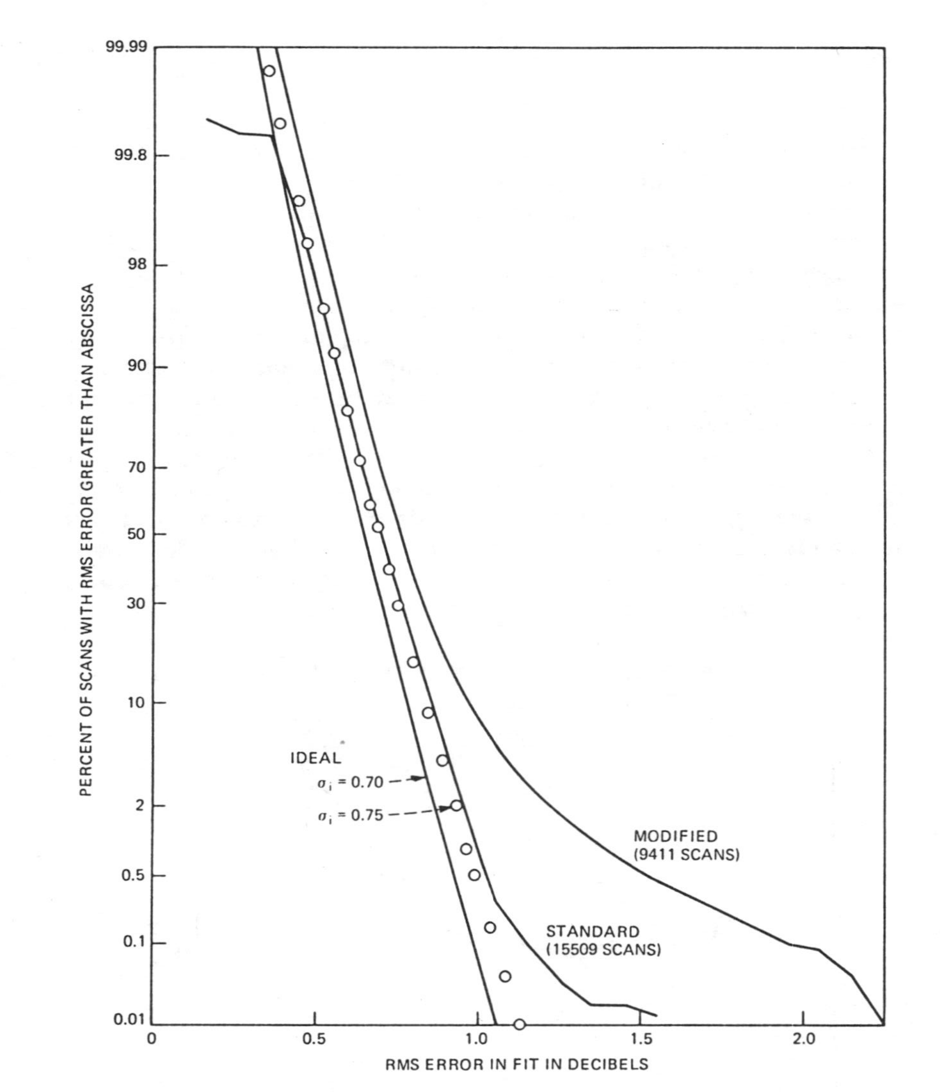

Fig. 16—Distribution of rms fit error for two scan subpopulations with fixed delay (6.3 ns) model.

consequence; they are distorted by quantization because one cannot associate any error with the 12 flat fades included in the data base. The deviation of the distributions from the ideal distribution at large errors is significant.

The large deviation of the modified fits in Fig. 16 reflects the failure of the fixed delay (6.3 ns) model to accurately fit the long delay fades. The tail deviation from ideal is modest down to about the 0.5 percent level, corresponding to a few tens of seconds per month. For compar-

* From the central limit theorem, we know that E^2_{rms} will be approximately Gaussian, as is χ^2, regardless of whether or not the measurement errors are precisely Gaussian.

† Note that although one cannot always reliably localize the values of the parameters in fitting with the simple three-path model (see discussions in Section 2.1 and Ref. 7), the error in the fit is always well defined.

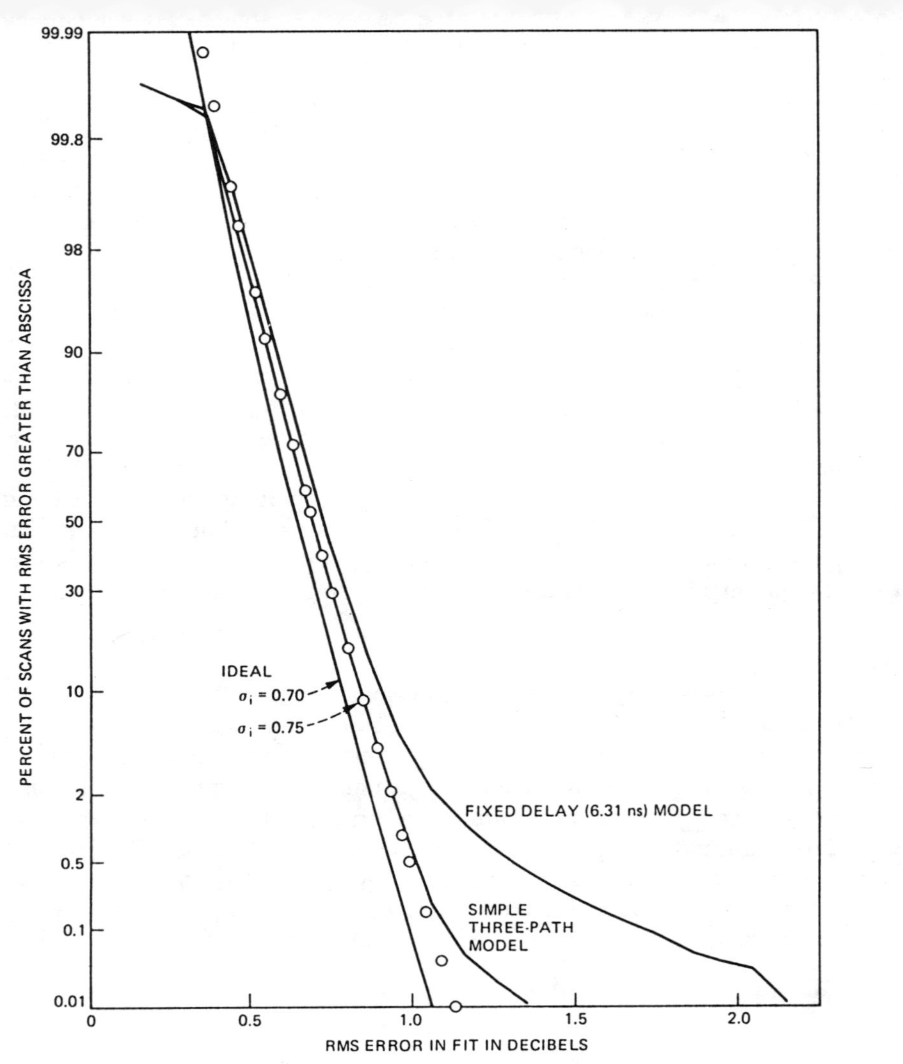

Fig. 17—Distribution of rms fit error for composite population with fixed delay and simple three-path models.

ison, we note that the rms error of the fit shown in Fig. 14a is 2.3 dB; this was the worst fit encountered for the fixed delay (6.3 ns) model. However, even in this case the model failure is hardly describable as severe. The model of the channel is depressed by 40 dB and has 9.5 dB of gain slope; the actual channel is depressed by 39 dB and has 11 dB of gain slope. Also, we note that the 6.3-ns delay model has the response minimum at about the same frequency as the best representation, the 26-ns delay model shown in Fig. 15.

The deviation of the tail of the error distribution for the three-path fade (Fig. 17) reflects the fact that there are fades that even this model has difficulty in fitting. An example of such a fade is shown in Fig. 18 along with the fit provided by the fixed delay (6.3 ns) model. The same rms error (1.6 dB) is obtained for all values of model delay between 0.05 and 9 ns; the fit degrades for larger delays. Either more than three rays are needed to describe the channel shape in Fig. 18, or the channel is so depressed that the amplitudes in the notch are distorted due to closeness to the noise level in the measuring equipment. The scan shown in Fig. 18 is one of three similar scans and has little statistical significance.

6.2 Maximum errors

Another type of error that can be used to judge the quality of the fit of the model to the channel is the worst-case error. That is, after fitting to each scan, one records the magnitude of the largest difference (in decibels) between the observed channel shape and the shape calculated from the model. The following paragraphs consider the distribution of these worst-case errors.

As in the preceding subsection, we can calculate an ideal distribution; however, the ideal distribution is not as realistic in this case since it is strongly dependent on the tails of the distributions of the individual measurement errors. We assume that each power measurement had

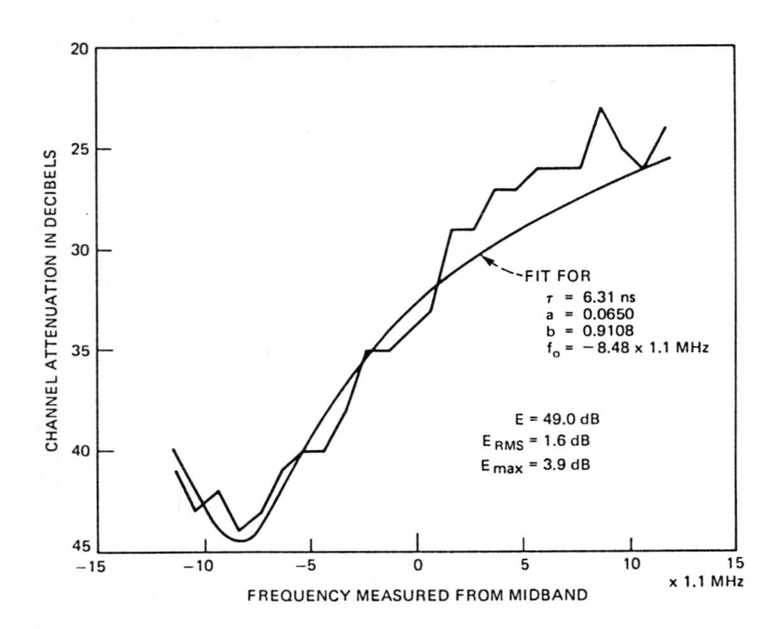

Fig. 18 –Severe fade observed on 22 June 1977 at 22 h, 29 m, 8.6 s.

an error in decibels that was Gaussian, with zero mean, a standard deviation of $\sqrt{20/23}\ \sigma_i$ to account for the three parameters estimated from the 23 observations per scan,* and that the errors are independent frequency to frequency. If the probability of any one measurement having an error less than x is denoted by $P_1(x)$, the probability that all 23 have values less than x is

$$P_{23}(x) = [P_1(x)]^{23}. \quad (24)$$

This is the probability that the maximum error is less than x, whereas we want the probability that it is greater than x which we denote $Q_{23}(x)$. It follows immediately from eq. (24) that

$$Q_{23}(x) = 1 - [P_1(x)]^{23}$$
$$= 1 - [1 - Q_1(x)]^{23}. \quad (25)$$

The distribution given by (25) is used as a reference in Figs. 19 and 20, which show the distribution of the maximum error for the same cases as in Figs. 16 and 17. Since the tails of these distributions are well behaved for larger errors, the distribution of the maximum errors is apparently dominated by the instrumentation noise. That is, if we use for the standard deviation of the measurement noise the value obtained from the mean of E_{rms} for one of these cases (as given in Section 6.1), the resulting worst-case error distribution calculated with eq. (25) will closely match the observed maximum error distribution.

VII. CONCLUSIONS

By analyzing the errors in fitting the observed channel characteristics in Section VI, we demonstrated that the simple three-path fade model is indistinguishable from a perfect model of a line-of-sight microwave radio channel.

The simple three-path model was used in Section V to characterize the channel delay difference. By two different methods, it was shown that, when there is 3 dB or more shape present in the channel, the average delay difference is between 5 and 8 ns. We developed a lower bound on the tails of the distribution of delay difference. From these results, which are shown in Fig. 13, we observe that a differential channel attenuation in-band of 3 dB or more may be due to delay differences as great as 43 ns. In another dimension, one would expect to see differential attenuation of 3 dB or more in-band due to delays greater than 20 ns for at least 70 seconds in a heavy fading month. This is comparable to the time the channel attenuation at a single frequency exceeds 40 dB.

* For comparisons with the three-path model, it is appropriate to use $\sqrt{19/23}\,\sigma_i$.

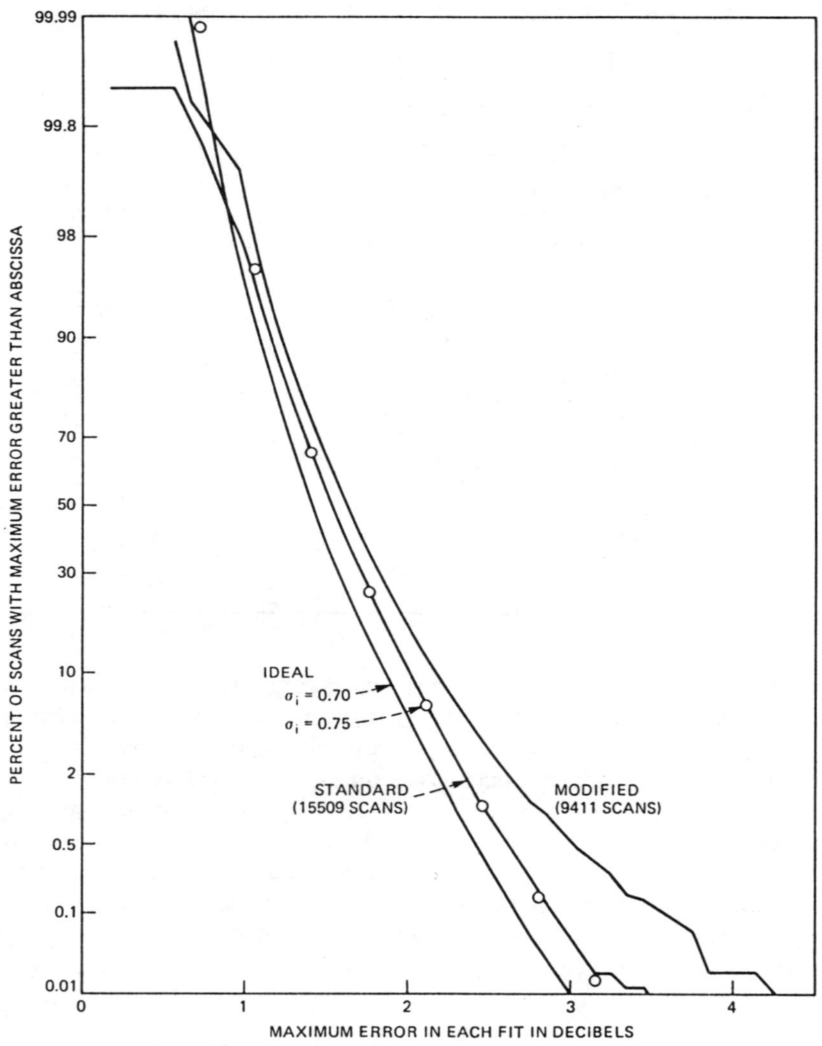

Fig. 19—Distribution of maximum (dB) fit error for two scan subpopulations with fixed delay (6.3 ns) model.

From the error analysis in Section VI, we also conclude that the fixed delay (6.3 ns) model is a very good approximation to the channel for all observed conditions. This conclusion is further substantiated by Figs. 14 and 18, which show the scans for which the fits with the fixed delay model exhibited the largest rms fit error (2.3 dB) and the largest maximum error (3.9 dB), respectively. The fixed delay model is preferable to the three-path model for channel modeling because it requires

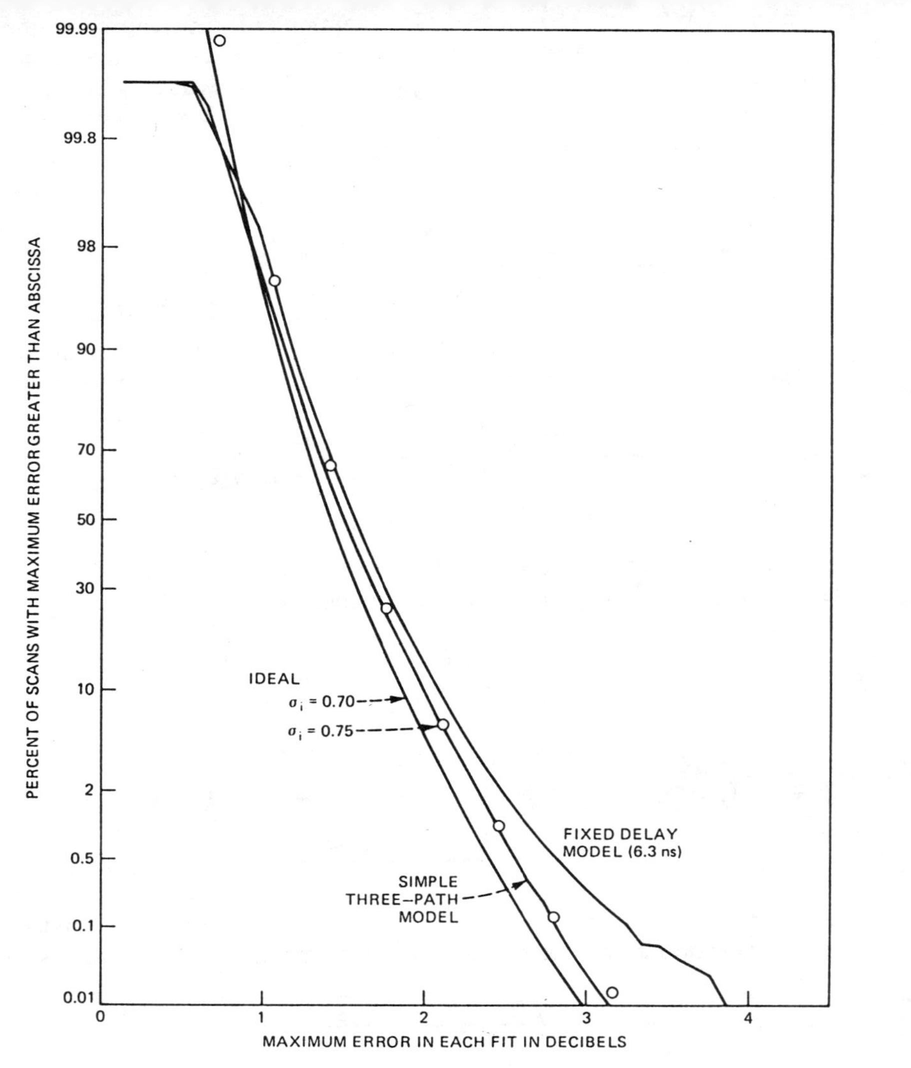

Fig. 20—Distribution of maximum (dB) fit error for composite population with fixed delay and simple three-path models.

only three parameters, and these can always be uniquely determined from a channel amplitude scan.

The statistics of the parameters of the fixed delay model as described in Section IV and shown in Figs. 8 to 12 provide the means of statistically generating all the channel conditions that one expects to see on a nominal hop channel operated at 6 GHz. If one determines, by laboratory test, the parameter values that will cause a particular error rate in a digital radio system, one can easily calculate the time during a heavy fading month that the error rate will equal or exceed this critical value. A companion paper describes the laboratory test and the required calculations.[8]

Future work will be directed toward verifying the model and model statistics with additional fading data obtained both at 6 GHz and at 4 GHz. Using coherent data obtained in 1973, it will be possible to determine the extent to which the channel is actually a minimum phase channel.

VIII. ACKNOWLEDGMENTS

This study would not have been possible without the contributions of many individuals. In particular, the radio equipment was installed, aligned, and maintained by R. A. Hohmann, C. W. Lundgren, and L. J. Morris with instrumentation designed by G. A. Zimmerman. The data processing expertise in setting up and calibrating the data base was provided by M. V. Pursley.

APPENDIX

Estimation of Parameters

The problem of estimating the parameters α, β, and f_o in Section III is equivalent to the problem of determining the first three terms in a subharmonic Fourier series expansion of a function in the frequency domain. Since such expansions are not standard, we provide a complete description of the methodology here.

From eqs. (4) and (9), we may express the weighted mean-square error between estimated and observed power as*

$$E = \frac{\sum C_n(Y_n - \alpha + \beta \cos(\omega_n - \omega_0)\tau)^2}{\sum C_n}. \tag{26}$$

For simplicity, we use a normalized weighting function, d_n, defined by

$$d_n = \frac{C_n}{\sum C_n} \tag{27}$$

so that

$$\sum d_n = 1. \tag{28}$$

In terms of the normalized weighting we may write (26) as

$$E = \sum d_n(Y_n - \alpha + \beta \cos(\omega_n - \omega_0)\tau)^2 \tag{29}$$

* Throughout this appendix, all summations are taken over all values of n corresponding to all frequencies observed in a scan.

or in expanded form as

$$E = \sum d_n Y_n^2 + \alpha^2 + \beta^2 \sum d_n \cos^2(\omega_n - \omega_0)\tau$$
$$+ 2\beta \sum d_n Y_n \cos(\omega_n - \omega_0)\tau$$
$$- 2\alpha\beta \sum d_n \cos(\omega_n - \omega_0)\tau - 2\alpha \sum d_n Y_n . \quad (30)$$

The error E is a minimum when α, β, and ω_0 are chosen so that the partial derivatives of E with respect to α, β, and ω_0 are all equal to zero. Setting the partial derivative of eq. (30) with respect to β equal to zero and solving for β gives

$$\beta = \frac{\alpha \sum d_n \cos(\omega_n - \omega_0)\tau - \sum d_n Y_n \cos(\omega_n - \omega_0)\tau}{\sum d_n \cos^2(\omega_n - \omega_0)\tau} . \quad (31)$$

Substituting (31) into eq. (30), we find E_β, the error minimized with respect to β, as

$$E_\beta = \sum d_n^2 Y_n^2 + \frac{1}{\sum d_n \cos^2(\omega_n - \omega_0)\tau}$$
$$\cdot \{\alpha^2[\sum d_n \cos^2(\omega_n - \omega_0)\tau - (\sum d_n \cos(\omega_n - \omega_0)\tau)^2]$$
$$- 2\alpha\, [(\sum d_n Y_n)(\sum d_n \cos^2(\omega_n - \omega_0)\tau)$$
$$- (\sum d_n \cos(\omega_n - \omega_0)\tau)(\sum d_n Y_n \cos(\omega_n - \omega_0)\tau)]$$
$$- (\sum d_n Y_n \cos(\omega_n - \omega_0)\tau)^2\} . \quad (32)$$

Minimizing this with respect to α requires that we set the partial derivative of E_β with respect to α equal to zero. This gives

$$\alpha = \frac{(\sum d_n Y_n)(\sum d_n \cos^2(\omega_n - \omega_0)\tau) - (\sum d_n \cos(\omega_n - \omega_0)\tau)(\sum d_n Y_n \cos(\omega_n - \omega_0)\tau)}{\sum d_n \cos^2(\omega_n - \omega_0)\tau - (\sum d_n \cos(\omega_n - \omega_0)\tau)^2} . \quad (33)$$

Substituting (33) into (32) gives $E_{\alpha\beta}$, the error minimized with respect to both α and β, as

$$E_{\beta\alpha} = \sum d_n Y_n^2 - \bar{Y}^2 - \frac{(\sum d_n (Y_n - \bar{Y})\cos(\omega_n - \omega_0)\tau)^2}{\sum d_n \cos^2(\omega_n - \omega_0)\tau - (\sum d_n \cos(\omega_n - \omega_0)\tau)^2} , \quad (34)$$

where

$$\bar{Y} = \sum d_n Y_n . \quad (35)$$

We note that we could have obtained this same expression by first minimizing with respect to α and then with respect to β; however, one obtains different but equivalent expressions for α and β, depending on the order of differentiation. We develop the alternative expressions for α and β in the following paragraphs.

Let us define some new quantities to simplify these expressions. Let the difference between the observed power and the weighted mean power in the band be designated by X_n; then

$$X_n = Y_n - \sum d_n Y_n = Y_n - \bar{Y} . \quad (36)$$

If we also define the quantities

$$X_c = \sum d_n X_n \cos \omega_n \tau , \quad (37)$$
$$X_s = \sum d_n X_n \sin \omega_n \tau , \quad (38)$$
$$D_a = \sum d_n \cos^2(\omega_n - \omega_0)\tau , \quad (39)$$
$$D_b = \sum d_n \cos(\omega_n - \omega_0)\tau , \quad (40)$$

we may rewrite α and β from eqs. (31) and (33) as

$$\alpha = \bar{Y} - \frac{[X_c \cos \omega_0\tau + X_s \sin \omega_0\tau]D_b}{D_a - D_b^2} \quad (41)$$

and

$$\beta = \frac{1}{D_a} \{(\alpha - \bar{Y})D_b - (X_c \cos \omega_0\tau + X_s \sin \omega_0\tau)\} . \quad (42)$$

Using (41) to eliminate α from (42), we obtain

$$\beta = -\frac{X_c \cos \omega_0\tau + X_s \sin \omega_0\tau}{D_a - D_b^2} . \quad (43)$$

We may use (43) in (41) to obtain

$$\alpha = \bar{Y} + \beta D_b . \quad (44)$$

Equations (43) and (44) are the estimators that would have been obtained if the order of taking partial derivatives in the preceding development had been reversed. It is apparent that, after one has estimated ω_0, one may estimate α and β by using either eqs. (41) and (42), (43) and (44), or eqs. (41) and (43).

The estimate of ω_0 that minimizes the weighted error is obtained by minimizing $E_{\beta\alpha}$ with respect to ω_0. Using eqs. (35) to (40) in eq. (34), we write

$$E_{\beta\alpha} = \sum d_n X_n^2 - \frac{[X_c \cos \omega_0\tau + X_s \sin \omega_0\tau]^2}{D_a - D_b^2} . \quad (45)$$

To see the explicit dependence of $E_{\beta\alpha}$ on ω_0, we define the following quantities

$$d_c = \sum d_n \cos^2 \omega_n \tau - (\sum d_n \cos \omega_n \tau)^2 , \quad (46)$$

$$d_s = \sum d_n \sin^2 \omega_n \tau - \left(\sum d_n \sin \omega_n \tau\right)^2, \tag{47}$$

$$d_{cs} = \sum d_n \cos \omega_n \tau \sin \omega_n \tau - \left(\sum d_n \cos \omega_n \tau\right)\left(\sum d_n \sin \omega_n \tau\right). \tag{48}$$

Substituting these into (45) gives

$$E_{\beta\alpha} = \sum d_n X_n^2 - \frac{[X_c \cos \omega_n \tau + X_s \sin \omega_0 \tau]^2}{d_c \cos^2 \omega_0 \tau + 2 d_{cs} \cos \omega_0 \tau \sin \omega_0 \tau + d_s \sin^2 \omega_0 \tau}. \tag{49}$$

Setting the partial derivative of $E_{\beta\alpha}$, as given by (49), equal to zero gives the estimator for ω_0 as

$$\omega_0 \tau = \mathrm{Tan}^{-1}\left[\frac{d_c X_s - d_{cs} X_c}{d_s X_c - d_{cs} X_s}\right]. \tag{50}$$

Obviously, two values of $\omega_0 \tau$ in the interval $(-\pi, \pi]$ will satisfy eq. (50). One of these, the principal value, lies in the interval $(-\pi/2, \pi/2]$, the other differs from the first by $\pm\pi$. We shall show that the two solutions are equivalent, but that our chosen solution is unique.

If we replace $\omega_0 \tau$ by $\omega_0 \tau \pm \pi$ in eqs. (39), (40), (43), (44), and (45), we see that D_b and β change sign and α and $E_{\beta\alpha}$ are unchanged. Since we want the solution with β greater than zero, we take the principal value solution to (50) if the resulting estimate of β is positive. Otherwise we add or subtract π to obtain a positive value for β and a value of $\omega_0 \tau$ in the appropriate interval.

While we could substitute the result of eq. (50) into (49) to obtain the minimum error, $E_{\alpha B \omega_0}$, it is more generally useful to evaluate $E_{\alpha\beta}$ for the optimum ω_0. This is especially true when we do not use the optimum ω_0, as given by eq. (50). The simplest form for $E_{\alpha\beta}$ is obtained by substituting (43) into (45) to give

$$E_{\alpha\beta} = \sum d_n X_n^2 - (D_a - D_b^2)\beta^2. \tag{51}$$

These equations were implemented, with the modifications described in Section 3.2, to obtain all the fits described in this paper.

REFERENCES

1. G. M. Babler, "A Study of Frequency Selective Fading for a Microwave Line-of-Sight Narrowband Radio Channel," B.S.T.J., *51*, No. 3 (March 1972), pp. 731–757.
2. G. M. Babler, "Selectively Faded Nondiversity and Space Diversity Narrowband Microwave Radio Channels," B.S.T.J., *52*, No. 2 (February 1973), pp. 239–261.
3. L. J. Greenstein, "A Multipath Fading Channel Model for Terrestrial Digital Radio," IEEE Trans. Commun. Tech., *COM-26*, No. 8 (August 1978), pp. 1247–1250.
4. R. L. Kaylor, "A Statistical Study of Selective Fading of Super-High Frequency Radio Signals," B.S.T.J., *32*, No. 5 (September 1953), pp. 1187–1202.
5. G. M. Babler, unpublished work.
6. W. C. Jakes, Jr., "An Approximate Method to Estimate An Upper Bound on the Effect of Multipath Delay Distortion on Digital Transmission," ICC '78 Conference Record, June 1978, *3*, pp. 47.1.1–5.
7. W. D. Rummler, "A Multipath Channel Model for Line-of-Sight Digital Radio Systems," ICC '78 Conference Record, June 1978, *3*, pp. 47.5.1–4.
8. C. W. Lundgren and W. D. Rummler, "Digital Radio Outage Due to Selective Fading—Observation vs Prediction from Laboratory Simulation," B.S.T.J., this issue, pp. 1073–1100.

A Polynomial Model for Multipath Fading Channel Responses

By L. J. GREENSTEIN and B. A. CZEKAJ

(Manuscript received January 10, 1980)

We present a channel model useful for analyzing the effects of multipath fading in digital radio systems. The frequency response of a fading channel is represented by the function $A_0 - \omega B_1 + j\omega A_1$, *where* ω $(=2\pi f)$ *is measured from the center of the channel, and* A_0, A_1, *and* B_1 *are variable coefficients that change slowly with time. The model consists of this function, the joint probability density function (pdf) for the three coefficients, and the average number of seconds per heavy-fading month for which this response applies. The model is derived from a large base of multipath fading data, obtained on a 26.4-mile path in Georgia in June 1977. It consists of nearly 25,000 recorded measurements of received power vs frequency in a 25.3-MHz bandwidth at 6 GHz. In this paper, we present the methods of data reduction and statistical analysis used to derive the model; describe some assessments of its validity; and discuss its limitations, virtues and possible uses. By all available measures, the model is highly accurate. It suffers from a potentially important phase ambiguity that can be resolved only via new, coherent measurements. The existing model should prove very useful in the design and planning of such measurements.*

I. INTRODUCTION

Multipath fading in microwave digital radio systems can be a major source of outage and, therefore, has been the subject of numerous recent investigations.[1-4] One important objective of current activity is to develop a statistical model of fading useful for estimating its effect on specific systems and for indicating possible methods of correction.

Two multipath fading characterizations have been proposed recently for terrestrial systems in the channelized common carrier bands. One represents the multipath fading frequency response as that due to an equivalent three-path medium.[5] The result is a three-parameter, complex gain function, where the joint probability distribution for the three parameters describes the fading statistically. The second characterization, which is the subject of this paper, expresses the multipath fading frequency response as a complex polynomial expanded about the channel center frequency.[6] Specifically, if

$$H_c(\omega) \triangleq \frac{\text{complex channel gain at any time}}{\text{complex channel gain during nonfading}}, \qquad (1)$$

where $\omega(=2\pi f)$ is measured from the center frequency, then the polynomial representation is

$$H_c(\omega) = A_0 + \sum_{n=1}^{N} (A_n + jB_n)(j\omega)^n, \qquad (2)$$

where the A- and B-coefficients vary slowly relative to the speeds of typical digital radio systems.

Note that $H_c(0)$ is just A_0, a real number, i.e., we arbitrarily (and with no loss in generality) specify the phase shift at the channel center frequency to be zero.

The polynomial function of eq. (2) can fit frequency responses of any shape with arbitrary accuracy merely by choosing N sufficiently large. Moreover, the difficulty of fitting coherently measured responses does not grow significantly with N since $H_c(\omega)$ is linearly related to the characterizing parameters (the A_n's and B_n's).

Several features make the polynomial representation attractive for multipath modeling. One is that it leads to simple methods for analyzing signal processing (note that $(j\omega)^n$ corresponds to the nth time derivative). A second is that it suggests a promising idea for adaptive equalization; specifically, the rational function $1/H_c(\omega)$ may be easy to realize adaptively when the complex zeros of $H_c(\omega)$, (2), have negative real parts. A third feature is parsimony, suggested by the conjecture[6] that a first-order function ($N = 1$) may be sufficiently accurate for terrestrial paths in the channelized bands below 15 GHz. In that case, the channel response function could be characterized by three linear coefficients (A_0, A_1, and B_1).

The work reported here has confirmed this conjecture for the data base used by Rummler.[5] The central activity has been to compute A_0, A_1, and B_1 from the recorded measurements and to analyze their joint statistics. The outcome is a model for the joint pdf, $p(A_0, A_1, B_1)$, that can be used to assess multipath effects in any digital radio system.

Section III presents a mathematical description of the new model. Section II gives an overview of the data reductions and analyses leading to this model, while Section IV gives an expanded discussion for readers interested in more details.

Section V discusses the accuracy of the derived model and presents supporting evidence for the steps taken to simplify it. Part of this

Reprinted with permission from *Bell System Technical Journal,* vol. 59, no. 7, pp. 1197–1225, Sept. 1980.

evidence is obtained by calculating outage times for numerous system approaches, using model versions of differing complexity. This exercise not only validates the simplest version but illustrates its important application to system studies. Section VI summarizes the uses and limitations of the model and cites the need for further measurements to reinforce and improve it.

II. DATA REDUCTIONS

2.1 The data base

The multipath fading data base is a set of measurements made on a 26.4-mile path in Palmetto, Georgia in June 1977.[5] During fading events, received power was measured at each of 24 frequencies distributed about 6034.2 MHz and spaced by 1.1 MHz. The measuring filter at each frequency had a 200-KHz bandwidth. The power samples were normalized by the unfaded received power level (as estimated from mid-day measurements), converted to decibels, quantized in 1-dB steps, and recorded on magnetic tape. Because of an equipment failure, data for the 19th frequency in the sequence (7.15 MHz above the center of the channel) were not obtained.

Each tape record therefore consists of 23 power response values, in quantized decibels, corresponding to a single measurement. No phase response data were recorded, resulting in unresolvable ambiguities in the reductions, as discussed later. Measurements were recorded at rates of either five per second or one per two seconds, depending on the speed of change in the fading response. The data base used here consists of 24,920 records spanning, roughly, 8400 seconds of fading.

2.2 The polynomial coefficients

The main reductions consist of (i) determining the coefficients in eq. (2) for each record in the data base and (ii) finding a mathematical description of the joint probability density function (pdf) of these coefficients. In this subsection, we outline the strategy used for determining the coefficients.

Each data record consists of the quantized values of $-10 \log_{10} |H_c|^2$ at 23 frequencies, where H_c is the complex channel gain.

We denote the recorded values by the set $\{P_i\}$, where i is the frequency index ($i = 1, 24$ excluding 19). The data reduction per record begins by converting each decibel quantity P_i into a power ratio $p_i (= 10^{-P_i/10})$, and fitting the sequence of p_i vs frequency with an Mth-order polynomial,

$$q(\omega) = D_0 + \omega D_1 + \cdots \omega^m D_m + \cdots \omega^M D_M. \tag{3}$$

In obtaining the coefficients ($D_0, \cdots D_M$), a form of least-squares optimization has been used, and M-values of 0, 2, and 4 have been tried. We have evaluated the errors between the p_i's and the fitting function $q(\omega)$ for each record and have studied the statistics of these errors over the ensemble of records that comprise the data base. Our finding is that, given the noise statistics and bandwidth of the measurements, the most suitable polynomial order is $M = 2$. For, when $M = 0$, the fitting function lacks the curvature needed to model the true fading response, while, when $M = 4$, the third- and fourth-order polynomial terms serve more to fit the measurement noise than to fit the underlying response. The curvature provided by $q(\omega)$ when $M = 2$ seems to be just right for the 25.3-MHz measurement bandwidth. For that case, we estimate the rms fractional difference between $q(\omega)$ and the true power response to be 5 percent at all frequencies.

Given a value for M and a procedure for obtaining the D_m's from the data records, we must choose the order (N) of the polynomial in eq. (2) and derive the A_n's and B_n's in that equation from the D_m's. Before deciding on N, we note that a power-gain function can be obtained from eq. (2) having the form

$$|H_c(\omega)|^2 = \hat{D}_0 + \omega \hat{D}_1 + \cdots \omega^m \hat{D}_m + \cdots \omega^{2N} \hat{D}_{2N} \tag{4}$$

where the $\hat{D}_m$'s are simply related to the A_n's and B_n's. The objective is to match this function to $q(\omega)$ in eq. (3).

One possibility is to choose $N = M/2$ and to find the A_n's and B_n's for which $\hat{D}_0 = D_0$, $\hat{D}_1 = D_1$, etc. The problem is that, if D_M happens to be negative, this important term cannot be matched using all real A_n's and B_n's. The approach we have used instead is to specify $N = M$, and to choose the $(2N + 1)$ A_n's and B_n's so that

$$\hat{D}_m = \begin{cases} D_m; & 0 \le m \le N \\ 0; & N < m \le 2N. \end{cases} \tag{5}$$

This procedure, however, is limited in at least one and sometimes two respects. The fundamental limitation concerns those A_n's and B_n's contained within the imaginary part of $H_c(\omega)$, *i.e.*, ($A_1, A_3 \cdots B_2, B_4 \cdots$). Given a valid solution to eq. (5), reversing all these coefficients in sign results in exactly the same function for $|H_c(\omega)|^2$. Thus, there are two possible solutions, each of which is equally valid given the data.

The polarity ambiguity in the subset ($A_1, A_3, \cdots B_2, B_4 \cdots$) is reflected as a polarity ambiguity in the phase response associated with $H_c(\omega)$, an inevitable consequence of noncoherent measurements. Our reduction approach is to assume equally likely polarities for any given fade. In computing A_n's and B_n's for each record, we have randomly selected the polarity in accordance with this assumption.

A second limitation on the application of eq. (5) arises when the $\hat{D}_m$'s for $n > N$ cannot all be forced to zero using real A_n's and B_n's. In performing reductions for $N = 2$, for example, a solution with real A_n's and B_n's is only possible when

$$\Delta = (D_2 - D_1^2/4D_0) \geq 0. \tag{6}$$

We will limit all remaining discussions to the case $N = M = 2$.

The condition $\Delta \geq 0$ corresponds roughly to $q(\omega)$ being concave over the measurement bandwidth. In such cases, $A_2 = B_2 = 0$, and the function for $H_c(\omega)$ reduces to a first-order polynomial. When $\Delta < 0$ ($q(\omega)$ more or less convex over the measurement bandwidth), no real combination of A_n's and B_n's will satisfy all of eq. (5). This condition is found to occur over 42 percent of the data base. Our strategy for these cases is to choose $(A_0, A_1, A_2, B_1, B_2)$ so that $(\hat{D}_0, \hat{D}_1, \hat{D}_2)$ matches (D_0, D_1, D_2); $\hat{D}_3 = 0$; and $\hat{D}_4 (= A_2^2 + B_2^2)$ has the smallest magnitude possible for real A_n's and B_n's. The result is a fourth-order component in $|H_c(\omega)|^2$, namely, $\omega^4 \hat{D}_4$, which we have found to be relatively small (in a sense that we define later).

The above strategy leads to the following set of solutions:

$$A_0 = \sqrt{D_0}; \tag{7a}$$

$$B_1 = -D_1/2A_0; \tag{7b}$$

$$A_1 = \begin{cases} \pm\sqrt{\Delta}; & \Delta \geq 0 \\ \pm\frac{1}{2}\sqrt{-B_1^2 + B_1\sqrt{B_1^2 - 8\Delta}}; & \Delta < 0 \end{cases} \tag{7c}$$

$$A_2 = \begin{cases} 0; & \Delta \geq 0 \\ \frac{A_1^2}{A_0}\left(1 + \frac{A_1^2}{B_1^2}\right); & \Delta < 0 \end{cases} \tag{7d}$$

and

$$B_2 = A_2 B_1 / A_1. \tag{7e}$$

In computing these coefficients, the polarity of A_1 is chosen randomly for each record.

The question arises of how the model should take account of the two distinct conditions on Δ. One idea is to treat all computed combinations of (A_0, A_1, B_1) as part of a single statistical population, without differentiating between the conditions $\Delta \geq 0$ and $\Delta < 0$, and to assume that $A_2 = B_2 = 0$ in *all* cases. The result would be a first-order polynomial for $H_c(\omega)$ defined by a single joint pdf of A_0, A_1 and B_1. This approach has been adopted for reasons of modeling simplicity, and is justified by data given in Section V.

2.3 The coefficient statistics

We begin the statistical modeling with the coefficient A_0, representing the complex channel gain at $\omega = 0$. The probability distribution of A_0 is found to be approximately log-normal, i.e., if

$$ADB \triangleq 20 \log_{10} A_0, \tag{8}$$

then the pdf of ADB is approximately Gaussian, with a mean (μ) of -21.39 dB and a standard deviation (Σ) of 6.562 dB. For convenience, we shall deal with ADB in terms of the standardized variable

$$a_o \triangleq (ADB - \mu)/\Sigma. \tag{9}$$

The precise pdf for a_o used in the model is presented in Section III.

The joint probability law for A_0, A_1, and B_1 can be represented by the joint pdf of A_1 and B_1, conditioned on a_o, times the pdf of a_o. Furthermore, system study results (Section V) indicate that A_1 and B_1, for given a_o, can be modeled as independent variables. Hence, we write

$$p(A_1, B_1 \mid a_o) = p_A(A_1 \mid a_o) p_B(B_1 \mid a_o). \tag{10}$$

To model $p_A(\,|\,)$ and $p_B(\,|\,)$, the entire population of data records was divided into 11 subpopulations, each corresponding to a specific interval of a_o. For example, all records with a_o between -2.25 and -1.75 constituted one such subpopulation, all those with a_o between -1.75 and -1.25 constituted another, and so on. Within each subpopulation, we computed the mean and standard deviation of A_1, as well as its probability distribution, and similarly for B_1.*

Both A_1 and B_1 are found to be essentially Gaussian in every subpopulation; in every case, moreover, each variable has a zero mean and a standard deviation that varies with the mid-value of a_o. Consequently, the joint pdf of A_1 and B_1, conditioned on a_o, reduces to a product of two Gaussian pdf's, each with a standard deviation that is a function of a_o. By finding suitable mathematical expressions for these functions, the statistical modeling of A_0, A_1, and B_1 is completed.

A final quantity to determine is the number of seconds, T_M, for which the multipath fading response applies in a heavy-fading month. The data base used here represents about 8400 seconds of fading which, Lundgren and Rummler have estimated, corresponds to two-thirds of a heavy-fading month for the measured path and frequency.[4] In attempting to relate this finding to other paths and frequencies, we have assumed that T_M is proportional to the multipath occurrence factor developed by Barnett.[7] This assumption permits the general expression for T_M given in the following section.

---

* The choice of 11 subpopulations and the particular a_o-intervals used is somewhat arbitrary. Our aim was to obtain both good resolution (narrow intervals of a_o) and high estimation accuracy (many samples per subpopulation).

III. THE MODEL

In this description, all numerical quantities in braces, { }, are data-derived constants that may vary with path length, antennas, locale, year, etc. It is hoped that the underlying model *structure*, consisting of the form of the transfer function and the functions for the coefficient pdf's, is generally applicable and that the constants alone might be subject to change. In any event, what follows is an estimated fading model based on reductions of the given data.

(*i*) The complex transfer function of a channel, normalized by its unfaded gain, is

$$H_c(\omega) = \begin{cases} 1 + j0 & \text{during non-fading periods;} \\ A_0 - \omega B_1 + j\omega A_1 & \text{during } T_M \text{ seconds per heavy-fading month.} \end{cases} \tag{11}$$

(*ii*) By assuming T_M to be proportional to the multipath occurrence factor,[7] we obtain

$$T_M = \{0.11\}\ cFd^3, \tag{12}$$

where c is the terrain factor, F is the system frequency in gigahertz, and d is the path length in miles.

(*iii*) The joint pdf of A_0, A_1, and B_1 (where A_0 is dimensionless and A_1 and B_1 are in units of seconds) can be represented by

$$p(a_o, A_1, B_1) = p_A(A_1 \mid a_o) p_B(B_1 \mid a_o) p_a(a_o), \tag{13}$$

where

$$a_o = \frac{20 \log_{10} A_o - \{-21.39\}}{\{6.562\}}. \tag{14}$$

(*iv*) The pdf of a_o is the nearly Gaussian function

$$p_a(a_o) = \frac{1}{\sqrt{2\pi}} \exp\left[-\frac{1}{2}[a_o + z(a_o)]^2\right] \cdot \left[1 + \frac{dz(a_o)}{da_o}\right], \tag{15}$$

where $z(a_o)$ is a small nonlinear term in a_o given by

$$z(a_o) = \{0.0742\} a_o^2 + \{0.0125\} a_o^3. \tag{16}$$

If this component were zero, $p_a(a_o)$ would be precisely Gaussian. Figure 1a shows $p_a(a_o)$ and compares it with the Gaussian pdf having the same mean (0.0) and variance (1.0). The difference seems small but has been found to be significant, as we will discuss in Section 5.2.

(*v*) The conditional pdf's of A_1 and B_1, $p_A(A_1 \mid a_o)$ and $p_B(B_1 \mid a_o)$, are both Gaussian with zero means and with standard deviations given by

$$\sigma_A = \text{Max}\{\{0.14\}, [\{0.309\} + \{0.13\} a_o]\} \times 10^{-9} \text{ s} \tag{17}$$

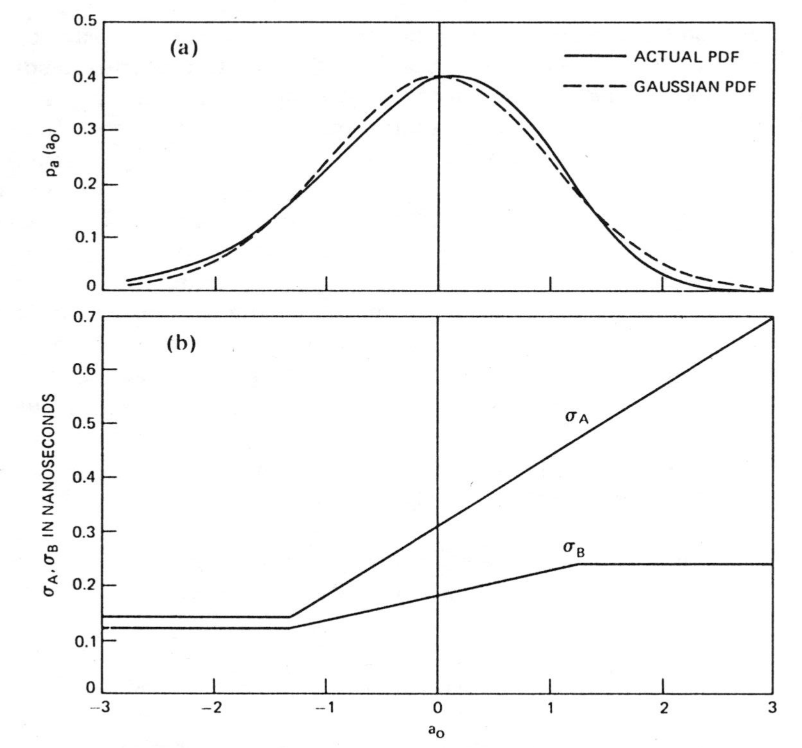

Fig. 1—Functions used in the multipath fading model: (a) the pdf of a_o, compared with a normal (Gaussian) pdf; and (b) the standard deviations of A_1 and B_1, conditioned on a_o.

and

$$\sigma_B = \text{Min}\{\{0.24\}, \text{Max}(\{0.12\}, [\{0.18\} + \{0.046\} a_o])\} \times 10^{-9} \text{ s}. \tag{18}$$

These functions, which model the variations of σ_A and σ_B with a_o, are shown graphically in Fig. 1b. They are derived by estimating σ_A and σ_B from finite numbers of records in nine a_o-intervals of finite width. The results are therefore quite approximate; our tests indicate, however, that they are not sensitive to the precise positions or widths of the a_o-intervals.

IV. DATA REDUCTIONS: EXPANDED DISCUSSION

This section presents mathematical, numerical, and graphical details underlying the data reductions and the construction of the model. The interested reader can thereby scrutinize the various stages of analysis and reasoning that led to the results in Section III. The less interested reader can, with no loss in continuity, proceed to Section V.

4.1 Data organization and notation used

The data base consisted originally of 24,920 records of decibel power gain vs frequency. An early decision was made to remove the first 100 records in the sequence, as they contained a number of normal (nonfading) responses. The model is derived from the resulting set of 24,820 records.

The second decision made was to create four subsets of data records, each containing about 2160 records distributed randomly over the total set (24,820 records). The aim was to verify the statistical regularity of the data base, i.e., to establish that no handful of records dominates the statistics and that individual subsets are representative of the data base as a whole. Having verified this regularity (by comparing results among the four subsets and between the four subsets and the total set), we were able to perform certain costly reductions using a relatively small set of records.

All computations in this study were performed using, as a data base, either the four individual subsets, the four subsets merged together (8640 records), or the total set of 24,820 records. In every case, we have found it helpful to regard the prevailing data base as an *ensemble* of records, each record being a sequence of 23 power gains. The notations, symbols, and definitions used in all computations are summarized in Table I, which should be consulted throughout subsequent discussions.

4.2 Deriving the polynomial coefficients

The process begins with 24,820 records of P_i vs i and ends with a population of coefficient sets, (A_0, A_1, B_1), whose reduction to a joint probability law (Sections III and 4.3) defines the fading model. Three stages of computation are involved here, namely, (*i*) adjusting the recorded P_i's to remove systematic calibration differences across the 23 measurement frequencies, (*ii*) determining, from these adjusted data, the optimal polynomial order and the resulting record-by-record coefficients of the power gain function, eq. (3), and (*iii*) determining, from these coefficients, the record-by-record coefficients of the complex gain function, eq. (2). We now discuss these three stages in turn.

4.2.1 Data adjustments

From physical considerations, we would expect the ensemble probability law for P_i to be the same for all i. Thus, the power gain at any given frequency should exhibit the same statistical behavior as that at any other frequency. This expectation is confirmed by the data, which show that the probability law for P_i is approximately Gaussian for all i. At the same time, however, the data show small but discernible variations of $\bar{P}_i$ and σ_i with i.

Table I—Symbols and Definitions

Symbol	Definition
i	Index of frequency ($i = 1$ for lowest frequency; $i = 24$ for highest frequency).
$\bar{x}_i$	Average of x_i over ensemble of records comprising data base (x_i can be any variable).
$\langle x_i \rangle$	Average of x_i over i for given record; $\langle x_i \rangle = \frac{1}{23} \sum_{i=1, \neq 19}^{24} x_i$.
R_i	Actual channel power gain, in dB, at ith frequency.
P_i	Recorded channel power gain, in dB, at ith frequency.
p_i	Power gain ratio at ith frequency; $p_i = 10^{P_i/10}$.
$q(\omega)$	Polynomial fitted to sequence of p_i's; $q(\omega) = \sum_{m=0}^{M} D_m \omega^m$.
Q_i	Decibel value of $q(\omega)$ at ith frequency (represents fitted approximation to P_i); $Q_i = 10 \log q(\omega_i) \equiv 10 \log q_i$.
σ_i^2	Ensemble variance of P_i; $\sigma_i^2 = \overline{P_i^2} - (\bar{P}_i)^2$.
e_i	Measurement error, $(P_i - R_i)$, at ith frequency; $\overline{e_i^2} = \sigma_{\text{meas}}^2$, all i.
E_i	Observed error, $(P_i - Q_i)$, at ith frequency.
ϵ_i	Modeling error, $(Q_i - R_i)$, at ith frequency.
E_{rms}^2	Mean-square observed error per record; $E_{\text{rms}}^2 = \langle E_i^2 \rangle$.
$\hat{E}_{\text{rms}}^2$	Approximation to E_{rms}^2 (quadratic in q_i); $\hat{E}_{\text{rms}}^2 = \left\langle \left\{ \frac{10}{\ln 10} \frac{p_i - q_i}{p_i} \right\} \right\rangle^2$.
E_B	Bias in observed errors per record; $E_B = \langle E_i \rangle$
$E_i^{(k)}$	kth root of kth central moment of E_i over ensemble; $E_i^{(k)} = \sqrt[k]{\overline{(E_i - \bar{E}_i)^k}}$.

These variations were computed for the total record set and for each of the four record subsets. Results are given in Fig. 2, where the vertical bars span values for the four subsets and the solid curves connect the values for the total set. Also given are results for the decibel value of $\bar{p}_i$. The variations for this quantity are virtually parallel to those for $\bar{P}_i$, a consequence of the fact that the pdf of P_i has roughly the same shape and variance for every i.

For all three quantities ($\bar{P}_i$, σ_i and $\bar{p}_i$), the consistency of results is evident among the four subsets and between those subsets and the total set. For each quantity, moreover, the average variation with i is more pronounced than the spread among the data subsets. These findings make clear that the variations with i are not statistical but, rather, the result of systematic calibration differences in the measurements. The variations with i of $\bar{P}_i$, in particular, can be explained in terms of such systematic differences. Comparable explanations for the variations of σ_i are not as forthcoming. Nevertheless, a decision was made to attribute the nonuniformities of both $\bar{P}_i$ and σ_i to systematic effects, and to compensate for them using simple adjustments of all the P_i's in the data base. The adjustments are of the form

$$P_i \Rightarrow \alpha_i P_i + \beta_i; \quad \text{each } i, \tag{19}$$

where the variations of α_i and β_i with i are chosen to serve two purposes. One is to render the ensemble means and variances of the adjusted P_i's uniform over i. The other is to conserve the "global" mean and variance ($\bar{P}$ and σ^2) of P_i, i.e., the mean and variance of the P_i's taken over the entire data base (24,820 records × 23 samples/record). It is easy to show that this is achieved by choosing α_i and β_i as

$$\alpha_i = \frac{\sigma}{\sigma_i}, \quad \beta_i = \bar{P} - \frac{\sigma}{\sigma_i}\bar{P}_i; \quad \text{each } i. \tag{20}$$

The variations of $\bar{P}_i$, σ_i, and $10 \log \bar{p}_i$ for the entire record set, when the data are adjusted according to (19) and (20), are given in Fig. 2 by the dashed curves. We see that, by making $\bar{P}_i$ and σ_i uniform over i, $10 \log \bar{p}_i$ has been made nearly uniform also, as we expect it to be from physical considerations. The α_i's and β_i's that produce these results are close to 1.0 and 0.0 dB, respectively, for all i.*

Though minor, the data adjustments have measurably reduced the curve-fitting errors associated with the next stage of reduction. This is made evident later.

4.2.2 *Coefficients of the power gain polynomial*

4.2.2.1 Least Squares Method. We wish to compute, for given M, those D_m's in eq. (3) that yield the best approximation to the channel power response in each record. Ideally, we would like to minimize some measure of $(Q_i - R_i)$ (see Table I), such as its mean-square average over i, $\langle \epsilon_i^2 \rangle$. Since the values of R_i are unknowable, the next best thing is to minimize some measure of $(P_i - Q_i)$, such as $\langle E_i^2 \rangle \equiv E_{\text{rms}}^2$. However, the Q_i's are logarithms of samples of $q(\omega)$, and so a precise least-squares minimization is not analytically tractable. Fortunately, so long as $|E_i| < 1$ dB for most i, E_{rms}^2 can be accurately approximated by $\hat{E}_{\text{rms}}^2$, as given in Table I. This quantity lends itself to least-squares minimization.

For a given M and a given record, the following approach is used to find $D_0, \cdots D_M$: (*i*) the 23 P_i's are converted to power ratios, p_i; (*ii*) the function $\hat{E}_{\text{rms}}^2$ is expressed in terms of the D_m's in eq. (3); (*iii*) the $M + 1$ derivatives of $\hat{E}_{\text{rms}}^2$ with respect to $D_0, \cdots D_M$ are computed and equated to zero; and (*iv*) the resulting $M + 1$ linear equations in $D_0, \cdots D_M$ are solved using matrix methods.

* The α_i's range from 0.945 to 1.029, and the β_i's range from −0.99 dB to 1.25 dB.

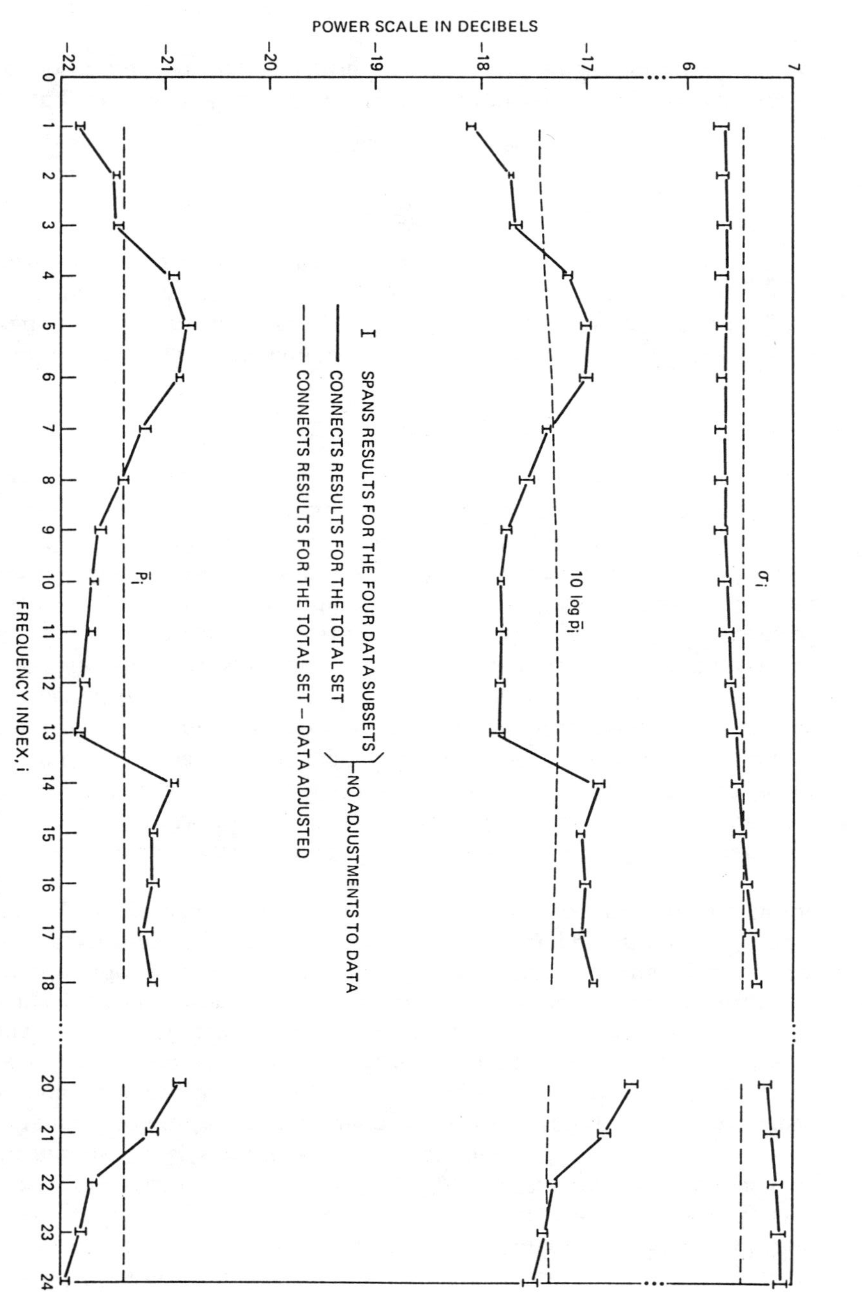

Fig. 2—Moments of the recorded power gains, as functions of frequency, before and after data adjustments.

As a refinement to this procedure, we add one more step: The derived D_m's are multiplied by a common factor chosen to force the bias error, $\langle E_i \rangle \equiv E_B$, to zero.* Thus, for example, if the initial solution for the D_m's leads to a bias error $E_B = X$, then adjusting each D_m by a multiplying factor $10^{X/10}$ will reduce E_B to zero. This step reduces E_{rms} and makes the Q_i's unbiased estimators of the P_i's.

Figure 3 shows two sample records for P_i, taken from the data base. In each case, the function $Q(\omega) = 10 \log q(\omega)$, as computed using the above procedure, is plotted for $M = 0$, 2, and 4. Note that, for $M = 0$, the fitting function lacks the curvature needed to accommodate typical fading patterns. For $M = 4$, on the other hand, the fitting function provides added curvature that may be artificial, i.e., it may serve to fit the measurement noise more than the underlying response. We now turn to a discussion of fitting errors and the most suitable choice for M.

4.2.2.2 Random Error Model. To begin, assume that 23 noisy samples of P_i are to be fitted by $Q(\omega) = 10 \log q(\omega)$. Assume, further, that the sample errors $e_i = (P_i - R_i)$ are independent, zero-mean Gaussian variables, each having an rms value σ_{meas}. Since the R_i's are unknowable, the e_i's cannot be observed directly; at best, their statistics can be inferred from an analysis of the observed errors, $E_i = P_i - Q_i$.

Specifically, suppose that M is a sufficiently high polynomial order that, in the absence of measurement noise, all significant variations of P_i with i could be accommodated by the fitting function $Q(\omega)$. In that case, all the E_i's would be due solely to measurement errors, the e_i's. To a first approximation, we could then say the following:[8] The E_i's are zero-mean Gaussian variables, each having an ensemble mean-square value

$$\overline{E_i^2} = \frac{22 - M}{23} \sigma_{meas}^2; \quad \text{each } i. \tag{21}$$

In addition, E_{rms}^2 is chi-square distributed, with $22 - M$ degrees of freedom and a mean given by eq. (21), i.e., $\overline{E_{rms}^2} = \overline{E_i^2}$. Finally, the errors of major interest, $\epsilon_i = (Q_i - R_i)$, are zero-mean Gaussian variables, each having a mean square value

$$\epsilon_i^2 \equiv \sigma_\epsilon^2 = \overline{e_i^2} - \overline{E_i^2}$$
$$= \frac{M + 1}{23} \sigma_{meas}^2; \quad \text{each } i. \tag{22}$$

The pertinence of this theory to our data reductions is as follows: If the observed E_i's are essentially zero-mean and Gauss-distributed across the data ensemble, each with the same $\overline{E_i^2}$; and if E_{rms}^2 is chi-square distributed with $22 - M$ degrees of freedom and a mean of $\overline{E_i^2}$, then the random-error model, as described above, can be assumed to apply. Accordingly, σ_{meas}^2 can then be estimated from measured error statistics using eq. (21), and the mean-square error between the fitted and true power gains can be estimated using eq. (22).

The reduced data show that, when $M = 0$, the populations of E_{rms}^2 and the E_i's do not have the properties of the random error model. This is to be expected, since a zeroth-order polynomial (horizontal line) cannot accommodate the true variations of faded power with frequency; hence, the E_i's are caused mostly by the inadequacy of the fitting function rather than by measurement noise.

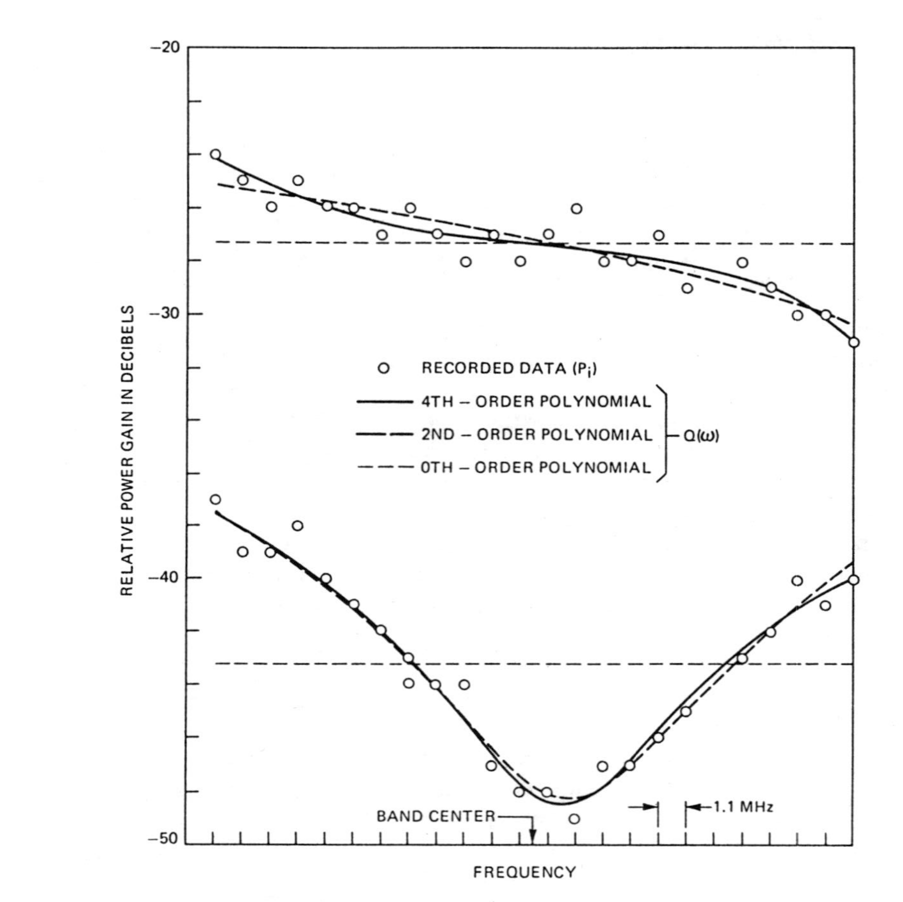

Fig. 3—Two examples of data records and their representations by zeroth-, second-, and fourth-order polynomials.

* Remember that the least-squares method is applied to $\hat{E}_{rms}^2$ rather than to the quantity of interest, E_{rms}^2. The possibility thus exists of a small bias error in the results, which this step removes.

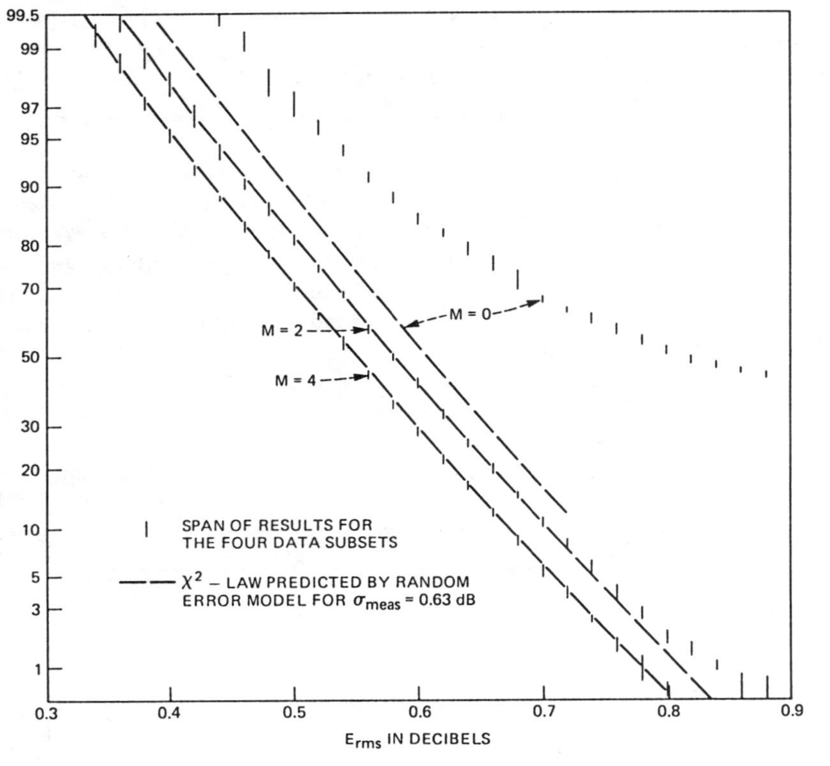

Fig. 5—Probability distributions for E_{rms}, using zeroth-, second-, and fourth-order polynomials, compared with chi-square distributions predicted by the random-error model.

For $M = 2$ and $M = 4$, however, the populations of E^2_{rms} and the E_i's do exhibit the properties of the random error model. For example, Fig. 4 shows variations of $\overline{E_i}$, $E_i^{(2)}$ and $E_i^{(4)}$ with i for $M = 2$ (see Table I). Each vertical bar spans the spread of values among the four data subsets, while the dashed lines give the values predicted by the random error model.

Figure 5 gives even stronger evidence for the random error model when $M \geq 2$. The vertical axis represents $\Pr\{E_{rms} > \text{abscissa}\}$, the vertical bars give the spread of this quantity among the four data subsets, and the dashed curves give the variations predicted by the random error model when $\sigma_{meas} = 0.63$ dB. For $M = 0$, the measured results compare very poorly with the predictions, for the reasons noted above, while for $M = 2$ and $M = 4$ they compare very well. Using eq. (22), the errors between Q_i and R_i are found to have the rms values $\sigma_\epsilon = 0.23$ dB for $M = 2$ and $\sigma_\epsilon = 0.29$ dB for $M = 4$.

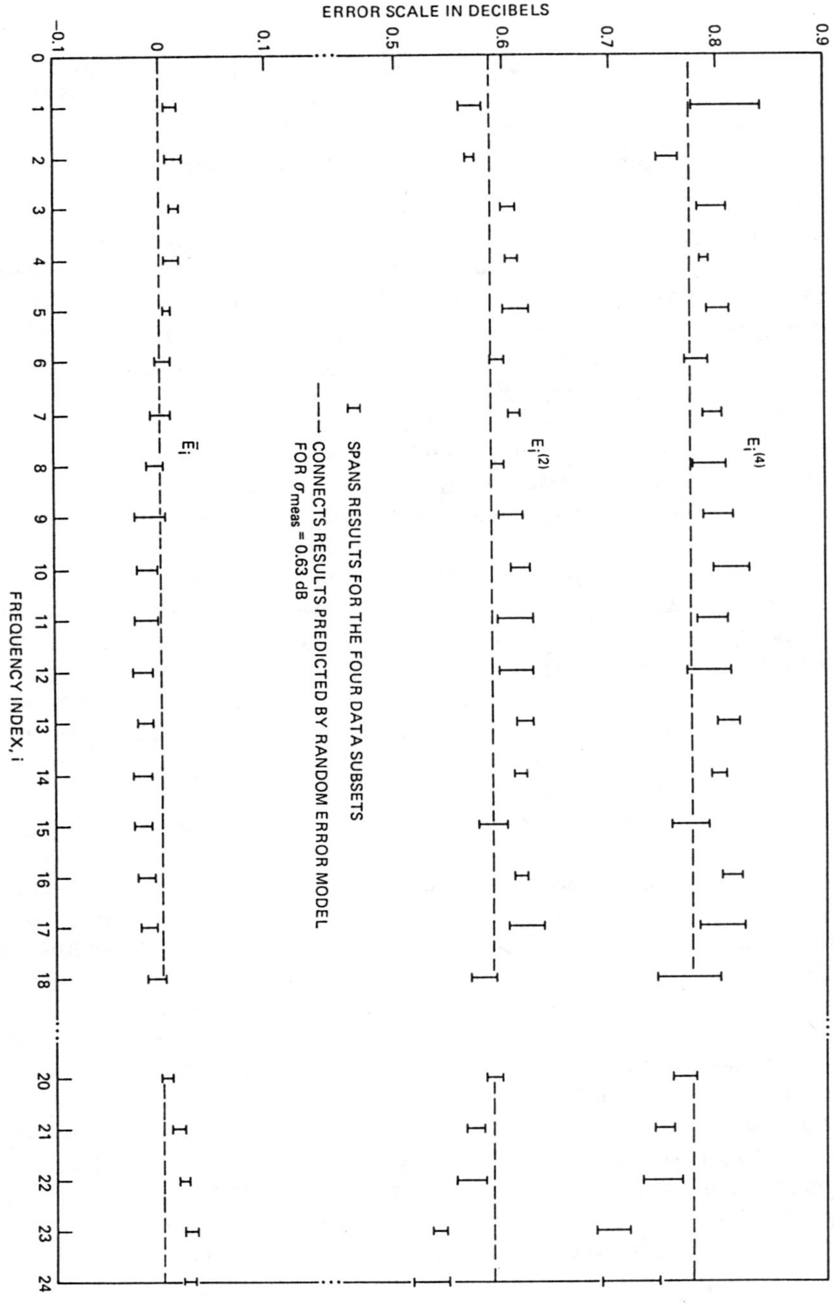

Fig. 4—Moments of observed fitting errors for $M = 2$, as functions of frequency, compared with values predicted by the random error model.

The rms modeling error, σ_ϵ, is smaller for $M = 2$ because only three polynomial coefficients are derived from 23 recorded samples, leaving 20 degrees of freedom for averaging out measurement noise. For $M = 4$, five polynomial coefficients are derived, leaving 18 degrees of freedom for noise averaging. We conclude that, for modeling purposes, M should be large enough to provide adequate curvature in $q(\omega)$ (as evidenced by E_i- and E^2_{rms}-populations that fit the random error model), but no larger (so as to minimize σ_ϵ). In view of this criterion, we have decided that $M = 2$ is the optimal polynomial order for the present data base.

Table II gives computed ensemble averages for $\hat{E}^2_{\text{rms}}$ (the statistic minimized by least-squares methods) and E^2_{rms} for $M = 0$, 2, and 4. The ensemble in this case was the four merged data subsets. The close agreements between $\overline{\hat{E}^2_{\text{rms}}}$ and $\overline{E^2_{\text{rms}}}$ for both values of M justify the approximation of E^2_{rms} by $\hat{E}^2_{\text{rms}}$ in the least-squares derivation of the D_m's.

The results in Table II also permit comparisons between the cases where the recorded data are adjusted, as described in Section 4.2.1, and not adjusted. The reduction in fitting error associated with these adjustments is small but distinct, indicating that the inferred calibration differences are real.

4.2.2.3 Coefficient Distributions. We have determined (D_0, D_1, D_2) for every record in the total set, using the least-squares fitting procedure described above. This effort has led to the set of cumulative probability distributions in Figs. 6, 7, and 8. In each case, the vertical bars span the results for the four data subsets, while the solid curve gives the result for the total set. In some cases, where the probability curve is steep and the spreads are small, the vertical bars can barely be distinguished. In all three figures, the results show good consistency among the four subsets and the total set.

The distribution for D_0, Fig. 6, is worth special note. From eq. (3), it is clear that D_0 can be interpreted as the power-gain ratio at the center of the channel ($\omega = 0$). Accordingly, it must always be positive, whereas D_1 and D_2 can have either polarity. Also, since the power-gain statistics should be the same at each frequency, the decibel value of D_0 should have the same distribution as do the P_i's. For that reason, the abscissa

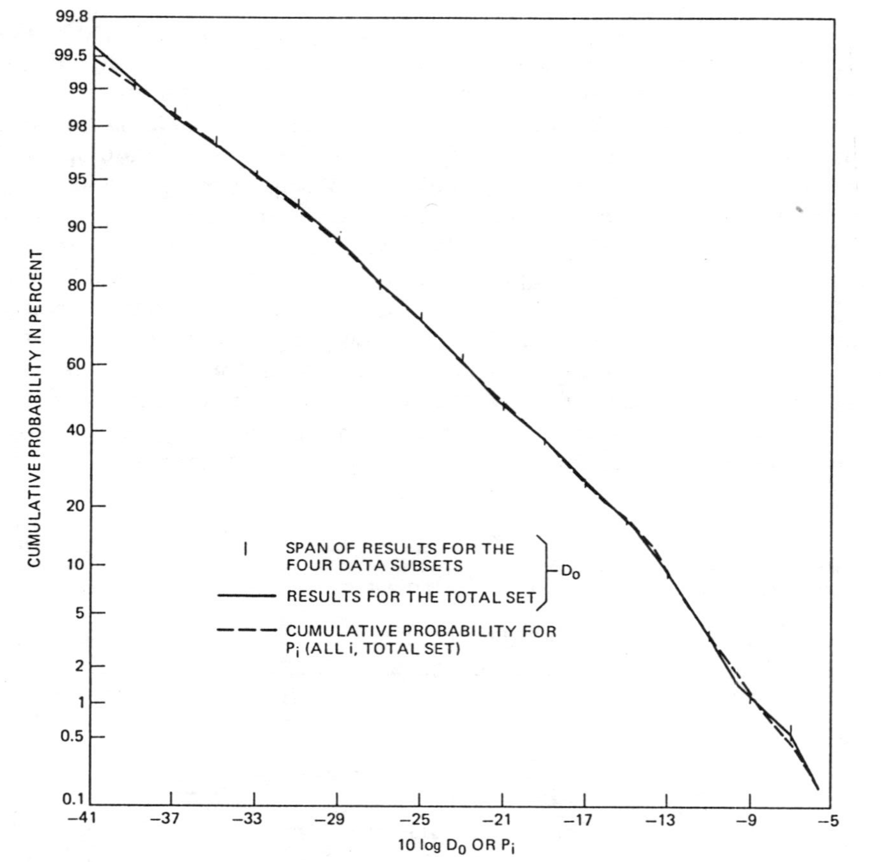

Fig. 6—Probability distributions for power-gain polynomial coefficient D_0 and recorded, adjusted dB power-gain P_i.

in Fig. 6 is scaled in units of 10 log D_0, and the distribution for all P_i's in the data base (24,820 records × 23 samples/record) is given as well (dashed curve). The agreement between the dashed and solid curves is excellent, as we should expect.

It should also be noted that a straight line on the probability paper used here would imply a Gaussian (or normal) probability law. Thus, 10 log D_0 is nearly normal, i.e., D_0 is approximately log-normal. Since $A_0 = \sqrt{D_0}$ (eq. (7a)), this means that A_0 is also approximately log-normal, as we have noted in Section III.

Table II—Error statistics for 8640 records

Measured Error Statistics	$M = 0$ Data Adjusted	$M = 0$ No Adjustment	$M = 2$ Data Adjusted	$M = 2$ No Adjustment	$M = 4$ Data Adjusted	$M = 4$ No Adjustment
$\overline{\hat{E}^2_{\text{rms}}}$ in dB²	—	—	0.361	0.486	0.314	0.373
$\overline{E^2_{\text{rms}}}$ in dB²	1.698	1.880	0.351	0.473	0.311	0.368

4.2.3 *Coefficients of the complex gain polynomial*

The method for computing the A_n's and B_n's from the D_m's for $M = 2$ was outlined in Section 2.2. The primary results are given by eqs. (6) and (7). The condition $\Delta \geq 0$ occurs over 58 percent of the data records, in which cases $|H_c(\omega)|^2$ can be matched to $q(\omega)$ without requiring a second-order coefficient (i.e., $A_2 = B_2 = 0$). Over the 42 percent of the records wherein $\Delta < 0$, however, $A_2, B_2 \neq 0$, and a small but nonzero fourth-order term, $(A_2^2 + B_2^2)\omega^4$, exists in $|H_c(\omega)|^2$. Although the method of solution is designed to minimize this term, it is necessary to evaluate its effect on the polynomial fitting.

We define the added distortion, δ, for a given record to be the rms decibel difference between $|H_c(\omega)|^2$ and $q(\omega)$, where the rms averaging is over i. Thus,

$$\delta = \sqrt{\left\langle 10 \log \left\{ \frac{q(\omega_i) + (A_2^2 + B_2^2)\omega_i^4}{q(\omega_i)} \right\} \right\rangle}. \tag{23}$$

Clearly, $\delta = 0$ for records in which $\Delta \geq 0$. We have computed a cumulative distribution for nonzero δ over 3600 records, specifically,

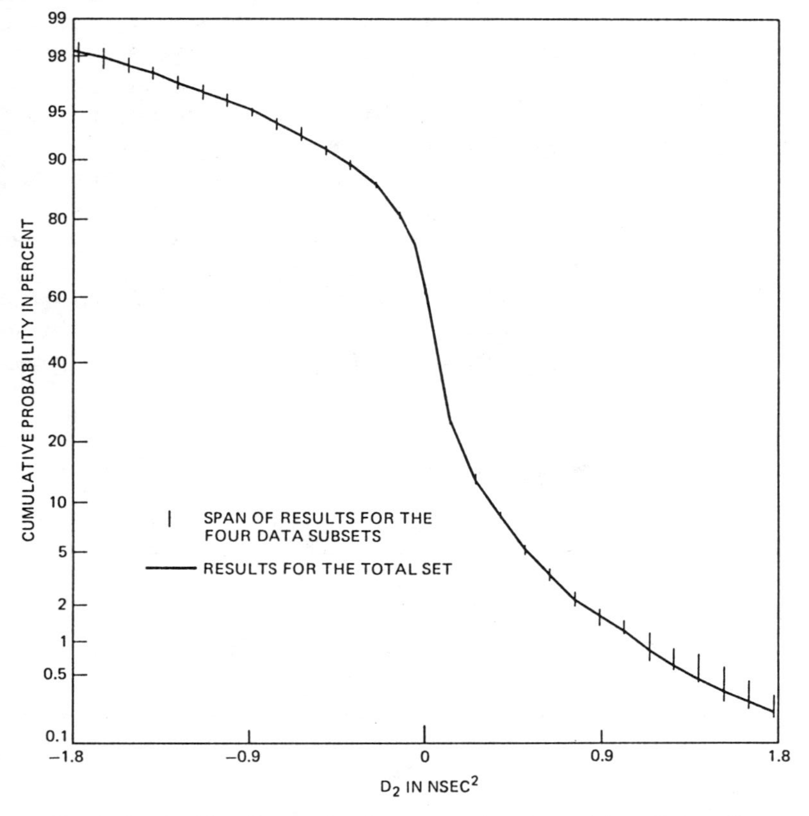

Fig. 7—Probability distribution for power-gain polynomial coefficient D_1.

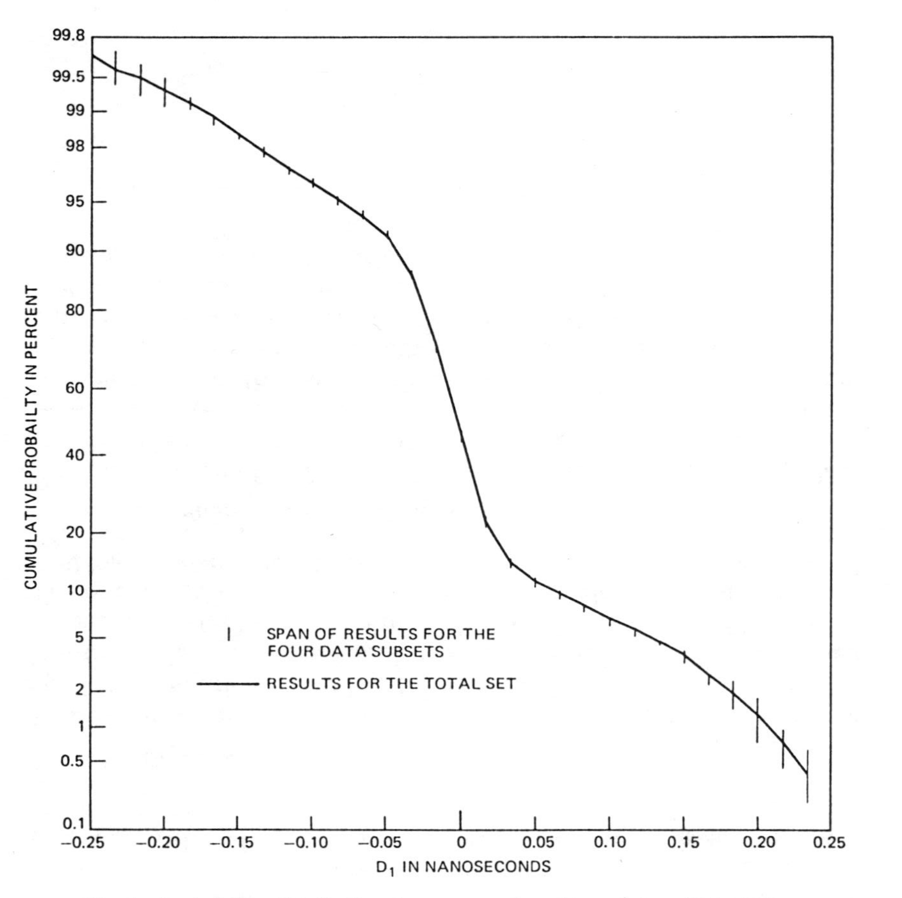

Fig. 8—Probability distribution for power-gain polynomial coefficient D_2.

those records within the four merged subsets where $\Delta < 0$. The result (Fig. 9) is that, for 99.4 percent of these records, δ is below 0.23 dB, i.e., smaller than the rms error due to noise. Over all records, moreover, δ is less than 0.52 dB.

Having established the accuracy of the procedure for computing the A_n's and B_n's, we have elected to ignore those nonzero values of A_2 and B_2 that arise in 42 percent of the records. This step permits a simpler model, as noted previously, at some cost in accuracy. Justification for it is given in Section V.

4.3 Deriving the coefficient statistics

Assume now that the set (A_0, A_1, B_1) is derived, record by record, for the total data base. Using this large population of coefficient sets, it should be possible to obtain a mathematical description for the pdf $p(A_0, A_1, B_1)$. The procedure used was outlined in Section 2.3, and the result was presented in Section III. In the following discussion, we present the evidence for (*i*) the pdf of a_o [eqs. (14) to (16)]; (*ii*) the pdf's of A_1 and B_1 conditioned on a_o; (*iii*) the functional variations of σ_A and σ_B with a_o [eqs. (17) and (18)]; and (*iv*) the statistical independence of A_1 and B_1 for given a_o.

4.3.1 PDF for a_o

The quantity a_o is a shifted, scaled version of 20 log A_0 which, from eq. (7a), is identical to 10 log D_0. The empirical distribution for this variate in Fig. 6 is represented in Fig. 10 by the circles. The dashed curve in Fig. 10 is based upon the pdf for a_o given by eqs. (15) and (16). We conclude, then, that our mathematical model for the distribution of 20 log A_0 is an accurate one.

4.3.2 Conditional PDF's for A_1 and B_1

To derive the pdf's of A_1 and B_1, conditioned on a_o, the 24,820 data records were grouped into 11 subpopulations, each corresponding to a particular range of a_o. For each of the central nine subpopulations, the a_o-range has a width of one-half a standard deviation, thereby balancing the objectives of good resolution (narrow a_o-range per subpopulation) and sample size sufficiency (many records per subpopulation).

Within each subpopulation, we computed the mean, standard deviation and cumulative distribution for A_1, and similarly for B_1. Since the polarity of A_1 was chosen randomly in each data record (a reaction to the absence of phase response data, as noted in Section 2.2), the mean for A_1 was very close to zero in every subpopulation.

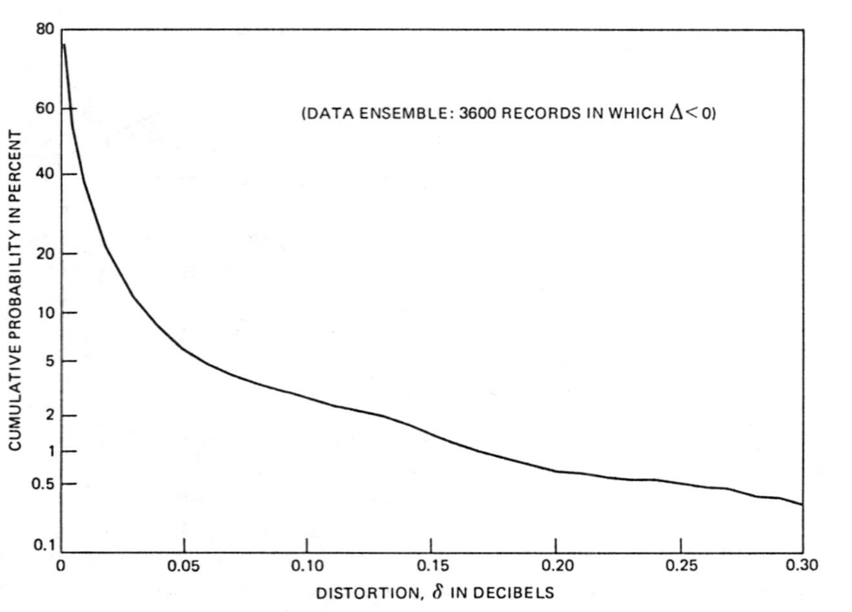

Fig. 9—Probability distribution for distortion term δ, caused by nonzero fourth-order coefficient in records for which $\Delta < 0$.

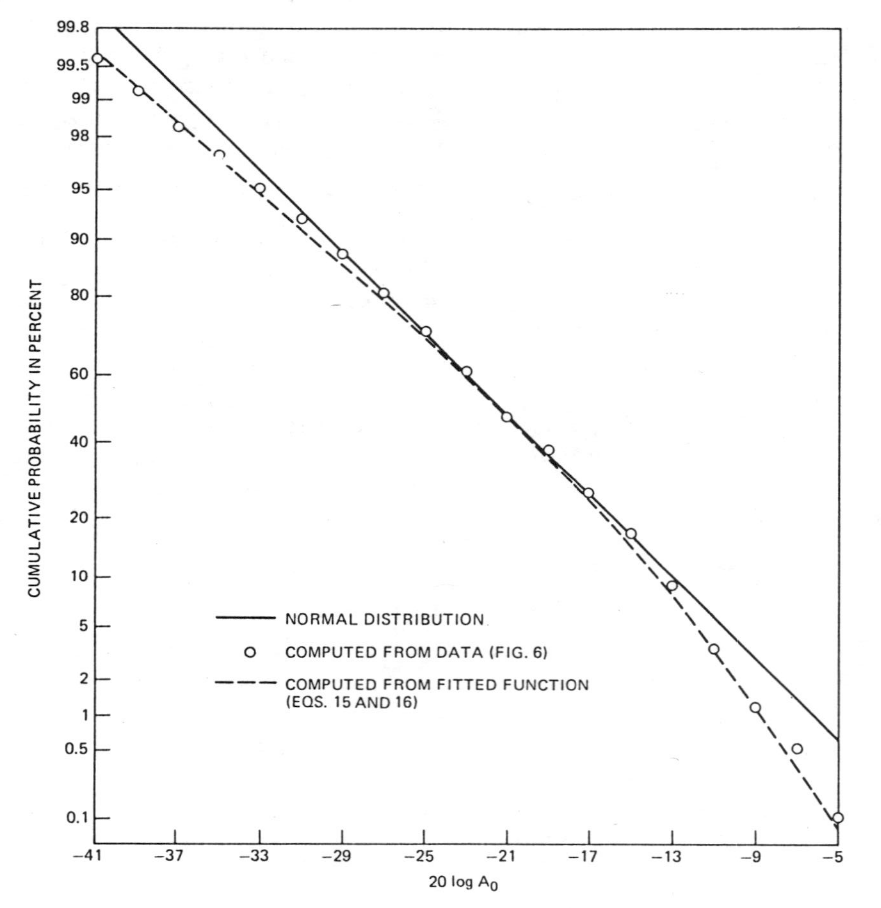

Fig. 10—Probability distributions (empirical and mathematical model) for complex gain polynomial coefficient A_0.

We also randomized the polarity of every reduced value of B_1 before analyzing the B_1 statistics in each subpopulation. The motive for this polarity randomization is that it simplifies the statistical modeling. Before randomization, the conditional mean of B_1 was not close to zero in every subpopulation, nor was its conditional distribution Gauss-like; *after* randomization, B_1 became a zero-mean Gauss-like variable in every subpopulation. Although this simplification entailed some doctoring of the derived B_1's, it should be noted that the polarity of B_1 is immaterial in any practical system context. For this quantity represents the slope, at band center, of the real part of $H_c(\omega)$, and whether the slope is upward to the right (B_1 negative) or upward to the left (B_1 positive) has no effect on either detection behavior or the realizability of equalizer circuits. The same cannot be said for the polarity of A_1, which can influence the stability of some equalizer designs.

Table III gives some pertinent data for the 11 subpopulations analyzed. For each one, identified by a number and an a_o-range, the table gives the number of records contained therein and the standard deviations for A_1 and B_1. Because of the polarity randomizations, the means for A_1 and B_1 are essentially zero in every case.

Some distributions for A_1/σ_A and B_1/σ_B are indicated in Figs. 11 and 12, respectively. Data are given for only five of the subpopulations, but all of the nine central subpopulations exhibit the Gauss-like distributions evident here. That is, A_1/σ_A and B_1/σ_B act, in every case, like zero-mean, unit-variance Gaussian variables. We can therefore invoke the following models:

$$p_A(A_1 \mid a_o) = \frac{1}{\sqrt{2\pi}\,\sigma_A(a_o)} \exp\left\{ -\frac{1}{2} \frac{A_1^2}{\sigma_A^2(a_o)} \right\} \tag{24}$$

and

$$p_B(B_1 \mid a_o) = \frac{1}{\sqrt{2\pi}\,\sigma_B(a_o)} \exp\left\{ -\frac{1}{2} \frac{B_1^2}{\sigma_B^2(a_o)} \right\}. \tag{25}$$

4.3.3 $\sigma_A(a_o)$ and $\sigma_B(a_o)$

Let us associate each σ_A in Table III with the midvalue of the a_o-range for which it is computed and do the same for each σ_B. (This procedure has no meaning, of course, for the semi-infinite ranges of the first and last subpopulations.) The result is that the circles and crosses in Fig. 13 can be regarded as samples of the functions $\sigma_A(a_o)$ and $\sigma_B(a_o)$, respectively. The solid curves provide accurate fits to these data that are described mathematically by eqs. (17) and (18).

Table III—Data for the 11 subpopulations

Subpop. Number	a_o-Interval	Population Size	σ_A (ns)	σ_B (ns)
1	< -2.25	515	0.154	0.126
2	$[-2.25, 1.75)$	660	0.153	0.125
3	$[-1.75, -1.25)$	1423	0.141	0.121
4	$[-1.25, -0.75)$	2878	0.176	0.140
5	$[-0.75, -0.25)$	4147	0.249	0.186
6	$[-0.25, 0.25]$	4925	0.278	0.175
7	$(0.25, 0.75]$	4340	0.370	0.185
8	$(0.75, 1.25]$	3522	0.440	0.239
9	$(1.25, 1.75]$	1933	0.503	0.233
10	$(1.75, 2.25]$	371	0.576	0.218
11	>2.25	106	0.656	0.158
		Total: 24,820		

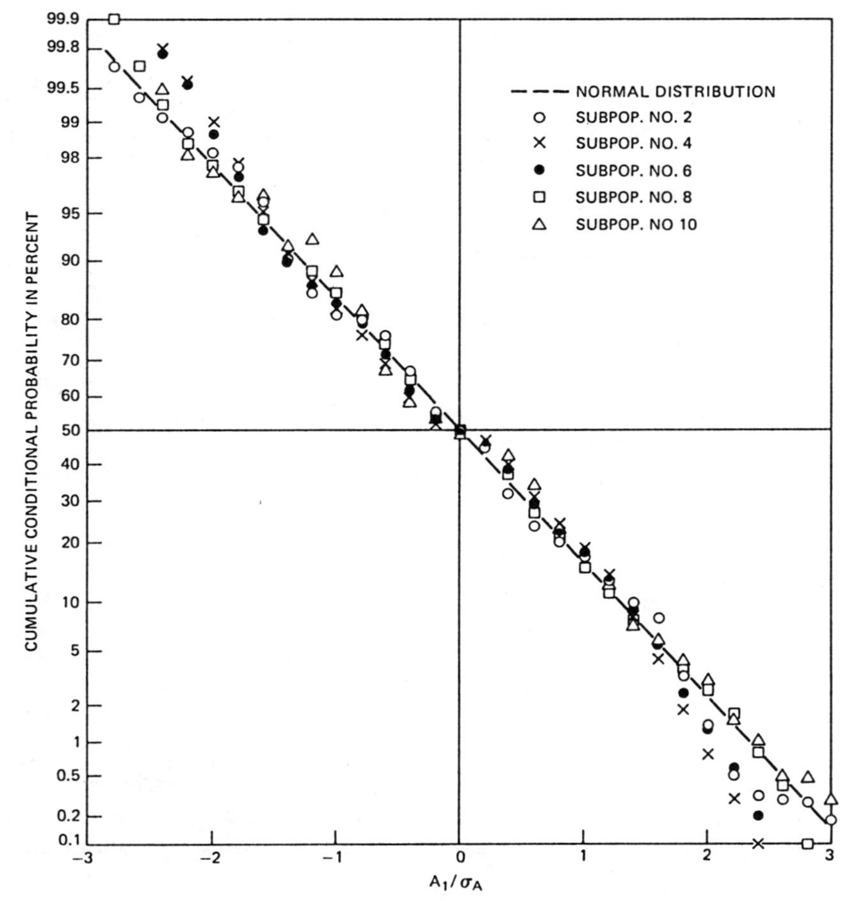

Fig. 11—Probability distributions for A_1/σ_A, conditioned on a_o, for five subpopulations.

4.3.4 Statistical independence of A_1 and B_1

Formal statistical testing has shown that A_1 and B_1 within each subpopulation are *not* mutually independent. We have assumed such independence anyway because of the resulting simplification of the model. The validity of this approach rests upon how it influences the outcome of system studies. The next section takes up this issue and gives justification for treating A_1 and B_1, conditioned as a_o, as independent variables.

V. ACCURACY OF THE MODEL

The multipath fading model presented here consists of (*i*) a polynomial representation for $H_c(\omega)$; (*ii*) a mathematical formula for the joint coefficient pdf; and (*iii*) a formula for T_M, the number of fading seconds per heavy-fading month. We now discuss the accuracy of each of these constituents.

5.1 The polynomial representation

The first step in using the polynomial representation consists of fitting $q(\omega)$, eq. (3), to the recorded power gain data. This was done, with $M = 2$, for each of 24,820 records. The error measured in each record is the rms decibel difference (E_{rms}) between the recorded and fitted power gains. Although the error between $q(\omega)$ and the true underlying response cannot be estimated from isolated records, its statistics can be inferred by analyzing the population of E_{rms} over thousands of records. Such an analysis* had led to the following conclusion: The decibel difference between the fitted and true power gains is approximately Gauss-distributed at each frequency, is due almost entirely to measurement noise (rather than inadequate curvature in the form of $q(\omega)$), and has an rms value of 0.23 dB. This corresponds to an rms error of 2.7 percent in the gain magnitude, $|H_c|$, which is sufficiently low for purposes of modeling.

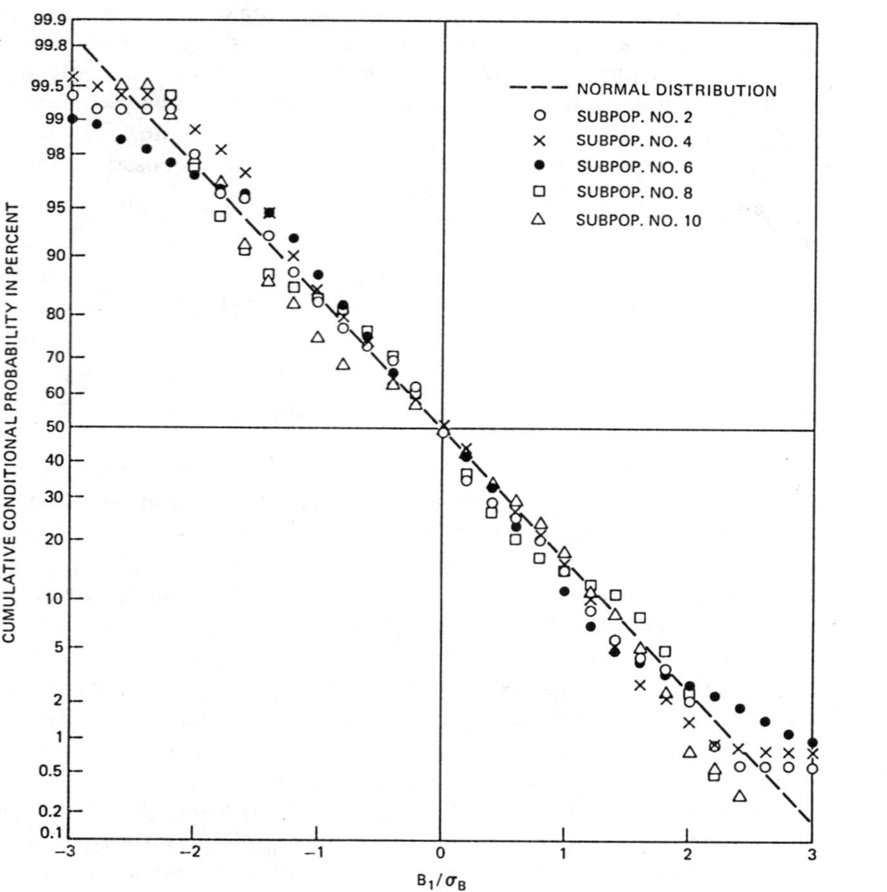

Fig. 12—Probability distributions for B_1/σ_B, conditioned on a_o, for five subpopulations.

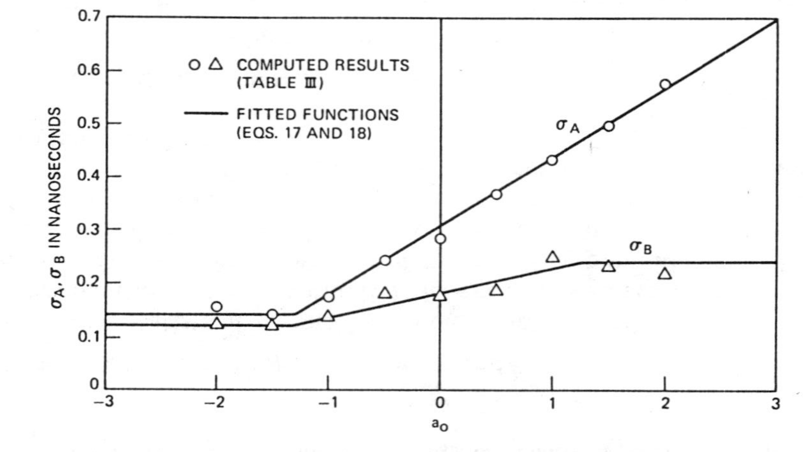

Fig. 13—Variations (empirical and mathematical model) of σ_A and σ_B with a_o.

The next step in the polynomial representation is to derive the A- and B-coefficients in eq. (2) from the D-coefficients in eq. (3). For 58 percent of the data records, coefficient sets (A_0, A_1, B_1) exist which provide a perfect match between $|H_c(\omega)|^2$ and $q(\omega)$. The one flaw in these cases is the unresolvable ambiguity (resulting from noncoherent measurements) in the polarity of A_1.

For the remaining 42 percent of the records, the method for choosing A's and B's leads, in addition, to nonzero A_2 and B_2 and a resulting fourth-order term in $|H_c(\omega)|^2$ not present in $q(\omega)$. The differences between $|H_c(\omega)|^2$ and $q(\omega)$ for these records were analyzed, and found to be minor (see Section 4.2.3 for more details).

Finally, all nonzero values computed for A_2 and B_2 have been discarded to simplify the model. To evaluate the validity of this step, we performed an outage analysis for several different modulations, specifically, 4-PSK, 8-PSK, 16-PSK and 16-QAM. For each modulation, cosine rolloff spectral shaping was assumed, and the rolloff factor (α) and channel bandwidth (W) were treated as variables. In every case,

* It is noteworthy that the distribution for E^2_{rms} derived here using a second-order $q(\omega)$ is nearly identical to that obtained for the three-path model of Rummler (Ref. 5). The slightly lower mean obtained here is due solely to the data adjustments described in Section IV.

we specified the symbol rate to be $W/(1 + \alpha)$, thereby matching the truncation bandwidth of the modulation to the channel bandwidth.

The outage analysis performed for each modulation used $W = 20$, 30, and 40 MHz and α-values between 0.1 and 1.0. For each of the resulting cases, the quantity "outage seconds/heavy-fading month" was estimated using the following procedure: Regions of the coefficient set $(A_0, A_1, B_1, A_2, B_2)$ for which the detected data eye is closed were derived analytically;* the number of data records for which $(A_0, \cdots B_2)$ lies within these so-called outage regions was counted; and the fraction of these records was multiplied by T_M in eq. (12). [We assumed $c = 1$ (average terrain), $F = 6$ (GHz), and $d = 26.4$ (miles), resulting in a value for T_M of 12,144 s].

This computation was done twice for each combination of modulation, W and α. In one computation, nonzero values of A_2 and B_2 were acknowledged in comparing each $(A_0, \cdots B_2)$ to the outage regions; in the other, A_2 and B_2 in each record were taken to be zero.

The numerical results span a range from 5 to 8000 seconds of outage. The results obtained by acknowledging nonzero (A_2, B_2) pairs and the results obtained by ignoring them differ by less than 10 percent for all cases treated. The simplification of omitting A_2 and B_2 from the model is thus found to be acceptable for purposes of making outage predictions.

5.2 The coefficient PDF

In deriving the mathematical result for $p(a_o, A_1, B_1)$, Section III, numerous approximations and assumptions were made, e.g., independence between A_1 and B_1 for given a_o; Gaussian conditional pdf's for A_1 and B_1; and the functions for $p_a(a_o)$, $\sigma_A(a_o)$, and $\sigma_B(a_o)$. The ultimate test of the result is whether it leads to the same performance predictions as those based upon the actual population of coefficient sets. Accordingly, outage seconds for the combinations of modulation, W, and α cited above were obtained using a separate method of calculation: In addition to counting data records over the coefficient outage regions, the model distribution $p(a_o, A_1, B_1)$ was integrated over these regions.

The main findings are as follows: When the outage calculated by counting data records is greater than 70 seconds per heavy-fading month, the outage calculated by integrating $p(a_o, A_1, B_1)$ differs by 16 percent or less; for outages less than 70 seconds, the difference is 50 percent or less. This is an excellent level of agreement for practical purposes.

The above findings vindicate the modeling assumption that A_1 and B_1 are independent for given a_o. It also confirms the accuracy of the functions used for $p_a(a_o)$, $\sigma_A(a_o)$, and $\sigma_B(a_o)$. It is noteworthy that, when outage is calculated using a purely Gaussian function for $p_a(a_o)$ [i.e., $z(a_o) = 0$ in eq. (16)], the agreement with counted data records is poor in several cases. In short, the refinement in $p_a(a_o)$ used here is important to the accurate use of the model in system outage predictions.

5.3 Expression for T_M

The expression given for T_M in Section III assumes only that this quantity is in a fixed proportion to the multipath occurrence factor of Barnett.[7] By computing the fading time and occurrence factor for the present data base, we have obtained the empirical results of eq. (12). This result is unverified, although recent studies have indicated that it is reliable to within a factor or two.[9]

There is further evidence supporting eq. (12) for T_M, as well as eq. (1) for $p_a(a_o)$: Using these two formulas, plus (14) and (16), we can show that deep fades observe the probability law

$$\Pr\{A_0 < L\} \doteq 1.7\, rL^2; \; (-10 \log L^2 > 26 \text{ dB}). \tag{26}$$

This law applies to heavy-fading months, and r is the multipath occurrence factor used to derive T_M. This result differs from that of Barnett[7] by the factor 1.7, which represents very good agreement.

VI. DISCUSSION AND CONCLUSION

We have presented a statistical model for describing the microwave channel distortions classified as multipath fading. The model is defined by the functions of eqs. (11) through (18) and by the 12 numerical constants contained therein. We do not know whether the functional structure is valid for other paths, microwave frequencies, seasons, etc., and, if it is, whether or how the numerical constants vary with these conditions. The way to resolve these questions is through more multipath experiments.

A major virtue of the polynomial model suggested by eq. (2) is the simplicity it affords in analyzing the response of a fading channel to arbitrary modulations. This is largely because the required polynomial order for $H_c(\omega)$ in channelized microwave systems is generally quite low. Ample support for this conclusion is given by the data reductions and system analysis results reported here.

Another useful feature of the polynomial model is that it suggests the form of an adaptive equalizer response, i.e., $1/H_c(\omega)$, which is easy to realize when the complex zeros of $H_c(\omega)$ have negative real parts. With this in mind, the polarity of A_1 becomes a very important

* The closed-eye criterion for "outage" leads to conservative performance measures but has the virtue of not depending on system fade margin or specified bit error rate; it also simplifies the outage analysis considerably.

issue. For when $A_1 > 0$, the realizable adaptive response $[A_o - \omega B_1 + j\omega | A_1 |]^{-1}$ is highly effective against multipath fading.[10] When $A_1 < 0$, however, such a response eliminates amplitude distortion at the cost of increased delay distortion. In lieu of explicative data, the present model assumes that $A_1 < 0$ precisely half the time. Whether this is true or not has important bearing on the achievement of successful equalizer approaches. Coherent multipath measurements would help to resolve the existing polarity uncertainty in the model.

The new model provides a useful starting point for the design of subsequent experiments. What it reveals about the depth, shape, and statistics of multipath fading responses can facilitate design choices for the channel probing signal, receiver processing, data acquisition and reduction strategies, and other experiment features.

REFERENCES

1. C. W. Anderson, S. Barber and R. Patel, "The Effect of Selective Fading on Digital Radio," Paper No. 33.5, ICC Conf. Record, June 1978.
2. L. J. Greenstein and V. K. Prabhu, "Analysis of Multipath Outage with Applications to 90-Mb/s PSK Systems at 6 and 11 GHz," IEEE Trans. Commun., *COM-27*, No. 1 (January 1979), pp. 68–75.
3. W. C. Jakes, Jr., "An Approximate Method to Estimate an Upper Bound on the Effect of Multipath Delay Distortion on Digital Transmission," IEEE Trans. on Comm., *COM-27*, No. 1 (January 1979), pp. 76–81.
4. C. W. Lundgren and W. D. Rummler, "Digital Radio Outage Due to Selective Fading—Observation vs. Prediction from Laboratory Simulation," B.S.T.J., *58*, No. 5 (May–June 1979), pp. 1073–1100.
5. W. D. Rummler, "A New Selective Fading Model: Application to Propagation Data," B.S.T.J., *58*, No. 5 (May–June 1979), pp. 1037–1071.
6. L. J. Greenstein, "A Multipath Fading Channel Model for Terrestrial Digital Radio Systems," IEEE Trans. on Commun., *COM-26*, No. 8 (August 1978), pp. 1247–1250.
7. W. T. Barnett, "Multipath Propagation at 4, 6 and 11 GHz," B.S.T.J., *51*, No. 2 (February 1972), pp. 321–361.
8. J. R. Green and D. Margerison, *Statistical Treatment of Experimental Data*, Elsevier, 1977; Chs 13 and 15.
9. W. D. Rummler, "Extensions of the Multipath Fading Channel Model," Paper No. 32.2, ICC Conf. Record, June 1979.
10. L. J. Greenstein and D. Vitello, "Digital Radio Receiver Responses for Combating Frequency-Selective Fading," IEEE Trans. on Commun., *COM-27*, No. 4 (April 1979), pp. 671–681.

ONE YEAR RESULTS OF SWEEP MEASUREMENTS OF A RADIO LINK

MARKUS LINIGER

Research and Development Department of the Swiss PTT
CH-3000 Bern 29

ABSTRACT

Sweep measurements over a bandwith of 40 MHz were carried out on a 111 km hop in the 11 GHz band in Switzerland, covering a period of one year with and without space diversity reception (in-phase combiner). The results are represented as the probability density function of polynomial approximation of the attenuation in decibels and are given for the copolarized and the crosspolarized RF-channel.

The description of the wave propagation is then combined with the performance (signature) of several digital radiolink equipments. This gives an estimation of the outage probability of a radio link.

1. INTRODUCTION

The transition of terrestrial radio links from analog FM-FDM to digital transmission has revealed the greater sensitivity of the new systems to selective multipath fading. Therefore, for the past three years the Research and Development Department of the Swiss PTT has conducted sweep measurements over a 40 MHz wide RF-channel over several hops of its existing radio-link network. The measurement system, the method of analysis, and first results have been given in (Ref. 1). This paper deals with results which were recorded during one year on a 111 km hop at 11 GHz. The propagation characteristics combined with the properties of the radio-link equipment permit to estimate the quality and availability of the link.

2. DESCRIPTION OF EXPERIMENT

The experiment was conducted on a hop of 111 km length and at 11.285 GHz. The transmitting antenna was a vertical polarized circular parabolic dish of 4.3 m diameter. Two offset-parabolic antennas (Type CM4671 by Thomson CSF) with a diameter of 4 m were used for reception. Their vertical separation was 11.8 m. This allowed to study the influence of space-diversity reception. The two received signals were added with an electronic RF-combiner (Ref. 2). During the measurements the diversity path was switched in and out at hourly intervals. In addition, provisions were made to receive also the cross-polarized signal, in order to study it and its influence at frequency separations between 0 and 40 MHz.

The measurement system consists of the RF-part of an FM receiver and transmitter and a link analyzer RM-4 (Wandel + Goltermann).

The three measured analog quantities, namely sweep voltage (corresponding to frequency), amplitude, and group delay are digitized and preprocessed by a computer (HP 2100). Two types of statistics are collected, longterm and short-term. Values for the longterm statistics are taken every 15 s and stored on magnetic tape. They consist of the results of an entire swept measurement as well as extreme and average values for the previous 15-second interval. If the variations of the measured values exceed given limits or if the amplitude falls below a given value, then the raw data are stored in addition to the longterm values. By means of this strategy a magnetic tape suffices for a recording period between 10 hours to 25 days depending on the fading activity.

3. MODELLING OF THE PROPAGATION CHARACTERISTICS

The characteristics of the propagation path are described by the following quantities, taken during periods of heavy fading:

- Amplitude frequency response A(f)
- Group delay D(f)
- Amplitude of the cross-polarized signal X(f)

The measurements were conducted with or without space-diversity reception. The amplitude and the group delay or alternatively the cross-polarized signal are examined in pairs for their correlation.

3.1 Amplitude frequency response A(f)

A total of 1.2 million sweeps without and 0.9 million with space diversity reception, taken during periods of heavy fading, were available for analysis. A polynomial of second degree namely, $A(f) = a_0 + a_1 f + a_2 f^2$, is fitted to the measured received levels in dB. The results are tripples of coefficients (a_0, a_1, a_2). The probability densities p(a_1, a_2, all a_0) are shown graphically in Fig. 1 and 2. Without diversity-protection (Fig. 1) large positive and negative linear distortions combined with negative parabolic distortions occur (① and ②), if a notch is situated above or below the measuring channel. If a notch falls within the channel, it produces positive parabolic distortion with only weak additional linear distortion ③. Space-diversity reception significantly reduces these distortions (Fig. 2).

Level statistics that would be measured with a single carrier can be obtained from the new measurements via the probability density

Reprinted from *IEEE Int. Conf. Comm.*, pp. 645-649, June 1983.

$p(a_0 \mid \text{all } a_1, a_2)$. From this the cumulative distribution function $P(a_0 < L)$, with L a threshold level, can be derived for the worst month, as shown in Figs. 19 and 20 for the cases without and with space-diversity protection. The ratio of the two distribution functions gives the improvement factor obtained with space diversity reception and an electronic inphase combiner.

The analysis of the densities $p(a_0, a_1, a_2)$ yields further results, if the logarithm of the marginal density functions

$$p(a_1 \mid a_0=\text{const, all } a_2),\ p(a_2 \mid a_0=\text{const, all } a_1) \quad (1)$$

are both approximated with Gaussian distributions. The resulting mean values $\overline{a}_1$ and $\overline{a}_2$ and the corresponding standard deviations σ_{a_1} and σ_{a_2} are plotted as functions of the coefficient a_0 of the fit polynomial (Figs. 3 and 4).

Without diversity reception the mean value $\overline{a}_1$ varies between -0.03 and +0.01 dB/MHz. For $a_0 < -5$dB the standard deviation σ_{a_1} remains practically constant at a value of 0.23 dB/MHz. Or stated differently, positive and negative linear distortions are to a first approximation equally likely and independent of the fade depth (at f=0).

The mean value $\overline{a}_2$ of the parabolic distortions varies for small fade depths around zero and increases for $a_0 < -25$ dB towards positive values (Fig. 4). The reason for this behavior is distortion caused by fading with a notch falling within the measuring bandwidth.

The application of the space-diversity system reduces the extent of the distortions approximately by a factor of 3.

3.2 Group delay D(f)

The evaluation of the group-delay measurement is restricted in this part of the paper to its frequency response alone, without taking into consideration the simultaneously measured amplitude response. It yields a first idea for the extent of the group-delay distortions due to multipath propagation effects. To describe the group delay, the polynomial $D(f) = d_0 + d_1 f + d_2 f^2$ is used.

Figures 5 and 6 illustrate the density functions of the recorded distortions. The extent of the distortions is detailed in Figs. 7, 8 and 9. Fig. 7 gives a plot of $p(d_0 \mid \text{all } d_1, \text{all } d_2)$, whereas Figs. 8 and 9 depict the mean value and the standard deviation of the coefficients d_1 and d_2. The behavior of the curves indicates a strong dominance of negative group-delay distortions, which means that non-minimum phase fading is infrequent. According to Fig. 9 positive parabolic group-delay distortions, d_2, are connected with negative values d_0 and vice versa. This behavior reinforces the above mentioned statement that a prevailing portion of all fading is minimum-phase fading.

Of particular interest is the correlation between non-minimum-phase fading and the fade depth a_0. Fig. 10 shows the density function $p(d_0, a_0)$ which is detailed numerically in Fig. 11. For $a_0 > -25$dB minimum-phase fading occurs almost exclusively. As soon as a_0 falls below -25 dB, the probability for non-minimum-phase fading increases.

The density function can be determined as

$$p(d_0) = \int p(a_0) \cdot p(d_0 \mid a_0 = \text{const.})\, da_0. \quad (2)$$

This gives an indication for the specifications of adaptive equalizers which, depending on the required quality of a radio-link connection, have to be able to equalize both classes of fadings.

3.3 Cross-polarized signal X(f)

The measurement of the cross-polarized signal yields information about the correlation between the copolarized useful signal and the cross-polarized interfering signal. Also important is the dispersion of the latter, which can be described by modeling the cross-polarized signal X(f) with a second-order polynomial similar to A(f). The probability density of the coefficients x_1 and x_2 of the fit polynomial $X(f) = x_0 + x_1 f + x_2 f^2$, which is shown in Fig. 12, indicates strong positive parabolic and small linear distortions. The probability density $p(x_0 - \overline{a}_0 \mid \text{all } x_1, \text{all } x_2)$ exhibits a near normal distribution around the average value of -33 dB with a standard deviation of 5 dB (Fig. 13). $\overline{a}_0$ is the received level during undisturbed propagation. The cross-polarized signal X(f) is not as distorted for $x_0 - \overline{a}_0 > -30$ dB as the copolarized one, A(f), and shows only similarly large distortions as A(f) for $x_0 - \overline{a}_0 < -35$ dB (Figs. 14 and 15).

The correlation between the useful signal A(f) and the interfering signal X(f) has to been evaluated in order to estimate the interference between a cross-polarized inband- or adjacent channel and the operating channel. Fig. 16 shows the probability density of the correlated occurrence of pairs of (x_0, a_0), from which a measure of the inband signal-to-interference ratio can be derived. The cross-polarized signal decreases proportionally to the useful signal for fading down to approximately -15 dB. The cross-polarized signal remains practically constant and is uncorrelated with the copolarized one for $a_0 < -20$ dB. Intersections at fixed levels of the copolarized signal yield the density functions $p(x_0 \mid a_0 = \text{const.})$, for each of which the average value and the standard deviation can be determined. Fig. 18 shows the function of the average $\overline{x}_0(a_0)$ and of $\overline{x}_0(a_0) \pm \sigma_{x_0}(a_0)$, which were obtained from the results of Fig. 16.

The influence of an adjacent cross-polarized signal (Δf = 30 MHz) is illustrated by the results of Fig. 17. This signal exhibits a slightly larger standard deviation and the average $\overline{x}_0(f_0 - 15\text{ MHz}, a_0(f_0 + 15\text{ MHz}))$ increases asymptotically towards a 5 dB larger value than that for the inband signal. This fact is also shown in Fig. 18.

4. CONNECTION BETWEEN PROPAGATION DATA AND PROPERTIES OF THE RADIO EQUIPMENT

The signature of a receiver specifies its sensitivity to multipath propagation effects. It can be transformed from the $(\lambda, \Delta f)$-space into the expanded polynomial space as was shown previously in (Ref. 1). The Boolean function $S(a_1, a_2)$ is equal to 1 for all cases of a degraded transmission channel (e.g. BER $> 10^{-6}$), otherwise S = 0.

The probability P_D, indicating a transmission system degraded on a particular hop by distortion, can be calculated as $P_D(BER > BER_0)$

$$P_D = \int_0^L p(a_0) \iint_{-\infty}^{+\infty} S(a_1,a_2)\ p(a_1,a_2|a_0=\text{const})\,da_1\, da_2\, da_0 \quad (3)$$

This value has to be augmented by the probability P_L, taking into account the falling of the level below a threshold level L, namely

$$P_L(P_{RX} < L) = \int_{-\infty}^{L} p(a_0|\text{all } a_1, \text{ all } a_2)\, da_0 \quad (4)$$

The probability P_L is equivalent to the well-known Rayleigh distribution obtained from single-carrier measurements and multiplied by a hop-dependent factor. It is plotted for the examined hop in Fig. 19 as curve 1. The application of space-diversity reception reduces this probability by the improvement factor I. Curve 1 in Fig. 20 gives directly the distribution function with diversity reception, as derived from the measurements.

In addition the same figures show curves representing the probabilities P_D for the signatures with Δf = 20 or 50 MHz and 20 lg λ = -5 to -25 dB (λ = 1-r, r = magnitude of the second vector in a two-ray model). The sum

$$P(L) = P_D(L) + P_L(L) \quad (5)$$

gives the probability of degraded transmission due to a low input level or to propagation distortions.

As a further influence the interference due to inband and adjacent channels has to be taken into account. A transmission system with a given S/N, e.g. 20 dB, is only feasible with the present antenna system in the experiment and in the presence of an inband interferer down to fading depths of approximately -20 dB. This event will occur with a probability of $3 \cdot 10^{-3}$ (Fig. 19) resp. $3 \cdot 10^{-4}$ (Fig. 20). For that purpose better antennas and eventually an adaptive cross polarization interference canceler are required.

The influence of the adjacent channel is approximately 5 dB larger than that of the inband channel. However, this is reduced due to filtering, e.g. by 15 dB, resulting in a fading margin of approx. 30 dB.

5. CONCLUSIONS

The propagation data presented were collected during an unfavorable month and on a hop with frequent multipath propagation fading. The combination of these results with the properties of the radio equipment gives an indication for its required signature and permits an estimation of the influence of inband- and adjacent-channel interferences. A transmission system exhibiting a signature 20 to 30 MHz wide and a value λ, where 20 lg $\lambda \approx$ -15 to -20 dB, should make possible a connection in accordance with CCIR recommendations.

6. ACKNOWLEDGEMENTS

The author would like to thank the personnel of the radio station Albis for taking care of the measuring and data collection equipment and Mr. Max Matter for the minute control and processing of the raw data. He also thanks Mr. Theo Formanek who developped the APL-programms for the threedimensional computer drawings.

REFERENCES

1. M. Liniger, "Sweep measurements of the transferfunction of a RF-channel and their representation by polynomials", Conference Record, IEEE International Conference on Communications (ICC'82), June 1982.
2. U. Gysel, "Elektronisches Raumdiversity System für Richtfunk-Anlagen", NTG Fachtagung in Munich, Germany, April 1980.

Markus I. Liniger, Ingenieur HTL 1965, Ingenieurschule Bern; Lizentiat (Mathematics, Physics, Astronomie), 1974, University of Berne, Switzerland; Research and Development Department of the Swiss PTT 1974-. Mr. Liniger is working on various electromagnetic wave propagation topics and is currently working on problems in line-of-sight microwave propagation and digital microwave radio systems. Member CCIR Commission 5 and 9.

DENSITY FUNCTION OF THE COEFFICIENTS OF THE FIT-POLYNOMIAL
ATTENUATION (F) = A0+A1×F+A2×F×F (1193785 SWEEPS)
5 DB>A0≥-45 DB
MARCH 81 - FEBRUARY 82 WITHOUT SPACE DIVERSITY

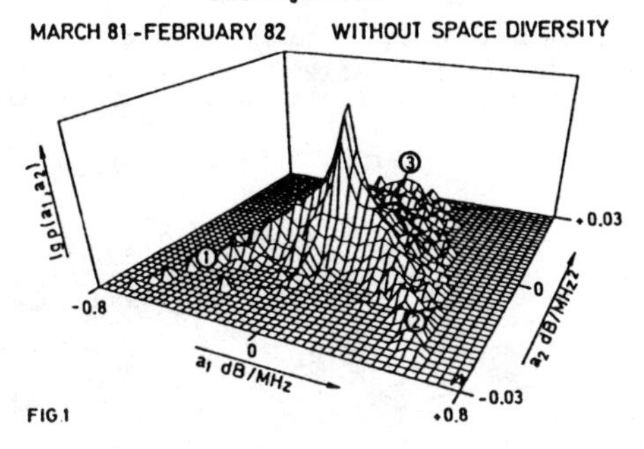

FIG 1

DENSITY FUNCTION OF THE COEFFICIENTS OF THE FIT-POLYNOMIAL
ATTENUATION (F) = A0+A1×F+A2×F×F (839610 SWEEPS)
5 DB>A0≥-45 DB
MAY 81 - FEBRUARY 82 WITH SPACE DIVERSITY

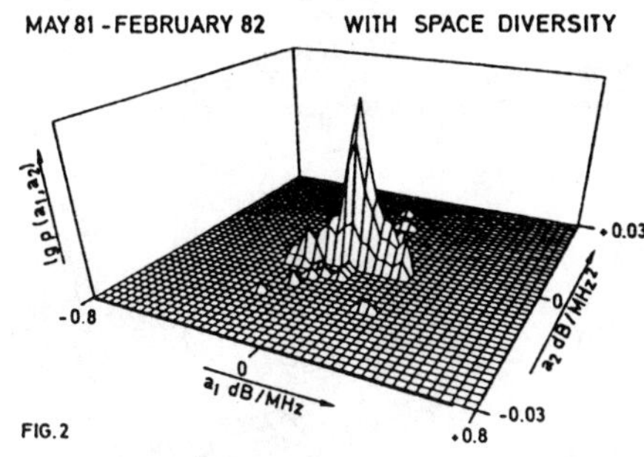

FIG. 2

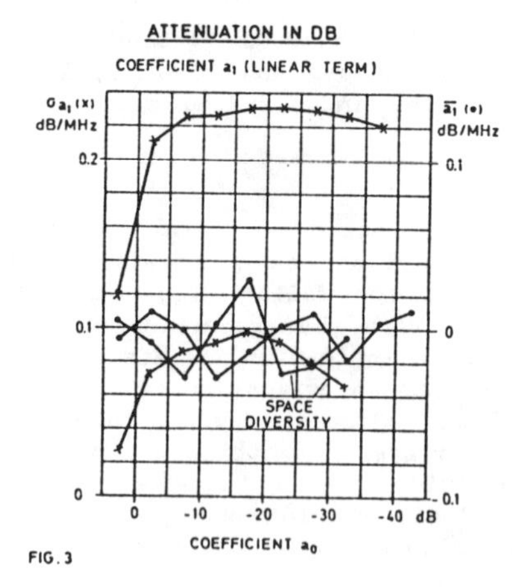

FIG. 3

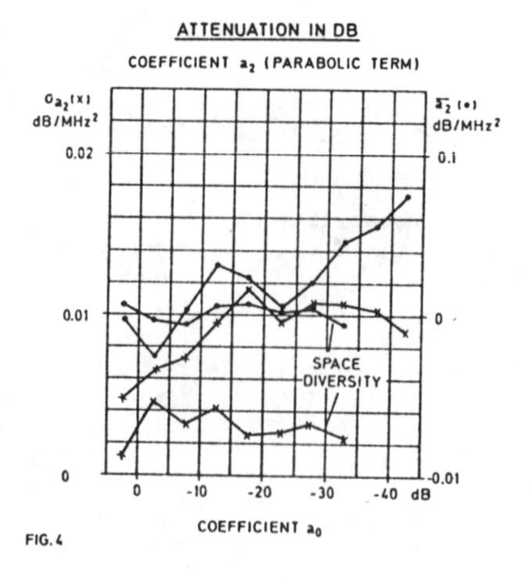

FIG. 4

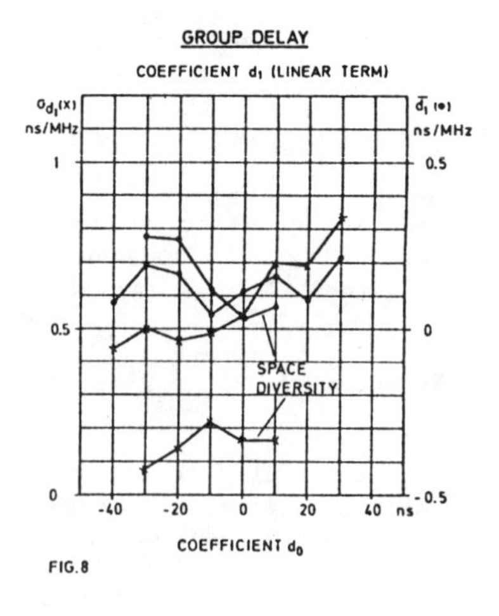

FIG. 8

DENSITY FUNCTION OF THE COEFFICIENTS OF THE FIT-POLYNOMIAL
GROUP DELAY (F) = $D_0 + D_1 \times F + D_2 \times F \times F$ (1101433 SWEEPS)
-55 NS $> D_0 \geq 55$ NS
MARCH 81 - FEBRUARY 82 WITHOUT DIVERSITY

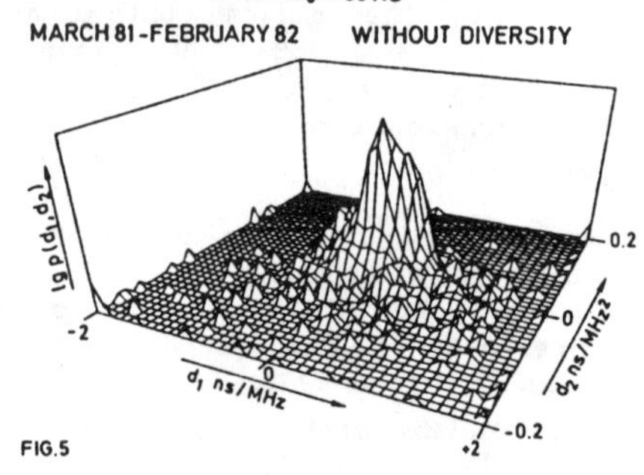

FIG.5

DENSITY FUNCTION OF THE COEFFICIENTS OF THE FIT-POLYNOMIAL
GROUP DELAY (F) = $D_0 + D_1 \times F + D_2 \times F \times F$ (538661 SWEEPS)
-55 NS $> D_0 \geq 55$ NS
MAY 81 - FEBRUARY 82 WITH SPACE DIVERSITY

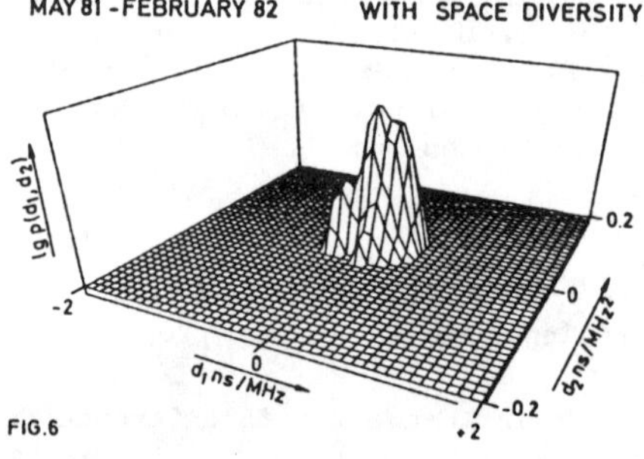

FIG.6

CORRELATION-MATRIX OF GROUP-DELAY VS ATTENUATION
VALUES AT F = 0

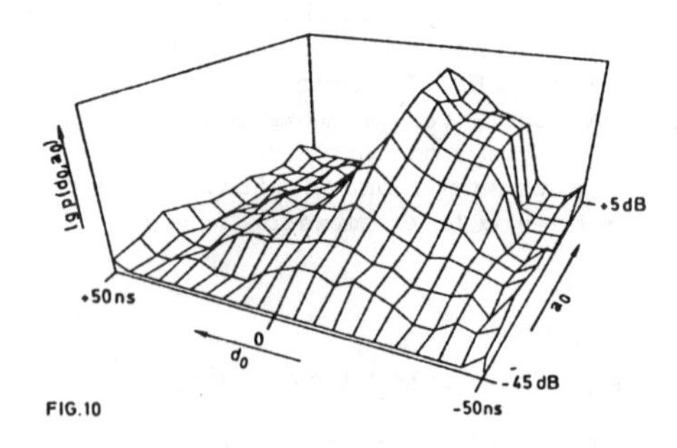

FIG.10

DENSITY FUNCTION OF THE COEFFICIENTS OF THE FIT-POLYNOMIAL
CROSS-POLAR-SIGNAL (F) = $X_0 + X_1 \times F + X_2 \times F \times F$
-30 DBM $> X_0 \geq -75$ DBM

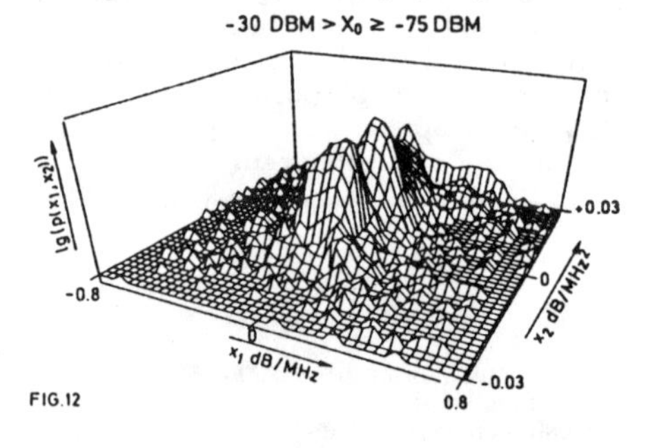

FIG.12

CORRELATION-MATRIX CO-VS CROSS-POLAR-SIGNAL
CO-POLAR RECEPTION WITHOUT SPACE-DIVERSITY

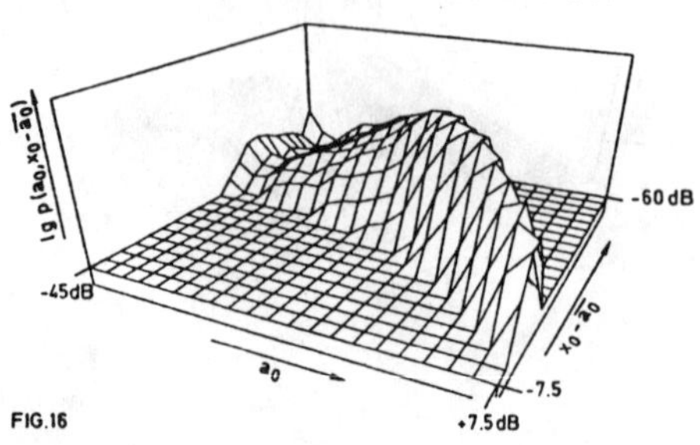

FIG.16

CORRELATION-MATRIX OF CO- VS CROSS-POLAR-SIGNAL
CO-POLAR-LEVEL AT CENTREFREQUENCY -15 MHZ
CROSS-POLAR-LEVEL AT CENTREFREQUENCY + 15 MHZ

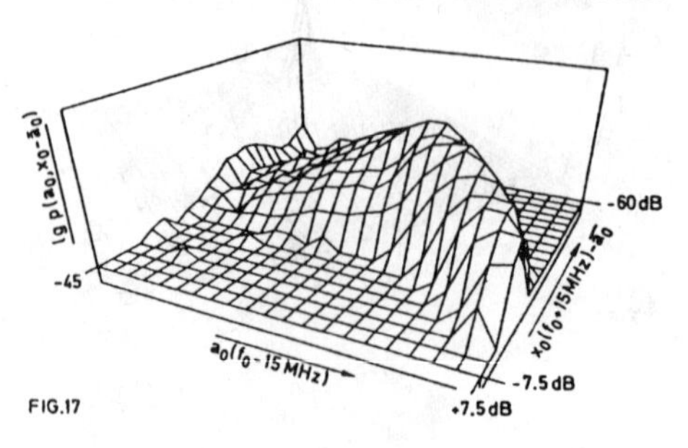

FIG.17

GROUP DELAY

DISTRIBUTION FUNCTION OF THE COEFFICIENT d_0 OF THE FIT-POLYNOMIAL $D(f) = d_0 + d_1 f + d_2 f^2$

d0 ns
30
20
10
0
-10
-20
-30
-40
SPACE DIVERSITY
0,01 0,1 1 5 10 30 50 70 90 95 99 99,9 %
P(d0 < L)

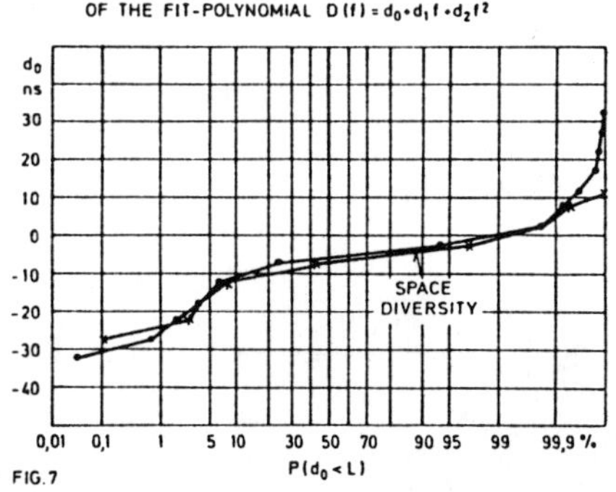

FIG.7

GROUP DELAY

COEFFICIENT d_2 (PARABOLIC TERM)

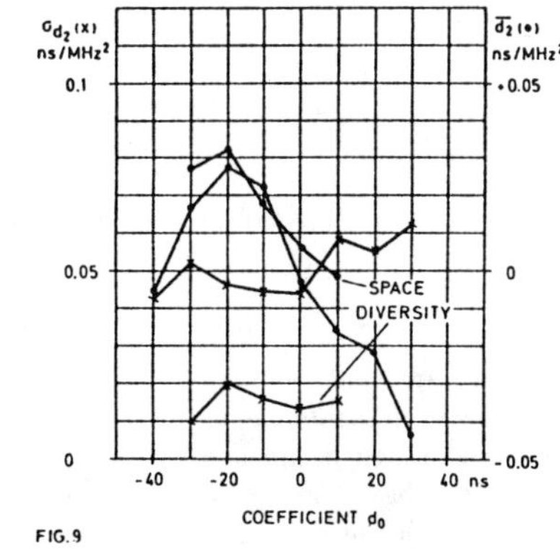

σd2(x) ns/MHz²
d̄2(•) ns/MHz²
0.1
0.05
0
+0.05
0
-0.05
SPACE DIVERSITY
-40 -20 0 20 40 ns
COEFFICIENT d0

FIG.9

CORRELATION GROUP-DELAY VS FADE DEPTH

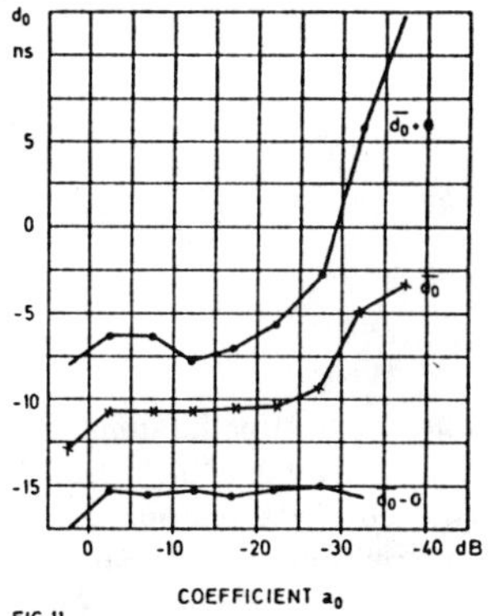

d0 ns
5
0
-5
-10
-15
d̄0+σ
d̄0
d̄0-σ
0 -10 -20 -30 -40 dB
COEFFICIENT a0

FIG.11

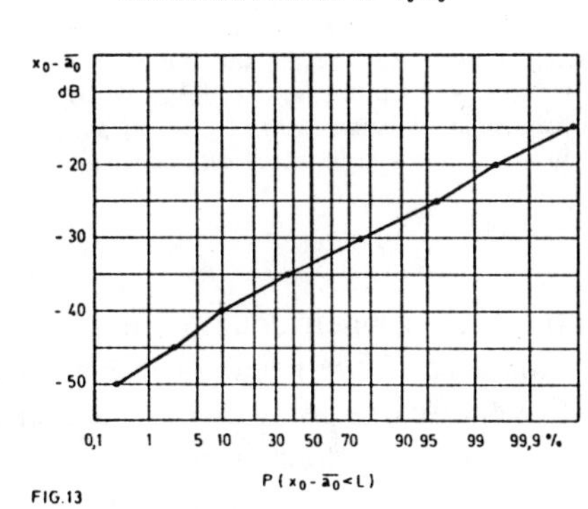

CROSS-POLAR-SIGNAL

DISTRIBUTION FUNCTION OF $x_0 - \bar{a}_0$

x0 - ā0 dB
-20
-30
-40
-50
0,1 1 5 10 30 50 70 90 95 99 99,9 %
P(x0 - ā0 < L)

FIG.13

CROSS-POLAR-SIGNAL

COEFFICIENT x_1 (LINEAR TERM)

σx1(x) dB/MHz
x̄1(•) dB/MHz
0.2
0.1
0
0.1
0
-0.1
-10 -20 -30 -40 dB
x0 - ā0

FIG.14

CROSS-POLAR-SIGNAL

COEFFICIENT x_2 (PARABOLIC TERM)

σx2(x) dB/MHz²
x̄2(•) dB/MHz²
0.02
0.01
0
0.01
0
-0.01
-10 -20 -30 -40 dB
x0 - ā0

FIG.15

CORRELATION CO-CROSSPOLAR SIGNAL

: CO-CHANNEL (f1 = f2)
: ADJACENT-CHANNEL (30 MHz)
x0(a0) - ā0 dB
-50
-40
-30
-20
-10
0
x0 - ā0 - σx0
x0 - ā0
x0 - ā0
x0 - ā0 + σx0
x0 - ā0 + σx0
XPD = 20 dB
0 -10 -20 -30 -40 dB
COEFFICIENT a0 OF COPOLAR SIGNAL

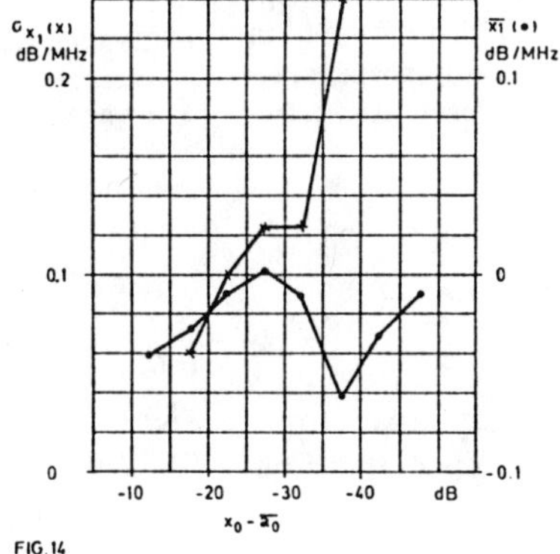

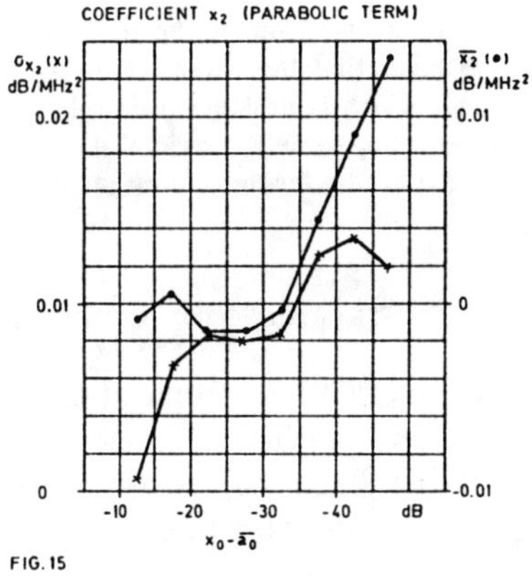

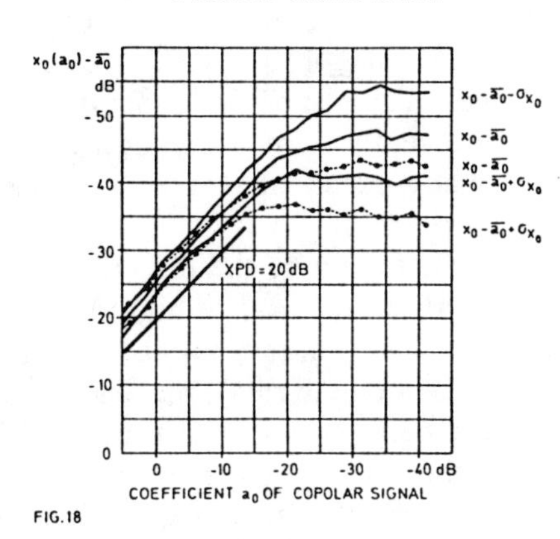

FIG.18

OUTAGE PROBABILITY VS FADE MARGIN

WITHOUT SPACE DIVERSITY

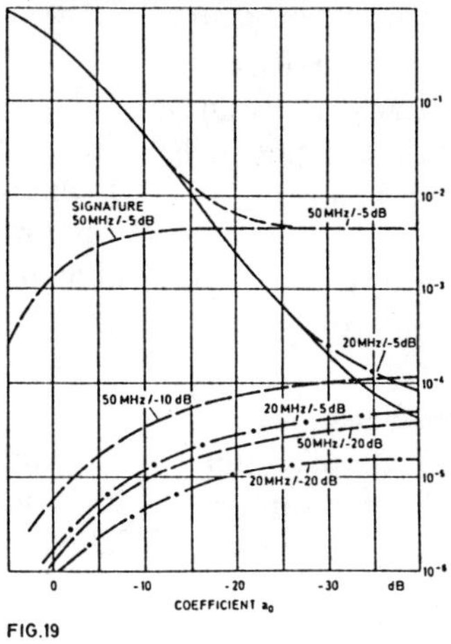

SIGNATURE 50MHz/-5dB
50MHz/-5dB
20MHz/-5dB
50MHz/-10dB
20MHz/-5dB
50MHz/-20dB
20MHz/-20dB
10⁻¹
10⁻²
10⁻³
10⁻⁴
10⁻⁵
10⁻⁶
0 -10 -20 -30 dB
COEFFICIENT a0

FIG.19

OUTAGE PROBABILITY VS FADE MARGIN

WITH SPACE DIVERSITY

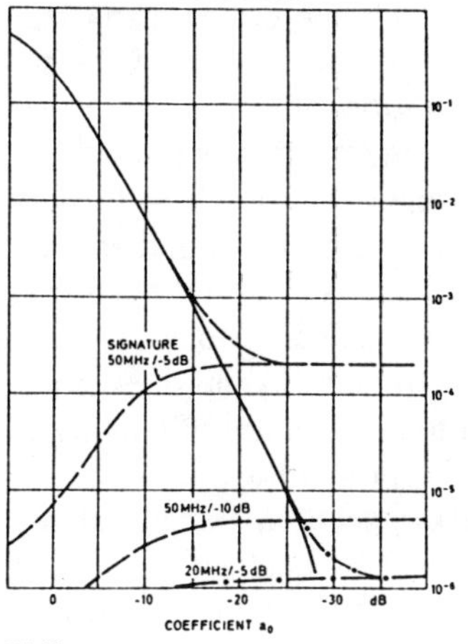

SIGNATURE 50MHz/-5dB
50MHz/-10dB
20MHz/-5dB
10⁻¹
10⁻²
10⁻³
10⁻⁴
10⁻⁵
10⁻⁶
0 -10 -20 -30 dB
COEFFICIENT a0

FIG.20

Selective Fading Radio Channels: Modeling and Prediction

JACQUES LAVERGNAT AND MICHEL SYLVAIN

Abstract—**Multipath fading is a major cause of impairment of high-bit-rate digital line-of-sight transmission. This paper presents a characterization of the multipath propagation channel which is independent of any equipment consideration. The method is based on a modeling of the channel transfer function over a bandwidth of a few tens of megahertz. It is shown that several simple three-parameter mathematical models give a good representation of the transfer function. The experimental statistics of these parameters have been established on three experimental links. With two of the mathematical models, the parameters' joint statistics can be expressed in terms of classical probability laws depending on two coefficients whose values reflect the channel selectivity. The last part of the paper describes the application of such a channel modeling to the prediction of the quality on a given hop. This assumes that the values of the statistical coefficients have been obtained, which is possible from relatively simple radio propagation measurements, namely, the joint distribution of the levels at two frequencies spaced a few tens of megahertz apart.**

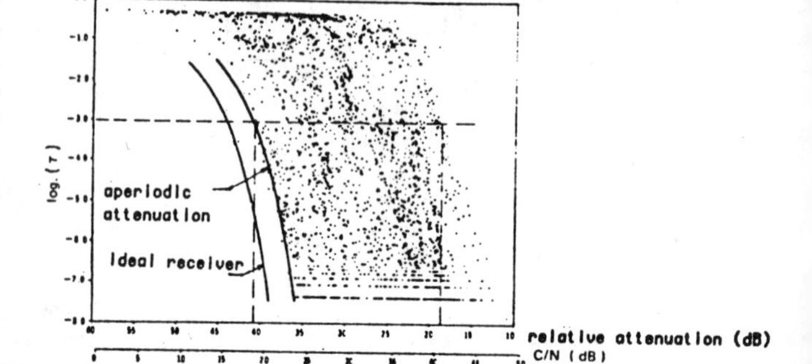

Fig. 1. Scatter diagram showing the relation between the log of the binary error rate and the carrier-to-noise ratio, or the attenuation at the center frequency. F = 4 GHz; rate = 45 Mbits/s; 8 PSK; D = 26 mi. The lower envelope shows the relation between the BER and an aperiodic attenuation, resulting in a flat fade margin of 41 dB (Giger and Barnett [9]).

I. Introduction: General Considerations

Multipath propagation (MP) due to "anomalous" refractive properties of the very low atmosphere has been known for a long time, since the first experiments devoted to this phenomenon were carried out around the 1950's. MP is a major source of impairment on line-of-sight microwave links. Because multipath fading is frequency selective, it generally severely affects high-bit-rate digital transmissions, which are currently planned for the communication networks. Therefore, MP and its effects on line-of-sight radio links have obviously been the subject of many studies for the past ten years. The ultimate object of the studies devoted to MP is the derivation of an accurate method to estimate the transmission quality of a link with specified characteristics such as frequency, hop length, slope of the link, heights of the antennas, climate, and so on.

As MP is a natural phenomenon, it is clear that it would be unreasonable to look for a pure deterministic model. The only relevant question is therefore the following: Within a given set of parameters (e.g., the above list) what is the probability (or equivalently, the fraction of time) that a specified transmission system will have sufficient quality? In fact, the answer to this question is not straightforward because it may decisively depend upon the system. This is essentially true when countermeasures are used. Moreover, in these cases we need additional information, such as the dynamical aspects of MP, which require more elaborate knowledge than is presently available. In front of such a "multiple" problem, the best way to tackle it seems to go from the simplest to the most complex. When the bandwidth of the radio system is narrow (typically, some megahertz) the statistical description of the level of the received field is sufficient, and many empirical laws have been proposed [1]. Whatever the provisions against these formulas may be, they lead to practical results which are not too bad.

The next level of difficulties is the introduction of medium-bandwidth systems ($\leqq 100$ MHz), discarding countermeasure devices. The rest of this paper deals with this problem. At this level, we are far from a complete knowledge, even if many excellent pioneer works exist [2]–[4].

Fig. 1 is a typical illustration of the difficulty in analyzing the influence of MP in the case of high-rate digital transmission. It appears that, contrary to the case of uniform attenuation (e.g., that due to rain), no functional relationship can be found between the bit error rate (BER) and the attenuation at the center frequency. Flat margin, relevant to uniform attenuation, directly gives the value of the attenuation beyond which the limit BER τ_0 is exceeded, and consequently, if the distribution of the level is known, it becomes a very simple task to get the fraction of time for which the quality ($\tau < \tau_0$) is good. The notion

Manuscript received August 7, 1986; revised December 16, 1986.
The authors are with CNET/PAB/RPE, 92131 Issy-les-Moulineaux, France.
IEEE Log Number 8613343.

Reprinted from *IEEE J. Selected Areas Comm.*, vol. SAC-5, no. 3, pp. 378–388, Apr. 1987.

of net margin[1] certainly originates from the convenience of flat margin. Fig. 1 clearly shows that this notion is only a global one, unable to represent the scattering of the experimental results. Net margin can sum up the quality of a link, but it is unlikely to obey simple laws versus propagation parameters.

In order to characterize the selectivity, two approaches are possible. Following the first one, one tries to find what characteristics of the transfer function are responsible for the errors and to establish accordingly a functional relationship between BER and some global parameters. For instance [5], in the case of 8-PSK modulation those parameters are the signal-to-noise ratio and the eye closure. This approach has, however, an important drawback: the effects of the propagation cannot be separated from those induced by the system.

The second approach, which avoids the previous drawback, is based upon modeling of the transfer function. This is the approach we have chosen, and the successive steps are the following.

1) Choose a mathematical form $H(\omega, \vec{p})$ for the model of the transfer function to be used. $\vec{p} = (p_1, p_2, \cdots, p_n)$ are the parameters of the model.

2) We determine the joint statistics of the model parameters $\Pr[\vec{p} = \vec{P}] = S(\vec{P}, \vec{s})$ where $\vec{s} = (s_1, \cdots, s_m)$ are statistical parameters (mean values, standard deviations, and so on).

3) At this stage, we make the hypothesis, which will be verified later, that the statistical law S is of the same type for all links. That is to say, for instance, that S is a normal law.

4) Ideally, we find a relationship between the statistical parameters and the environmental characteristics of the link. $\vec{s} = F(\vec{q})$ where $\vec{q} = (q_1, q_2, \cdots, q_k)$ are pertinent physical quantities to be defined.

This last step, however, assumes that measurements of the transfer function over a medium bandwidth were performed on a large number of links. This is not the case, and it will unlikely be so in the near future when considering the complexity and difficulties of these measurements. However, let us notice that once a statistical model is established, any partial information concerning a link, such as the distribution of levels at a fixed frequency or the joint distribution of levels at two frequencies, could be used to infer some knowledge about the statistical parameters $\vec{s}$.

This method works fairly well when data at two fixed frequencies, or at worst at one frequency, are available. It must be underlined that the method we propose does not suppress the need for a large campaign of collecting data, but only alleviates it.

The paper is divided into three parts. In the first one, mathematical representation models are presented and discussed. The second one is devoted to the properties of statistical modeling. The last part deals with the problem of estimating the statistical parameters on a given link.

[1]Net margin is the value of the attenuation at fixed frequency which is exceeded during the same fraction of time as a given BER (often 10^{-3}).

II. Mathematical Representation of the Transfer Function

A. Generalities

Let us just recall that the MP channel is changing slowly enough (a few milliseconds) as compared to the propagation time of a signal ($\simeq 100\ \mu$s) to be considered as a filter uniquely characterized by its transfer function.

Indeed, we are mainly interested in the normalized transfer function

$$F_r(\omega) = \frac{F_{\mathrm{MP}}(\omega)}{F_0(\omega)} \tag{1}$$

where $F_{\mathrm{MP}}(\omega)$ is the transfer function of the MP channel and $F_0(\omega)$ is the transfer function of the reference channel.

As measurements do not give absolute propagation time and phase, we have to model

$$H(\omega) = F_r(\omega) \exp j(\omega t_0 - \phi_0) = G(\omega)\, e^{-j\phi(\omega)}, \tag{2}$$

ϕ_0 and t_0 being two arbitrary constants.

Let us note that we have experimental access to $G(\omega)$ and $T(\omega) = (d\phi/d\omega)$. We are now faced with the choice of a model $H(\omega, \vec{p})$ which must be good enough to be used in place of (2). The only physical model to describe the multipath propagation channel is the multiray model, largely described in the literature. But this model is not simple (too many parameters are involved) and is tedious (time consuming) to manipulate, and the derivation of the actual values of its parameters from data on a limited bandwidth may be doubtful.

Thus, all authors have used [2]–[4] simpler mathematical models. These models generally belong to one of two families: polynomial expansions or ray models.

It must be immediately pointed out that these models are convenient for representation only, and it would be illusive to search for any relationship between environmental parameters and their original parameters $\vec{p}$.

Of course, we wish the modeled transfer function to be similar to the experimental one. To evaluate this similarity, we can use the distance between curves, measured, for instance, by the least mean-square (LMS) residual error or by the largest deviation. This criterion may be criticized because it is well known that the BER is a very sensitive quantity and that even if two transfer functions are close to one another the BER's they induce may differ by orders of magnitude. In fact, it has been shown statistically [7] that the two approaches based upon either the distance between curves or the distance between BER's lead to the same results.

Besides, a good model has necessarily, for practical reasons, a small number of parameters.

B. Polynomial Expansion

The most natural mathematical model is undoubtedly a polynomial expansion of the transfer function. Nevertheless, different expansions may be chosen; for instance, the gain curve $G(\omega)$ and the delay distortion curve $T(\omega)$ are

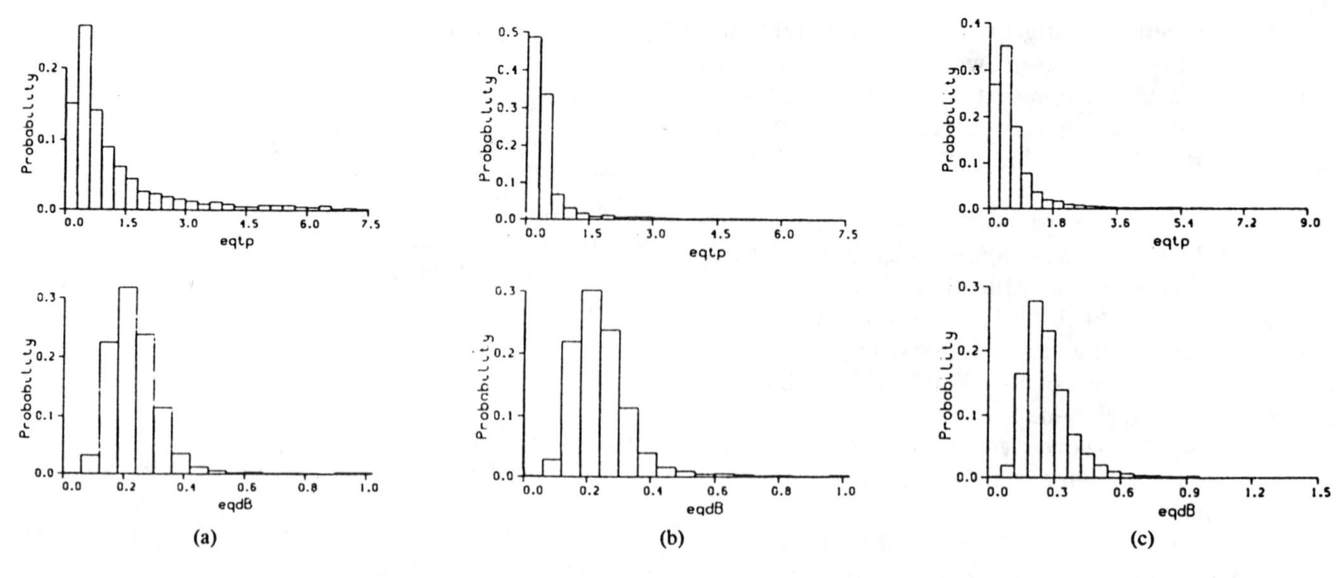

Fig. 2. Residual mean-square errors on amplitude (eqdB) in decibels and delay distortion (eqtp) in nanoseconds. These results were obtained over 100 000 transfer functions gathered in June–July 1982 on the link Marcheville–Viabon near Paris. (a) Complex polynomial expansion of first degree. (b) Rummler's model. (c) Normalized two-ray model.

expanded separately either by a Taylor technique or by an LMS estimation. However, inside this class of models, using a complex polynomial expansion as proposed first by Greenstein [3] provides a good and simple model. This model, limited to the first-degree complex polynomial, reads

$$H(\Omega) = R_0 + (R_1 + jX_1)\,\Omega \tag{3}$$

where $\Omega = \omega - \omega_c$ is the frequency measured from the center of the frequency range. The gain curve is written

$$\begin{aligned} G^2(\Omega) &= R_0^2 + 2R_0R_1\Omega + (R_1^2 + X_1^2)\,\Omega^2 \\ &= D_0 + D_1\Omega + D_2\Omega^2. \end{aligned} \tag{4}$$

This model depends upon three real parameters R_0, R_1, and X_1, the estimation of which is a little bit complex [3], [7], but it is possible by means of a weighted regression on the gain curve (4) to get satisfactory results. The experimental delay distortion curve only allows the determination of two ambiguous signs when trying to compute R_0, R_1, X_1 from the estimated values of D_0, D_1, D_2. It is of interest to recall [7] that this method of estimation induces in fact two different procedures, depending upon the sign of the discriminant $\Delta = D_2 - (D_1^2/4D_0)$. This could result in some undesirable effects on the parameters' statistics.

Fig. 2(a) gives a typical distribution of residual errors which was obtained during the modeling over 55 MHz of data gathered on the Marcheville–Viabon hop near Paris, France. Therefore, it appears that such a model is well suited to represent the multipath transfer function on a medium bandwidth. Let us recall that 0.5 dB is the accuracy with which the gain curve was measured.

C. Rummler's Model

This model is one of the ray models. It can be derived from the physical N-ray model using some simplifying assumptions [2], [6]. But due to this simplification, the remaining rays have no physical reality.

The general two-ray model, which is a four-parameter one, reads

$$H(\omega) = a(1 - be^{j(\omega - \omega_0)\tau}). \tag{5}$$

Rummler proposed reducing by 1 the number of parameters in fixing τ to a value depending on the bandwidth Δf: $\tau_0 = (1/6\Delta f)$. This important point, noted by Rummler himself, has unfortunately often been forgotten by later authors.

The power curve and the delay distortion curve are given in this model by

$$P(\omega) = \alpha - \beta \cos(\omega - \omega_0)\,\tau_0 \tag{6}$$

$$T(\omega) = \tau_0(\tfrac{1}{2} + a^2(b^2 - 1)/2P(\omega)) \tag{7}$$

with

$$\alpha = a^2(1 + b^2)$$

and

$$\beta = 2a^2b. \tag{8}$$

The three free parameters to be estimated are thus a, b, and ω_0. Although the weighted LMS estimation of α, β, and ω_0 is nearly straightforward, the derivation of a and b from (8) results again in two procedures due to the constraint $\alpha > \beta$ [6]. The delay distortion curve is employed to discriminate between the minimum- and the nonminimum-phase transfer function.

Fig. 2(b), similarly to Fig. 2(a), shows the quality of the modeling. The gain curves are as well represented as with Greenstein's model, and the delay distortion curves are even better represented.

D. Normalized Two-Ray Model

It has been shown [6] that instead of fixing τ, it is as efficient to impose $a = 1$, resulting in the normalized two-ray model invoked by some authors [4], [8]. With this model, the gain is written

$$P(\omega) = 1 + b^2 - 2b \cos(\omega\tau + \phi). \quad (9)$$

There are again three parameters b, ϕ, and τ.

This model sets a difficult problem because the required estimation is a nonlinear one relative to the three parameters. A direct computation does not work for all transfer functions; we are thus obliged to use an exhaustive method, which is too time consuming.

Noting that τ is expected to be small, i.e, on the order of a few nanoseconds, another method consists of estimating b, ϕ, and τ from an identification of (4) with the limited expansion of (9). Details of this procedure are explained in [7]. Therefore, the parameters of the normalized two-ray model appear to be a simple transformation of the parameters of Greenstein's model.

We would like to emphasize again that, as for the other models, the existence of a double procedure may induce some distortion in the statistics of the original parameters, a point to which we will come back later.

The quality of the representation by the normalized two-ray model is illustrated in Fig. 2(c).

In conclusion, transfer functions over medium bandwidths ($\leqq 100$ MHz) causing multipath propagation are well represented by at least three mathematical models: a complex polynomial of degree 1, Rummler's model, and the normalized two-ray model.

This assertion has been verified over all the links from which data were available to the authors. Moreover, as the different estimation procedures all give a similar quality of representation, we have chosen the methods which result in the most convenient statistics. This will be the subject of the following section.

III. Statistical Modeling of the Transfer Function

A. Generalities

Once multipath transfer functions have been collected on an experimental link, it is possible to compute for each of them estimated values of the model parameters (p_1, p_2, p_3) and to obtain experimental histograms of these parameters.

The next step is to find a relevant mathematical formulation describing these histograms which uses as small a number of coefficients as possible (these coefficients will hereafter be called statistical parameters to distinguish them from the model parameters). To be fully significant, of course, these results have to be general enough and not specific to a particular link. To establish such general behavior, we need data from many different and representative experimental links. Unfortunately, propagation experiments involving transfer function measurements are rather uncommon, and we had to restrict our analysis to only three links, the characteristics of which are given in Table I. Data from the two French experiments were available to us; as for the Atlanta–Palmetto link, we used the detailed statistical results previously published [3], [10]. Although our database is restricted to only three links, we think that they are different enough to allow significant results to be gained.

TABLE I
Characteristics of Experimental Links

Hop	length	frequency	bandwidth	period of measure
Marcheville-Viabon (France)	37 km	11 GHz	55MHz	June-July 1982
Atlanta-Palmetto (USA)	42 2 km	6 GHz	25 3MHz	June 1977
Lannion-Roc Tredudon (France)	50 km	11 GHz	56.5MHz	January 1979

The model is intended to represent the propagation channel only during multipath events, but it is not easy to set a limit between very faint events and a quiet situation. The criterion we adopted was to retain those transfer functions which presented an attenuation greater than a given threshold at at least one of the frequencies within the band of interest. One is easily convinced that the choice of the threshold value S has no damaging consequence as long as all transfer functions likely to impair the transmission are kept in the model. But the resulting joint probability distribution of the p_i parameters evidently depends on that choice. In practice, a value of S around 10 dB seems to be a good tradeoff. In some experiments (for instance, on the Atlanta–Palmetto link), recorded data were selected according to another criterion; one therefore needs to be cautious when comparing data from different sources.

It results from the previous discussion that a first statistical parameter to consider is the percentage of the time during which the model applies (i.e., during which the threshold condition is satisfied), or occurrence.

A last point must be kept in mind when analyzing the obtained statistics. We have mentioned (Section II) that, whatever the model, the parameters are estimated according to a double procedure, leading to two subpopulations of transfer functions. Unfortunately, these two subpopulations do not exhibit the same parameter statistics [7], as shown by the examples of Fig. 3. The relative importance of these two classes of functions (which can change from one link to another) may therefore influence the overall statistics.

To give a mathematical formulation of the experimental parameter statistics is a very important point because the simpler this formulation, the easier its use in a predictive method. We try, of course, to give this formulation in terms of classical probability distributions. It is most convenient to formulate the joint probability of the three model parameters. This can be practically achieved if at least one of them is statistically independent from the others. Otherwise, one is generally reduced to favoring one

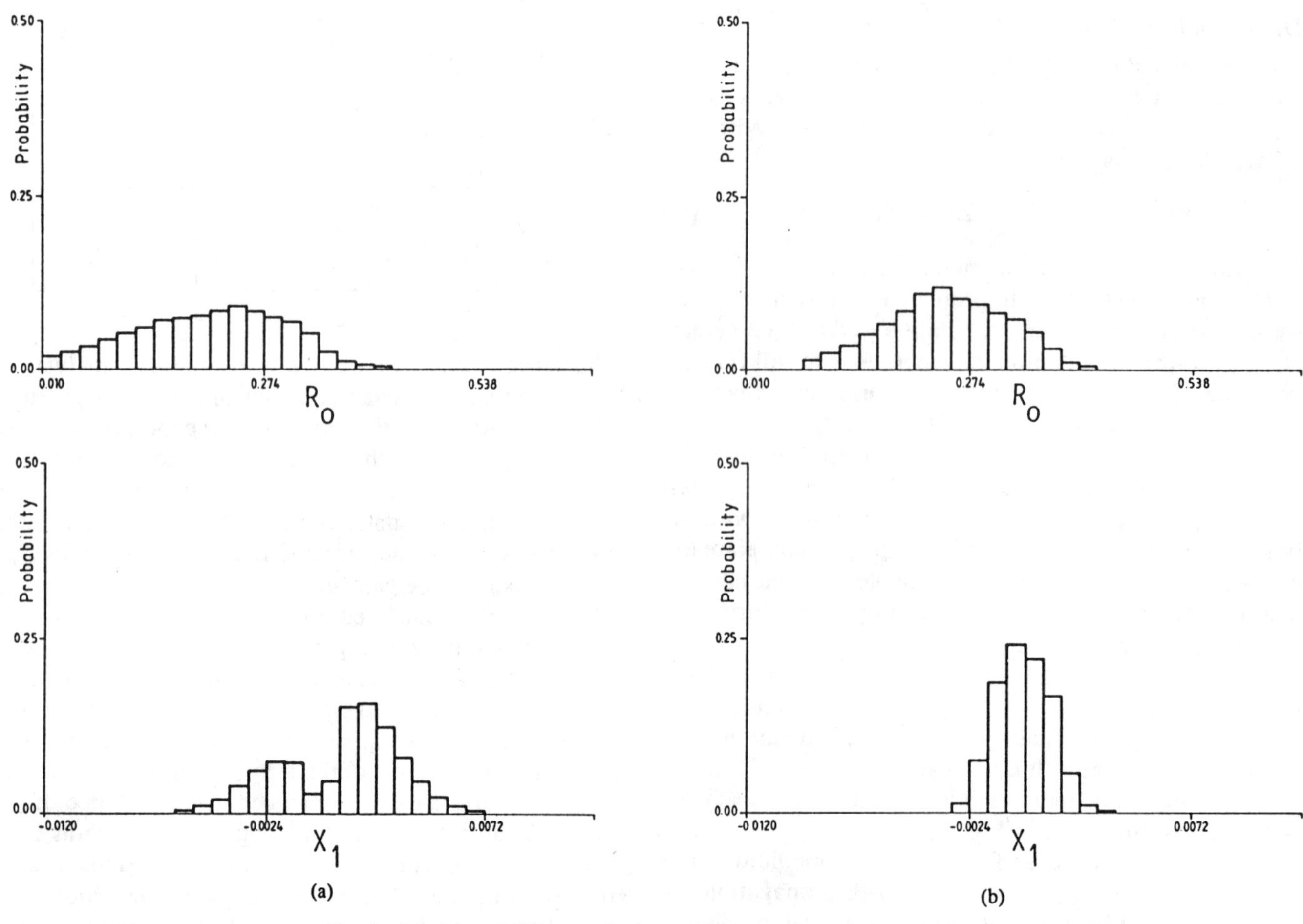

Fig. 3. Examples of differences in the model parameters statistics according to the parameter estimation procedure. (Greenstein's model. (a) $\Delta > 0$. (b) $\Delta < 0$.)

of the parameters and to giving its probability distribution and the conditional probabilities of the two others.

The parameters we have considered in Section II are those which appear in the expression of the modeled complex transfer function. They are not necessarily those leading to the most convenient probability laws.

The range of variation of the model parameters is limited, both by intrinsic constraints (such as $\alpha > \beta$ in the Rummler model) and by the threshold condition. On the other hand, we try to represent their probability distributions with the help of theoretical laws whose range of definition is generally unlimited. As a consequence, if $\mathfrak{D}$ is the domain of variation of the vector of parameters $\vec{p}$, and $f(\vec{p})$ is the mathematical expression used to represent the probability distribution of the model,

$$\Pr(\vec{p}) = \begin{cases} \dfrac{f(\vec{p})}{\int_{\mathfrak{D}} f(\vec{p})\, dp} & \text{if } \vec{p} \in \mathfrak{D} \\ 0 & \text{if } \vec{p} \notin \mathfrak{D} \end{cases} \quad (10)$$

$$\vec{p} = (p_1, p_2, p_3).$$

It must be emphasized that f just gives a mathematical description of the empirical distribution and must not be considered as an underlying true statistical structure. Therefore, the goodness of fit of the experimental distribution by (10) cannot be tested by a statistical test, which cannot be expected to be satisfied. To evaluate the agreement of the mathematical approximation with the data, we then compute a distance between the experimental distribution function and the distribution function of a random sample from the mathematical law. We think that the most convenient distance to use is the χ^2 distance [12], which is written here

$$d = \sum_{i=1}^{N} |f_i - \hat{f}_i| \quad (11)$$

where $i = 1$ to N are the classes of the histogram, and f_i and $\hat{f}_i$ are the probability of class i from the theoretical and the experimental distributions. d varies from 0, when both distributions are identical, to a maximum of 2 when they are completely disconnected.

B. Results

A detailed analysis of the statistics of the model parameters has been given in previous papers [7], [11]. We present here only the main results. Of prime importance is the fact that for a given mathematical model, we have found that the representation of the parameters' distribution function has the same mathematical structure for all links, the differences affecting only the numerical values of some coefficients.

1) Greenstein's Model: In the scope of this model, we did not find a statistically independent parameter and could not express conveniently the joint probability. A good representation of the experimental statistics consists of a lognormal law for $P_0(R_0)$ and normal laws with zero means and standard deviations increasing with R_0 for the conditional probabilities $P_1(R_1/R_0)$ and $P_2(X_1/R_0)$.

2) Normalized Two-Ray Model: The parameters' statistics depend markedly on the estimation procedure used. The best one consists of an identification with a polynomial expansion, which leads to a unimodal distribution for τ [7].

The main interest of this model then resides in the complete statistical independence of its three parameters:

$$f_{2N}(b, \tau, \phi) = P_\tau(\tau)\, P_b(b)\, P_\phi(\phi). \tag{12}$$

τ follows closely a gamma law, with two parameters a and λ,

$$P_\tau(\tau) = \frac{a\lambda}{\Gamma(\lambda)}\, e^{-a\tau} \tau^{\lambda-1}. \tag{13}$$

Estimation of the values of a and λ by a minimum-likelihood algorithm is tedious, but they can be computed immediately from the average and variance of τ:

$$\bar{\tau} = \lambda/a \qquad \mathrm{Var}(\tau) = \lambda/a^2. \tag{14}$$

P_b can be approximated by a uniform distribution from $b_{\min}$ (a value related to the threshold S) to 1. P_ϕ is uniform on $(-\pi, +\pi)$, apart from the threshold condition.

To check the accuracy of this mathematical approximation, we computed the χ^2 distance between the experimental distribution and a random sample, obeying the theoretical law. The results are (Marcheville–Viabon link, $S = 10$ dB)

- $d_\tau = 0.211$ (histogram of 30 classes, sample of size 20 000)
- $d_b = 0.378$
- $d_{b\tau} = 0.439$ for the double distribution of b and τ (histogram of 16 × 16 classes, random sample of size 50 000).

We see that the uniform law does not describe very accurately the distribution of b, producing the same distance on the double (b, τ) probability. On the other hand, the formulation is very pleasant, with only two statistical parameters to describe the selectivity of the channel (for instance, λ and a or λ and τ). The values of these parameters for the three experimental links are given in Table II.

3) Rummler's Model: In this model, parameter ω_0 (the notch frequency) is statistically independent from the two others. We thus have

$$f_R(a, b, \omega_0) = P_{\omega 0}(\omega_0)\, P_{a,b}(a, b). \tag{15}$$

Apart from the threshold effect, $P_{\omega 0}$ is a uniform distribution over $(-\pi, +\pi)$.

Instead of a and b, Rummler [10] used the two derived parameters

$$A' = -20 \log a' \qquad B' = -20 \log (1 - b')$$

with

$$a' = a \text{ and } b' = b \quad \text{if } b < 1$$
$$a' = ab \text{ and } b' = 1/b \quad \text{if } b > 1. \tag{16}$$

He gives

$$P_{A',B'} = P_{B'}(B')\, P_{A'/B'}(A'/B') \tag{17}$$

with a power law for $P_{B'}$, a normal law with a constant standard deviation, and a mean increasing with B' for $P_{A'/B'}$.

We found that a simpler expression for the joint probability distribution is obtained when using

$$A = 10 \log \alpha = 10 \log (a^2(1 + b^2))$$
$$B = 10 \log \beta = 10 \log (2a^2 b). \tag{18}$$

$P_{A,B}(A, B)$ can then be represented by a bivariate normal distribution:

$$P_{A,B} = \frac{1}{2\pi\sigma_A\sigma_B\sqrt{1-\rho^2}} \exp\left\{\frac{-1}{2(1-\rho^2)} \cdot \left[\frac{(A-m_A)^2}{\sigma_{A^2}} - 2\rho\frac{(A-m_A)(B-m_B)}{\sigma_A\sigma_B} + \frac{(B-m_B)^2}{\sigma_B^2}\right]\right\} \tag{19}$$

and depends on five parameters m_A, m_B, σ_A, σ_B, and ρ.

As the domain $\mathfrak{D}$ (10) has a limit which is not analytically simple, these statistical parameters cannot be estimated from the experimental moments and have to be computed by an iterative procedure. The results obtained on some data from the experimental links are given in Table III.

For the Viabon–Marcheville data, an evaluation of the χ^2 distance from the mathematical representation leads to

TABLE II
NORMALIZED TWO-RAY MODEL. STATISTICAL PARAMETERS OBTAINED ON EXPERIMENTAL LINKS

Hop	Model Parameters		
	τ (ns)	λ	a
Marcheville-Viabon	0.485	1.66	3.42
Lannion_Roc Treduon	0.474	2.12	4.47
Atlanta-Palmetto	0.279	1.34	4.79

TABLE III
RUMMLER'S MODEL. STATISTICAL PARAMETERS OBTAINED ON EXPERIMENTAL LINKS

Hop	Model Parameters				
	m_A	σ_A	m_B	σ_B	ρ
Marcheville-Viabon	-7.25	6.5	-5.5	6.5	0.45
Lannion-Roc Tredudon	-8.5	9	-3	8.5	0.75
Atlanta-Palmetto	-24.	7.5	-14.5	7.5	0.

$$\begin{cases} d_A = 0.128 \\ d_B = 0.146 \\ d_{A,B} = 0.149, \end{cases}$$ conditions similar to that for the normalized two-ray model

indicating a good agreement.

We thus have a statistical model describing nicely the data, but with five statistical parameters to specify the bivariate normal when we need only two parameters to specify the probability distribution of the parameters of the normalized two-ray model.

In fact, all statistical models represent the same phenomenon and should be equivalent. It should then be possible to express all of them with the same minimum number of statistical parameters—the "degree of freedom" of the problem. We therefore tried to find some relationship between the statistical parameters of Rummler's model. We proceeded empirically, using several data samples from the three experimental links, and found that the relations (20) are approximately satisfied:

$$\sigma_A = \sigma_B = 7.5 \text{ dB}$$
$$m_B = 0.6 m_A. \tag{20}$$

The representation of the data is not so good, the χ^2 distance becoming

$$\begin{cases} d_A = 0.152 \\ d_B = 0.247 \\ d_{AB} = 0.224, \end{cases}$$

but we get a more tractable model, and the data representation remains quite acceptable.

IV. Application to the Prediction of the Quality

A. Generalities

We have shown in the previous section that with each given mathematical model representing multipath transfer functions is associated a joint probability distribution with a specific mathematical structure. Moreover, in the case of two of these models, namely, Rummler's model and the normalized two-ray model, this probability distribution is completely specified by the values of two statistical coefficients. We derived these coefficients on a few experimental links from transfer functions measurements.

A true prediction model should ideally allow us to derive the values of these parameters as well as the occurrence (time percentage of applicability) of the mathematical model from the characteristics of the link under consideration. At the present time, the physical behavior of the propagation channel is not sufficiently understood to derive these relations theoretically, and transfer function data are too scarce to derive them experimentally. The best we can do, therefore, is to find a method to evaluate these three parameters on a given link from experiments simpler than transfer function measurements. The idea is the following one. Once the transfer function statistical model is completely specified, the channel is, from a radioelectrical point of view, perfectly known, and such information as the distribution of the field level at a fixed frequency or at several fixed frequencies can be derived. If this particular information varies enough with the values of the statistical parameters, we can hope to estimate the latter from the former. We explore this possibility in the present section.

B. Estimation of the Statistical Parameters from the Level Distribution at a Fixed Frequency

Many investigations have been concerned with the field level at a fixed frequency, leading to empirical prediction rules [1]. Moreover, this measurement is the simplest one to implement. It would thus be quite desirable to take it as a basis for an estimation procedure of the statistical parameters.

A little thought shows that there is some chance of success only in the context of Rummler's model. In the case of the normalized two-ray model, we have seen that all the information about the selectivity of the channel is concentrated on the behavior of τ; but at a given frequency ω, whatever the value of τ, the total phase $(\omega\tau + \phi)$ has the same distribution due to the freedom on ϕ, and thus, no information concerning τ can be inferred from the level at only one frequency.

Starting with Rummler's model, we simulated by random choice (with a sample size of 50 000) the distribution of the level[2] at the center frequency for several values of the statistical parameters, m_A varying from -5 to -20 dB (with a step of 5 dB) and ρ varying from 0 to 1 (with a step of 0.1). As our transfer function model is restricted by the threshold condition, we can simulate the fixed frequency level distribution only for attenuations greater than this threshold S. As a consequence, all simulated curves originate from the point of abscissa S decibels and ordinate 100 percent. On the other hand, it is known that at large attenuations the slope of the distribution curve tends towards one decade of probability for 10 dB of attenuation, all the curves becoming parallel to each other. Significant differences between the curves corresponding to different values of the statistical parameters thus appear principally in the range of attenuations between 10 dB and

[2]The level distribution curve must be understood as the curve whose ordinate gives the time percentage during which the attenuation is greater than that indicated by the abscissa.

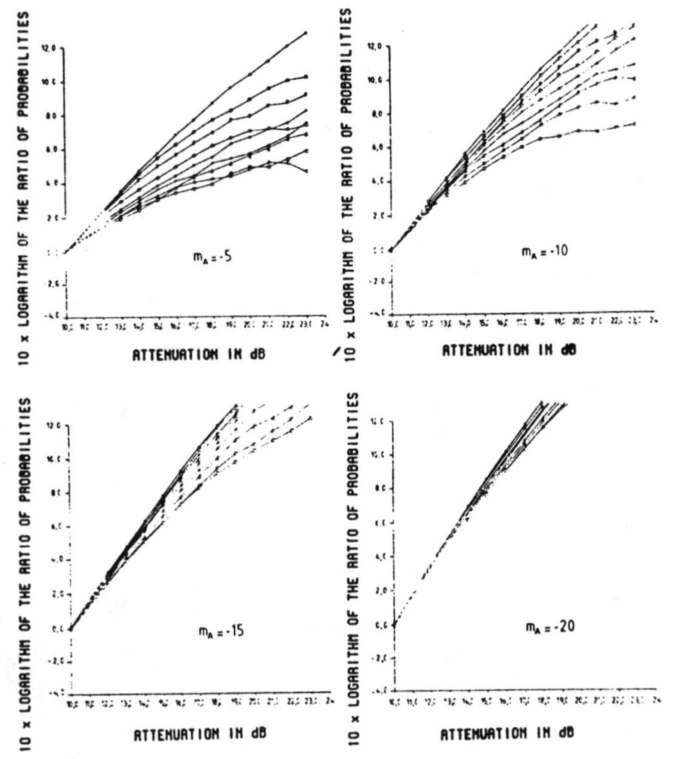

Fig. 4. Simulated curves of the normalized fixed frequency level distribution.

about 20 dB where these curves are not parallel. These differences are nevertheless not very important and do not appear clearly on the distribution curves themselves. To make them more apparent, we eliminated the general slope of all curves by using their difference in probability (on a logarithmic scale) with a reference curve, which was the one having a slope of 5 dB per decade of probability. We thus obtained the curves given on Fig. 4.

To obtain the values of the statistical parameters for a particular link from the experimental curve recorded on it,

1) we normalize the experimental curve as indicated above, and

2) we compare the experimental normalized curve to the simulated ones and look for the most similar one.

Examples of experimental normalized curves are given in Fig. 5.

At this point, let us make some comments. The parameter estimation is based on comparisons of curves. This method is not very precise for two main reasons.

1) As can be seen in Fig. 4, some of the simulated curves happen to be very close, and the result is thus somewhat ambiguous.

2) In the examples of Fig. 5, the experimental curves have the same shape as the simulated ones, making possible an identification. However, it occurs that the experimental curve has a different shape, and the parameter estimation is then very unreliable.

In the absence of data, it is tempting to use a fixed frequency prediction formula [13]. The concavity of the normalized curve is then opposite to that of the model because these prediction formulas are not well suited to attenuations less than 20 dB.

C. Computation of the Outage Time; Comparison with Experiment

Up to here, we have developed a modeling of the propagation channel independent from any transmission equipment.[3] The object of a statistical model is, however, to allow an evaluation of the quality of transmission of a link. This evaluation is done by computing the outage time, generally defined as the time when the BER is greater than a given limit (usually 10^{-3}). If T is the period of interest (for instance, an average month, or the worst month), the outage time can be expressed as

$$T_0 = T \cdot P_r(M) \cdot P_r(\text{BER} > 10^{-3} \mid M) \quad (21)$$

where $P_r(M)$ is the occurrence (fraction of time during which the model is applicable) and $P_r(\text{BER} > 10^{-3} \mid M)$ is the probability that a transfer function given by the model implies BER $> 10^{-3}$.

If the threshold has been correctly chosen, all the transfer functions likely to affect the link are taken into account by the model and

$$P_r(\text{BER} > 10^{-3} \mid M) = P_r(\text{BER} > 10^{-3})/P_r(M). \quad (22)$$

1) Evaluation of $P_r(BER > 10^{-3} \mid M)$: In order to compute this probability, we first need the signature of the transmission system. For a completely specified system, the signature is the part of the space of the representation model parameters leading to bad transmission (BER $> 10^{-3}$). If we call φ this domain,

$$P_r(\text{BER} > 10^{-3} \mid M) = \int_{\varphi} P_r(p_1, p_2, p_3)\, d\boldsymbol{P} \quad (23)$$

with $P_r(\vec{p})$, the probability law of the model parameters, given by (10).

In practice, it is difficult to describe analytically the domain φ, as well as the domain $\mathfrak{D}$ considered in the definition of $P_r(\vec{p})$; thus, (23) is computed by a Monte Carlo method. N triplets of the parameters of the model are drawn at random, according to the probability law of the model; if N' of them lead to a point inside the signature, we have

$$P_r(\text{BER} > 10^{-3} \mid M) = N'/N. \quad (24)$$

The only difficulty is in choosing the value of the sample size N in order to have a good estimate: it has been done empirically by comparing the result to that obtained when doubling N; a sample size of about 10 000 leads to satisfactory results.

2) Evaluation of $P_r(M)$: $P_r(M)$ is the probability that the threshold condition is satisfied, i.e., the probability of an attenuation greater than S (10 dB) on a bandwidth Δf

[3]Except for the bandwidth, and considering, strictly speaking, the antenna directivity as part of the channel.

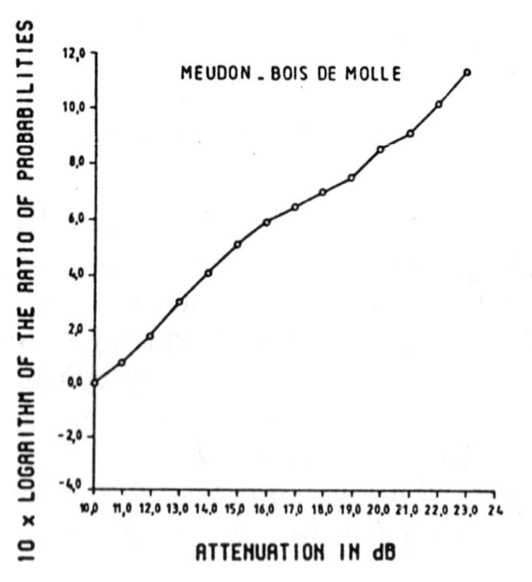

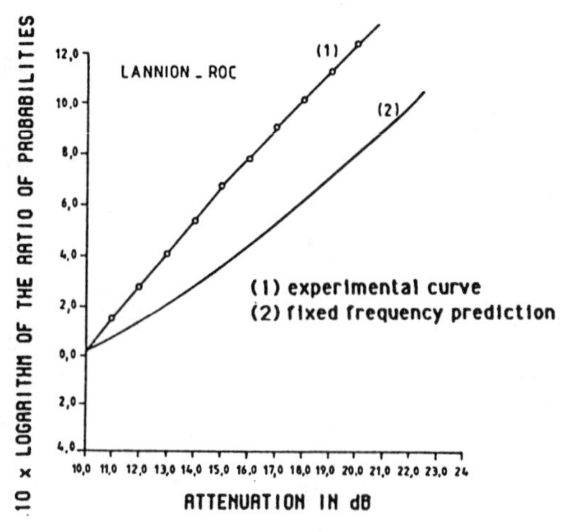

Fig. 5. Experimental normalized curve (fixed frequency level distribution).

(55 MHz). It is not known *a priori*, but can easily be derived if we know the transfer function statistical model and the experimental fixed frequency level distribution.

Let us assign the fixed frequency distribution to the center frequency. It is clear that each time the attenuation at the center frequency is greater than S the corresponding transfer function is taken into account by the model, but the reverse is not true. We can then simulate a sample of transfer functions obeying the model probability law and determine the fraction of them r for which the attenuation at the center frequency is greater than S. We have

$$P_r(M) = \frac{P_f(S)}{r} \tag{25}$$

with $P_f(S)$ being the probability of an attenuation greater than S at one frequency. Of course, the value of r depends on the values of the two other statistical parameters of the model.

3) Comparison with Experiment: To compare the results of the proposed method with experiment, we need a link for which we have both the level distribution at one frequency and the outage time.

A first comparison was done with data recorded from March to June 1982 on the Lannion–Roc link [14]. Table IV gives the details of the outage time computation and the comparison with the experimental result. There is an agreement within a factor of 2.

Other comparisons were made for some links of the French operational network. Results are given in Table V. In that case, we did not have the outage time (BER > 10^{-3}), but the time requirement for switching to the rescue channel (BER > 10^{-6}) (it does not change the method, but only the signature to be used). Besides, the BER was recorded at the end of a multihop link, when our method has to be applied to each hop separately; we had therefore to distribute the errors among the hops, which induces some uncertainties. However, the results are quite good for two of the links. The large discrepancy observed on the third one (Nancy–Reims) is attributed to a long-duration ground duct, inducing an atypical shape of the level distribution curve and a very unreliable estimation of the statistical parameters.

TABLE IV
STEPS OF CALCULUS OF PREDICTION FOR A TEST LINK AND COMPARISON TO EXPERIMENT

Hop: Lannion-Roc Treduder	Flat fade margin: 43dB	Period: March-June 1982	
Step 1	Identification of statistical parameters	$m_A = -15$ dB	$\rho = 0.3$
Step 2	Computation of r	r = 0.92	
Step 3	Computation of $P_r(M)$	10^{-2}	
Step 4	Computation of $P_r(\text{ber} > 10^{-3}/M)$	0.023	
Step 5	Computed outage time	2.3 10^{-4}	
Experimental outage time = 1.5 10^{-4}			

D. Estimation of the Statistical Parameters from the Level Distribution at Two Frequencies

If the level distribution at a fixed frequency allows a rough determination of the statistical parameters of Rummler's model, we must admit it is merely makeshift. Although fixed frequency level distribution is the only widely available data, it is worthwhile to look for more elaborate data leading to a more precise estimation of the statistical parameters. The next degree of complexity, yet liable to experiment, is the joint distribution of levels at two frequencies. The second frequency brings all the more information as it is decorrelated from the first one. This is why we chose two frequencies 50 MHz apart, near both ends of the frequency band of interest.

The method is similar to that using the one-frequency level distribution. We have simulated the two-frequency level distributions (with a random size of 50 000) for both Rummler's model (with m_A from -5 to -25 dB with a step of 1 dB and ρ from 0 to 0.9 with a step of 0.1) and the normalized two-ray model (with $\bar{\tau}$ from 0.1 to 1 ns

TABLE V
COMPARISONS OF PREDICTION AND MEASUREMENT OF OUTAGE TIME ON OPERATION LINKS

Link	Paris-Rouen	Poitiers-Limoges	Nancy-Reims
Period	24-10-84/13-11-84	24-9-84/3-10-84	8-11-84/5-12-84
ber>10^{-6} (prevision)	2.2 10^{-3}	8.35 10^{-4}	3.3 10^{-3}
ber>10^{-6} (measurement)	4 10^{-3}	8 10^{-4}	1.1 10^{-4}
$\tau = \frac{\text{measured outage time}}{\text{expected outage time}}$	1.8	0.96	0.03

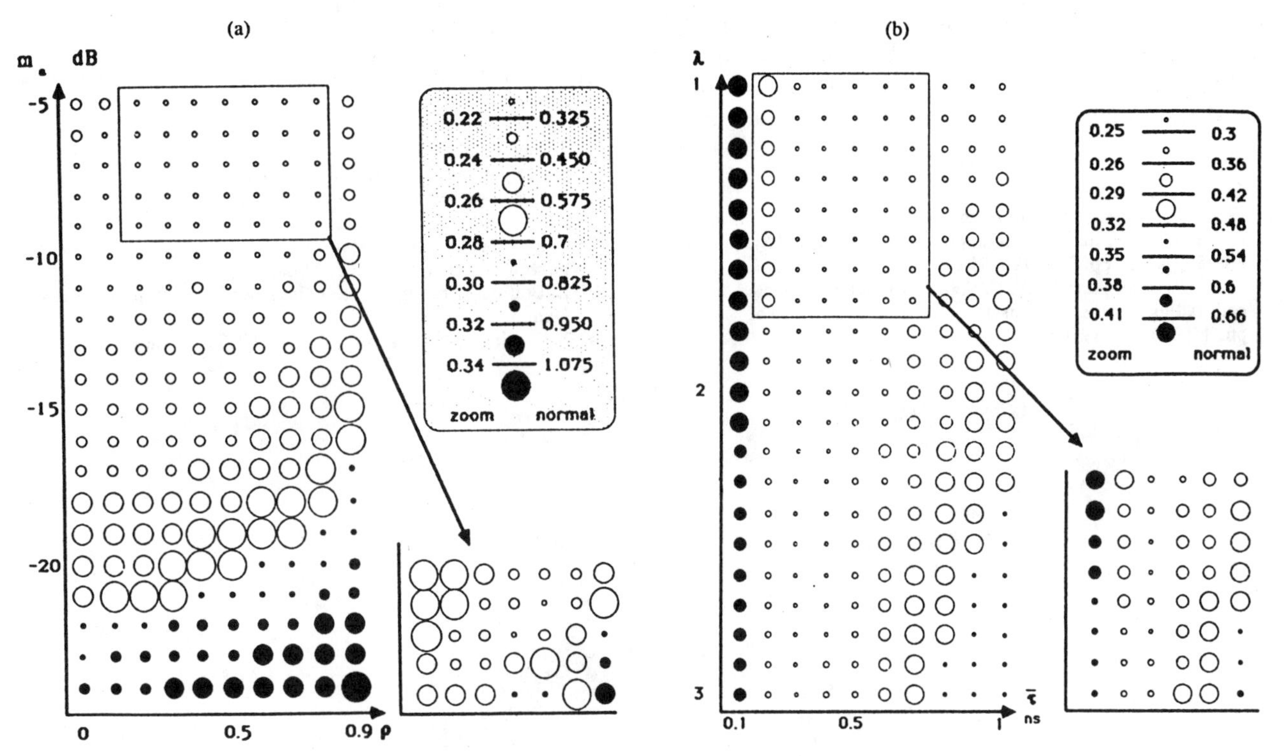

Fig. 6. χ^2 distance between the set of simulated joint distribution of the level at two frequencies, 50 MHz apart, and the experimental one. (a) This distance is represented versus the two parameters m_A and ρ of Rummler's model. (b) This distance is represented versus the two parameters $\bar{\tau}$ and λ of the normalized two-ray model.

with a step of 0.1 ns and λ from 1 to 3 with a step of 0.1). The range of variation of the parameters has been chosen to cover all the values obtained by direct estimation in Section III.

The experimental distribution obtained on the Marcheville–Viabon hop was then compared to the simulated ones by computing in each case the χ^2 distance, and the statistical parameters are estimated by minimizing this distance.

In the case of Rummler's model [Fig. 6(a)], we thus obtain

$$m_A = -6.5 \qquad \rho = 0.55.$$

In the case of the normalized two-ray model [Fig. 6(b)], we get

$$\bar{\tau} = 0.4 \qquad \lambda = 1.3.$$

In both cases, the estimation is without ambiguity and leads to values close to those obtained directly from the parameters' joint distribution. We thus have a simple and efficient method to characterize the multipath selectivity on a given link.

Unfortunately, the lack of experimental data prevents us from confirming this method by applying it to links which were not used to build it.

V. CONCLUSION

We have described in this paper a prediction method for the quality of a microwave link affected by MP. This method is based on a mathematical modeling of the propagation channel which applied fairly well to all links on which it was tested.

A good representation of the transfer function on a bandwidth of a few tens of megahertz is available with three-parameter mathematical functions. The statistical behavior of the multipath channel is then given by three statistical parameters: the first one, called occurrence, is the percentage of time during which the model applies; the two others specify the joint probability distribution of the mathematical model parameters and describe the selectivity of the channel.

We have shown that the values of these parameters can be roughly estimated from the level distribution at one frequency and more safely from the joint distribution of the levels at two frequencies.

It is now very desirable to develop measurements of the two-frequency level distribution on a large number of links in order to establish the relations between the values of the statistical parameters and the characteristics of the link, which remains the ultimate aim of MP studies.

References

[1] CCIR, Vol. V, Rep. 338.

[2] W. D. Rummler, "Time and frequency domain representation of multipath fading on line-of-sight microwave paths," *Bell Syst. Tech. J.*, vol. 59, no. 5, pp. 763-796, 1980.

[3] L. J. Greenstein and B. A. Czekaj, "A polynomial model for multipath fading channel responses," *Bell Syst. Tech. J.*, vol. 59, no. 7, pp. 1197-1225, 1980.

[4] M. H. Meyers, "Multipath fading characteristics of broadband radio channels," *IEEE Global Telecommun. Conf.*, Atlanta, GA, 1984, pp. 4.5.1.1-4.5.1.6.

[5] S. Mayrargue, "Caractérisation paramétrique de la fonction de transfert d'un canal radioélectrique en vue de l'évaluation du taux d'erreur," *Ann. Télécommun.*, vol. 40, no. 11-12, pp. 626-633, 1985.

[6] J. Lavergnat and M. Sylvain, "Analyse théorique d'un canal de propagation en présence de trajets multiples," *Ann. Télécommun.*, vol. 40, no. 11-12, pp. 572-583, 1985.

[7] M. Sylvain and J. Lavergnat, "Modelling the transfer function in medium bandwidth radio channels during multipath propagation," *Ann. Telecommun.*, vol. 40, no. 11-12, pp. 584-604, 1985.

[8] W. C. Jakes, Jr., "An approximate method to estimate an upper bound of the effect of multipath delay distortion on digital transmission," *IEEE Trans. Commun.*, vol. COM-27, pp. 76-81, 1979.

[9] A. J. Giger and W. T. Barnett, "Effects of multipath propagation on digital radio," *IEEE Trans. Commun.*, vol. COM-29, pp. 1345-1352, 1981.

[10] W. D. Rummler, "A new selective fading model: Application to propagation data," *Bell Syst. Tech. J.*, vol. 58, no. 5, pp. 1037-1071, 1979.

[11] J. Lavergnat and M. Sylvain, "Statistiques de la fonction de transfert par trajets multiples pour des largeurs de bande moyennes. Application à la prévision de la qualité," *Ann. Télécommun.*, vol. 40, no. 11-12, pp. 604-616, 1985.

[12] J. L. Chandon and S. Pinson, *Analyse Typologique: Théorie et Applications.* Paris, France: Masson, 1981.

[13] L. Boithias, *Propagation des Ondes Radioélectriques.* Paris, France: Dunod, 1983.

[14] J. C. Pinault, CNET/NT/LAB/MER/177, France, Note, 1984.

Part IV
Receiver Techniques

Receiver Techniques for Microwave Digital Radio

J. K. CHAMBERLAIN, F. M. CLAYTON, HIKMET SARI, AND PATRICK VANDAMME

Editor's Note: This tutorial article was originally published in the November 1986 issue of the IEEE COMMUNICATIONS MAGAZINE, as part of the Special Series on Microwave Digital Radio. It has been updated by the authors and edited for inclusion in this book.

INTRODUCTION

PREVIOUS leadoff tutorials in this book [1]–[3] have discussed methods of modulation and demodulation, and the problems associated with abnormal propagation conditions. This article describes the receiver techniques that have been developed to deal with linear signal distortions and interferences arising from anomalous propagation.

The receiver function is in many ways the most fundamental aspect of communication. In a digital radio system, it may be separated into three tasks, all of which are essential for accurate recovery of the transmitted data. The first is to demodulate the received signal, and requires the synchronization of a local reference (in both frequency and phase) to the carrier used in the modulation process in the transmitter. The second is to extract symbol timing so that the demodulated signal can be sampled in synchronism with the transmitter clock. The third function is to estimate the values of transmitted data symbols.

No matter how effective a digital radio receiver is in performing these tasks under normal propagation conditions, there can be major difficulties associated with its operation in the presence of multipath fading [3]. For the small fraction of the time that any given link is affected by such propagation anomalies, system availability may be drastically reduced unless appropriate countermeasures are employed.

In the following sections, various adaptive receiver techniques that have been proposed to combat multipath effects are considered. Equalization of channel distortion by frequency-domain and time-domain circuitry is covered first, followed by a discussion of carrier and symbol synchronization subsystems. Space and frequency diversity are treated next, and the final section deals with methods for adaptive cancellation of cross-polarized interference. A generalized block diagram of a receiver in which these techniques are implemented is shown in Fig. 1, with fixed filters, amplifiers, downconverters, etc., omitted to emphasize the circuits of interest.

ADAPTIVE CHANNEL EQUALIZATION

The first defense that suggests itself against the degrading effects of anomalous propagation is adaptive equalization—equalization arranged to adjust itself as the degradations themselves vary. The desired channel characteristics (amplitude and group-delay) will probably have been specified in the frequency domain. One class of equalizer, then, attempts to maintain them in their undegraded frequency-domain form. However, the proximate cause of channel failure is really propagation-induced intersymbol interference, which is a time-domain effect; time-domain signal processing is therefore, in a sense, the most natural technique on which to base equalization. The following sections give some account of these two broad classes of equalizers, noting, as a measure of their effectiveness, what modification they make to two-path fading "signatures"—the contour lines that separate regions of acceptable and unacceptable performance in the parameter space of the two-ray fading model [3].

Frequency-Domain Equalization

Basic Principle—If the channel equalizing function is realized at the intermediate frequency (IF) as an analog network, its transfer function will depend on a few (usually one or two) variable parameters. Since information concerning the channel group-delay response is not directly available, the equalizer is adjusted so as to minimize distortion in the overall channel amplitude-frequency response within the signal bandwidth.

Control is achieved by monitoring the output power spectrum at two or three frequencies, using a set of narrow-band filters, and comparing the measured powers with each other or with predetermined undistorted levels. A general block diagram applicable to frequency-domain equalization is given in Fig. 2(a).

From Slope to Notch Correction—The general nature of anomalous propagation [3] is such that, in the frequency domain, multipath distortion most often appears as "slope" asymmetries in the radio channel response. Consequently, the first type of frequency-domain equalizer to be developed [4]–[6] was a linear amplitude-slope equalizer, often referred to simply as a *slope equalizer.* Its function is to introduce an amplitude tilt correction which restores equality to the spectral density measured at the passband edges, as shown in Fig. 2(b). A slope equalizer is primarily able to compensate for frequency-selective fades where any attenuation "notch" lies outside the passband. It can match the amplitude characteristics that these produce; uncompensated group-delay distortions, being more concentrated around the notch frequency, are small and contribute only weakly to system performance degradation. Note that, since information on group-delay distortion is generally not obtained in the spectrum monitoring

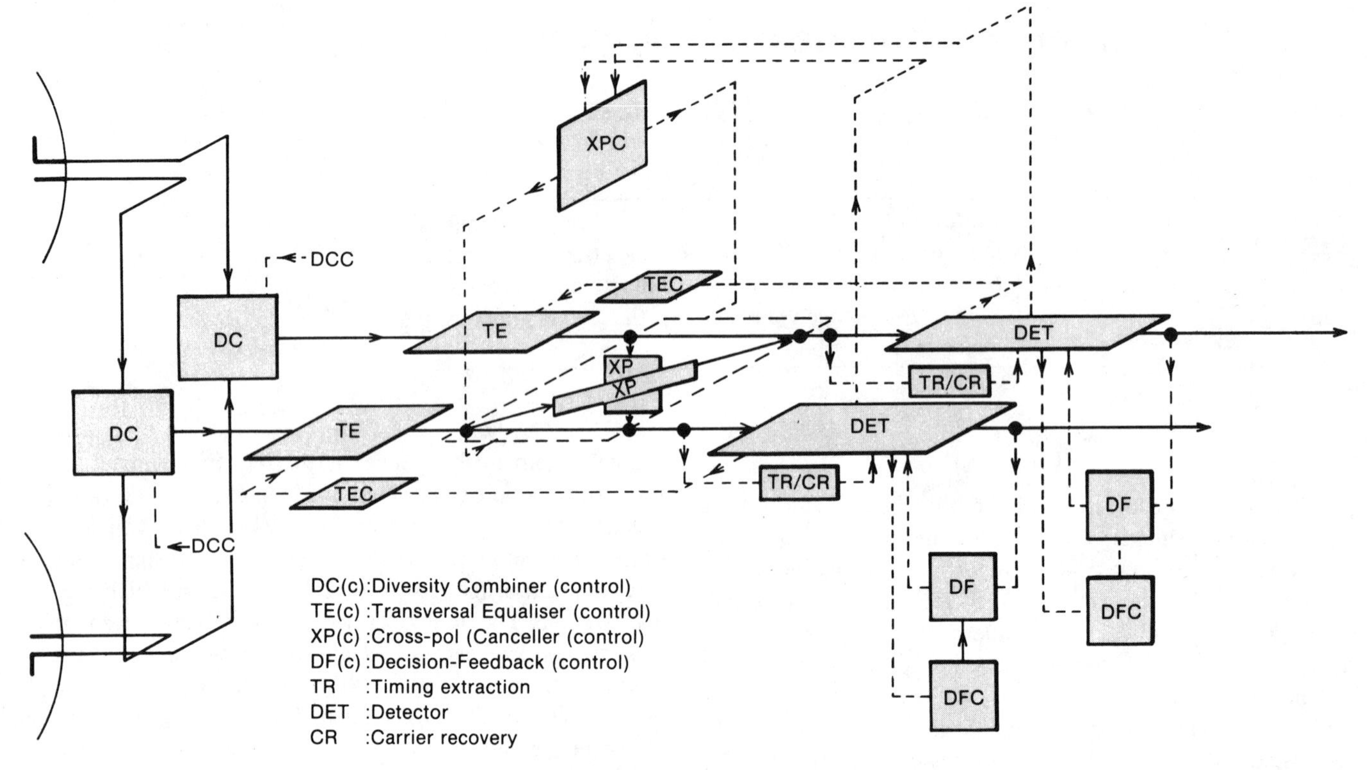

Fig. 1. Generalized Receiver with Adaptive Techniques.

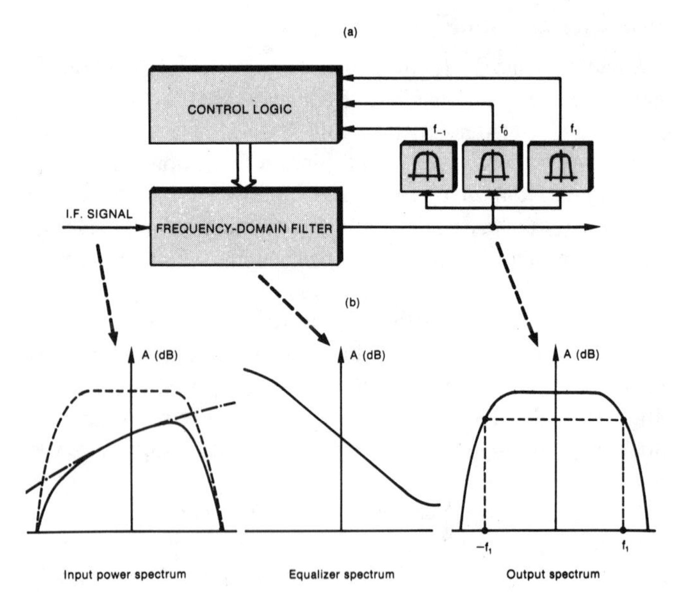

Fig. 2. Frequency-Domain Equalization.
a) Block diagram
b) Example of spectrum correction with a slope equalizer.

process, slope equalizers are usually designed with flat group-delay characteristics.

In contrast, little performance improvement is achieved for fades producing an in-band notch, since in that case a linear amplitude slope is a poor approximation to the distortions actually experienced by the channel. Consequently, the main effect of slope equalizers on system signatures is to reduce their frequency width.

Another class of frequency-domain equalizers attempted to produce transfer functions that approximate the inverse of channel characteristics conforming to a two-ray propagation model [3]. They consist of a resonator filter whose sharpness factor and center frequency are controlled to track the fade notch; hence, the generic name *notch equalizer* [7]. Such circuits always exhibit a concave group-delay characteristic. As a consequence, they tend to produce significant reductions in signal distortion when the channel experiences minimum-phase fading, but double the group-delay distortions for nonminimum-phase fading. System signatures for minimum-phase fades can be considerably reduced, while those for nonminimum-phase fades are roughly equivalent to what is obtained with a slope equalizer: they are somewhat narrowed, but otherwise generally not improved.

Performance in the Field—Frequency-domain equalizers are generally considered to be simple, low-cost countermeasures for selective fading, but their performance as observed in field-trial experiments is rather variable. Outage reduction factors of between 1.5 and 5 have been reported for channels operated with adaptive equalization only, but much larger factors (up to several hundred) can be obtained when both equalization and space diversity are used. The approximate nature of the equalization means that some residual distortion is generally unavoidable; nevertheless, the technique is sometimes adequate for systems with moderately complex modulations (9-QPRS, 8-PSK, 16-QAM). However, in bandwidth-efficient systems such as 64-QAM or 256-QAM, its role is more usually supplementary to the time-domain equalization discussed in the next section.

Time-Domain Equalization

Basic Principle—The basic principle of time-domain equalization can best be visualized by considering the overall channel impulse response (defined to include the filtering in the modulation and demodulation processes). Digital communication systems are commonly designed to make this response satisfy Nyquist's first criterion—having regular zero-crossings spaced at multiples of the symbol interval—so that at the receiver, the transmitted data symbols can be detected without any mutual interference. Multipath propagation distorts the impulse response, introducing intersymbol interference and hence reducing the system margin against thermal noise, interference, and imperfections.

In QAM systems (including QPSK), two separate data systems are transmitted orthogonally on a single carrier: one stream (usually designated $I(t)$) modulates cos $\omega_c t$, the other ($Q(t)$) modulates sin $\omega_c t$, and the two modulated signals are added [2]. Ideally, each data stream appears at its own baseband detector in the receiver without pulse distortion and without additive interference from the other stream. As a consequence, signal distortions can be conveniently divided into two classes. One is *in-phase* distortion, which results from the perturbations of the amplitude and group-delay responses that have even symmetry about the channel's center frequency and produce pulse distortions within each data stream separately. The other is *quadrature* distortion, which results from asymmetries in the channel responses that produce *cross-talk (cross-rail interference)* between the two streams.

To illustrate in-phase and quadrature distortions in a simple way, suppose that the undistorted signal at a given baseband detector is a rectangular pulse, and that two-ray fading occurs to produce a second received pulse, generally of different amplitude and RF phase, that is offset in time by a small fraction of the symbol period. The situation shown in Fig. 3(a) then obtains: the in-phase pulse shows partial cancellation between the delayed and undelayed pulses, and, except for the special case of the second ray arriving precisely in phase or in antiphase, a large quadrature distortion component is generated.

In practice, this idealized response is modified in two ways: common types of demodulator act so as to eliminate present-symbol quadrature cross-talk at the decision instant (Fig. 3(b)); and practical filtering smooths the pulse shapes. To illustrate these effects, the responses of Figs. 3(c)–3(f) were calculated for a QPSK system, with equipment imperfections ignored. They show the distortion produced when a multipath notch of fixed depth is offset by various amounts from the channel center frequency [3]. In each case, the dark circles represent the decision instant for which the present-symbol cross-talk is zero. The crosses, which are separated from the decision instant by multiples of the symbol period, indicate the interference that would be received from previous pulses (samples to the right, called *postcursors*) and future pulses (samples to the left, called *precursors*).

From this discussion it is clear that, to be of value, a time-domain equalizer must be able to compensate for both in-phase and quadrature distortion. A typical scheme to achieve this is shown in its baseband form in Fig. 4, where the interconnected adaptive networks are all transversal filters with a tapped delay-line structure. (For more details, see [8].) Conceptually, the total equalizer may conveniently be divided into two parts. The feedforward (or *nonrecursive*) section consists of the $F_{ij}(\omega)$ responses. The feedback section comprises the $G_{ij}(\omega)$ responses, and may be arranged as shown to take its input alternatively from before or after the decision device. Thus, with the general structure of Fig. 4, either linear or nonlinear operation is possible; their relative merits will be considered next.

Linear Equalizers—Although recursive linear equalizers are used in some digital radio systems [9], this discussion will be confined to the most common time-domain equalizer structure—the nonrecursive linear equalizer [10]–[13]. This type can be implemented either in the IF stage of the receiver [10], or at baseband. IF implementation has advantages with respect to carrier recovery (see below), but baseband equalizers are sometimes preferred for technological reasons, and additionally provide compensation for asymmetrical modem imperfections. A baseband nonrecursive linear equalizer has the structure of Fig. 4 without the feedback section.

The capability of a linear equalizer to reduce intersymbol interference is determined by the number of delay-line taps and the position of the reference tap.* For digital radio applications, typically 5 or 7 taps are used, with the center tap chosen as reference so as to give equal effectiveness against distortion from both minimum- and nonminimum-phase fading. As a result of their ability to compensate for group-delay distortions, linear transversal equalizers at this level of complexity provide a somewhat better performance than frequency-domain equalizers—in a 140-Mb/s 16-QAM system, a two-path fade with 6.3 ns delay and notch depth approaching 20 dB can be accommodated. However, there are two basic limitations to what these equalizers can achieve in the presence of deep fades. First, optimal filtering for severely distorted channel transfer functions can only be achieved by using many taps; and second, in attempting to contrive the necessary equalization, the control algorithms may produce gain compensations that lead to a significant enhancement of receiver noise or adjacent-channel interference. Thus, this type of equalizer is generally most effective in combination with space diversity (see the section titled "Synergy of Adaptive Equalization and Diversity Combining").

Decision Feedback Equalizers—This nonlinear equalization technique is based on direct cancellation of the intersymbol interference from previously detected symbols, and consequently is always realized as a baseband subsystem. A decision feedback equalizer has the structure shown in Fig. 4, arranged so that the inputs to the filters G_{ij} are regenerated data symbols (switch in position 1). All the filters are optimized concurrently. The linear feed-forward section of the equalizer attempts to compensate only for precursor intersymbol interference. Postcursor interference, including any associated with the operation of the feed-forward section, is cancelled in the feedback part, and, since this latter cancellation is noise-free, effective equalization is possible even for fades that produce an infinite in-band notch. The main

* At any given sampling instant, the symbol being decided on at the detector is also in transit on the tapped delay line. The *reference tap* is the one at which this symbol is located.

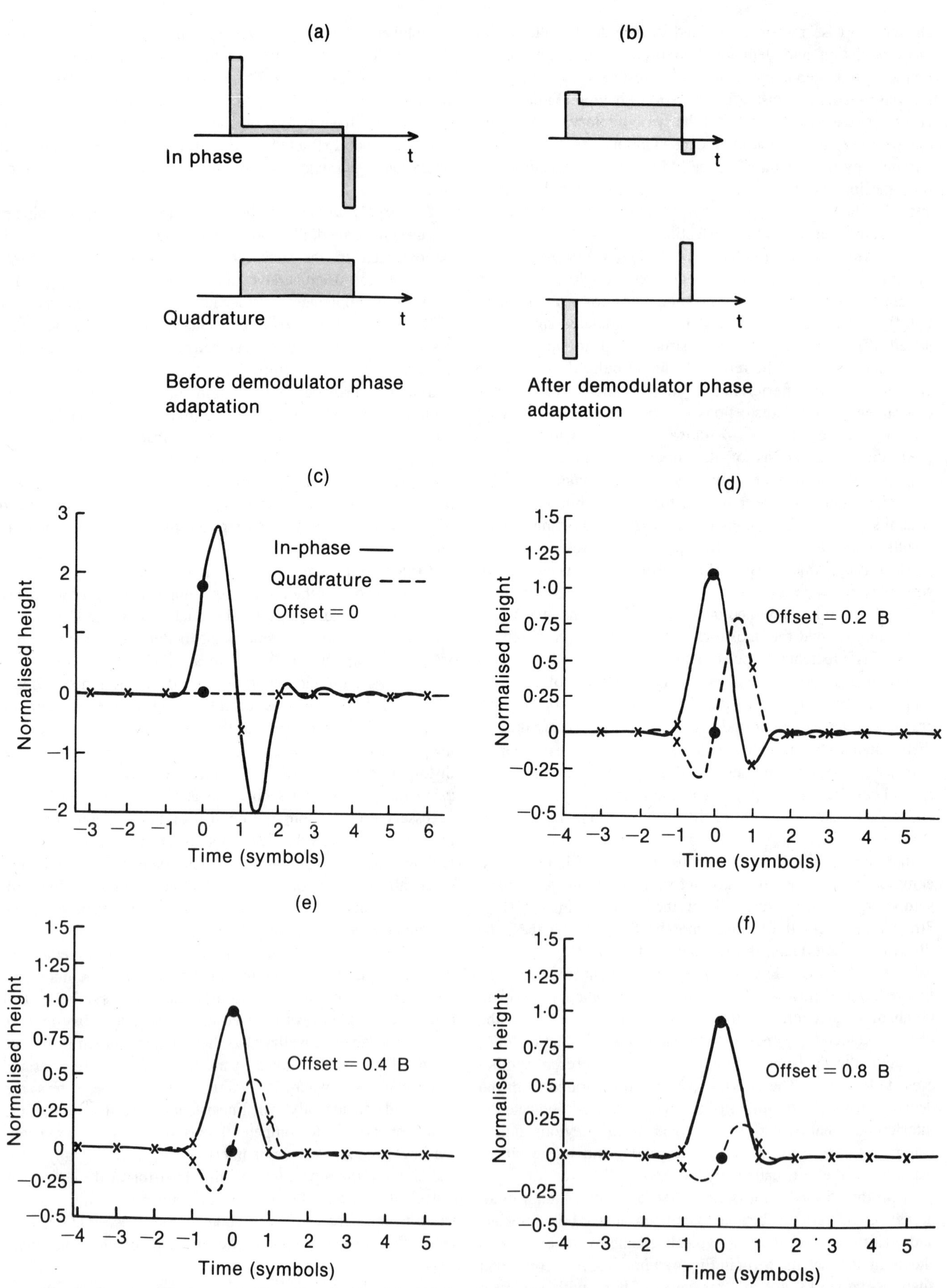

Fig. 3. Receiver Response to Two-Ray Fading (the "Offset" is the Notch Frequency Relative to the Channel Center and B is the Reciprocal of the Symbol Period).

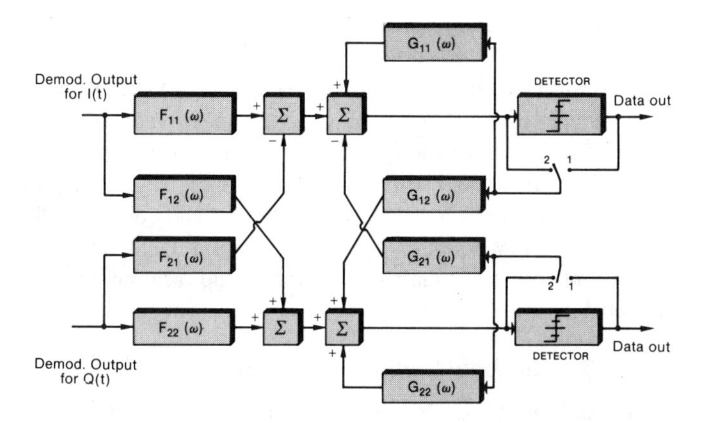

Fig. 4. Generalized Block Diagram of Baseband Equalizers for QAM Systems. (Switch in Position 2 for Recursive Linear Equalizers and Position 1 for Decision Feedback Equalizers.)

drawback is that any regeneration errors give rise to doubled intersymbol interference and the possibility of error propagation; fortunately, this has been found to have only a marginal effect on overall performance.

The decision feedback equalizers used in digital radio systems typically comprise one or two feedback taps and two or three feedforward taps. This combination gives good performance against minimum-phase fades, but in nonminimum-phase fading, when precursor intersymbol interference dominates, its behavior is similar to that of a linear equalizer. Thus, the potential improvement to be gained from using a nonlinear equalizer depends on the relative probability of occurrence of minimum- and nonminimum-phase fades. Measurements indicate that for noncatastrophic fades minimum-phase conditions occur more frequently. However, further field results are needed to establish the extent to which the theoretical superiority of nonlinear equalization is realized in practice.

Zero-Forcing Criterion vs. Minimum Mean-Square Error Criterion—So far, this discussion of time-domain equalizers has concentrated on their structure as determining theoretical performance. How far this potential can be realized in practice depends critically on the means devised to adapt the equalizer's characteristics to those of the time-varying channel. It should be emphasized that, in comparison with the rapid signaling rates in high-capacity digital radio systems, the time variations of the channel occur very slowly and consequently can be tracked by simple adaptation algorithms.

The first method of control to come into common use was based on cancellation of intersymbol interference—called "zero-forcing." Each adaptive tap in the equalizer cancels one sample of the channel impulse response, and the reference tap provides the wanted main sample. However, convergence of the control algorithm is not guaranteed if the data eye pattern at the equalizer input is closed, and no account is taken of additive noise or possibly enhanced interference contributions from symbols lying outside the span of the equalizer.

Better performance can be achieved if the tap gains are adjusted to minimize the mean square error at the equalizer output, defined as the sum of the intersymbol interference and noise powers. This forms the basis for the "stochastic gradient" algorithm (also known as "least mean squares"), developed in the early 1970s. Furthermore, since the mean square error is a quadratic function of the tap gains, convergence is always guaranteed, independent of the level of noise or interference.

Despite the superior performance of the stochastic gradient algorithm, the zero-forcing technique is almost universally used in microwave digital radio applications. This is certainly due to the simplicity of its implementation. The algorithm control signals are obtained by correlating the signs of the detected data symbols with the sign of the instantaneous error signal at the equalizer output [14]. Thus, all the required signals are available at the equalizer output, and tap-gain updating circuitry can consist of a few logic gates. In contrast, the control signal in the stochastic gradient algorithm is obtained by correlating the output error signal with the signal values present in the equalizer delay line. The process may be simplified by considering only the signs of the signals involved, but still requires more complex circuitry.

It must also be pointed out that the reduction of system outage is not simply determined by the form of the equalizer, but depends, too, on the robustness of the timing and carrier synchronization schemes, and how they interact with equalizer operation. These issues are considered next.

Synchronization

Before symbol regeneration can take place, two essential functions must be performed by the receiver in a coherent digital communication system: synchronization of the carrier wave (*carrier recovery*) and symbol timing (*timing recovery*). For correct performance, the locally generated reference waveforms must be accurately phase-locked to those produced in the transmitter. Both recovery circuits must extract their references from the received signal and, to avoid excessive outage, should be robust to the distortions introduced by multipath fading. If synchronization is lost, then, when conditions improve, reacquisition should be as rapid as possible over the full range of frequency uncertainty in the incoming signals.

Adaptive Timing Recovery

Uncertainty in the frequency of the timing clock can be kept rather small, typically of the order of one part in 10^6. However, as well as maintaining a low value of steady state phase jitter during normal propagation conditions, the timing recovery circuit must be robust to selective fades, and must track as necessary to ensure a near-optimum sampling instant.

Most equalizers used in digital radio systems are "synchronous"—that is to say they have symbol-period tap spacing—and it is well-known that their performance is very dependent on the sampling instant used [15], [16]. Transversal equalizers have transfer functions that are periodic in the frequency domain, and, in synchronous equalizers, the period is equal to the Nyquist bandwidth (the reciprocal of the symbol period). On the other hand, radio systems commonly employ Nyquist filtering with between 30 and 50 percent of excess bandwidth. Consequently, a synchronous equalizer cannot make independent corrections in both the Nyquist bandwidth and the excess bandwidth, and so exact equalization of this type of channel is generally impossible. To see how the residual distortion is influenced by shifts in the symbol timing phase, note that, because of the sampled detection process, the equalizer effectively operates on the folded channel spectrum, whose shape depends on the phase of the sampling clock [15], [16]. An improper sampling instant may lead to a folded spectrum with deep depressions or even spectral nulls, and this considerably limits the performance obtainable from a synchronous equalizer.

Fractionally-spaced equalizers [17] are in principle able to overcome this difficulty. For example, consider a system designed with a raised-cosine spectral rolloff [2], in which case the signal bandwidth never exceeds twice the Nyquist bandwidth. An equalizer with half-symbol-period tap-spacing synthesizes a transfer function with a period equal to twice the Nyquist bandwidth, and is insensitive to changes in the sampling instant. However, control of these equalizers presents various problems which are not yet fully resolved.

To optimize the performance of synchronous equalizers, an appropriate sampling instant must be selected. Various schemes have been proposed for tracking its optimum position during fading events; maximization of the sampled signal energy, tracking of the zero-crossings, and minimization of the output mean-square-error are among the more widely used. If applied at the equalizer output, most of these lead to almost the same near-optimal sampling instant. In addition, it seems that using the equalized signal in this way tends to improve the general robustness of timing recovery schemes in the presence of selective fading.

Robust Carrier Recovery

From power-efficiency considerations, it is desirable that the reference carrier needed for coherent demodulation should be extracted from the modulated signal available at the receiver. The frequency uncertainty of this carrier is determined by the stability of the microwave local oscillators used for IF/RF up- and down-conversion, and is more than an order of magnitude larger than that of the symbol timing clock. To acquire lock over a relatively large frequency range, digital radio receivers employ acquisition-aiding techniques such as frequency sweeping. This consists in applying an additional periodic signal to the VCO of the carrier-recovery loop, until the carrier is acquired. The mean acquisition time is then determined by the frequency of the sweeping signal.

The other crucial aspect of carrier recovery is steady-state performance. As the complexity of high-level modulation techniques increases, so does their sensitivity to phase jitter, and the carrier-recovery loop must provide correspondingly tighter control. For normal propagation conditions, this is achieved by choosing the loop parameters appropriately. However, any signal distortions induced by multipath fading will increase the loop noise spectral density and its noise bandwidth, and will reduce the phase detector gain.

As in timing recovery, to improve robustness against selective fading, the signals used for carrier recovery should be derived from the output of any adaptive channel equalization. This presents no problems for IF equalizers, but means that, for baseband implementation, the equalizer (or its feedforward part in a decision-feedback scheme) will be inside the carrier-recovery loop. Excessive delays must then be avoided to prevent consequent degradations in loop performance. Care also needs to be taken to minimize any interaction between the equalizer adaptation algorithm and control of the carrier-recovery loop; selective correlation techniques and constraint of the equalizer reference tap are commonly-adopted measures.

Time-Domain Equalization in Practice

It is difficult to make a reliable estimate of the performance improvement that would result from adding time-domain equalization to a digital radio system which is otherwise unprotected against multipath propagation. This is because most published data from field-trial experiments refer to systems already equipped with space diversity. Indeed, equalization was first seen as a supplementary countermeasure to be used on paths where space diversity was not powerful enough to ensure that performance requirements were met; it is only in recent years that the idea of using time-domain equalization as a self-sufficient technique for medium-length paths has been considered.

The available data [9], [18] indicate that the outage-time reduction factor for linear equalizers does not exceed 5. In general, this is considerably lower than the reduction factors suggested by current theoretical prediction methods. However, these usually rely on data from system signatures measured with static channel characteristics at high signal-to-noise ratios, which, it is now commonly agreed, do not fully represent the effects of fading on practical microwave links. The following paragraphs discuss some other important factors that need to be taken into account.

First, a large body of propagation data indicates that deep selective fading is often accompanied by a significant flat fade (sometimes referred to as the *median signal depression*). This reduction in received signal power, when combined with enhancement of noise and interference as a result of linear equalization, can produce unacceptable bit-error ratios, even if intersymbol interference has been cancelled. It can therefore contribute significantly to system outage, particularly on paths with a limited fade margin.

Another way in which system performance may be degraded is through *hysteresis* effects in the receiver synchronization processes. If the propagation-induced distortion increases beyond the compensation ability of the equalizer, the system bit-error ratio may degrade to the point where the carrier recovery circuit loses synchronism, which in turn may cause the equalizer control algorithm to diverge. Such severe propagation events are usually of short duration. However, the time necessary for the system to recover satisfactory transmission quality will depend on the re-synchronization ability of the various adaptive circuits, and may contribute significantly to the total system outage. The robustness of a system in these conditions can be characterized by two ''synchronization signatures.'' One corresponds to values of the fade parameters for which carrier synchronization loss occurs, the other to the

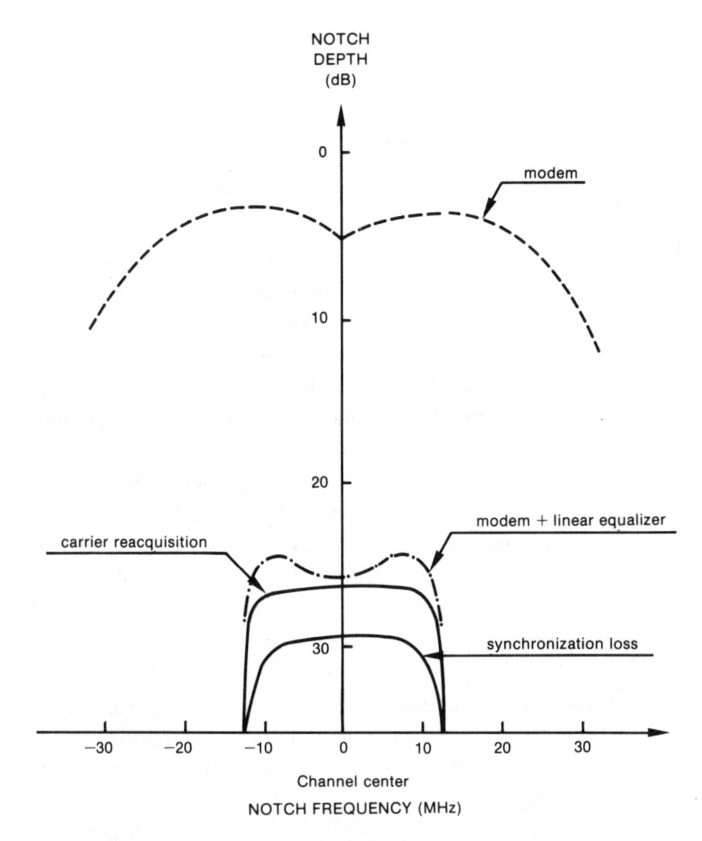

Fig. 5. Example Set of Signatures for a 140-Mb/s 16-QAM System. (Dashed Curves Represent Conventional Signatures Without and with Linear Equalizers.)

values for which carrier acquisition is possible again. An example of such a pair of signatures for a 140-Mb/s 16-QAM system is given in Fig. 5. In a well-designed system, these two signatures should be as nearly identical as possible; to minimize any hysteresis effects, the equalizer control has to be constrained so that its convergence is assured even when carrier synchronization is lost [9], [19].

Finally, the dynamic behavior of the channel may also be an important influence on the equalizer improvement factor. Maximum rates of variation are usually considered to be 100 dB/s for a fade notch deepening at a fixed frequency, and 100 MHz/s for a notch moving across the band. For high-capacity digital radio systems, these figures still represent a slowly-varying channel, capable of being adequately tracked. However, difficulties may arise when the channel experiences nonminimum-phase fading, particularly if this succeeds minimum-phase fading via an ''infinite-notch'' transition, resulting in a rapid polarity inversion of the group-delay characteristics [20]. Both equalizer control and timing recovery circuits have difficulty maintaining lock through such events.

Diversity Reception

The Field Available for Reception

A previous tutorial in this book [3] has described the digital radio channel, and has explained how its characteristics may be distorted or otherwise modified in the process of propagation through the atmosphere. In particular, atmospheric inhomogeneities, or reflections from terrain or water, can give rise to interference patterns in the vicinity of the receiving station. Since the surfaces and atmospheric layers that cause them are comparatively uniform in the horizontal sense, these patterns have a predominantly vertical variation; and, like all interference effects that depend on differential path lengths, they vary with frequency. Taking this simplified view, and assuming that relative signal delay due to differential feeder lengths is separately accounted for, we may therefore regard a receiving antenna as drawing its signal from a space of only two dimensions—one of frequency, and the other corresponding to height. At unfavorable positions in this space, where gradients of amplitude and phase are high, signal distortion may be severe.

Selecting a Better Signal

The most obvious way to safeguard signals against destruction by propagation effects is to have available alternative segments in the height-frequency space, selecting at any moment whichever offers the best signal. This is one form of diversity protection that has long been employed in analog radio systems, with the choice between alternatives usually being made on the basis of detected signal strength. For digital systems, however, the correlation between signal strength and the primary performance measure, bit-error ratio, is not good; this quantity itself is therefore normally measured and used as a criterion for switching.

The alternatives from which the choice is made may employ the same band of frequencies (*channel*) but different heights, or different channels with the same height. The first arrangement, called *selection space diversity*, requires at least one additional antenna, spaced vertically from the main antenna by a distance that can range from several meters to a few tens of meters; this represents a significant addition to the cost of the installation.

The second arrangement—*frequency diversity*—avoids this expense, but requires additional receivers (and corresponding transmitters) at the appropriate alternative frequencies; these may be used to protect more than one primary channel in an ''m for n'' arrangement. A disadvantage of this form of diversity protection is that it reduces the number of active channels that can be accommodated in a given frequency band, though this may be offset by integrating it with the function of protection switching to alternative channels on equipment failure.

Both forms of selection diversity are reported to give substantial reductions in propagation-induced outage time—in several cases between one and two orders of magnitude [21]–[23]. An extreme form of frequency diversity that switches between widely separated frequency bands gives comparable improvements [24].

Continuous Combining: Cooperation, not Competition

It will sometimes happen that samples of the signal field for a particular channel taken by antennas at different heights will both exhibit the same fading degradations. At such times, no means of improvement is available; but this is rarely the case because correlation between field samples is low if the antennas are well separated. More commonly—apart from median-level-depression effects associated with radio-refractive-index anomalies—each component frequency has a usable level on at least one antenna. It must then generally be possible in principle to obtain a better signal for the regenerator by judiciously combining the separate samples, rather than merely choosing between them as alternatives as in selection space diversity.

If there were sufficient information available about the paths taken by the signal, an optimum channel could be synthesized from an appropriate frequency-dependent blending of the channels associated with the individual antennas. Again in principle, such information could be obtained to some approximation through the operation of adaptive equalizers on the channels. This degree of sophistication has not yet been reached; so far, the development of digital-radio diversity combining has been limited to schemes for processing two diversity signals with at most two adjustable variables—their relative attenuation and phase shift.

Maximum-power combining, carried over from analog practice, was the earliest method employed in digital radio systems, and is still common. This method relies on the fact that at its maximum value—achieved by adjustment of the relative phase of the component signals—the combined signal power has a stationary point with respect to small perturbations of that phase [25]. Perturbations are impressed on one of the components at the receiver input, and the detected combined power is used to optimize the setting of the combiner phase shifter in a feedback loop. Field results for outage times achievable with this method of combining show reduction factors of between a half and one-and-a-half orders of magnitude relative to nondiversity [26].

Effective as these results show the method to be, it operates on a criterion that generally fails to minimize propagation-induced channel distortion, and to that extent it is suboptimal. A number of more recent schemes, implemented and proposed, address this minimization directly.

An early approach [27] consisted in adjusting the relative phase of the main and diversity signals in accordance with the amplitude of certain frequency components of the signals detected before and after combining. In terms of the simplest two-ray model of multipath fading, this strategy results in mutual cancellation of the echo signals entering the main and diversity antennas, and hence, in undistorted combination of the main rays at the combiner output. Useful improvements relative to maximum-power combining have been claimed for this method, but in its basic form it has two deficiencies. First, if a fading situation, however mild, produces fields at the two antennas that are similar over the frequency range of the channel, the control strategy will ensure that not only the echo signals, but also the wanted main signals, will mutually cancel, leaving no useful resultant. This has to be prevented by reverting to maximum-power combining under such circumstances. Second, the control strategy loses some of its effectiveness if fading is not of the simple two-ray type.

As high-level quadrature-amplitude-modulation methods (16-QAM, 64-QAM, 256-QAM) come into use, it becomes increasingly important to avoid signal distortions which collapse the multilevel eye patterns. Thus, it is natural that the next phase of combiner development should be concerned with maintaining the integrity of channel characteristics during multipath fading. Control is achieved by methods similar to those employed for frequency-domain equalization—detecting the power present in two or more sub-bands of the combined channel and adjusting the relative phase of the main and diversity signals according to the result of the measurement. Initially [25], [28], this has been done simply so as to force first-order symmetry and thereby minimize damaging quadrature crosstalk. The problem of signal loss when both antennas receive similarly faded signals remains, and is overcome by reverting to maximum-power combining, as before. Diversity improvement factors of more than an order of magnitude have been reported [28].

All the schemes described so far have availed themselves of only one variable to optimize combining—relative phase of the main and diversity signals. Interesting proposals have recently been made [29], [30] for including the adjustment of relative signal amplitudes in the combining control process as well. This not only allows more general forms of multipath distortion to be corrected, thereby easing the task of any subsequent adaptive equalizer, but also, by taking account of channel noise, overcomes the signal-loss difficulties mentioned earlier. No field results have yet been reported for this type of scheme.

Synergy of Adaptive Equalization and Diversity Combining

On many paths, propagation degradations are so severe for some types of digital signal that neither diversity nor adaptive equalization, used alone, is sufficient to keep outage time within acceptable limits. In combination, however, these two countermeasures are nearly always adequate, it being found that their combined improvement factor is usually higher than the product of the individual factors: the two act together synergistically. Various reasons have been proposed for this effect [26, Sec. 4.3]; they mostly depend on the idea that a diversity combiner conditions the signal that the equalizer must handle in a way that facilitates its mode of action.

Cancellation of Cross-Polarization Interference

Depolarization in Anomalous Propagation: The Need for Countermeasures

In considering the use of co-frequency dual-polarization operation to double the capacity of digital radio systems, it is important to recognize that it is not sufficient simply to require high cross-polarization discrimination (XPD) from the antennas in normal propagation conditions; degraded cross-polarization behavior of the radio channel during multipath fading may also severely limit system performance [31]–[33]. The most serious deterioration of XPD is usually associated with deep fading of the co-polar signal, and can be ascribed to two distinct types of mechanism. Scattering, either from refractive inhomogeneities or from the ground or water surface along the path, produces a cross-polarized signal that is essentially independent of the cross-polar properties of the antennas [31], [34]. Potentially more damaging effects arise from interactions between rays (either direct or reflected) arriving at off-axis angles, and then the detailed shape of the antenna cross-polar radiation pattern is important [35], [36]. In either case, it seems that antenna design and path planning cannot always be relied upon to eliminate the problem; some form of adaptive cancellation of the unwanted cross-polarized signal is required for difficult paths.

Cross-Polarization Interference Cancellation: The Basic Principles

The possibility of adaptive cross-polar interference cancellation depends on two conditions: the ability to make an accurate estimate of the parameters of the interfering signal (size, distortion, and so on), and the availability of a signal from the orthogonal receiver that can be processed to provide the cancelling waveform (Fig. 6). Although, in principle, cancellation can take place at RF, IF, or baseband, it is difficult to achieve more than simple amplitude and phase adjustment with any accuracy at RF, and consequently the more sophisticated cancellers tend to operate at IF or baseband. The IF choice offers several advantages from a practical point of view [37]: it is relatively easy to incorporate into existing equipment; components need only cover a narrow fractional bandwidth; and only a single transversal filter is required (per signal) for cross-polar interference cancellation. However, some sophistication [38] is necessary to allow asynchronous operation of local oscillators in the receivers associated with orthogonally-polarized signals.

Two levels of cancellation complexity can be identified. One is *simple cancellation*, which relies on the assumption that the interfering signal is an essentially undistorted replica of the orthogonally-polarized transmission; the other is *full cancellation*, which attempts to take account of the fact that the interference may have undergone frequency-selective distortion in the depolarization process. In simple cancellation, the canceller transfer functions $K_{ij}(\omega)$ in Fig. 6 reduce to nondispersive (complex) gains; in full cancellation, these functions are in general dispersive. Both implementation and control are obviously more complex in the latter type of canceller. This is offset, however, by its potential ability to deal with a much wider class of depolarization events, particularly those associated with frequency-selective propagation; expected improvements in overall system performance are therefore larger.

The mode of operation of an interference canceller is perhaps best explained as a generalization of the channel equalization described earlier. Thus, the canceller transfer functions $K_{ij}(\omega)$ in Fig. 6 must be chosen so that the output signals $O_1(\omega)$ and $O_2(\omega)$ are, respectively, good approximations to the original transmitted signals $V_1(\omega)$ and $V_2(\omega)$, on the basis of a specified error criterion (for example, least-mean-square or zero-forcing [39]). As the bottom matrix in the figure shows, there are enough degrees of freedom in specifying the $K_{ij}(\omega)$ responses to achieve this result, at least in principle.

Distortion of the wanted transmission as a result of multipath propagation may cause practical problems in the cancellation process. Consequently, the use of adaptive

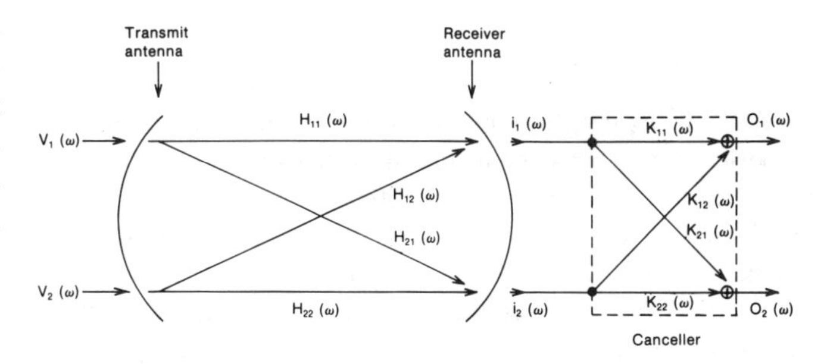

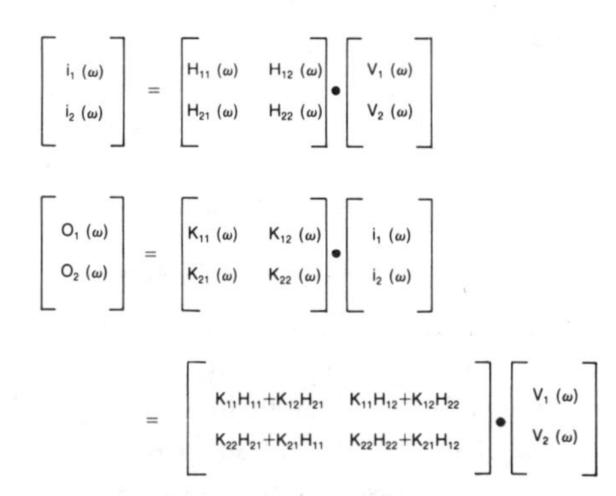

$$= \begin{bmatrix} K_{11}H_{11}+K_{12}H_{21} & K_{11}H_{12}+K_{12}H_{22} \\ K_{22}H_{21}+K_{21}H_{11} & K_{22}H_{22}+K_{21}H_{12} \end{bmatrix} \bullet \begin{bmatrix} V_1(\omega) \\ V_2(\omega) \end{bmatrix}$$

Fig. 6. Cross-Polarization Coupling and Cancellation.

channel equalization in combination with an interference canceller is now being suggested as the preferred method of producing an adequately-corrected signal in all but the most severe fading conditions.

Control: The Key to Effectiveness

Pilot Tone Techniques—Initial ideas for canceller control [40], [41] relied on the assumed transmission of "beacon signals," that is, pilot tones that could be extracted in the receiver and used to decide the relative amplitude and phase of the signals combined to achieve cancellation. This principle is still finding application in systems with appropriately designed modems [42], and appears to provide useful cancellation even in the presence of significant linear distortion.

Correlation Techniques—The use of correlation as a canceller control technique appeared in an early proposal for cancellation of nondispersive cross-polar interference [37]. Subsequently, in considering control algorithms for frequency selective cancellers implemented as cross-connected transversal filters, it became clear that there are significant similarities between these algorithms [39] and the techniques of channel equalization discussed earlier. These ideas have been brought together in [43]. This article describes a transversal equalizer in which the distortion-correction and interference-cancellation functions are combined, and whose control is achieved through any of the methods (for example, zero forcing, stochastic gradient algorithm) appropriate to this class of linear filter. The power and flexibility of such an equalizer/interference canceller combination is readily apparent, and, following successful practical implementations of IF designs [43], [44], it may well find wide application in future high-speed digital radio systems.

Field Experience: Does Cross-Polarization Interference Cancellation Work?

There is, as yet, little published information on operational experience with cross-polar interference cancellers. However, two trials have been reported that indicate the order of performance improvement that can be obtained. Tests of a simple canceller on a 90-Mb/s, 6-GHz, 8-PSK system over a 35.4-mile path in the southern USA [45] showed an outage improvement factor of 2.5; and recent experiments with more sophisticated cancellers in a 200-Mb/s, multi-carrier, 16-QAM system on a difficult over-water path in Japan [44] produced outage improvement factors of 3 and 5, depending on the number of transversal filter taps used. Clearly, more work in this area is needed, and may be expected, as the deployment of dual-polar systems to double digital radio capacity becomes more widespread.

Conclusion

In this article, we have described some of the adaptive receiver techniques that have been devised to combat the damaging effects of anomalous propagation in digital radio systems. The variety and sophistication of these techniques continue to increase rapidly, as evidenced by recent papers on angle-diversity reception [46], [47] and on more general schemes for cross-polar interference cancellation [48], as well as digital implementation of the processes involved [49], [50]. The tutorial article that leads off the next part of this book [51] discusses how propagation models can be combined with characterizations of receivers, as they may be assisted by such countermeasures, to provide quantitative predictions of system performance. The leadoff articles for the final part [52] will make clear the importance to future systems of the adaptive techniques we have discussed here.

References

[1] D. P. Taylor and P. R. Hartmann, "Telecommunications by microwave digital radio," Part I of this book. [An earlier version was published in *IEEE Commun. Mag.*, vol. 24, no. 8, Aug. 1986, pp. 11–16.]

[2] T. Noguchi, Y. Daido, and J. A. Nossek, "Modulation techniques for microwave digital radio," Part II of this book. [An earlier version was published in *IEEE Commun. Mag.*, vol. 24, no. 11, Nov. 1986, pp. 21–30.]

[3] W. D. Rummler, R. P. Coutts, and M. Liniger, "Multipath fading channel models for microwave digital radio," Part III of this book. [An earlier version was published in *IEEE Commun. Mag.*, vol. 24, no. 11, Nov. 1986, pp. 30–42.]

[4] C. Anderson, J. Barber, and R. Patel, "The effect of selective fading on digital radio," *IEEE Trans. Comm.*, vol. COM-27, no. 12, pp. 1870–1876, Dec. 1979.

[5] P. Hartmann and E. Allen, "An adaptive equalizer for correction of multipath distortion in a 90 Mbit/s-8PSK system," in *Proc. Int. Conf. Comm.*, Boston, MA. paper 5.6, June 1979.

[6] T. Giuffrida, "Measurements of the effects of propagation on digital radio systems equipped with space diversity and adaptive equalisation," in *Proc. Int. Conf. Comm.*, Boston, MA, paper 48.1, June 1979.

[7] S. Komaki, I. Horikawa, K. Morita, and Y. Okamoto, "Characteristics of a high-capacity 16 QAM digital radio system in multipath fading," *IEEE Trans. Comm.*, vol. COM-27, no. 12, pp. 1854–61, Dec. 1979.

[8] S. U. H. Qureshi, "Adaptive equalization," *IEEE Commun. Mag.*, vol. 20, no. 2, pp. 1–16, Mar. 1983.

[9] G. Bonnerot, M. Daout, and Ch. Lerouge, "Fading resistant 140 Mbit/s long haul system design," *GLOBECOM '83 Conf. Rec.*, vol. 1., San Diego, CA, pp. 302–306, Nov./Dec. 1983.

[10] S. Takenaka, K. Ogawa, E. Fukuda, and H. Nakamura, "A transversal fading equalizer for a 16 QAM microwave digital radio," in *ICC '81 Conf. Rec.*, Denver, CO, paper 46.2, June 1981.

[11] C. P. Bates and M. A. Skinner, "Impact of technology on high-capacity digital radio systems," in *ICC '83, Conf. Rec.*, vol. 3., Boston, MA, pp. 1467–1471, June 1983.

[12] T. Noguchi, T. Ryu, Y. Koizumi, S. Mizoguchi, and M. Yoshimoto, "6 GHz 135 Mbps digital radio system with 64 QAM modulation," in *ICC '83 Conf. Rec.*, vol. 3., Boston, MA, pp. 1472–1477, June 1983.

[13] J. D. McNichol, S. Barber, and F. Rivest, "Design and application of the RD-4A and RD-6A 64QAM digital radio systems," in *ICC '84 Conf. Rec.*, vol. 2, Amsterdam, The Netherlands, pp. 646–652, 1984.

[14] R. W. Lucky, "Automatic equalization for digital communications," *Bell Syst. Tech. J.*, vol. 44, pp. 547–588, Apr. 1965.

[15] D. L. Lyon, "Timing recovery in synchronous equalized data communication," *IEEE Trans. Comm.*, vol. COM-23, no. 2, pp. 269–274, Feb. 1975.

[16] J. Mazo, "Optimum timing phase for an infinite equalizer," *Bell Syst. Tech. J.*, vol. 54, no. 1, pp. 189–291, Jan. 1975.

[17] G. Ungerboeck, "Fractional tap-spacing equalizer and consequences for clock recovery in data modems," *IEEE Trans. Comm.*, vol. COM-24, no. 8, pp. 856–864, Aug. 1976.

[18] G. Fenderson and S. Shepard, "Adaptive transversal equalizer for 90 Mbit/s 16 QAM systems in the presence of multipath propagation," in *Proc. Int. Conf. Comm.*, Boston, MA, June 1983.

[19] J. Nossek, G. Sebald, and W. Grafinger, "Adaptive time-domain equalization of high-capacity digital radio systems," in *Proc. IEE Telecom. Trans. Conf.*, London, Mar. 1985.

[20] A. Leclert, and P. Vandamme, "Decision feedback equalization of dispersive radio channels," *IEEE Trans. Comm.*, vol. COM-33, no. 7, pp. 676–684, July 1985.

[21] A. J. Giger, and W. T. Barnett, "Effects of multipath propagation on digital radio," *IEEE Trans. Comm.*, vol. COM-29, no. 9, pp. 1345–1351, Sept. 1981.

[22] D. R. Smith, and J. J. Cormack, "Improvement in digital radio due to space diversity and adaptive equalization," *GLOBECOM '84 Conf. Rec.*, Atlanta, GA, pp. 1486–1491, Nov. 1984.

[23] P. L. Dirner, and S. H. Lin, "Measured frequency diversity improvement for digital radio," *IEEE Trans. Comm.*, vol. COM-33, no. 1, pp. 106–109, Jan. 1985.

[24] F. Fabbri, "Field tests on high capacity digital radio systems," *Alta Frequenza*, vol. LIII, no. 1, pp. 6–15, Jan. 1984.

[25] P. D. Karabinis, "Maximum-power and amplitude-equalizing algorithms for phase control in space diversity combining," *Bell Syst. Tech. J.*, vol. 62, no. 1, pp. 63–89, Jan. 1983.

[26] Recommendations and Reports of the CCIR, 1986. Report 784-2: "Effects of propagation on the design and operation of line-of-sight radio-relay systems."

[27] S. Komaki, Y. Okamoto, and K. Tajima, "Performance of 16-QAM digital radio system using new space diversity," in *Int. Conf. Comm.*, Seattle, WA, Paper 52.2, 1980.

[28] S. Komaki, K. Tajima, and Y. Okamoto, "A minimum dispersion combiner for high capacity digital microwave radio," *IEEE Trans. Comm.*, vol. COM-32, no. 4, pp. 419–428, Apr. 1984.

[29] Y. S. Yeh and L. J. Greenstein, "A new approach to space diversity combining in microwave digital radio," *AT&T Tech. J.*, vol. 64, no. 4, pp. 885–905, Apr. 1985.

[30] J. Henriksson, "Decision directed diversity combiners for digital radio links," *Acta Polytech. Scandinavica*, no. 54, 1984.

[31] S. H. Lin, "Impact of microwave depolarization during multipath fading on digital radio performance," *Bell Syst. Tech. J.*, vol. 56, no. 5, pp. 645–674, May 1977.

[32] P. M. Cronin, "Dual-polarized digital radio operation in a fading environment," in *Int. Conf. Comm.*, Seattle, WA, Paper 52.1, 1980.

[33] S. Barber, "Cofrequency cross-polarized operation of a 91 Mb/s digital radio," in *Int. Conf. Comm.*, Denver, CO, Paper 46.6, 1981.

[34] T. O. Mottl, "Dual-polarized channel outages during multipath fading," *Bell Syst. Tech. J.*, vol. 56, no. 5, pp. 675–701, May 1977.

[35] K. Morita, "Fluctuation of cross-polarization discrimination due to fading," *Rev. Elec. Comm. Lab.*, Japan, vol. 19, pp. 549–552, 1971.

[36] R. L. Olsen, "Cross-polarization during clear-air conditions on terrestrial links: a review," *Radio Sci.*, vol. 16, no. 5, pp. 631–647, Sept. 1981.

[37] P. W. Huish and J. H. A. McKeown, "Interference cancellation in terrestrial digital microwave radio-relay systems," *IEE Colloquium on "Terrestrial digital microwave systems,"* Digest no. 1983/14, contribution no. 2.

[38] T. Ryu, M. Tahara, and T. Noguchi, ''IF band cross-polarization canceller,'' in *Int. Conf. Comm.*, Amsterdam, Netherlands, pp. 1442–1446, 1984.
[39] M. Kavehrad, ''Adaptive cross-polarization interference cancellation for dual-polarized M-QAM signals,'' in *Int. Conf. Comm.*, Boston, MA, Paper 29.7, 1983.
[40] T. S. Chu, ''Restoring the orthogonality of two polarizations in radio communication systems, Part 1,'' *Bell Syst. Tech. J.*, vol. 50, no. 9, pp. 3063–3069, Nov. 1971.
[41] T. S. Chu, ''Restoring the orthogonality of two polarizations in radio communication systems, Part II,'' *Bell Syst. Tech. J.*, vol. 52, no. 3, pp. 319-327, Mar. 1973.
[42] Y. Aono, Y. Daido, S. Takenaka, and H. Nakamura, ''Cross polarization interference canceller for high-capacity digital radio systems,'' in *Int. Conf. Comm.*, Chicago, IL, Paper 39.5, 1985.
[43] J. Namiki and S. Takahara, ''Adaptive receiver for cross-polarized digital transmission,'' in *Int. Conf. Comm.*, Denver, CO, Paper 39.6, 1981.
[44] M. Araki, H. Ichikawa, and A. Hasimoto, ''100km overwater span digital radio system,'' in *Int. Conf. Comm.*, Chicago, IL, Paper 15.5, 1985.
[45] K. T. Wu and T. S. Giuffrida, ''Feasibility study of an interference canceller for co-channel cross-polarization operation of digital radio,'' in *Int. Conf. Comm.*, Philadelphia, PA, Paper 2B.7, 1982.
[46] E. H. Lin, A. J. Giger, and G. D. Alley, ''Angle diversity on line-of-sight microwave paths using dual-beam dish antennas,'' in *Conf. Rec., ICC '87*, pp. 831–841.
[47] P. Balaban, E. A. Sweedyk, and G. S. Axeling, ''Angle diversity with two antennas: model and experimental results,'' in *Conf. Rec., ICC '87*, pp. 846–852, June 1987.
[48] B. Lankl, ''Cross polarization interference canceller for QAM digital radio systems with asynchronous clock and carrier signals,'' in *Conf. Rec., GLOBECOM '86*, pp. 523-529, Dec. 1986.
[49] B. Bacetti, S. Bellini, G. Filberti, and G. Tartara, ''Full digital adaptive equalization in 64-QAM radio systems,'' *IEEE J. Sel. Areas Comm.*, vol. SAC-5, no. 3, Apr. 1987, pp. 457–465.
[50] H. Matsue, H. Ohtsuka, and T. Murase, ''Digitalized cross-polarization interference canceller for multilevel digital radio,'' *IEEE J. Sel. Areas Comm.*, vol. SAC-5, no. 3, Apr. 1987, pp. 493-501.
[51] L. J. Greenstein and M. Shafi, ''Outage calculation methods for microwave digital radio,'' Part V of this book. [An earlier version was published in *IEEE Commun. Mag.*, vol. 25, no. 2, Feb. 1987, pp. 30–39.]
[52] H. Yamamoto, K. Kohiyama, O. Kurita, M. H. Meyers, V. K. Prabhu, G. Hart, and J. A. Steinkamp, ''Future trends in microwave digital radio,'' Part VI of this book. [Three mini-papers, introduced by Yamamoto, giving views from Asia, North America, and Europe. Earlier versions were published in a three-part article in *IEEE Commun. Mag.*, vol. 25, no. 2, Feb. 1987, pp. 40–52.]

ADAPTIVE EQUALIZATION

SHAHID QURESHI

High speed data transmission over voice band telephone lines.

INTRODUCTION

THE rapidly rising need for higher speed data transmission to furnish computer communications has been met primarily by utilizing the widespread network of voice-bandwidth channels developed for voice communications. A modulator-demodulator (MODEM) is required to carry digital signals over these analog passband (nominally 300 to 3000 Hz) channels by translating binary data to voice-frequency signals and back (Fig. 1).

Real analog channels reproduce at their output a transformed and corrupted version of the input waveform. Statistical corruption of the waveform may be additive and/or multiplicative, because of possible background thermal noise, impulse noise and fades. Examples of deterministic (although not necessarily known) transformations performed by the channel are frequency translation, nonlinear or harmonic distortion and time dispersion.

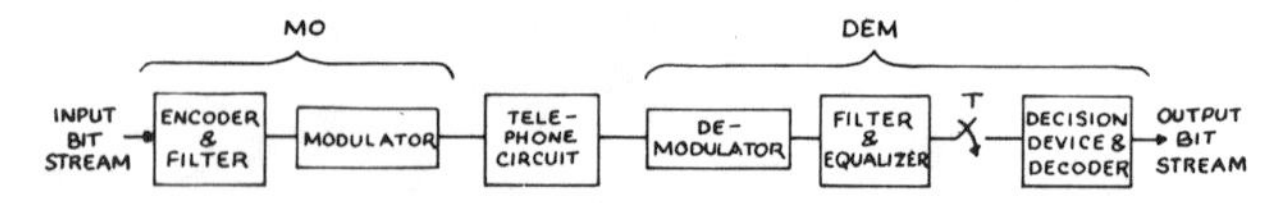

Fig. 1. Data transmission system.

In telephone lines, time dispersion results from the deviation of the channel frequency response from the ideal characteristics of constant amplitude and linear phase (or constant delay). The idea of equalization is simply to compensate for nonideal characteristics by additional filtering, and dates back to the use of loading coils to improve the characteristics of telephone cables for voice transmission.

A modem transmitter collects an integral number of bits of data at a time and encodes them into symbols for transmission at the signaling rate. In pulse amplitude modulation, each signal is a pulse whose amplitude level is determined by the symbol, e.g., amplitudes of −3, −1, 1, and 3 for quaternary transmission. In efficient digital communication systems the effect of each symbol transmitted over a time dispersive channel extends beyond the time interval used to represent that symbol. The distortion caused by the resulting overlap of received symbols is called intersymbol interference (ISI) [1]. This distortion is one of the major obstacles to reliable high-speed data transmission over low-background-noise channels of limited bandwidth.

It was recognized early in the quest for high speed data transmission that rather precise compensation, or equalization, is required to reduce the intersymbol interference introduced by the channel. In addition, in most practical situations the channel characteristics are not known beforehand. For medium-speed (up to 2400 b/s) modems it is usually adequate to design and use a compromise (or statistical) equalizer which compensates for the average of the range of expected channel amplitude and delay characteristics. However, the variation in the characteristics within a class of channels, as in the lines found in the switched telephone network, is large enough so that automatic adaptive equalization is used nearly universally for speeds higher than 2400 b/s. Even 2400 b/s modems now often incorporate this feature.

INTERSYMBOL INTERFERENCE

Intersymbol interference arises in all pulse-modulation systems, including frequency-shift keying (FSK), phase-shift keying (PSK) and quadrature amplitude modulation (QAM). However, its effect can be most easily described for a baseband pulse-amplitude modulation (PAM) system. A model of such a PAM communication system is shown in Fig. 2. A baseband equivalent model such as this can be derived for any linear modulation scheme. In this model the "channel" includes the effects of the transmitter filter, the modulator, the transmission medium and the demodulator.

A symbol x_m, one of L discrete amplitude levels, is transmitted at instant mT through the channel, where T seconds is the signaling interval. The channel impulse response h(t) is shown in Fig. 3. The received signal r(t) is the superposition of the impulse responses of the channel to each transmitted symbol and additive white Gaussian noise n(t):

$$r(t) = \sum_j x_j \, h(t - jT) + n(t).$$

If we sample the received signal at instant $kT + t_0$, where t_0

Reprinted from *IEEE Comm. Mag.*, vol. 20, no. 2, pp. 9–16, Mar. 1982.

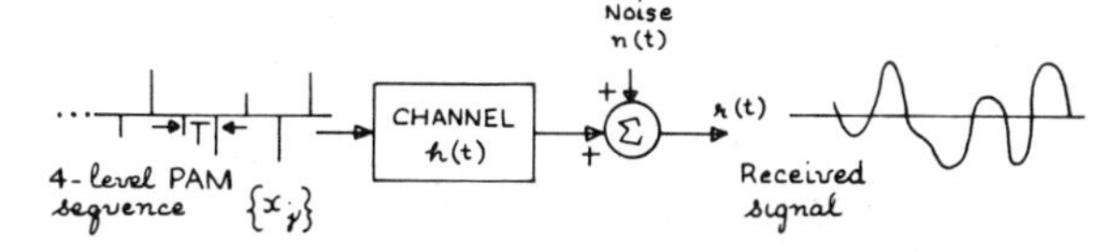

Fig. 2. Baseband PAM system mode.

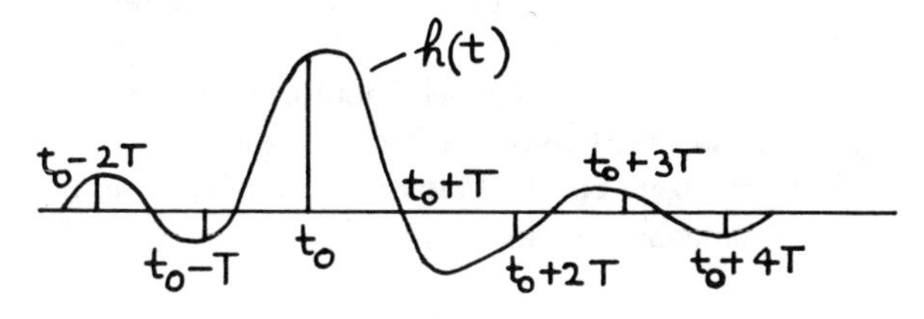

Fig. 3. Channel impulse response.

accounts for the channel delay and sampler phase, we obtain

$$r(t_O + kT) = x_k\, h(t_O) + \sum_{j \neq k} x_j\, h(t_O + kT - jT) + n(t_O + kT).$$

The first term on the right is the desired signal since it can be used to identify the transmitted amplitude level. The last term is the additive noise, while the middle sum is the interference from neighboring symbols. Each interference term is proportional to a sample of the channel impulse response, $h(t_O + iT)$, spaced a multiple iT of symbol intervals T away from t_O as shown in Fig. 3. The ISI is zero if and only if $h(t_O + iT) = 0$, $i \neq 0$; that is, if the channel impulse response has zero crossings at T-spaced intervals.

When the impulse response has such uniformly-spaced zero crossings, it is said to satisfy Nyquist's first criterion. In frequency domain terms, this condition is equivalent to

$$H'(f) = \text{constant for } |f| \leqslant 1/2T.$$

H(f) is the channel frequency response and H'(f) is the "folded" (aliased or overlapped) channel spectral response after symbol-rate sampling. The band $|f| \leqslant 1/2T$ is commonly referred to as the Nyquist or minimum bandwidth. When $H(f) = 0$ for $|f| > 1/T$ (the channel has no response beyond twice the Nyquist bandwidth), the *folded* response H'(f) has the simple form

$$H'(f) = H(f) + H(f - 1/T),\ 0 \leqslant f \leqslant 1/T.$$

Figures 4 (a) and (d) show the amplitude response of two linear-phase lowpass filters: one an ideal filter with Nyquist bandwidth and the other with odd (or vestigial) symmetry around 1/2T Hz. As illustrated in Fig. 4 (b) and (e), the folded spectrum of each filter satisfies Nyquist's first criterion. One class of linear-phase filters, commonly referred to in the literature [1], is the raised-cosine family with cosine rolloff around 1/2T Hz.

In practice, the effect of ISI can be seen from a trace of the received signal on an oscilloscope with its time base synchronized to the symbol rate. Figure 5 shows a trace (eye pattern) for a two-level or binary PAM system. If the channel satisfies the zero ISI condition, there are only two distinct levels at the sampling time t_O. The eye is then fully open and

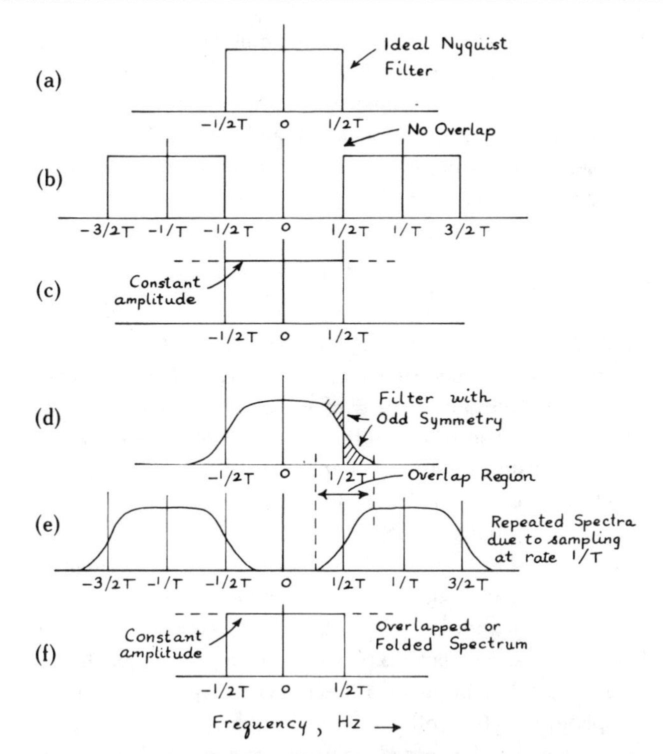

Fig. 4. Linear phase filters which satisfy Nyquist's first criterion.

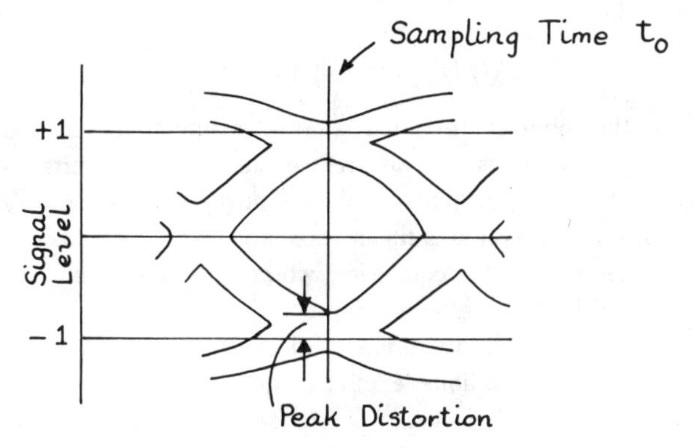

Fig. 5. Binary eye pattern.

the peak distortion is zero. Peak distortion (Fig. 5) is the ISI that occurs when the data pattern is such that all intersymbol interference terms add to produce the maximum deviation from the desired signal at the sampling time.

The purpose of an equalizer, placed in the path of the received signal, is to reduce the ISI as much as possible to maximize the probability of correct decisions.

LINEAR TRANSVERSAL EQUALIZERS

Among the many structures used for equalization the simplest is the transversal (*tapped delay-line* or *nonrecursive*) equalizer shown in Fig. 6. In such an equalizer the current and past values $r(t - nT)$ of the received signal are linearly weighted by equalizer coefficients (*tap gains*) c_n and summed to produce the output. If the delays and tap-gain multipliers are analog, the continuous output of the equalizer z(t) is sampled at the symbol rate and the samples go to the decision device. In the now universally used digital implementation, samples of the received signal at the symbol rate are stored in a digital shift register (or memory), and the equalizer output samples (sums of products) $z(t_O + kT)$ or z_k are computed digitally, once per symbol, according to

$$z_k = \sum_{n=0}^{N-1} c_n \, r(t_O + kT - nt)$$

where N is the number of equalizer coefficients.

The equalizer coefficients, c_n, $n = 0, 1, ..., N-1$ may be chosen to force the samples of the combined channel and equalizer impulse response to zero at all but one of the N T-spaced instants in the span of the equalizer. This is shown graphically in Fig. 7. Such an equalizer is called a zero-forcing (ZF) equalizer [2].

If we let the number of coefficients of a ZF equalizer increase without bound, we would obtain an infinite-length equalizer with zero ISI at its output. The frequency response C(f) of such an equalizer is periodic, with a period equal to the symbol rate 1/T because of the T second tap spacing. After sampling, the effect of the channel on the received signal is determined by the *folded* frequency response H′ (f). The combined response of the channel, in tandem with the equalizer, must satisfy the zero ISI condition or Nyquist's first criterion,

$$C(f)\, H'(f) = 1, \ |f| \leqslant 1/2T.$$

From the above expression we see that an infinite-length zero-ISI equalizer is simply an inverse filter, which inverts the *folded* frequency response of the channel. A finite-length ZF equalizer approximates this inverse and so may excessively enhance noise at frequencies where the folded channel spectrum has high attenuation.

Clearly, the ZF criterion neglects the effect of noise altogether. Also, a finite-length ZF equalizer is guaranteed to minimize the peak distortion or worst-case ISI only if the peak distortion before equalization is less than 100 percent; i.e., if a binary eye is initially open. However, at high speeds on bad channels this condition is often not met.

The least mean-square (LMS) equalizer [1] is more robust. Here the equalizer coefficients are chosen to minimize the mean-square error—the sum of squares of all the ISI terms plus the noise power at the output of the equalizer. Therefore, the LMS equalizer maximizes the signal-to-distortion ratio at the equalizer output within the constraints of the equalizer length and delay.

The delay introduced by the equalizer depends on the position of the main or reference tap of the equalizer. Typically, the tap gain corresponding to the main tap is the largest.

If the values of the channel impulse response at the sampling instants are known, the N coefficients of the ZF and the LMS equalizers can be obtained by solving a set of N linear simultaneous equations for each case.

AUTOMATIC SYNTHESIS

Before regular data transmission begins, automatic synthesis of the ZF or LMS equalizers for unknown channels, which involves the iterative solution of one of the above-mentioned sets of simultaneous equations, should be carried out during a training period.

Most current high-speed modems use LMS equalizers because they are more robust and superior to the ZF equalizers in their convergence properties. In the remainder of this article we shall restrict our attention to LMS equalizers.

During the training period, a known signal is transmitted and a synchronized version of this signal is generated in the receiver to acquire information about the channel characteristics. The training signal may consist of periodic isolated pulses or a continuous sequence with a broad, even spectrum such as the widely used maximum-length shift-register or pseudo-noise (PN) sequence [1,3]. The latter has the advantage of much greater average power, and hence a larger received signal-to-noise ratio (SNR) for the same peak transmitted power. The training sequence must be at least as long as the length of the equalizer so that the transmitted signal spectrum is adequately dense in the channel bandwidth to be equalized.

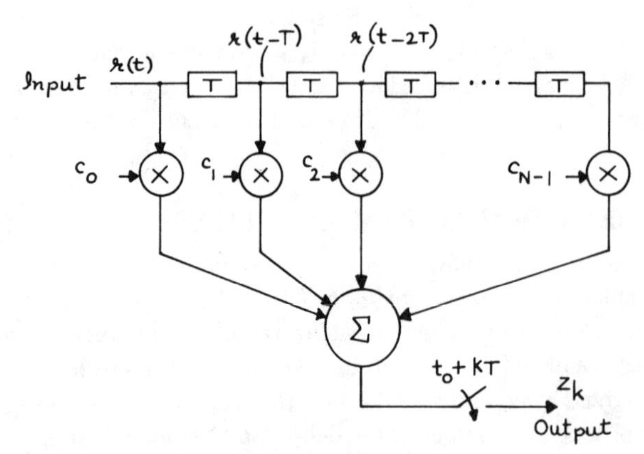

Fig. 6. Linear transversal equalizer.

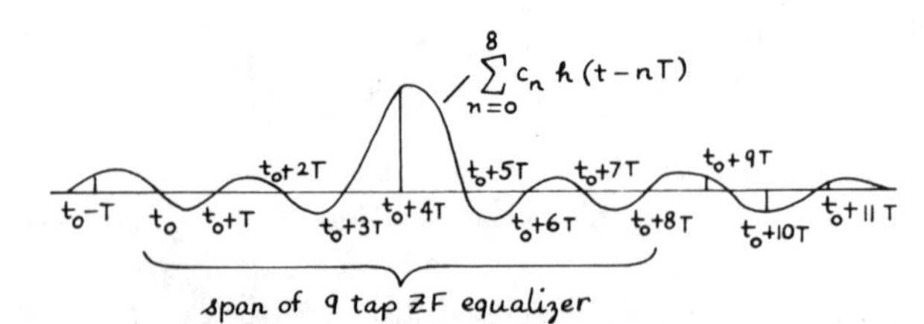

Fig. 7. Combined impulse response of a channel and zero-forcing equalizer in tandem.

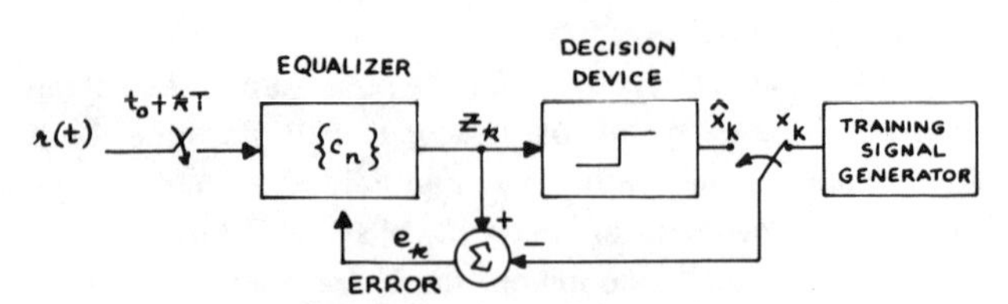

Fig. 8. Automatic adaptive equalizer.

Given a synchronized version of the known training signal, a sequence of error signals $e_k = z_k - x_k$ can be computed at the equalizer output (Fig. 8), and used to adjust the equalizer coefficients to reduce the sum of the squared errors. The most popular equalizer adjustment method involves updates to each tap gain during each symbol interval. Iterative solution of the coefficients of the equalizer is possible because the mean-square error (MSE) is a quadratic function of the coefficients. The MSE may be envisioned as an N-dimensional paraboloid (punch bowl) with a bottom or minimum. The adjustment to each tap gain is in a direction opposite to an estimate of the gradient of the MSE with respect to that tap gain. The idea is to move the set of equalizer coefficients closer to the unique optimum set corresponding to the minimum MSE. This symbol-by-symbol procedure is commonly referred to as the continual or stochastic update method because, instead of the true gradient of the mean-square error,

$$\partial E[e_k^2] / \partial c_n(k),$$

a noisy but unbiased estimate

$$\partial e_k^2 / \partial c_n(k) = 2 e_k r(t_O + kT - nT)$$

is used.

Thus, the tap gains are updated according to

$$c_n(k+1) = c_n(k) - \Delta e_k r(t_O + kT - nT), n = 0, 1, \ldots, N-1,$$

where $c_n(k)$ is the nth tap gain at time k, e_k is the error signal and Δ is a positive adaptation constant or step size.

EQUALIZER CONVERGENCE

The convergence behavior of the stochastic update method is hard to analyze. However, for a small step size and a large number of iterations, the behavior is similar to the steepest-descent algorithm, which uses the actual gradient rather than a noisy estimate.

Here we list some general convergence properties: (a) fastest convergence (or shortest settling time) is obtained when the (folded) power spectrum of the symbol-rate sampled equalizer input is flat, and when the step size Δ is chosen to be the inverse of the product of the received signal power and the number of equalizer coefficients; (b) the larger the variation in the above-mentioned folded power spectrum, the smaller the step size must be, and therefore the slower the rate of convergence; (c) for systems where sampling causes aliasing (channel foldover or spectral overlap), the convergence rate is affected by the channel delay characteristics and the sampler phase, because they affect the aliasing. This will be explained more fully later.

ADAPTIVE EQUALIZATION

After the initial training period, the coefficients of an adaptive equalizer may be continually adjusted in a *decision-directed* manner. In this mode the error signal $e_k = z_k - \hat{x}_k$ is derived from the final (not necessarily correct) receiver estimate $\{\hat{x}_k\}$ of the transmitted sequence $\{x_k\}$. In normal operation the receiver decisions are correct with high probability, so that the error estimates are correct often enough to allow the adaptive equalizer to maintain precise equalization. Moreover, a decision-directed adaptive equalizer can track slow variations in the channel characteristics or linear perturbations in the receiver front end, such as slow jitter in the sampler phase.

The larger the step size, the faster the equalizer tracking capability. However, a compromise must be made between fast tracking and the excess mean-square error of the equalizer. The excess MSE is that part of the error power in excess of the minimum attainable MSE (with tap gains frozen at their optimum settings). This excess MSE, caused by tap gains wandering around the optimum settings, is directly proportional to the number of equalizer coefficients, the step size and the channel noise power. The step size that provides the fastest convergence results in a mean-square error is, on the average, 3 dB worse than the minimum achievable MSE. In practice, the value of the step size is selected for fast convergence during the training period and then reduced for fine tuning during the steady-state operation (or data mode).

EQUALIZERS FOR QAM SYSTEMS

So far we have only discussed equalizers for a baseband PAM system. Modern high-speed modems almost universally use phase-shift keying (PSK) for lower speeds, e.g., 2400 to 4800 b/s, and combined phase and amplitude modulation or, equivalently, quadrature amplitude modulation (QAM) [1], for higher speeds, e.g., 4800 to 9600 or even 14,400 b/s. QAM is as efficient in bits/second per Hz as vestigial— or single-sideband modulation—yet enables a coherent carrier to be derived and phase jitter to be tracked using easily implemented decision-directed carrier recovery techniques.

Figure 9 shows a generic QAM system, which may also be used to implement PSK or combined amplitude and phase modulation. Two double-sideband suppressed-carrier AM signals are superimposed on each other at the transmitter and separated at the receiver, using quadrature or orthogonal carriers for modulation and demodulation. It is convenient to represent the *in-phase* and *quadrature* channel lowpass filter output signals in Fig. 9 by $y_r(t)$ and $y_i(t)$, as the real and imaginary parts of a complex-valued signal $y(t)$. (Note that the signals are real, but it will be convenient to use complex notation.)

The baseband equalizer [4], with complex coefficients c_n, operates on samples of this complex signal $y(t)$ and produces complex equalized samples $z(k) = z_r(k) + j z_i(k)$, as shown in Fig. 10. This figure illustrates more concretely the concept of

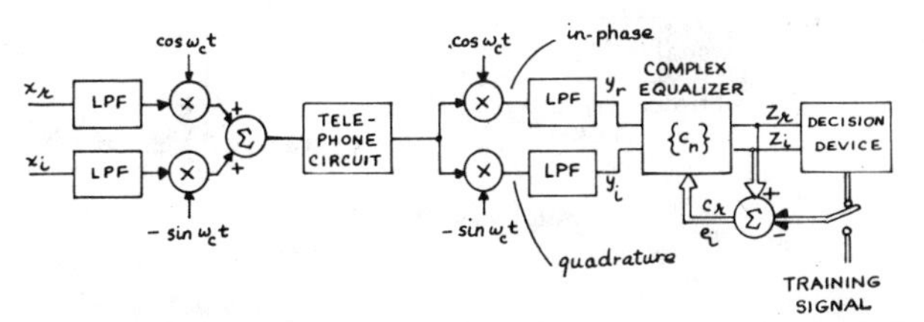

Fig. 9. QAM system with baseband complex adaptive equalizer.

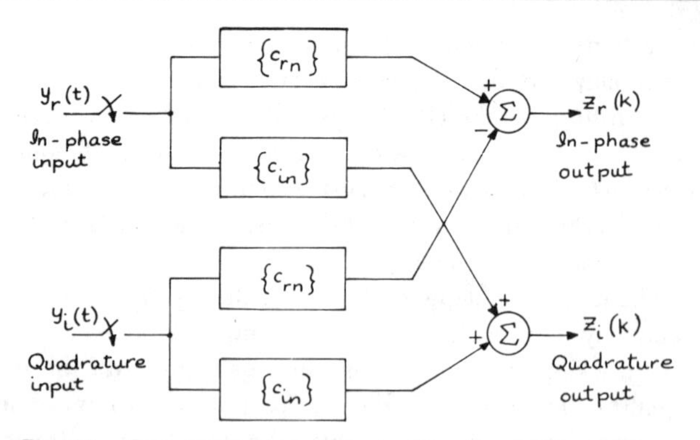

Fig. 10. Complex transversal equalizer for QAM modems.

a complex equalizer as a set of four real transversal filters (with cross-coupling) for two inputs and two outputs. While the *real* coefficients c_{rn}, $n = 0, \ldots, N-1$, help to combat the intersymbol interference in the in-phase and quadrature channels, the *imaginary* coefficients c_{in}, $n = 0, \ldots, N-1$, counteract the cross interference between the two channels. The latter may be caused by asymmetry in the channel characteristics around the carrier frequency.

The coefficients are adjusted to minimize the mean of the squared magnitude of the complex error signal, $e(k) = e_r(k) + j\,e_i(k)$, where e_r and e_i are the differences between z_r and z_i, and their desired values. The update method is similar to the one used for the PAM equalizer except that all variables are complex-valued;

$$c_n\,(k+1) = c_n(k) - \Delta\; e_k\; y^*(t_O + kT - nT),$$
$$n = 0, 1, \ldots, N-1,$$

where y^* is the complex conjugate of y. Again, the use of complex notation allows the writing of this single concise equation, rather than two separate equations involving four real multiplications, which is what really has to be implemented.

The complex equalizer can also be used at passband [5] to equalize the received signal before demodulation as shown in Fig. 11. Here the received signal is split into its in-phase and quadrature components by a pair of so-called phase-splitting filters, with identical amplitude responses and phase responses that differ by 90°. The complex passband signal at the output of these filters is sampled at the symbol rate and applied to the equalizer delay line in the same way as at baseband. The complex output of the equalizer is demodulated, via multiplication by a complex exponential as shown in Fig. 11, before decisions are made and the complex error computed. Further, the error signal is remodulated before it is used in the equalizer adjustment algorithm. The main advantage of implementing the equalizer in the passband is that the error signal can be fed back for phase correction without delay, thus enabling fast phase jitter to be tracked more effectively. The same advantage can be attained with a baseband equalizer by putting a jitter tracking loop after the equalizer.

DECISION-FEEDBACK EQUALIZERS

We have discussed different placements and adjustment methods for the equalizer, but the basic equalizer structure has remained a linear and nonrecursive filter. A simple nonlinear equalizer [6], which is particularly useful for channels with severe amplitude distortion, uses decision feedback to cancel the interference from symbols which have already been detected. Figure 12 shows such a decision-feedback equalizer (DFE). The equalized signal is the sum of the outputs of the forward and feedback parts of the equalizer. The forward part is like the linear transversal equalizer discussed earlier. Decisions made on the equalized signal are fed back via a second transversal filter. The basic idea is that if the value of the symbols already detected are known (past decisions are assumed to be correct), then the ISI contributed by these symbols can be canceled exactly, by subtracting past symbol values with appropriate weighting from the equalizer output. The weights are samples of the tail of the system impulse response including the channel and the forward part of the equalizer.

The forward and feedback coefficients may be adjusted simultaneously to minimize the mean squared error. The update equation for the forward coefficients is the same as for the linear equalizer. The feedback coefficients are adjusted according to

$$b_m\,(k+1) = b_m\,(k) - \Delta\; e_k\; d_{k-m},\; m = 1, \ldots, M,$$

where d_k is the kth symbol decision, $b_m(k)$ is the mth feedback coefficient at time k and there are M feedback coefficients in all. The optimum LMS settings of b_m, $m = 1, \ldots, M$, are

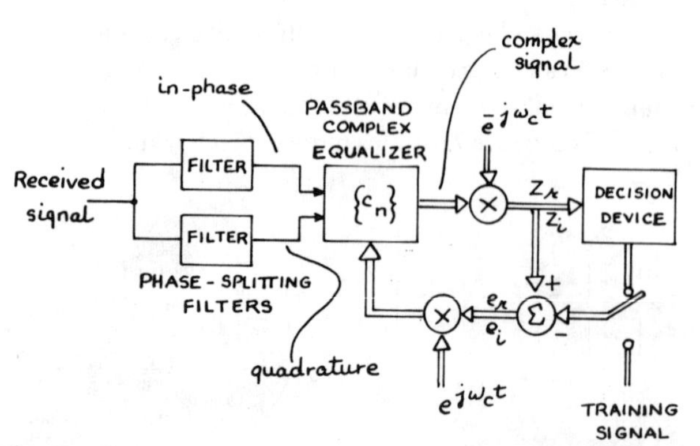

Fig. 11. Passband complex adaptive equalizer for QAM system.

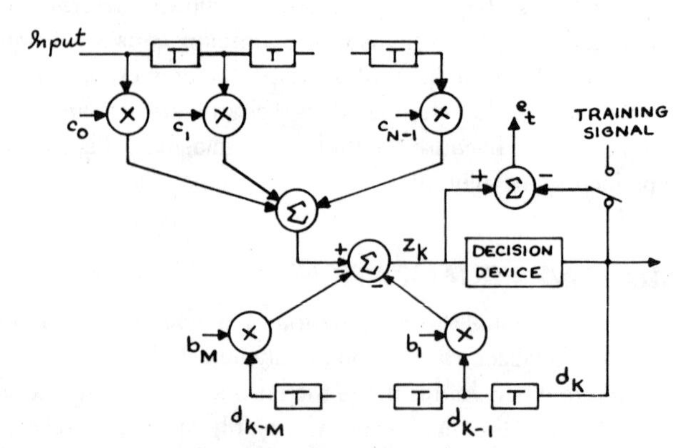

Fig. 12. Decision-feedback equalizer.

those that reduce the ISI to zero, within the span of the feedback part, in a manner similar to a ZF equalizer. Note that since the output of the feedback section of the DFE is a weighted sum of noise-free past decisions, the feedback coefficients play no part in determining the noise power at the equalizer output.

Given the same number of overall coefficients, does a DFE achieve less mean squared error than a linear equalizer? There is no definite answer to this question. The performance of each type of equalizer is influenced by the particular channel characteristics and sampler phase, as well as the actual number of coefficients and the position of the reference or main tap of the equalizer. However, the DFE can compensate for amplitude distortion without as much noise enhancement as a linear equalizer. The DFE performance is less sensitive to the sampler phase.

An intuitive explanation for these advantages is as follows: The coefficients of a linear transversal equalizer are selected to force the combined channel and equalizer impulse response to approximate a unit pulse. In a DFE, the ability of the feedback section to cancel the ISI, because of a number of the past symbols, allows more freedom in the choice of the coefficients of the forward section. The combined impulse response of the channel and the forward section may have nonzero samples following the main pulse. That is, the forward section of a DFE need not approximate the inverse of the channel characteristics, and so avoids excessive noise enhancement and sensitivity to sampler phase.

When a particular incorrect decision is fed back, the DFE output reflects this error during the next few symbols because the incorrect decision traverses the feedback delay line. Thus, there is a greater likelihood of more incorrect decisions following the first one, i.e., error propagation. Fortunately, the error propagation in a DFE is not catastrophic. On typical channels, errors occur in short bursts that degrade performance only slightly.

FRACTIONALLY-SPACED EQUALIZERS

A fractionally-spaced transversal equalizer [7,8] is shown in Fig. 13. The delay line taps of such an equalizer are spaced at an interval τ which is less than, or a fraction of, the symbol interval T. The tap spacing τ is typically selected such that the bandwidth occupied by the signal at the equalizer input is $|f| < 1/2\tau$, i.e., τ-spaced sampling satisfies the sampling theorem. In an analog implementation, there is no other restriction on τ, and the output of the equalizer can be sampled at the symbol rate. In a digital implementation τ must be KT/M, where K and M are integers and M > K. In practice, it is more convenient to choose $\tau = T/M$, where M is a small integer, e.g. The received signal is sampled and shifted into the equalizer delay line at a rate 2/T and one output is produced each symbol interval (for every 2 input samples):

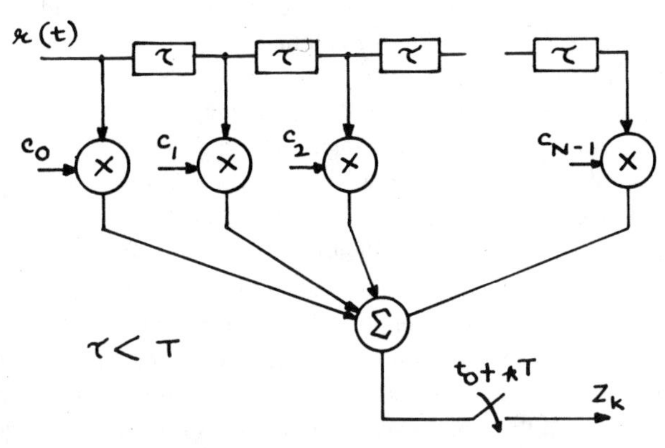

Fig. 13. Fractionally-spaced equalizer.

$$z_k = \sum_{n=0}^{N-1} c_n\, r(t_O + kT - nT/2).$$

The coefficients of a T/2 equalizer may be updated once per symbol based on the error computed for that symbol, according to

$$c_n(k+1) = c_n(k) - \Delta\, e_k\, r(t_O + kT - nT/2), \qquad n = 0, 1, \ldots, N-1.$$

One important property of a fractionally-spaced equalizer (FSE) is the insensitivity of its performance to the choice of sampler phase. This distinction between the conventional T-spaced and fractionally-spaced equalizers can be heuristically explained as follows: First, symbol-rate sampling at the input to a T equalizer causes spectral overlap or aliasing, as explained in connection with Fig. 4. When the phases of the overlapping components match they add constructively, and when the phases are 180° apart they add destructively, which results in the cancellation or reduction of amplitude as shown in Fig. 14. Variation in the sampler phase or timing instant corresponds to a variable delay in the signal path; a linear phase component with variable slope is added to the signal spectrum. Thus, changes in the sampler phase strongly influence the effects of aliasing; i.e., they influence the amplitude and delay characteristics in the spectral overlap region of the sampled equalizer input. The minimum MSE achieved by the T equalizer is, therefore, a function of the sampler phase. In particular, when the sampler phase causes cancellation of the band-edge ($|f| = 1/2T$ Hz) components, the equalizer cannot manipulate the null into a flat spectrum at all, or at least without significant noise enhancement (if the null is a depression rather than a total null).

In contrast, there is no spectral overlap at the input to an FSE. Therefore, such an equalizer can adjust the channel spectrum (amplitude and phase) at the two band-edge regions before symbol-rate sampling (and spectral overlap) at the equalizer output. Thus, the sensitivity of the minimum MSE, achieved with a fractionally-spaced equalizer with respect to the sampler phase, is typically far smaller than with a T equalizer.

Another point of view is as follows: It has been shown that the optimum receive filter in a linear modulation system is the cascade of a filter matched to the actual channel, with a transversal T-spaced equalizer. The fractionally-spaced equalizer, by virtue of its sampling rate, can synthesize the best combination of the characteristics of an adaptive matched filter and a T-spaced equalizer, within the constraints of its length and delay. A T-spaced equalizer, with symbol-rate sampling at its input, cannot perform matched filtering. An FSE can effectively compensate for more severe delay distortion and deal with amplitude distortion with less noise enhancement than a T equalizer.

Comparison of the performance of T and T/2 equalizers for QAM systems operating over representative voice-grade telephone circuits has shown the following additional properties: (a) a T/2 equalizer with the same number of coefficients (half the time span) performs almost as well or better than a T equalizer; (b) a pre-equalizer receive shaping filter is not required with a T/2 equalizer; (c) for channels with severe band-edge delay distortion, the T equalizer performs noticeably worse than a T/2 equalizer regardless of the choice of sampler phase.

OTHER APPLICATIONS

In this section we briefly mention applications of automatic or adaptive equalization in areas other than telephone-line modems.

One such application is generalized automatic channel equalization, where the entire bandwidth of the channel is to be equalized without regard to the modulation scheme or transmission rate to be used on the channel. The tap spacing and input sample rate are selected to satisfy the sampling theorem, and the equalizer output is produced at the same rate. During the training mode a known signal is transmitted, which covers the bandwidth to be equalized. The difference between the equalizer output and a synchronized reference training signal is the error signal. The tap gains are adjusted to minimize the mean-square error in a manner similar to that used for an automatic equalizer for synchronous data transmission.

On telephone line circuits the primary cause of intersymbol interference is linear distortion because of imperfect amplitude and group-delay characteristics. In radio and undersea channels, ISI is due to multipath transmission, which may be viewed as transmission through a group of channels with different delays. Adaptive equalizers are capable of correcting for ISI due to multipath in the same way as ISI from linear distortion. One special requirement of equalizers intended for use over radio channels is that they be able to track the time varying channel characteristics typically encountered. The convergence rate of the algorithm employed then becomes important during normal data transmission rather than just during the training period.

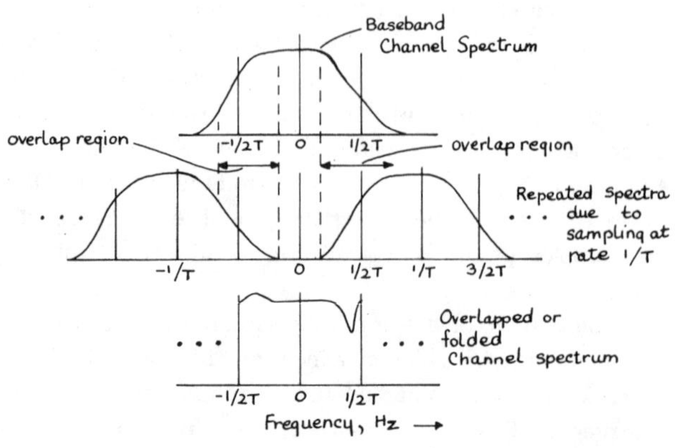

Fig. 14. Spectral overlap at the input to a T equalizer.

Experimental use of fixed transversal equalizers has also been made in digital magnetic recording systems. The recording method employed in such a case must be linear instead of the saturated magnetization normally used. Having linearized the "channel," equalization can be employed to combat intersymbol interference at increased recording densities, using a higher symbol rate or multilevel coding.

Two related areas where the techniques developed for adaptive equalization find application are adaptive filtering for cancellation of noise or an interfering signal, and adaptive channel modeling or identification for echo cancellation in communication circuits for speech or data transmission.

IMPLEMENTATION APPROACHES

One may divide the methods of implementing adaptive equalizers into the following general categories: analog, hardwired digital and programmable digital.

Analog adaptive equalizers, with inductor-capacitor (LC) tapped delay lines and switched ladder attenuators as tap gains, were among the first implementations. The switched attenuators later gave way to field-effect transistors as the variable gain elements. Analog equalizers were soon replaced by digitally implemented equalizers for reduced size and increased accuracy. Recently, however, there is renewed interest in large-scale integrated (LSI) analog implementations based on the charge-coupled device (CCD) technology. Here the equalizer input is sampled but not quantized. The sampled analog values are stored and transferred as charge packets. The variable tap gains are typically stored in digital memory locations and the multiplications between the analog sample values and the digital tap gains take place in analog fashion, as via multiplying digital-to-analog converters. This technology is still in infancy and has yet to find its way into practice. However, it has significant potential in applications where the symbol rates are high enough to make digital implementations impractical or very costly.

The most widespread technology of the last decade for adaptive equalizer implementation may be classified as hardwired digital technology. In such implementations the equalizer input is made available in sampled and quantized form suitable for storage in digital shift registers. The variable tap gains are also stored in shift registers and the formation and accumulation of products takes place in logic circuits connected to perform digital arithmetic. This class of implementations is characterized by the fact that the circuitry is hardwired for the sole purpose of performing the adaptive equalization function with a predetermined structure. Examples include the early units based on metal-oxide semiconductor (MOS) shift registers and transistor-transistor logic (TTL) circuits. Later implementations were based on MOS

LSI circuits with dramatic savings in space, power dissipation and cost.

The most recent trend in implementing adaptive equalizers is toward programmable digital signal processors. Here, the equalization function is performed in a series of steps or instructions in a microprocessor or a digital computation structure specially configured to efficiently perform the type of digital arithmetic (e.g., multiply and accumulate) required in digital signal processing. The same hardware can then be time-shared to perform functions such as filtering, modulation and demodulation in a modem. Perhaps the greatest advantage of programmable digital technology is its flexibility, which permits sophisticated equalizer structures and training procedures to be implemented with ease.

CONCLUSION

This paper serves as a broad-brush introduction to a rich and mature field. Interested readers will find a wealth of information in the brief list of references given here. More comprehensive lists of references are available in [6, 9 and 10]. Despite the maturity of the field, adaptive equalization is still an area of active interest; a good example is fractionally-spaced and fast-training equalizers and their implementation.

REFERENCES

[1] R. W. Lucky, J. Salz and E. J. Weldon, Jr., *Principles of Data Communication*, New York: McGraw-Hill, 1968.

[2] R. W. Lucky, "Automatic equalization for digital communication," *Bell Syst. Tech. J.*, April 1965, pp. 547-588.

[3] K. H. Mueller and D. A. Spaulding, "Cyclic equalization—a new rapidly converging equalization technique for synchronous data communication," *Bell Syst. Tech. J.*, February 1975, pp. 369-406.

[4] J. G. Proakis and J. H. Miller, "An adaptive receiver for digital signaling through channels with intersymbol interference," *IEEE Trans. Inform. Theory*, July 1969, pp. 484-497.

[5] D. D. Falconer, "Jointly adaptive equalization and carrier recovery in two-dimensional digital communication systems," *Bell Syst. Tech. J.*, March 1976, pp. 317-334.

[6] C. A. Belfiore and J. H. Park, Jr., "Decision feedback equalization," *Proc. IEEE*, August 1979, pp. 1143-1156.

[7] G. Ungerboeck, "Fractional tap-spacing equalizer and consequences for clock recovery in data modems," *IEEE Trans. Commun.*, August 1976, pp. 856-864.

[8] S. U. H. Qureshi and G. D. Forney, Jr., "Performance and properties of a T/2 equalizer," *Nat. Telecomm. Conf. Record*, December 1977.

[9] R. W. Lucky, "Survey of communication theory literature: 1968-1973," *IEEE Trans. Inform. Theory*, November 1973, pp. 725-739.

[10] J. G. Proakis, "Advances in equalization for intersymbol interference," *Advances in Communication Systems*, vol. 4, New York: Academic Press, 1975.

[11] L. E. Franks, ed., *Data Communication*, Dowden, Hutchinson and Ross, 1974.

A TRANSVERSAL FADING EQUALIZER FOR A 16-QAM MICROWAVE DIGITAL RADIO

S. TAKENAKA†, E. FUKUDA†, H. NAKAMURA† and K. OGAWA††

†Fujitsu Laboratories Ltd. ††Fujitsu Ltd.

1015 Kamikodanaka, Nakahara-ku, Kawasaki, 211, JAPAN

ABSTRACT

This paper proposes a transversal adaptive equalizer for a 200 Mbps 16-QAM digital radio to equalize waveform distortion caused by multipath fading. The technique of transversal adaptive equalization is very effective against multipath fading: (1) with long delay differences (2) deviated from a two-ray path model, and (3) with non-minimum phase response.

The performance of a transversal equalizer was investigated theoretically and experimentally. The capability of the equalizer was calculated by computer simulation with the number of taps as the parameter. The hardware of a 5-tap transversal equalizer was implemented and the experiment was conducted under a two-ray multipath model. Multipath channel distortion was equalized up to an inband linear amplitude dispersion of 17 dB for a short delay difference (2 ns), and up to 7.5 dB for a long delay difference (20 ns). These values agree well with the calculated values.

It is also shown that a transversal equalizer, used in combination with a simple frequency domain adaptive equalizer, greatly improves performance of the system against multipath fading.

1. INTRODUCTION

A 200 Mbps 16-QAM digital radio system [1] is now under development to replace the conventional analog FM in microwave trunk transmission systems.

In designing digital radio systems, special consideration has to be given to the effect of multipath fading on the microwave propagation path, especially in a long-haul 16-QAM system. To combat multipath fading, space diversity and adaptive equalization techniques are indispensable. Several such techniques have been developed, like inphase combining space diversities, interfering ray cancellation space diversities[2] and simple frequency domain adaptive equalizers.[1],[3] In the following cases, however, these techniques cannot always provide sufficient improvement:
(1) Fading with long delay differences.
(2) Channel responses which deviate from those of the two-ray path model.
(3) Fading with non-minimum phase response which gives the delay inversion characteristics.
The time domain adaptive equalizer, a transversal equalizer, is expected very effective in equalizing these various types of distortion.[4]

This paper proposes a transversal adaptive equalizer for the 16-QAM digital radio system to equalize waveform distortions caused by multipath propagation.

The performance of a transversal equalizer was investigated, theoretically and experimentally. The capability of the equalizer to equalize distortion induced by two-ray model fading was calculated by computer simulation with the number of taps as the parameters. Calculation was also done for residual distortion caused by a frequency domain adaptive equalizer[1] A transversal adaptive equalizer with 5 taps was developed and its performance was evaluated. Experimental results confirm validity of the simulation results and the effectiveness of implementing a transversal fading equalizer for the 16-QAM system.

2. COMPUTER SIMULATION OF A TRANSVERSAL EQUALIZER[5]

2.1 Channel model

Fig. 1 shows the channel model used for simulation representing the main function of the 16-QAM system. System parameters of the channel model are listed in Table 1. The signal path shown in Fig. 1 is represented in complex form. Spectrum shaping of 50% Nyquist roll-off is performed by a carrier frequency bandpass filter. The baseband transversal equalizer is two dimensional, with inphase and quadrature components. These are equal numbers of forward and backward taps. The control algorithm of automatic equalization is a mean-square-error method which performs well in fast convergence and has no divergence. The transmission path frequency responses are those of the multipath fading generated on the microwave propagation path or the residual responses of pre-equalizers (frequency domain adaptive equalizers).[1] In calculating intersymbol interference, fifty symbols before and after the center symbol were observed.

Peak intersymbol interference (peak distortion[5]) which directly relates to eye closure was used to evaluate residual intersymbol interference after equalization. Peak intersymbol interference D_p is defined as follows.

$$D_p = \sum_{n \neq 0} |h(nT)| \qquad (1)$$

where $h(t)$: impulse response of the system
nT : sampling time.

Reprinted from *IEEE 17th Int. Conf. Comm.*, vol. 3, pp. 46.2.1–46.3.5, June 1981.

For the allowable value of peak intersymbol interference, 0.5 was selected. This is because in the 16-QAM system, system degradation caused by distortion other than fading is estimated 6 dB which coresponds to an eye closure of 50%.

2.2 Transmission path frequency response

It is known through field experiments that there are two types of multipath fading, which can be represented in a two-ray multipath model.[6] One is generated by the interfering ray refracted in the atmospheric layer and features a short time delay difference between the direct ray and the interfering ray. The other is refracted on the sea-surface or in the ground and features a rather long time delay difference between the two rays.

The two-ray path model was used to simulate multipath fading response and typical delay differences of 2 and 4 ns were chosen for the former case and 10 and 20 ns for the latter case. Fig. 2 shows the frequency response of two-ray path model fading. Amplitude response $H(\omega)$ and delay response $\tau(\omega)$ are given by equations (2) and (3).

$$H(\omega) = \sqrt{1 + \rho^2 + 2\rho \cos(\omega\tau + \phi)} \quad (2)$$

$$\tau(\omega) = \frac{d\theta}{d\omega} = \frac{\rho\tau\{1 + \cos(\omega\tau + \phi)\}}{1 + \rho^2 + 2\rho \cos(\omega\tau + \phi)} \quad (3)$$

where ρ: amplitude ratio of interference to direct rays
τ: delay difference between the two rays
ϕ: phase difference between the two rays.

Residual distortion characteristics caused by a frequency domain adaptive equalizer (a pre-equalizer in the IF band) before the transversal equalizer are given for transmission path responses. The frequency domain adaptive equalizer uses a variable resonator which is controlled to compensate for multipath fading dispersion by information extracted from the signal power spectrum at three different frequencies. Residual distortion occurs when the center frequency of the resonator deviates from the notch frequency of fading. Typical residual responses are shown in Fig. 3, when $\rho = 0.97$ and $\tau = 2$ ns.

2.3 Simulation results

The performance of a transversal equalizer represented in ρ-τ relation curves with the number of taps as the parameter is shown in Fig. 4. The chosen notch frequency of fading is 15 MHz from the center frequency, where much waveform distortion occurs and where equalization is most difficult. Regions where $D_p \geqq 0.5$ are above the ρ-τ curves, when system outage occurs. The dashed line depicts the performance of a 16-QAM system without an equalizer.
The 5-tap equalizer can considerably improve the effects of fading in spite of the small number of taps.

Fig. 5 shows the performance of a transversal equalizer for the residual distortion caused by the frequency domain adaptive equalizer shown in Fig. 3. Residual curves ① and ② are equalized with only a few taps. Curve ③ which has abrupt changes is harder to equalize; it requires 10 taps for equalization.

Then, if the residual shown by curve ③ is suppressed, transversal equalizers with only a few taps combined with frequency domain adaptive equalizers provide considerable improvement against multipath fading.

3. HARDWARE IMPLEMENTATION OF THE 5-TAP TRANSVERSAL EQUALIZER

The 5-tap transversal equalizer for the 200 Mbps 16-QAM system is realized as shown in Fig. 6.

Configuration of the transversal adaptive equalizer with 5 taps is shown in Fig. 7.
The signal is delayed one symbol time in the delay circuit. The variable tap coefficients are made with IC mixers, which are frequency-compensated within a deviation of 1 dB in the range from 5 kHz to 30 MHz. ECL ICs are used in the logic circuits. They make decisions on signals and construct error signals to control each tap.

The control algorithm for adaptation is a simple zero-forcing method which is easy to construct. The integrators which determine the response speed of the equalizer are RC low pass filters with an RC time constant of 1 millisecond to catch up the fading rate (100 dB/second).

4. RESULTS OF EXPERIMENTS

The experimental test setup is shown in Fig. 8. A 16-QAM modem operating at a 140 MHz carrier frequency and with a bit rate of 200 Mbps was used. The reference carrier for demodulation was a recovered carrier. A hardwired carrier was provided in case the carrier recovery circuit locked out. Two-ray model fading was generated in the fading simulator.

In evaluating performance of the equalizer, amplitude ratio ρ was measured when bit error rate Pe exceeds 10^{-4}due to fading in the absence of thermal noise.

Fig. 9 shows the relation between fading notch frequency f_N and amplitude ratio ρ for a delay difference of $\tau = 2$ ns. In Fig. 9, the circles (o) show the results with the hard-wired carrier and the crosses (x) show the results with the recovered carrier. Performance with the recovered carrier was inferior because the carrier recovery circuit started to lock out. Performance with the hard-wired carrier represents that of the equalizer itself. The equalizer can equalize up to $\rho = 0.92$ of $\tau = 2$ ns fading. Fig. 10 shows eye diagrams before and after equalization for fading of $\rho = 0.82$ and $\tau = 2$ ns.

Experiments were also conducted for $\tau = 4$, 10, and 20 ns. Fig. 11 shows the ability of the equalizer to equalize two-ray multipath propagation distortion. Simulation results are shown by double circles (◎). These agree well with the results of the experiments.

Bit error rate Pe relative to carrier phase error $\Delta\theta c$ was measured in the presence of additive noise and without fading as shown in Fig. 12. Degradation with the equalizer (solid line) was much less than without it. Degradation of Pe was two or three times the optimum and corresponds to an S/N degradation of 0.3 dB when $\Delta\theta c$ equals 10 degrees. S/N degradation without the equalizer was 6 dB when $\Delta\theta c$ equals 10 degrees.

5. CONCLUSION

Performance of a transversal equalizer to equalize multipath fading in a 16-WAM microwave digital radio system was investigated theoretically and experimentally. The results are as follows.

(1) The equalizer capability relative to the number of taps was clarified.

(2) A transversal equalizer with only a few taps provides remarkable improvement, when combined with a simple frequency domain adaptive equalizer.

(3) The 5-tap transversal equalizer can equalize multipath fading of $\rho \leq 0.92$ for $\tau = 2$ ns, and $\rho \leq 0.4$ for $\tau = 20$ ns. These values correspond to inband linear amplitude dispersions of 17 dB and 7.5 dB, respectively.

(4) Carrier phase error is compensated by the equalizer. S/N degradation due to phase error of 10 degrees was within 0.3 dB, compared with 6 dB or more without an equalizer.

The transversal adaptive equalizer is expected to provide great improvement against various types of multipath channel distortions.

ACKNOWLEDGEMENT

The authors wish to thank Dr. H. Yamamoto and Mr. K. Morita, NTT, for many useful discussions and suggestions. They would also like to thank Messrs. T. Tsuda, Fujitsu Labs., and J. Dodo and H. Kurematsu, Fujitsu Ltd. for their guidance and encouragement.

REFERENCES

[1] I. Horikawa, Y. Okamoto, and K. Morita, "Characteristics of a High Capacity 16-QAM Digital Radio on a Multipath Fading Channel", ICC'79, 1979, PP.48.4.1-48.4.6.

[2] S. Komaki, Y. Okamoto, and K. Tajima, "Performance of 16-QAM Digital Radio System using New Space Diversity", ICC'80, 1980, PP.52.2.1-52.2.6.

[3] P. Hartmann and B. Bynum, "Adaptive Equalization for Digital Microwave Radio Systems", ICC'80, 1980, PP.8.5.1-8.5.6.

[4] M. Araki and T. Murase, "Effects of a Transversal Equalizer for a 16-QAM Microwave System", Paper of Tech. Group, IECE of Japan, CS80-125, Sept. 1980.

[5] R. W. Lucky, J. Salz, and E. J. Weldon, "Principle of Data Communication", McGraw-Hill, 1968.

[6] W. C. Jakes, "An Approximate Method to Estimate an Upper Bound on the Effect of Multipath Delay Distortion on Digital Transmission", IEEE Trans. Communications, vol. COM-27, No.1, Jan. 1979, PP. 76 - 81.

Table 1 System Parameters of Channel Model

Modulation	16 QAM
Demodulation	Coherent Detection
Symbol Rate	50 MB
Carrier Freq.	140 MHz
Filtering	50% Roll-Off

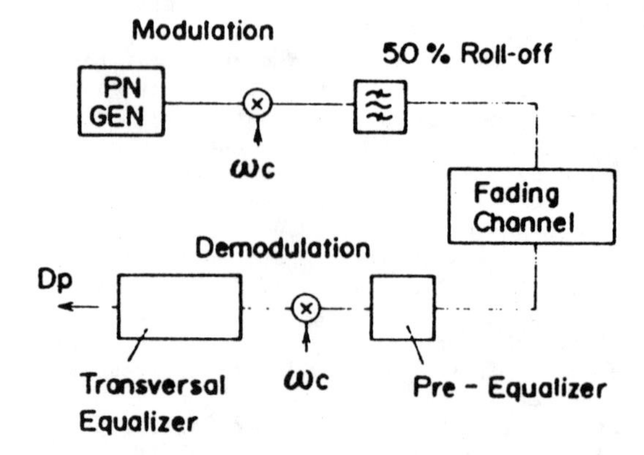

Fig. 1 Channel Model Used for Simulation

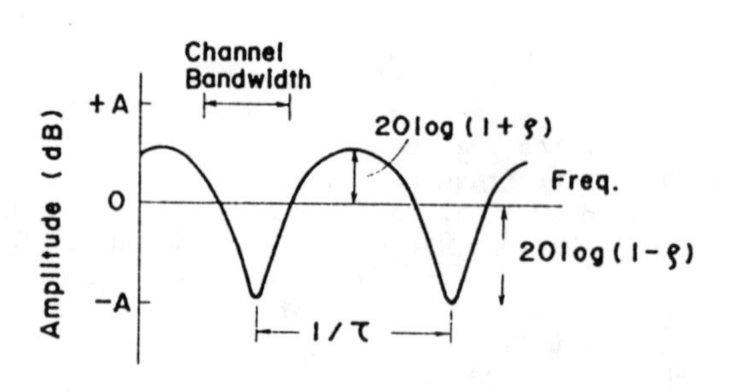

Fig. 2 Frequency Response of Fading

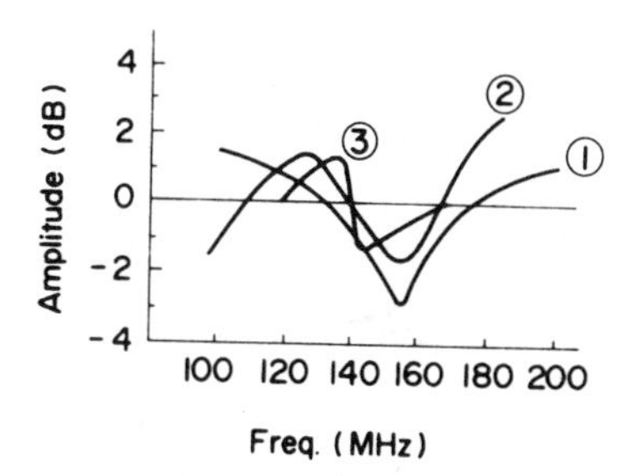

Fig. 3 Residual Response of a Frequency Domain Adaptive Equalizer

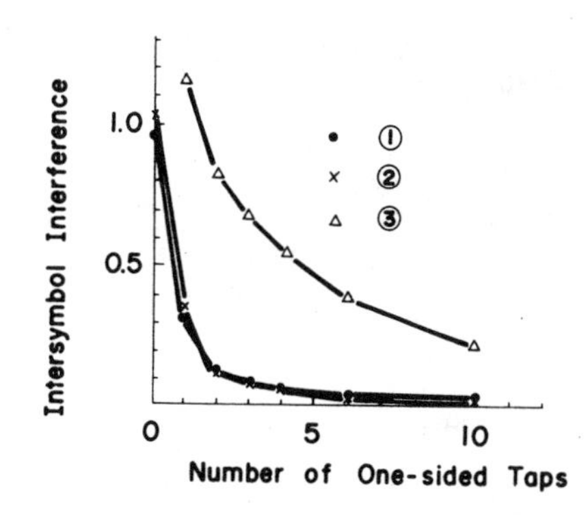

Fig. 5 Number of Taps vs Intersymbol Interference

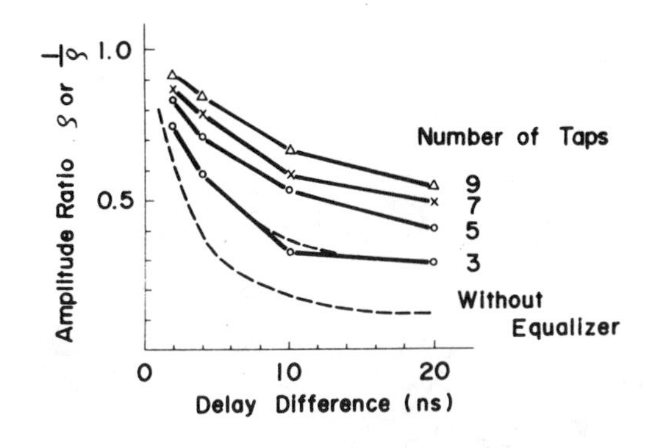

Fig. 4 Performance of the Equalizer with the number of taps as the parameter (D_p = 0.5)

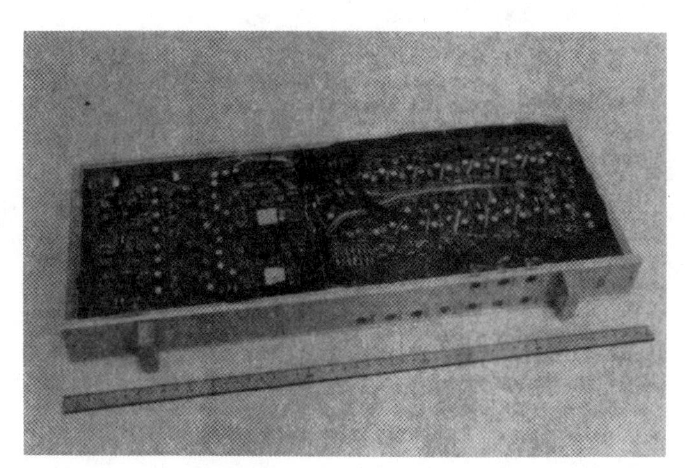

Fig. 6 Photo of the Equalizer

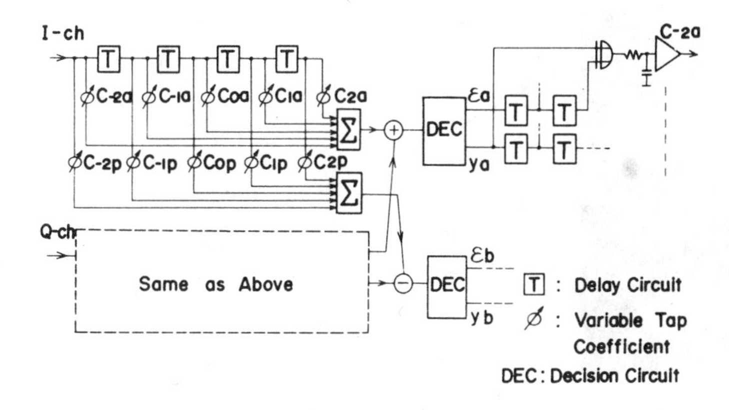

Fig. 7 Configuration of the Transversal Adaptive Equalizer (5 Taps)

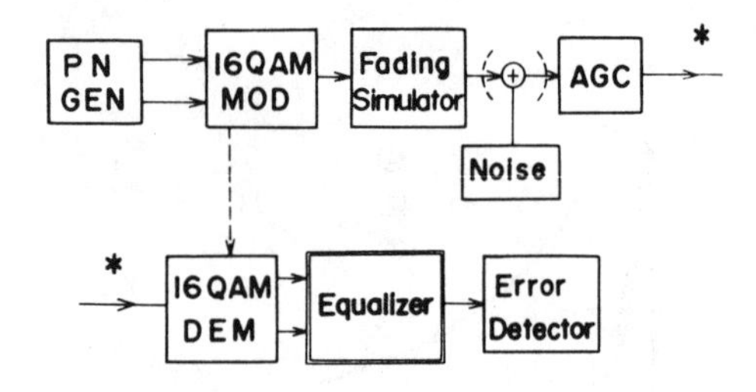

Fig. 8 Experimental Test Setup

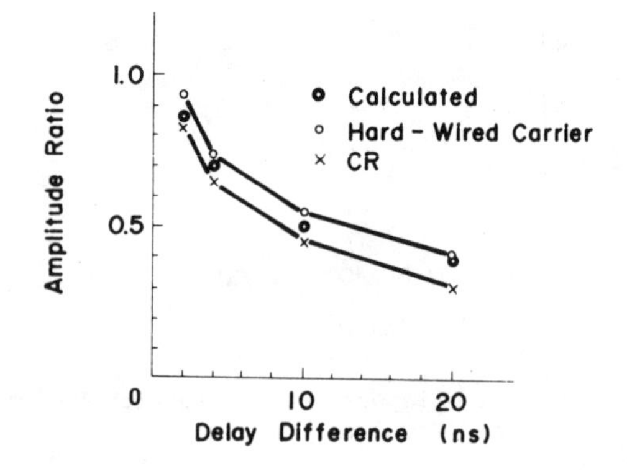

Fig. 11 Performance of 5-Tap Equalizer ($Pe = 10^{-4}$)

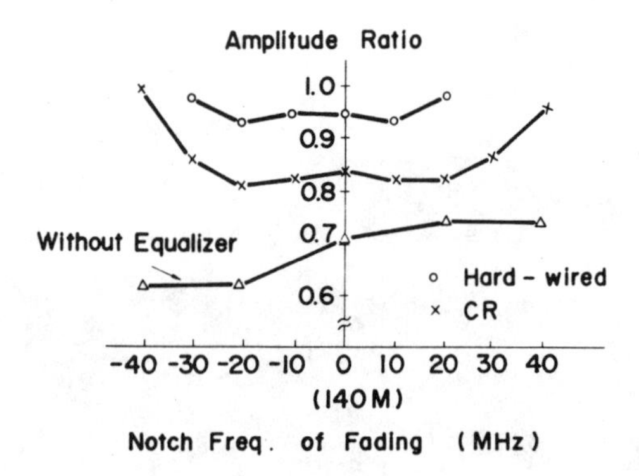

Fig. 9 Notch Freq. vs Amplitude Ratio (τ = 2 ns, $Pe = 10^{-4}$)

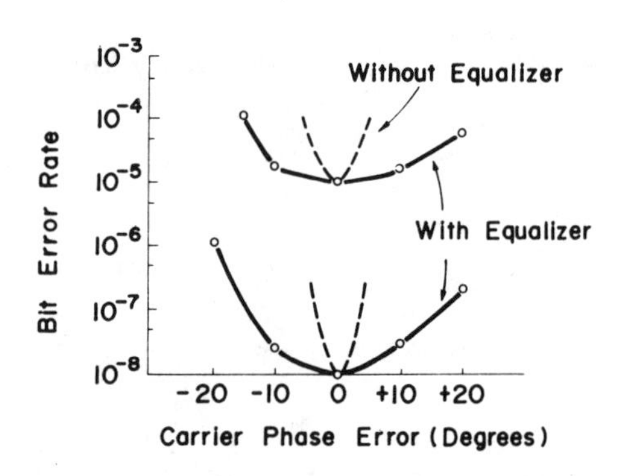

Fig. 12 Carrier Phase Error vs Bit Error Rate

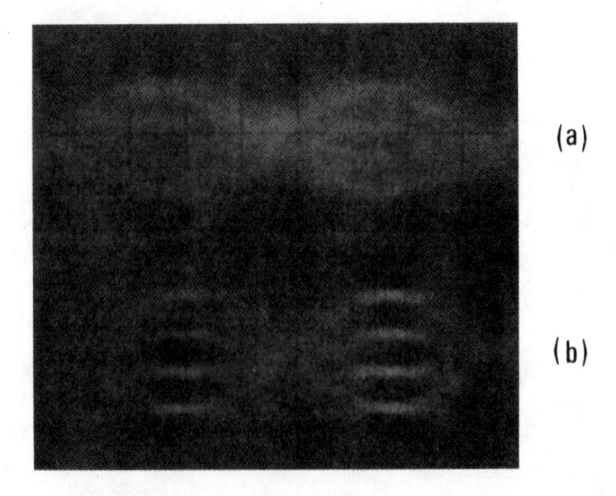

Fig. 10 Eye Diagrams. (a) Unequalized, (b) Equalized, Horizontal: 5 ns/div.

A Minimum Dispersion Combiner for High Capacity Digital Microwave Radio

SHOZO KOMAKI, MEMBER, IEEE, KOJIRO TAJIMA, MEMBER, IEEE, AND YOSHIHARU OKAMOTO, MEMBER, IEEE

***Abstract*—A new minimum-dispersion (MID) combiner, which reduces multipath degradation in a high capacity digital microwave radio, is proposed. A high capacity digital microwave radio is inherently very sensitive to waveform distortion caused by multipath in-band delay dispersion and in-band amplitude dispersion. To minimize the in-band dispersion, the combined-signal spectrum shape from the two antennas is monitored before and after a small change takes place in the combining phase. The phase shifter rotates in the direction of the flatter of the spectrum shapes, either the one before or the one after monitoring.**

Performance evaluations through simulation calculation and theoretical estimation using in-band amplitude dispersion probability density are given. More improvement can be obtained when the MID combiner is used instead of the maximum power (MAP) combiner currently in use. Laboratory and field experiments, using a 200 Mbit/s 16-QAM signal, verify these analyses and show an additional outage reduction factor of more than 5.

I. Introduction

COUNTERMEASURES for multipath fading are urgently needed in developing an economical digital microwave radio (DMR), designed to meet the growing digital transmission demands. As a countermeasure, space diversity and an adaptive equalizer have been introduced into the recent DMR [1]-[5].

The conventional space diversity combining method utilizes the MAP (maximum power) combiner, which combines two received signals on a cophase basis [6], [7] in order to maximize the level of the combined signal. As such, the MAP combiner design does not provide for minimizing in-band amplitude and delay dispersions, and its performance is not sufficient for use in a high capacity DMR.

On the other hand, another combining method employing the MID (minimum dispersion) combiner [8], [9] can suppress in-band dispersion. The MID combiner monitors the combined signal spectrum shape and combines two received signals to reduce spectrum shape distortion.

This paper describes the theoretical analysis for the operation of the MID combiner, its performance and outage probability reduction factors. It also describes the performance evaluation derived from simulation experiments and a field test.

In this analysis, it is assumed that multipath fading is manifested by a two-ray model, and that the interfering rays received by secondary antennas have about the same delay times.

II. MID Combiner Operation Principle

A. Operation

Multipath fading is caused by destructive interference between rays arriving at a receiving antenna via different paths, such as the one taken by direct rays and the interfering rays which are refracted in an atmospheric layer or reflected off the earth's surface, as shown in Fig. 1. The difference in the delay time between the interfering ray and the direct ray is τ_1 and the amplitude is r_1, which is normalized by the direct ray amplitude A_1. The received signal R_1 is given by

$$R_1 = A_1(1 + r_1 e^{j\omega\tau_1}) = A_1\sqrt{1 + r_1^2 + 2r_1 \cos \omega\tau_1}\, e^{j\theta_1} \tag{1}$$

where

$$\theta_1 = \tan^{-1}\left[\frac{r_1 \sin \omega\tau_1}{1 + r_1 \cos \omega\tau_1}\right] \tag{2}$$

and ω denotes angular frequency. The interfering ray has a different phase $\omega\tau_1$ from that of the direct ray. The phase difference between them increases in proportion to frequency increases. Thus, received signal level R_1 varies according to changes in the frequency. This results in amplitude frequency response degradation.

A secondary antenna, which is separated vertically from the main antenna, is used in space diversity reception.

This secondary signal R_2 is given by

$$R_2 = A_2(1 + r_2 e^{j\omega\tau_2}) = A_2\sqrt{1 + r_2^2 + 2r_2 \cos \omega\tau_2}\, e^{j\theta_2} \tag{3}$$

where

$$\theta_2 = \tan^{-1}\left[\frac{r_2 \sin \omega\tau_2}{1 + r_2 \cos \omega\tau_2}\right]. \tag{4}$$

The combined signal R_c is then

$$R_c = R_1 + R_2 e^{j\phi} \tag{5}$$

where ϕ denotes the phase shift. Then, the combined signal power $P_c(\omega)$ is given by

$$\begin{aligned} P_c(\omega) &= |R_c|^2 \\ &= A_1^2(1 + r_1^2 + 2r_1 \cos \omega\tau_1) \\ &\quad + A_2^2(1 + r_2^2 + 2r_2 \cos \omega\tau_2) + 2A_1A_2 \\ &\quad \cdot \{\cos \phi + r_1 \cos (\omega\tau_1 - \phi) + r_2 \cos (\omega\tau_2 + \phi) \\ &\quad + r_1 r_2 \cos (\omega\Delta\tau + \phi)\} \end{aligned} \tag{6}$$

where $\Delta\tau$ denotes the delay difference as follows.

$$\Delta\tau = \tau_2 - \tau_1. \tag{7}$$

A conventional MAP combiner combines the received signals so that R_1 and R_2 are cophase at the center frequency.

Paper approved by the Editor for Radio Communication of the IEEE Communications Society for publication after presentation at the International Conference on Communications, Seattle, WA, June 1980. Manuscript received July 24, 1981; revised October 5, 1983.

The authors are with the Yokosuka Electrical Communication Laboratory, Nippon Telegraph and Telephone Public Corporation, Yokosuka-shi, Kanagawak-en, 238-03, Japan.

Reprinted from *IEEE Trans. Comm.*, vol. COM-32, no. 4, pp. 419–428, Apr. 1984.

This combiner chooses a ϕ, as follows.

$$\phi_{MAP} = \theta_1 - \theta_2 = \tan^{-1}\left[\frac{r_1 \sin \omega_c\tau_1}{1 + r_1 \cos \omega_c\tau_1}\right] - \tan^{-1}\left[\frac{r_2 \sin \omega_c\tau_2}{1 + R_2 \cos \omega_c\tau_2}\right]$$

$$= \tan^{-1}\left[(r_1 \sin \omega_c\tau_1 - r_2 \sin \omega_c\tau_2 - r_1r_2 \cdot \sin \omega_c \Delta\tau)/(1 + r_1 \cos \omega_c\tau_1 + r_2 \cos \omega_c\tau_2 + r_1r_2 \cos \omega_c\Delta\tau)\right] \quad (8)$$

where ω_c is the angular center frequency for the passband.

Another combining method exists which uses the cancellation (CNCL) combiner. It minimizes the amplitude of the sum of the interfering rays, and its detailed operation is related in [8]. It combines two interfering rays in antiphase, as shown in Fig. 2, and the phase shift is chosen as follows.

$$\phi_{CNCL} = \pi - (\omega_c\tau_2 - \omega_c\tau_1) = \pi - \omega_c\Delta\tau. \quad (9)$$

A MID combiner combines the received signal so that dispersion is minimized at the passband. In other words, it minimizes the amplitude ratio between the sum of direct rays A_1 and A_2 and the sum of interfering rays B_1 and B_2, because the in-band dispersion is minimized when the amplitude ratio is minimized. Therefore, the MID combiner does not function in the same way as the CNCL combiner, because the CNCL combiner minimizes the amplitude of the sum of the two interfering rays B_1 and B_2, while ignoring the amplitude of the sum of the two direct rays A_1 and A_2. When the interfering rays B_1 and B_2 are equal in amplitude, both of these combining methods work the same. This point will be addressed later. A MID combiner chooses the phase shift as follows. The introduction for this equation is shown in the Appendix. There is no notch in the combined signal in the passband, i.e., when $(P_r(\omega_+) - P_r(\omega_c))\cdot(P_r(\omega_-) - P_r(\omega_c)) \leq 0$,

$$\phi_{MID} = \tan^{-1}\left[\frac{-XY - Z\sqrt{Y^2 + Z^2 - X^2}}{XZ + Y\sqrt{Y^2 + Z^2 - X^2}}\right]$$

or

$$\pi - \tan^{-1}\left[\frac{-XY + Z\sqrt{Y^2 + Z^2 - X^2}}{-XZ + Y\sqrt{Y^2 + Z^2 - X^2}}\right]; \quad \text{when } X^2 < Y^2 + Z^2$$

$$\phi_{MID} = \tan^{-1}\left[-\frac{r_1 \cos \omega_c\tau_1 \sin \frac{\Delta\omega\tau_1}{2} - r_2 \cos \omega_c\tau \sin \frac{\Delta\omega\tau_2}{2} + r_1r_2 \sin \omega_c\Delta\tau \sin \frac{\Delta\omega\Delta\tau}{2}}{r_1 \sin \omega_c\tau_1 \sin \frac{\Delta\omega\tau_2}{2} + r_2 \sin \omega_c\tau \sin \frac{\Delta\omega\tau_2}{2} + r_1r_2 \cos \omega_c\Delta\tau \sin \frac{\Delta\omega\Delta\tau}{2}}\right]; \quad \text{when } X^2 > Y^2 + Z^2 \quad (10)$$

where X, Y, and Z are given by (A.3) $\Delta\omega$ is the bandwidth in angular frequency, and

$$\omega_+ = \omega_c + \frac{\Delta\omega}{2}$$

$$\omega_- = \omega_c - \frac{\Delta\omega}{2}. \quad (11)$$

On the other hand, when a notch exists in the passband, i.e., when $(P_r(\omega_+) - P_r(\omega_c))\cdot(P_r(\omega_-) - P_r(\omega_c)) > 0$,

$$\phi_{MID} = \tan^{-1}\left[\frac{-xy - z\sqrt{y^2 + z^2 - x^2}}{xz + y\sqrt{y^2 + z^2 - x^2}}\right]$$

or

$$\pi - \tan^{-1}\left[\frac{-xy + z\sqrt{y^2 + z^2 - x^2}}{-xz + y\sqrt{y^2 + z^2 - x^2}}\right]; \quad \text{when } x^2 < y^2 + z^2$$

$$\phi_{MID} = \tan^{-1}\left[-\frac{r_1\left(1 - \cos\frac{\Delta\omega\tau_1}{2}\right)\sin \omega_c\tau_1 - r_2\left(1 - \cos\frac{\Delta\omega\tau_2}{2}\right)\cdot \sin \omega_c\tau_2 - r_1r_2\left(1 - \cos\frac{\Delta\omega\Delta\tau}{2}\right)\cos \omega_c\Delta\tau}{-r_1\left(1 - \cos\frac{\Delta\omega\tau_1}{2}\right)\cos \omega_c\tau_1 - r_2\left(1 - \cos\frac{\Delta\omega\tau_2}{2}\right)\cdot \cos \omega_c\tau_2 - r_1r_2\left(1 - \cos\frac{\Delta\omega\Delta\tau}{2}\right)\sin \omega_c\Delta\tau}\right];$$

$$\text{when } x^2 > y^2 + z^2 \quad (12)$$

where x, y, and z are defined by (A.9). Except for the special case of $A_1r_1 = A_2r_2$, phase shifts ϕ_{MAP}, ϕ_{CNCL}, and ϕ_{MID} differ from each other, as shown in (8)-(10) and (12).

Fig. 3 shows the ϕ_{MAP}, ϕ_{CNCL}, and ϕ_{MID} for various r_1, r_2, and $\Delta\tau$ values. ϕ_{MAP} is not always the same as ϕ_{MID}, because the correlation between ϕ_{MAP} and ϕ_{MID} is low, as seen in Fig. 3(a). On the other hand, ϕ_{MID} is nearly, but not always, equal to ϕ_{CNCL}, as shown in Fig. 3(b).

In the case of $A_1r_1 = A_2r_2$, the CNCL combiner operates exactly the same as the MID combiner. The interfering ray can be cancelled out completely, and the frequency response becomes flat.

An example of frequency characteristics created by multipath fading is shown in Fig. 4. Here, dotted lines show the frequency characteristics corresponding to the two received signals before combining. The MID combiner is seen to give the best performance among these three combining methods.

B. Control Circuit

An example of the MID combiner configuration is provided in Fig. 5. A combined signal spectrum shape is monitored at several frequencies by narrow-band filters and detectors. Then the phase shifter is rotated to an arbitrary direction, and the

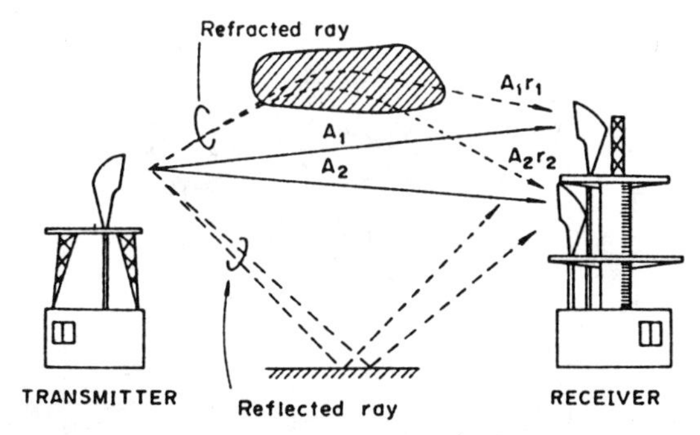

Fig. 1. Multipath fading model.

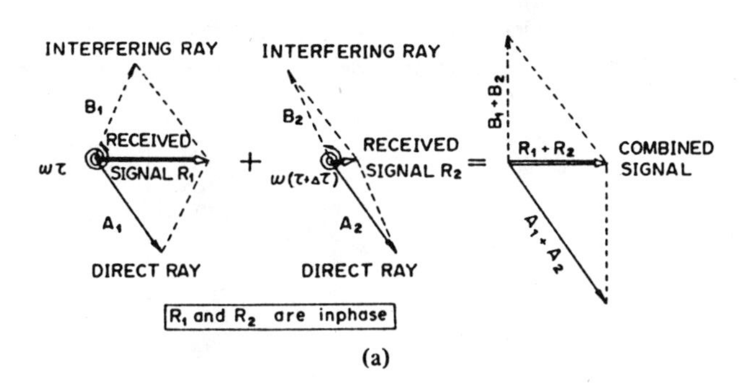

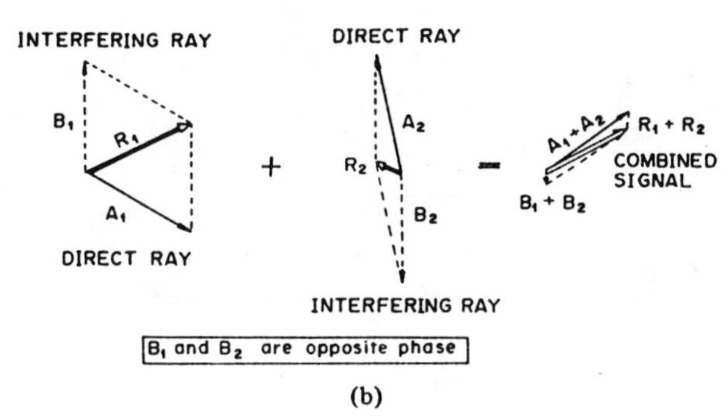

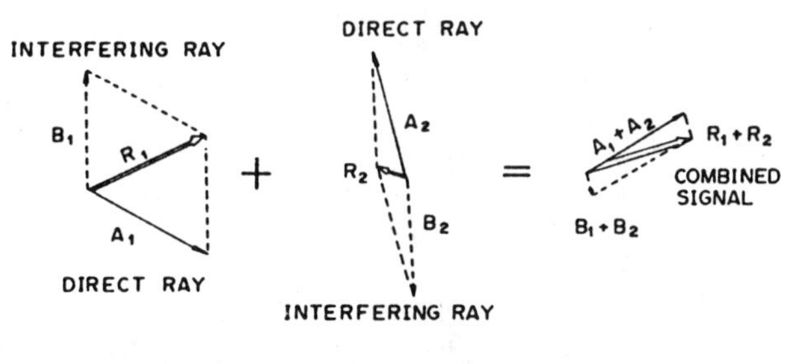

Fig. 2. Vector diagrams for various combiners. (a) MAP combiner. (b) CNCL combiner. (c) MID combiner.

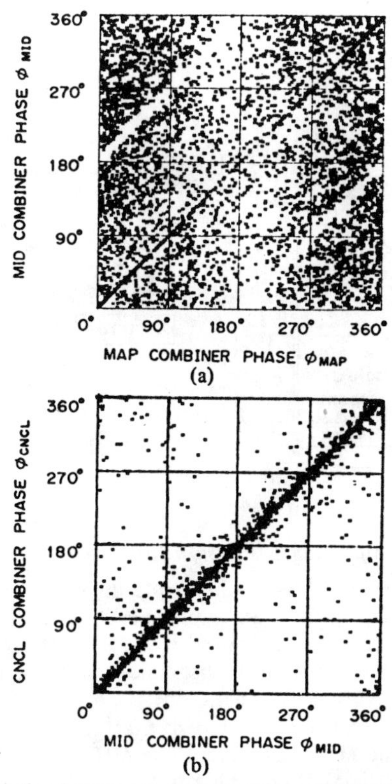

Fig. 3. Correlation between various combiner phase shifts. (a) Correlation between ϕ_{MAP} and ϕ_{MID}. (b) Correlation between ϕ_{MID} and ϕ_{CNCL}.

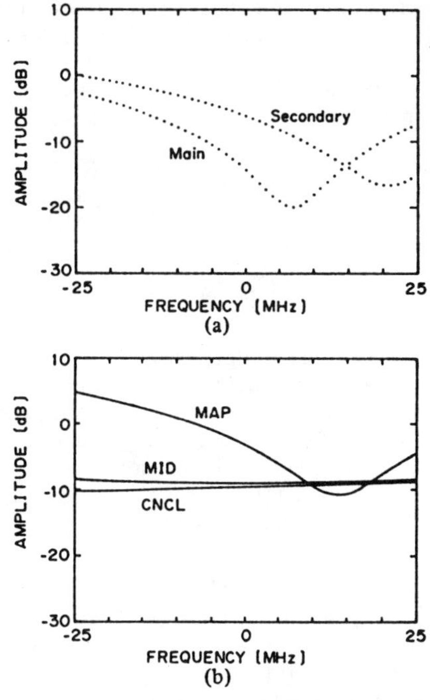

Fig. 4. Frequency response for several kinds of combining methods. (a) Single receptions. (b) Various combiners.

combined signal spectrum shape is monitored again. From these two spectrum shapes, two peak-to-peak in-band amplitude dispersions are calculated. After comparing these two in-band amplitude dispersions, the phase shifter is rotated in the direction in which in-band amplitude dispersion decreases. In this way, the in-band dispersion is minimized. A microcomputer can be used for this purpose.

III. Performance Simulation Calculations

For fading, it was assumed that the interfering signal amplitudes vary at random, and that their distributions are uniform. Their correlation is denoted by K_r.

Then, computer simulations are performed under the following conditions:

$$\begin{cases} p(r_1) = p(r_2) = [0, 1] \\ \overline{r_1 \cdot r_2} = K_r^2 \\ A_1 = A_2 = 1 \\ \tau_1 \text{ is fixed} \\ 0 < \omega\Delta\tau \leqq 2\pi. \end{cases} \tag{13}$$

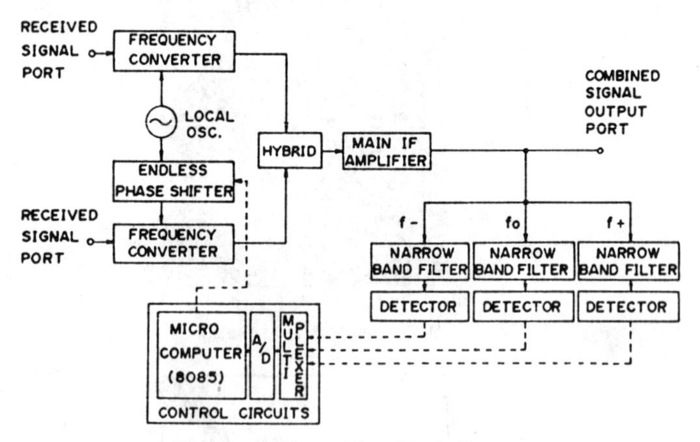

Fig. 5. MID combiner block diagram.

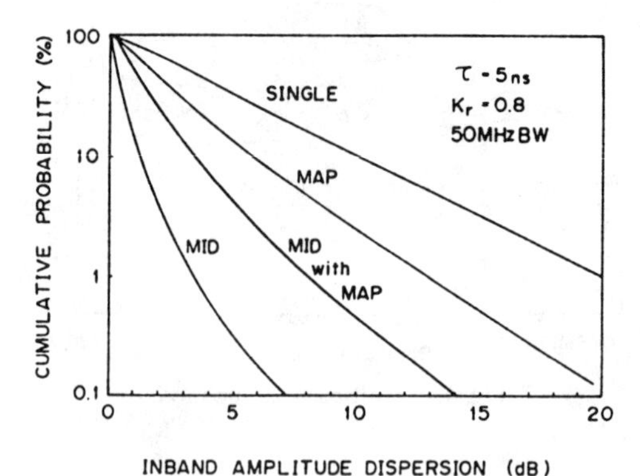

Fig. 6. Cumulative in-band amplitude dispersion probability example.

A. In-Band Amplitude Dispersion Improvement

The degree of in-band amplitude dispersion improvement, used in estimating the outage probability reduction factors, is calculated in this section. As seen in Fig. 6, the MID combiner improves in-band amplitude dispersion, and this improvement is relatively greater than that achieved by the MAP combiner. Therefore, it can be concluded that the MID combiner is superior to the MAP combiner. The MID-with-MAP combiner is inferior to the MID combiner in the aspect of in-band amplitude dispersion, because the MID-with-MAP combiner operates the same as the MAP combiner when the combined signal is low.

The MAP combiner performance decreases when the delay difference between the direct ray and the interfering ray increases. Fig. 7 shows calculated in-band amplitude dispersion improvements versus delay difference in the MID and MAP combiners. In this figure, improvement factors are defined as follows.

$$I_{MID} = P_{single}(D \geqq 5 \text{ dB})/P_{MID}(D \geqq 5 \text{ dB})$$
$$I_{MAP} = P_{single}(D \geqq 5 \text{ dB})/P_{MAP}(D \geqq 5 \text{ dB}) \quad (14)$$

where D denotes in-band amplitude dispersion and P_{single}, P_{MID}, and P_{MAP} are cumulative in-band amplitude dispersion probabilities. This figure shows that the MID combiner improvement factor does not decrease rapidly in comparison with the MAP combiner. It can be concluded that the MID combiner is most suited for paths with a long delay difference between the direct and interfering ray.

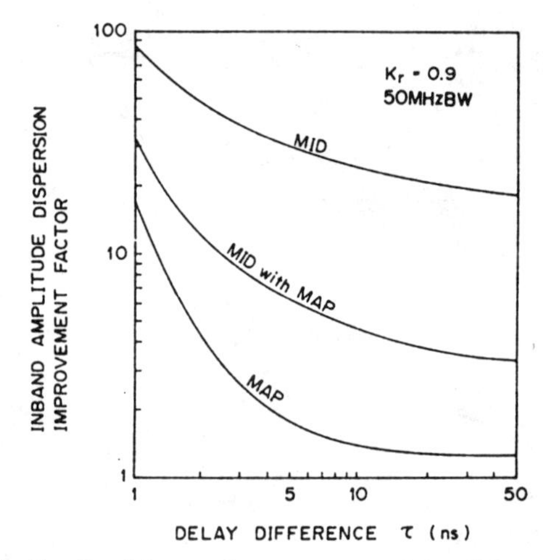

Fig. 7. Combiner performance versus delay difference.

B. Performance of Infering Rays with Unequal Amplitudes

When there are several interfering rays having small delay differences, the overall interfering ray amplitude varies randomly during fades. In these fades, the condition $r_1 = r_2$ does not hold true. However, MID combiner reception is still good during these fades.

The calculated results are shown in Fig. 8. The correlation coefficient between the interfering ray amplitudes is denoted by K_r. This figure shows that minimum dispersion combiner performance does not decrease, even when interfering rays are unequal. This point is clarified by the following explanation.

1) When $r_1 \neq r_2$, the interfering rays cannot be cancelled out However, both r_1 and r_2 cannot be in unity at the same time, so at least one of the received signals has a nearly flat frequency response.

2) When $r_1 = r_2 = 1$, the interfering rays are cancelled out, even if large in-band amplitude dispersion is present.

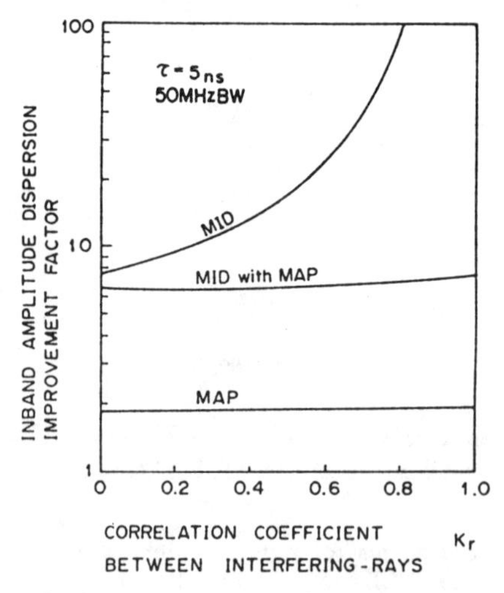

Fig. 8. Performance when interfering rays are not equal.

C. Signal Loss Prevention

If phase difference $\omega\tau_1$ is the same as $\omega\tau_2$, the direct rays received by the main and secondary antennas are simultaneously cancelled out when the interfering rays are cancelled out. Therefore, the combined signal loss increases in MID combiner reception. This signal loss is largest when the interfering ray amplitude of the signal received by the main antenna is equal to that of the signal received by the secondary antenna ($K_r = 1$). Fig. 9 shows the attenuation distributions for single, MAP combiner, and MID combiner reception, when $K_r = 1$.

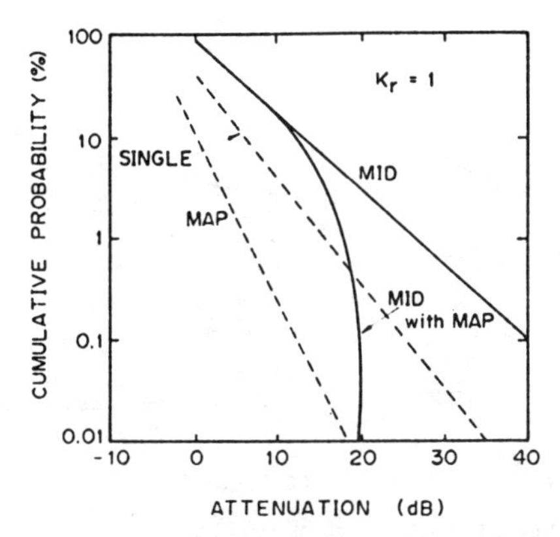

Fig. 9. Cumulative attenuation probability.

The MID combiner attenuation is larger than that of the MAP combiner, as mentioned above. Since received signal loss increases the circuit outage caused by thermal noise and interference, it is necessary for the MID combiner to prevent signal loss. Combined signal level monitoring is used for this purpose. The MID-with-MAP combiner operates the same as the MID combiner, when the combined signal attenuation is smaller than 20 dB, and operates the same as the MAP combiner, when the combined signal attenuation is larger. Attenuation probability is improved by using the level recovery algorithm, as shown in Fig. 9 (MID with MAP).

IV. Theoretical Analysis for Outage Probability Reduction Factor

Circuit outage due to multipath fading can be estimated by the probability that in-band amplitude dispersion exceeds the value D_0 [10]. This value D_0 is about 5 dB for 16 QAM, about 10 dB for QPSK, and 2 dB for 64 QAM. In this section, the outage probability reduction factors for the various combiners are discussed with this value in mind.

A. *In-Band Amplitude Dispersion Probability*

If delay differences τ_1 and τ_2 are smaller than Nyquist duration T, in-band amplitude dispersion can be approximated by the ratio between two received signal powers measured at the extreme edges of the passband. Received signal probability density can be represented by a gamma distribution:

$$f(X) = \frac{\lambda^\lambda}{\Gamma(\lambda)} X^{\lambda-1} e^{-\lambda X} \tag{15}$$

where

$$\Gamma(\lambda) = \int_0^\infty t^{\lambda-1} \cdot e^{-\lambda}\, dt \tag{16}$$

and λ is the parameter showing fading depth. Actual fading observation has shown that $\lambda = 1$ for single reception and $\lambda = 2$ for dual diversity reception [11].

Cumulative probability of power ratio between the received signals $Z = X_1/X_2$ becomes

$$P(Z) = \frac{1}{B(\lambda)} \int_0^{\sin^2\theta/2} x^{\lambda-1}(1-x)^{\lambda-1}\, dx \tag{17}$$

where

$$B(\lambda) = \Gamma(\lambda)^2/\Gamma(2\lambda)$$

$$\theta = \tan^{-1}\left(\sqrt{\frac{4(1-\rho)Z}{(1-Z)^2}}\right) \tag{18}$$

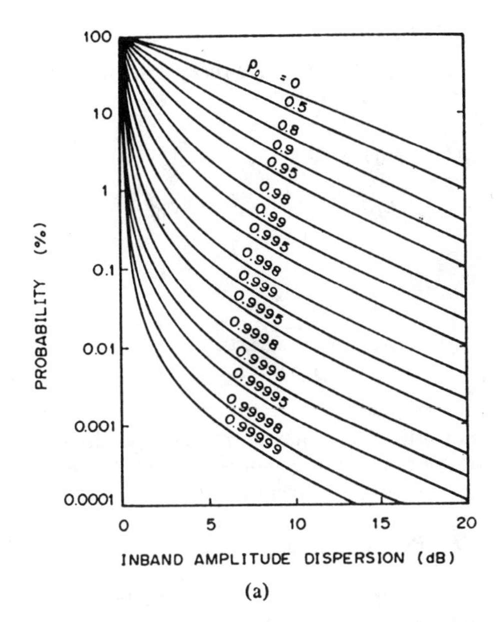

(a)

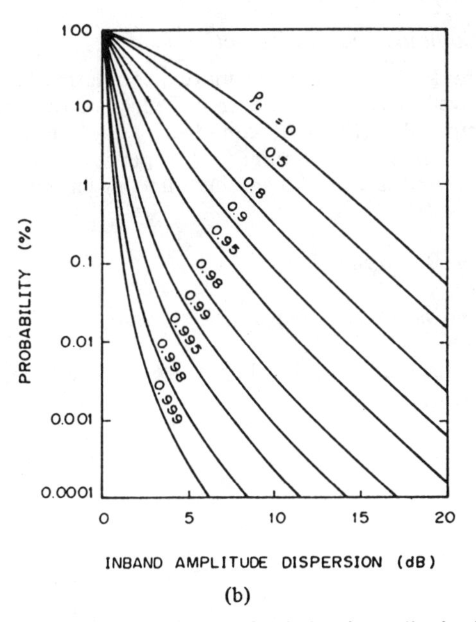

(b)

Fig. 10. Cumulative distribution for in-band amplitude dispersion. (a) Single reception. (b) Dual diversity.

$\rho(\Delta\omega)$ mutual correlation coefficient between two received signal powers, monitored at the extreme edges of passband $\Delta\omega$.

The in-band amplitude dispersion is defined by the absolute value $|Z|$. Then the cumulative probability for in-band amplitude dispersion D becomes as follows.

$$\begin{cases} P_s(D) = 2\alpha; & \lambda = 1 \text{ (single reception)} \\ P_{SD}(D) = 6\alpha^2 - 4\alpha^3; & \lambda = 2 \text{ (dual diversity)} \end{cases} \tag{19}$$

where

$$\alpha = \frac{1}{2}\left\{1 + \frac{1-D}{\sqrt{(1+D)^2 - 4\rho D}}\right\}. \tag{20}$$

Some calculated distribution examples are shown in Fig. 10.

B. MAP Combiner Outage Reduction Factor

Outage probabilities for single reception and MAP combiner can be approximated as follows.

$$P_0 = P_S(D > D_0) = 2\alpha_0$$
$$P_{MAP} = P_{SD}(D > D_0) = 6\alpha_0^2 - 4\alpha_0^3 \quad (21)$$

where

$$\alpha_0 = \frac{1}{2}\left\{1 + \frac{1 - D_0}{\sqrt{(1 + D_0)^2 - 4\rho_0 D_0}}\right\} \quad (22)$$

and ρ_0 denotes the frequency correlation coefficient for single or MAP combiner reception. Therefore, the outage probability reduction factor can be shown as follows.

$$I_{MAP} = P_0/P_{MAP} = 1/(3\alpha_0 - 2\alpha_0^2). \quad (23)$$

Fig. 11 shows several examples of calculated MAP combiner outage reduction factors.

C. MID Combiner Outage Reduction Factor

For the MID combiner, the mutual correlation between the two received signals at the extreme edges of the passband is high compared to the MAP combiner, because the MID combiner minimizes in-band amplitude dispersion. From field experiments conducted in Japan, the following empirical formula was obtained:

$$\rho_{MID} = 0.463\rho_0 + 0.537. \quad (24)$$

Fig. 12 shows the field experimental results and the empirical formula.

Then, the outage probability can be estimated by the following equation.

$$P_{MID} = 6\alpha_{MID}^2 - 4\alpha_{MID}^3 \quad (25)$$

where

$$\alpha_{MID} = \frac{1}{2}\left\{1 + \frac{1 - D_0}{\sqrt{(1 + D_0)^2 - 4\rho_{MID} D_0}}\right\}. \quad (26)$$

From (21) and (26), the outage reduction factor can be shown as follows.

$$I_{MID} = P_0/P_{MID} = \alpha_0/(3\alpha_{MID}^2 - 2\alpha_{MID}^3). \quad (27)$$

Fig. 13 gives some examples of the calculated results.

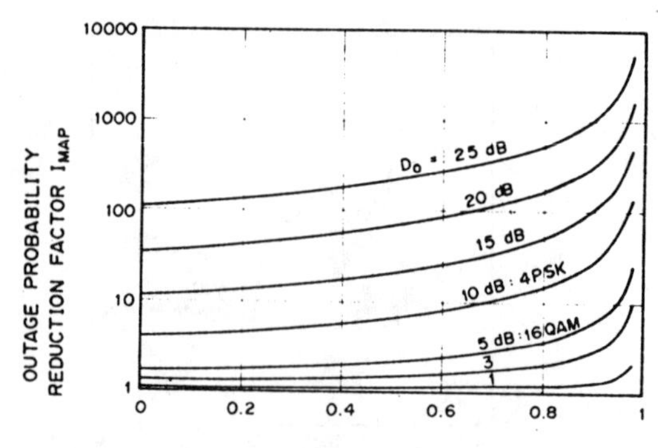

Fig. 11. Outage reduction by MAP combiner.

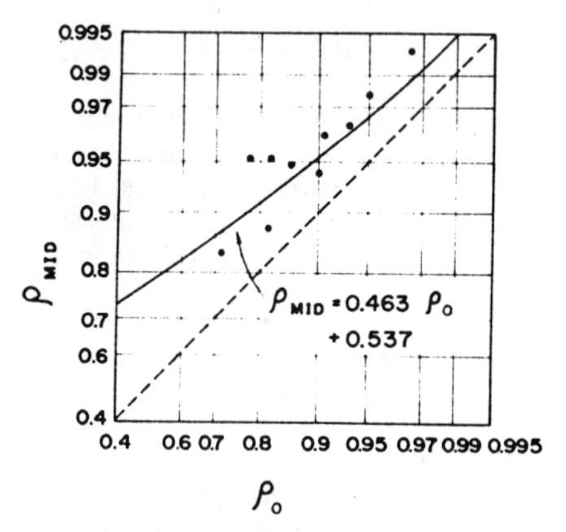

Fig. 12. Relation between ρ_{MAP} and ρ_{MID}.

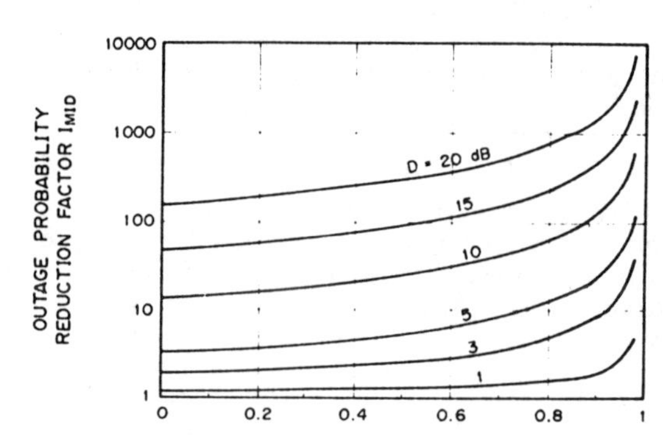

Fig. 13. Outage reduction by MID combiner.

V. Experimental Results

A. Description of the Experimental System

To verify MID combiner performance under multipath fading conditions, simulation experiments and a field test were carried out. A 200-Mbit/s 16-QAM signal was used. The spectrum was shaped by a 50 percent rolloff filter. The fading simulator used in this experiment generated a two-ray fading model, with uniform amplitude distributions r_1 and r_2, and uniform phase distribution for ϕ_1 and ϕ_2. The delay difference was easily varied by changing the delay line length.

The MID combiner configuration is shown in Fig. 14. A microcomputer (8085) was used in the control circuit for control algorithm flexibility. The combiner was installed in a test 16-QAM digital radio repeater. The combiner circuits are very compact and do not increase the repeater size. The combined signal is demodulated to four 50 Mbit/s streams. Errors are counted every 0.2 s and received power levels are monitored by AGC voltage. In-band amplitude dispersion was measured at three frequencies, 115, 140, and 165 MHz. The measured flat fade margin for the system was about 40 dB.

B. Simulation Experiments

The results obtained from the simulation experiments are shown in Figs. 15–18. The time series data are shown in Fig. 15.

The outage probability improvement for the MID combiner, observed during the simulation experiments, is five times greater than that for the MAP combiner.

The measured in-band amplitude dispersion cumulative probability is shown in Fig. 16. From the experiment, it was confirmed that the MID combiner can improve in-band amplitude dispersion.

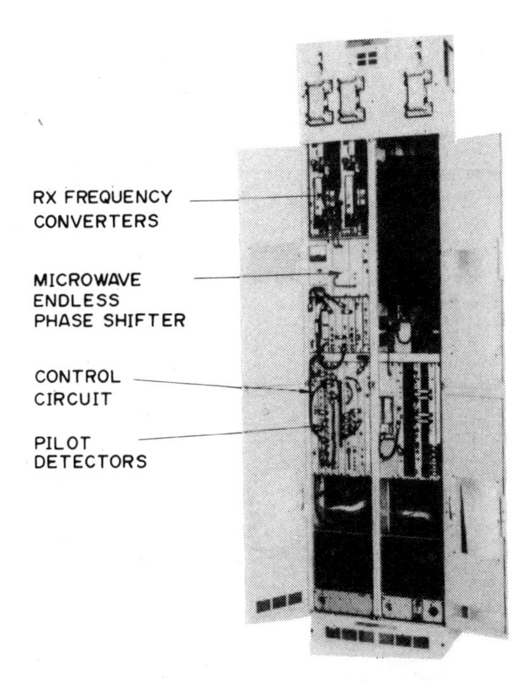

Fig. 14. MID combiner installed in 16 QAM digital radio repeater.

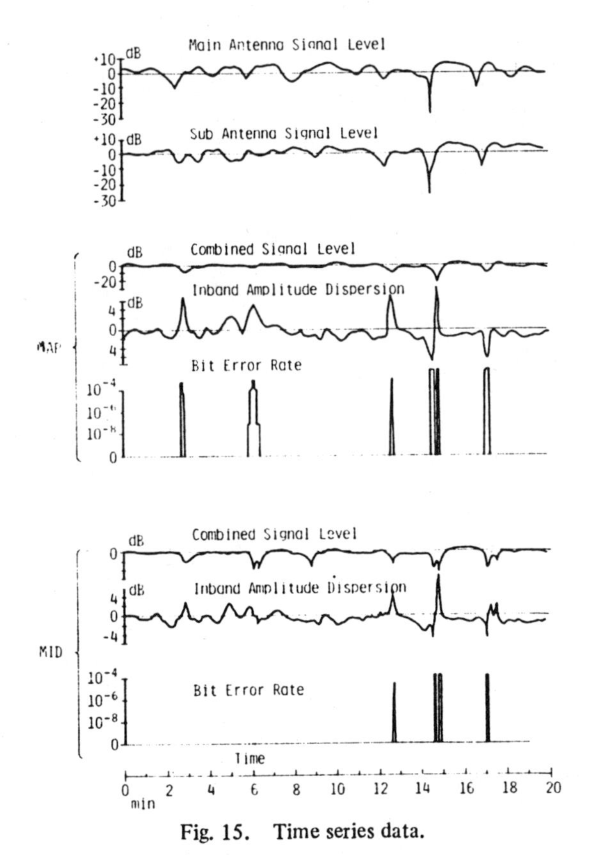

Fig. 15. Time series data.

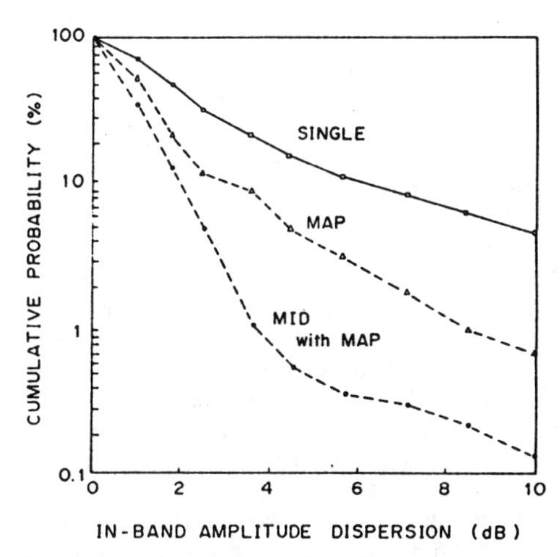

Fig. 16. Measured cumulative in-band amplitude dispersion distribution.

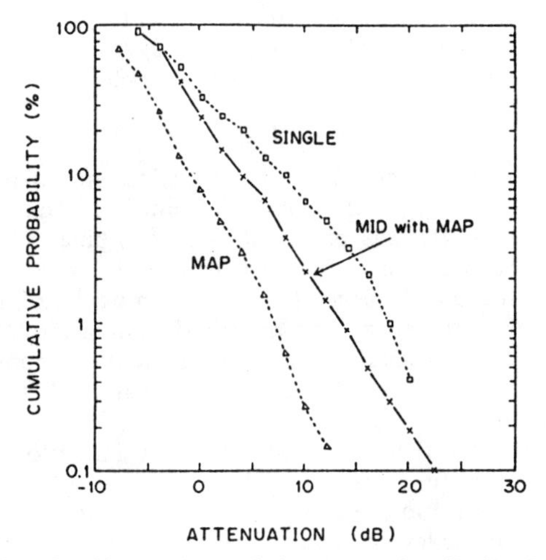

Fig. 17. Measured cumulative attenuation distribution.

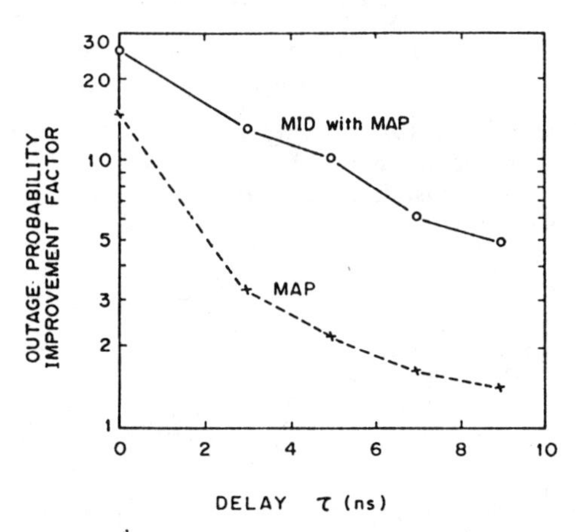

Fig. 18. Measured outage reduction factor.

The fade depth cumulative probability is shown in Fig. 17. It can be seen that the MID-with-MAP combiner fade depth is between the single and MAP combiner reception.

The outage probability improvement factor, shown in Fig. 18, is associated with the delay change. In the case of a small delay, the MID combiner will have nearly the same improvement factor as the MAP combiner. In the case of a larger delay, the MID combiner improvement factor will not decrease rapidly. The MAP combiner improvement factor, however, decreases rapidly. Thus, the MID combiner displays greater potential in large delay differences rather than in small delay differences.

C. Field Experiments

The field experiments were conducted from August to October 1980 over one of the worst propagation paths in FM routes existing in Japan [12]. This path extends 53 km over water between Muroran and Kakkumi on the island of Hokkaido. The power ratio of the direct ray to the ray reflected from the surface of the sea was about 6 dB. The delay difference between the two rays was 4.3 ns. The equipment configuration was almost the same as in the simulation experiments. A transmitter, operating at 1.5 W output power, and horn reflector antennas, having a 42.5 dB gain, were used. The flat fade margin was more than 34 dB.

Fig. 19 shows the cumulative in-band amplitude dispersion probability for various configurations. The new MID combiner reduces the dispersion probability to less than 1/5 of that gained by the conventional MAP combiner at 5 dB. This reduction factor is the same as that observed for outage probability reduction, as shown in Fig. 20. It has been confirmed by the field experiment that the MID combiner is superior to the MAP combiner currently in use.

The MID combiner is able to suppress the amplitude of the interfering ray after combining the signals, so the polarity inversion effect in the delay characteristic is suppressed. This polarity inversion is known as the nonminimum phase fading condition and occurs when the interfering ray amplitude exceeds that of the direct ray. It degrades the performance of the dynamic equalizer. Since the MID combiner suppresses the nonminimum phase fading condition, dynamic equalizer potential as well as its synergetic effects are enhanced.

VI. Conclusion

As an effective countermeasure against multipath fading on microwave propagation paths, a new MID (minimum dispersion) combiner was proposed. This combiner reduces the outage probability in high speed digital microwave radio systems which use high spectrum efficiency modulation methods. From analysis based on the multipath fading model, outage probability and in-band dispersion are reduced considerably compared with those of the conventional MAP (maximum power) combiner.

To verify the theoretical analysis, both simulation and field experiments were carried out. These tests show that the MID combiner has an outage probability improvement factor of 5. Combined signal level loss can be reduced through the dual use of the MID and MAP combining methods.

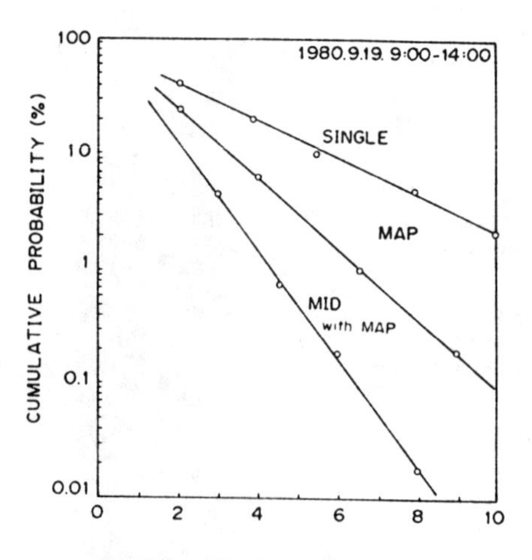

Fig. 19. Cumulative in-band amplitude dispersion distribution observed in the field experiments.

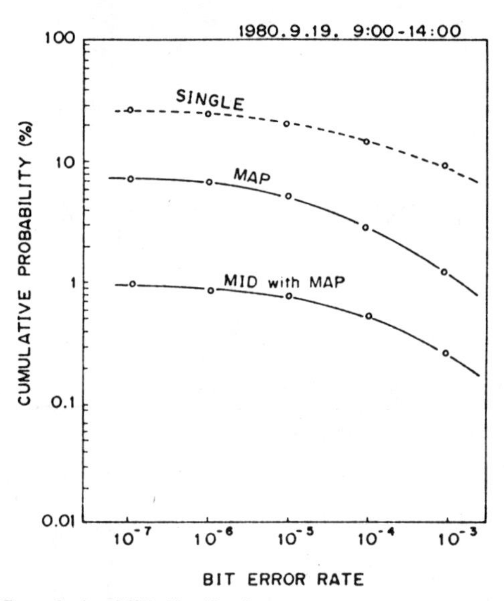

Fig. 20. Cumulative BER distribution observed in the field experiments.

Appendix

Phase Shift to Minimize the In-Band Amplitude Dispersion

1) In case of $(P_r(\omega_+) - P_r(\omega_c))\cdot(P_r(\omega_-) - P_r(\omega_c)) \leq 0$, that is, when the combined signal frequency response has no notch in the passband, in-band amplitude dispersion can be defined as follows.

$$D = P_r(\omega_+) - P_r(\omega_-). \tag{A.1}$$

Then, the in-band dispersion is minimized, when $D = 0$ or $\partial D/\partial\phi = 0$. From (5),

$$\begin{aligned} D = {} & A_1{}^2(1 + r_1{}^2 + 2r_1 \cos \omega_+\tau_1) + A_2{}^2(1 + r_2{}^2 \\ & + 2r_2 \cos \omega_+\tau_2) - A_1{}^2(1 + r_1{}^2 + 2r_1 \cos \omega_-\tau_1) \\ & - A_2{}^2(1 + r_2{}^2 + 2r_2 \cos \omega_-\tau_2) + 2A_1A_2\{\cos\phi \\ & + r_1 \cos(\omega_+\tau_1 - \phi) + r_2 \cos(\omega_+\tau_2 + \phi) \\ & + r_1r_2 \cos(\omega_+\Delta\tau + \phi)\} = X + Y \sin\phi + Z\cos\phi \end{aligned} \tag{A.2}$$

where

ω_+, ω_- angular frequency at the extreme edge of the passband

τ_1, τ_2 delay differences between direct rays and interfering rays

$$\Delta\omega = \omega_+ - \omega_-$$

$$\Delta\tau = \tau_2 - \tau_1$$

$$X = -4 \cdot \left\{A_1{}^2 r_1 \sin\omega_c\tau_1 \sin\frac{\Delta\omega\tau_1}{2} + A_2{}^2 r_2 \sin\omega_c\tau_2 \cdot \sin\frac{\Delta\omega\tau_2}{2}\right\}$$

$$Y = -4A_1A_2\left\{-r_1\cos\omega_c\tau_1 \sin\frac{\Delta\omega\tau_1}{2} + r_2\cos\omega_c\tau_2 \cdot \sin\frac{\Delta\omega\tau_2}{2} + r_1r_2\cos\omega_c\Delta\tau \cdot \sin\frac{\Delta\omega\Delta\tau}{2}\right\}$$

$$Z = -4A_1A_2 \left\{ r_1 \sin \omega_c\tau_1 \sin \frac{\Delta\omega\tau_1}{2} + r_2 \sin \omega_c\tau_2 \cdot \sin \frac{\Delta\omega\tau_2}{2} + r_1r_2 \sin \omega_c\Delta\tau \cdot \sin \frac{\Delta\omega\Delta\tau}{2} \right\} \quad (A.3)$$

Equation (A.3) and $D = 0$ lead to

$$\phi = \begin{cases} \tan^{-1}\left[\dfrac{-XY - Z\sqrt{Y^2+Z^2+X^2}}{XZ + Y\sqrt{Y^2+Z^2-X^2}}\right] & \text{or} \\ \pi - \tan^{-1}\left[\dfrac{-XY + Z\sqrt{Y^2+Z^2-X^2}}{-XZ + Y\sqrt{Y^2+Z^2-X^2}}\right]; & \\ \text{when } X^2 \leqq Y^2 + Z^2. \end{cases} \quad (A.4)$$

This equation holds true when $X^2 \leqq Y^2 + Z^2$. This condition denotes that D reaches zero.

On the other hand, in case of $X^2 > Y^2 + Z^2$, D does not reach zero. In this case, D is positive or negative, so the D is minimized at $\partial D/\partial\phi = 0$. From (5)

$$\frac{\partial D}{\partial\phi} = \frac{\partial P_r(\omega_+)}{\partial\phi} - \frac{\partial P_r(\omega_-)}{\partial\phi} = 4A_1A_2 \left\{ \left(r_1 \cos \omega_c\tau_1 \cdot \sin \frac{\Delta\omega\tau_1}{2} - r_2 \cos \omega_c\tau_2 \sin \frac{\Delta\omega\tau_2}{2} + r_1r_2 \sin \omega_c\Delta\tau \cdot \sin \frac{\Delta\omega\Delta\tau}{2} \right) \cos\phi + r_1 \sin \omega_c\tau_1 \sin \frac{\Delta\omega\tau_1}{2} + r_2 \cdot \sin \omega_c\tau_2 \sin \frac{\Delta\omega\tau_2}{2} + r_1r_2 \cos \omega_c\Delta\tau \cdot \sin \frac{\Delta\omega\Delta\tau}{2} \sin\phi \right\}. \quad (A.5)$$

Then, $\partial D/\partial\phi = 0$ leads to

$$\phi = \tan^{-1}\left[-\frac{r_1 \cos \omega_c\tau_1 \sin \frac{\Delta\omega\tau_1}{2} - r_2 \cos \omega_c\tau_2 \sin \frac{\Delta\omega\tau_2}{2} + r_1r_2 \sin \omega_c\Delta\tau \cdot \sin \frac{\Delta\omega\Delta\tau}{2}}{r_1 \sin \omega_c\tau_1 \sin \frac{\Delta\omega\tau_1}{2} + r_2 \sin \omega_c\tau_2 \sin \frac{\Delta\omega\tau_2}{2} + r_1r_2 \cos \omega_c\Delta\tau \cdot \sin \frac{\Delta\omega\Delta\tau}{2}}\right] \quad \text{when } X^2 > Y^2 + Z^2. \quad (A.6)$$

2) In case of $(P_r(\omega_+) - P_r(\omega_c))\cdot(P_r(\omega_-) - P_r(\omega_c)) \geqq 0$, that is, when the combined signal frequency response has a notch or convex area in the passband, in-band amplitude dispersion can be defined as follows.

$$D = P_r(\omega_+) + P_r(\omega_-) - 2P_r(\omega_c). \quad (A.7)$$

From (5)

$$\begin{aligned} D &= A_1^2(1 + r_1^2 + 2r_1 \cos \omega_+\tau_1) + A_2^2(1 + r_2^2 + 2r_2 \cdot \cos \omega_+\tau_2) + A_1^2(1 + r_1^2 + 2r_1 \cos \omega_-\tau_1) \\ &\quad + A_2^2(1 + r_2^2 + 2r_2 \cos \omega_-\tau_2) - 2A_1^2(1 + r_1^2 + 2r_1 \cos \omega_c\tau_1) - 2A_2^2(1 + r_2^2 + 2r_2 \cos \omega_c\tau_2) \\ &\quad + 2A_1A_2 \{\cos\phi + r_1 \cos(\omega_+\tau_1 - \phi) + r_2 \cos(\omega_+\tau_2 + \phi) + r_1r_2 \cos(\omega_+\Delta\tau + \phi)\} + 2A_1A_2 \{\cos\phi \\ &\quad + r_1 \cos(\omega_-\tau_1 - \phi) + r_2 \cos(\omega_-\tau_2 + \phi) + r_1r_2 \cos(\omega_-\Delta\tau + \phi)\} - 4A_1A_2 \{\cos\phi + r_1 \\ &\quad \cdot \cos(\omega_c\tau_1 - \phi) + r_2 \cos(\omega_c\tau_2 + \phi) + r_1r_2 \cos(\omega_c\Delta\tau + \phi)\} \\ &= x + y \sin\phi + z \cos\phi \end{aligned} \quad (A.8)$$

where

$$x = -4 \cdot \left\{ A_1^2 r_1 \left(1 - \cos \frac{\Delta\omega\tau_1}{2}\right) \cos \omega_c\tau_1 + A_2^2 r_2 \left(1 - \cos \frac{\Delta\omega\tau_2}{2}\right) \cos \omega_c\tau_2 \right\}$$

$$y = -4A_1A_2 \left\{ r_1 \left(1 - \cos \frac{\Delta\omega\tau_1}{2}\right) \sin \omega_c\tau_1 - r_2 \left(1 - \cos \frac{\Delta\omega\tau_2}{2}\right) \sin \omega_c\tau_2 - r_1r_2 \cdot \left(1 - \cos \frac{\Delta\omega\Delta\tau}{2}\right) \sin \omega_c\Delta\tau \right\}$$

$$z = -4A_1A_2 \left\{ r_1 \left(1 - \cos \frac{\Delta\omega\tau_1}{2}\right) \cos \omega_c\tau_1 + r_2 \left(1 - \cos \frac{\Delta\omega\tau_2}{2}\right) \cos \omega_c\tau_2 + r_1r_2 \left(1 - \cos \frac{\Delta\omega\Delta\tau}{2}\right) \sin \omega_c\Delta\tau \right\}. \quad (A.9)$$

Equation (A.9) and $D = 0$ leads to

$$\phi = \begin{cases} \tan^{-1}\left[\dfrac{-xy - z\sqrt{y^2+z^2-x^2}}{xz + y\sqrt{y^2+z^2-x^2}}\right] & \text{or} \\ \pi - \tan^{-1}\left[\dfrac{-xy + z\sqrt{y^2+z^2-x^2}}{-xz + y\sqrt{y^2+z^2-x^2}}\right]; & \\ \text{when } x^2 \leqq y^2 + z^2. \end{cases} \quad (A.10)$$

On the other hand, in case of $x^2 > y^2 + z^2$, D does not reach zero. In this case, D is minimized at $\partial D/\partial\phi = 0$. From

(5)

$$\begin{aligned}\frac{\partial D}{\partial \phi} &= \frac{\partial P_r(\omega_+)}{\partial \phi} + \frac{\partial P_r(\omega_-)}{\partial \phi} - 2\frac{\partial P_r(\omega_c)}{\partial \phi} \\ &= -4A_1A_2\left\{ r_1\left(1 - \cos\frac{\Delta\omega\tau_1}{2}\right)\sin\omega_c\tau_1 \right. \\ &\quad - r_2\left(1 - \cos\frac{\Delta\omega\tau_2}{2}\right)\sin\omega_c\tau_2 \\ &\quad \left. - r_1r_2\left(1 - \cos\frac{\Delta\omega\Delta\tau}{2}\right)\cos\omega_c\Delta\tau\right\}\cos\phi \\ &\quad - 4A_1A_2\left\{ -r_1\left(1 - \cos\frac{\Delta\omega\tau_1}{2}\right)\cos\omega_c\tau_1 \right. \\ &\quad - r_2\left(1 - \cos\frac{\Delta\omega\tau_2}{2}\right)\cos\omega_c\tau_2 \\ &\quad \left. - r_1r_2\left(1 - \cos\frac{\Delta\omega\Delta\tau}{2}\right)\sin\omega_c\Delta\tau\right\}\sin\phi. \qquad \text{(A.11)}\end{aligned}$$

Then, $\partial D/\partial\phi = 0$ leads to

$$\phi = \tan^{-1}\left[-\frac{r_1\left(1-\cos\frac{\Delta\omega\tau_1}{2}\right)\sin\omega_c\tau_1 - r_2\left(1-\cos\frac{\Delta\omega\tau_2}{2}\right)\cdot\sin\omega_c\tau_2 - r_1r_2\left(1-\cos\frac{\Delta\omega\Delta\tau}{2}\right)\cos\omega_c\Delta\tau}{-r_1\left(1-\cos\frac{\Delta\omega\tau_1}{2}\right)\cos\omega_c\tau_1 - r_2\left(1-\cos\frac{\Delta\omega\tau_2}{2}\right)\cdot\cos\omega_c\tau_2 + r_1r_2\left(1-\cos\frac{\Delta\omega\Delta\tau}{2}\right)\sin\omega_c\Delta\tau}\right]. \qquad \text{(A.12)}$$

ACKNOWLEDGMENT

The authors wish to thank Dr. H. Yamamoto and K. Morita for many useful discussions and suggestions.

REFERENCES

[1] P. R. Hartmann, "A 90 Mb/s digital transmission system at 11 GHz using 8 PSK modulation," in *Proc. Int. Conf. Commun.*, 1976, pp. 188, 193.

[2] I. Godier, "DRS 8 digital radio for long haul transmission," in *Proc. Int. Conf. Commun.*, June 1979, p. 102.

[3] W. T. Barnett, "Measured performance of a high capacity 6 GHz digital radio system," in *Proc. Int. Conf. Commun.*, 1978, p. 47.4.1.

[4] I. Horikawa *et al.*, "Design and performance of a 200 Mbit/s 16 QAM digital radio system," *IEEE Trans. Commun.*, vol. COM-27, Dec. 1979.

[5] T. S. Giuffrida, "Measurements of the effects of propagation on digital radio systems equipped with space diversity and adaptive equalizer," in *Proc. Int. Conf. Commun.*, June 1979, p. 48.1.1.

[6] A. Vigants, "Space diversity engineering," *Bell Syst. Tech. J.*, vol. 54, Jan. 1975.

[7] H. Makino and K. Morita, "Design of space diversity receiving and transmitting systems for line-of-sight microwave links," *IEEE Trans. Commun. Technol.*, vol. COM-15, p. 603, Aug. 1967.

[8] S. Komaki, K. Okamoto, and K. Tajima, "Performance of 16-QAM digital radio system using new space diversity," in *Proc. Int. Conf. Commun.*, 1980, p. 52.2.1, or U.S. Patent 4 326 294.

[9] Y. Y. Wang, "Simulation and measured performance of a space diversity combiner for 6 GHz digital radio," *IEEE Trans. Commun.*, vol. COM-27, p. 1896, Dec. 1979, or W. T. Barnett, U.S. Patent 4 261 056.

[10] S. Komaki, I. Horikawa, K. Morita, and Y. Okamoto, "Characteristics of a high capacity 16 QAM digital radio system in multipath fading," *IEEE Trans. Commun.*, vol. COM-27, p. 1854, Dec. 1979.

[11] S. Sakagami and H. Hosoya, "Some experimental results on in-band amplitude dispersion and a method for estimating in-band linear amplitude dispersion," *IEEE Trans. Commun.*, vol. COM-30, pp. 1875–1888, Aug. 1982.

[12] T. Murase, K. Morita, and S. Komaki, "Correction techniques for multipath distortion in a high capacity 16-QAM system," in *Proc. Int. Conf. Commun.*, 1981, p. 46.1.1.

A New Approach to Space Diversity Combining in Microwave Digital Radio

By Y. S. YEH and L. J. GREENSTEIN*

(Manuscript received May 30, 1984)

In this paper we describe a new approach to dual-channel space diversity combining in microwave digital radio. This approach features (1) adaptive control of the relative amplitudes and phases of the two branch gains; and (2) a search strategy, based on noncoherent spectrum measurements at the combiner output, that simultaneously accounts for both dispersion and noise. Computer programs have been developed to simulate the search process and to analyze the resulting performance. Eight representative channel response pairs are postulated and performance results are presented for each. They show that the scheme provides a high degree of channel equalization over bandwidths up to at least 40 MHz, and that, in receivers not using adaptive equalizers, it offers major improvements in detection performance over selection diversity.

I. INTRODUCTION

In a terrestrial digital radio link, frequency selective fading caused by multipath propagation presents the major threat to system availability. Efforts to reduce channel dispersion, and thus to increase availability, typically center on the use of adaptive equalization and/or dual-branch space diversity.[1-11]

Most conventional space diversity schemes use either selection switching or so-called "in-phase" combining of the diversity branches. The latter approach concentrates on maximizing the combiner output power rather than on minimizing channel dispersion. Recent work, however, has dealt with "out-of-phase" combining, which reduces output dispersion by suitably adjusting the relative phases between the two branches.[4,8] This approach can completely eliminate dispersion for certain two-path propagation situations, but not under more general and realistic conditions.

In this paper, we describe a combiner in which the relative phases *and* amplitudes of the two branches are controlled. With little increase in complexity, this approach allows the effects of both dispersion and noise to be jointly minimized. We shall consider this type of combining within the context of M-level Quadrature Amplitude Modulation (M-QAM) systems.

Two specific approaches for finding the best amplitude and phase adjustments are described in Section II. One in particular, based on noncoherent spectrum measurements at the combiner output, is identified for further study. The simulation and analysis of this scheme are discussed in Section III, and its performance for several postulated dual-channel response pairs is assessed in Section IV. Comparisons with other diversity and nondiversity approaches are also given and attest to the effectiveness of the new scheme.

II. THE DIVERSITY COMBINER

2.1 *Rationale*

The idea of using both amplitude and phase adjustments in space diversity combining, while not entirely new, has yet to be fully understood and optimally exploited. We now illustrate the potential power of this form of combining under quite general circumstances. To do so, we will invoke some recent work on the modeling and analysis of multipath fading responses.

Let $H_1(f)$ and $H_2(f)$ be the complex frequency responses of a fading channel as viewed by two vertically spaced receiver antennas. Under nonfading conditions, these functions are flat with frequency at unity amplitude ($|H_1(f)| = |H_2(f)| = 1$). In all that follows, we measure f from the center of the radio channel, so that $H_1(f)$ and $H_2(f)$ are baseband functions. Moreover, we are interested in their variations over just the interval $[-W/2, W/2]$, where W is the channel bandwidth in hertz. Since multipath fading arises from a finite number of discrete propagation paths, we can present $H_1(f)$ and $H_2(f)$ in the following general forms:

$$H_1(f) = \sum_{k=1}^{K_1} R_{1k}\exp\{-j(\omega\tau_{1k} - \theta_{1k})\} \quad (1)$$

and

$$H_2(f) = \sum_{k=1}^{K_2} R_{2k}\exp\{-j(\omega\tau_{2k} - \theta_{2k})\}. \quad (2)$$

* Authors are employees of AT&T Bell Laboratories.

Reprinted with permission from *AT&T Technical Journal,* vol. 64, no. 4, pp. 885–895, April 1985.

In (1), K_1 is the number of paths and τ_{1k}, R_{1k}, and θ_{1k} are the time delay, amplitude, and phase, respectively, associated with the kth path. Similar definitions apply to K_2, τ_{2k}, R_{2k}, and θ_{2k} in $H_2(f)$.

Typically, multipath propagation on microwave radio links exhibits delay spreads on the order of 10 ns or less, i.e., the largest and smallest τ values differ by amounts small compared to $1/W$, where $W \leq 40$ MHz in the common carrier bands. This observation inspired earlier efforts to approximate fading channel responses using low-order polynomials in $j\omega$.[12]

To be concrete, let $H_1(f)$ and $H_2(f)$ be represented by the infinite power series

$$H_1(f) = e^{-j\omega t_1}[A_1 + j\omega B_1 + (j\omega)^2 C_1 + \cdots] \tag{3}$$

and

$$H_2(f) = e^{-j\omega t_2}[A_2 + j\omega B_2 + (j\omega)^2 C_2 + \cdots], \tag{4}$$

where t_1 and t_2 are arbitrary, and the A's, B's, and so on are complex coefficients. Using the power series expansion for e^{jx}, they can be easily related to the parameters of the functions (1) and (2), e.g.,

$$A_1 = \sum_{k=1}^{K_1} R_{1k} e^{j\theta_{1k}}; \qquad B_1 = \sum_{k=1}^{K_1} R_{1k} e^{j\theta_{1k}}(t_1 - \tau_{1k}), \tag{5}$$

and so on.

The essence of *first-order* polynomial fitting is this: Given $H_1(f)$, a value for t_1 can usually be found such that $(A_1 + j\omega B_1)$ is the dominant part of (3) over $[-W/2, W/2]$, all higher-order terms in $j\omega$ being small, in some sense. Similarly, a value for t_2 can usually be found that does the same for $H_2(f)$, (4). That such first-order polynomial fitting is reasonable to do in common carrier channels has been supported by theory,[13] noncoherently measured data,[12] and (more recently) coherently measured data.[14]

Now suppose that a space diversity combiner were used having an adjustable time delay (τ) and complex gain (β) in the second branch. The composite channel response, as viewed at the combiner output, would then be

$$H(f) = H_1(f) + \beta e^{-j\omega\tau} H_2(f). \tag{6}$$

If τ and β were adaptively adjusted to be

$$\tau = t_1 - t_2 \quad \text{and} \quad \beta = -B_1/B_2, \tag{7}$$

we could then write [see (3) and (4)]

$$H(f) = [A_1 - A_2 B_1/B_2] + \left\{ \begin{array}{l} \text{second- and} \\ \text{higher-order} \\ \text{terms in } j\omega \end{array} \right\}. \tag{8}$$

Thus, by proper choice of delay and gain in one branch, the channel response could be made dispersionless except for small higher-order terms in $j\omega$. This is a quite general result for the channels of interest and shows the power of the combining approach under study. If β were a phase-only factor $[\beta = \exp(j\phi)]$, such a strong reduction in dispersion would only be possible in those fortuitous situations where $|B_1| = |B_2|$.

2.2 The combiner structure

The above discussion suggests both a particular structure, (6), for the combiner and particular solutions, (7), for the variable parameters. The discussion was intended, however, to provide insight rather than to identify a serious design approach. For one thing, a variable time delay would be difficult to implement and would offer little incremental benefit in most cases. In addition, the solutions of (7) do not properly address either the residual dispersion (i.e., higher orders in $j\omega$) or the receiver noise. If, for example, A_1/B_1 and A_2/B_2 happened by chance to be close in value, the first term in (8) would be severely weakened in the process of eliminating the $j\omega$ term; this would enhance the effects of both the remaining dispersion and the noise.

Accordingly, we propose a combiner in which the relative amplitudes and phases in the two diversity branches are adaptively adjusted, but not the delays. Thus,

$$H(f) = \beta_1 H_1(f) + \beta_2 H_2(f), \tag{9}$$

as indicated in Fig. 1, where β_1 and β_2 are adapted gains. (Because the dominant thermal noise is introduced before the combiner, several ways of adapting β_1 and β_2 would, in theory, yield equivalent performance. For example, one gain could be held fixed, or adapted in amplitude only, with the other being adapted in both amplitude and phase; or both gains could be adapted in both amplitude and phase. Each of these approaches would permit the adaptation of the *relative* complex branch gains, which is all that matters.) Moreover, we propose the use of control strategies that take proper account of both dispersion and noise.

2.3 Control strategies

We shall discuss two distinct approaches for controlling the gain pair (β_1, β_2) in (9). The first approximates the theoretically best way to do combining when there is no post-combiner equalization. The second approach, thought suboptimal, has features that make it attractive both with and without post-combiner equalization. The new scheme reported here incorporates the second approach, whose performance we will compare with that of the first.

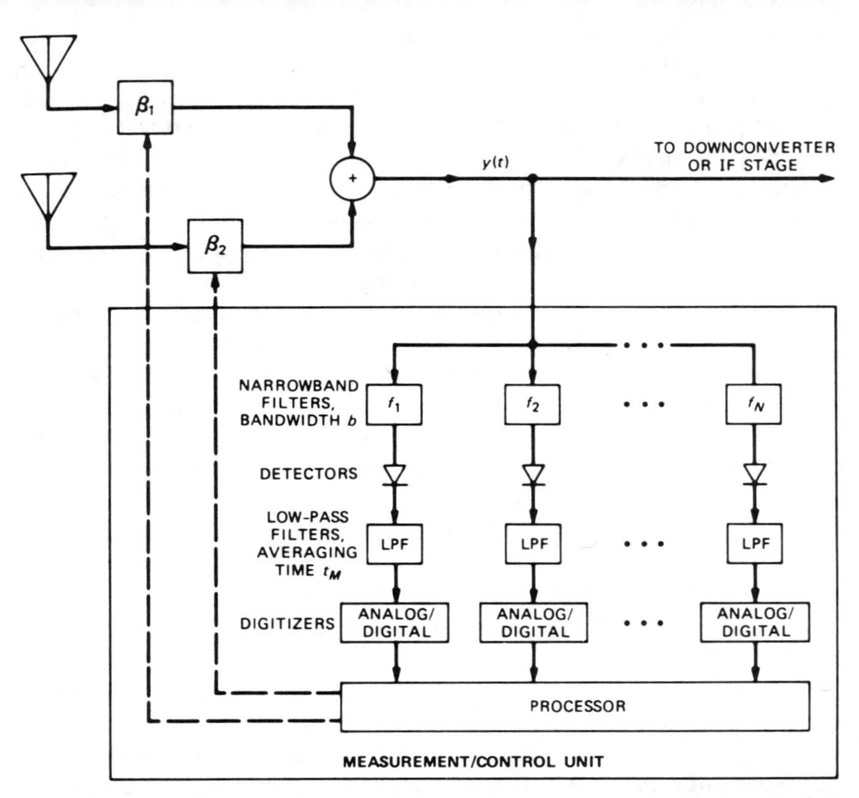

Fig. 1—Block diagram of the space diversity combiner. In this scheme, noncoherent spectral measurements are made on the combiner output at three or more inband frequencies, and the control measure formed from them is used to adjust the complex gains β_1 and β_2. The combining is shown at Radiofrequency (RF) but could be at Intermediate Frequency (IF) instead.

Approach 1: Assume for now that there is no post-combiner equalization. In that case, the detection bit error rate is virtually minimized by choosing (β_1, β_2) to maximize the ratio of sampled signal to root mean squared (rms) distortion at the receiver output. By "sampled signal," we mean the half-distance between signal levels, as sampled in every period at the in-phase and quadrature detectors; by "rms distortion," we mean the rms sum of thermal noise and Intersymbol Interference (ISI) sampled at the detectors. This signal-to-distortion ratio [defined formally by (26) and expressed by (28), below] depends on $H_1(f)$, $H_2(f)$, β_1, β_2, and other factors; most important, it is convex in β_1 and β_2. The result is that there is a unique value of relative complex gain, β_1/β_2, that maximizes this ratio. We shall regard as *optimal* any gain pair (β_1, β_2) that exhibits this maximizing relative gain.

There is a practical way to realize optimal values for β_1 and β_2, namely, gradient search methods using data decisions.[15,16] The action is similar to that of an adaptive transversal equalizer, with β_1 and β_2 taking the place of the optimized tap gains. We shall refer to this or any decision-directed scheme that optimizes (β_1, β_2), in the sense defined above, as Approach 1.

Approach 2: A potential liability of Approach 1 is that it relies on having accurate data decisions, a condition that may not always exist (e.g., during recovery from severe fades). The scheme to be reported here is based on a different strategy, which we designate as Approach 2. It consists of (1) performing certain noncoherent spectral measurements on the combiner output; and (2) sequentially searching over (β_1, β_2) so as to maximize a certain quantity [eq. (18), below] computed from these measurements. By adapting β_1 and β_2 in this way, a close approximation to the "optimal" condition defined above can be achieved, as we show later. That is, Approach 2 should yield near-optimal detection performance for receivers having no post-combiner equalization. For cases where such equalization *is* used, Approach 2 would serve a different purpose—reducing the signal dispersion as seen by the equalizer input, thereby simplifying the requirements on equalizer design (e.g., number of taps) and improving convergence speed. Thus, Approach 2 has the twin virtues of not relying on data decisions and having utility both with and without post-combiner equalization.

To describe Approach 2, we refer to the block diagram of Fig. 1. The combiner output signal, $y(t)$, has a power spectrum density given by

$$S_y(f) = \underbrace{S(f)\,|H(f)|^2}_{\text{Signal}} + \underbrace{N_o(|\beta_1|^2 + |\beta_2|^2)\,|H_R(f)|^2}_{\text{Noise}}, \tag{10}$$

where N_o is the power spectrum density of the receiver input noise,* $H_R(f)$ represents whatever receiver selectivity precedes the combiner output, and $S(f)$ is the spectral density of the signal (excluding channel and combiner effects). More specifically,

$$S(f) = S_o\,|H_T(f)|^2\,|H_R(f)|^2, \tag{11}$$

where $|H_T(f)|^2$ represents the spectral shaping in the transmitter and S_o is a spectral density scale factor.

All functions and parameters in (10) and (11) are design-specified except $|\beta_1|$ and $|\beta_2|$, which are controlled by the combiner circuitry, and $|H(f)|$, which must be measured in real time. Our scheme estimates $|H(f)|$ at N evenly spaced frequencies (N odd) within the

* Though not made explicit by the figure, we assume there is sufficient front-end amplification that the combiner gains β_1 and β_2 have no effect on receiver noise figure.

channel bandwidth by estimating the corresponding values of $S_y(f)$ [see eq. (10)]. Based on these estimates, β_1 and β_2 are adjusted to maximize a computed performance measure, Y, which we introduce shortly. Before doing so, we define the following:

$$\Delta f \triangleq \text{Spacing between estimates of } |H(f)|, \tag{12}$$

where $(N-1)\Delta f \leq W$;

$$H_n \triangleq |H(f_n)|, \qquad f_n = n\Delta f\,(n = 0, \pm 1, \cdots \pm (N-1)/2); \tag{13}$$

$$\bar{H} \triangleq \operatorname*{Ave}_n\{H_n\} = \text{“Average Signal Gain”}; \tag{14}$$

$$X_{sig} \triangleq \operatorname*{Ave}_n\{S(f_n)\}(\bar{H})^2 W = \text{“Signal Power”}; \tag{15}$$

$$X_{dis} \triangleq \operatorname*{Ave}_n\{S(f_n)(H_n - \bar{H})^2\} W = \text{“Distortion Power”}; \tag{16}$$

$$X_{noise} \triangleq N_o(|\beta_1|^2 + |\beta_2|^2)\operatorname*{Ave}_n\{|H_R(f_n)|^2\} W = \text{“Noise Power”}; \tag{17}$$

where W and T are the channel bandwidth and digital symbol period, respectively. All quantities in these equations are known a priori except the H_n values, which are measured.

We now define the performance measure to be computed and maximized, namely,

$$Y \triangleq X_{sig}/(X_{dis} + X_{noise}). \tag{18}$$

This ratio is an approximation, computed from noncoherent spectral measurements, of the detector output signal-to-distortion ratio defined by (26), below. It is an apt measure to maximize in its own right, for the following reasons: In typical digital radio links, noise will not be a serious factor unless $H_1(f)$ and $H_2(f)$ are strongly faded. Therefore, maximizing Y will, in most cases, amount to minimizing the ratio X_{dis}/X_{sig}, which is a measure of the dispersion in $H(f)$. Including X_{noise}, however, safeguards against minimizing this ratio at an undue cost in signal (X_{sig}) and thus seriously degrading the signal-to-noise ratio.

The control strategy is therefore as follows (see Fig. 1): At the combiner output, a parallel bank of envelope detectors is used to estimate $|H(f)|$ at N frequencies. The spectral samples are digitized and applied to a microprocessor, which computes Y. This measure drives the search over $|\beta_1|$ (or $|\beta_2|$) and $\phi = \text{Arg}\{\beta_2\}$, i.e., these quantities are adjusted so as to maximize Y. Typically, they are adjusted iteratively, e.g., ϕ is changed in 0.1-radian step until a local maximum is found; then $|\beta_1|$ or $|\beta_2|$ is changed from 1 in steps of 0.1 until a maximum is found; and this process repeats, possibly using smaller steps in successive rounds, until Y can no longer be increased by varying either β or ϕ. If each measurement (i.e., set of estimates of $|H(f)|$ at N frequencies) and computation for Y takes t_M seconds, and N_A steps are needed to find (β_1, β_2), then the "solution time" of the combiner will be about $N_A t_M$. This number should be small compared to 1 second to achieve timely adaptation to multipath fades.

2.4 Measurements

We now discuss the scheme for measuring the set of H_n's in (13). In our simulations (Section III) we treat only the case where $N = 3$ and $\Delta f = W/2$ (i.e., three samples, taken at the channel edges and center). In practice, the outer samples would probably be closer-in so as to minimize errors from adjacent-channel interference. Also, higher values of N (e.g., $N = 5$) might be worthwhile.

To see how accurate estimates of H_n might be obtained, let $G(f)$ represent a low-pass power gain function with bandwidth $b/2 \ll W$. We envision the measurement of H_n as involving a bandpass filter with power response $G(f - n\Delta f)$ followed by envelope detection and t_M-second averaging of the detector output (Fig. 1). Referring to eq. (10), the average power at the output of the bandpass filter will be

$$\bar{P}_n = \int \{S(f)|H(f)|^2 + N_o(|\beta_1|^2 + |\beta_2|^2)|H_R(f)|^2\} G(f - n\Delta f)df, \tag{19}$$

where $S(f)$ is defined by (11). Assuming a square-law detector, the time-averaged detector output will be

$$\langle \bar{P}_n \rangle = \bar{P}_n + \begin{Bmatrix} \text{Fluctuation Noise;} \\ \text{Variance} \sim \bar{P}_n/bt_M \end{Bmatrix}. \tag{20}$$

We now define two constants related to the system design functions, namely,

$$\eta_n \triangleq N_o \int |H_R(f)|^2 G(f - n\Delta f)df \tag{21}$$

and

$$\zeta_n \triangleq \int S(f)G(f - n\Delta f)df. \tag{22}$$

Based on (19) and the fact that $H(f)$ changes little over the bandwidth b, a microprocessor can estimate H_n using the formula

$$\hat{H}_n = \sqrt{\frac{\langle \bar{P}_n \rangle - \eta_n[|\beta_1|^2 + |\beta_2|^2]}{\zeta_n}}, \tag{23}$$

where $\langle \bar{P}_n \rangle$ is measured in real time; $|\beta_1|$ and $|\beta_2|$ are controlled parameters of known value; and η_n and ζ_n are predetermined constants.

Equation (23) shows how $\hat{H}_n$ is computed in terms of measured or known quantities. To see what this computed number represents, we insert (19) through (22) into (23) and obtain

$$\hat{H}_n = \sqrt{\frac{\int S(f)G(f - n\Delta f)|H(f)|^2 df}{\int S(f)G(f - n\Delta f) df} + \left\{\begin{matrix}\text{Term due solely to} \\ \text{fluctuation noise}\end{matrix}\right\}}. \quad (24)$$

We can now cite choices for b and t_M that lead to accurate and sufficiently rapid estimations of H_n. As b gets very small, the first term under the radical sign in (24) approaches H_n^2, so that the major inaccuracy in $\hat{H}_n$ is due to fluctuation noise. To be more precise, the first term is close to H_n^2 so long as $|H(f)|^2$ changes little over the passband of $G(f - n\Delta f)$. Since we are considering propagation media with delay spreads of just a few nanoseconds, the design rule $b \leq 2$ MHz should permit more than adequate resolution in this regard. To achieve low mean-square fluctuation noise as well [second term in (24)], the condition $bt_M \geq 4000$ should be satisfied [see (20)]. Thus, with $b = 2$ MHz and $t_M = 2$ ms, H_n can be approximated with high accuracy by the quantity $\hat{H}_n$. Moreover, this design choice would permit numerous iterations of the search over β_1 and β_2 before the medium response changes appreciably.

III. PERFORMANCE STUDY

3.1 *General*

We have written a set of computer programs to simulate the behavior of the combiner scheme described above (Approach 2) and to analyze its performance and that of other receiver techniques. Each simulation is done for a specific pair of fading functions, $H_1(f)$ and $H_2(f)$, and for a specific value of Carrier-to-Noise Ratio (CNR). What is simulated is the sequential search over β_1 and β_2, as performed by a receiver in real time to maximize the computed measure Y [see eq. (18)].

The analysis programs compute a detection performance measure for a nonequalized receiver using the (β_1, β_2) pairs derived in the simulations for Approach 2. The same measure is also computed for the cases of *optimal* combining (Approach 1) and no combining (non-diversity). Also, the analysis programs examine the signal dispersion at the combiner output for these various cases, which has relevance to receivers with post-combiner equalization.

3.2 *Response pairs studied*

We have specified eight distinct pairs of $H_1(f)$ and $H_2(f)$ for purposes of study. These pairs are collectively representative of what

Table I—Dual-channel response pairs studied

CASE	PLOTS OF $\|H_1(f)\|, \|H_2(f)\|$ (IN dB)	SPACE DIVERSITY CHANNEL	PATH NUMBER, k	τ_k (IN ns)	$R_k \exp(j\theta_k)$
1	$\|H\|$ IN dB; 10, 0, −10; $\|H_2\|$, $\|H_1\|$; f IN MHz: −10, 0, 10	1	1	-4.0	1.0 + j0.1
			2	+1.0	-0.5 + j0.9
		2	1	-2.0	1.1 + j0.4
			2	+4.0	0.9 − j0.4
2	10, 0, −10; $\|H_2\|$, $\|H_1\|$; −10, 0, 10	1	1	-4.0	0.2 + j0.1
		2	1	-5.0	-1.1 + j0.6
			2	+3.0	0.5 + j1.1
3	10, 0, −10; $\|H_2\|$, $\|H_1\|$; −10, 0, 10	1	1	-1.0	1.0 + j0.1
			2	+2.0	-0.9 + j0.5
		2	1	-2.0	1.1 + j0.9
			2	+2.0	-1.0 + j0.5
4	10, 0, −10; $\|H_2\|$, $\|H_1\|$; −10, 0, 10	1	1	-1.0	1.0 + j0.0
			2	+2.0	-0.9 + j0.0
		2	1	-2.0	1.1 + j0.0
			2	+2.0	-0.5 + j0.0

might arise in actual radio links using space diversity. In each case, $H_1(f)$ corresponds to a one-, two- or three-path medium, and similarly for $H_2(f)$. [In terms of (1) and (2), $K_1 = 1$, 2, or 3 in each case, and similarly for K_2.] The corresponding time delays and complex gains are summarized in Table I. Also shown for each case are graphs of $|H_1(f)|$ and $|H_2(f)|$, in decibels, over a 40-MHz bandwidth.

Table I—(Cont.) Dual-channel response pairs studied

CASE	PLOTS OF $\lvert H_1(f)\rvert$, $\lvert H_2(f)\rvert$ (IN dB)	SPACE DIVERSITY CHANNEL	PATH NUMBER, k	τ_k (IN ns)	$R_k \exp(j\theta_k)$
5	10, 0, -10; -10, 0, 10; $\lvert H_2\rvert$, $\lvert H_1\rvert$	1	1	-2.0	1.0 + j0.0
			2	+2.0	-0.872 j0.223
		2	1	1.0	0.54 + j0.84
			2	+3.0	0.66 j0.60
6	10, 0, -10; -10, 0, 10; $\lvert H_2\rvert$, $\lvert H_1\rvert$	1	1	-3.0	1.0 + j0.0
			2	+3.0	-0.5 + j0.0
		2	1	-3.0	-0.068 - j0.998
			2	+1.0	0.48 + j0.76
7	10, 0, -10; -10, 0, 10; $\lvert H_1\rvert$, $\lvert H_2\rvert$	1	1	-3.0	0.0 j1.0
			2	0.0	0.5 + j0.5
			3	+2.0	1.0 + j0.0
		2	1	3.0	0.0 j1.0
			2	0.0	0.5 + j0.5
			3	+2.0	0.5 + j0.0
8	10, 0, -10; -10, 0, 10; $\lvert H_2\rvert$, $\lvert H_1\rvert$	1	1	-3.0	1.0 + j0.5
			2	-1.0	0.2 + j0.7
			3	+2.0	0.5 - j1.0
		2	1	2.0	1.0 j0.2
			2	+1.0	1.0 j0.2
			3	+2.0	0.5 + j0.5

Because of additive noise, the receiver performance for a given response pair would be affected by any amplitude scaling of $(H_1(f), H_2(f))$. We permit the possibility of such a scaling, for each case in Table I, by including it in the carrier-to-noise ratio parameter discussed below.

ADAPTIVE RECEIVER FOR CROSS-POLARIZED DIGITAL TRANSMISSION

Junji NAMIKI* & Shigeru TAKAHARA**

C & C Systems Research Labs.*, Microwave & Satellite Comm. Division**
Nippon Electric Co., Ltd.
1-1 Miyazaki Yonchome, Takatsu-ku, Kawasaki-City
Kanagawa Pref. 213 Japan

Abstract

This paper introduces a new kind of depolarization equalizer, based on the conventional adaptive linear equalization, and shows that the capacity of a given frequency allocation can be doubled, while satisfying the outage probability objective. Crosstalk elimination for entangled depolarization due to a non regenerative repeater, due to a selective fading channel and due to timing differences between orthogonal data are simulated. The superiority of this proposed equalizer is proven in comparison with existing cancelation systems.

Finally, more efficient utilization of available bandwidth in 4 GHz band is proposed by using the co-channel allocation plan with this new equalizer. More than 1800 channels can be carried with feasible modems, in 20 MHz bandwidth where even 1152 channels are very hard to carry for 32 QAM single modem.

1. Introduction

Microwave transmission is making rapid progress in both terrestrial and satellite communications. Demand for radio communications is expected to grow further from now on, according to expansion of mobile communication services or satellite data networks. Therefore, the need for the currently utilized frequency band reuse is increasingly raised, along with the exploitation of the quasi-millimeter wave and even higher frequency bands.

Cross-polarization techniques which can share one radio band with two orthogonal channels among facing antenna sections are very attractive for efficient bandwith utilization. As weather factors, such as rainfall, cause cross-polarization crosstalk, it is not sufficient to merely improve antenna and power feed system for keeping channels stable. Therefore, many adaptive compensation circuits for cross-polarization deterioration have been reported. [1], [2], [3], [4] They, however, do not consider the entangled combination of signal distortion with cross-polarized crosstalk, which appears in fading channels or in non-regenerative repeating systems. Accordingly, this entangled crosstalk is not cancelled by former methods. Moreover, if two cross-polarized signals are used by two independent stations, so as to double the capacity of a conventional SCPC satellite system, carrier frequency and clock timing differences between two orthogonal data items must be considered. A new kind of depolarization equalizer is introduced, which can solve these problems. This equalizer is based on conventional linear equalization techniques, so crosstalk can be cancelled by use of two orthogonal complex signals in any section among baseband, intermediate frequency band and radio frequency band. This paper proposes a co-channel allocation plan with this depolarization equalizer in a high capacity long-haul digital microwave transmission system. The paper also shows that 1800 channel per 20 MHz long-haul transmission in the 4 GHz band is sufficiently feasible.

2. Depolarization equalizer

2.1 Co-channel duad modem and depolarization equalizer

A typical depolarization model for a regenerative repeating system is depicted in Fig. 1, which can be seen in a conventional terrestrial digital microwave system.

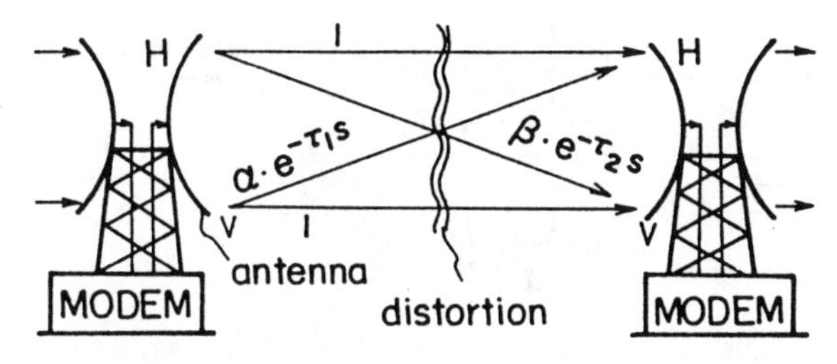

Figure 1 Depolarization model in terrestrial microwave transmission

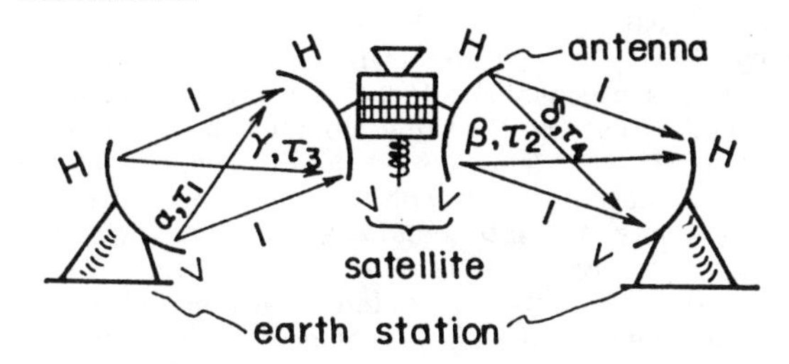

Figure 2 Depolarization model for a non-regenerative repeating system

The model for a non-regenerative repeating system, such as satellite links, is depicted in Fig. 2. Provided a coherent detection is adopted in a receiver for demodulation, equivalent baseband models for Figs. 1 and 2 are depicted in Figs. 3, and 4, respectively, where α, β, γ and δ are complex constants expressing cross-polarization crosstalk, τ_i ; ith path (antenna feed system) difference in seconds, H; horizontally polarized complex signal. V ; vertically polarized complex signal. ω_h and ω_v; carrier frequencies of H and V. "Distortions" stand for filtering operation or fading channel characteristics. In general, these distortions can be distributed over every path with different appearances from each other. These models can be applied to the circular polarization systems with the same expressions.

If an interleave allocation plan is adopted, two adjacently cross-polarized signals V_1 and V_2 facing the desirered signal H are considered as the cross-polarization crosstalk sources. It is disadvantageous for crosstalk cancellation in the IF or baseband section. A parallel

Reprinted from *IEEE 17th Int. Conf. Comm.*, pp. 46.3.1-46.3.5, June 1981.

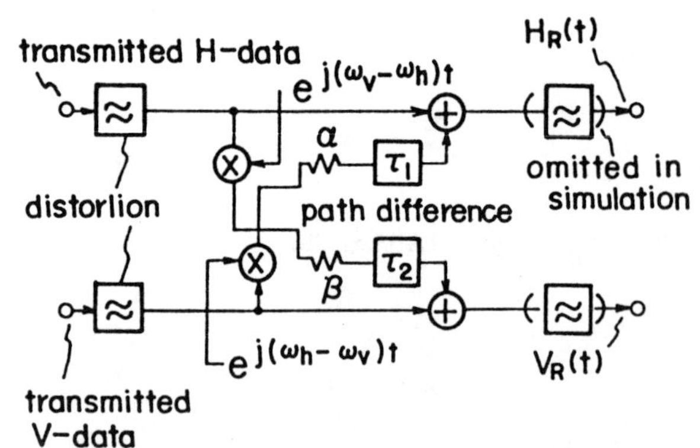

Figure 3 Equivalent baseband model for Fig. 1

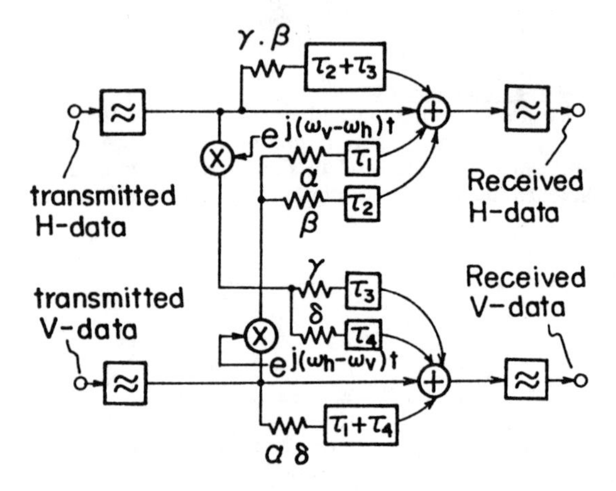

Figure 4 Equivalent baseband model for Fig. 2

data transmission system sharing the same carrier frequency band by corss-polarization techniques is considered in a co-channel allocation plan. It transmits two independent data by the same modulation at the same transmission rate. It demodulates the received signals by each coherent detector and each suitable sample timing. Namely two same modulators and demodulators face each other by way of cross-polarized antenna system. These sets of modulators and demodulators are called a co-channel duad modem (CDM) hereafter (see Fig.7). Of course it produces many advantages for demodulation, in that CDM has a synchronized carrier commonly as well as clock. An $H_R(t)$, one CDM output, which is coherently detected for H, is expressed simply by taking note of only facing cross-polarized singal, V, as follows.

$$
\begin{aligned}
H_R(t) = & \sum a_n \cdot h(t-nT) \\
& + \alpha \cdot e^{-j\omega_v \tau_1} \sum b_n \cdot v(t-nT-\tau_1)\; e^{-j(\omega_h-\omega_v)t} \\
& + \beta e^{-j\omega_v \tau_2} \sum b_n \cdot v(t-nT-\tau_2)\; e^{-j(\omega_h-\omega_v)t} \\
& + \beta\cdot\gamma\, e^{-j\omega_v(\tau_2+\tau_3)} \sum a_n \cdot h(t-nT-\tau_2-\tau_3) \qquad (1)
\end{aligned}
$$

Similarly,

$$
\begin{aligned}
V_R(t) = & \sum b_n \cdot v(t-nT) \\
& + \gamma e^{-j\omega_h \tau_3} \sum a_n \cdot h(t-nT-\tau_3) \cdot e^{-j(\omega_v-\omega_h)t} \\
& + \delta e^{-j\omega_h \tau_4} \sum a_n \cdot h(t-nT-\tau_4) \cdot e^{-j(\omega_v-\omega_h)t} \\
& + \alpha\cdot\delta e^{-j\omega_h(\tau_1+\tau_4)} \sum b_n \cdot v(t-nT-\tau_1-\tau_4) \qquad (2)
\end{aligned}
$$

, where h(t) and v(t) are impulse responses, and a_n and b_n are complex digital data to be sent for H and V, respectively.

Provided

$$\frac{|\omega_h-\omega_v|}{2\pi} \ll \frac{1}{T} \qquad (3)$$

terms which include $e^{-j(\omega_h-\omega_v)t}$ in Eq.(1) and Eq.(2) are considered as complex constants, in comparison with equalizer tap-weight update speed, as mentioned later. Moreover, a set of Λ_l values satisfying

$$
\begin{aligned}
\sum a_n \cdot h(t-nT) + \alpha' \sum a_n \cdot h(t-nT-\tau) \\
\approx \sum_l \Lambda_l \sum_n a_n \cdot h(t-nT-lT) \qquad (4)
\end{aligned}
$$

can be found. Therefore, $H_R(t)$ can be rewritten as follows.

$$
\begin{aligned}
H_R(t) = & \sum \Lambda_l \sum a_n \cdot h(t-nT-lT) \\
& + \sum \Lambda_l' \sum b_n \cdot v(t-nT-lT) \qquad (5)
\end{aligned}
$$

$V_R(t)$, also, can be rewritten similarly. From this expression, the equalizer for the entangled depolarization at least must have the structure corresponding to the above mentioned Eq.(5). In the next section, an equalizer having the above mentioned structure is introduced.

2.2 Problems for a pair tap crosstalk cancellation

The equivalent baseband models of previously reported crosstalk cancellation systems [1], [2], [4], are depicted commonly in Fig. 5. Due to lack of a straight path weighting circuit, this structure is not sufficient to cancel the heavy crosstalk which occurse when XPD reduces to even 6 dB. For strict crosstalk cancellation, the structure depicted in Fig. 6 is required. No matter

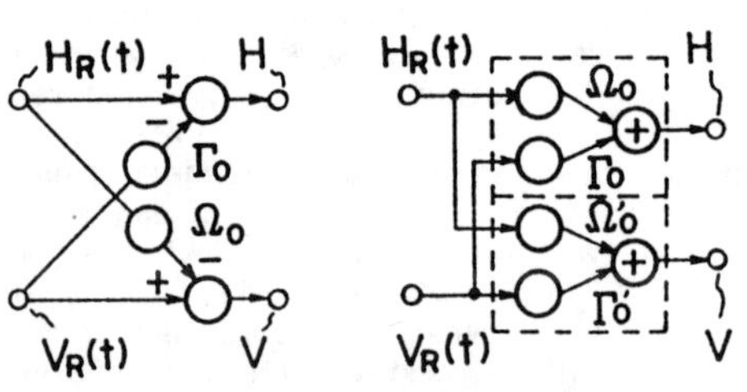

Figure 5 Simple crosstalk canceller

Figure 6 One pair top crosstalk canceller

how heavy the crosstalk may be, it is cancelled completely by this canceller, as long as the crosstalk does not include the signal distortion or timing difference between two orthogonal bits of data. Provided a couple of data items, $H_R(t)$ and $V_R(t)$, supplied to the canceller are sampled at a suitable timing for each signal, the crosstalk component in $H_R(t)$ becomes intolerably different from $V_R(t)$ itself. Consequently, cancellation ability decreases markedly, as the timing difference becomes larger. Therefore, another kind of structure becomes required for the general crosstalk cancellation.

2.3 Proposal on a new depolarization equalizer

An equivalent baseband model for the new depolarization equalizer is depicted in Fig. 7. In this figure, two orthogonal received data bits, $H_R(t)$ and $V_R(t)$, are supplied to a couple of input terminals, respectively.

Since the equalizer is controlled by sampled values, all equalizer signals can be treated as sampled values. Equalizer output samples are given as

$$Y_n = \sum_i \Omega_i \cdot H_{n-i} + \sum_i \Gamma_i \cdot V_{n-i} \tag{6}$$

, where H_n and V_n are samples of the incoming signal to the equalizer. As mentioned later, it is noteworthy for Y_n to have the same expression as that of a depolarization signal with distortion. To obtain both equalized orthogonal data bits, a couple of equalizers are required, which have the same structures, such as Fig. 7, except for

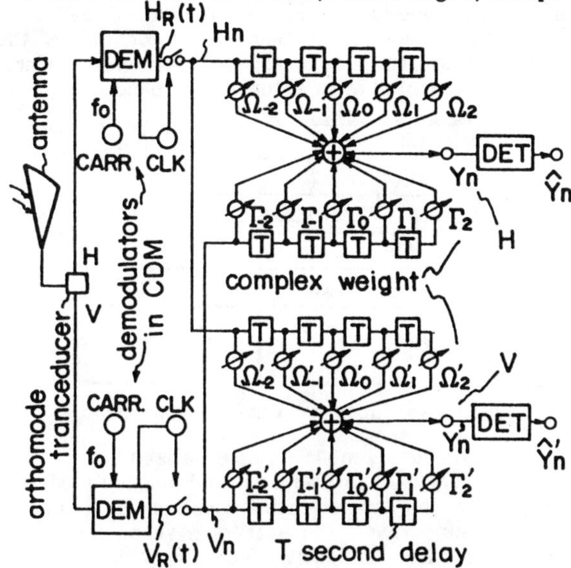

Figure 7 5 pair tap depolarization equalizers

inital values of two center tap weights. H_n and V_n are supplied as the demodulated sample value output of the co-channel duad modem (CDM). Consequently, they are sampled at a suitable timing for each modem in general and supplied to a couple of equalizers, simultaneously. The initial vaue $\Omega_0^{(0)}$ of an upper equalizer in Fig. 7 is set to a value of 1 and set to zeros for all other tap weights. The qualized $H_R(t)$ appears at an upper output. For a lower equalizer in Fig. 7, a value of 1 is assigned to only $\Gamma'^{(0)}$ with zeros for all other tap weights. The equalized $V_R(t)$ appears at a lower output.

Each tap weight is updated, according to the following iterative formulas.

$$\Omega_i^{(n+1)} = \Omega_i^{(n)} - k \cdot H^*_{n-i} \cdot (Y_n - \hat{Y}_n) \tag{7}$$

$$\Gamma_i^{(n+1)} = \Gamma_i^{(n)} - k \cdot V^*_{n-i} \cdot (Y_n - \hat{Y}_n) \tag{8}$$

for getting equalized $H_R(t)$

$$\Omega'^{(n+1)}_i = \Omega'^{(n)}_i - k \cdot H^*_{n-i} \cdot (Y'_n - \hat{Y}'_n) \tag{9}$$

$$\Gamma'^{(n+1)}_i = \Gamma'^{(n)}_i - k \cdot V^*_{n-i} \cdot (Y'_n - \hat{Y}'_n) \tag{10}$$

for getting equalized $V_R(t)$

,where k is a positive adaptation constant, $\hat{Y}_n$ is the decision data for Y_n and the asterisk indicates complex conjugate. These are applied expressions from well known iterative formulas based on mean square criteria, whose convergency is strictly guaranteed. [5] [6] Rightly, many other iterative formulas which are popular in linear adaptive equalization, can be applicable such as zero-forcing method, based on worst case criteria.

It is easy to expand the equalizer structure in Fig. 7 to a double sampling adaptive equalizer, whose delay components have T/2 second delay, and whose equalizability is insensitive to sample timing. Moreover a decision feedback equalization algorithm is also applicable. [7] The equalizer structure based on this algorithm is depicted in Fig. 8. This structure has a

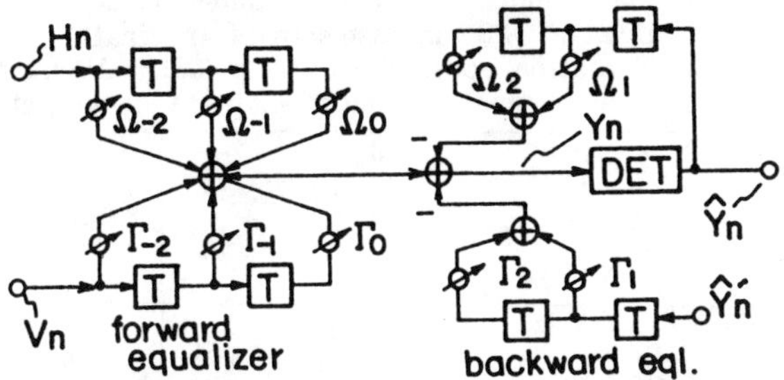

Figure 8 Decision feedback equalizer for cross-polarized signal (Yn side only shown)

remarkable feature, which reproduces the post-cursor from the past data, and subtracts it from the forword equalizer output. So, the noise component included in the past data is not enlarged. This is very favorable for equalizing a deep selective fading characteristic, which often occurs in a terrestrial microwave transmission channel. To avoid expression confusion, only the linear equalizer is studied from now on.

The joint operation between carrier recovery circuit and equalizer has been studied for a long time in a data modem. No intolerable problems are reported for a two dimensional (complex) signal.

3. Fundamental simulation on equalizer performance

3.1 Timing differences absorption

The relatione between crosstalk suppression and the number of pair taps is shown in Fig. 9 by simulating operations of 50 % rall-off shaping 16 QAM-CDM with 10 dB XPD. It can be seen that even T/8 timing difference markedly influences one-tap equalizer equalizability. This figure shows, however, the timing difference are absorbed merely by increasing the number of taps.

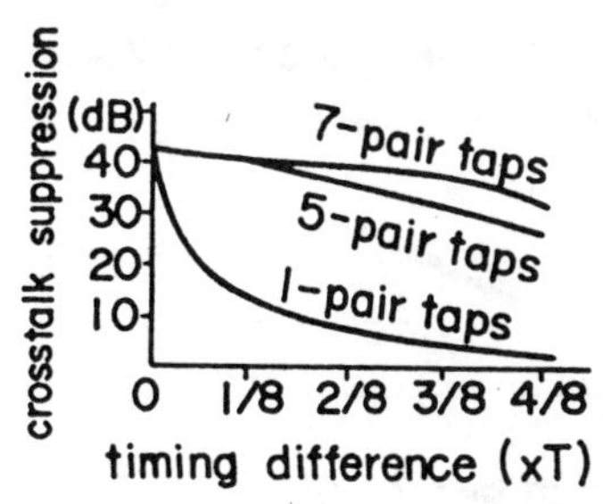

Figure 9 Timing difference absorption due to multi pair taps

3.2 Equalizability for depolarization with distortion

The equalizability for entangled depolarization due to non-regenerative repeating is very interesting. QPSK-CDM with 50 % roll-off Nyquist filtering was chosen as data transmission format, since multi-level QAM is not used in a non-regenerative repeating system. A couple of QPSK signals are transmitted through the channel depicted in Fig. 2 (or 4). Channel parameter values in Fig. 4 are chosen as follows α, β, γ, δ =0.2 (14 dB XPD).

$\tau_1, \tau_2, \tau_3, \tau_4 = T/8$. $\omega_h - \omega_v = 0$. 7.7 dB inband amplitude slope was chosen as the channel distortion. Figure 10 depicts QPSK signal points, scatter diagram, distorted by the amplitude slope. The scatter diagram depicted in Fig. 11 is of received QPSK with crosspolarization interference in addition to the distortion. Two similer received signals, such as in Fig. 11, are supplied to a couple of pair input terminals of equalizers. Equalization processes are depicted in the form of a relation between the number of control times (abscisa) and sampled values of QPSK inphase signal (ordinate) in Fig. 12. The adaptation constant is set to 0.01. About 400 times tap weight control yields complete convergency.

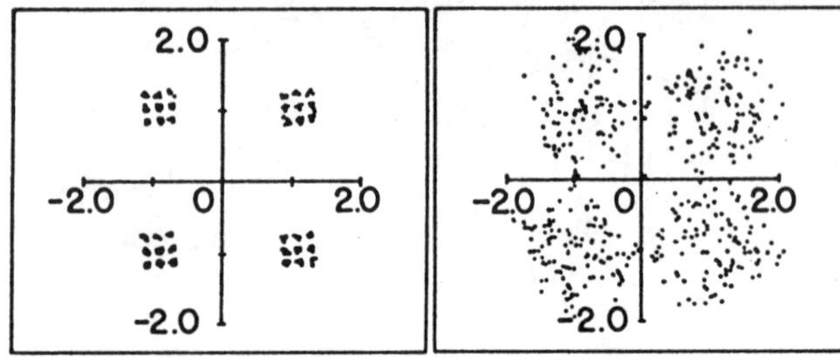

Figure 10 QPSK distorted by 7.7 dB amplitude slope

Figure 11 Received QPSK with crosstalk and distortion

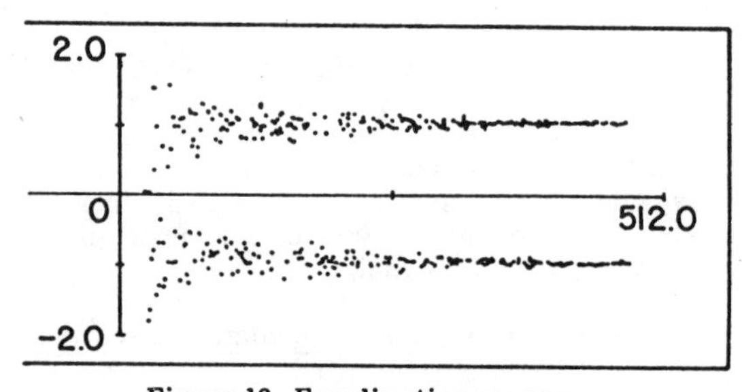

Figure 12 Equalization process

The relation between the crosstalk suppression and the number of pair taps is depicted in Fig. 13, which shows that even interference purely due to crosstalk cannot be eliminated by the pair tap equalizer.

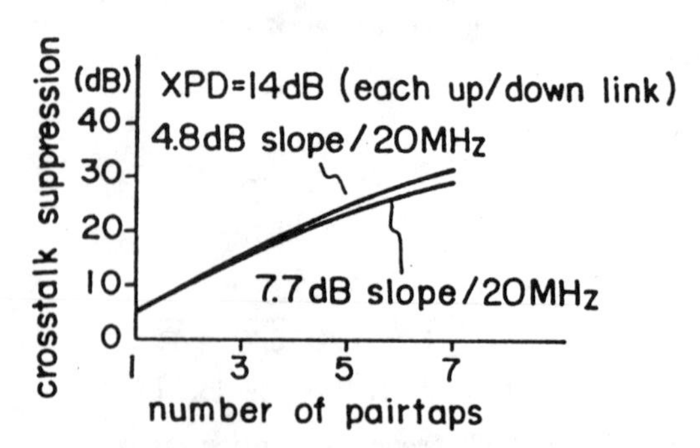

Figure 13 Equalizability for combination with crosstalk and distortion

4. More Channels per 20 MHz in 4 GHz Band

4.1 FCC provision for 4 GHz band

The FCC (Federal Communication Commission) provides that, for digital transmission in the 4 GHz band, the equipment to be used for voice transmission shall be capable of satisfactory operation within 20 MHz authorized bandwidth to allow encoding at least 1152 voice channels. This provision is very severe in modem manufacturing, even in comparison with other frequency bands. To satify this, 32 QAM with sharp roll-off shaping filters is required, at least. FCC, however, shows another provision, wherely the required minimum number of channels may be reduced by a 1/N factor, provided that N transmitters may be operated satisfactorily within an authorized bandwidth. This description suggests an adoptation of a cross-polarized configuration to meet the above capacity requirements.

Table 1 shows conbinations with channel capacity and shaping filter roll-off per 20 MHz for 32 QAM single transmitter and 16 QAM co-channel duad modem (CDM). This table clearly asserts a 16 QAM-CDM is very attractive for equipment feasibility. Hereafter, the 16 QAM-CDM system toreances are considered under the conditions expected in an actual 4 GHz band.

System	channels/20MHz	roll-off/20MHz
16QAM-CDM	1344 (672x2)	81 %
	1536 (672x2+96x2)	60 %
	1728 (672x2+96x4)	41 %
	2016 (672x3)	21 %
32QAM single modem	1152	27.8%

Table 1 Available systems in 4GHz band

4.2 Critical channel model to be equalized

It is very hard to find a critical channel model to be equalized by the depolarization equalizer, so as to satisfy an outage objective. The authors have not seen such a model, even for a selective fading channel without the cross-polarization crosstalk. In practice, other supports for fading deterioration can be considered, such as space diversity and IF section amplitude equalizer employed in Bell's AR-6A. Since heavy duct fading is characterized chiefly by a 2-path model, only duct fading was considered, which has relatively a small path difference and causes deep fading. The outage is assumed to occur on the whole when a fading notch having a deeper depth than some criterion appears inband. Therefore, outage probability P_{out} can be roughly estimated by using statistical characteristics of notch depth and of its appearance. Moreover, P_{out} can be simplified as follows.

$$P_{out} = P_D \cdot P_\rho \cdot P_R \qquad (11)$$

where P_D: Probability that a fade produces a minimum (notch) in a band B Hz.

$$P_D = B\overline{\tau} \text{ [8]} \qquad (12)$$

$$\overline{\tau} = \mu \cdot 3.7(D/20)^3 \times 10^{-9} \text{ (sec) [9]} \qquad (13)$$

$$(\mu = 0{,}07)$$

P_ρ : Probability that a notch depth (fade level) reduces to L (fade depth = -20 Log L db: L=1-r, where r is amplitude ratio,

$$P_\rho = L^2 \text{ [9]} \qquad (14)$$

P_R = multipath occurrence probability

$$P_R = C\left(\frac{f_c}{4}\right) \cdot D^3 \times 10^{-5} \text{ [9]} \qquad (15)$$

where terrain factor c=1 for average terrain, f_c is

the carrier frequency in GHz, D is hop length in miles.

From (11), critical notch depth L becomes equal to 0. 032, under the following conditions that f_c = 4 GHz, D=26.4 miles, B=20 MHz and P_{out} = 0.01%/2500 km (CCIR Rec.). Concerning the worst cross-polarization discrimination (XPD) in 4 GHz, Barnett's (Bell Telephone Lab.) literature [10] is applicable. The worst XPD can be expected to be greater than 6 dB for heavy rainfall or during multipath fading. According to the above-mentioned considerations, a critical channel model is introduced where XPD=4.4 dB and L=0.032.

4.3 Simulation on the critical channel model

Figure 4 was chosen as the depolarization channel model for 4 GHz terrestrial digital transmission. Channel parameters are selected as follows. τ_1, τ_2=T/8, (ω_h-ω_v)=0, XPD=20 dB~4.4 dB (α, β =0.1~0.6). Values L and $\bar{\tau}$ in the previous section determine the maximum distortion to be equalized, whose frequency characteristic is depicted in Fig. 14.

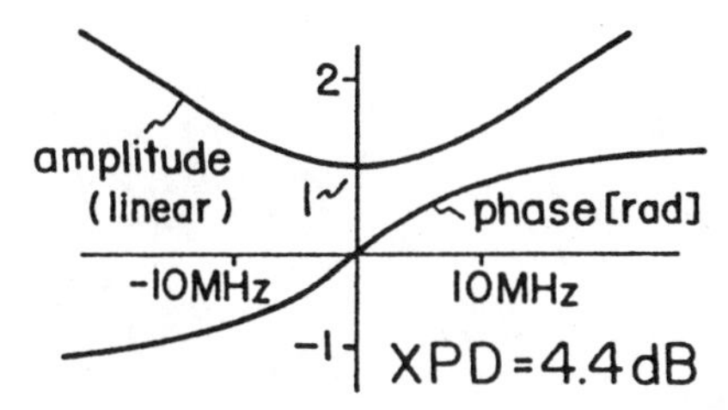

Figure 14 Maximum distortion to be equalized

Figure 15 shows 16 QAM signal points deteriorated by the above mentioned distortion. 16 QAM signal points, received through the Fig. 4 channel model, are shown in Fig. 16.

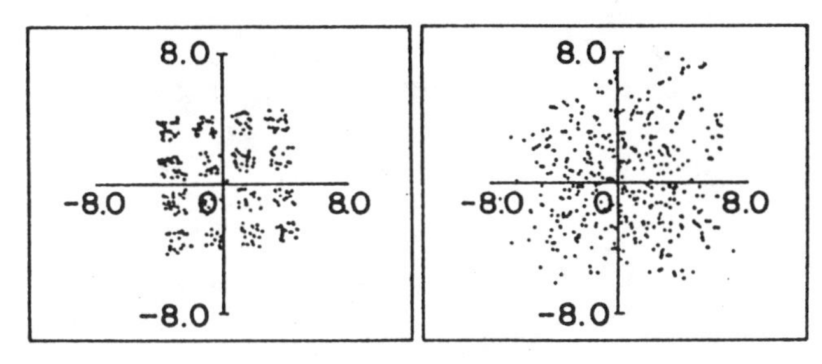

Figure 15 16 QAM signal points deteriorated by channel distortion

Figure 16 Received 16 QAM with crosstalk and distortion

16QAM CDM		XPD db , linear expression in(α)					
		20.0db (0.1)	14.0db (0.2)	10.5db (0.3)	8.0db (0.4)	6.0db (0.5)	4.4db (0.6)
Number of pair-taps	7	0.0052	0.0098	0.0093	0.0135	0.0144	0.0271
	5	0.0279	0.0192	0.0290	0.0448	0.0388	0.0603
	3	0.1029	0.0646	0.0843	0.1230	0.1003	0.1475

Table 2 The equalized signal rms deviation normalized by a signal point distance

Equalized signal rms deviations, normalized by signal point distance, are shown in Table 2. Since a very good equalizability can be obtained by 7 pair taps, it is sufficiently expected that even more severe channel conditions can be equalized.

It is expected that the co-channel duad modem (CDM) with depolarization equalizer will yield a highly efficient digital transmission system in the 4 GHz band, though many unknow problems will have to be disposed of.

5. Conclusion

For eliminating the entangled depolarization crosstalk, a new depolarization equalizer was proposed. Co-channel duad modem (CDM) was introduced with the above mentioned equalizer, so that highly efficient digital transmission in the microwave band may be attainable. In principle, any depolarization can be equalized completely, as long as a sufficient number of taps can be prepared. Though considerations for the critical channel model to be equalized are not sufficient, the CDM prospect could be found based on simulation results. Since an adaptive transversal equalizer for 200Mbit 16 QAM is under developmental research in NTT, Japan, the depolarization equalizer seems to be quite feasible. (NTT=Nippon Telephone and Telegram Public corporation.)

Acknowledgement

The authors are grateful to Mr. Y. MATSUO, a research manager for useful discussions, to Mr. S. YOKOYAMA, an assistant general manager for his encouragement and also to many members of the Communication Research Labs., C & C Systems Research Laboratories in NEC.

References

[1] Lin-shan Lee: "New Automatic Polarization Chanelling Control for Multiplestation Satellite Communication Systems" ICC'78 pp43.3.1-43.3.5

[2] D.F. DIFONZO, W.S. TRACHTMAN, A.E. WILLIAMS: Adaptive Polarization Control for Satellite Frequency Reuse Systems" COMSAT REVIEW pp253-283, 19

[3] P. Monsen: "Digital Communications Receiver for Dual Input Signal" United States Patent Appl. No. 712, 147, 1976

[4] W.J. Weber III: "A Decision-Directed Network for Dual-Polarization Crosstalk Cancellation" ICC'79 pp40.4.1-40.4.7 , 1979

[5] P.S. Tong, B. Liu: "Automatic Time Domain Equalization in the Presence of Noise" Proc. of National Electrics Conference Vol. 23 pp262-266, 1967

[6] L.E. Davissen, M. Schwaltz: "Analysis of a Decision-Directed Receiver with Unknown Prior" IEEE trans. vol. IT-16, pp270-276 May 1970

[7] K. Watanabe, K. Inoue, Y. Sato: "A 4800 BPS Microprocessor Data Modem" ICC'77 pp47.6.252-47.6.256

[8] W.D. Rummler: "A Multipath Channel Model for Line-of-Sight Digital Radio Systems" ICC'78 pp47.5.1-47.5.4

[9] V.K. Prabhu, L.J. Geenstein: "Analysis of Multipath Outage with Applications to 90 MB/S PSK Sysems at 6 and 11 GHz" ICC'78 pp47.2.1-47.2.5

[10] W.T. Barnett: "Deterioration of Cross-Polarization Disrimination During Rain and Multipath Fading at 4 GHz" IEEE Int. Conf. Commun. (Inst Electr. Electron Eng), 1974 pp12D-1-12D-4.

Part V
Performance Calculations

Outage Calculation Methods for Microwave Digital Radio

LARRY J. GREENSTEIN AND MANSOOR SHAFI

Editor's Note: This tutorial article was originally published in the February 1987 issue of the IEEE COMMUNICATIONS MAGAZINE, as part of the Special Series on Microwave Digital Radio. It has been updated by the authors and edited for inclusion in this book.

INTRODUCTION

PREVIOUS tutorial articles in this book [1]–[4] have discussed the rapid growth, sophisticated methods, and special technical problems of this dynamic field. All have emphasized the importance of multipath fading as an obstacle to link performance, and [3] dealt specifically with multipath fading channel models. In this paper, we show how such models can be used in estimating the reliability of digital radio links.

To be specific, the subject of this paper is the prediction of link *outage time*. We will define this quantity and show that its estimation plays a pivotal role on several levels: In the initial design of a radio system, in evaluating specific hardware realizations, and in comparing alternative design approaches. We aim to impart an understanding of the various estimation methods that exist and their virtues and limitations.

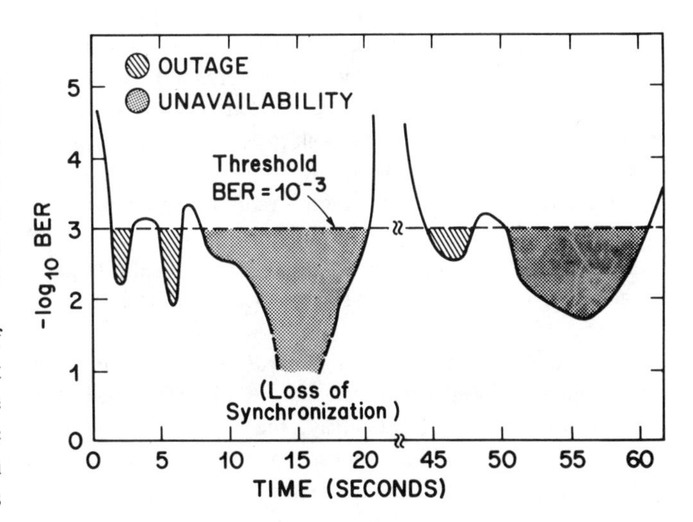

Fig. 1. Bit error ratio history for a particular link, showing outage times and unavailability times. The threshold BER is 10^{-3}.

UNAVAILABILITY AND "OUTAGE" TIME

According to CCIR/CCITT recommendations [5], a radio link is deemed to be *unavailable* if, in at least one direction of transmission, either of two conditions occur for at least ten consecutive seconds: (i) The digital transmission is interrupted (e.g., alignment or timing is lost); or (ii) the bit error ratio exceeds 10^{-3}. Time periods in which the bit error ratio exceeds 10^{-3} for fewer than ten seconds are regarded as periods of unacceptable performance, or *outage*, and *outage time* is the accumulated seconds for all outage events in a given time period, say, a year.* This distinction between unavailability and outage is depicted in Fig. 1.

Radio unavailability is usually associated with long-lasting events like rain (primarily in bands above 10 GHz), equipment failures, and propagation anomalies such as ducting. Outages, on the other hand, tend to be associated with brief events like multipath fading. This explains the important connection between channel fading models and estimations of outage time.

A multipath-related component of outage time is the time it takes a receiver to recover synchronization after losing it because of high bit error ratios. This "hysteresis" effect is ignored in most existing methods of outage time estimation, the implicit assumption being that such periods are relatively short.

The objectives for outage time can be quite stringent. The CCIR allocation followed by European nations is 20 seconds per heavy-fading (or worst-case) month on a 40-km path. On its long-haul routes in the United States, AT&T has an objective of 10 seconds per year for such paths. These are, indeed, very stringent objectives, considering that typical radio paths can experience thousands of seconds of multipath fading in a heavy-fading month. The resulting demands on system design are great, particularly as designers reach for higher capacities by using multilevel modulations and co-channel dual-polarized transmissions.

PREDICTION METHODS I: GENERAL APPROACH BASED ON OUTAGE REGIONS

The most reliable way to assess performance for a given radio system on a given path is to install the system on that path and measure its outage time. This is also the most expensive way. Moreover, it is quite time-consuming, since one is obliged to wait until nature provides enough fading

* This simplified view of outage time will serve the purpose of the present article. However, a more precise definition for periods of unacceptable performance takes into account such measures as number of error-free seconds in a month, number of severely errored seconds in a month, number of degraded minutes in a month, and residual bit error ratio (see [5] for details). These measures reflect the growing importance of data as well as digitized voice in digital radio transmission.

events to be statistically reliable, and then hope that the measurement system works properly at the most interesting times. Finally, this approach cannot be readily used in the all-important design stages, wherein alternative approaches are evaluated and compared.

For these reasons and more, digital radio practitioners have resorted to obtaining statistical models for the multipath channel response [3] and performing system studies—in the laboratory and/or on the computer—using these models. In this way, for example, the same representative ensemble of channel responses can be applied to a wide variety of modulations, adaptive techniques, and design parameters without actually building anything (for example, in the design stage, using computer methods); or they can be applied to one or more hardware realizations (for example, in the evaluation/selection stage, using laboratory measurements). Moreover, arbitrary changes can be made in the statistics governing the ensemble of channel responses, both to accommodate different radio paths and to permit sensitivity studies.

The methods for evaluating digital radio performance seem to be many and varied, and the literature of the past several years is filled with examples. However, most of the popular methods use the following four steps (see Table I):

TABLE I
OUTAGE CALCULATION PROCEDURE

STEP 1: CHOOSE A PERFORMANCE THRESHOLD CRITERION
- Bit Error Ratio Exceeds Threshold Value (Typically, 10^{-3})
- Data Eye Closes
- Signal-to-(ISI + Noise) Ratio at Detector Exceeds Threshold Value

STEP 2: SPECIFY A STATISTICAL CHANNEL MODEL
- Two-Path Model
- Simplified Three-Path Model
- General Three-Path Model
- Others

STEP 3: FIND THE PARAMETER REGION, Ω, OVER WHICH THE PERFORMANCE THRESHOLD IS EXCEEDED
- Laboratory Measurements
- Mathematical Analysis
- Computer Simulation

STEP 4: FIND THE FRACTION OF FADE RESPONSES FOR WHICH THE MODEL PARAMETERS LIE WITHIN Ω
- Integrate the Joint PDF of the Model Parameters Over Ω.
- Generate Many Sets of Model Parameters, Using Monte Carlo Simulations, and Count the Fraction of Sets Within Ω.

RESULT: CONDITIONAL PROBABILITY (GIVEN FADING) OF PERFORMANCE THRESHOLD BEING EXCEEDED. MULTIPLY BY TOTAL NUMBER OF FADING SECONDS (MODEL PARAMETER T_f) TO CONVERT TO OUTAGE TIME.

1) Choose A Performance Threshold Criterion—The most commonly used performance threshold criterion is that the short-term bit error ratio (say, the one-second average, denoted by BER) should not exceed some value (say, 10^{-3}). A less-used performance threshold criterion is that the data eye, as seen at the detector, should not close. This criterion has been used in some analytical studies, both for convenience [6] and because the closed-eye probability is closely related to the outage time in receivers dominated by multipath distortion [7]. Finally, some studies use a specially defined signal-to-distortion ratio at the detector as a performance measure, and the performance threshold criterion is that this ratio must exceed some value [8]–[13]. The reason for doing this, as we will see later, is that the signal-to-distortion ratio in question relates to a simple and tight upperbound on BER.

2) Specify A Statistical Channel Model—In most digital radio studies, the model for the complex channel transfer function begins with the same representation. Stating it somewhat formally,

$$H(f)=\begin{cases} 1+j0 & \text{During Normal Propagation} \\ F(f;\ \{c\}) & \text{During Multipath Fading} \\ & (T_f \text{ Seconds/Year}) \end{cases}$$

$$-W/2 \leq f \leq W/2 \tag{1}$$

where W is the channel bandwidth; f is measured from the center of the channel; and $F(\cdot)$ is a complex function of f containing a finite set of parameters (the c's) that change randomly with time. Because the c's change very slowly compared to the signaling rate used over the channel, $F(\cdot)$ can be regarded as a quasi-static function of f whose parameters are random variables. The statistical model then consists of (i) the form of the function, $F(f;\ \{c\})$; (ii) the joint conditional probability density function (pdf) of its random parameters; and (iii) a number or formula for the effective number of multipath fading seconds per year, T_f.* Typically, T_f is chosen so that the yearly (or monthly) distribution of power fading at a single frequency, as derived using (i), (ii), and (iii), is consistent with known results for single-frequency fading [14].

[Note that we are not discussing models for *multiple* channels, that is, the channel models associated with space diversity reception and/or dually polarized transmission. Such models have been reported, for example those in [12], [13], [15], but the present discussion is confined to single-polarization nondiversity channels.]

The primary distinction among models described in the literature resides in the function (hereafter denoted simply by $F(f)$) used to characterize the complex response over the channel bandwidth. Two of the most popular models use a

* At the discretion of the user, T_f can be made to represent the number of fading seconds per heavy-fading month rather than per year. In temperate northern climates, it is usual to assume three heavy-fading months per year (June, July, and August) and to multiply the outage time for such a month by three to estimate yearly outage time.

function of the form

$$F(f) = a[1 - b \exp(-j2\pi(f - f_0)\tau)]. \quad (2)$$

In the so-called *two-path model* [16], [17], there is no amplitude scale factor (equivalently, $a = 1$) and b, f_0, and τ are the variable parameters. In the so-called *simplified three-path model [18], [19],* τ is a parameter related to the channel bandwidth whose value is set at 6.3 ns; and a, b, and f_0 are the variable parameters. For purposes of discussing either model, we recast the parameter b in terms of the decibel value of maximum fade depth,

$$B \triangleq \begin{cases} -20 \log_{10}(1-b) & \text{if } b<1 \\ -20 \log_{10}(1-1/b) & \text{if } b>1. \end{cases} \quad (3)$$

Also, the a-parameter in the simplified three-path model is recast in terms of its decibel value,

$$A \triangleq -20 \log_{10} a. \quad (4)$$

Confusion often arises from the observation that $F(f)$ in (2) has the appearance of a two-path (*not* a three-path) response. However, it can be envisioned as arising from three paths (or propagation rays), wherein two of the paths are so close in time delay that their composite response is a constant (a) over the channel bandwidth; and frequency selectivity is attributed to a third path at relative delay τ. This useful picture gives the simplified three-path model its name. On the other hand, radio links with terrain reflections (from either land or sea) can have a more physically motivated three-path situation. In such cases, a so-called *general three-path model* of the type reported in [20], [21] may be appropriate. For this model, we denote the path gains by a_0, a_1, and a_2, and the relative delays between the primary and two secondary paths by τ_1 and τ_2.

Other models are discussed in [3]. For the purposes of this paper, however, the above descriptions of the two-path, simplified three-path, and general three-path models will suffice.

3) Find The Parameter Region, Ω, Over Which The Performance Threshold Is Exceeded—For concreteness in discussing Ω, we will specify the receiver performance threshold to be BER $= 10^{-3}$ and assume that the simplified three-path model is being used. Extensions to other performance thresholds and channel models will be obvious.

Suppose now that one has a specific radio link design, including power levels and receiver noise figure. For that design, there will be some regions of the parameter space (a, b, f_0) over which BER $> 10^{-3}$. That region of (a, b, f_0)-space is what we mean by Ω, which can thus be regarded also as an "outage region." This seeming abstraction can be made more visual by fixing the value of a (or A in (4)); finding the resulting outage region within the (b, f_0)-plane (or (B, f_0)-plane, using (3)); and repeating this process for numerous values of A. The resulting set of two-dimensional outage regions characterizes the outage performance of the radio equipment. Note that this characterization does *not* depend on the joint parameter statistics, i.e., on the probability density function $p(a, b, f_0)$; it is a function solely of the radio design. This is why the outage region in parameter space (Ω) has value in its own right: Without resort to model parameter statistics, which might be unknown or uncertain for a particular path, competing radio designs can be compared by viewing the relative sizes of their outage regions [22]–[24].

How to find Ω? There are three distinct approaches that can be used, each of which has been extensively reported in the literature. They are (i) laboratory measurements; (ii) mathematical analysis; and (iii) computer simulation. In any of these approaches, the aim is to find the model parameter combinations for which the performance threshold is just met. Collectively, these combinations define the boundaries of the outage region.

The laboratory measurement approach is the most reliable, as it makes no idealized assumptions about how the radio equipment works. However, it requires the development of laboratory circuitry to emulate the fading channel [25], and it is limited to radio hardware already built. The use of mathematical analysis permits a wide variety of design approaches to be considered, but it can run into tractability problems in dealing with nonlinearities and other imperfections. The computer simulation approach, in which random data streams and their passage through the radio link are emulated via numerical methods, permits greater tractability; the cost, however, can be very long computer running times. In short, all three approaches have their virtues and their limitations.

4) Find The Fraction of Fade Responses For Which The Model Parameters Lie Within Ω—The final step is performed in one of two ways. The investigator either (i) integrates the joint parameter pdf over the outage region, Ω; or (ii) generates, via Monte Carlo simulation, thousands of parameter sets (for example, $\{a, b, f_0\}$) and counts the fraction of sets that lie within Ω. The result, in either case, is the probability that the performance threshold is exceeded, conditioned on the fact that fading is occurring. This is called the *conditional outage probability,* P_0. To convert to outage seconds per year, this quantity is multiplied by T_f in (1). Thus, the outage region Ω, which characterizes just the radio design, is coupled with the model characteristics $p(a, b, f_0)$ and T_f, which characterize just the channel, to produce estimates of outage time.

The choice between integrating the parameter pdf over Ω or simulating the parameter sets and counting those within Ω is a matter of judgment and taste. Each approach has its advocates, as the literature shows. The integration approach invariably requires numerical methods and, to ensure computational accuracy, these methods must be carefully considered. The simulation approach is in this respect simpler, since virtually any pdf can be accommodated by Monte Carlo methods. However, the attainment of statistical accuracy can place heavy demands on the number of simulation trials.

Treating The Effects of Diversity

Space and frequency diversity are both powerful and widely used countermeasures to multipath fading. For analog radio systems, improvements in outage performance resulting from using these techniques can be estimated from the empirical relationships of Vigants [26], which are based on single-frequency fading laws. For digital radio, however, outage calculation methods that include diversity effects lack completeness and maturity.

To evaluate links equipped with space diversity, appropriate dual channel models can be combined with Monte Carlo simulations of the model parameters to assess specific designs [11]. The traditional approach of first estimating outage time without space diversity, and then applying an improvement (reduction) factor, is another possibility. However, improvement factors based on single-frequency fading laws are bound to be too conservative for dispersion-dominated systems. On the other hand, improvement factors derived for digital radio channels would have to account for the difference between selection diversity and combining diversity, differences among types of combining diversity, and other design details. In short, the problem is a complex one.

Similar comments pertain to frequency diversity. The results of Vigants for single-frequency fading show the improvement factor to be proportional to the frequency separation of the working and protection channels. For dispersion-dominated digital radio links, however, this finding may be inappropriate. This is suggested by the recent findings of Lee and Lin [27], which show that switching to an adjacent channel is better than switching to channels two or more bandwidths away. In any case, frequency diversity improvement factors may depend on the details of the protection switching algorithm, and data on methods for dealing with the numerous possible strategies are not easily obtained.

Now there is a new category of diversity techniques, referred to as *horizontal diversity* [28], *angle diversity* [29], [30], and *aperture diversity* [31], that has created considerable interest in the field of microwave digital radio. These techniques exploit the fact that two antenna beams originating from the same height but having different patterns can, between them, virtually eliminate the occurrence of frequency-selective (or spatially-selective) reception nulls. Although current efforts to understand and capitalize on this phenomenon are intense, matters are not yet at the point where statistical models—and corresponding outage calculation methods—are at hand.

For all of the above reasons, the general treatment of diversity in estimating digital radio outage remains an open issue. We will not explore it further in this paper, except to note its importance to the evaluation of future sophisticated systems.

Historical Review

Today, most predictions of digital radio outage time are based on the notion of parameter outage regions, as discussed above, or use variations on that general approach. Some important methods, on the other hand, are alternatives to it. To set the stage for discussing several methods in detail, we present here some historical perspective. The demands of brevity limit us to citing but a fraction of the many valuable contributions to this subject.

The importance of multipath fading effects on microwave digital radio links was not widely appreciated until the mid-1970's, and so the history of published outage prediction methods appears to have a beginning as recent as 1978. That year, at the International Conference on Communications (ICC) in Toronto, three papers were presented on the subject [32]–[34]. We shall begin with the two that were published (in expanded form) the following January in the IEEE Transactions on Communications; the third will be discussed later in connection with equipment signatures.

The studies published by Greenstein and Prabhu [35] and by Jakes [36] used a common multipath model agreed upon by the three investigators. In the absence of extensive measured data, they hypothesized a two-path situation in which the direct path has a fixed gain, the second path has an exponentially-distributed gain, and its delay is statistically independent of the gain and is also exponentially distributed. In short, they drew on the sparse measured results available at that time to compose an early version of the two-path model in common use today [16], [17].

Greenstein and Prabhu defined outage as occurring whenever BER exceeds 10^{-6}, and they studied various modulations for transmitting 90 Mb/s at both 6 and 11 GHz. Jakes defined outage as occurring whenever the peak-to-peak envelope delay distortion within the signal band exceeds some modulation-related constant times the symbol period, and he studied several modulations and bit rates. In both analytical studies, the assumed joint pdf of the secondary path gain and delay was integrated over the derived outage region. The results showed the serious inadequacy of digital radio transmission without some form of multipath equalization, even for short-haul systems, if stringent reliability objectives are to be met.

In the meantime, Rummler was completing the development of his simplified three-path model and, in fact, had reported it at the same ICC in Toronto [37]. An expanded version of this paper appeared in the *Bell System Technical Journal* in May–June 1979 [18], along with a paper by Lundgren and Rummler that applied the new three-path model to outage predictions [38]. In this study, the performance threshold criterion was BER $= 10^{-3}$, and the outage region was derived via laboratory measurements of a particular unequalized 6-GHz radio modem. The estimates of outage obtained from laboratory measurements were found to be in good agreement with field measurements made at the same time.

Lundgren and Rummler presented their outage region, initially, in the following way: For a given f_0 (see (2)), there is a critical A for each B (see (4) and (3)) beyond which outage occurs. The set of curves of critical A versus B (one curve for each of several values of f_0) defines the outage region in (A, B, f_0)-space. Integrating the joint pdf of A and B over the exterior of each curve, and suitably summing over f_0, yields the conditional outage probability (or outage time, if multiplication by T_f is included).

For the radio system they were studying, Lundgren and Rummler found that outage time was dominated by frequency-selective fading (multipath distortion) rather than by total signal fading relative to noise. Accordingly, they removed the parameter A from consideration and obtained a single plot of critical B versus f_0. The outage region corresponds to the exterior of this curve, which was found to have the general shape of a "W." (Later on, Rummler was to refer to such outage region boundaries as *W-curves* [39]; they are also related to the *M-curves* discussed shortly.)

The paper by Lundgren and Rummler was among the first to report a strong correspondence between in-band selectivity (the dB difference between maximum and minimum channel response values) and BER in dispersion-limited receivers. The correspondence suggests a simple link between the probability distribution of this dispersion measure and the curve of outage probability vs. threshold BER. This link has been pursued by investigators in Japan [20], [40], [41] and elsewhere, and we will return to it later in connection with the linear amplitude dispersion (LAD) method.

The W-curves (critical B versus f_0) mentioned above bring us to the subject of *M*-curves and signatures. Much credit for introducing these ideas goes to Emshwiller, who reported a new method for estimating outage time in his 1978 ICC paper in Toronto [34]. He assumed a two-path model, with the delayed path having a gain $b < 1$ and the delay, τ, being a random variable of arbitrary distribution. Using computer simulations of an 8-PSK modem, he obtained curves corresponding, essentially, to critical values of $(1 - b)$ versus notch frequency, f_0, with τ as a parameter. These curves tend to be *M*-shaped, as seen in Fig. 2(a), and the outage region corresponds to their interiors. Recasting the ordinate as B, (3), leads to the W-curve shown in Fig. 2b. Both kinds of curves are generally called *signatures*.

The width of the signature curve is not usually much larger than the channel bandwidth, reflecting the fact that the most damaging notches are ones that occur in-band. The dip at the band center signifies that slightly off-centered notches can be worse than centered ones because they introduce more cross-rail interference. Emshwiller was able to show that the conditional outage probability due to multipath distortion, computed using a two-path model, is proportional to the mean-square value of the τ-distribution.

This important scaling relationship for distortion-dominated receivers was separately derived by Campbell and Coutts [16], who showed additionally that the outage time scales with the square of the symbol rate $(1/T)$. This knowledge permits the radio system investigator to measure a single signature curve (that is, for any combination of τ and symbol rate); integrate the appropriate pdf over the interior of that curve (as if τ were fixed); and scale the result by a simple rule to obtain outage for any $\overline{\tau^2}$ and $1/T$.

The utility of a single signature curve in assessing outage, at least for distortion-dominated receivers, applies equally to the simplified three-path model. In this case, τ in (2) is fixed at 6.3 ns and a is a variable; however, the signal scaling by a is unimportant when noise is not the primary cause of outage. Hence, a single curve of critical B vs. f_0, obtained via either measurement or computation but in the absence of noise, is sufficient; conditional outage probability is found by just integrating the joint pdf of B and f_0 over the outage region defined by this curve.

The simplicity of the signature approach to outage time calculations for distortion-dominated receivers is thus made

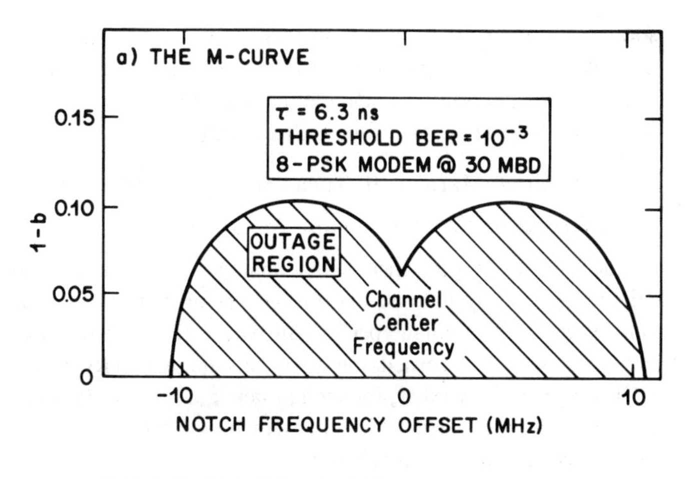

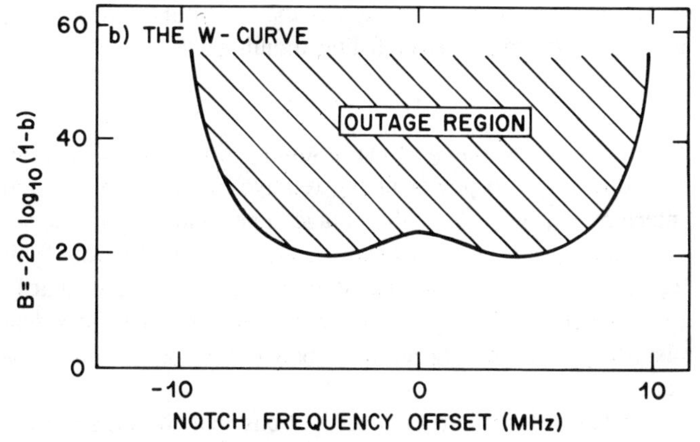

Fig. 2. M- and W-curves for an 8-PSK modem and a delay parameter of 6.3 ns.

clear. For such receivers, the outage region, Ω, is defined by a two-dimensional contour (the signature), and can be derived and integrated over with relative ease.

In recent years, significant improvements in receiver equalization have gone far to reduce the impact of multipath distortion relative to that of noise. It is noteworthy, then, that signatures can also be useful when noise *is* an important cause of outage. An example is the more general signature method based on the simple three-path model that was developed early by Lundgren and Rummler [38] and which has been used subsequently by Rummler and others [39]–[42]. As this method demonstrates, however, the inclusion of noise effects requires bringing in a third model parameter (the scaling factor a, in this case), with a major increase in the required amount of measurement and computation.

More recently, a general signature method applicable to the two-path model was developed by Meyers [43]. The method permits additive noise effects to be included and uses the Gauss-Quadrature rule to permit rapid integrations over the parameter outage region.

Despite these possibilities for including thermal noise, there has been a need to enlarge on the body of outage prediction methods so that additional factors can be taken into account. They include the widespread use of space and frequency diversity; the possible use of co-channel dual polarization; and the prominence on many links of interference from various sources. Among the responses to this need are the method of composite fade margin [39], [44], the method of linear amplitude dispersion [20], and the method of model parameter simulation [8], [10]. We discuss these next.

Prediction Methods II: Three Specific Approaches

The Method of Composite Fade Margin (CFM)

The outage performance of an analog radio link can be related to the thermal noise margin, also referred to as the *flat fade margin*. The outage performance of a digital radio link, however, is not governed solely by this quantity. Rummler [39], [44] has shown that, for such systems, a *composite fade margin* can be defined which includes the effects of thermal noise, dispersion, and interference.

The starting point is to express the total outage time for a given path during a heavy-fading month as

$$T_{out} = T_N + T_D \tag{5}$$

where T_N is the outage time assuming no signal distortion, i.e., due solely to power fading relative to thermal noise and interference; and T_D is the outage time assuming no noise, i.e., due solely to dispersion caused by multipath. This equation reflects the fact that some outages are due primarily to signal power fading and some are due primarily to signal distortion, with in-between cases being more or less insignificant.

We know from earlier investigations that the outage time component T_N can be expressed as

$$T_N = rL_f^2 T_0 \tag{6}$$

where r is a *multipath occurrence factor* that has been related to carrier frequency, path length and terrain type [14], [45]; T_0 is the total number of seconds in a month; and L_f^2 is the amount of single-frequency fading (expressed as a relative power gain) for which thermal noise plus interference would bring BER to its specified threshold value. If we assume a threshold value of 10^{-n}, then L_f^2 is the solution to

$$\frac{L_f^2 C}{I+N} = CNR_n \tag{7}$$

where C, I, and N are the average powers of the signal, interference, and noise, respectively, at the receiver input; and CNR_n is that input signal-to-(interference plus noise) ratio for which BER $= 10^{-n}$. Note that, in this particular formulation, the interference is implicitly assumed not to fade. However, a simple modification can be made to account both for interferers that fade when the signal does and for interferers that do not.

The solution for L_f^2 in (7) is

$$L_f^2 = \frac{CNR_n}{(C/I)} + \frac{CNR_n}{(C/N)}. \tag{8}$$

The reciprocal of the second term on the right is the flat fade margin in the absence of interference ($I = 0$). We denote the dB value of this quantity by *FFM*, and so

$$L_f^2(I=0) = 10^{-FFM/10}. \tag{9}$$

With interference included, the flat fade margin is denoted by *FFM′* and, using the above relationships, we can express it by

$$L_f^2(I \neq 0) = 10^{-FFM'/10} = 10^{-FFM/10} + 10^{-(\{C/I\} - \{CNR_n\})/10} \tag{10}$$

where $\{X\}$ denotes X in dB. Combining (10) with (6) yields T_N in terms of *FFM*, $\{C/I\}$ and $\{CNR_n\}$.

The remaining task is to obtain T_D, the outage time due solely to multipath dispersion. This quantity can be estimated by the various signature methods described earlier for noiseless receivers, or it can be measured in the laboratory or on radio paths. It can also be estimated using the method of linear amplitude dispersion discussed in the next subsection. In any event, T_D can be cast in terms similar to (6) and (9) by defining the *dispersive fade margin* (*DFM*) to be the dB quantity that satisfies

$$T_D = r10^{-DFM/10} T_0. \tag{11}$$

The quantity *DFM*, in comparison with *FFM*, tells the designer how large the transmitter power should be before diminishing returns set in. For example, increasing *FFM* to values larger than *DFM* + 6 dB will do little to reduce total outage time.

To bring all of the above ideas together, the composite fade margin (*CFM*) is defined to be the dB quantity that satisfies

$$T_{out} = r10^{-CFM/10} T_0. \tag{12}$$

Combining (10)–(12) with (6), *CFM* can be expressed in terms of *FFM*, *DFM*, $\{C/I\}$ and $\{CNR_n\}$ by

$$10^{-CFM/10} = 10^{-FFM/10} + 10^{-(\{C/I\} - \{CNR_n\})/10} + 10^{-DFM/10}. \tag{13}$$

The Method of Linear Amplitude Dispersion (LAD)

Various reports [20], [38], [40], [41] have suggested that BER can be closely related to either peak-to-peak amplitude dispersion (the dB amplitude range of the propagation response within the channel bandwidth) or linear amplitude dispersion (the dB difference between response amplitudes at two frequencies, usually chosen to be near the upper and lower band edges). Let Z denote the (positive) dB difference between the response amplitudes at f_1 and f_2; and let Z_0 denote the value of Z beyond which a given radio equipment generally exceeds the performance threshold. If one can determine the probability that Z exceeds Z_0 during multipath fading (with noise absent), then the conditional probability of outage due solely to dispersion can be estimated. This probability can be converted to the outage time component T_D in (5) by multiplying it by T_f, the time-scaling factor of the channel model used.

One formulation for the probability distribution of Z, developed by Sakagami and Hosoya [20], invokes the general three-path model cited previously and assumes that, over the population of all fading events, the channel *power* responses at any two frequencies are correlated first-order Gamma variates [21]. This assumption can be made consistent with known statistical results for single-frequency fading by specifying $T_f = rT_0$, where r and T_0 are introduced in (6).

Denoting the power responses at two frequencies f_1 and f_2 by $P(f_1)$ and $P(f_2)$, with $P(f_1)$ taken to be the larger, we can write the equation pair

$$z \triangleq P(f_1)/P(f_2); \qquad Z \triangleq 10 \log_{10} z. \tag{14}$$

To find the cumulative distribution for z, it is only necessary to know the correlation coefficient, $\rho_{\Delta f}$, between the first-order Gamma variates $P(f_1)$ and $P(f_2)$. For nondiversity reception, the distribution can be shown to be [46]

$$F(z) = 1 - ((z-1)/\sqrt{(z+1)^2 - 4\rho_{\Delta f} z}). \tag{15}$$

In light of previous discussion, we can then express T_D as

$$T_D = rT_0 F(z = 10^{Z_0/10}). \tag{16}$$

The problem thus reduces to finding $\rho_{\Delta f}$. This can be done if the statistical means and variances of the model parameters (a_0, a_1, a_2, τ_1 and τ_2) are known or can be estimated. Alternatively, if single-frequency data at frequencies f_1 and f_2 are available, the quantity $\rho_{\Delta f}$ can be calculated directly. A typical value for $\rho_{\Delta f}$ for $f_2 - f_1 = 35$ MHz is 0.99. In any case, $\rho_{\Delta f}$ must be accurately determined for the LAD method to work, because $F(z)$ is highly sensitive to it. For example, $F(10)$ is roughly ten times higher for $\rho_{\Delta f} = 0.99$ than it is for $\rho_{\Delta f} = 0.999$. This sensitivity is an important consideration in using the LAD method.

In summary, the LAD method requires determining a threshold amplitude difference, Z_0, for a particular radio equipment; and a correlation coefficient, $\rho_{\Delta f}$, for the radio link over which it operates. It is then possible to estimate the conditional probability of outage due to dispersion. Multiplying this quantity by $T_f = rT_0$ will yield the component T_D in (5). The formulation for the component T_N can then follow the same lines described previously.

The Method of Model Parameter Simulation

One version of the outage time prediction method described earlier avoids the explicit determination of parameter outage regions. It consists of generating, via Monte Carlo simulations, an adequate number (perhaps thousands) of sets of model parameters; computing, for each such parameter set, a performance measure; and counting the fraction of trials for which some threshold is violated (say, BER $> 10^{-n}$). That fraction estimates the conditional outage probability (P_0) and multiplication by T_f in (1) yields outage time.

This method was first developed and used for digital radio outage calculations by Foschini and Salz [8] and has been used in numerous subsequent outage studies. These include assessments of finite-tap and infinite-tap equalizers [9], [10], space diversity combining [11], and dual-polarization reception [12], [13]. Thus, the method has broad applicability. Moreover, it requires little more than a way to generate model parameters in accordance with the model statistics and a way to estimate the performance measure. Monte Carlo methods for generating the model parameters are well known, so we will not discuss them here.

To estimate BER, Foschini and Salz derived a simple and tight upperbound formula,

$$BER \leq 2 \exp(-\rho/2) \tag{17}$$

where ρ is a particular ratio of signal to distortion for the design samples at the detector. The "signal" part of ρ is the square of the half-distance between points in the sampled-signal constellation; and the "distortion" part is the mean square sum of intersymbol interference and thermal noise at the sample times. For M-level quadrature amplitude modulation (M-QAM), it is easy to show that

$$\rho = \frac{3}{M-1} \Gamma \tag{18}$$

where Γ is a readily derived function of the channel response and the transmitter and receiver filtering [10]. If BER $= 10^{-n}$ is the specified performance threshold, the requirement on Γ to avoid outage is

$$\Gamma \geq \frac{2(M-1)}{3} \ell n \, (2 \times 10^n) \equiv \Gamma_0(M, n). \tag{19}$$

For $n = 4$, for example, the decibel values of Γ_0 for $M = 16$, 64, and 256 are 20.0 dB, 26.2 dB, and 32.3 dB, respectively. Because of the bounding relationship given by (17), these requirements may be high by 1–2 dB in most cases.

A study using this simulation method thus reduces to computing Γ for many Monte Carlo trials, and estimating P_0 as the fraction of trials for which $\Gamma < \Gamma_0$. The resulting simplicity permits a wide range of design approaches to be evaluated and compared. On the negative side, the method does not readily address the full range of possible link impairments (nonlinearities, phase jitter, etc.). However, interference effects could be accommodated by expanding the definition and analysis of ρ, for example, along the lines described in [13].

APPLICATIONS

There are numerous modes of application for the outage calculation methods we have discussed here. The following examples illustrate two distinct types of application and the relevant method of calculation for each.

Example 1—A 16-QAM 4-GHz system is to operate, without space diversity, on a 40-km path over normal terrain (surface roughness $= 20$ m) in a maritime subtropical region of the USA. The interference on that path is dominated by a satellite downlink signal, with received power 63 dB below that of the unfaded desired signal; the unfaded C/N at the receiver is 60 dB; the estimated correlation coefficient between faded powers at frequencies 8 MHz above and below the channel center is 0.995; and the modem is sufficiently equalized that BER does not exceed its threshold of 10^{-4} unless the linear amplitude dispersion between those two frequencies exceeds 10 dB. The aim is to estimate the outage time in a year, using the LAD method to account for dispersive effects.

First, we derive T_N in (5). To estimate the multipath occurrence factor, we use [45] and obtain $r \cong 0.161$. The

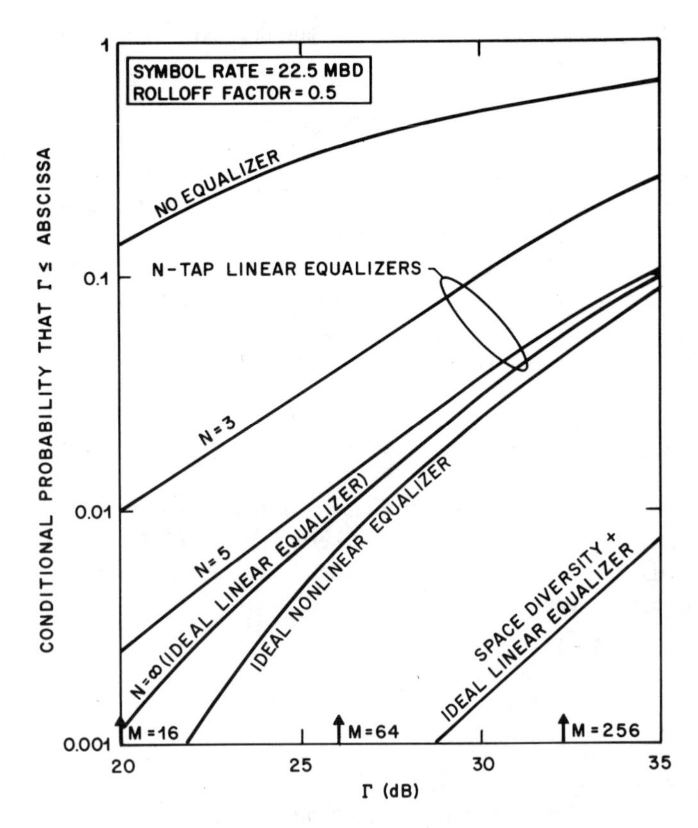

Fig. 3. Study results using the method of model parameter simulation. The markers shown for $M = 16, 64$ and 256 indicate Γ_0 for a threshold bit error ratio of 10^{-4}.

appropriate CNR_n in (7), for $n = 4$ and 16-QAM, and allowing for modem impairments, is estimated to be 20 dB. Given that C/N is 60 dB, the flat fade margin, FFM, is thus 40 dB. Given that C/I is 63 dB, FFM' is found from (11) to be 38.23 dB. The corresponding T_N, using (6) with $r = 0.161$, is 63.6 seconds per heavy-fading month.

Second, we derive T_D in (5). Since we are using the LAD method, the effective number of fading seconds per heavy-fading month can be estimated from $T_f = rT_0 \cong 424{,}500$ seconds. From (15), (16), and the given values for Z_0 (10 dB) and $\rho_{\Delta f}(0.995)$, the conditional outage probability due to dispersion is calculated to be 0.00123. Multiplying by T_f yields $T_D = 523.1$ seconds per heavy-fading month.

Finally, the estimated outage time per heavy-fading month is $T_N + T_D \cong 586.7$ seconds. Assuming three heavy fading months, we obtain $T_{out} \cong 1{,}760$ seconds per year. Note that this design would not satisfy stringent long haul objectives, and so the use of space diversity would be essential.

Example 2—In the 6-GHz band in the USA, digital radio systems typically use M-level QAM, symbol rates of 22.5 Mbd, and cosine rolloff shaping with a rolloff factor near 0.5. It is of interest to estimate conditional outage probabilities for such systems over a wide range of possible designs, including different approaches to equalization and space diversity, and different numbers of modulation levels.

Three published theoretical studies, each using the model parameter simulation method and the simplified three-path model, have addressed these issues [9]–[11]. Fig. 3 shows probability distributions for Γ, as defined by (18), for a variety of designs. The dB value of Γ that must be exceeded when the bit error ratio threshold is 10^{-4} is marked on the abscissa for each of three values of M. Each such Γ corresponds to Γ_0 in (19) for $n = 4$. For each M, the ordinate associated with the threshold Γ is the conditional outage probability (i.e., $P_0 = Pr\{\Gamma < \Gamma_0\}$).

The top three of the central curves show the theoretical capabilities of linear equalizers with 3, 5, and an infinite number of T-spaced taps [9]. The lowest of the central curves, labeled "Ideal Nonlinear Equalizer," is an optimistic bound on what the best possible receiver could accomplish [10]. The potential improvements on linear equalization are seen to be small, at least for the fading statistics assumed in this study.

The result of space diversity plus equalization (lowest curve) is dramatically better than that for equalization alone [11]. This improvement is consistent with observations from field measurements. However, the improvement obtained with equalization alone over the case of no equalization (highest curve) is much larger in the figure than what has been measured. One possible reason is that the bounding relationship given by (17) is particularly pessimistic for unequalized receivers, working best for cases where the residual distortion has Gauss-like statistics. Probably more important, however, is the fact that the method of outage prediction ignores such real-life effects as channel dynamics and receiver hysteresis. We address these and other issues in the following concluding remarks.

Conclusion

As our discussions have made clear, a reliable outage calculation method would be one that accounts for both noise (power fading) and intersymbol interference (dispersion); includes the effects of both external interference and equipment imperfections; and is general enough to accommodate both space diversity reception and dual-polar operation. We have also seen that, while each of the published methods satisfies some of these requirements, none satisfies all. Moreover, most of the existing methods do not take account of the time variation properties of fading channels or the receiver hysteresis effects associated with recovery from loss of synchronization, although recent efforts are aimed at correcting this deficit [43]. The possible offsetting effects of diversity methods on hysteresis also need to be better understood.

An essential part of realizing the advances called for above is obtaining improved channel models. More accurate statistical characterizations, inclusion of time-variation properties, and "portability" of models from one path to another would enhance the prediction accuracy of methods that use these models. Expanding the models in frequency extent would also be helpful, as this would facilitate the prediction of outage time reductions using frequency diversity.

In conclusion, the need for reliable predictions of outage time continues to challenge system analysts, system designers, and route engineers. Much remains to be done in this important aspect of microwave digital radio.

References

[1] D. P. Taylor and P. R. Hartmann, "Telecommunications by microwave digital radio," Part I of this book. [An earlier version was

published in *IEEE Commun. Mag.*, vol. 24, no. 8, Aug. 1986, pp. 11-16.]

[2] T. Noguchi, Y. Daido, and J. A. Nossek, "Modulation techniques for microwave digital radio," Part II of this book [An earlier version was published in *IEEE Commun. Mag.*, vol. 24, no. 11, Sept. 1986, pp. 21-30.]

[3] W. D. Rummler, R. P. Coutts, and M. Liniger, "Multipath fading channel models for microwave digital radio," Part III of this book. [An earlier version was published in *IEEE Commun. Mag.*, vol. 24, no. 11, Nov. 1986, pp. 30-42.]

[4] J. K. Chamberlain, F. M. Clayton, H. Sari, and P. Vandamme, "Receiver techniques for microwave digital radio," Part IV of this book. [An earlier version was published in *IEEE Commun. Mag.*, vol. 24, no. 11, Nov. 1986, pp. 43-54.]

[5] Recommendations and Reports of the CCIR, Study Group 9, 16th Plenary Assembly, Dubrovnik, Yugoslavia, 1986.

[6] L. J. Greenstein and B. A. Czekaj-Augun, "Performance comparisons among digital radio techniques subjected to multipath fading," *IEEE Trans. Commun.*, vol. COM-30, no. 5, May 1982, pp. 1184-1197.

[7] O. Andrisano, "The combined effects of noise and multipath propagation in multilevel PSK radio links," *IEEE Trans. Commun.*, vol. COM-32, no. 4, Apr. 1984, pp. 411-418.

[8] G. J. Foschini and J. Salz, "Digital communications over fading radio channels," *Bell Syst. Tech. J.*, vol. 62, no. 2, part 1, Feb. 1983, pp. 429-456.

[9] N. Amitay and L. J. Greenstein, "Multipath outage performance of digital radio receivers using finite-tap adaptive equalizers," *IEEE Trans. Commun.*, vol. COM-32, no. 5, May 1984, pp. 597-608.

[10] W. C. Wong and L. J. Greenstein, "Multipath fading models and adaptive equalizers in microwave digital radio," *IEEE Trans. Commun.*, vol. COM-32, no. 8, Aug. 1984, pp. 928-934.

[11] L. J. Greenstein and Y. S. Yeh, "A simulation study of space diversity and adaptive equalization in microwave digital radio," *AT&T Tech. J.*, vol. 64, no. 4, Apr. 1985.

[12] N. Amitay and J. Saltz, "Linear equalization theory in digital data transmission over dually-polarized fading radio channels," *AT&T Tech. J.*, vol. 63, no. 10, part 1, Dec. 1984, pp. 2215-2259.

[13] L. J. Greenstein, "Analysis/simulation study of cross-polarization cancellation in dual-polarization digital radio," *AT&T Tech. J.*, vol. 64, no. 10, part 1, Dec. 1985.

[14] CCIR Rpt. 338-5, "Propagation data and prediction methods required for line-of-sight radio relay systems," Study Group 5, 16th Plenary Assembly, Dubrovnik, Yugoslavia, 1986.

[15] W. D. Rummler, "A statistical model of fading on a space diversity radio channel," *Bell Syst. Tech. J.*, vol. 61, no. 9, Nov. 1982, pp. 2185-2219.

[16] J. C. Campbell and R. P. Coutts, "Outage prediction of digital radio systems," *Elec. Lett.*, vol. 18, no. 25/26, Dec. 1982.

[17] M. H. Meyers, "Multipath fading characteristics of broadband radio channels," in *Conf. Rec.*, GLOBECOM '84, Paper 45.1, Nov. 1984.

[18] W. D. Rummler, "A new selective fading model: Application to propagation data," *Bell. Syst. Tech. J.*, vol. 58, no. 5, May-June 1979, pp. 1037-1071.

[19] W. D. Rummler, "More on the multipath fading channel model," *IEEE Trans. Commun.*, vol. COM-29, no. 3, Mar. 1981, pp. 346-352.

[20] S. Sakagami and Y. Hosoya, "Some experimental results on in-band amplitude dispersion and a method for estimating in-band linear amplitude dispersion," *IEEE Trans. Commun.*, vol. COM-30, no. 8, Aug. 1982, pp. 1875-1888.

[21] M. Shafi, "Statistical analysis/simulation of a three ray model for multipath fading with applications to outage prediction," *IEEE J. Sel. Areas Comm.*, vol. SAC-5, no. 3, Apr. 1987, pp. 389-401.

[22] M. Shafi and D. J. Moore, "Adaptive equalizer improvements for 16 QAM and 64 QAM digital radio," in *Conf. Record, ICC '84*, pp. 998-1002, May 1984.

[23] K. Metzger and R. Valentin, "An analysis of the sensitivity of digital modulation techniques to frequency-selective fading," *IEEE Trans. Commun.*, vol. COM-33, no. 9, Sept. 1985, pp. 986-992.

[24] Y. Daido, E. Fukuda, and Y. Takeda, "Theoretical evaluation of signatures and CNR penalties caused by modern impairments in multilevel QAM digital radio systems," *IEEE Trans. Commun.*, vol. COM-34, no. 7, July 1986, pp. 654-661.

[25] A. J. Rustako, Jr., C. B. Woodworth, R. S. Roman, and H. H. Hoffman, "A laboratory simulation facility for multipath fading microwave radio channels," *AT&T Tech. J.*, vol. 64, no. 10, Dec. 1985, pp. 2281-2317.

[26] A. Vigants, "Space-diversity engineering," *Bell. Syst. Tech. J.*, vol. 54, no. 1, Jan. 1975, pp. 103-142.

[27] T. C. Lee and S. H. Lin, "More on frequency diversity for digital radio," in *Conf. Rec.*, GLOBECOM '85, Paper 36.7, Dec. 1985.

[28] M. F. Gardina and S. H. Lin, "Measured performance of horizontal space diversity on a microwave radio path," in *Conf. Rec.*, GLOBECOM '85, Paper 36.6, Dec. 1985.

[29] E. H. Lin, A. J. Giger, and G. D. Alley, "Angle diversity on line-of-sight microwave paths using dual-beam dish antennas," in *Conf. Rec., ICC '87*, pp. 831-841, June 1987.

[30] P. Balaban, E. A. Sweedyk, and G. S. Axeling, "Angle diversity with two antennas: model and experimental results," in *Conf. Rec., ICC '87*, pp. 846-852, June 1987.

[31] P. M. Dekan, J. H. Berg, and M. Evans, "Aperture diversity using similar antennas," in *Conf. Rec., ICC '87*, pp. 842-845, June 1987.

[32] W. C. Jakes, Jr., "An approximate method to estimate an upper bound on the effect of multipath delay distortion on digital transmission," in *Conf. Rec., ICC '78*, Paper 47.1, June 1978.

[33] V. K. Prabhu and L. J. Greenstein, "Analysis of multipath outage with applications to 90 Mb/s PSK systems at 6 and 11 GHz," in *Conf. Rec., ICC '78*, Paper 47.2, June 1978.

[34] M. Emshwiller, "Characterization of the performance of PSK digital radio transmission in the presence of multipath fading," in *Conf. Rec., ICC '78*, Paper 47.3, June 1978.

[35] L. J. Greenstein and V. K. Prabhu, "Analysis of multipath outage with applications to 90 Mb/s PSK systems at 6 and 11 GHz," *IEEE Trans. Commun.*, vol. COM-27, no. 1, Jan. 1979, pp. 68-75.

[36] W. C. Jakes, Jr., "An approximate method to estimate an upper bound on the effect of multipath delay distortion on digital transmission," *IEEE Trans. Commun.*, vol. COM-27, no. 1, Jan. 1979, pp. 76-81.

[37] W. D. Rummler, "A multipath channel model for line-of-sight digital radio systems," *Conf. Rec., ICC '78*, Paper 47.5, June 1978.

[38] C. W. Lundgren and W. D. Rummler, "Digital radio outage due to selective fading-observation vs. prediction from laboratory simulation," *Bell Syst. Tech. J.*, vol. 58, no. 5, May-June 1979, pp. 1073-1100.

[39] W. D. Rummler, "A simplified method for the laboratory determination of multipath outage of digital radios in the presence of thermal noise," *IEEE Trans. Commun.*, vol. COM-30, no. 3, Mar. 1982, pp. 487-494.

[40] S. Komaki, *et al*, "Characteristics of a high capacity 16 QAM digital radio system in multipath fading," *IEEE Trans. Commun.*, vol. COM-27, no. 12, Dec. 1979, pp. 1854-1861.

[41] Y. Serizawa and S. Takeshita, "A simplified method for prediction of multipath fading outage of digital radio," *IEEE Trans. Commun.*, vol. COM-31, no. 8, Aug. 1983, pp. 1017-21.

[42] G. H. Niezgoda, D. P. Taylor, and M. Shafi, "Further results on the dispersive fade performance of a 49 QPRS 90 Mb/s digital radio," *Conf. Rec., ICC '84*, pp. 653-656, May 1984.

[43] M. H. Meyers, "Multipath fading outage estimates incorporating path and equipment characteristics," in *Conf. Rec., GLOBECOM '84*, Paper 45.2, Nov. 1984.

[44] W. D. Rummler, "A comparison of calculated and observed performance of digital radio in the presence of interference," *IEEE Trans. Commun.*, vol. COM-30, no. 7, July 1982, pp. 1693-1700.

[45] W. T. Barnett, "Multipath propagation at 4, 6 and 11 GHz," *Bell Syst. Tech. J.*, vol. 51, no. 2, Feb. 1972, pp. 321-361.

[46] K. Morita, T. Murase, and S. Komaki, "Design considerations for 4/5/6 L-D1 digital radio system," *Rev. Elect. Commun. Lab.*, vol. 30, no. 5, 1982, pp. 846-858.

[47] A. L. Martin, "Digital microwave radio-a new system measurement technique," in *Conf. Rec., ICC '86*, pp. 472-476, June 1986.

Multipath Propagation at 4, 6, and 11 GHz

By W. T. BARNETT

(Manuscript received August 24, 1971)

Signals at 4, 6, and 11 GHz, transmitted over a 28.5-mile radio relay path in Ohio, were continuously monitored during the late summer of 1966. Previous publications have reported on the observed 4- and 6-GHz multipath fading statistics, and on the improvements available with space or frequency diversity. This paper presents data for the 11-GHz transmission, and, in combination with the earlier results, establishes an empirical frequency dependence for the amplitude statistics.

A general treatment of the relationships between the factors underlying multipath propagation is intractable. However, based on the results in this and other papers, a general relationship is given for the probability of deep multipath fading which is linear in frequency, cubic in path length, and varies with meteorological-geographical factors.

Temporal aspects of the Ohio data were also investigated at all frequencies, utilizing both a 1-hour and a 1-day clock time interval. It was found that the multipath fade time statistic can be described by a single parameter for either interval. A subset of the multipath fading hours was also analyzed using a 1-minute clock interval, with the result that the difference between the minute median fade and the hourly median fade is frequency independent, and normally distributed with a standard deviation of 5.5 dB.

I. INTRODUCTION

Although it is a relatively rare phenomenon, multipath propagation constitutes a fundamental limitation to the performance of microwave radio systems. During a period of multipath propagation, the narrowband output from a single receiving antenna can be reduced to equipment noise levels for seconds at a time. Corrective measures such as frequency diversity or space diversity then must be introduced to provide satisfactory commercial operation.[1-3] Propagation data required for economical system design and detailed performance estimates were not available prior to 1966. To fill this need an extensive experimental program was undertaken on a typical radio relay path in Ohio. Previous studies[1,2,4] have reported on the amplitude statistics obtaining during multipath propagation at 4 and 6 GHz, both with and without frequency or space diversity. Data were also obtained for a single frequency in the 11-GHz band. The multipath fading data for this signal have now been analyzed and statistics for the total time faded (P), the number of fades (N), their average duration ($\bar{t}$), and the fade duration distribution are presented in Section IV as functions of fade depth.

Multipath propagation is by its very nature dependent upon the operating microwave frequency; the variation of the fading characteristics with frequency has been considered by many investigators[5,6] with controversial results.* This is not surprising considering the time-variant, nonstationary behavior of the phenomena. However, the data obtained in Ohio were extensive enough to give statistical stability which, with the 11:6:4 frequency sampling, allows a meaningful comparison in Section V of P, N, and $\bar{t}$ as functions of frequency.

It is clear that a great deal is known about one path in Ohio. Generalization of these results to other paths requires an underlying theory. The experimental data show that P, N, and $\bar{t}$ can be quite closely represented by simple, one-term algebraic functions of fade depth. This agrees with predictions by S. H. Lin[7] based on analysis of a simple and plausible analytic model for multipath fading. It is therefore reasonable to assume that the variation of P, N, and $\bar{t}$ with fade depth for all paths subject to multipath fading will have the same functional dependence as did the Ohio path. A general formulation which includes the most important path parameters is proposed in Section VI for the coefficient in the equation relating the total time faded and the fade depth during the so-called worst fading month. This estimate provides necessary information for microwave radio system design in the continental U.S.A.

The intensity of multipath fading varies greatly, even during the normally active summer months; during some days there will be exten-

* Reference 5 gives many references on multipath fading investigations.

Reprinted with permission from *Bell System Technical Journal,* vol. 51, no. 2, pp. 321–361, Feb. 1972.

sive multipath fading, while on others there will be none. Statistics for time bases shorter than a month—or the entire 68-day period for the test reported here—are also of interest. The time faded characteristic was studied for the 4-, 6-, and 11-GHz signals on both a daily (24-hour) and an hourly basis. Section VII concludes with a study of minute-by-minute variations within an hour for a subset of the multipath fading hours.

All the experimental results mentioned in the preceding paragraphs were obtained from a data base comprising all the time intervals with deep multipath fading.[2] In sum, these intervals were about 15 percent of the total measurement time. The P, N, and $\bar{t}$ statistics for the remaining 85 percent of the time are given in Section VIII for typical 4- and 6-GHz signals. The 11-GHz data for this interval were not included because of the difficulty in identifying rain attenuation data; meteorological measurements were not made in conjunction with this experiment.

II. SUMMARY

Highlights of the results detailed in Sections IV thru VIII are given in this section. A few definitions are needed first:

L: Normalized algebraic value of envelope voltage (fade depth in dB $= -20 \log L$)
P: Fraction of time T that the envelope voltage is $\leqq L$
N: Number of fades (during T) of the envelope voltage below L
t: Duration of a fade below L in seconds ($\bar{t}$ = average duration)
f: Frequency in GHz
D: Path length in miles

The major results are:

(*i*) The 11-GHz amplitude statistics for the data base interval (T) of 5.26×10^6 seconds and for fade depths exceeding 15 dB are $P = 0.69L^2$, $N = 12{,}300L$, $\bar{t} = 330L$. Also $t/\bar{t}$ is log-normal and independent of L with 1 percent of the fades at any level longer than ten times the average.

(*ii*) The P and N statistics for the 4- , 6- , and 11-GHz data are, within experimental error, linear functions of frequency given by $P = 0.078fL^2$ and $N = 1000fL$. The comparable $\bar{t}$ statistic is given by $\bar{t} = 410L$.

(*iii*) An empirical estimate of P for the worst fading month is

$$P = rL^2, \qquad L \leqq 0.1$$

where r is defined as the multipath occurrence factor and is given by

$$r = c\left(\frac{f}{4}\right)(D^3)(10^{-5})$$

with

$$c = \begin{cases} 1 & \text{average terrain} \\ 4 & \text{over-water and Gulf Coast} \\ 0.25 & \text{mountains and dry climate.} \end{cases}$$

(*iv*) Of the days in the 1966 Ohio data base, about 12 had more fading than the average while 54 had less. The worst day contained about 48 percent of the total fade time at or below 40 dB while the worst hour contained some 20 percent.

(*v*) The simple model, $P = aL^2$, can be used to characterize shorter periods with multipath fading.* The cumulative empirical probability distribution (c.e.p.d.) with $a \equiv a_d$ is for daily fading

$$\Pr\,(a_d \geqq A) \cong \exp\,[-1.2\sqrt{A(4/f)}]$$

and for the hourly fading with $a \equiv a_h$

$$\Pr\,(a_h \geqq A) \cong \exp\,[-0.7\sqrt{A(4/f)}].$$

The hourly median fade depth value exceeded by 1 percent of the hours is 18 dB below free space.

(*vi*) The random variable defined as the difference between the median for a minute in a fading hour and the median for the entire hour was found to be normally distributed with zero mean and a standard deviation of 5.5 ± 1.5 dB.

III. EXPERIMENTAL DESCRIPTION[2]

The data presented were obtained by the MIDAS† measuring equipment at West Unity, Ohio. The basic data consist of measurements of the received envelope voltages of standard TD-2 (4 GHz), TH (6 GHz), and TL (11 GHz) signals; Table I is a list of the center frequencies of each channel. A functional block diagram is shown on Fig. 1. The 4-GHz and 6-GHz channels were standard in-service FM radio channels with nominally constant transmitted power (±0.5 dB). The 11-GHz

* The change in the coefficient from r to a is made to clearly differentiate between the total measurement period and the daily (or hourly) epoch.
† An acronym for Multiple Input Data Acquisition System.

Table I—Radio Channels Measured at West Unity, Ohio

Channel No.*	Frequency (MHz)	Antenna	Polarization
4-7	3750	Horn Reflector	V
4-1	3770		H
4-8	3830		V
4-2	3850		H
4-9	3910		V
4-11	4070		V
4-6	4170		H
6-11	5945.2		H
6-13	6004.5		H
6-14	6034.2		V
6-15	6063.8		H
6-17	6123.1		H
6-18	6152.8		V
11-1	10995		V

* The 4-X channels correspond to standard TD-2 radio system signals; 6-X corresponds to TH; 11-1 corresponds to TL.

channel was added especially for the test program and was unmodulated, with the RF equipment housed in an outdoor cabinet.

West Unity, Ohio, was chosen as the site for this experiment because it is part of a major cross-country route in an area known to suffer multipath fading. The hop monitored was of typical length—28.5

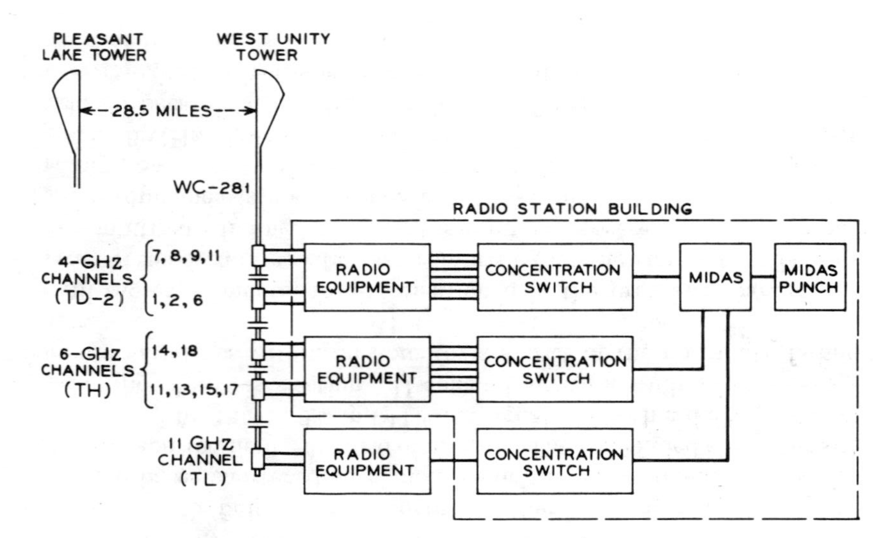

Fig. 1—1966 experimental layout, Pleasant Lake–West Unity.

miles—with negligible ground reflections. The path clearance was adequate even for the extreme of equivalent earth radius (k) equal to two-thirds, as shown on the path profile in Fig. 2. It is believed that this path is typical of those inland paths subject to multipath fading conditions.

The MIDAS equipment sampled each signal five times per second, converted each measurement to a decibel scale, and recorded the data in digital form for subsequent computer processing (in the absence of fading the recording rate was less than the sample rate). Further equipment details are given in Ref. 2.

The data were obtained during the period from 00:28 on July 22 to 08:38 on September 28, 1966. The total elapsed time was 5.9×10^6 seconds of which 5.26×10^6 seconds was selected for the data base; the balance was unusable mainly because of maintenance of the radio equipment or MIDAS. Within the data base, 7.8×10^5 seconds contained all the multipath fading in excess of approximately 10 dB. The balance of the time, 4.48×10^6 seconds, was categorized as nonfading time.

A natural epoch for multipath fading is the 24-hour period from noon to noon. It was convenient to number these periods from 1 to 69 starting at noon on July 21 and ending at noon on September 28. Here the

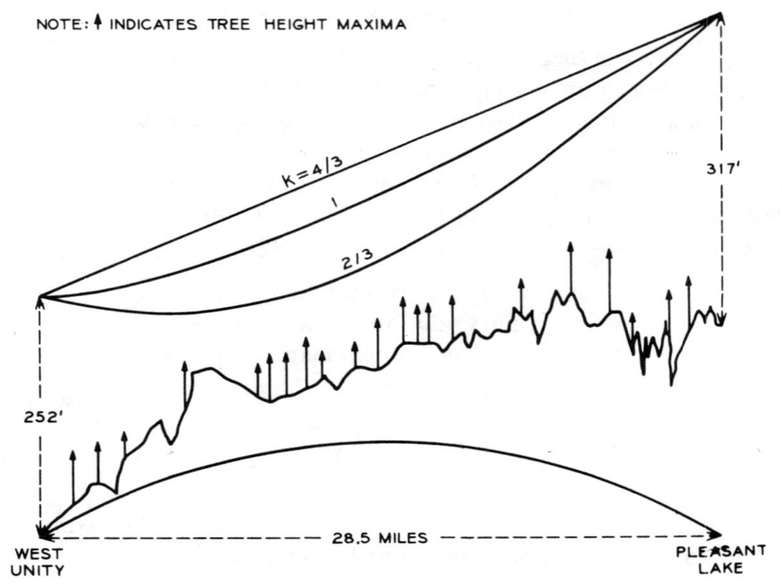

Fig. 2—West Unity–Pleasant Lake path profile.

missing end periods from 12:00, July 21 to 00:28, July 22 and 08:38 to 12:00 on September 28 have been assumed negligible. Most of the multipath fading was found to occur in the period between midnight and 9 A.M. as will be discussed later. These latter time periods were, for all practical purposes, subject to continuous measurement for 66 of the 69 periods. Thus, we reduce the multipath fading data base to 66 nine-hour periods. These were used for channel characterization and for investigating the daily and hourly statistical properties of multipath fading.

All fading distributions will be given in terms of the received voltage relative to the midday normal in dB. The rms variation in the dB reference level was estimated as ± 0.8 dB.[2]

IV. 11-GHz MULTIPATH RESULTS

The 11-GHz data were analyzed in terms of the statistical properties previously reported for the 4- and 6-GHz data.[2,4] These were (*i*) the fraction (P) of 5.26×10^6 seconds that the signal was faded below a given level L, (*ii*) the number of fades (N) below L, (*iii*) the average duration in seconds ($\bar{t}$) of fades below L, and (*iv*) the fade duration distribution. The data were carefully inspected to insure that only multipath fading was included and that rain fading was excluded. This was done by inspection of signal level vs time plots with the determination made by the frequency of the fading and by comparison with the 4- and 6-GHz data. As in the case of the 4- and 6-GHz data, we were most interested in fades greater than 15 dB. However, reliable data for the 11-GHz signal were limited to fade depths of 35 dB because the reference level of received signal strength was 5–10 dB lower than that for the 4- and 6-GHz signals.

The data for the fractional fade time are given in Fig. 3. They are adequately represented by a straight line whose equation is $P = 0.69L^2$. The data for the number of fades are given in Fig. 4 along with the fitted line $N = 12{,}300L$. The data for the average fade duration are obtained from the ratio of the total time faded to the number of fades and are given in Fig. 5 along with the fitted line $\bar{t} = 330L$. These variations of P, N, and $\bar{t}$ with L are in agreement with those previously found for the more extensive 4- and 6-GHz data and are as predicted from a mathematical model of the multipath fading process.[7]

The probability that a fade of depth $-20 \log L$ dB lasts longer than t seconds, i.e., the fade duration distribution, can be estimated by dividing the number of fades of depth L and duration t seconds or longer by the total number of fades of depth L. A normalization is made

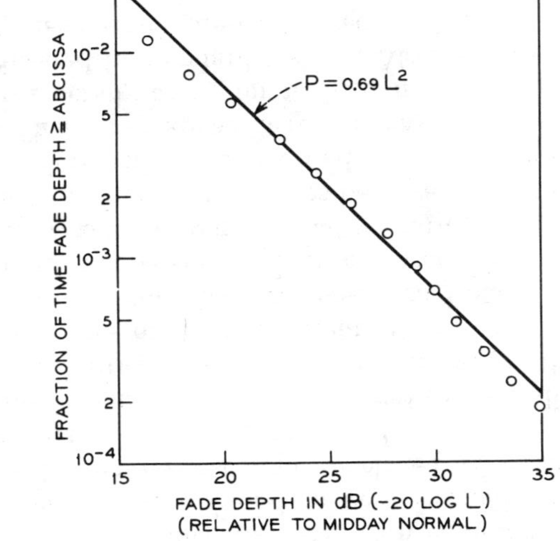

Fig. 3—11-GHz fade depth distribution, 1966 West Unity.

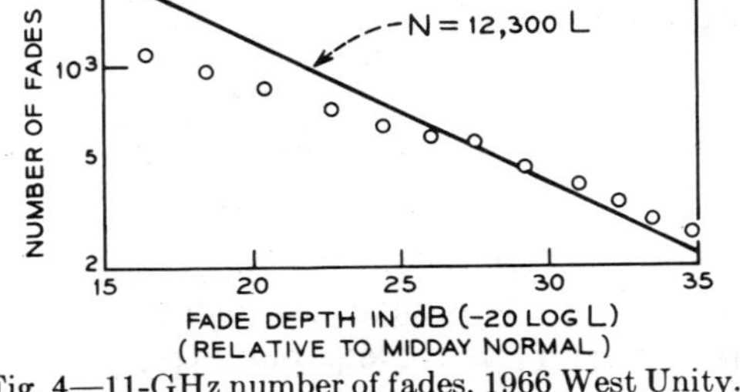

Fig. 4—11-GHz number of fades, 1966 West Unity.

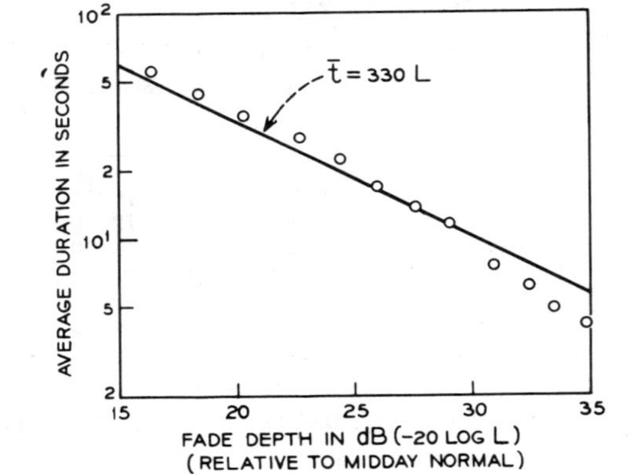

Fig. 5—11-GHz average fade duration, 1966 West Unity.

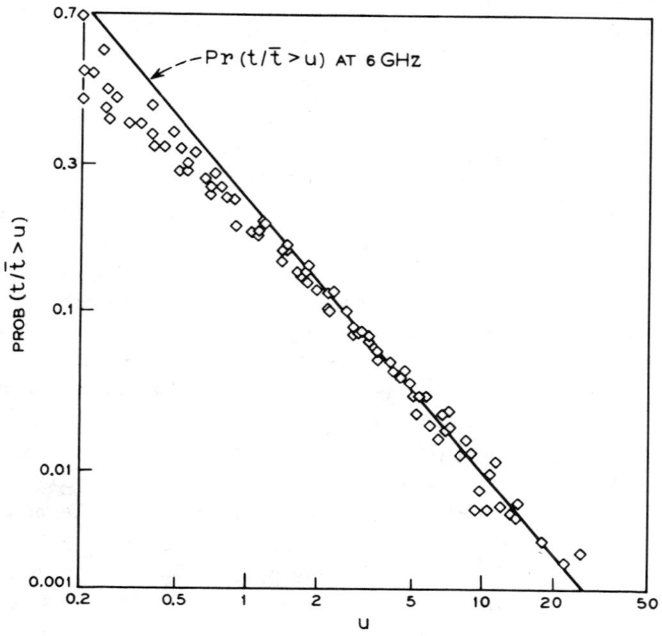

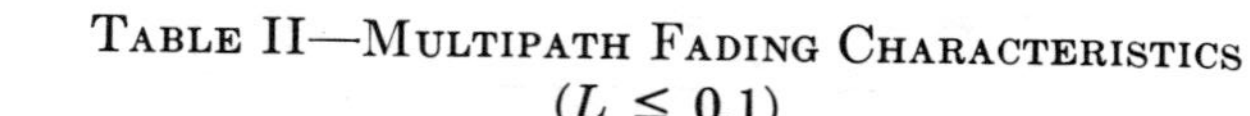

Fig. 6—11-GHz fade duration distribution: probability that the fade duration, normalized to its mean for a given fade depth, is longer than a given number. Data pooled for all fade depths greater than 10 dB.

with respect to the average fade duration. The 11-GHz data are plotted on Fig. 6, using a normal probability scale, for all fades $\geqq$10 dB. The data indicate that $t/\bar{t}$ is independent of L and that the probability is approximately log normal with 1 percent of the fades being longer than ten times the average fade duration. The line on Fig. 6, taken from Fig. 40 of Ref. 4, represents the fade duration distribution for the corresponding 6-GHz data. Thus, the fade duration distributions, when properly normalized, appear to be invariant with frequency.

V. MULTIPATH EFFECTS AS A FUNCTION OF FREQUENCY

The 11-GHz results of Section IV can be combined with those previously obtained for 4 and 6 GHz[2,4] to obtain an estimate of the variation of the characteristics with microwave frequency. This treatment is valid because all the data were obtained under identical conditions: same path, same antennas,* and same time period.

* The different beamwidths of the horn reflector for the three frequencies play a minor role because the variations in angle-of-arrival of the multipath components are generally less than the smallest beamwidth, which is ±0.6 degree at 11 GHz.

Table II—Multipath Fading Characteristics ($L \leqq 0.1$)

Freq (GHz)	P	N	$\bar{t}$
4	$0.25L^2$	$3670L$	$408L$
6	$0.53L^2$	$6410L$	$490L$
11	$0.69L^2$	$12300L$	$330L$

Table II summarizes the 4-, 6-, and 11-GHz results. The tabulated coefficients incorporate the effects of the environment and frequency. Plotting them versus frequency (as in Fig. 7) allows us to observe that the N and P coefficients increase, within experimental error, linearly with f while $\bar{t}$ is longer at 6 GHz and shorter at 11 GHz with respect to 4 GHz. Based upon these data, an approximation that $\bar{t}$ is independent of f is reasonable. The functional dependence is described by:

$$P = 0.078fL^2, \quad (1)$$

$$N = 1000fL, \quad (2)$$

$$\bar{t} = 410L, \quad (3)$$

with f in GHz.

The deviation of the P and N coefficients of Table II from these empirical equations is less than ±1 dB which is within the bounds of experimental error.[2] The $\bar{t}$ coefficients agree with equation (3) within ±2 dB. This is satisfactory since the $\bar{t}$ data were originally obtained as the ratio of the P and N data at each fade level; ±1 dB variation each in P and N corresponds to ±2 dB variation in $\bar{t}$.

The multiple transmission paths which give rise to the fading effects are generated by irregularities in the refractivity gradient in the volume defined by the beamwidths of the two antennas. As the relative path lengths vary with time the composite received signal may fade due to destructive interference (or be enhanced by constructive interference). It is easy to see that a given change in relative path length will cause more signal variations at higher frequencies because of the proportionally larger phase variations; we have found that the effect in Ohio in 1966 was linear. There is no apparent reason why this variation with frequency does not generally apply for multipath fading for a normal overland path engineered in standard fashion. Also, a linear variation of P with frequency has been theoretically predicted by C. L. Ruthroff[6] from a careful analysis of a simple physical model of multipath fading.*

* The results discussed here predate Ruthroff's analysis.

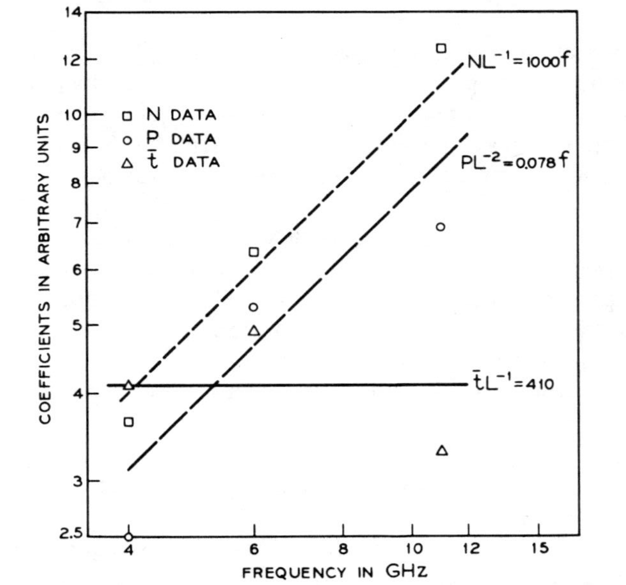

Fig. 7—Coefficients of P, N, and $\bar{t}$ characteristics versus frequency.

VI. OCCURRENCE OF MULTIPATH FADING

6.1 *General*

It is well known that the time (probability) distribution of the envelope of a microwave signal subject to multipath fading depends upon path length, path geometry, terrain clearance, type of terrain, and meteorological conditions in a complex manner. A general treatment of these relationships is intractable. However, based on the results discussed in previous sections and in other papers, an engineering estimate (incorporating the most important factors) of the fade depth distribution can be made for typical microwave paths for the heavy fading time of the year, i.e., the so-called worst month fading. In the results that follow adequate path clearance and negligible ground reflections are assumed.

6.2 *Relation to the Rayleigh Distribution*

Quite often in propagation studies it is assumed that the probability distribution of the envelope (v) of the received signal is given by the Rayleigh formula

$$\begin{aligned} \Pr(v < L) &= 1 - e^{-L^2} \\ &\cong L^2 \quad \text{for} \quad L < 0.1. \end{aligned} \tag{4}$$

One physical basis of this distribution is the limiting case of the envelope of an infinitely large number of equal amplitude signals of the same frequency, but random phase. Since this is a good approximation in many situations, e.g., tropospheric and mobile radio propagation, this distribution has seen much use. In the case of line-of-sight microwave radio, this is not a good assumption and the distribution is not directly applicable. From Table II the results for the fade depth distribution P vary as L^2 but with different coefficients.* The coefficient is generally not fixed, but depends upon the time base of the data, and upon the particular path parameters. The path parameters can be incorporated in the coefficient by expressing the multipath fade depth distribution as

$$\Pr(v < L) = rL^2 \qquad L < 0.1 \tag{5}$$

where r is defined as the multipath occurrence factor; $r = 1$ is appropriate to the Rayleigh distribution.

6.3 *Path Parameters*

As discussed in Section V, r is directly proportional to frequency; terrain and distance effects have to be incorporated. An engineering estimate for r can be given as a product of three terms†

$$r = c\left(\frac{f}{4}\right)D^3 10^{-5} \tag{6}$$

where: f is frequency in GHz,
D is the path length in miles,

$$c = \begin{cases} 1 & \text{average terrain} \\ 4 & \text{over-water and Gulf Coast} \\ 0.25 & \text{mountains and dry climate.} \end{cases}$$

The terrain effects and the distance dependence are based on applicable (albeit meager) Bell System data, most of which was acquired at 4 GHz on paths of 20–40 miles length. The plot given on Fig. 8 extends beyond this range. Indeed it can be argued that the curves should become parallel to the abscissa as D decreases (no multipath fading for paths sufficiently short[6]) and parallel to the ordinate (saturation) as D increases.

---

* An analysis of a mathematical model for multipath fading shows that the deep fade region of the distribution will be proportional to L^2 under very general conditions (Ref. 7).

† This empirical result for r is partially supported by British data as reported by K. W. Pearson[8] and is similar to a concise result reported by S. Yonezawa and N. Tanaka.[9]

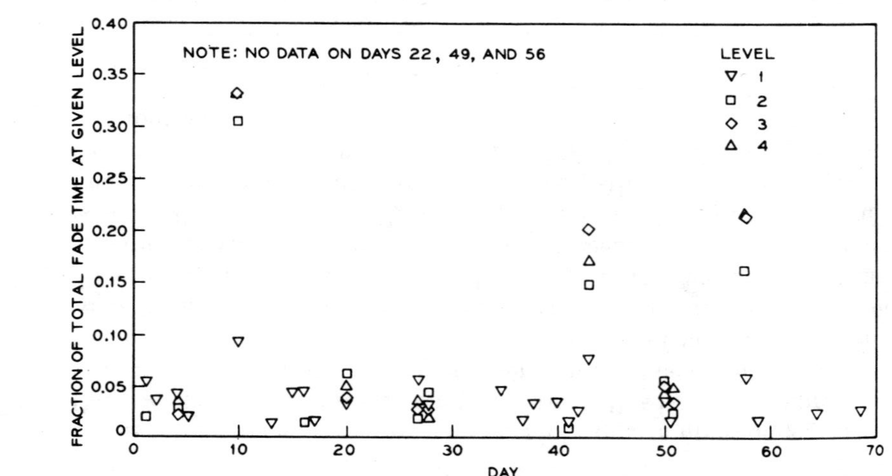

Fig. 8—Worst month multipath fading: $P = rL^2 = c(f/4)\,D^3L^2$.

The plotted values are certainly upper bounds for either extreme. The fD^3 dependence has been theoretically obtained by Ruthroff.[6]

The engineering estimate, equation (6), indicates that on a path of above average length, maintenance of the per-hop fading outage usually obtaining requires compensation for the additional free-space loss ($\propto D^2$) and for increased multipath ($\propto D^3$), which combine to impose a D^5 (15 dB/octave) length dependence.

VII. TIME CONCENTRATION OF DEEP MULTIPATH FADING

7.1 *Introduction*

The results and estimates already given utilize the entire data base, thus averaging temporal effects. It is well established that multipath fading occurs most often at night, with a few nights experiencing considerably more fading than most of the others. Describing this variability statistically is the objective here. We consider the fade time statistic for hourly and for daily periods and the median fade depth during an hour or a minute.

The analysis includes data from four fade depth values,* 9.8 dB, 20.4 dB, 31 dB, and 40.1 dB (henceforth labeled as levels 1 through 4). At each fade depth and for each analysis period the fade time for the seven 4-GHz channels was arithmetically averaged, as was that for the six 6-GHz channels. The fade time for the 11-GHz channel was used

* The unusual numbers are the result of quantization and calibration.[2]

TABLE III—FADE TIME DATA
(Seconds at or Below Given Fade Depth)

Freq Band (GHz)	Fade Depth 1 (9.8 dB)	2 (20.4 dB)	3 (31 dB)	4 (40.1 dB)
4	148,427	13,771	1329	135
6	259,933	27,503	2562	312
11	243,977	32,232	2982	*

* No data was obtained at 11 GHz for fade depth 4; see Section IV for further details.

directly. The resulting data will be referred to as the 4-, 6-, and 11-GHz fade times respectively. The fade time totals for the entire test period (5.26×10^6 seconds) are given in Table III.

7.2 *Distribution by Days—Rank Order Data*

The fade times for fade depths 1–4 were separately compiled for each of the 66 noon-to-noon periods. As expected there is considerable variation. As an example, Fig. 9 shows a plot of the 6-GHz fade time versus day number. Here the value plotted is the ratio of the fade time for the day to the total fade time, given in Table III, for a fixed fade depth. Much of the deep fading (levels 2, 3, 4) occurred on days 10,

NOTE: NO DATA ON DAYS 22, 49, AND 56
LEVEL ▽ 1 □ 2 ◇ 3 △ 4
FRACTION OF TOTAL FADE TIME AT GIVEN LEVEL
DAY

Fig. 9—Daily variation of multipath fading at 6 GHz, 1966 West Unity.

43, and 58 while the fading time for level 1 was more widely distributed.

The days were separately rank ordered for each frequency and each level with the variable again being the fraction of the total fade time; the results are given on Figs. 10a–c.* A few observations from these plots:

(*i*) The worst day fraction increases with level.
(*ii*) The data for level 1 do not fall off as rapidly with rank order as for levels 2–4.
(*iii*) Long tails in the rank order are prevalent.

Some of the more pertinent statistics are summarized in Table IV.

As already noted from Fig. 9, the bulk of the deep fading occurred on three days (10, 43, 58). The fraction of the total fading at the sample levels summed for these three days ranges from 0.55 to 0.74. Day 10 was the worst day in all cases. It appears that if a day suffered extensive 20-dB fading it also suffered 30- and 40-dB fading, but this indicator is not valid for 10-dB fading. In fact, about two-thirds of the days had 10-dB fading while only one-third had some 40-dB fading.

The statistical worst night is of particular interest. Figure 11 is based upon the observation that the worst day fraction increases with fade depth. The data points are fairly consistent except for levels 3 and 4 at 6 GHz which, for some unknown reason, do not show the expected increase relative to level 2. The line on Fig. 11 can be used as an estimate of the worst day fraction as a function of level. This estimate predicts that for systems with 40-dB fade margins the worst day will have 48 percent of the total fading within the worst month.†

A different perspective on the daily fading time can be obtained from Figs. 12a–c, which replot the rank order data on a logarithmic scale which has the effect of emphasizing the tail behavior. Generally, the tail is longer for lesser fade depths. It is interesting to compare these data with the result that would obtain for a uniform fade time distribution: a horizontal line at 0.015 (1/66). This line intercepts the level 2, 3, 4 data in the range of 10–15 days which means that this number of days had more fading than the average for the entire period while some 51–56 days have less fading. We shall return to the daily data in a later section where we shall see that they can be reduced to a more

* The data were plotted for all the days such that the cumulative sum of the plotted fade times just exceeded 99 percent of the total; note change of scale at rank order day 5.

† Here we take our statistics as representative of the worst month, the argument being that our results for a late summer—early fall period are generally comparable to the so-called worst fading month in a year.

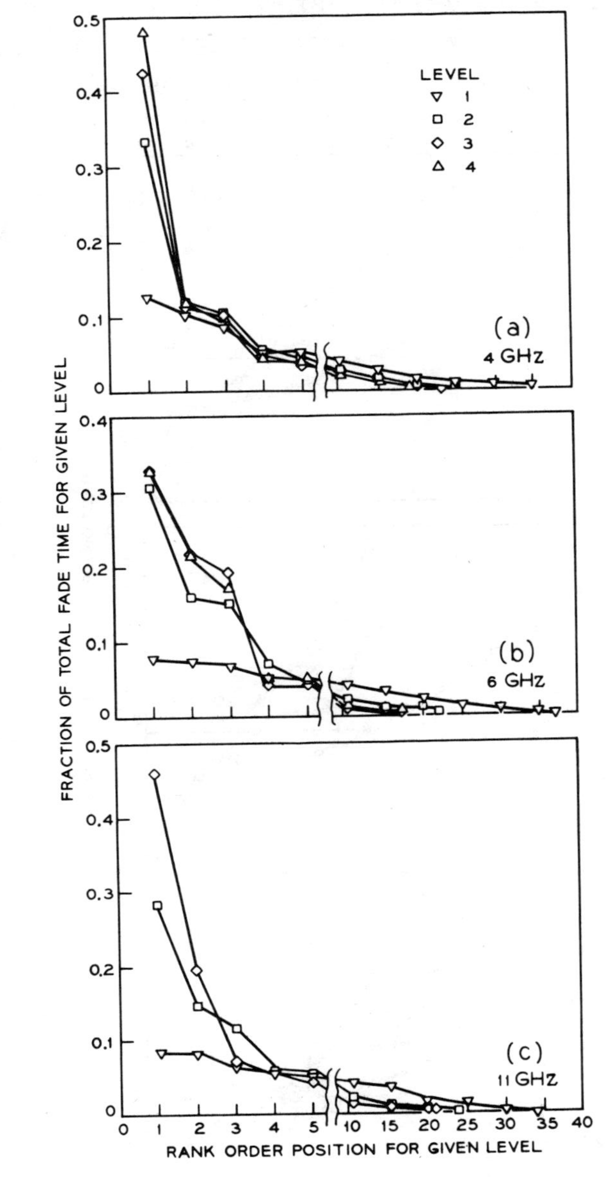

Fig. 10a—Rank order of 4-GHz daily fade times, 1966 West Unity.

Fig. 10b—Rank order of 6-GHz daily fade times, 1966 West Unity.

Fig. 10c—Rank order of 11-GHz daily fade times, 1966 West Unity.

TABLE IV—DAILY FADE TIME STATISTICS

Freq (GHz)	Level	Number of Days With Fade Time >0	Fraction of Total Fade Time: Worst Day	Fraction of Total Fade Time: Sum of 3 Worst Days	Number of Days to Give 0.99 of Total
4	1	46	0.125	0.31	35
	2	36	0.33	0.56	25
	3	30	0.43	0.64	23
	4	26	0.48	0.70	19
6	1	46	0.077	0.22	37
	2	35	0.30	0.61	22
	3	27	0.33	0.73	17
	4	24	0.33	0.71	16
11	1	43	0.083	0.22	34
	2	35	0.29	0.55	24
	3	30	0.47	0.74	21

meaningful form given the appropriate statistical treatment and mathematical modeling.

7.3 *Distribution by Hours—Rank Order Data*

The preceding treatment on daily fade time is repeated here for hourly fade time. This fade time is expressed as a fraction of all time during the entire measurement period as given in Table III. Of course, greater scatter can be expected in the hourly data than in the daily data.

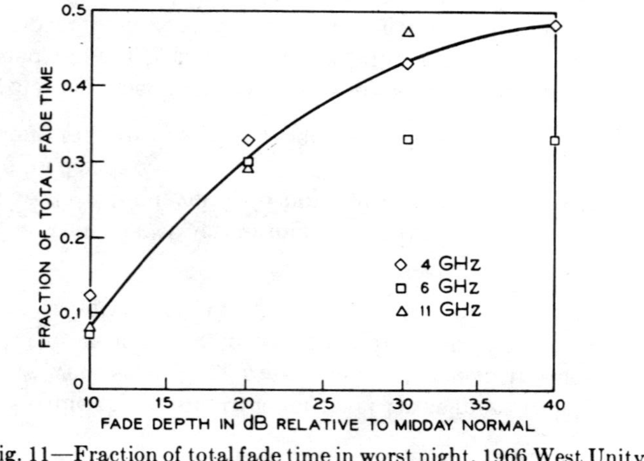

Fig. 11—Fraction of total fade time in worst night, 1966 West Unity.

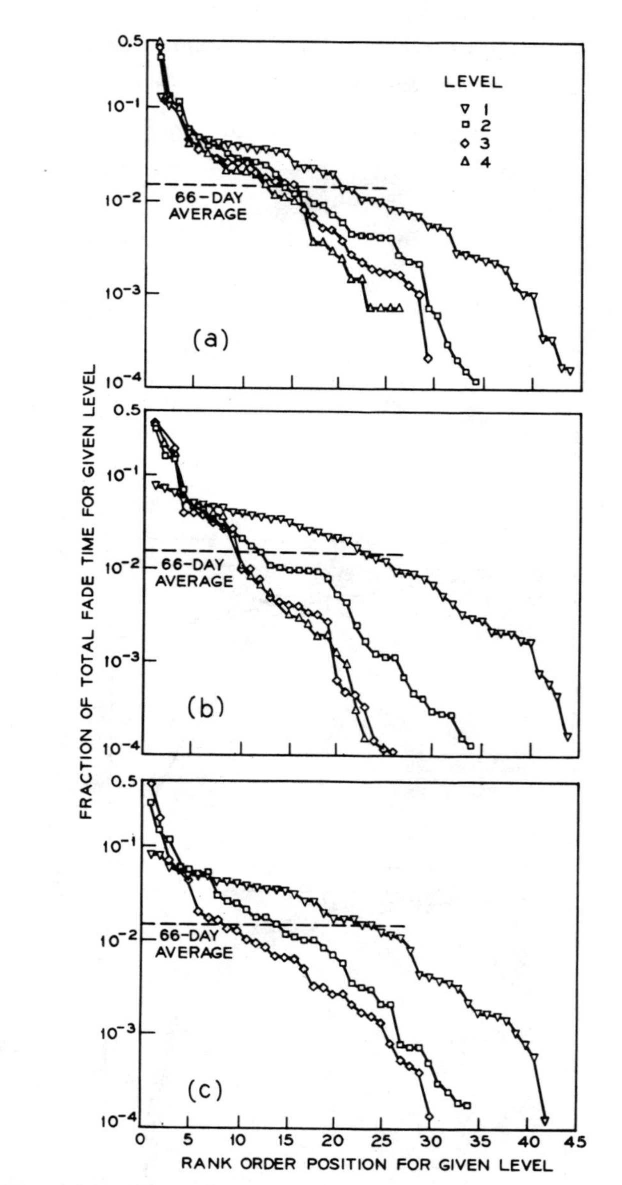

Fig. 12a—4-GHz rank order of daily fade times, 1966 West Unity.

Fig. 12b—6-GHz rank order of daily fade times, 1966 West Unity.

Fig. 12c—11-GHz rank order of daily fade times, 1966 West Unity.

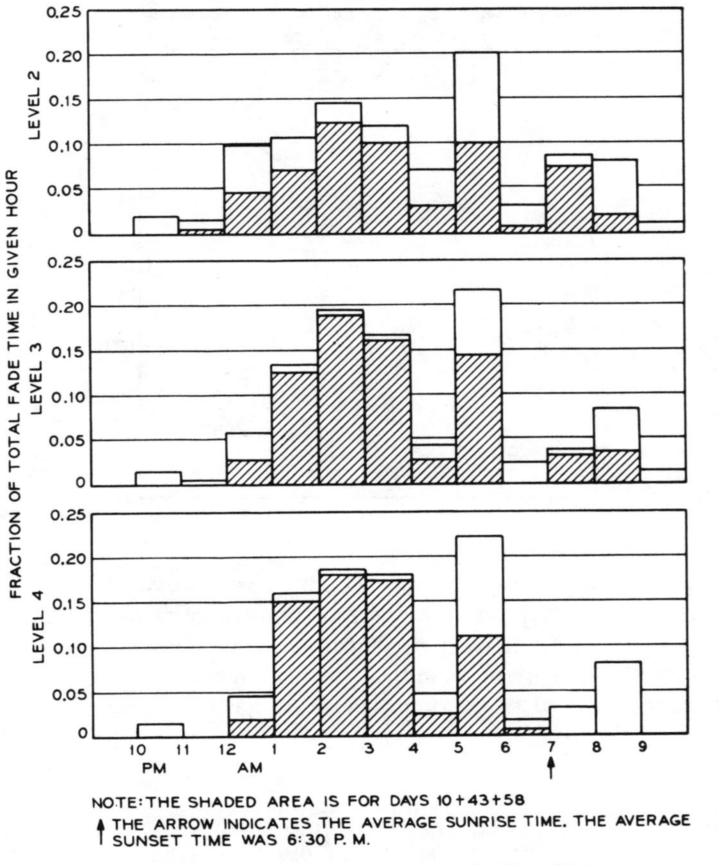

Fig. 13—6-GHz hour-of-day ranking, 1966 West Unity.

Figure 13 shows the distribution of fading for levels 2, 3, and 4 for the 6-GHz channels as a function of the hour of the day. Deep fading was generally within a 9-hour period between 12 P.M. and 9 A.M. The hours were rank ordered by level within a particular frequency band as shown on Figs. 14a–c. The general observations that can be made are similar to the "days" case:

(*i*) The worst night fraction increases with fade depth.
(*ii*) The level 1 fraction does not fall off very rapidly.
(*iii*) Long tails are even more prevalent than for daily fading.

Some of the pertinent statistics are summarized in Table V.

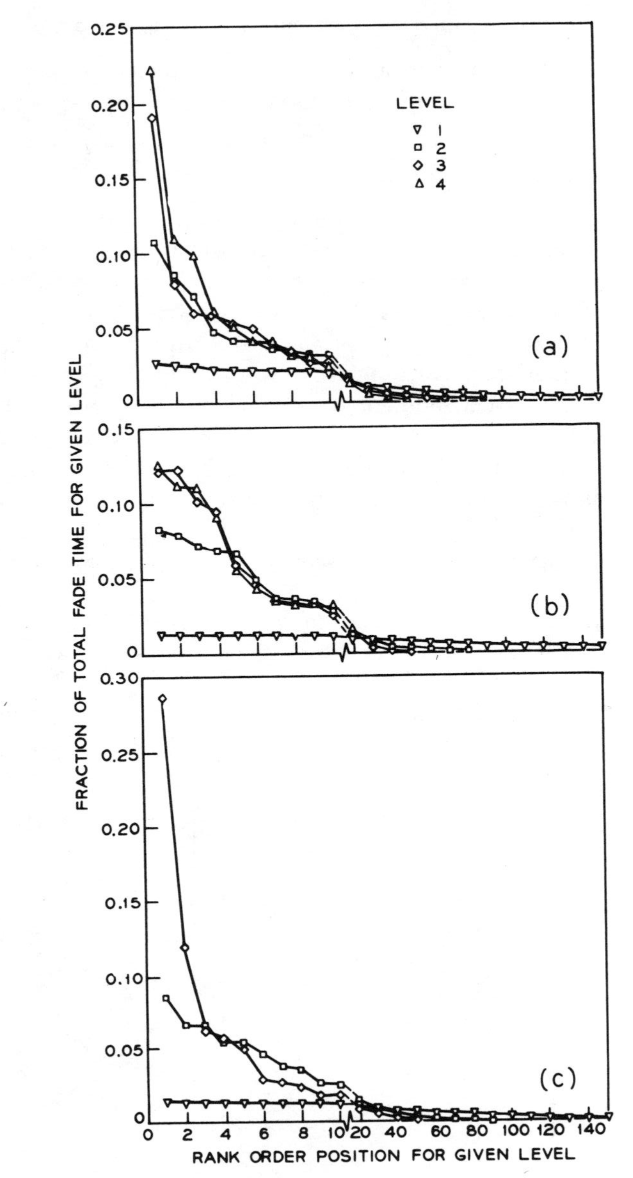

Fig. 14a—4-GHz rank order of hourly fade times, 1966 West Unity.

Fig. 14b—6-GHz rank order of hourly fade times, 1966 West Unity.

Fig. 14c—11-GHz rank order of hourly fade times, 1966 West Unity.

TABLE V—HOURLY FADE TIME STATISTICS

Freq (GHz)	Level	Number of Hours	Fraction of Total Fade Time		Number of Hours to Give 0.50 0.90 0.99 of Total Fade Time		
			Worst Hour	10 Worst Hours	0.50	0.90	0.99
4	1	220	0.027	0.226	33	103	163
	2	117	0.107	0.525	10	48	80
	3	88	0.192	0.621	7	36	69
	4	61	0.222	0.699	5	24	47
6	1	259	0.014	0.128	50	138	206
	2	123	0.083	0.559	9	39	83
	3	78	0.123	0.681	5	24	49
	4	56	0.126	0.672	6	22	43
11	1	230	0.015	0.136	48	127	189
	2	121	0.085	0.495	11	46	88
	3	70	0.286	0.694	4	28	53

The worst hour for each transmission band is plotted versus fade depth in Fig. 15. The data spread is greater than for the days case (Fig. 11) with 6 GHz again exhibiting the least variation. The line on Fig. 15 can be used as an estimate of the worst hour fraction as a function of level. Thus, the worst day (Fig. 11) and worst hour (Fig. 15) estimates

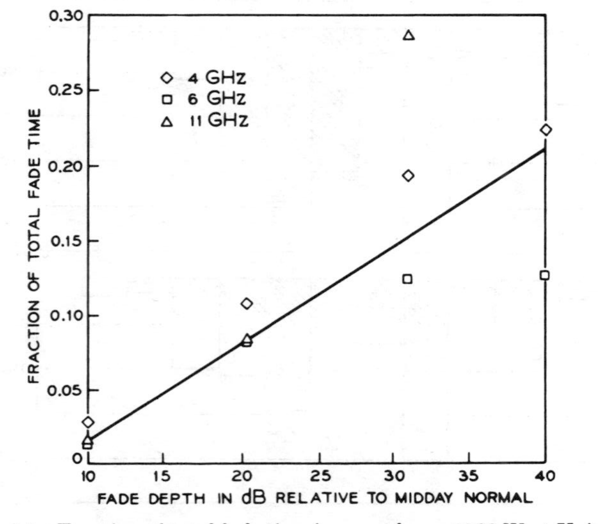

Fig. 15—Fraction of total fade time in worst hour, 1966 West Unity.

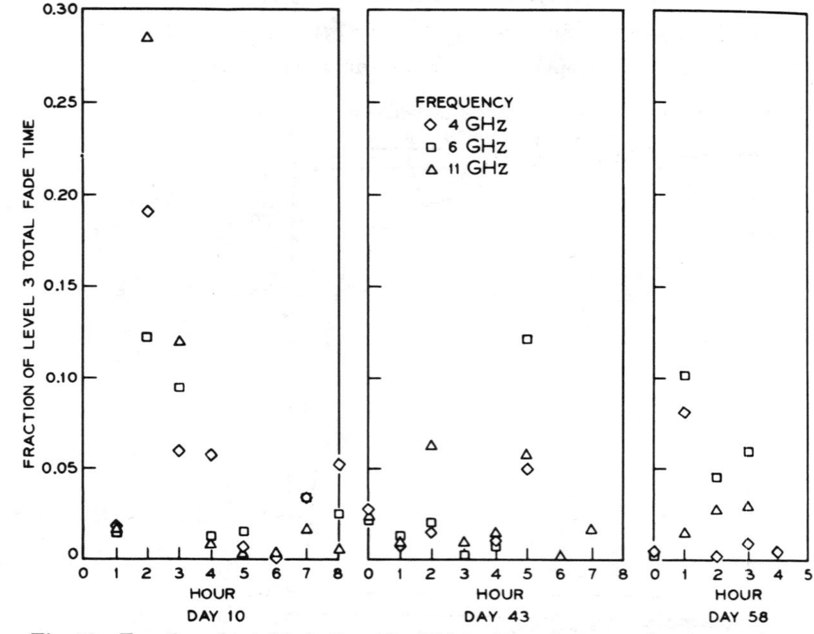

Fig. 16—Fraction of total fade time: level 3 hourly variation on three worst days.

for 40 dB predict 48 percent of the worst month multipath in a single day with 21 percent in the worst hour.

Days 10, 43, and 58 merit special study since they contain a majority of the deep fade time. The hourly variation in fade time for level 3 is given on Fig. 16. It is obvious from these data that there is no fixed relation between the frequency bands on an hourly time scale.* The hour from 2 A.M. to 3 A.M. on day 10 was the worst hour with the fractional fade time ranking with frequency as 11–4–6. However, the hour from 5 A.M. to 6 A.M. on day 43 was also a bad one with the fractional fade time ranking with frequency as 6–11–4. On day 58 the hour from 1 A.M. to 2 A.M., which was also outstanding, had the frequency order 6–4–11. However, the overall statistics show that fading severity increases with frequency.

7.4 *Hourly Median for a 4-GHz Channel*

The data reported in previous sections were in terms of the fraction of time that some fixed fade depth was exceeded; a reversal of these

* This conclusion does not change if absolute fade time is used instead of fractional fade time.

roles is equally valid. The variable examined in this section is the fade depth exceeded for a total of 30 minutes in an hour (hourly median). Figure 17 shows a rank order of the hourly median data for one of the 4-GHz channels as obtained directly from the experimental data for each hour. This particular channel is considered typical. The worst hourly median was 20.5 dB below free space and some 10 hours had hourly medians in excess of 15 dB. The general tendency is quite regular and shows a slowly decreasing median value with 120 hours experiencing hourly median fades in excess of 5 dB.

7.5 *Analytic Model for Hourly Median*

The single-channel fade depth statistics have a common characteristic: the fractional probability that the signal v is at or below L is proportional to L^2 (see Table II). Lin[7] has shown that this is a general property of fading signals under very general conditions, i.e.,

$$P \equiv \Pr(v \leqq L) = aL^2 \equiv \frac{t_L}{T} \qquad L \leqq 0.1 \tag{7}$$

where a is an environmental constant and t_L/T is the fractional fade time for the time period T.

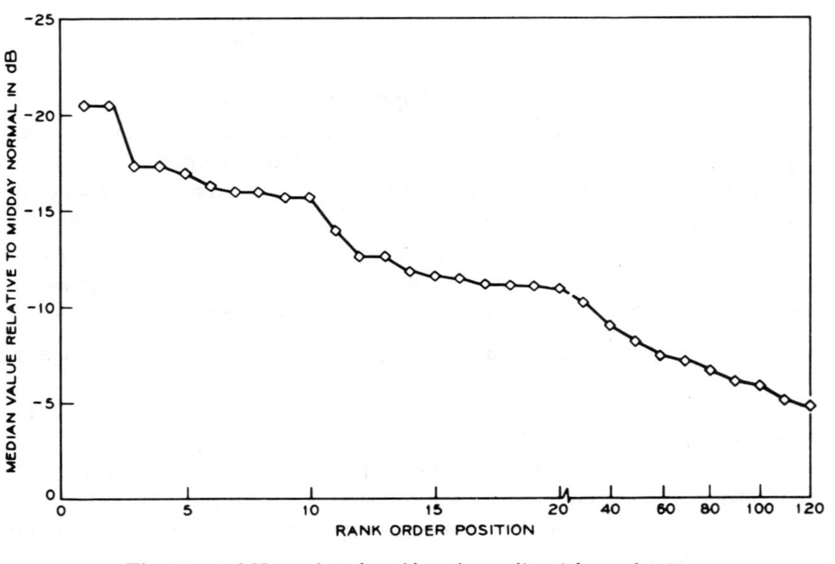

Fig. 17—4-GHz rank order of hourly median (channel 4-7).

This formula will be used here with the following modification for analytical simplicity,

$$P = \Pr(v \leqq L) = \begin{cases} aL^2 = \dfrac{t_L}{T} & 0 \leqq L^2 \leqq \dfrac{1}{a} \\ 1 & L^2 \geqq \dfrac{1}{a} \end{cases}. \tag{8}$$

For this simple model the median value, L_m, is given by

$$L_m^2 = \frac{1}{2a} = \frac{L^2 T}{2t_L}. \tag{9}$$

This relation can be used to calculate values of L_m from the 4-GHz hourly rank order data of Fig. 14a. The results for levels 2 and 3 are shown on Fig. 18 along with the 4-GHz hourly median data from Fig. 17. There is good agreement for the first 20 rank order days. Level 3 predicts a worst hour median 2.5 dB higher and level 2 predicts a worst hour median 1 dB lower than the Fig. 17 data.

The calculated results roll off faster below 10 dB than the Fig. 17 data, which means that the aL^2 model does not hold when the hourly median is less than 10 dB. This is to be expected because the aL^2 model applies for multipath fading while the Fig. 17 data contains a con-

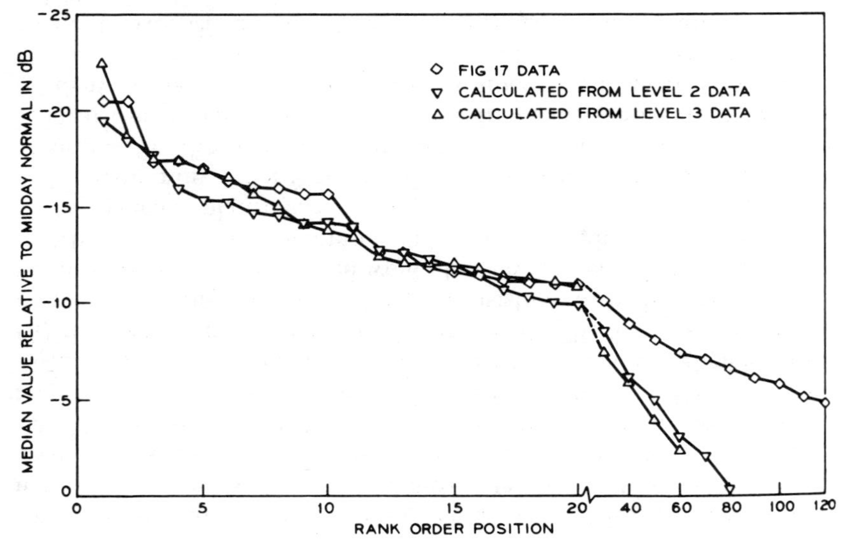

Fig. 18—Comparison of 4-GHz hourly median data of Fig. 17 with calculated values.

siderable number of hours during which the signal is depressed with little multipath fading. In any case, the analytic model (8) is adequate for the higher values of the hourly median which is the region of greatest interest. This model will now be applied to the nightly and hourly rank order data presented in Figs. 10 and 14.

7.6 *Empirical Probability Distribution of Daily Fade Time*

The rank order data (Section 7.2) can be used to estimate the probability distribution for the daily fade time by plotting the value of the ith ordered sample versus the probability estimate $(N)^{-1}(i - 0.5)$, defined as the cumulative empirical probability distribution (c.e.p.d.).[10] The random variable t_L is defined as the total amount of time during the 9-hour period for which the signal level is less than or equal to L.* The rank order daily fade time data (Section 7.2) are samples of t_L, with t_{Li} the ith rank ordered sample value. Thus the c.e.p.d. is:

$$P_{Li} \equiv \Pr(t_L > t_{Li}) = \frac{i - 0.5}{N_L} \tag{10}$$

where N = number of sample values.

Repeating equation (7) in a form consistent with the above definitions gives

$$\Pr(v_i \leqq L) = a_i L^2 = \frac{t_{Li}}{T_d} \tag{11}$$

where v_i is the envelope voltage during the ith interval,
a_i is the environmental constant during the ith interval,
T_d = 9 hours.

Combining (10) with (11) gives

$$P_{Li} = \Pr\left(\frac{t_L}{L^2 T_d} \geqq \frac{t_{Li}}{L^2 T_d}\right) = \Pr(a_d \geqq a_i). \tag{12}$$

Thus the c.e.p.d. for t_L is identical to that for the random variable a_d, the daily environmental constant.

In the calculation of P_{Li} for levels 2–4 the values used for N_L will be those given in Table IV. At level 2 there were 36 days with non-zero fade time at 4 GHz and 35 at both 6 and 11 GHz. If the aL^2 model is interpreted in a deterministic sense then all days with level 2 fade time should have level 3 fade time; yet there were only 30 such days at 4 GHz. There is no inconsistency because the aL^2 model is statistical so that not all level 2 fades also generate level 3 fades; thus the 30 samples are used to construct a c.e.p.d. which can be compared to that obtained for the 36 samples at level 2. The corresponding procedure was followed for level 4 at 4 GHz and for levels 2–4 for 6 and 11 GHz. Two basic assumptions are made: (*i*) 0.2-second sampling has a negligible effect; (*ii*) the samples at any level are independent. The first assumption will be justified if the level 4 results are consistent with the level 2 results because the sampling interval would have a greater effect on the level 4 results. The second assumption only requires independence from day to day which is plausible.

The daily rank order fade time data have been plotted on Figs. 19a–c according to (12). The probability scale is exponential and the abscissa is logarithmic. The data for all three frequencies appear to be independent of level and approximately linear with increasing scatter above 70 percent. The conclusion is that the aL^2 representation is adequate over the 20–40 dB fade depth range for daily fading.

In Section V we examined the frequency dependence of the environmental constant. Utilizing that relation, and normalizing to 4 GHz, equation (12) becomes:

$$P_{Li} = \Pr\left[\frac{a_d}{\left(\frac{f}{4}\right)} \geqq \frac{a_i}{\left(\frac{f}{4}\right)}\right]. \tag{13}$$

The level 2 data for 4, 6, and 11 GHz has been plotted in Fig. 20 according to (13). The reduced data are consistent for the three frequencies; a straight line whose equation is

$$\Pr\left(\frac{t}{L^2 T_d} = a_d \geqq A\right) = \exp\left(-1.2\sqrt{A\left(\frac{4}{f}\right)}\right) \tag{14}$$

provides a good fit (± 2 dB) to the data below 0.9. Similar results are obtained for levels 3 and 4 but with increased scatter.

Figure 20 indicates that the environmental parameter a_d is linearly dependent on frequency on a day-to-day statistical basis for multipath fading. This is a stronger result than that of Section V, where the linear frequency dependence was found valid for the measurement period taken as a whole. The net result of this analysis is that the daily fade time for a day picked at random can be calculated statistically.

The result, (14), can be checked against the results given in Table II for the entire measurement period in the following manner. Equation (11) gives, for the ith fading day out of N,

* The 9-hour period was chosen because most of the daily fading occurred between 12 P.M. and 9 A.M.

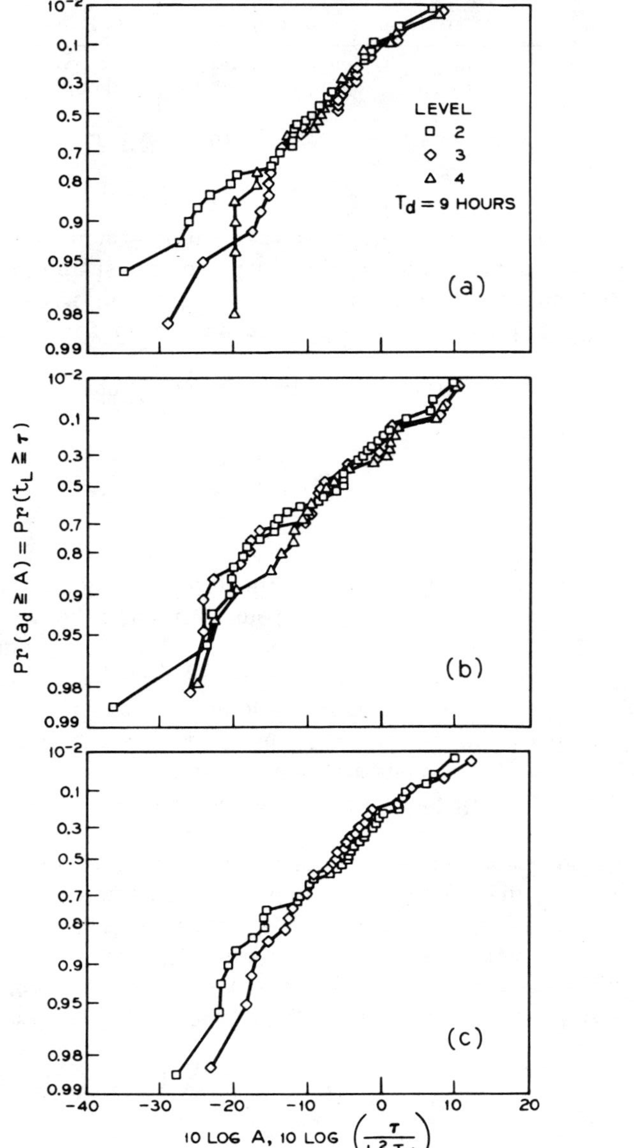

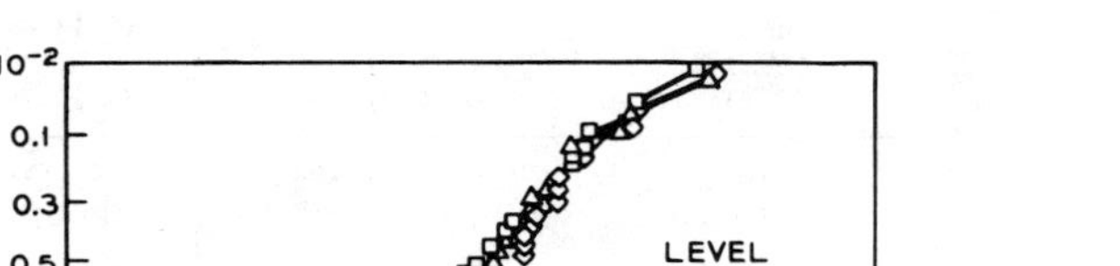

Fig. 19a—4-GHz daily fade time, 1966 West Unity, cumulative empirical probability distribution.
Fig. 19b—6-GHz daily fade time, 1966 West Unity, cumulative empirical probability distribution.
Fig. 19c—11-GHz daily fade time, 1966 West Unity, cumulative empirical probability distribution.

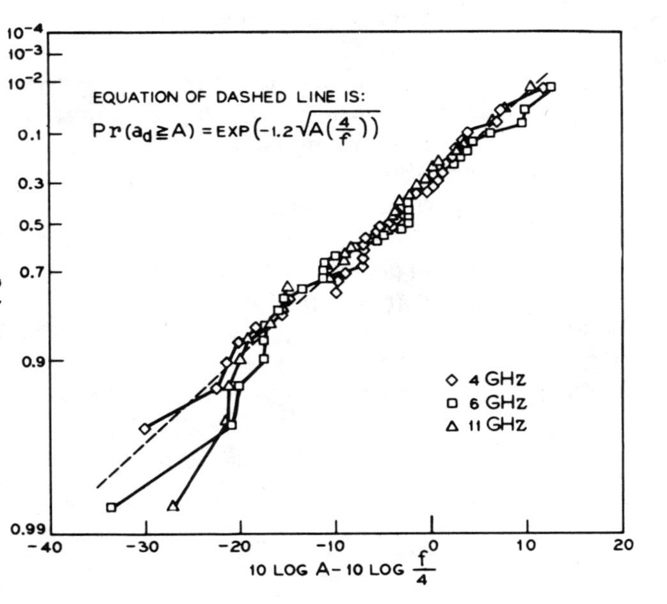

Fig. 20—Daily fade time for level 2, cumulative empirical probability distribution. 4, 6, and 11 GHz, 1966 West Unity.

$$\Pr(v_i \leq L) = a_i L^2 = \frac{t_{Li}}{T_d}. \tag{11}$$

The fractional fade time accumulated over the N days is [equation (5)]

$$\Pr(v \leq L) = r_N L^2 = \frac{\sum_{i=1}^{N} t_{Li}}{NT_d} = \frac{\sum_{i=1}^{N} a_i L^2 T_d}{NT_d}. \tag{15}$$

Thus

$$r_N = \frac{\sum_{i=1}^{N} a_i}{N} \tag{16}$$

so that r_N is the average value of the a_i's which in turn can be calculated from

$$P(a_d \leq A) = 1 - \exp\left(-1.2\sqrt{A\left(\frac{4}{f}\right)}\right). \tag{17}$$

Thus

$$r_N = \int_0^\infty ap(a)\,da = \int_0^\infty a\frac{dP}{dA}\,da \tag{18}$$

$$= 1.4\left(\frac{f}{4}\right).$$

To convert from N periods of 9 hours each to the entire measurement period of 5.26×10^6 seconds the above result must be multiplied by $[(N)(32,\ 400)/5.26 \times 10^6]$. Substitution of the number of days with nonzero level 2 fade time (Table IV) gives the results shown in Table VI. The coefficients obtained from the daily fade times are in fair agreement with the overall coefficients which is a reassuring check on the consistency of the results.

As a digression it is to be noted that the usual Rayleigh assumption for modeling the propagation medium corresponds to $A = 1$. Equation (18) shows that the average value of a_d corresponds to $A = 1.4$. It appears that the Rayleigh assumption is reasonable on the average but it should be recognized that some 30 percent of the days will have greater fading.

The calculation of the daily median is the last topic in this section. As noted in Section 7.5, the median value L_m for the aL^2 distribution model is given as

$$L_m^2 = \frac{1}{2a} \tag{19}$$

or

$$20 \log L_m = -10 \log a - 3 \text{ dB}. \tag{20}$$

Values for $20 \log L_m$ can be read off directly from Fig. 20, e.g., at 4 GHz the 90-percent point is —8 dB relative to midday normal, while the 1-percent point is -14 dB. This calculation is valid only for median values less than some -10 dB because as the value of a gets small the calculated median values will be much too high. This occurs because the range of validity of the aL^2 representation certainly does not extend above -10 dB relative to midday normal. As a matter of fact, the daily median is uninteresting and is included here only for completeness. The next section will take up the matter of the hourly variation for which the median calculation is more meaningful.

TABLE VI—FADE TIME COEFFICIENT OF L^2

Freq (GHz)	Calculated from Daily Fade Time	Measured (Table II)
4	0.3	0.25
6	0.45	0.53
11	0.82	0.69

7.7 *Empirical Probability Distribution of Hourly Fade Time*

The treatment of the daily fade time in Section 7.6 will be applied to the hourly fade time in this section. As in Section 7.6, we define*

- t_L total time during an hour for which the signal level is less than or equal to L,
- t_{Li} ith rank ordered sample value,
- N_L number of samples,
- v_i envelope voltage during ith hour,
- a_i environmental constant for the ith hour,
- T_h one hour (3600 seconds).

The cumulative empirical probability distribution for the hourly data is constructed according to (see Section 7.6)

$$P_{Li} \equiv \frac{i - 0.5}{N_L} = \Pr\left(\frac{t_L}{L^2T_h} \geqq \frac{t_{Li}}{L^2T_h}\right) = \Pr\,(a_h \geqq a_i) \tag{21}$$

with

$$\Pr\,(v_i \leqq L) = a_iL^2 = \frac{t_{Li}}{T_h}. \tag{22}$$

The hourly rank order data on Figs. 14a–c are replotted on Figs. 21a–c according to equation (21). The probability scale is exponential and the abscissa is logarithmic. The 4-GHz results on Fig. 21a are consistent with less than 3 dB scatter from 0.8 to 0.01 and increasing scatter for smaller data values. The cutoff value imposed by the 0.2-second sampling rate is -22.2 dB for level 2, -11.6 dB for level 3, and -2.5 dB for level 4. Since the 4- and 6-GHz data is averaged for 7 and 6 channels respectively, the actual cutoff point is some 8 dB lower. In any case increased scatter is to be expected for smaller sample values.

The 6-GHz results on Fig. 21b are consistent for levels 2 and 3 but the level 4 data is offset. If all the sample hours had the same amount of fade time at a given level then the c.e.p.d. would be a vertical line on Fig. 21b. One possible explanation, therefore, is that the level 4 hours

* The hourly data utilizes similar notation to that for the daily data.

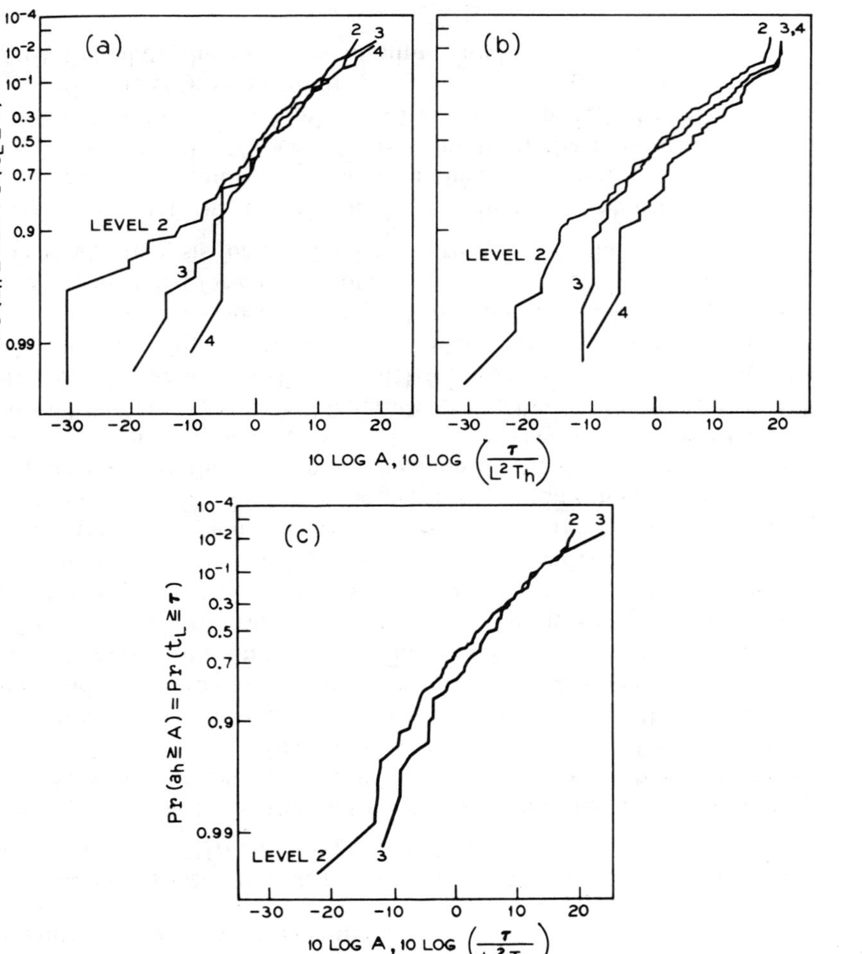

Fig. 21a—4-GHz hourly fade time, 1966 West Unity, cumulative empirical probability distribution.

Fig. 21b—6-GHz hourly fade time, 1966 West Unity, cumulative empirical probability distribution.

Fig. 21c—11-GHz hourly fade time, 1966 West Unity, cumulative empirical probability distribution.

at 6 GHz tended to be more alike than the level 2 and level 3 hours. This behavior was also noted in conjunction with Figs. 14b and 15. We assume that the 6-GHz hourly data for level 4 is atypical.

The 11-GHz results on Fig. 21c are reasonably consistent. Since

there was only one 11-GHz channel, the effect of the 0.2-second cutoff is clearly discernible.

The level 2 data from Figs. 21a–c is cross-plotted on Fig. 22 where the frequency has been normalized to 4 GHz. Thus, assuming that the level 2 data is typical, it is found that the distribution of the hourly environmental constant a_h for hours containing level 2 fades is approximately given by

$$P(a_h \leqq A_h) = 1 - \exp\left(-0.7[A(4/f)]^{1/2}\right). \tag{23}$$

The square-root function in the exponent was arbitrarily chosen to agree with the result for the daily data, e.g., (14). A slightly larger value than 0.5 would give a better fit for the smaller sample values but this was considered unimportant.

From equation (23), the 50-percent point for 4 GHz falls at $A_h = 1$, with the 99-percent point at $A_h = 30$. This means that for a fading hour the level 2 fade time will exceed 1080 seconds with 1 percent probability.

The hourly median can now be obtained based on the aL^2 model (see Section 7.3):

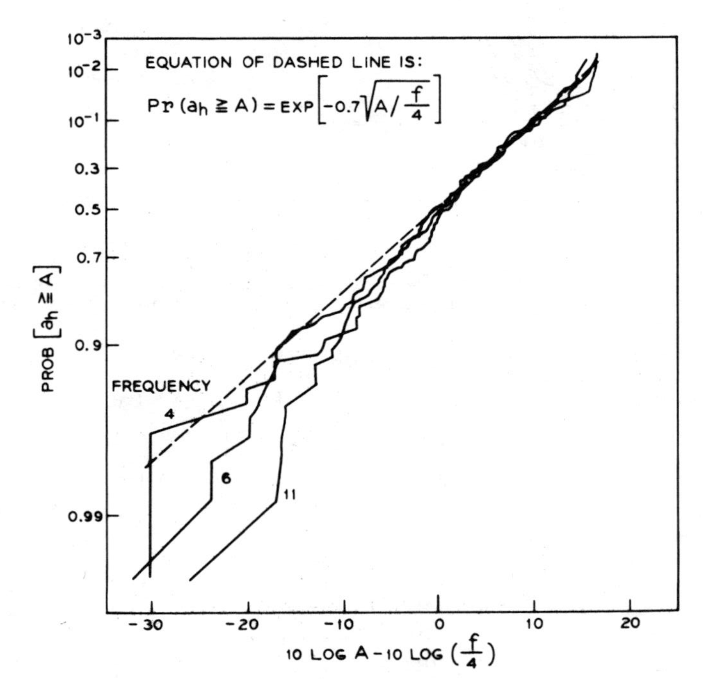

Fig. 22—Hourly fade time for level 2, cumulative empirical probability distribution, 4, 6, and 11 GHz, 1966 West Unity.

$$20 \log L_{mh} = -10 \log a_h - 3 \text{ dB}. \qquad (24)$$

Values for L_{mh} can be obtained from Fig. 22 using (24). For example, at 4 GHz the 1-percent median is −18 dB relative to midday normal. The actual maximum data point shown, however, falls at 10 log A = 17 dB which gives a median of −20 dB. This is in good agreement with the minimum median of −20.5 dB for the data given on Fig. 17 for one of the 4-GHz channels. This points up the problems of using a best fit line to estimate tail probabilities. Within such limitations it appears that the simple aL^2 model for the hourly and daily variations of multipath fade time is adequate.

7.8 *Empirical Probability Distribution of the Median of the Fade Depth Distribution for a Minute in a Fading Hour*

In preceding sections, the multipath fading data have been examined on a daily basis (Sections 7.2 and 7.6) and an hourly basis (Sections 7.3, 7.4, 7.5, and 7.7). Finer scale variations also are of interest. The sampling rate for a single radio channel varies from 0.2 second to 30 seconds depending on the amount of activity. This suggests that the smallest consecutive time interval that can be used in the construction of fade depth distributions is one minute. The measurement technique guarantees that if the 30-second rate is being used then the difference between any two 30-second samples is less than 2 dB.

The previous section (7.7) gave an estimate of the probability distribution of the hourly median fade depth of a fading hour. It is logical then to consider the median of the fade depth distribution for each minute within a clock hour. One channel in each of the three bands, 4, 6, and 11 GHz, was selected for study during five hours with multipath activity. The hourly medians in dB for each combination are given in Table VII. Four of the hours selected were drawn from among the ten having the most fading, with one lesser fading hour (day 10, 5–6 A.M.) included for comparison.

The data analysis for the five hours proceeds as follows:

(*i*) Construct the experimental fade depth distribution for each minute within the hour and for the entire hour.

(*ii*) Estimate the 50-percent dB point from the fade depth distribution for: (a) each minute within the hour: m_i dB, $1 \leq i \leq 60$; (b) the entire hour: h dB.

(*iii*) Calculate the difference in minute and hour medians:

$$d_i = h - m_i \text{ dB}. \qquad (25)$$

(*iv*) Rank order the d_i values from largest to smallest (i is then

TABLE VII—MEDIAN VALUES OF THE HOURLY FADE DEPTH DISTRIBUTION

Day	Hour	4 GHz	6 GHz	11 GHz
10	2–3 A.M	(1)* −20.5 dB	(2) −23.5 dB	(1) −27.5 dB
	3–4 A.M.	(3) −17.4 dB	(4) −21.0 dB	(2) −22.7 dB
	5–6 A.M.	(25) −11.5 dB	(17) −13.0 dB	(20) −14.2 dB
28	0–1 A.M.	(7) −16.4 dB	(10) −16.2 dB	(7) −17.4 dB
43	5–6 A.M.	(6) −17.5 dB	(1) −22.8 dB	(3) −21.0 dB

* The circled numbers give the hourly rank order position of the fade time at or below level 3 (−31.0 dB) in the hour.

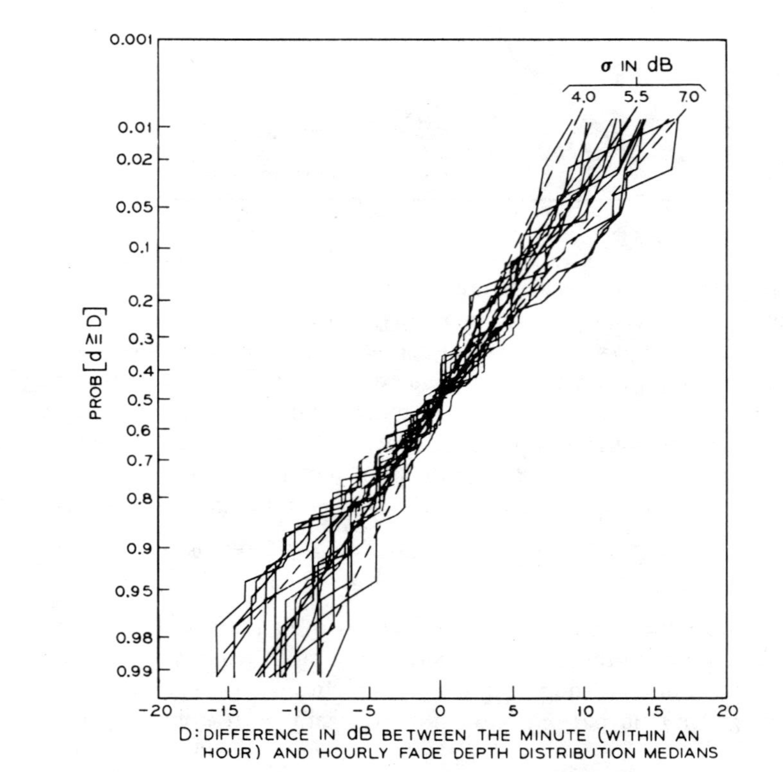

Fig. 23—Cumulative empirical probability distribution for the difference between the minute and hourly fade depth distribution medians. Data samples of five hours for 4, 6, and 11 GHz.

redefined as the rank order index with $i = 1$ for the worst 50-percent minute median fade value as normalized to the hourly median).

(v) Construct the cumulative empirical probability distribution for d, that is,

$$\Pr [d \geqq d_i] \cong \frac{i - 0.5}{60}. \tag{26}$$

The c.e.p.d. for d is plotted on Fig. 23 for all five hours and the three radio channels. This single plot suffices because there is no consistent difference between the different hours for a particular channel or between the different channels in a particular hour. As expected, the 50-percent point falls at the 0-dB difference point (within ±1 dB). The entire set of data appears to be normal with a mean of 0 dB and a standard deviation of 5.5 ± 1.5 dB. It can be seen that, for a multipath fading hour, the minute medians vary considerably as compared to the hourly median. This is not surprising since the average duration of a multipath fade varies from 4 seconds at a −40-dB fade depth to 40 seconds at a −20-dB fade depth.[4]

To recapitulate, the hourly median can be estimated from Fig. 22 using equation (24) and the difference in the hourly and minute median calculated using a normal distribution with a mean of 0 dB and σ = 5.5 dB.

VIII. AMPLITUDE STATISTICS FOR ENTIRE TEST PERIOD

8.1 *Introduction*

The effects of multipath propagation are most important in the deep fade region, because the received signal can be rendered unusable. The signal statistics for shallow fade depths also are of interest if only because the signal amplitude resides in this range for the vast majority of time. At West Unity an elapsed time of 5.26×10^6 seconds (T_o) was the total data base; of this total 0.78×10^6 seconds (T_A) contained all the deep multipath fading and was subjected to detailed analysis.[1,2,4] In this section, statistics for the remaining 4.48×10^6 seconds (T_B) will be presented for two 4-GHz and two 6-GHz channels. The data for the 11-GHz channel was not included in this analysis because of the difficulty of separating out the effects of rain attenuation.

8.2 *Fade Depth Distribution*

The fade depth distributions for 4 and 6 GHz are given on Figs. 24 and 25, respectively, for the three time bases T_A, T_B, and $T_o = T_A + T_B$.

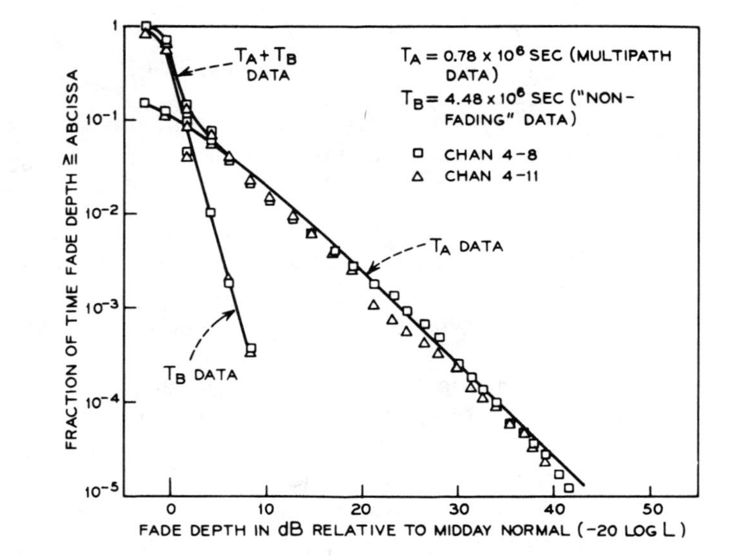

Fig. 24—4-GHz fade depth distribution for the entire test period, 1966 West Unity.

The lines on the figures are smoothed through the data, with the deep fade equations $0.25L^2$ and $0.53L^2$ (as given in Table II) used below −20 dB for 4 and 6 GHz respectively. As expected, the T_B data dominates the total distribution above the 10-percent point.

It should not be inferred from these results that there was zero probability of upfades above +3 dB. The equipment was designed to give this value whenever the signal level was in excess thereof.

The data for the fade depths less than 20 dB have been replotted on Figs. 26 and 27 on a normal probability scale where each set of data has been normalized to its own time base, e.g., the data for the multipath period are expressed as a fraction of 0.78×10^6 seconds (T_A). The data are given for only one of the channels in each band since the two channels have almost identical statistics in this fade depth range (see Figs. 24 and 25).

The plots show that neither the data for the total measurement period of 5.26×10^6 seconds ($T_A + T_B$) nor for the "nonfading" period of 4.48×10^6 seconds (T_B) are lognormal. During normal daytime periods of transmission on a single hop when the atmosphere is well mixed the envelope voltage scintillates and has a lognormal distribution with a standard deviation less than 1 dB. The T_B data is drawn from

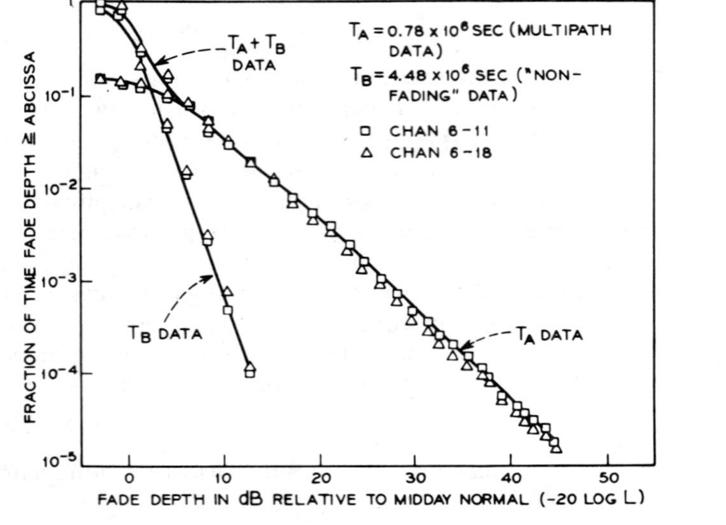

Fig. 25—6-GHz fade depth distribution for the entire test period, 1966 West Unity.

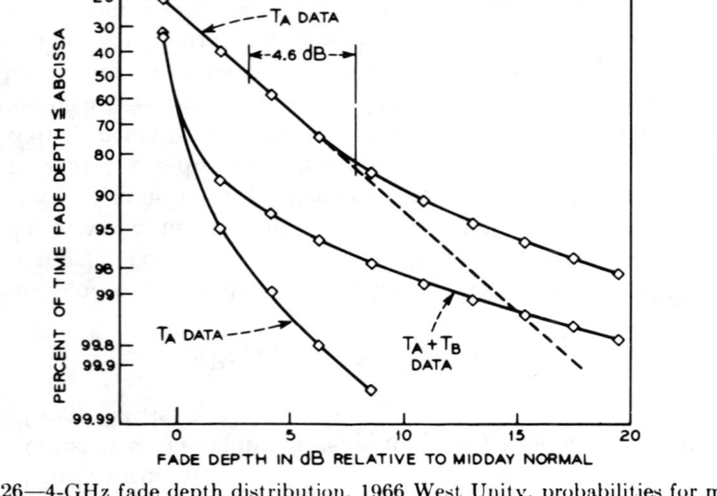

Fig. 26—4-GHz fade depth distribution, 1966 West Unity, probabilities for measurement intervals T_A (0.78 × 10⁶ seconds), T_B (4.48 × 10⁶ seconds), and $T_A + T_B$ (5.26 × 10⁶ seconds).

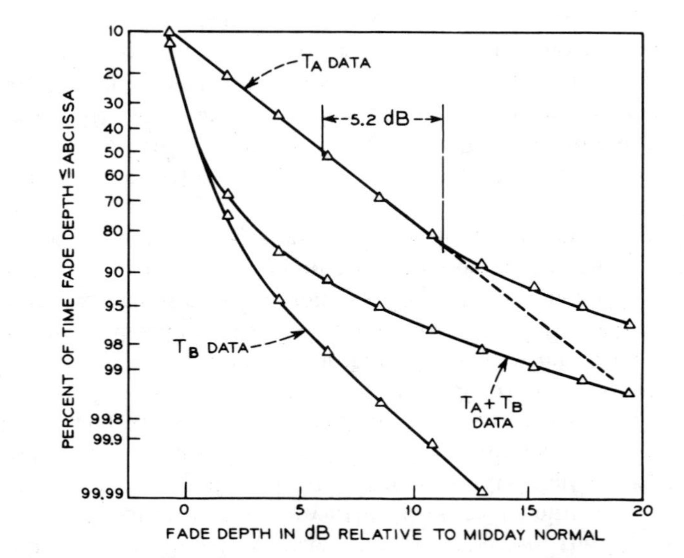

Fig. 27—6-GHz fade depth distribution, 1966 West Unity, probabilities for measurement intervals T_A (0.78 × 10⁶ seconds), T_B (4.48 × 10⁶ seconds), and $T_A + T_B$ (5.26 × 10⁶ seconds).

a mixture of such periods and others with mild fading. This mixture, coupled with the approximately 2-dB quantizing intervals, makes it difficult to draw definitive conclusions from either the T_B or the $(T_B + T_A)$ data in the central part of the distribution.

The T_A data are approximately lognormal over the central 80 percent of the distribution, with the characteristics given in Table VIII. As can be seen from Figs. 26 and 27, the lognormal characteristic is useless for predicting the deep fade behavior. This seems to be a common finding; an observable which can be modeled as having multiplicative components is usually lognormal near its median. However, a more sophisticated model is needed for calculation of the tails of the distribution.[7]

Table VIII—Characteristics of Shallow Fades During Periods Inclusive of All Deep Multipath Fades

Characteristic	4 GHz	6 GHz
50% point	3.1 dB	6.0 dB
σ	4.6 dB	5.2 dB

8.3 *Number of Fades and Average Fade Durations*

Data on the number of fades and the average fade duration were also obtained for a 4-GHz and a 6-GHz radio channel, as shown on Figs. 28a–b and 29a–b respectively. The number of fades occurring during the deep fade total time (T_A) first increases and then decreases as the fade depth increases below 0 dB. The line through the deep fade region, $3670L$ for 4 GHz on Fig. 28a and $6410L$ for 6 GHz on Fig. 29a, are the least squares fitted lines to the data for all the channels in the separate bands.[4] The data for the balance of the measurement time (T_B) varies more rapidly as a function of fade depth, i.e., approximately a factor of 100 from 0 to -10 dB. Of course, the T_B data has many more fades at 0 dB fade than the T_A data. Note that the deep fade fitted line would overestimate the number of fades by a factor of 2 at a -10-dB fade depth but would be quite adequate for prediction at 0 dB fade depth.

The average fade duration at any fade depth is obtained from the ratio of the total time at or below the fade depth to the number of fades of this depth. Values for this variable have been obtained from the data for each of the three time bases—T_A, T_B, and $T_A + T_B$—as shown on Figs. 28b and 29b for 4 and 6 GHz respectively. The lines $408L$ (4 GHz) and $490L$ (6 GHz) have been obtained for the deep fade

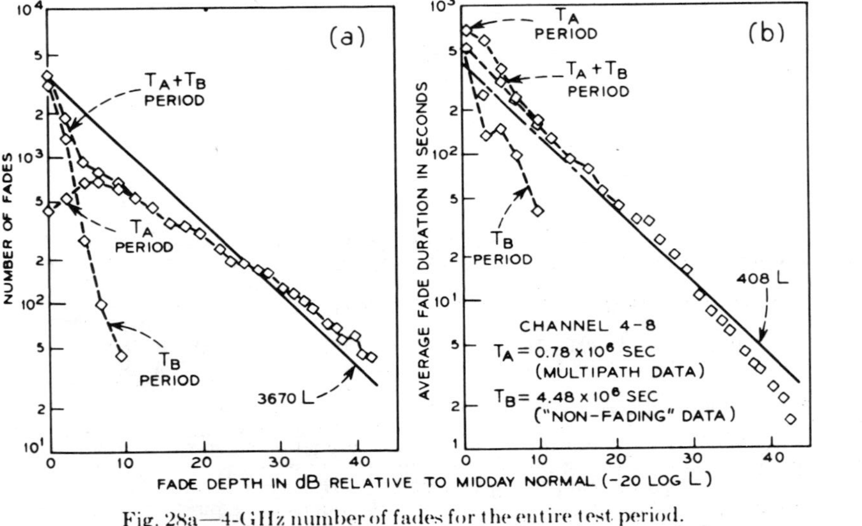

Fig. 28a—4-GHz number of fades for the entire test period.
Fig. 28b—4-GHz average fade duration for the entire test period.

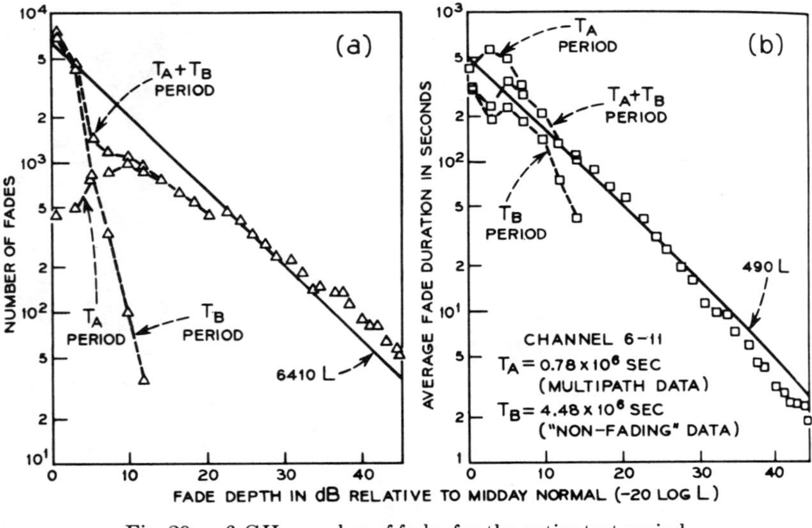

Fig. 29a—6-GHz number of fades for the entire test period.
Fig. 29b—6-GHz average fade duration for the entire test period.

data. However, these deep fade lines, extended to 0 dB, are a good representation of the data for the entire dB range. This is further evidence in support of Lin's finding that the average fade duration is less sensitive to the fading conditions than is either the number of fades or the fade depth distribution.

IX. ACKNOWLEDGMENTS

The author is indebted to many of his colleagues. The experimental data was obtained by MIDAS which was the creation of G. A. Zimmerman. The computer data processing capability was provided by C. H. Menzel. The data tabulation and plots were done by Miss E. J. Emer and Miss P. L. Russell. The interest and support of E. E. Muller and K. Bullington were invaluable.

REFERENCES

1. Vigants, A., "The Number of Fades in Space-Diversity Reception," B.S.T.J., *49*, No. 7 (September 1970), pp. 1513–1530.
2. Barnett, W. T., "Microwave Line-of-Sight Propagation With and Without Frequency Diversity," B.S.T.J., *49*, No. 8 (October 1970), pp. 1827–1871.
3. Chen, W. Y. S., "Estimated Outage in Long-Haul Radio Relay Systems with Protection Switching," B.S.T.J., *50*, No. 4 (April 1971), pp. 1455–1485.
4. Vigants, A., "Number and Duration of Fades at 6 and 4 GHz," B.S.T.J., *50*, No. 3 (March 1971), pp. 815–841.

5. Beckmann, P., and Spizzichino, A., *The Scattering of Electromagnetic Waves from Rough Surfaces*, New York: Pergamon Press, 1963, pp. 355–367.
6. Ruthroff, C. L., "Multiple-Path Fading on Line-of-Sight Microwave Radio Systems as a Function of Path Length and Frequency," B.S.T.J., *50*, No. 7 (September 1971), pp. 2375–2398.
7. Lin, S. H., "Statistical Behavior of a Fading Signal," B.S.T.J., *50*, No. 10 (December 1971), pp. 3211–3270.
8. Pearson, K. W., "Method for the Prediction of the Fading Performance of a Multisection Microwave Link," Proc. IEE, *112*, No. 7 (July 1965), pp. 1291–1300.
9. Yonezawa, S., and Tanaka, N., *Microwave Communication*, Tokyo: Maruzen Co., Ltd., 1965, pp. 25–60.
10. Wilk, M. B., and Gnanadesikan, R., "Probability Plotting Methods for the Analysis of Data," Biometrika, *55*, No. 1 (1968), pp. 1–17.

CHARACTERIZATION OF THE PERFORMANCE OF PSK DIGITAL RADIO TRANSMISSION IN THE PRESENCE OF MULTIPATH FADING

M. Emshwiller
Bell Laboratories
North Andover, Massachusetts

ABSTRACT

A method is presented for calculating the fraction of time that the transmission performance over a single radio hop would be unacceptable due to multipath fading. Analysis based on the two ray fading model allows the development of a system signature that provides a meaningful and quantitative comparison between systems. These system signatures are discussed with respect to their properties, including how to measure them. Under certain assumptions about the statistics of fades a closed form expression can result for the outage time experienced due to multipath fading.

This work has been motivated by the need to predict digital radio transmission performance, but the approach is general and applicable to both analog (FM, SSB) and digital (PSK, FSK, QAM, etc.) systems.

The signatures and predicted results are shown for a few different systems. These results indicate that unprotected wideband PSK systems will suffer bit error rates in excess of 10^{-3} for a substantial fraction of time due to shaped fades.

INTRODUCTION

Fading on line-of-sight radio hops has been the subject of considerable experimental and analytical study.[1-3] Most explanations assume that there are multiple paths which give rise to interference which is frequency dependent. This is easily expressed in Equation 1 where L and ϕ are the received amplitude and phase.

$$Le^{i\phi} = 1 + K_1 e^{i\omega\tau_1} + K_2 e^{i\omega\tau_2} + \ldots \qquad (1)$$

The direct ray is normalized to unity and the K's and τ's are amplitude and delay for the interfering rays. This has a reasonable physical explanation due to the layered variable index of refraction (at microwave frequencies) of the atmosphere, particularly during the 'fading months'.[4]

A difficulty arises in measuring the interfering rays. If they are known, it is a straight forward and unambiguous calculation to describe the amplitude and phase functions. In the lab or on the computer the interfering rays are easily generated, but in the field only the magnitude of L is readily measured and this information is insufficient to unambiguously define the interfering rays.

To reduce the ambiguity and facilitate calculation the general multipath case of equation 1 is simplified to a two ray model which still permits a good approximation to most shaped fades. This simplified model allows the development of a signature that characterizes a system and allows quantitative comparisons with other systems. This signature is a new and useful measure for predicting system behavior.

CALCULATION PROCEDURE

The calculation of unacceptable performance outage can be formulated into a four step procedure. This procedure is implicit in performance calculations done by others, but is explicitly delineated here so that focus can be placed on each step. Only the effect of the path is treated, the availability of protection channels or the failure of equipment is not considered.

The result will be a predicted fraction of time that a particular radio hop will be considered to have unacceptable performance. In specific terms unacceptable performance may mean a switch is requested, the channel noise is considered excessive, or some other criteria such as a certain bit error rate has been exceeded.

These four steps are:

1) Define the propagation model in terms of a set of parameters. This is equivalent to choosing the fading model. The number of parameters is the degrees of freedom allowed by the model.

2) Establish the performance characteristics of the system with respect to the values of the parameters. We are interested in a go-no-go decision about system performance based on values of the various parameters. There will be a region or regions, not necessarily contiguous, where the system performance is considered unacceptable. These regions constitute a 'signature' of the particular system. To show the nature and significance of this signature is the main purpose of this paper. These regions of poor performance can be determined by a variety of ways such as; measuring the system in the laboratory, simulating the system on the computer, or by analysis.

3) Develop the statistics of the values of the parameters by measuring actual propagation conditions over a period of time. These statistics will allow the empirical establishment of a probability density function (pdf) over the space of the parameters.

4) Integrate the pdf given by the statistics (step 3) over the regions of the signature (step 2). If both are empirical tables then the only recourse is to do a numerical integration, but if the regions can be described by a analytic curve and the pdf can be approximated by an analytic function then a closed form integration may be possible.

FADING MODELS

a. One Parameter Model

One of the simplest models that can be formulated is to assume all fading is flat across the radio

Reprinted from *IEEE Int. Conf. Comm.*, vol. 3, pp. 47.3.1–47.3.6, June 1978.

channel. This requires only one parameter to describe the fade, namely the depth of the fade relative to the normal received signal. This model has been used historically to predict system behavior. The system signature consists merely of determining the flat fade margin, i.e., how small a signal can be received before thermal (front end) noise is sufficient to cause the system to become unacceptable. The gathering of fading statistics is relatively simple as it is only necessary to measure the level of a single tone[1]. In the case of low index FM as is used in most line-of-sight radio telephony the measure of the carrier power is a good approximation to a single tone.

The advantages are the simplicity of the model, the ease of characterizing the system, (step 2) and the ease of measuring the fade statistics (step 3). The disadvantage is that no allowance is made for shaped (frequency selective) fades and it is known that shaped fades will cause impairments even if the thermal noise limit has not been exceeded. The inability to include these shaped fades leads to optimistic (small) predictions of outage time, especially for digital systems.

b. Two Parameter Model

The next level of sophistication in modelling radio fades is to assume there is a second path that interferes destructively with the normal or direct ray. The resultant phasor diagram is shown in Figure 1. The received amplitude and phase are given by L and ϕ respectively. The direct ray is normalized to unity amplitude and zero phase and the interfering ray is of magnitude K with a phase angle given by the product of frequency and the relative delay between the two rays. If the frequency range is over a narrow bandwidth the resultant will trace a small arc on the phasor plane and along that arc L and ϕ will vary with frequency. This model has been used extensively by many authors.[1-3, 5-7]

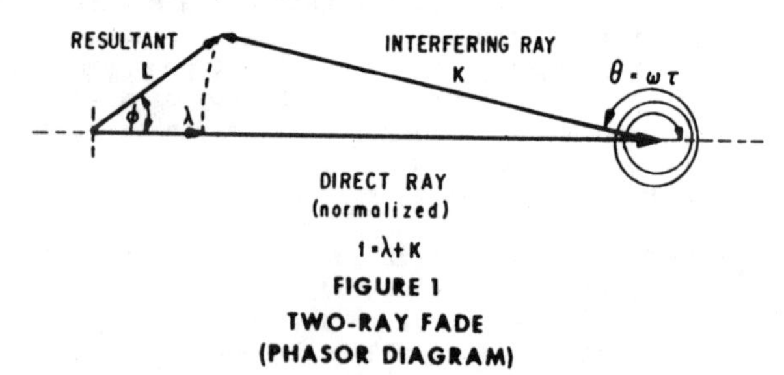

FIGURE 1
TWO-RAY FADE
(PHASOR DIAGRAM)

There are two parameters available using this model, the relative ray delay τ, and K, the amplitude of the interfering ray. The resultant phasor can be written as

$$L\,e^{i\phi} = 1 + Ke^{i\omega\tau} \quad (2)$$

$$L = (1 + K^2 + 2K\cos\omega\tau)^{\frac{1}{2}} \quad (3)$$

$$\phi = \text{Arctan}\,\frac{K\sin\omega\tau}{1 + K\cos\omega\tau} \quad (4)$$

The interference will be destructive and lead to the deepest fade (smallest L) when

$$\omega_b\tau = 2\pi(n - 1/2) \quad (5)$$

The frequency defined by ω_b (radians per second) occurs at the bottom of the fade and is referred to as the notch frequency. The offset frequency ω_o is the difference between the notch and the center frequency ω_c of the radio channel, i.e.

$$\omega_o = \omega_b - \omega_c \quad (6)$$

Figure 2 illustrates these values on a plot of the magnitude of L versus frequency. Using equations 5 and 6 one can write

$$\tau = \frac{2\pi(n-\frac{1}{2})}{\omega_c(1+\frac{\omega_o}{\omega_c})} \quad (7)$$

Typically, $\omega_o/\omega_c \ll 1$, e.g. 10/4000, so that τ has a coarse or gross value related to the integer n and center frequency ω_c, while offset frequency ω_o will be related to small variations in delay.

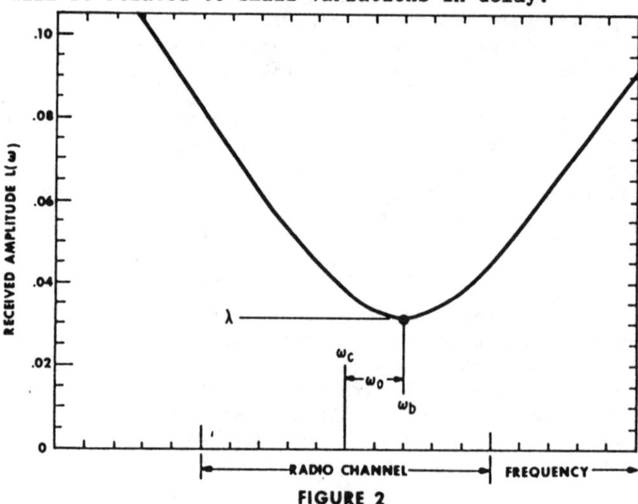

FIGURE 2
TWO-RAY FADE MINIMUM λ OCCURS AT ω_b, OFFSET ω_o FROM CENTER ω_c

If a fade has a minimum at some particular offset in a given channel then the ray delay is restricted to a discrete set of values as given by the integer n in equation (7). On the other hand, the depth of the fade can assume any arbitrary value. Define the minimum value of L (at the notch) to be λ, then

$$\lambda = 1 - K \quad (8)$$

Therefore with two parameters, λ (or K) and τ, the two ray fading model can be completely specified. It is not difficult to set up in the laboratory a two-ray fading simulator and measure the system behavior versus the two parameters. Computer simulation and analytical studies[5] can also proceed handily using this two ray model.

Since there are only two parameters it is possible to plot the regions of outage on a two dimensional plane. Figure 3 shows typical results (not to scale) derived by measuring or calculating a full set of the values of λ and τ that will cause unacceptable performance, indicated by the shaded areas. In these regions the delay is such that a minimum fade (notch) is in or near the radio channel and the fade is sufficiently deep to cause the channel performance to become unacceptable. The nature of the shape, width, and height of the regions, individually and collectively, constitute the signature of the particular system being measured. In Figure 3 the width of the individual regions is grossly exaggerated for illustrative clarity.

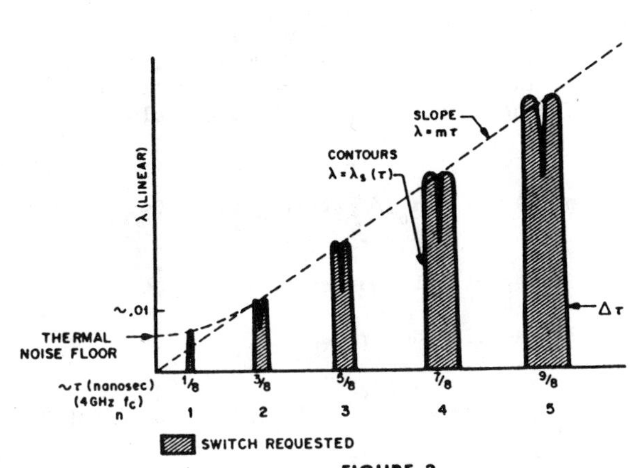

FIGURE 3
SWITCH REQUEST CONTOURS
(WIDTH GREATLY EXAGGERATED)

The nature of this signature is not obvious and requires some study to grasp its significance. First the occurrence of the discrete regions is a consequence of the fact that the second ray must have a delay such that destructive interference occurs at the receiving antenna. This implies that the path length is an odd number of half-wavelengths longer. In equation 7 this is indicated by the integer values of n. In Figure 3 the value of n is listed under the abscissa as well as the delay time (in nanoseconds) if the radio channel center frequency is 4 GHz. The vertical scale is linear as would be necessary for the integration procedure.

The measurement (or calculation) of the shape and size of an individual region is done in the following manner. The gross delay is selected so that the fade minimum occurs in or near the radio channel. This means that the value of the integer n has been selected in equation 7. The delay is 'fine tuned' to move the minimum to the offset frequency desired. The depth of fade can be changed until λ decreases far enough to cause the system performance to become unacceptable. This defines the performance threshold (λ_s) for that particular offset frequency at that particular gross delay for that particular system. A deeper fade will cause unacceptable performance but a lesser fade will be acceptable.

This procedure is illustrated in Figure 4 where the locus of points defined by the change of λ versus the change in offset frequency defines an individual region as shown by the dashed curve. If the offset is large, i.e. the notch is well away from the radio channel, then no amount of fading will cause unacceptable performance.

The performance threshold can be defined in terms of noise in a measurement slot (FM radio) or in terms of bit error rate (digital transmission) or any other criteria suitable to the particular system.

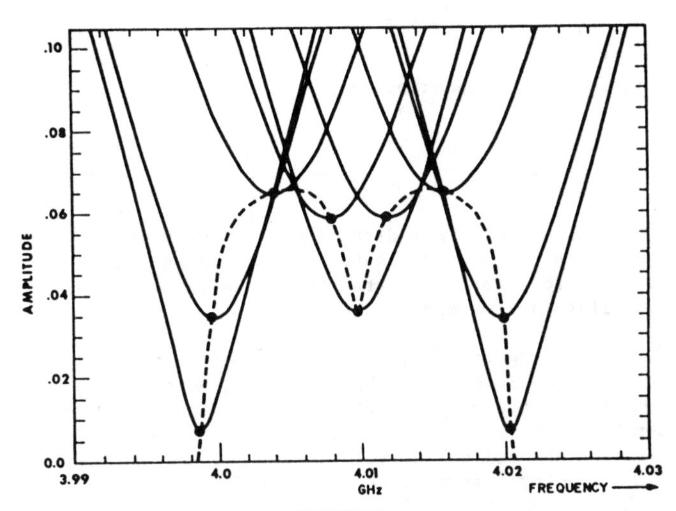

FIGURE 4
DASHED CURVE DEFINED AS LOCUS OF MINIMUMS OF FADES CAUSING UNACCEPTABLE PERFORMANCE

Several such regions for an 8-PSK system are plotted for various values of gross delay (different n) in Figure 5. Note that the maximum offset frequency is essentially constant and that the height of the curves increases linearly with delay.

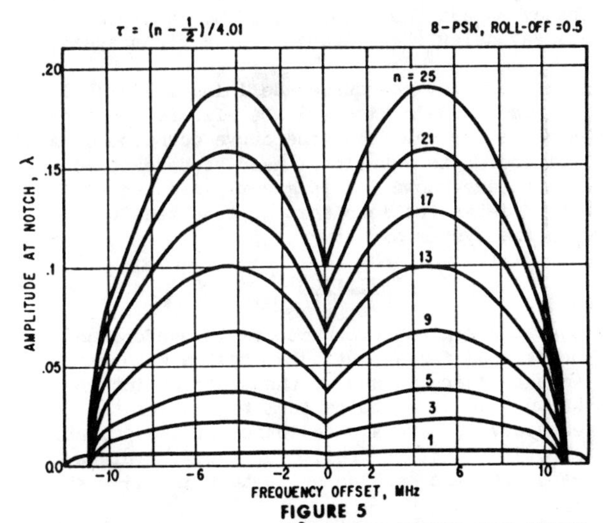

FIGURE 5
CONTOURS OF BER = 10^{-3} vs FADE NOTCH AND DELAY

This linear relationship can be substantiated algebraically as follows. Expand L in a power series in radian frequency about ω_b, the notch frequency. The first term in the series is λ and the first derivative of L with respect to ω will be zero. The second derivative will give

$$(d^2L/d\omega^2)/\lambda = (1-\lambda)(\tau^2/\lambda^2) \qquad (9)$$

The division on the left by λ is to normalize the response, i.e. the absolute amplitude is not important if the detector is insensitive to amplitude. If the relationship between λ and τ is linear so that

$$\lambda = m\tau \qquad (10)$$

and $1-\lambda$ is approximated as 1 then the normalized shape of the amplitude versus frequency is independent of τ, i.e.

$$(d^2L/d\omega^2)/\lambda \equiv 1/_{m}2 \qquad (11)$$

and to a good approximation

$$L/\lambda \equiv 1 + \frac{(\omega-\omega_b)^2}{2} \left(\frac{1}{m^2}\right) + \ldots \qquad (12)$$

This normalized shape determines the intersymbol interference which dominates the thermal noise effects for the case where the shaped fade is above the flat fade margin.

A similar argument holds for the envelope delay which is defined as the first derivative of the phase. If the envelope delay is expanded in a power series then its first derivative is zero and the second derivative (parabolic delay distortion) is dominated by a term of the form $2(\tau/\lambda)^3$ or

$$d^2(d\phi/d\omega)/d\omega^2 \propto 2(\tau/\lambda)^3 = 2/m^3 \qquad (13)$$

Hence the shape of the envelope delay with frequency is independent of λ to the first order, providing λ is much less than one and is linear with τ.

The important conclusion is that for a receiver with an amplitude insensitive phase (or frequency) detector, and carrier phase and envelope delay adjustments and for fades above thermal noise the performance is determined solely by the offset frequency ω_o and the slope, m, of notch depth versus ray delay (Equation 10) for a given system.

This implies that the curves defining individual regions as shown in Figure 3 are all similar except for a scale factor. Only one curve corresponding to one value of n in equation 7 need be determined and all others are known. This greatly reduces the number of determinations that must be made to develop the system signature.

STATISTICS

Now that the regions of unacceptable performance can be defined for a particular system with respect to two parameters λ, and τ, the areas of integration are known. In order to perform the integration, it is necessary to have a description of the probability distribution function (pdf) over the λ, τ plane, or at least over the portion of it that includes the M-shaped curves.

In order to facilitate the integration the following assumptions are made:

A) The pdf is separable into a function of λ times a function of τ, i.e. $p(\lambda,\tau)=p_\lambda(\lambda)p_\tau(\tau)$

B) The pdf of λ is a constant given by 4r where r is the fade occurrence factor given by Barnett.[1] This factor includes distance, frequency, and terrain dependence and varies with the particular radio hop. Note that the cumulative probability $P(\lambda<\lambda_o)$ is not the same as the probability $P(L<L_o)$. The former refers to the value of L at the notch ($L=\lambda$) while the latter is measured at a fixed frequency. The relationship has been studied by the author in an unpublished work.[6]

C) The pdf of τ is a slowly varying function with a characteristic time τ_c. The value of τ_c is left as a parameter to fit measured fading data. A general form would be

$$p_\tau(\tau) = f(\tau/\tau_c)/\int_0^\infty f(\tau/\tau_c)d\tau \qquad 0<\tau<\infty \qquad (14a)$$

Some particular cases that have been used are:

Exponential decay[7]

$$p_\tau(\tau) = \frac{1}{\tau_c} e^{-\tau/\tau_c} \qquad (14b)$$

Gaussian[8]

$$p_\tau(\tau) = \frac{2}{\tau_c\sqrt{2\pi}} e^{-\tau^2/2\tau_c^2} \qquad (14c)$$

It is expected that τ_c would have a dependence on distance similar to Ruthroff's[4] relationship of maximum delay to distance.

INTEGRATION

Given a signature of a system (a la Figure 3) and with the above assumptions made for the statistics it is possible to derive a closed form solution to the integration (step 4).

From Figure 3 it is obvious that the integration will be the sum of a number of integrals such as

$$\text{F.T.} = \sum_{n=1}^{\infty} \int_{\tau_n-\Delta\tau_n/2}^{\tau_n+\Delta\tau_n/2} \int_0^{\lambda_s} p(\lambda,\tau) \, d\lambda \, d\tau \qquad (15)$$

where F.T. stands for the fraction of time that the performance is unacceptable. The regions can be approximated by assuming a rectangular shape with width given by (derived from Equation 7)

$$\Delta\tau_n = \tau_n \frac{\Delta f}{f_c} \qquad (16)$$

where Δf is the bandwidth of the curves as shown in Figure 5. The mean height of the curves is given by

$$\lambda_n = m\,\tau_n \qquad (17)$$

where mean height is defined as that height that gives the same area as an exact integral under the curve. Both of these factors vary as τ and hence the area of each separate region will vary as τ^2.

From the statistical assumptions the integrand can be written

$$p(\lambda, \tau) = (4r)\, p_\tau(\tau) \qquad (18)$$

Let $\lambda_s = \lambda_m$ and use equations 16 and 17 to define the limits. Neglect the change in $p(\tau)$ across each narrow region and the integration yields a summation

$$\text{F. T.} = 4rm\frac{\Delta f}{f_c}\sum \tau_n^{\ 2}\, p_\tau(\tau_n) \qquad (19)$$

The exact integral has been approximated with a summation of integrals over rectangular areas each of which has height $m\tau_n$, width Δf, and constant pdf given by $p_\tau(\tau)$ evaluated at τ_n for that region. For a slowly varying pdf and narrow regions this is a good approximation.

This summation can be approximated by assuming a continuous function made up of intervals of duration $\frac{1}{f_c}$ and whose mean value of the nth interval is $p_\tau(\tau_n)$ so that the summation is replaced by an integration

$$\text{F. T.} \doteq \frac{4rm\Delta f}{f_c}\int_0^\infty \tau^2\, p_\tau(\tau)d\tau \qquad (20)$$

which yields

$$\text{F. T.} = 4rm\Delta f\ \tau_c^{\ 2}\ \langle(\frac{\tau}{\tau_c})^2\rangle \qquad (21)$$

The expectation value given by $\langle\mu^2\rangle$ (where $\mu = \tau/\tau_c$) is a number depending upon the functional form of $p_\tau(\tau)$ and the definition of τ_c. For the cases cited in Equation 14 the values are

Exponential decay	$\langle\mu^2\rangle = 2$	(22a)
Gaussian	$\langle\mu^2\rangle = 1$	(22b)

If the signature of a system were to be represented by a single number it is the product $m\Delta f$. If m is increased (steeper slope on Figure 3) or Δf is increased (wider curves in Figure 5) the fraction of time will increase proportionately. Both factors are dependent upon system design choices such as baud rate, filter roll-off, modulation format, etc.

EXAMPLES

Equation 21 gives the expected outage time under the assumptions given above. Some examples of individual regions measured on different systems are shown in composite Figure 6. From a single contour the values of m and Δf can be established. Listed in Table I are the values for the curves shown in Figure 6.

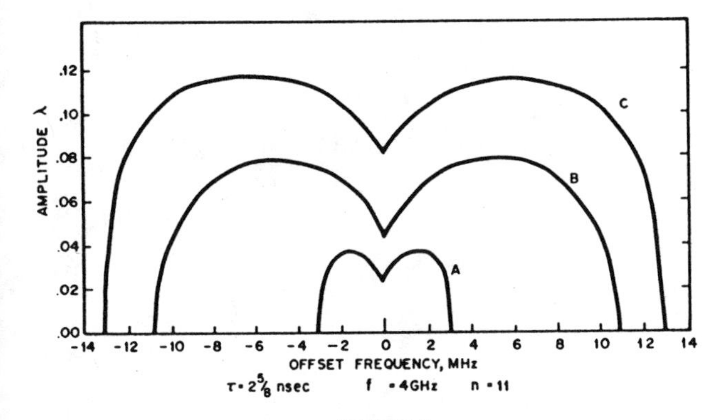

FIGURE 6
REPRESENTATIVE REGIONS FOR DIFFERENT SYSTEMS

TABLE I

	System	m	Δf	Criteria
A)	4 GHz TD-2 FM radio*	.014	.006	55 dBrnc0
B)	Computer Simulation 8-PSK 4 GHz 45 Mb/s	.027	.022	BER = 10^{-3}
C)	Experimental 8-PSK 4 GHz 45 Mb/s	.043	.026	BER = 10^{-3}

The dimensions of m are normalized amplitude per nanosecond and Δf is in GHz. If τ_c is given in nanoseconds for Equation 21 then the dimensions are compatible.

The computer simulation model (8-PSK) has perfect Nyquist filter roll-off (50%) characteristics and the ability to find the ideal timing and phase adjustments so it is not surprising that the simulated model performs better than a real system.

The value of the fractional time of outage depends not only on the system but on the value of the characteristic time τ_c of the ray delay distribution. In Figure 7 the outage is plotted as a function of the characteristic time for the different systems. The value of r is taken to be .176 corresponding to a 26 mile hop over average terrain at 4 GHz. A factor of 2 is included by assuming the exponential decay model of Equation 14b and 22a. Also shown for reference is the fractional times predicted by the flat fade (one parameter) model for different fade margins (L_o). The fractional time predicted by that model is[1]

$$\text{F.T.} = rL_o^{\ 2} \qquad (23)$$

Note that the digital systems have a much greater fractional time of unacceptable performance predicted than the FM system and that the shaped fades will cause more poor performance than flat fades unless the characteristic time is quite short. Also indicated on Figure 7 is a point marked J that is predicted by Jakes[7] indicating $\tau_c = 0.3$ ns.

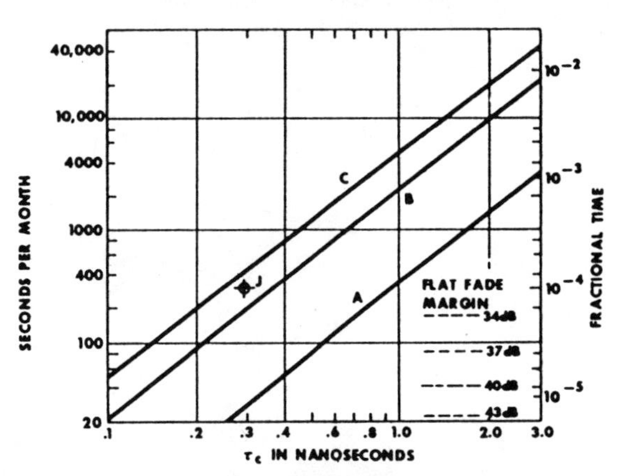

FIGURE 7
FRACTIONAL TIME OF UNACCEPTABLE PERFORMANCE

*Measurements on the FM system were made by S. H. Lee (BTL internal publication).

Note that performance is unacceptable for 300 seconds per month instead of the 45 seconds predicted by a flat fade model for a system with a 40 dB margin. Such a digital system has an outage time commensurate with a hypothetical system with a 31 dB flat fade margin. This clearly indicates that the conventional concept of fade margin breaks down for this case. Curve A shows that for τ_c = 0.3 ns the effects of shaped fades on FM radio is small.

CONCLUSIONS

A method has been presented for the calculation of the fractional time that a line of sight radio hop will have unacceptable performance. Heretofore most calculations of performance have considered the fading to be flat with a decision dependent only on one system parameter, - the fade margin set by thermal noise. This results in an overly optimistic prediction of performance, especially for systems sensitive to transmission path distortions.

The two ray model allows two parameters to be chosen to fit the fading statistics. It also provides for a signature that can easily be determined in the laboratory for any particular system. The properties of the signature allow for simple and direct comparison between systems as well as a calculation of the contribution of selective fading to system outage.

Under certain reasonable assumptions about the statistics of fades a closed form expression results from the integration to find the fractional time that a radio hop will have unacceptable performance.

REFERENCES

1. W. T. Barnett, "Multipath Propogation at 4, 6 and 11 GHz," B.S.T.J. Vol. 51, #2, February, 1972, Pg. 321.

2. S. H. Lin, "Statistical Behavior of a Fading Signal," B.S.T.J., Vol. 50, #10, December, 1971, Pg. 3211.

3. G. M. Babler, "A Study of Frequency Selective Fading for a Microwave Line-of-Sight Narrowband Radio Channel," B.S.T.J., Vol. 51, #3, March, 1972, Pg. 731.

4. C. L. Ruthroff, "Multi-Path Fading on Line-of-Sight Microwave Radio Systems as a Function of Path Length and Frequency," B.S.T.J. Vol. 50, #7, September, 1971, Pg. 2375.

5. V. K. Prabhu and L. J. Greenstein, "Analysis of Multipath Outage with Application to 90 Mb/s PSK Systems at 6 and 11 GHz," IEEE International Communications Conference, ICC 1978.

6. M. Emshwiller, BTL Internal Memorandum.

7. W. C. Jakes, Jr. "An Approximate Method to Estimate the Effect of Multipath Delay Distortion on Digital Transmission." IEEE International Communications Conference, ICC 1978.

8. H. Yamamoto and S. Nakamura, "Waveform Distortion and Interchannel Interference Due to Frequency Selective Fading in Microwave PCM Systems," Rev. Elec. Comm. Lab. Vol. 17, #3-4, March-April, 1969, p. 173.

Digital Radio Outage Due to Selective Fading—Observation vs Prediction From Laboratory Simulation

By C. W. LUNDGREN and W. D. RUMMLER

(Manuscript received July 7, 1978)

A statistical model (introduced in a companion paper) of fading on a radio path is used with laboratory measurements on a digital radio system to estimate the outage due to multipath fading, where outage is the time that the bit error rate (BER) exceeds a threshold. Over the range of BER of interest (10^{-6} to 10^{-3}), the calculated outage agrees favorably with the outage observed during the period for which the fading model was developed. It is further shown that the calculated outage, when scaled to a heavy fading month on the basis of single-frequency, time-faded statistics, agrees equally well with the outage observed on the same path during a heavy fading month. The agreement between measured and predicted outage substantiates the selective fading model. The prescribed laboratory measurements characterize the sensitivity of the radio system to selective fading. Thus, the methodology provides a useful basis for comparing the outage of alternative realizations of digital radio systems.

I. INTRODUCTION

Present interest in using high-speed common carrier digital radio[1–5] has precipitated a need for estimating the performance of such systems during periods of selective (multipath) fading. This paper describes a method of characterizing a digital radio system in the laboratory which allows the outage to be accurately estimated. For a digital radio system, outage requirements are stated in terms of the number of seconds in a time period (usually a heavy fading month) during which the bit error rate (BER) may exceed a specified level; typically, 10^{-3} or 10^{-4} is appropriate to voice circuit application.

The method is based upon a statistical channel model[6] developed from measurements on an unprotected 26.4-mile hop in the 6-GHz band in Palmetto, Georgia in 1977 using a general trade 8-PSK digital radio system as a channel measuring probe. The modeled fading occurrences were scaled to the basis of a heavy fading month using the occurrence of time faded below a level at a single frequency as the means of calibration. The bit error rate performance of the digital radio system was measured during the time period used for channel modeling and for an extended period corresponding to a heavy fading month. This same radio system was later subjected to a measurement program in the laboratory using a multipath simulator which provides a circuit realization of the fading model. The measured results are used with the channel model to determine the occurrence of channel conditions which will cause the BER to exceed a given threshold. Comparisons on the basis of the modeling period and a heavy fading month show good agreement between calculated and observed outages for BERs between 10^{-6} and 10^{-3}.

The properties of the fixed-delay channel model are reviewed briefly in Section II as a basis for describing the measurements and for the subsequent outage calculations. This three-parameter channel model is realized in the laboratory by an IF fade simulator. The simulator and its use in obtaining the necessary laboratory data are described in Section III. The procedures to be followed in calculating outage times for a given BER are described in Section IV. Calculated and observed outage times are compared in Section V. Conclusions are provided in Section VI.

II. MODEL DESCRIPTION—METHODOLOGY

It has been demonstrated[6] that the complex voltage transfer function of a line-of-sight microwave radio channel is well modeled by the function

$$H(\omega) = a\,[1 - b e^{-j(\omega-\omega_0)\tau}] \tag{1}$$

with τ fixed. A 6-GHz channel (30-MHz bandwidth) has been characterized statistically by the model with $\tau = 6.3$ ns. Such a channel has a power transfer function given by

$$|H(\omega)|^2 = a^2[1 + b^2 - 2b\cos(\omega - \omega_0)\tau] \tag{2}$$

and an envelope delay distortion function, i.e., the derivative of the phase of $H(\omega)$ with respect to ω, given by

$$D(\omega) = \frac{b\tau(\cos(\omega - \omega_0)\tau - b)}{1 + b^2 - 2b\cos(\omega - \omega_0)\tau}. \tag{3}$$

In the following paragraphs, we summarize the properties of the model, the statistics of the model parameters, and the measurement objectives.

Reprinted with permission from *Bell System Technical Journal,* vol. 58, no. 5, pp. 1073–1102, May/June 1979.

2.1 *Fixed delay model*

A plot of the attenuation produced by the fixed delay model of eq. (1) is shown in Fig. 1. Since the delay is fixed at 6.3 ns, the spacing between nulls of the response, 158.4 MHz, is much larger than the channel bandwidth. The parameters a and b control the depth and shape of the simulated fade, respectively. The parameter $f_o(=\omega_0/2\pi)$ determines the position of the fade minimum or notch. Both the notch frequency, f_o, and the response frequency, f, are measured from the center of the 30-MHz channel for convenience.

The model function of eq. (1) may be interpreted as the response of a channel which provides a direct transmission path with amplitude a and a second path providing a relative amplitude b at a delay of 6.3 ns and with a phase of $\omega_0\tau + \pi$ (independently controllable) at the center frequency of the channel. This interpretation is represented in Fig. 2 by a phasor diagram at $\omega = 0$, the center frequency of the channel. Varying the frequency, ω, over the channel bandwidth (30 MHz) moves the angle of the interfering ray through an arc of about 60 degrees ($2\pi \times 30$ MHz $\times$ 6.3 ns $\approx \pi/3$), centered at the position shown. This diagram is useful for understanding the fade simulation; it also provides an alternate means of describing the position of the notch. The notch position may be specified by its frequency, f_o, or by ϕ, the angle of the interfering path at the center of the channel.

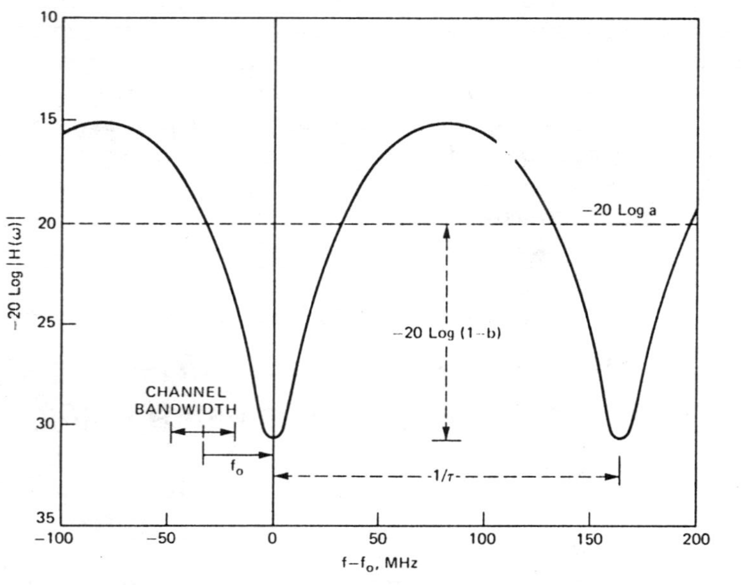

Fig. 1—Attenuation of channel model function, $H(\omega) = a[1 - b\exp(-j(\omega - \omega_0)\tau)]$, for $\tau = 6.3$ ns, $a = 0.1$, $b = 0.7$.

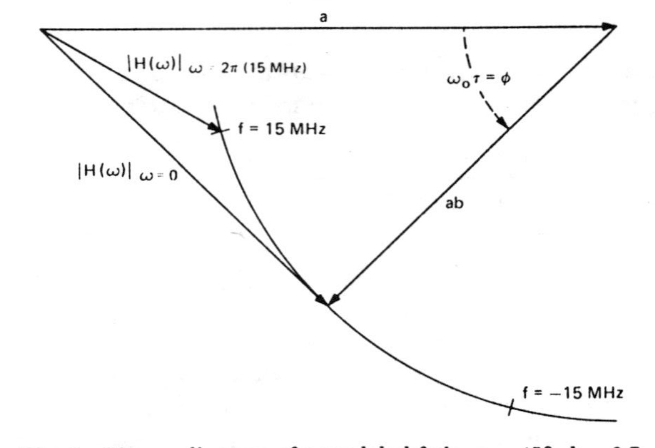

Fig. 2—Phasor diagram of a modeled fade, $\phi = 45°$, $b = 0.7$.

From Figs. 1 and 2, or eq. (1), it may be seen that varying a changes the overall level and varying b changes the shapeliness of the modeled fade. Furthermore, if the minimum is within the 30-MHz channel bandwidth ($|\phi| < 30°$), the fixed delay model can generate notches with a wide range of levels and notch widths. With the minimum out of band, it can generate a wide range of combinations of levels, slopes, and curvatures within the channel bandwidth.

2.2 *Model statistics*

The statistics of the model parameters were obtained from a selected data base during which heavy fading activity was observed.[6] The distribution of b is best described in terms of $B = -20\log(1 - b)$. Figure 3 shows the distribution of B and the least-squares straight line fit to the distribution over the region where it best represents selective fading—between B values of 3 and 23 dB. The channel is described by B greater than 23 dB for less than 0.15 percent of the observed time which makes the distribution less certain beyond this point. At the other extreme, during the periods of time when there is little or no selective fading, the channel is characterized by values of B less than 3 dB. Thus, the fitted line represents a lower bound on the distribution for B less than 3 dB. Since the fitted line has an intercept of 5400 seconds, we may model the fraction of 5400 seconds during which B exceeded a value X by the probability distribution

$$P(B > X) = e^{-X/3.8}. \tag{4}$$

Thus the probability of finding a value of B between X and $X + dX$ is

$$p_B(X)\,dX = \frac{dX}{3.8}\,e^{-X/3.8}. \tag{5}$$

The distribution of a is lognormal with a standard deviation of 5 dB and a mean that is dependent on B (or b). Hence, the probability that

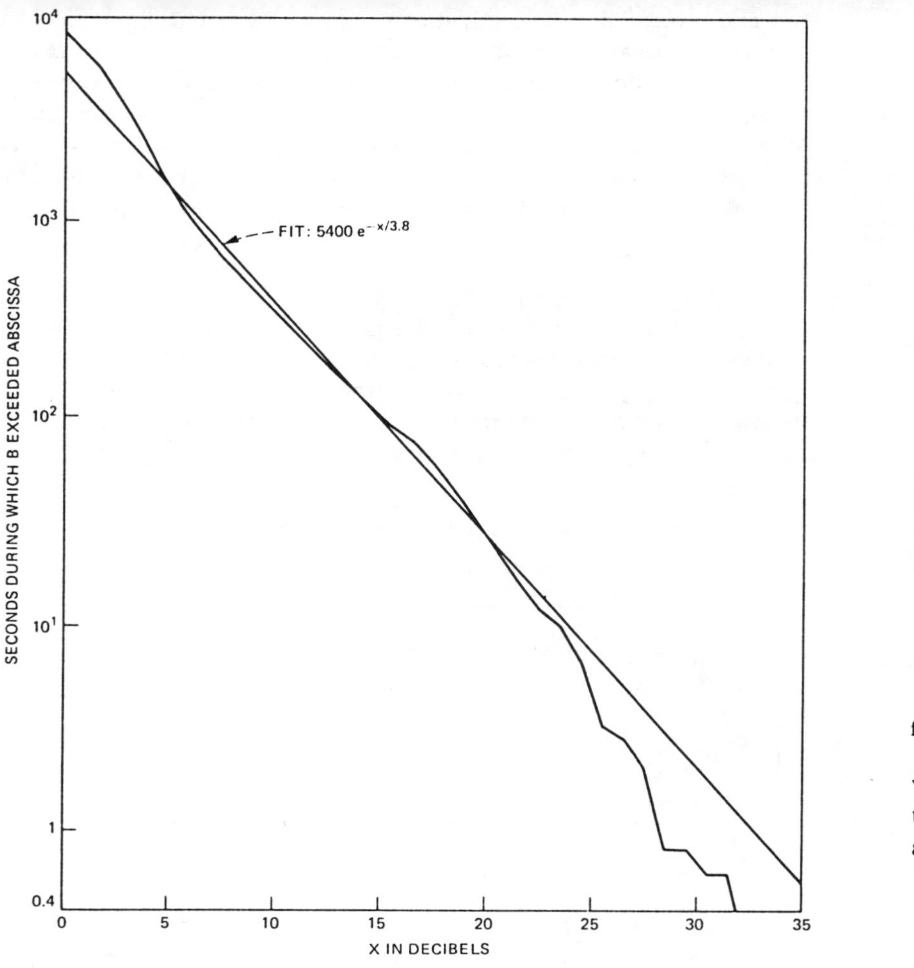

Fig. 3—Distribution of B for model data base period.

$A = -20 \log a$ has a value between Y and $Y + dY$ is given by

$$p_A(Y)\, dY = \frac{dY}{5\sqrt{2\pi}} e^{-[Y - A_0(B)]^2/50}. \qquad (6)$$

The relationship between A_0, the mean of the distribution, and B is given in Fig. 4.

The distribution of f_o is found to be independent of A and B. It is usually simpler to work with ϕ rather than f_o. The two variables are simply related in that ϕ is defined on the interval $(-\pi, \pi)$ and a 2.5-degree change in ϕ corresponds to a 1.1-MHz change in f_o. For the fixed delay model, the variable ϕ has been found to have a probability density function that can be described as uniform at two levels, with

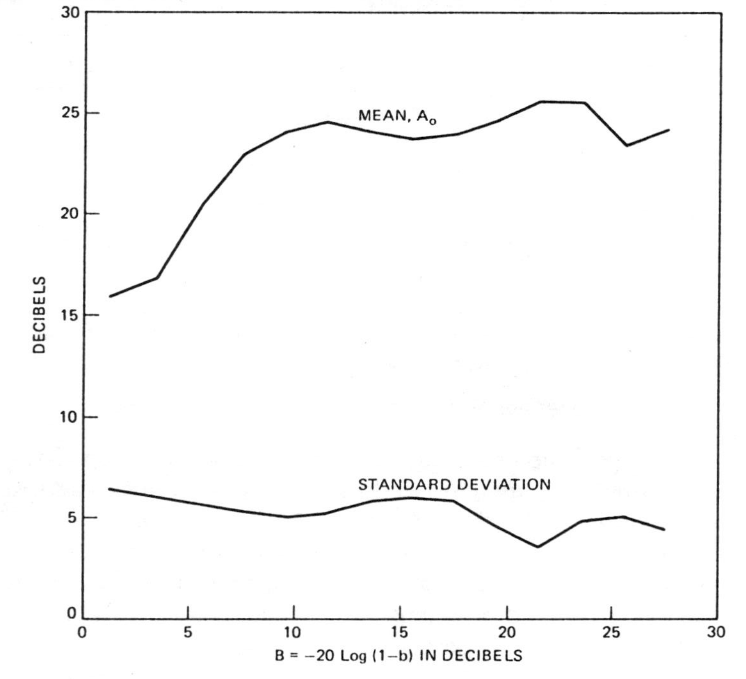

Fig. 4—Mean and standard deviation of the distribution of $A = -20 \log a$ as a function of B.

values less than $\pi/2$ being five times more likely than values greater than $\pi/2$. Thus, we have the probability density function per degree as:

$$p_\phi(\phi) = \begin{cases} \dfrac{1}{216} & |\phi| < 90^\circ \\ \dfrac{1}{1080} & 90^\circ \le |\phi| \le 180. \end{cases} \qquad (7)$$

The functions in eqs. (5) to (7) can be used to determine the probability of finding a, b, and f_o in some region of a-b-f_o space. This probability can be converted to number of seconds in the observation period by multiplying by 5400 seconds. To convert this probability to the number of seconds in a month requires scaling the data base. The scaling may be obtained from Fig. 5 which shows, for several frequencies in the band, the time during the model data base period that the channel was faded below a given level. Distributions are shown for average power in the band and for power at selected frequencies at the center and near the edges of the radio channel. (Frequencies indicated are at IF where the center frequency is at 70 MHz.) For the path used,

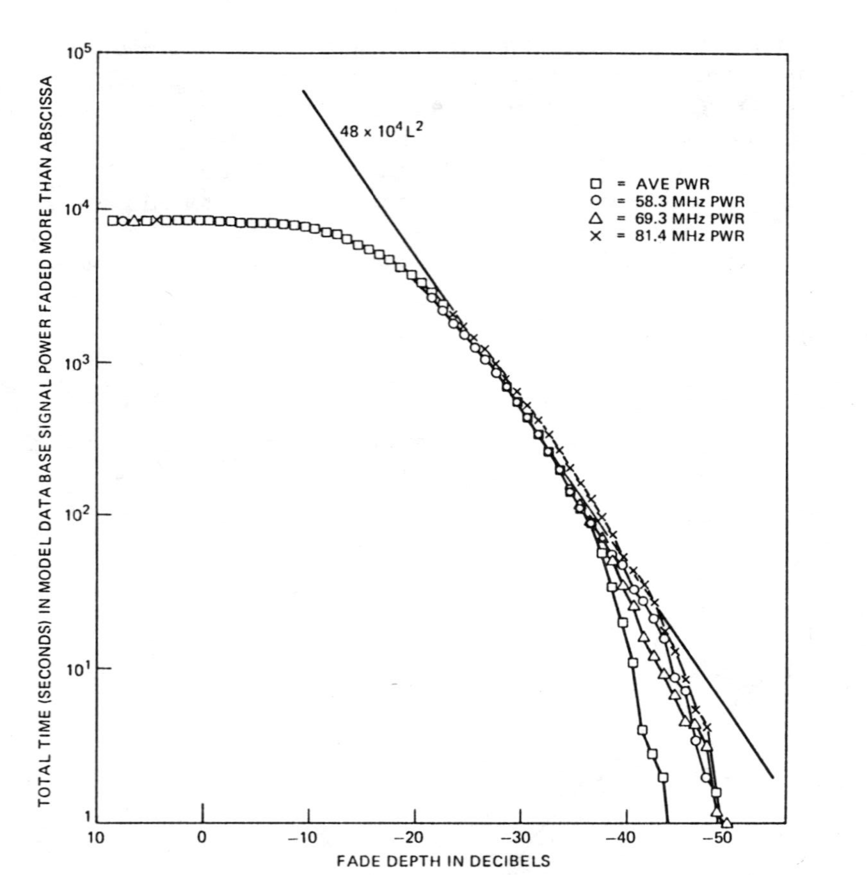

Fig. 5—Amplitude distributions for model data base period.

one expects the received voltage (single frequency) relative to midday average to be less than L for $72.5 \times 10^4\, L^2$ seconds in a month.[7] For the data base used, the fading is best described by $48 \times 10^4\, L^2$ hence, the data represent ⅔ of a fading month. To obtain outage on a seconds-per-heavy-fading-month basis, the probabilities calculated with eqs. (5) to (7) must be multiplied by 5400×1.5 or 8100 seconds.

2.3 *Outage estimation*

The fixed delay model described above can be simulated with an equivalent circuit laboratory measurement to determine the equipment response to multipath fading. Conceptually, one determines critical values of A and B for which a specified error rate is achieved for each fade notch position. In practice, it is difficult to maintain a constant BER; it is more expedient to fix b and vary the carrier-to-noise ratio (a) while plotting the BER. From the resulting curves, one may compute critical contours of A and B for each prescribed notch location and BER. Using eqs. (5) and (6), the probability that A and B lie on the high error rate side of a given critical contour may be calculated.

By repeating this calculation for a uniform set of notch positions and using (7) to determine the probability weighting given to each and summing, one may estimate the probability of all selective fades that produce a BER exceeding the prescribed one. Multiplying this probability by 5400 gives the outage time expected over the data base period; multiplying by 8100 gives the expected outage time per heavy fading month.

The following section describes the laboratory measurement; Section IV describes the reduction of the measured curves and parameters to outage times.

III. LABORATORY MEASUREMENTS

Figure 6 illustrates stressing of a digital radio system by means of an IF fade simulator. The simulator, which is inserted after linear IF preamplification but before any high-gain amplification, shapes both the desired signal and the effective received noise. It is necessary to operate the simulator at a sufficiently large input carrier-to-noise ratio that the concomitantly shaped noise at its output remains a negligible contributor to degraded system performance throughout the operating range of interest.

Within its restricted frequency range of operation, the IF simulator is adjusted to achieve those specific shapes implied by Fig. 2. Although the measurements could have been made using an RF fade simulator, the choice of an IF simulator was based primarily upon considerations of signal and noise levels, and the repeatability of adjustments. The following section describes an IF shape-stressing measurement in the minimum detail necessary to qualify the data collected.

3.1 *Representative IF two-path fade stressing measurement*

The block diagram of Fig. 7 illustrates an arrangement employing an IF fade simulator and an IF flat noise source. A pseudo-random test

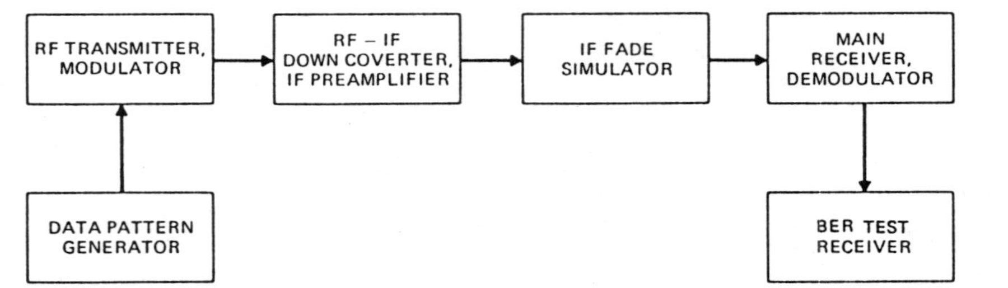

Fig. 6—IF fade stressing.

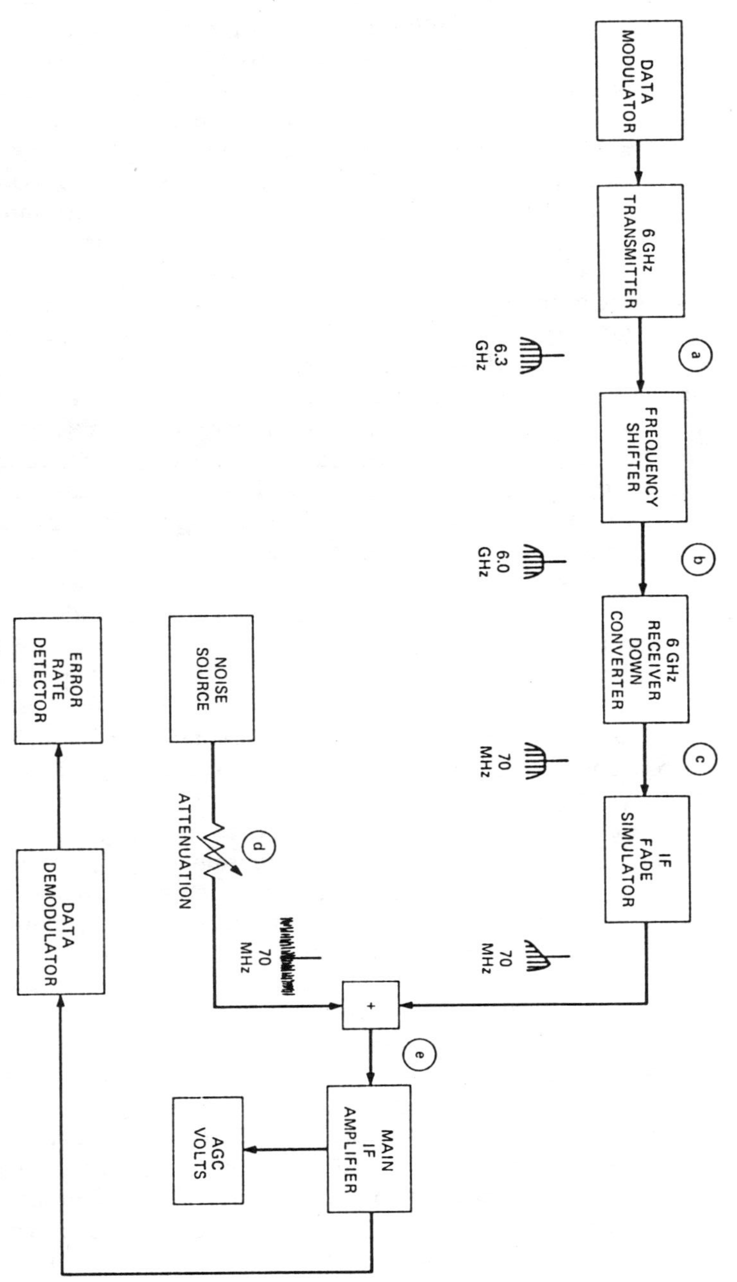

Fig. 7—Laboratory test arrangement.

pattern modulates the 6-GHz radio transmitter whose output is nominally 5 watts (ⓐ, in Fig. 7). The output spectrum is usually shaped by a bandpass filter following the microwave power amplifier to comply with FCC regulations.

To enable back-to-back operation of the transmitter and receiver of a single repeater which normally operate on different radio channels, a radio test translator was employed. The translator output power was approximately −30 dBm (adjustable, at ⓑ) to simulate the unfaded received signal level (RSL) observed typically in the field.

Assuming a linear RF-IF conversion gain of 20 dB, the signal power at the input (ⓒ) to the IF fade simulator is −10 dBm. The simulator incorporates low-noise linear amplification. A reference insertion loss for the main unfaded ray is 10 dB, including the output power summer. Hence the maximum desired signal power at the input to the main IF amplifier (ⓔ) is −20 dBm.

Assuming a 30-MHz receiver noise bandwidth and a current-art receiving system noise figure of 5 dB, the total system noise power is approximately −95 dBm, referred to the receiver's input port. This results in a flat receiver noise contribution of −85 dBm at input ⓔ to the main IF amplifier. Consequently, the maximum attainable carrier to simulator-shaped RF noise ratio is 10 log $(C_0/N_{rf}) = -20 - (-85) =$ 65 dB. The noise contributed by the fade simulator amplifiers must not exceed −100 dBm, to be negligible.

Flat IF noise much larger than the unwanted and shaped system noise is added artificially at ⓔ and is adjusted in magnitude by a calibrated attenuator ⓓ to superpose thermal noise degradations upon the simulated selective fading degradations of the desired signal. One would ideally measure the added IF noise power in the final predetection bandwidth of the digital radio system, or twice the Nyquist bandwidth. It is more convenient in the laboratory to reference carrier-to-noise ratios to the output of the main IF amplifier by using the precalibrated AGC voltage (assuming that wideband AGC detection is employed), to measure both the unshaped signal and noise powers. The carrier-to-noise ratio at the detector would be higher—by the ratio of the system noise bandwidths that would be measured at the respective points.

The noise source output in Fig. 7 may be adjusted so that an attenuator setting of 0 dB ⓓ results in a noise power equivalent to that of the unfaded signal power (the attenuator is then calibrated directly in uncorrected C_0/N_{if}, in dB). As the IF fade simulator is readjusted to achieve different prescribed fade shapes, its mean insertion loss may also change. The change in insertion loss is determined by monitoring the change in signal power at ⓔ; the same loss increment (dB) must be added to the noise attenuator ⓓ to reestablish the 0 dB reference.

3.2 IF two-path fade simulator

Figure 8 illustrates splitting the desired IF signal into an arbitrarily phased, adjustable "main" component and a "delayed" component fixed in delay (τ ns) but adjustable in magnitude. The main component is further resolved into orthogonal components (inset to Fig. 8) using wideband networks exhibiting flat gain and well-behaved delay. A particular sum vector is constructed by adjusting the orthogonal components to establish a simulated fade notch frequency; in practice, the phase sense of 0- and 90-degree components are independently reversible, as indicated by the switches in the figure, for complete flexibility in notch frequency selection.

The 6.3-ns fixed delay added to the delayed path imparts a phase shift of 159 degrees at the 70-MHz IF center frequency. This is shown built out to 225 degrees, relative to the 0-degree transmission path, using a 66-degree wideband network of the same type. The delayed vector is fixed in direction opposite the midrange position of the adjustable main vector, corresponding to a channel-centered fade ($\phi = 0$ degrees).

Since $1/\tau = 158.4$ MHz, a change of 1 degree in direction of the main vector corresponds to a frequency displacement of the fade notch location of 0.44 MHz. For $\phi = -45$ degrees, the notch location is displaced 19.8 MHz below the channel center ($f_o = -19.8$ MHz). The magnitude of the delayed component is then adjusted to achieve the desired notch depth.

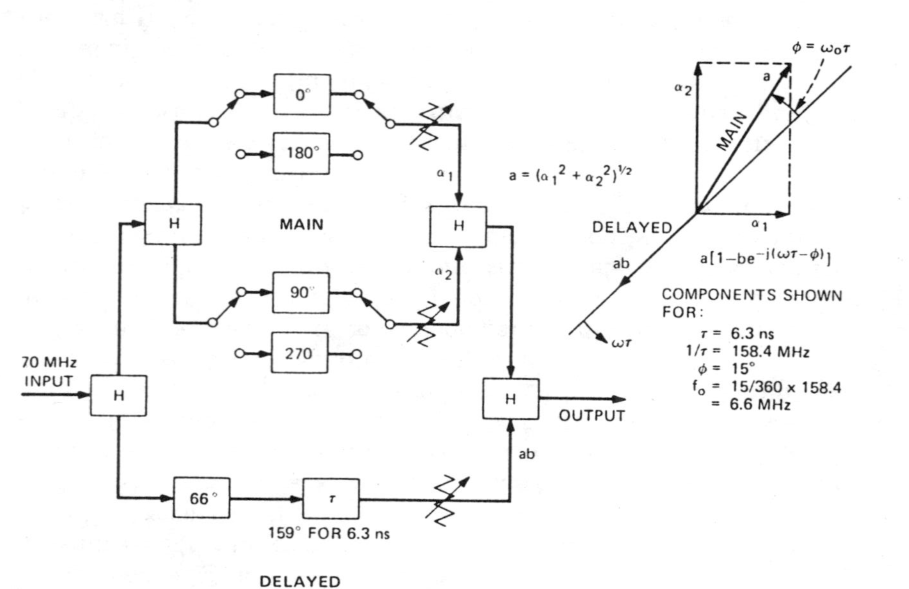

Fig. 8—IF fade simulator—conceptual block diagram.

3.3 Digital radio performance stressed by in-band selectivity and thermal noise

The radio equipment was measured at uniformly spaced notch frequencies separated by 4.4 MHz ($\Delta\phi = 10$ degrees). To fully characterize a period of variation in ϕ, or f_o, one would need to make 36 sets of measurements. *Ideally,* half may be omitted because of symmetry. For given values of A and B, the same error rate should obtain for a notch at a given frequency displacement above or below the channel band center. Variations in B ought not to have a significant effect for $|\phi|$ greater than 90 degrees. It was determined that detailed measurements were required for nine different values of f_o to characterize the digital radio tested.

Using a wideband RF fade simulator in the field, the digital radio performance for out-of-channel notch locations was relatively independent of whether minimum or nonminimum-phase fade simulations were employed. The nonminimum phase fade is modeled by eq. (1) with the sign of the phase term reversed. This leaves the amplitude [eq. (2)] unchanged, but reverses the sign of the envelope delay distortion [eq. (3)]. We conclude that the minimum phase channel model is sufficiently general for use in simulating the channel and in estimating performance.

The IF fade simulator was adjusted for each notch frequency, and the depth of notch was varied by adjusting the magnitude of the delayed component. Then various amounts of IF thermal noise were added. Figure 9 typifies the performance data collected. BERs are plotted *versus* the uncorrected IF carrier-to-noise ratio (C/N_{if}), for a constant fade notch offset from midchannel ($f_o = -19.8$ MHz, $\tau = 6.3$ ns). Each curve corresponds to a different notch depth ($B = -20 \log(1-b)$ dB), and hence a different amplitude and delay shape in the radio channel. Each curve is also identified with an in-band selectivity, defined as the difference between the maximum and minimum attenuation present in the (25.3-MHz) channel bandwidth. The lower-left "baseline" curve presents the unshaped signal, flat fading performance obtained by adding only IF thermal noise. This curve was verified (without the added IF noise) by attenuating the received RF input signal in the back-to-back configuration.

Consideration was given to matching the order of measurements to characteristics of the particular digital radio tested. For example, considerable scattering of data at low error rates can result from synchronizations involving different reference carrier phases. The authors elected to perform several synchronizations while observing the BER for each phase, and then chose that phase giving the worst performance.* Synchronization was accomplished at the low error rate

* Because the phase information in the measured system was Gray coded and digital access was on a per-rail basis, one rail had twice the BER of the other two. All measurements in the field and in the laboratory were made on this worst-case rail.

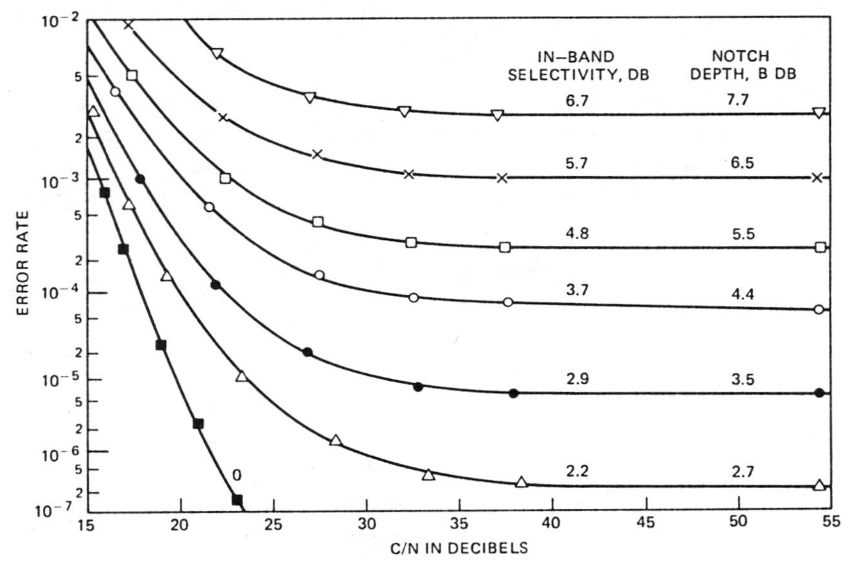

Fig. 9—High-speed digital radio IF dispersive fade simulations, $\tau = 6.3$ *ns*, $f_0 = -19.8$ MHz.

(bottom) of each curve, and this phase relationship was maintained for all data points obtained for each curve.

From the baseline curve of Fig. 9, a BER $= 1 \times 10^{-6}$ obtains for $10 \log(C/N_{if}) \doteq 21.5$ dB. For the digital radio system installed on the instrumented hop and reported in the figure, the measured flat fade margin for a threshold BER $= 1 \times 10^{-6}$ was 40.5 dB This leads to an unfaded IF carrier-to-noise ratio $10 \log(C_0/N_{if}) \doteq 21.5 + 40.5 = 62$ dB.

From the baseline curve for a BER $= 1 \times 10^{-7}$, note that insertion of a fade whose notch depth is 6.5 dB results in four orders of magnitude degradation in BER performance; equivalently, an in-band selectivity of only 5.7 dB in 25.3 MHz results in a BER $> 1 \times 10^{-3}$.

The asymptotic regions in Fig. 9, corresponding to high values of C/N_{if}, are not normally presented in characterizations of this type; however, system outage depends primarily upon the performance in these asymptotic regions. Thus, under typical fading conditions, the transmitted carrier power might be increased at will without improving the BER significantly. The effects of decreasing the carrier power are discussed in Section 4.4.

A family of curves like those shown in Fig. 9 was obtained (but are not given here) for each of nine uniformly spaced frequency offsets below midchannel to characterize the digital radio system sufficiently for the prediction of outage. A number of spot checks were also made using both RF and IF fade simulators at symmetrical positive and negative offset frequencies, to establish that acceptable symmetry existed.

IV. CALCULATION OF OUTAGE

This section describes four methods of calculating outage. The derivation of the critical curves of A and B, which provide the basis for making and understanding these calculations, is given in Section 4.1. In Section 4.2 the detailed calculation of outage from the critical A-B curves is described. It is shown in Section 4.3 that for the present system this method may be greatly simplified by calculating only selectivity-caused outage (i.e., neglecting thermal noise). Section 4.4 presents an approximate method of accounting for the effects of thermal noise. Section 4.5 provides a basis for estimating the selectivity-caused outage from a single measurement.

4.1 Derivation of critical characteristics

To calculate the outage for a fixed bit error rate, one must first obtain the critical curves of A and B at each simulated value of f_o, the notch position. Thus, from Fig. 9 which corresponds to $f_o = -19.8$ MHz (or $\phi = -45°$), we obtain six points on the critical curve of A and B for a BER of 10^{-3}, one point from each of the six curves which cross the critical BER. The value of B is obtained from the value of b since

$$B = -20 \log(1 - b). \tag{8}$$

For the curve in Fig. 9 corresponding to $B = 4.4$ dB, we obtain the corresponding critical value of A for a BER of 10^{-3} from the value of carrier-to-noise ratio, which is 20.2 dB where this curve crosses the 10^{-3} BER line. Since the carrier-to-noise ratio is 62 dB when the channel is unfaded, the 20.2 dB value corresponds to a relative average power loss of 41.8 dB,

$$L_s = 62 - 20.2 = 41.8 \text{ dB}. \tag{9}$$

Without loss of generality, we assume that the PSK signal has a rectangular spectrum of width f_b; consequently, the relative power transmitted by the model is obtained from eq. (2) as*

$$P_{av} = \frac{1}{2\pi f_b} \int_{-\pi f_b}^{\pi f_b} |H(\omega)|^2 \, d\omega$$

$$= a^2 \left\{ 1 + b^2 - 2b \cos 2\pi f_o \tau \left(\frac{\sin \pi f_b \tau}{\pi f_b \tau} \right) \right\}. \tag{10}$$

* The calculated result is not critically dependent on the flatness of the signal spectrum or the spectral width chosen. We have used for f_b a value of 25.3 MHz as representing the effective width of the signal.

Defining a correction term by

$$C = -10 \log\left\{1 + b^2 - 2b \cos 2\pi f_o\tau\left(\frac{\sin \pi f_b\tau}{\pi f_b\tau}\right)\right\}, \tag{11}$$

we may express the signal loss as

$$L_s = -10 \log P_{av} = A + C. \tag{12}$$

Thus, we obtain the critical value of A as

$$A = L_s - C. \tag{13}$$

For $B = 4.4$ dB ($b = 0.4$) and $f_o = -19.8$ MHz, we find $C = 2.06$ dB and the critical value of A is $41.8 - 2.1 = 39.7$ dB.

Carrying out these calculations for the six curves in Fig. 9, one can generate the critical curve of A and B for $f_o = -19.8$ MHz and a BER of 10^{-3}. The curve is shown in Fig. 10 along with the critical curves for several other values of the BER. A complete set of curves must be generated for all values of f_o.

The curves in Fig. 10 are typical of the critical curves obtained for $|f_o| \leq 33$ MHz. The intercept with the A-axis represents the flat fade margin for the given BER; this margin is independent of notch position. The intercept of a critical contour with the B-axis represents the shape, or relative fade depth, margin for the given notch position. For values of B to the right of this intercept, the critical value of BER cannot be obtained at any carrier-to-noise ratio for the given notch position.

4.2 Outage calculation—detailed method

The probability, P_o, of finding A and B outside all critical contours may be written with eqs. (5), (6), and (7) as

$$P_o = \int_{-\pi}^{\pi} p_\phi(\phi) P_c(\phi)\, d\phi, \tag{14}$$

where

$$P_c(\phi) = \int_0^\infty \int_{A_c(X)}^\infty p_A(Y) p_B(X)\, dY\, dX, \tag{15}$$

and $A_c(X)$ is the functional relation of the critical values of A to B (or X), for B less than B_c, the B-axis intercept, and for a given BER and ϕ value.* Since measurements were made for a uniformly spaced set of notch positions with spacing $\Delta\phi = 10°$, we may approximate (14) by

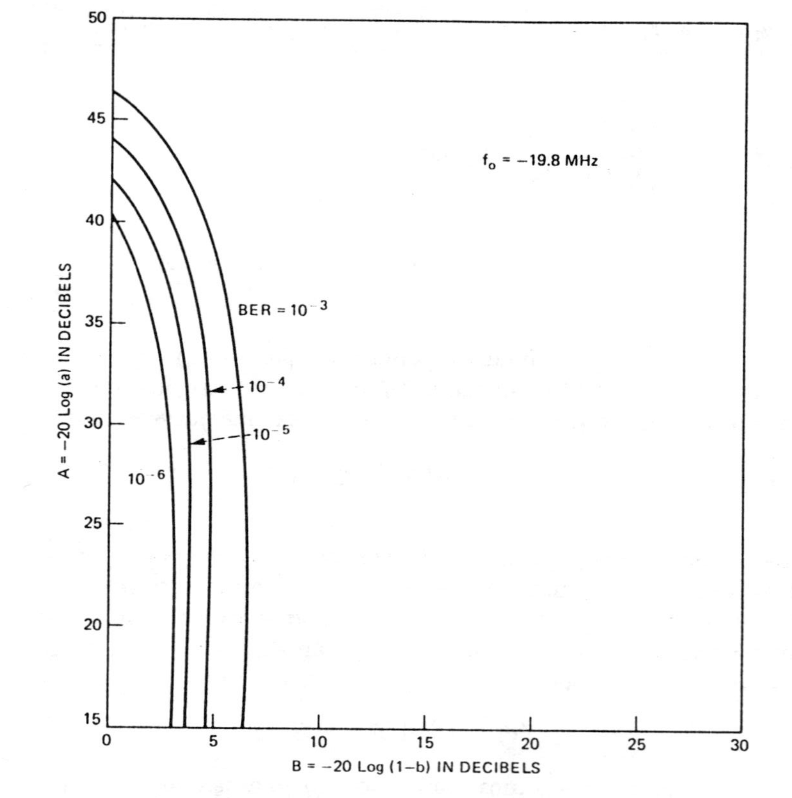

Fig. 10—Critical curves of A and B for $f_o = -19.8$ MHz.

$$P_o = \Delta\phi \sum_{\text{All } \phi_i} p_\phi(\phi_i) P_c(\phi_i). \tag{16}$$

To illustrate the calculation of outage probability with eqs. (15) and (16), we shall calculate the term in the summation of eq. (16) corresponding to a BER of 10^{-3} and $\phi_i = -45°$ (or $f_o = -19.8$ MHz). From Fig. 11, which is taken from Fig. 10, we note that the double integral in eq. (15) may be broken into integrations over two regions. Thus

$$P_c(\phi_i) = \int_{B_c}^\infty \int_{-\infty}^\infty p_A(Y) p_B(X)\, dY\, dX + \int_0^{B_c} \int_{A_c(X)}^\infty p_A(Y) p_B(X)\, dY\, dX, \tag{17}$$

where the two double integrals correspond to integrations over Regions 1 and 2, respectively, in Fig. 11. Outage due to the occurrence of A and B in Region 1 may be described as outage due only to shape or

* The dependence of the function $A_c(X)$ and the asymptote B_c on BER and ϕ is not explicitly denoted to keep notation simple.

selectivity. In Region 2, outage is due to the combined effects of signal loss and selectivity.

Using eqs. (5) and (6), the integral over Region 1 is obtained as $e^{-B_c/3.8}$. The contribution due to thermal noise and shape (Region 2) is slightly more complicated. Dividing the interval 0 to B_c in Fig. 11 into N subintervals, as shown in Fig. 12, the probability of being in Region 2 is the sum of the probabilities for each subinterval. Thus eq. (17) becomes

$$P_c(\phi_i) = e^{-B_c/3.8} + \sum_{k=1}^{N} [e^{-B_{k-1}/3.8} - e^{-B_k/3.8}] P_g\left(\frac{A_k - A_0(B_k)}{5}\right), \quad (18)$$

where

$$P_g(X) = \frac{1}{\sqrt{2\pi}} \int_X^\infty e^{-x^2/2}\, dx. \quad (19)$$

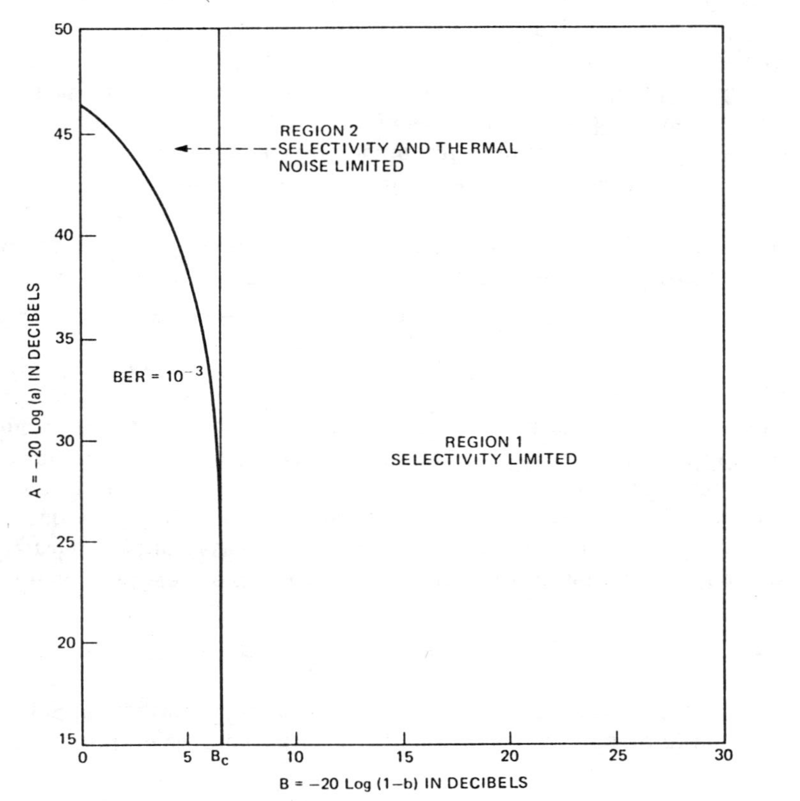

Fig. 11—Classification of outage with respect to critical curve for BER = 10^{-3}, f_0 = 19.8 MHz.

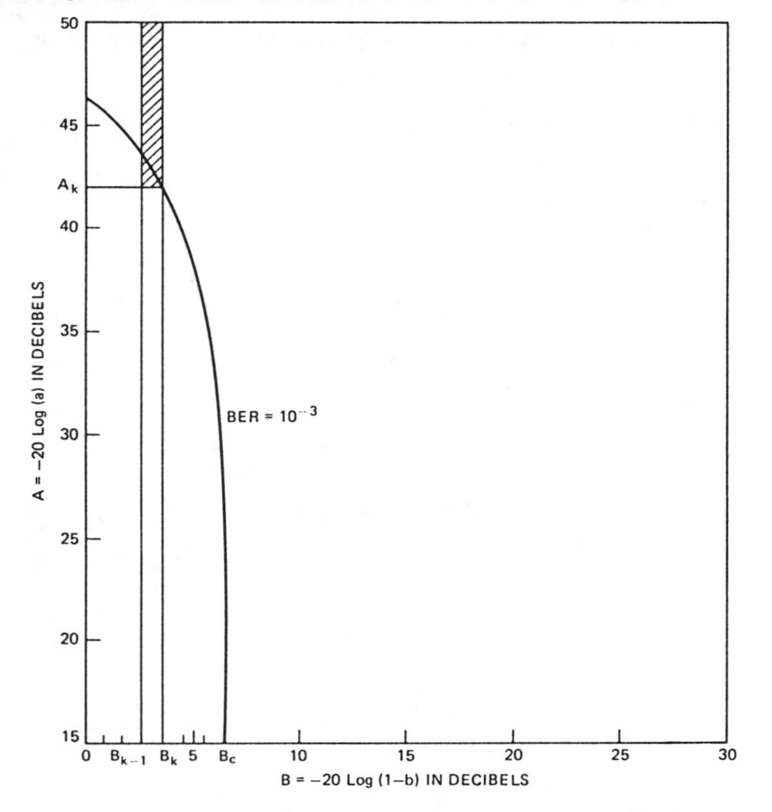

Fig. 12—Outage calculation for an incremental interval.

Evaluating the two components of eq. (18) from Fig. 11, we find

$$P_c(-45°) = 0.181 + 0.003 = 0.184. \quad (20)$$

This calculation was performed for 10 values of ϕ_i from −5 to −85 degrees in 10-degree steps. Using these results in eq. (16) and multiplying by two to account for positive values of ϕ_i which are assumed to contribute equally, we find the probability, P_o for a BER of 10^{-3} as

$$P_o = 0.0996.$$

The expected outage for the data base period is, then,

$$T_o = 5400 \times 0.0996 = 538 \text{ seconds}. \quad (21)$$

4.3 Outage calculation—selectivity only

It is apparent from eq. (20) that most of the outage for the system under study is caused by selectivity, fades characterized by A and B values in Region 1. From eqs. (14) and (17), we may express P_{os}, the probability of outage due to selectivity, as

$$P_{os} = \int_{-\pi}^{\pi} \int_{B_c}^{\infty} p_\phi(\phi) p_B(X) \, dX \, d\phi. \quad (22)$$

For the system studied for a BER of 10^{-3}, a finite B_c is obtained only for $|\phi_i| < 90$ degrees. Hence, we may use eq. (7) to simplify (22)*

$$P_{os} = \frac{2\Delta\phi}{216} \sum_{i=1}^{10} e^{-B_c(\phi_i)/3.8}. \quad (23)$$

From eq. (23) we see that the outage due only to selectivity depends on the relationship between B_c, the asymptote of critical B values, and the notch angle or notch frequency. Figure 13 shows the relationship between B_c and the notch frequency for four values of BER. It is apparent from eq. (22) that the outage probability is the probability of finding B and f_o values in the region above this curve. Such curves, therefore, provide a useful basis for evaluating the selectivity outage of a digital radio system.

4.4 Outage calculation—approximate method

For a radio system sensitive to both thermal noise and selectivity, the calculation of Section 4.3 is inadequate and that of Section 4.2 is unduly cumbersome.

To illustrate a simpler, but generally applicable, method and at the same time to provide a useful incidental result, let us evaluate the effect of reducing the transmitted power by 10 dB. For the reduced power system, the carrier-to-noise ratio would be 52 dB for the unfaded channel, and the critical curves of A and B would be shifted by 10 dB. Figure 14 shows the critical curve of A and B for a 10^{-3} BER and $f_o = -19.8$ MHz with an overplot of the conditional mean of the distribution of A. The dotted curves on Fig. 14 represent 2-sigma intervals on either side of the mean. From the properties of the Gaussian distribution, one may determine that more than 95 percent of the values of A and B will lie between these two dotted curves. We designate as A_m and B_m the coordinates of the intersection of two curves: the critical $A - B$ curve and the conditional mean curve. Then approximating the critical curve of A and B with a straight line segment tangent at (A_m, B_m), with slope s, we may approximate the probability of outage by integrating the probabilities over the region to the right of the tangent line. Using eqs. (15) and (16), we obtain

$$P_o = \Delta\phi \sum_{\text{All } \phi_i} p_\phi(\phi_i) \int_0^{\infty} \int_{A_m+s(B-B_m)}^{\infty} p_A(Y) p_B(X) dY \, dX. \quad (24)$$

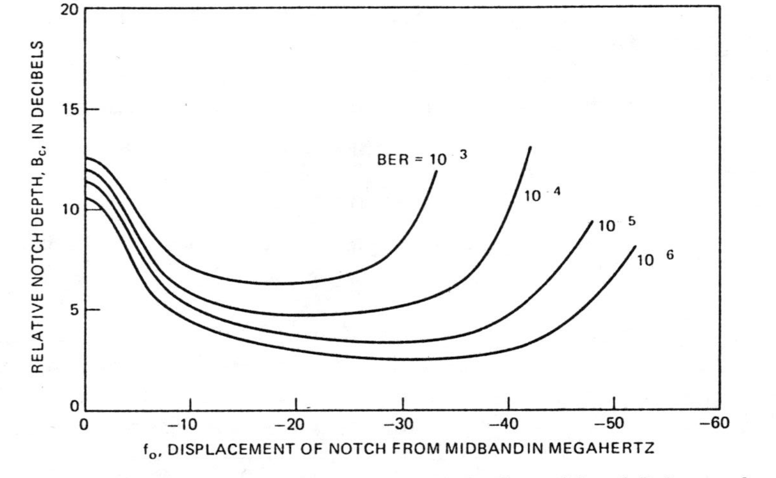

Fig. 13—Asymptotic performance curves. Locus of values of f_o and B that produce a fixed BER at high carrier-to-noise ratio.

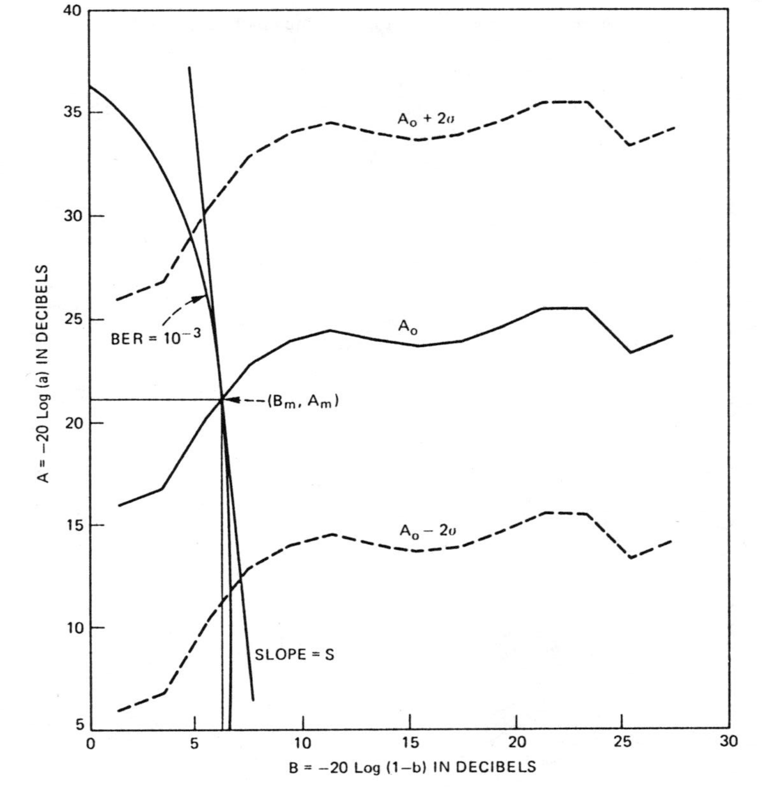

Fig. 14—Approximate outage calculation for 10 dB less transmitted signal.

* The factor of two is required in eq. (23) because the indicated summation corresponds to an integration only over negative notch frequencies ($\phi_i < 0$).

Interchanging the order of integration and ignoring* the B dependence of $A_0(B)$, this becomes

$$P_o = \Delta\phi \sum_{\text{All } \phi_i} p_\phi(\phi_i)e^{-B_m/3.8}e^{0.866/s^2}. \tag{25}$$

Evaluating eq. (25) for a 10^{-3} BER and multiplying the result by 5400 gives an outage estimate for the data base period of 602 seconds. Recalculating the total outage time at a 10^{-3} BER using the method of Section 4.2 [eqs. (16) and (18)] gives 636 seconds, which verifies the accuracy of the approximate method. The estimate of 636 seconds was calculated as an upper bound; the 602 seconds calculated using (25) tend to be a lower bound. We conclude that backing off transmitted power by 10 dB would increase the outage by about 12 percent (538 to 602).

4.5 A further simplification

In this section, we show that the outage due to selectivity can be estimated approximately for a given BER from a determination of the in-band selectivity required (with the notch out-of-band) to produce that BER. Such a measurement may provide a useful approximation for any digital system using quadrature modulation components;[9] however, we provide a justification based on the performance of the system at hand. In-band selectivity is defined as the difference between the maximum and minimum attenuation present in the (25.3-MHz) channel bandwidth.

Since the in-band selectivity is a constant for any of the curves shown in Fig. 9, one can use Fig. 9 to plot the asymptotic BER against in-band selectivity for $f_o = -19.8$ MHz. Such a plot was generated for each notch position measured to produce the family of curves shown in Fig. 15. Note that, except for notch positions near the band center, the BER is uniquely related to the in-band selectivity. Neglecting the in-band notches, we find that a 10^{-3} BER corresponds to an in-band selectivity of 5.5 dB.

If we use the original model of eq. (2) to determine the values of B that will produce an in-band selectivity of 5.5 dB for a number of different notch positions, we would generate Fig. 16. It is apparent that for this system there is a good correspondence between the curves of asymptotic performance (Fig. 13) and the curves of constant in-band selectivity (Fig. 16).

To reinforce this conclusion, we provide Figs. 17, 18, and 19. Figure 17 shows the locus of in-band selectivity in a 25.3-MHz band corre-

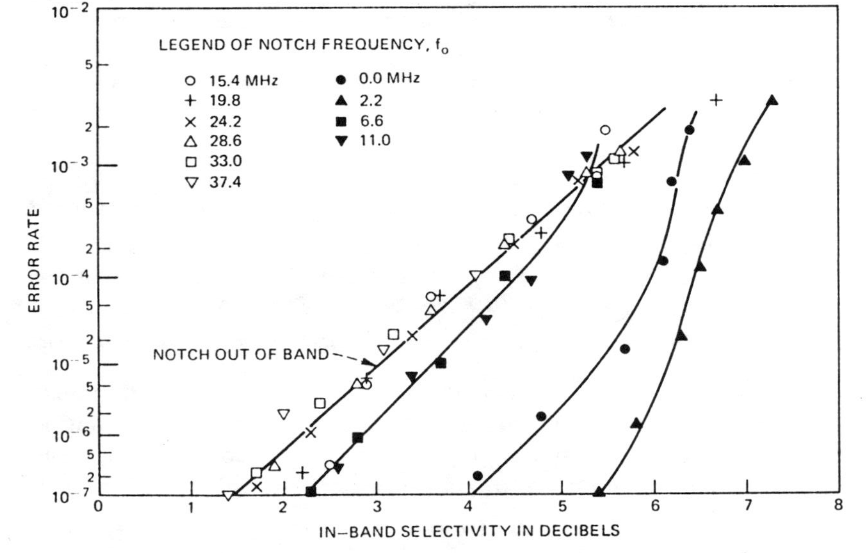

Fig. 15—Measured asymptotic bit error rate vs peak-to-peak amplitude difference in a 25.3-MHz band.

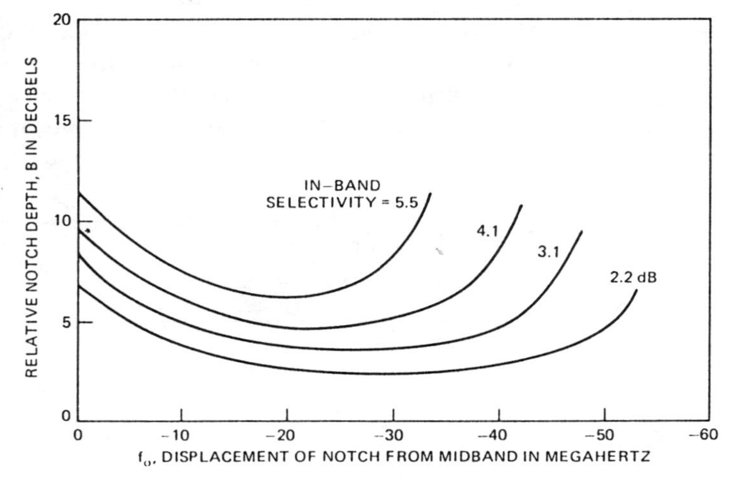

Fig. 16—Locus of B and f_o for modeled fades that have fixed peak-to-peak amplitude in a 25.3-MHz band.

sponding to each of the curves of constant BER in Fig. 13. That is, for each BER and each value of notch position, f_o, we have plotted the peak-to-peak amplitude difference in the band for the corresponding value of B_c, the asymptotic critical value of B. Figure 18 shows a similar set of curves with the peak-to-peak delay distortion in a 25.3-

* Including the effect of the slope of $A_0(B)$ at $B = B_m$ gives the same symbolic result with s interpreted as the algebraic sum of the slope of the tangent and dA_0/dB evaluated at $B = B_m$.

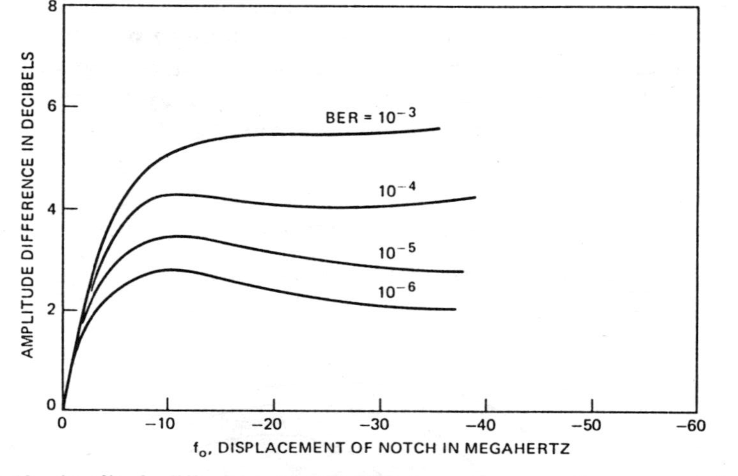

Fig. 17—In-band selectivity (in 25.3-MHz bandwidth) corresponding to asymptotic critical values of notch depth (B_c) for several values of BER.

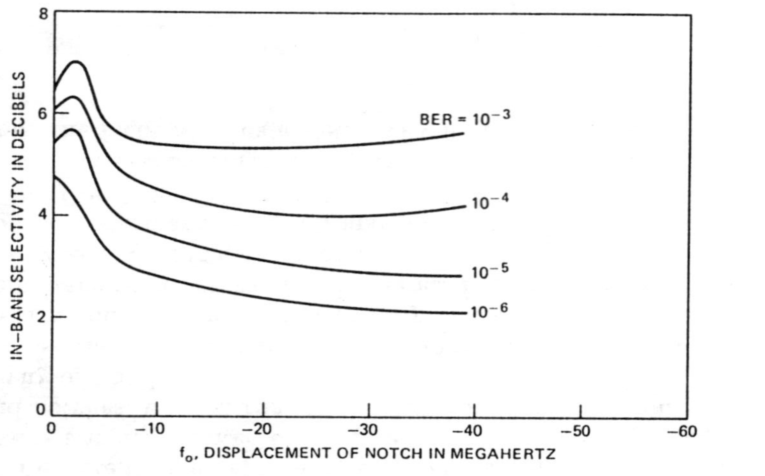

Fig. 18—Peak-to-peak envelope delay distortion in 25.3-MHz bandwidth corresponding to asymptotic critical values of notch depth (B_c) for several values of BER.

MHz band as the ordinate. Similarly, Fig. 19 has as the ordinate the "slope," or amplitude difference at a separation of 25.3 MHz. It is again clear from these three figures that the in-band selectivity is the relevant channel impairment giving rise to errors. We see from Fig. 18 that, for out-of-band notches, high BERs are obtained with very small values of peak-to-peak delay distortion, and from Fig. 19 that for in-band notches high BERs are obtained for very small values (zero at mid-band) of slope.

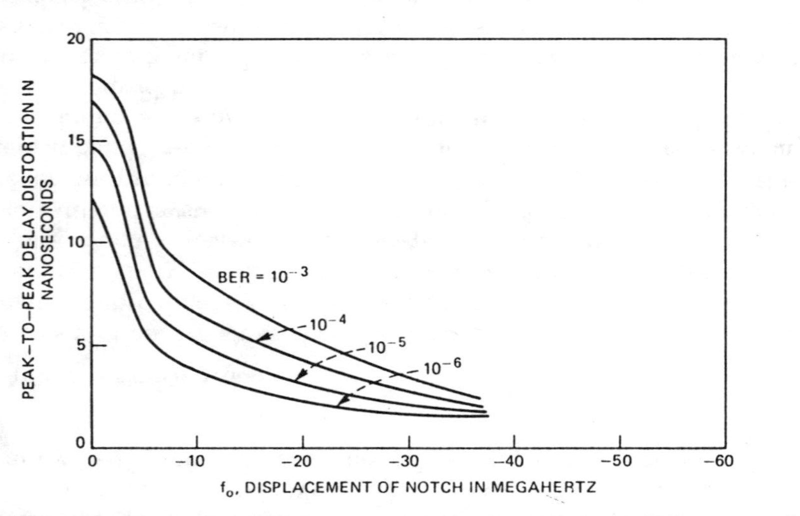

Fig. 19—Amplitude difference at a 25.3-MHz frequency separation corresponding to asymptotic critical values of notch depth (B_c) for several values of BER.

The model data base was analyzed to determine the time during which the in-band selectivity in a band of 25.3 MHz exceeded a given value. Figure 20 presents this distribution for in-band selectivity as calculated from the modeled fits. Figure 20 can be used directly in conjunction with Fig. 15 to calculate the outage times for the model data base.* For instance, from Fig. 15 we note that 5.5 dB of selectivity corresponds to a 10^{-3} BER. We use Fig. 20 to determine that 5.5 dB was exceeded for 520 seconds.

V. COMPARISONS OF CALCULATED AND OBSERVED OUTAGES

Using the methods of Sections 4.2 to 4.5, outage times were calculated for bit error rates of 10^{-3} to 10^{-6} for both the model data base period and for a heavy fading month, by multiplying the outage probabilities by 5400 and 8100, respectively.

5.1 Model data base period

Calculated and observed† outages for the model data base period are shown in Table I. In general, comparing the calculated results with observed results, we see that the outage is underestimated at high BERs and overestimated at low BERs. Any estimation procedure based on the current modeled state of the channel will tend to underestimate

* In practice, one would use a single measurement of in-band selectivity. For instance, in Fig. 9 one would take the 5.7-dB value corresponding to the curve asymptotic at a 10^{-3} BER.

† Because of quantization, the outage times observed from the field experiment correspond to bit error rates of 1.26×10^{-3}, 1.57×10^{-4}, 0.981×10^{-5}, and 1.19×10^{-6}.

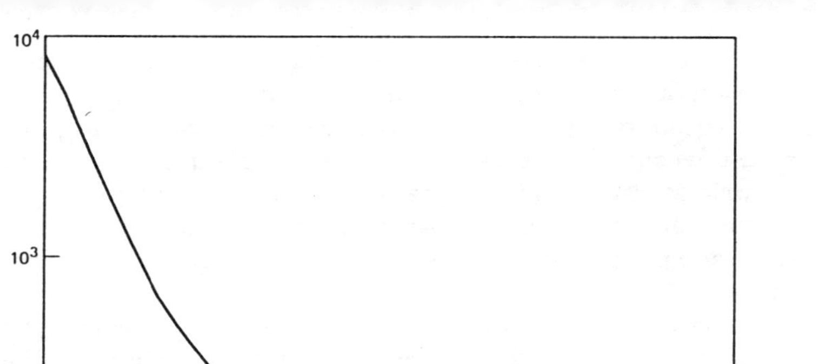

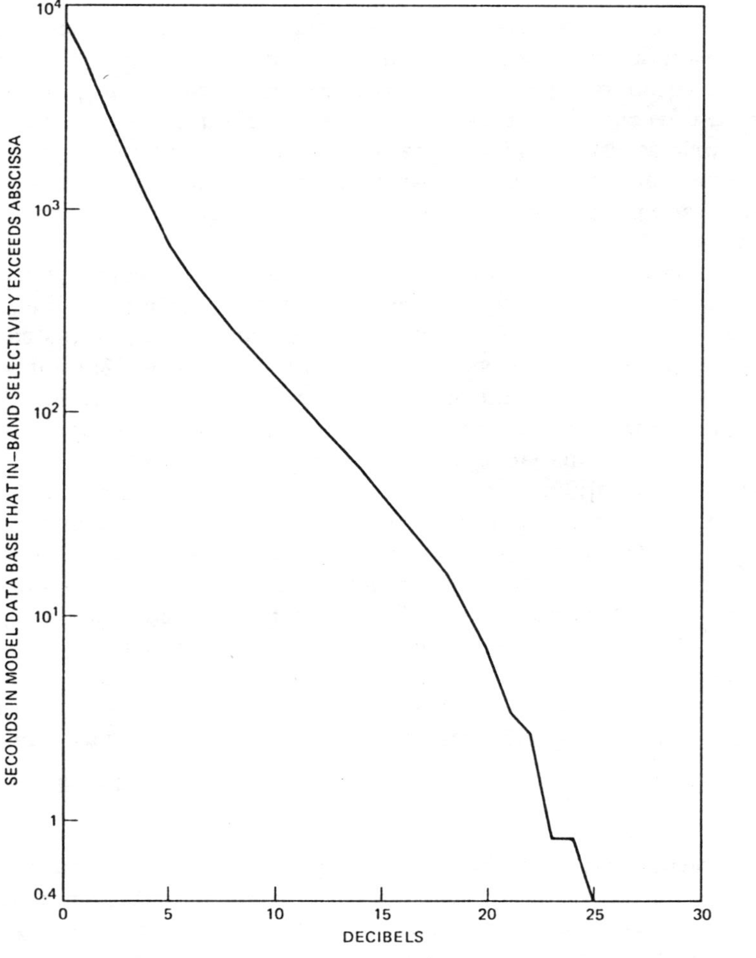

Fig. 20—Distribution of in-band selectivity (25.3-MHz bandwidth) for model data base.

Table I—Outage in modeling data base period (seconds)

BER =	10^{-3}	10^{-4}	10^{-5}	10^{-6}
Observed	636	903	1191	1487
Detailed calculation (Section 4.2)	538	960	1430	1860
Approximate calculation (Section 4.4)	527	950	1420	1830
Asymptotic calculation (Section 4.3)	527	950	1420	1830
Selectivity calculation (Section 4.5)	510	900	1570	2730

outage at high BERs because of hysteresis effects in the radio receiving equipment. That is, when the channel condition becomes sufficiently severe, the bit error rate becomes high enough (on the order of 10^{-3}) that the timing and/or phase of the radio system breaks lock. If the channel impairment becomes less severe, the BER will not improve until the system resynchronizes. The hysteresis is important at the 10^{-3} BER, since a significant fraction of the events that cause 10^{-3} BER will cause the system to break lock.

One would expect to overestimate the outage at low BERs because of the method of taking data. Recall that, in measuring the curves in Fig. 9, it was found that the BER depended on the phase to which the system had locked. The recorded performance represented the worst-phase condition. At a 10^{-6} BER, the best phase produces a BER that is about ⅓ that produced by the worst phase; the difference in BER from worst to best phase at a 10^{-3} BER is negligible. Hence, one would expect outage to be overestimated significantly at low bit error rates.

In comparing the outage calculated from in-band selectivity (Section 4.5) to the outage observed, we find that the overestimation of outage at low BERs is more severe than with the other methods. This is due to the greater sensitivity of the differential selectivity method to the bias induced by choosing the worst-case phase. For instance, comparing calculations at a 10^{-6} BER, we find that Fig. 20 is steeper for amplitude differences near 2 dB than is Fig. 3 near B values of 3.5 dB. (Figure 9 verifies the appropriateness of this comparison). More generally, one expects the method based on in-band selectivity to overestimate the outage because the method is based on notches out of band. From Fig. 15, it is apparent that, for a given ΔA, some scans will not have the BER specified.

We conclude that, although calculation of outage from sensitivity to in-band selectivity provides quick estimates, they are less accurate. The calculation requires knowledge of the distribution of in-band selectivity over a specified bandwidth. These statistics are neither simple nor generally available. It has been shown,[10] for instance, that slope statistics have a nontrivial dependence upon the measurement bandwidth.

It is clear that the calculations based on selectivity (Sections 4.3 and 4.5) agree for the system studied here because that system has very little outage due to thermal noise limitations, and because it is sensitive primarily to in-band amplitude excursions. The extent to which these statements are true for other systems is currently unknown.

5.2 Outage on a monthly basis

The results in Table I may be put on the basis of a heavy fading month by increasing them by a factor of 1.5, as discussed in Section 2.2. The resulting outages (including the scaled observed outage) are compared with the outage observed in a one-month period[8] in Table II. We see that the outage times observed in the total one-month period agree well with the values obtained by scaling the observed

Table II—Outage in a heavy fading month (seconds)

BER =	10^{-3}	10^{-4}	10^{-5}	10^{-6}
Observed (Ref. 8)	1000	1320	2100	2900
Scaled observation from Table I	955	1350	1790	2230
Calculation (Section 4.2–4.4)	800	1430	2140	2760
Selectivity calculation (Section 4.5)	770	1350	2350	4100

outage for the data base period used in modeling, except for the slight divergence appearing at low BERs. This divergence should not be unexpected for this equipment. As may be seen in Fig. 15, a 10^{-6} BER may be caused by differential amplitude selectivity in band of 2 dB. Such modest amounts of selectivity may be expected to occur sometimes in the presence of very moderate selective fading. The modeling data base was constructed by selecting only periods of significant selective fading. This reinforces the comments made in conjunction with Fig. 3, namely, that the model distribution of B represents a lower bound for small values of B which can contribute to outage at the 10^{-6} BER level.

VI. CONCLUSIONS

We have demonstrated the validity of a technique for estimating the unprotected outage of a digital radio system due to selective fading on a particular hop in the 6-GHz common carrier band. The technique required field measurements to statistically characterize the parameters of a model of propagation on the hop. It also requires performance data obtained in the laboratory for the radio system by stressing it with a two-path fade simulator with a differential path delay of 6.3 ns, corresponding to the fixed delay channel model. Since the radio path on which these measurements were made has a length close to the average for the Bell System long haul radio network and has an average incidence of fading activity, the channel model is representative of a typical path. At the very least, the technique provides a basis for determining the relative merits of various digital radio systems operating without benefit of space diversity.

For the system under test, outage was calculated by four different methods. Because this system was selectivity-limited rather than noise-limited, all four methods predicted approximately the same outage as that summarized in Table I; however, the method based on in-band selectivity is more severely biased at low BERs. The method based on asymptotic performance and that based on in-band selectivity can only be used to estimate outage due to selectivity. If the transmitted power of the system under test were reduced by 10 dB, both of the other two methods, the detailed and the approximate method, would predict an increase in outage time of about 12 percent.

VII. ACKNOWLEDGMENTS

The conclusions of this effort depend upon data collected on a 6-GHz digital radio hop installed by R. A. Hohmann and L. J. Morris, using instrumentation designed by G. A. Zimmerman. M. V. Pursley's assistance in processing the data was invaluable. Consistent laboratory data used to close the loop reflect contributions by T. J. West and G. B. Thomas to the methodology of selective fade simulation at IF, and A. E. Resch who performed the measurements.

REFERENCES

1. E. Takeuchi and P. Tobey, "A 6 GHz Radio for Telephony Applications," Conference Record ICC 1976, Vol. II, June 1976, p. 18–27.
2. P. R. Hartman and J. A. Crossett, "A 90 MBS Digital Transmission System at 11 GHz using 8 PSK Modulation," Conference Record ICC 1976, Vol. II, June 1976, p. 18–8.
3. A. J. Giger and T. L. Osborne, "3A-RDS 11 GHz Digital Radio System," Conference Record ICC 1976, Vol. II, June 1976, p. 18–1.
4. I. Godier, "DRS-8 Digital Radio for Long-Haul Transmission," Convention Record ICC 1977, Vol. 1, June 1977, p. 102.
5. W. A. H. Wood, "Modulation and Filtering Techniques for 3 Bits/Hertz Operation in the 6 GHz Frequency Band," Convention Record ICC 1977, Vol I, June 1977, p. 97.
6. W. D. Rummler, "A New Selective Fading Model: Application to Propagation Data," B.S.T.J, this issue, pp. 1037–1071.
7. W. T. Barnett, "Multipath Propagation at 4, 6, and 11 GHz," B.S.T.J., *51*, No. 2 (February 1972), pp. 321–361.
8. W. T. Barnett, "Measured Performance of a High Capacity 6 GHz Digital Radio System," Conference Record ICC 1978, Vol. III, June 1978, pp 47.4.1–47.4.6.
9. J. R. Gray and T. J. West, private communication.
10. G. M. Babler, "A Study of Frequency Selective Fading for a Microwave Line-of-Sight Narrowband Radio Channel," B.S.T.J., *51*, No. 3 (March 1972), pp. 731–757.

MULTIPATH FADING OUTAGE ESTIMATES INCORPORATING PATH AND EQUIPMENT CHARACTERISTICS

M. H. Meyers

AT&T Bell Laboratories
North Andover, Massachusetts 01845

ABSTRACT

A method of predicting digital radio outage due to multipath fading which incorporates both path and equipment characteristics is presented. The basic channel model is a 2-ray path with the delay between the rays characterized by an exponential distribution whose mean is hop dependent. Recent work by Campbell and Coutts has demonstrated the ability of the channel model to predict outage of digital radio systems. Our result extends the limits of application of their formula and can be considered a generalization of their expression for outage. By using a Gauss-quadrature formula to evaluate the equation for outage, a simple hop-dependent expression is found for the outage time due to multipath fading. Our formula gives outage time directly and relates it intrinsically to laboratory measurement of the radio system's fading signatures taken near the characteristic mean delay of the channel. Comparison of predicted outages with actual field-measured performance attests to the utility of the method. Of particular importance is the ability to differentiate between and assess the impact of channel-dependent parameters upon system performance.

INTRODUCTION

In this paper, we present an analysis technique for predicting digital radio outage due to multipath fading that incorporates both path variations and equipment characteristics. In the next section, we present our channel model for dispersive multipath fading. Mathematically, it is a 2-ray model characterized by an exponential distribution for the delay difference between the two rays. The mean of the exponential delay distribution is directly related to the severity of the dispersive fading with larger mean delays corresponding to narrower, and less easily equalized, inband notches (for a fixed fade depth). The multipath model is a modification of the one proposed by Emshwiller (Ref. 1). The model can also be obtained by slightly changing the multipath channel model proposed by Jakes (Ref. 2). In the form we use here, it was the basis for the important work done by Campbell and Coutts (Ref. 3) who provided an expression for calculating the outage due to dispersive fading.

In this model for multipath fading, a radio channel is completely characterized by the mean channel delay and the amount of fading time. The former is important because measurements have shown that all paths do not fade similarly in terms of dispersion. The latter is significant because the same hop can experience large seasonal and year-to-year variations in the amount of fading time. A method for obtaining these key multipath parameters of the fading process is contained in Ref. 4.

Once the channel has been characterized, a method of analyzing the overall digital radio system is presented in this paper. By using a Gauss-quadrature formula to evaluate the equation for outage, a simple hop-dependent expression is found for the outage time due to multipath fading. Our formula gives outage time directly and relates it intrinsically to laboratory measurement of the radio system's fading signatures taken near the characteristic mean delay of the channel. Our result generalizes the expression found by Campbell and Coutts (Ref. 3).

In this paper, we combine the channel characterization with our outage expression and compare outage predictions with measured performance observed during past experiments. Good agreement is obtained between predicted and measured outages.

MULTIPATH FADING MODEL

The model we will use is mathematically given by the transfer function (Refs. 3, 4)

$$1 - b\,e^{-j2\pi(f-f_o)\tau} \tag{1}$$

where τ is the delay between the primary and secondary rays, f_o is the frequency location of the notch minimum, and b is the magnitude of the interfering ray with

$$B = -20\log(1-b), \tag{2}$$

the depth of the notch minimum in dB. Statistically, the multipath fading model is described by

1) an exponential distribution with mean τ_o for the delay

$$p_\tau(\tau) = \frac{1}{\tau_o} e^{-\tau/\tau_o}, \tag{3}$$

2) a uniform distribution over the frequency range $(-1/2\,\tau,\ 1/2\,\tau)$ for the notch location

$$p_{f_o}(f_o) = \begin{cases} \tau & |f_o| < \dfrac{1}{2\tau} \\ 0 & \text{elsewhere}, \end{cases} \tag{4}$$

3) and a uniform distribution for the interfering ray magnitude over the range (0, 1)

$$p_b(b) = \begin{cases} 1 & 0 \le b \le 1 \\ 0 & \text{elsewhere}, \end{cases} \tag{5}$$

so that the fade depth distribution (in dB) is given by

$$p_B(B) = \left(\frac{\ln 10}{20}\right) 10^{-B/20} \quad B \ge 0. \tag{6}$$

Reprinted from *IEEE 3rd Global Telecomm. Conf.*, vol. 3, pp. 1466–1470, November 1984.

CALCULATING OUTAGE FOR MULTIPATH FADING

The time (in seconds) spent in outage, where the bit error rate (BER) is above 10^{-3}, can be written as (Ref. 3)

$$\text{outage} = T_f \cdot P_o \tag{7}$$

where T_f is the amount of time (in seconds) with multipath fading and P_o is the conditional probability of outage given multipath fading occurred. We next concern ourselves with finding an easily evaluated expression for P_o, the conditional probability of outage given multipath fading.

In its most general form, the probability of outage is given by

$$P_o = \iiint p_{B,f,\tau}(B, f, \tau)\, dB\, df\, d\tau \tag{8}$$

where the triple integration is over the space of (τ, f_o, b) parameters such that outage occurs. We can rewrite this expression as

$$P_o = \int \left\{ \iint p_{B,f|\tau}(B, f|\tau = \tau^*)\, dB\, df \right\} p_\tau(\tau^*) d\tau^*. \tag{9}$$

The bracketed term is equal to the probability of outage given that the delay random variable equals a particular value τ^*.

The next step in evaluating the expression for probability of outage is to use a Gauss-quadrature formula to express the integration over the exponential delay random variable as the weighted sum of conditional probabilities

$$P_o = \sum_{i=1}^{N} \lambda_i \left\{ \int_{f=-\frac{1}{2\tau}}^{\frac{1}{2\tau}} \int_{B=B_{\tau_i}(f)}^{\infty} p_{B,f|\tau}(B, f|\tau = \tau_i)\, dB\, df \right\} \tag{10}$$

where the $\{\lambda_i\}$ are weights, the $\{\tau_i\}$ are the nodes of the Gauss-quadrature for mean delay τ_o, and $B_{\tau_i}(f_o)$ is the fade depth at frequency f_o which causes a 10^{-3} BER. This fade depth is also a function of the system flat fade margin (S/N).

A general description of Gauss-quadrature integration can be found in Ref. 5, while the appendix of Ref. 6 gives specific implementation details for computing the weights and nodes. For our purposes here, we find that the nodes and weights corresponding to an exponential distribution are commonly tabulated in mathematical handbooks (Ref. 7, Page 923, also known as Gauss-Laguerre integration). Table 1 shows the nodes and weights for N = 1, 3, 5, 7, and 9. It can be shown that very few terms are needed for Eq. 10 to converge.

In Eq. 10, we recognize that the conditional probability

$$\int_{f=-\frac{1}{2\tau}}^{\frac{1}{2\tau}} \int_{B=B_{\tau_i}(f)}^{\infty} p_{B,f|\tau}(B, f \mid \tau = \tau_i)\, dB\, df \tag{11}$$

is the integral, over fade depth B and notch frequency f_o, of the equipment fade signature (Ref. 1) taken at a particular delay τ_i. Therefore, all that remains is to evaluate the weighted area below the equipment fade signature when τ equals τ_i. The integral over the equipment fade signature is approximated by the sum

$$\tau_i (\Delta f) \sum_j 10^{-B_{\tau_i}(f_j)/20} \tag{12}$$

where

$$\Delta f = |f_{i+1} - f_i| \tag{13}$$

is the spacing between evaluated points on the equipment fade signature. The factor of τ_i is due to the unformly distributed notch frequency location which can occur anywhere between $-1/2\tau_i$ and $1/2\tau_i$.

If we assume the fading is equally likely to be minimum phase or nonminimum phase, the expression for the conditional probability of outage becomes

$$P_o = \Delta f \sum_{i=1}^{N} \lambda_i \tau_i \left\{ \sum_j \frac{1}{2} 10^{-B_{\tau_{i,min}}(f_j)} + \frac{1}{2} 10^{-B_{\tau_{i,non}}(f_j)} \right\} \tag{14}$$

where

$$B_{\tau_{i,min}}(f)$$

is the minimum phase equipment fading signature, and

$$B_{\tau_{i,non}}(f)$$

is the nonminimum phase equipment fading signature.

ABSCISSAS AND WEIGHT FACTORS FOR LAGUERRE INTEGRATION

Abscissas – α_i (Zeros of Laguerre Polynomials)

Weight Factors – λ_i

n	α_i	λ_i
$n=2$	0.58578 64376 27	(-1)8.53553 390593
	3.41421 35623 73	(-1)1.46446 609407
$n=3$	0.41577 45567 83	(-1)7.11093 009929
	2.29428 03602 79	(-1)2.78517 733569
	6.28994 50829 37	(-2)1.03892 565016
$n=4$	0.32254 76896 19	(-1)6.03154 104342
	1.74576 11011 58	(-1)3.57418 692438
	4.53662 02969 21	(-2)3.88879 085150
	9.39507 09123 01	(-4)5.39294 705561
$n=5$	0.26356 03197 18	(-1)5.21755 610583
	1.41340 30591 07	(-1)3.98666 811083
	3.59642 57710 41	(-2)7.59424 496817
	7.08581 00058 59	(-3)3.61175 867992
	12.64080 08442 76	(-5)2.33699 723858
$n=6$	0.22284 66041 79	(-1)4.58964 673950
	1.18893 21016 73	(-1)4.17000 830772
	2.99273 63260 59	(-1)1.13373 382074
	5.77514 35691 05	(-2)1.03991 974531
	9.83746 74183 83	(-4)2.61017 202815
	15.98287 39806 02	(-7)8.98547 906430
$n=7$	0.19304 36765 60	(-1)4.09318 951701
	1.02666 48953 39	(-1)4.21831 277862
	2.56787 67449 51	(-1)1.47126 348658
	4.90035 30845 26	(-2)2.06335 144687
	8.18215 34445 63	(-3)1.07401 014328
	12.73418 02917 98	(-5)1.58654 643486
	19.39572 78622 63	(-8)3.17031 547900
$n=8$	0.17027 96323 05	(-1)3.69188 589342
	0.90370 17767 99	(-1)4.18786 780814
	2.25108 66298 66	(-1)1.75794 986637
	4.26670 01702 88	(-2)3.33434 922612
	7.04590 54023 93	(-3)2.79453 623523
	10.75851 60101 81	(-5)9.07650 877336
	15.74067 86412 78	(-7)8.48574 671627
	22.86313 17368 89	(-9)1.04800 117487
$n=9$	0.15232 22277 32	(- 1)3.36126 421798
	0.80722 00227 42	(- 1)4.11213 980424
	2.00513 51556 19	(- 1)1.99287 525371
	3.78347 39733 31	(- 2)4.74605 627657
	6.20495 67778 77	(- 3)5.59962 661079
	9.37298 52516 88	(- 4)3.05249 767093
	13.46623 69110 92	(- 6)6.59212 302608
	18.83359 77889 92	(- 8)4.11076 933035
	26.37407 18909 27	(-11)3.29087 403035

TABLE I GQR NODES (α_i) AND WEIGHTS (λ_i)

It is convenient to define an equipment-dependent variable *ro* or relative outage which results in

$$P_o = \frac{(\Delta f)}{2} \sum_{i=1}^{N} \lambda_i \tau_i \, ro\,(\tau_i) \tag{15}$$

where we have defined the relative outage, $ro(\tau)$, as

$$ro\,(\tau_i) = \sum_j \left\{ 10^{-B_{\tau_i, min}(f_j)/20} + 10^{-B_{\tau_i, non}(f_j)/20} \right\} \tag{16}$$

which is the sum of both minimum and nonminimum phase terms.

The *ro*, or relative outage, is a measure which can be used to compare the relative merits of various systems. Most important, the *ro* is directly related to the amount of outage time in seconds. Combining Eqs. 7 and 15, we find the outage in seconds for a hop is given by

$$\text{outage} = \frac{T_f \, \Delta f}{2} \sum_{i=1}^{N} \lambda_i \tau_i \, ro(\tau_i) \text{ (seconds)}. \tag{17}$$

Equation 17 gives the outage in seconds due to multipath fading. The expression is both hop dependent (τ_o, T_f) and equipment dependent ($ro\,(\tau_i)$). The structure of the outage expression directly shows how improvements in the equipment fade signatures reduce outage. Assume, for example, we have a digital radio system characterized by equipment signatures $B_{\tau_i}^{(1)}$ (f) (for simplicity, assume the minimum and nonminimum phase signatures are identical). Now, let there be another digital radio system characterized by equipment signatures $B_{\tau_i}^{(2)}$ (f) which are the same width as Case (1) but differ in height by a fixed ΔB dB, i.e.,

$$B_{\tau_i}^{(2)}(f) = B_{\tau_i}^{(1)}(f) + \Delta B. \tag{18}$$

The change in outage can be computed by combining Eqs. 16-18 so that

$$\text{outage (Case 2)} = \frac{T_f\,(\Delta f)}{2} \sum_{i=1}^{N} \lambda_i \tau_i \, ro^{(2)}(\tau_i)$$

$$= \frac{T_f\,(\Delta f)}{2} \sum_{i=1}^{N} \lambda_i \tau_i \left\{ \sum_j 10^{-B_{\tau_i}^{(2)}(f_j)} \right\}$$

$$= \frac{T_f\,(\Delta f)}{2} \sum_{i=1}^{N} \lambda_i \tau_i \left\{ \sum_j 10^{-\Delta B/20}\, 10^{-B_{\tau_i}^{(1)}(f_j)/20} \right\}$$

$$= 10^{-\Delta B/20} \cdot \text{outage (Case 1)}. \tag{19}$$

This illuminates the fundamental relationship between the equipment signature and outage performance. Equation 19 shows that, assuming the widths of the equipment signatures are equal, an improvement of ΔB dB in the equipment signatures produces a $10^{-\Delta B/20}$ reduction in outage time. Expressing the ratio of outage times in dB, each dB of improvement in the signatures produces a 0.5-dB reduction in outage.

We note that one must compare the signatures around the τ_o characteristic of a hop and that the improvement may change from one delay to another in accordance with how the equipment signatures vary. For example, equipment signatures taken at 6.3 ns may not be accurate for a 0.5-ns hop.

We next discuss the sensitivity of the outage time result and the number of terms needed to evaluate the sum of Eq. 17. Since each $ro(\tau)$ requires the evaluation of a number of equipment fading signatures at specific values of τ, the fewer terms necessary for convergence of the result, the less effort will be required to characterize digital radio systems. Referring to Table 1 which contains a list of the nodes and weights for the Gauss-quadrature of an exponential function, we next substitute these values into Eq. 17. With the definition

$$\tau_i = \alpha_i \tau_o, \tag{20}$$

we can write the outage equation as

$$\text{outage} = \frac{T_f(\Delta f)\,\tau_o}{2} \sum_{i=1}^{N} (\lambda_i \alpha_i)\, ro\,(\tau_i) \tag{21}$$

$$= T_f\,(\Delta f/2)\,\tau_o\, ro_{eff}\,(\tau_o)$$

where we have defined the *effective relative outage*

$$ro_{eff}\,(\tau_o) = \sum_{i=1}^{N} (\lambda_i \alpha_i)\, ro\,(\tau_i)\ . \tag{22}$$

Experience has shown that Eq. 22 converges quickly. For a 5-point quadrature

$$\text{outage} = \frac{T_f(\Delta f)\,\tau_o}{2} \left\{ \begin{aligned} &(0.521)\, ro\,(0.131\,\tau_o) \\ &+ (0.398)\, ro\,(0.706\,\tau_o) \\ &+ (0.076)\, ro\,(1.79\,\tau_o) \\ &+ (0.0036)\, ro\,(3.54\,\tau_o) \\ &+ (0.000023)\, ro\,(6.32\,\tau_o) \end{aligned} \right\} \tag{23}$$

$$= \frac{T_f\,(\Delta f)\,\tau_o}{2} \left\{ \begin{aligned} &(0.521)\, ro\,(0.131\,\tau_o) \\ &+ (0.398)\, ro\,(0.706\,\tau_o) \\ &+ (0.076)\, ro\,(1.79\,\tau_o) \\ &+ (0.0036)\, ro\,(3.54\,\tau_o) \end{aligned} \right.$$

where the last expression was obtained by dropping the fifth term in the sum which can be shown to have no measurable effect upon the final result.

It has been found that N = 5 is adequate for evaluating Eq. 21. The outage is dominated by the middle two of the four remaining terms in Eq. 23. Typically, the first and last terms contribute less than 10 percent of the total outage calculation. Equation 23 is the expression we will use in the next sections to evaluate the performance of digital radio systems. In the rest of this paper, we will evaluate the relative outages at the equipment fade signature points corresponding to Δf = 4 MHz. This value will be the basis for calculating $ro\,(\tau_o)$ and $ro_{eff}\,(\tau_o)$.

It is interesting to consider the simplest case for evaluating Eq. 21 where N = 1. The outage expression reduces to the particularly simple, but less accurate,

$$\text{outage} = \frac{T_f\,(\Delta f)\,\tau_o}{2}\, ro\,(\tau_o). \tag{24}$$

We have found Eq. 24 to be as much as 60-percent optimistic in estimating outage. Equation 24 is similar to Eq. 21 except that the effective relative outage $ro_{eff}\,(\tau_o)$ has been approximated by the relative outage evaluated at the mean delay, i.e., $ro\,(\tau_o)$. Of particular interest here is the form of the outage relation with respect to the channel mean delay τ_o.

Emshwiller's (Ref. 1) outage prediction, with his slightly different model, predicted that outage increases as the square

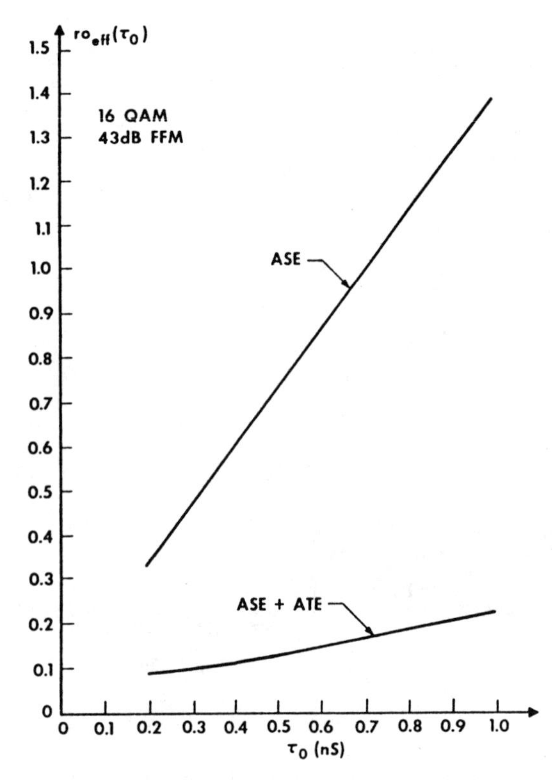

FIGURE 1. ro_{eff} (τ_0) VS. τ_0

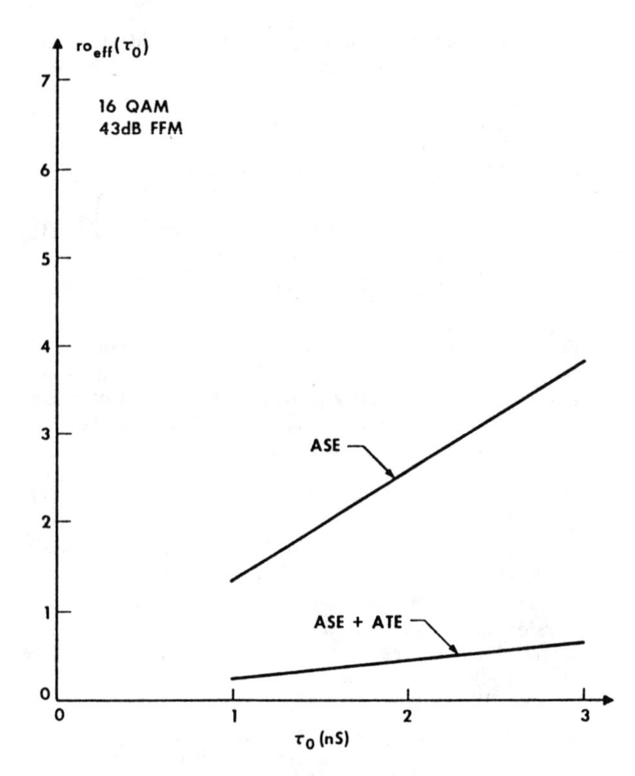

FIGURE 2. ro_{eff} (τ_0) VS. τ_0

of the channel mean delay τ_o. More recently, Campbell and Coutts also obtained an expression (Ref. 3) for outage which varied as τ_o^2. The τ_o^2 proportionately will hold if and only if ro_{eff} (τ) is proportional to τ in the region of interest, primarily from 0.7 τ_o to 1.8 τ_o. Common to the analyses of both Emshwiller and Campbell and Coutts, was the assumption that the equipment fading signatures could be scaled from one delay to another simply as

$$\Delta B = 20 \log \left(\frac{\tau_1}{\tau_2}\right). \tag{25}$$

With this assumption, the exponential distribution for delay τ in Eq. 9 could be integrated in closed form with no need for the Gauss-quadrature technique used here. In both cases, the τ_o^2 proportionality of outage was a consequence of Eq. 25. Interestingly, the simple formula is accurate for the specific digital radio systems evaluated by Emshwiller and Coutts, et al. It is the advent of more sophisticated equalization and higher level modulations which require a more general analysis. Our result for outage is valid even when equipment fade signatures cannot be scaled according to Eq. 25.

Figures 1 and 2 are plots of the calculated effective relative outage for two 22-Mbaud, 16-QAM systems, one with an adaptive slope equalizer (ASE), the other equipped with an ASE plus an adaptive transversal equalizer (ATE). The variation of relative outage with channel mean delay is clearly seen to be a function of the equalization. For both systems, as the delay becomes smaller, the effective relative outage converges to the value corresponding to that associated with the flat fade margin of the system. Seen as a function of the mean channel delay, the relative performance of the two systems is seen to vary. Therefore, the relative merits of two systems must be seen in the context of the specific channel (i.e., τ_o) under consideration.

We note that the flat and dispersive components of the fading are simultaneously evaluated in the term "effective relative outage." Accurate evaluation of the outage due to simultaneous flat and dispersive fading is integral in this method of analysis which treats both forms of degradation together.

COMPARISON OF OUTAGE PREDICTION WITH MEASURED RADIO PERFORMANCE

An important test of the theory developed in this paper is how well it agrees with field experience on hops that have been monitored in the past. The expression for outage, with ΔF = 4 MHz and τ_o in ns, is given by

$$\text{outage} = \frac{T_f \, \tau_o \, ro_{eff}(\tau_o)}{500}. \tag{26}$$

In order to predict the outage of a digital radio system, we must know how susceptible a particular system is to multipath fading as a function of the channel model mean delay. This information is contained in the term $ro_{eff}(\tau_o)$ which is computed from the equipment fade signatures. Ideally, the required equipment signatures should be measured in the laboratory with the appropriate channel delays. The individual signatures are then integrated and added with the weights given in Eq. 22. An unfaded S/N of 60 dB (43-dB flat fade margin) characterizes the thermal sensitivity of the system.

We next compare predicted and observed outage performance for a 6-GHz digital radio system (Ref. 8) monitored in Palmetto in 1982. Employing the channel identification algorithm of Ref. 4 to extract the mean delay and fading time, we find that for the 1982 measurement period, the channel is characterized by τ_o = 0.5 ns and a total fading time of 1.33 × 10^6. Combining these parameters with the equipment signatures of an ASE equipped system, we predict 852 seconds of outage compared to the observed 680 seconds. The difference is approximately 25 percent or 0.97 dB.

Besides the measurement period discussed above, we have evaluated other experiments to validate our outage prediction technique. Of a total of five experiments, only one measurement period differed from the prediction by more than 47 percent and the average error was 20 percent.

A variable which is beyond our control and remains unknown for the fading periods measured is the ratio of minimum to nonminimum phase fades. We have assumed a 50/50 split probabilistically. Deviations from this ratio could account for 1- to 2-dB changes in the outage predictions.

ACKNOWLEDGEMENTS

Many people have contributed in numerous ways which have served to clarify and illuminate the theoretical and practical aspects of the fade model and outage analysis. For their insights, interest, and support, C. P. Bates, W. J. Schwarz, J. J. Kenny, G. Turner, R. R. Grady, and R. L. Lahlum are gratefully acknowledged.

REFERENCES

1. M. Emshwiller, "Characterization of the Performance of PSK Digital Radio Transmission in the Presence of Multipath Fading," ICC'78.

2. W. C. Jakes, Jr., "An Approximate Method to Estimate an Upper Bound on the Effect of Multipath Delay Distortion on Digital Transmission," IEEE Trans. on Communications, Vol. COM-27, No. 1, January 1979, pp. 76-81.

3. J. C. Campbell and R. P. Coutts, "Outage Prediction of Digital Radio Systems," Electronics Letters, December 1982, Vol. 18, No. 25/26.

4. M. H. Meyers, "Multipath Fading Characteristics of Broadband Radio Channels," Globecom '84, this volume.

5. W. Gautchii, "On the Construction of Gaussian Qaudrature Rules from Modified Moments," Math. Comp., No. 24 (April 1970), pp. 245-260.

6. M. H. Meyers, "Computing the Distribution of a Random Variable via Gaussian Quadrature Rules," B.S.T.J., Vol. 61, No. 9, November 1982, pp. 2245-2261.

7. M. Abramowitz and I. A. Stegun, "Handbook of Mathematical Functions," NBS Applied Math Series, No. 55, December 1972.

8. A. Ranade and P. E. Greenfield, "An Improved Method of Digital Radio Characterization from Field Measurements," ICC'83.

Digital Communications Over Fading Radio Channels

By G. J. FOSCHINI and J. SALZ

(Manuscript received August 19, 1982)

A major contribution to system outage in a terrestrial digital radio channel is deep fading of the frequency transfer characteristic, which in addition to causing a precipitous drop in received signal-to-noise ratio (s/n) also causes signal dispersion that can result in severe intersymbol interference. Because the temporal variation of the channel is slow compared to the signaling rate, the information theoretic channel capacity and the "Efficiency Index" in bits/cycle—a figure-of-merit we use for the communication techniques considered—can be viewed as random processes. Starting from an established mathematical model characterizing fading channels (derived from extensive measurements), we estimate the probability distribution of channel capacity and the distributions of efficiency indices for different communications techniques. The repertoire of communication methods considered involves quadrature amplitude modulation with adaptive linear and decision feedback equalization, and maximum likelihood sequence estimation. For specific outage objectives the maximum number of bits per cycle achievable by each technique is estimated. The sensitivity of the distributions to bit-error-rate objective and unfaded s/n is assessed. For certain desired operating points the efficacy of adaptive equalization is demonstrated. There are some operating points where adaptive equalization alone is not adequate and therefore space diversity should be considered. An estimate of the effect of frequency diversity is also included.

I. INTRODUCTION AND SUMMARY

Fading of terrestrial digital radio channels owing to multipath reception is a prime cause of system outage. For a specific hop a mathematical model of these fades has been developed by W. D. Rummler[1,2] from extensive measurements of the channel frequency power transfer characteristic over time. The radio channel has a time-varying frequency characteristic, with additive Gaussian noise; however, the temporal variations are sufficiently slow in comparison to the data symbol rate that the characteristics can be represented as a random ensemble of static frequency power transfer functions. The presence of additive noise implies that each member of the ensemble is limited to a maximum rate of transmission of data, depending on the communication method. For each specific communication technique, the stochastic nature of the channel makes it meaningful to consider the probability distribution of data rates that can be supported at a certain bit-error-rate objective.

The purpose of this paper is to explore the relative performance of various communication techniques employing quadrature amplitude modulation (QAM), distinguished by the type of equalization method used. These techniques include variants of linear equalization, decision feedback equalization, and maximum likelihood sequence estimation (MLSE). For these methods a unified set of Chernoff bounds on the probability of error is obtained. Given a communication method, a channel impulse response, an error-rate objective, a received unfaded channel s/n, a channel bandwidth, and a signaling rate we use the Chernoff bounds to estimate the maximum number of bits per cycle of bandwidth (not necessarily integer-valued) for which the constraints are met. By computing the maximum number of bits per cycle supported by each member of a large representative population of channels, we obtain the cumulative probability distribution function. One can use the cumulative distribution curve to determine the probability of outage at a prescribed bit rate.

The information theory bound on the number of bits per cycle attainable is also derived. In a sequence of plots we compare the different schemes with each other and with the information theory limit.

If $F(r)$ is the probability distribution function of data rates associated with a communication method and we set an outage objective, ξ, then the value r_ξ for which $F(r_\xi) = \xi$ represents the maximum data rate at which it is possible to transmit and meet the outage objective. We present and discuss these distributions in the context of desired long-haul and short-haul outage objectives and rates associated with the digital hierarchy constraints. The efficacy of adaptive equalization is established. The advantage of decision feedback and MLSE over

Reprinted with permission from *Bell System Technical Journal,* vol. 62, no. 2, pp. 429–456, Feb. 1983.

optimum linear equalization is not very substantial. There are some desired operating points for which space diversity should be considered.

For a fair comparison of different communication techniques, the transmitter filter shape must be optimized for a fixed transmitter power. We found the performance to be insensitive to whether or not the transmitter filter is optimized and we provide a theoretical guideline to indicate when this optimization becomes significant.

Our results indicate that optimized equalizer structures yield data rates only a few bits/cycle below channel capacity. It therefore appears that higher dimensional constellations[3] spanning two to four symbol intervals could go a long way toward obtaining that which can be expected practically. Although we did not analyze higher dimensional signal design or optimize the constellation in QAM, it is reasonable to expect that these techniques can offer at most an equivalent few dB increase in s/n. Another method of achieving coding gain "of the order of 3-4 dB" is described in Ref. 4. Moreover, the real limitations on the selection of signal points in a practical system will most likely arise from the nonlinear operation of radio frequency (RF) power amplifiers rather than from s/n limitations.

We argue the merits of adaptive transversal equalization and provide numerical support for our claims. This is not to say that fixed or even adjustable bump and/or slope equalizers in the frequency domain could not provide adequate performance in some cases. However, fluctuating (and sometimes nonminimum) phase distortion associated with fading and other linear filters admits robust and stable compensation via adaptive transversal filters. These structures with adjustable taps can automatically equalize any phase characteristic without noise enhancement and therefore are natural candidates in these applications, especially at a high number of levels where even small amounts of phase distortion can degrade system performance.

Our analysis was carried out with ideal models and an infinite number of taps. The actual number of taps needed in any application would be determined from experiments and/or more detailed analysis.

II. THE EQUALIZED QAM SYSTEM—IDEALIZED MODEL

The use of equalizers to mitigate the effects of intersymbol interference and noise in voiceband data transmission is by now standard practice. We are thus naturally led to consider the application of these techniques in digital data transmission over the radio channel where slowly varying, frequency-selective fading is the predominant impairment. Here we review and derive the applicable mathematical theory that will be used in the sequel to evaluate the system performance indices.

To focus on basics and avoid extensive numerical analysis, we consider idealized equalizer models represented as transversal filters with an infinite number of taps. Tap adjustment algorithms are well established and our formulas are derived under the assumption that the taps have converged to their optimum values.

Our analyses are based on the digital communications model depicted in Fig. 1. To appreciate the applicability and generality of this baseband model to digital radio communications, we observe that, for any bandpass linear channel, the output waveform, when the input is any linearly modulated data wave, can be represented as

$$s(t) = \mathrm{Re}\left\{\sum_n \tilde{a}_n \tilde{h}(t - nT + t_0)\exp[i(2\pi f_0 t + \theta)]\right\},$$

where $\mathrm{Re}\{\cdot\}$ stands for the "real part." The data symbols $\{\tilde{a}_n\}$ transmitted at T-second intervals, are statistically independent and, in two-dimensional modulation systems such as QAM, they assume complex values. The overall equivalent baseband impulse response, $\tilde{h}(t)$, is also complex-valued. The real part represents the in-phase response, while the imaginary part is the quadrature component. The frequency, f_0, is the carrier frequency, θ is the carrier phase, and, t_0 is the timing phase. Ideal demodulation with a known carrier frequency f_0 and carrier phase θ implies a translation of the received bandpass signal to baseband. The real part of the resulting complex signal represents the in-phase modulation, while the imaginary part is the quadrature modulation. This then is the rationale, in addition to economics of notation, for using the complex baseband model depicted in Fig. 1.

We restrict our treatment to ideal Nyquist systems with no excess bandwidth. This permits less cumbersome calculations without loss of physical insights. We also derive our formulas by assuming flat transmitting filters and prove later that in-band optimum shaping yields imperceptible additional benefits. Also neglected is adjacent channel interference, as ideal bandlimiting eliminates this problem.

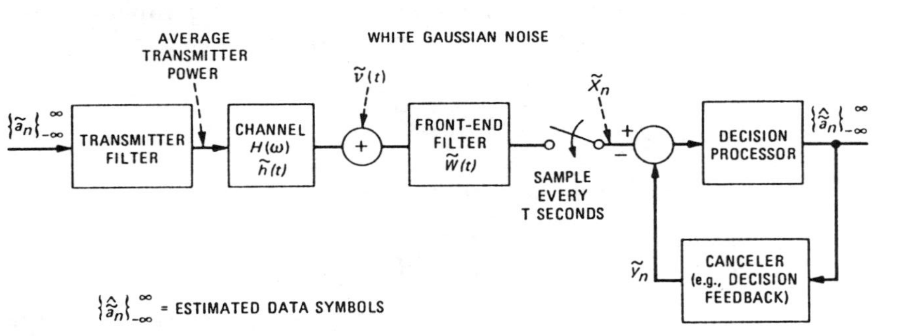

Fig. 1—Complex baseband model for QAM data transmission.

We now return to Fig. 1 and discuss the various functions and notations indicated. Without loss of generality we assume that the complex data symbols, $\{\tilde{a}_n = a_n + ib_n\}_{-\infty}^{\infty}$, take on values on a set of positive and negative odd integers with equal probability. Accordingly,

$$E\tilde{a}_n = 0 \quad \text{and} \quad E|\tilde{a}_n|^2 = 2\frac{L^2 - 1}{3},$$

where $E(\cdot)$ denotes mathematical expectation and L (even) is the maximum number of data levels assumed by a_n and b_n. Thus, in QAM L^2 data points are available for conveying information and the source therefore generates

$$R = \frac{\log_2 L^2}{T} \text{ bits/sec.} \tag{1}$$

For a given channel bandwidth, $\mathscr{W}$, the efficiency index is defined as

$$I = R/\mathscr{W} = \frac{2\log_2 L}{\mathscr{W}T} \text{ bits/cycle.} \tag{2}$$

As we shall see, the relationship among P_e- probability of error, s/n, $\mathscr{W}$, T and $H(\omega)$-channel frequency characteristics is rather complicated. The determination of the relationship for different communication techniques is our chief task in the sequel.

From a mathematical point of view, the fading radio channel is characterized by a slowly varying linear distorting filter whose base-band equivalent complex impulse response is the Fourier transform of the transfer function $H(\omega)$, shifted to zero frequency:

$$\tilde{h}(t) = h_1(t) + ih_2(t) = \int_{-2\pi\mathscr{W}}^{2\pi\mathscr{W}} H(\omega)e^{i\omega t}\frac{d\omega}{2\pi}.$$

At the receiver the added complex noise process, $\tilde{v}(t) = v_1(t) + iv_2(t)$, is assumed to be white Gaussian with $v_1(t)$ independent of $v_2(t)$ and each possessing a double-sided spectral density, $N_0/2$. So,

$$E|\tilde{v}(t)|^2 = Ev_1^2(t) + Ev_2^2(t)$$
$$= N_0\delta(0),$$

where $\delta(\tau)$ is the Dirac delta function. The average transmitted signal power, P_0, for a flat transmitting filter can easily be calculated. However, for our purposes a more relevant quantity is the received, unfaded signal power

$$P = K^2P_0 = 2\frac{L^2 - 1}{3}\frac{K^2}{T^2},$$

where K is a constant that includes the effects of amplifiers, antennas, and the unfaded channel loss. Also, the added average noise power in the Nyquist band, $\mathscr{W} = 1/2T$, is

$$P_v = \frac{N_0}{T}.$$

Thus the unfaded received s/n, a most important system parameter, is

$$s/n = \rho = 2\frac{L^2 - 1}{3}\frac{K^2}{N_0}\frac{1}{T}. \tag{3}$$

The receiver structures under consideration consist of a perfect demodulator followed by a front-end filter possessing the complex impulse response $\tilde{W}(t)$, a sampler, a decision device, and a canceler. The design of an optimum receiver entails the selection of $\tilde{W}(t)$ and the canceler for a particular channel characteristic. Since the channel characteristics are usually unknown to the receiver, these components must be determined adaptively.

To understand the function of the canceler, consider the signal sample at the output of filter $\tilde{W}(t)$,

$$\tilde{x}_n = \sum_{k=-\infty}^{\infty} \tilde{r}_k\tilde{a}_{n-k} + \tilde{z}_n, \quad -\infty \le n \le \infty, \tag{4}$$

where $\tilde{r}_k = \tilde{r}(kT + t_0)$ is the overall complex-sampled system impulse response and $\tilde{z}_n$ is the Gaussian noise output sample. Ideally the canceler strives to synthesize the value

$$\tilde{y}_n = \sum_{k\in S} \tilde{r}_k\tilde{a}_{n-k} \tag{5}$$

and to subtract it from (4) where the set of integers S is defined as $\{k \in S: k = -N_1 \cdots -1, 1 \cdots N_2\}$. The canceler's ability to synthesize these values presumes that some past $(k = 1, \cdots, N_2)$ and/or future $(k = -1, \cdots, -N_1)$ transmitted data symbols are perfectly detected and, moreover, that the set of complex numbers, $\tilde{r}_k$, are adaptively estimated.

The front-end filter, $\tilde{W}(t)$, is usually determined adaptively by minimizing the mean-squared error (MSE) between the sample, $\tilde{x}_n - \tilde{y}_n$, and the expected data symbol $\tilde{a}_n$:

$$MSE[N_1, N_2, \tilde{W}(t)] = E|\tilde{x}_n - \tilde{y}_n - \tilde{a}_n|^2, \tag{6}$$

and the optimum filter, $\tilde{W}(t)$, is chosen to achieve

$$(MSE)_0 = \min_{\tilde{W}(t)} MSE[N_1, N_2, \tilde{W}(t)] = MSE[N_1, N_2, \tilde{W}_0(t)]. \tag{7}$$

Since (6) is a quadratic functional of $\tilde{W}(t)$, a unique minimum can always be found. It is standard to represent the linear filter $\tilde{W}(t)$ by a transversal structure and in practice the search for the minimum is accomplished by varying the taps of this filter until a minimum of the time average of the squared error is found. Clearly, to realize such a minimization procedure, estimates of the transmitted data symbols must be used.

III. SYSTEM PERFORMANCE—GENERAL

To get at the efficiency index of a system, the error rate as a function of data rate for any choice of the canceler set, $\{S\}$, and front-end filter, $\tilde{W}(t)$, must be explicitly expressed. Unfortunately, exact relationships are not mathematically tractable for the simplest of systems and so we must employ upperbounds. Fortunately, for the systems under consideration, it is possible to obtain exponentially tight inequalities.

With this approach in mind, note that after perfect cancellation, the decision variable, from (4) and (5) becomes

$$\begin{aligned} s_n &= \tilde{x}_n - \tilde{y}_n \\ &= \tilde{r}_0\tilde{a}_n + \sum_{k \notin J} \tilde{r}_k\tilde{a}_{n-k} + \tilde{z}_n, \end{aligned} \tag{8}$$

where now the set J is $S \cup 0$, $\{k \in J: k = -N_1 \cdots 0 \cdots N_1\}$. Decisions in QAM are made on the real part of s_n and, separately, on the imaginary part of s_n. Simple calculations give

$$Re(s_n) = \mu_0 a_n - v_0 b_n + \sum_{k \notin J} (\mu_k a_{n-k} - v_k b_{n-k}) + z_{n1},$$

and

$$Im(s_n) = \mu_0 b_n + v_0 a_n + \sum_{k \notin J} (\mu_k b_{n-k} + v_k a_{n-k}) + z_{n2}, \tag{9}$$

where

$$\tilde{r}_k = \mu_k + iv_k,$$

and

$$\tilde{z}_k = z_{k1} + iz_{k2} = \int_{-2\pi W}^{2\pi W} \tilde{v}(t)\tilde{W}(t)dt.$$

For an L-level system, slicing levels are placed at $0 \pm 2\mu_0 \cdots \pm \mu_0(L-2)$ and compared with the received samples $Re(s_n)$ and $Im(s_n)$. An error occurs whenever the noise plus intersymbol interference (in-phase and quadrature) exceed in magnitude the distance from the transmitted level to the nearest decision threshold, μ_0. However, the outside two levels can be in error in one direction only.

Now denote the event of an error committed in the "real" rail by E_r and in the "imaginary" rail by E_i. Then the probability of system error, P_e, is the probability of either (or both) E_r or E_i occurring,

$$P_e = P(E_r \cup E_i) \leq P(E_r) + P(E_i), \tag{10}$$

where

$$P(E_r) = \left(1 - \frac{1}{L}\right) \Pr\left[\left| z_{n1} - \sum_{k \notin J} (\mu_k a_{n-k} - v_k b_{n-k}) + v_0 b_n \right| \geq \mu_0\right]$$

and

$$P(E_i) = \left(1 - \frac{1}{L}\right) \Pr\left[\left| z_{n2} - \sum_{k \notin J} (\mu_k G_{n-k} + v_k a_{n-k}) - v_0 a_0 \right| \geq \mu_0\right]. \tag{11}$$

Because of symmetry, $P(E_r) = P(E_i) = P(E)$, and so we only need to upperbound $P(E)$.

We adopt a bounding procedure introduced by B. Saltzberg[5] to analyze the error rate in an unequalized baseband system. We have extended Saltzberg's approach to our systems and it can be shown that

$$P(E; A, B, \delta) \leq 2 \exp\left\{ \frac{\left[\mu_0 - (L-1)\left(\sum_{k \in A} |\mu_k| + |v_k| + \delta v_0\right)\right]^2}{2\left\{\sigma_{z_1}^2 + \frac{L^2 - 1}{3}\left[\sum_{k \in B} \mu_k^2 + v_k^2 + (1-\delta)v_0^2\right]\right\}} \right\}. \tag{12}$$

The set of integers A and B form a partition on the set of integers not included in J. That is,

$$A \cup B = \Omega = \{k: k \notin J\}$$

and

$$A \cap B = \phi.$$

The variable $\delta = 1$ *or* 0, and

$$\sigma_{z_1}^2 = \frac{N_0}{2} \int_{-\infty}^{\infty} |\tilde{W}(t)|^2 dt.$$

The sharpest upperbound is obtained by minimizing (12) with respect to the sets A, B, and δ. Algorithms for carrying out this minimization can be devised readily.

IV. SYSTEM PERFORMANCE

4.1 Discussion

Equation (12) is a rather general upperbound on the error rate for any passband linear data transmission system and it will now be specialized to include the effects of the different choices of equalizers.Before proceeding with the detailed numerical analysis, we need to make a connection between the mean-squared error (MSE), which is minimized by equalizers, and the system probability of error, which, ideally, should be the quantity minimized.

A straightforward but tedious approach for getting at the error rate might be to first determine the filter, $\tilde{W}(t)$, which minimizes the MSE for any particular equalization scheme, calculate the overall resulting impulse response, and then use eq. (12) to upperbound the error rate. This approach can be circumvented by exploiting the explicit relationship between the minimum MSE and the value of the overall impulse response at $t = t_0$ when the optimum filter, $\tilde{W}_0(t)$, is used.

The optimum structure of the minimum mean-squared error receiver can be shown to consist[6] of a matched filter in cascade with a transversal filter combined with a linear intersymbol interference canceler. The implication of this structure is that the resulting overall system transfer function is a real function of frequency. Or, the complex-sampled impulse response, $\{\mu_k + iv_k\}_{-\infty}^{\infty}$, must be a real number at $k = 0$, which results in $v_0 = 0$. This follows from the Fourier Transform representation of $\tilde{r}(t)$, from which we see that at $t = 0$ the integrand is real and nonnegative. Indeed the overall phase characteristic has been removed by the matched filter (without enhancing the noise*). Using the fact that $v_0 = 0$ and careful numerical analysis of the available channel characteristics, our calculations showed that for all practical purposes the bound (12) becomes

$$P(E, S) \leq 2 \exp\left\{\frac{-\mu_0^2}{2\left[\sigma_{z_1}^2 + \sigma^2(L) \sum_{k \notin S} (\mu_k^2 + v_k^2)\right]}\right\}, \quad (13)$$

where we set $\sigma^2(L) = \dfrac{L^2 - 1}{3}$.

As will become apparent, the argument of the exponential function in (13) can be directly related to the minimum mean-squared error.

Towards this end we recall a well-known[6] result that states that the best achievable MSE has the simple representation,

$$(MSE)_0 = 2\sigma^2(L)(1 - \mu_0), \quad (14)$$

where μ_0 is the sample at $t = t_0$ at the output of the optimum filter. Also, when the optimum filter, $\tilde{W}_0(t)$, is used, a straightforward calculation of the resulting MSE gives

$$(MSE)_0 = 2\sigma^2(L)(1 - \mu_0)^2 + 2\sigma^2(L) \sum_{k \notin S} (\mu_k^2 + v_k^2) + 2\sigma_{z_1}^2. \quad (15)$$

Relationships (14) and (15) make it possible to write (13) as

$$P(E; S) \leq 2 \exp\left\{-\frac{1}{(MSE)_0}\left[1 - \frac{(MSE)_0}{2\sigma^2(L)}\right]^2\right\}$$

$$\sim 2\exp - \frac{1}{(MSE_0)} \quad \text{for} \quad N_0 \to 0, \quad (16)$$

relating error rate and minimum MSE. This is an extremely useful inequality since (MSE_0) as a function of channel characteristics is often explicitly known for different equalizer structures.

It is also interesting to note (this has been pointed out before[8]) that the filter, $\tilde{W}(t)$, that minimizes MSE also minimizes the upperbound on P_e. This is true because the same quadratic functionals in $\tilde{W}(t)$ are involved in the optimization of both expressions.

We are now in a position to specialize our formulas to the various equalizer structures under investigation.

The six examples that follow do not require the knowledge of channel phase characteristics to compute performance. Implicit in each of these schemes is the complete removal of phase distortion, which can be accomplished without noise enhancement. Only a magnitude characterization of the channel transfer response was available at the time the work reported here was done. While departure from flatness of the magnitude fundamentally affects performance, theoretically, departure of phase from linear has no effect on attainable performance. Therefore, the lack of phase characterization of the channels was not an obstacle to our study. However, a complex characterization of the channel would be useful in determining the minimum number of required taps in the designs of the equalizers.

4.2 Pure phase equalization

In this particular equalizer, $N_1 = N_2 = 0$ (where N_1 and N_2 are the lengths of the precursive and postcursive cancelers, respectively). We choose $\tilde{W}(t)$ so that

* A fractionally spaced ($T/2$) transversal filter can automatically synthesize any matched filter and thus eliminate phase distortion and also compensate for timing phase (see Ref. 7.)

$$W(\omega) = e^{-i\phi(\omega)}, |\omega| \leq \frac{\pi}{T}$$

$$= 0, |\omega| > \frac{\pi}{T},$$

where $W(\omega)$ is the Fourier transform of $\tilde{W}(t)$ and $\phi(\omega)$ is the channel phase characteristic. For this choice of filter, only the magnitude of the channel transfer function enters into the computation of the bound, as shown in eq. (13).

Using the well-known Poisson sum formula along with some algebra, it is possible to write (13) more explicitly, i.e.,

$$P_e \leq 2 \exp\left\{-\frac{\rho}{2\sigma^2(L)} \frac{\langle H \rangle^2}{1 + \rho \, \langle (H - \langle H \rangle)^2 \rangle}\right\}, \tag{17}$$

where we used the shorthand notation

$$\langle \cdot \rangle = \frac{T}{2\pi} \int_{-\pi/T}^{\pi/T} [\cdot] d\omega$$

and H is used in place of $|H(\omega)|$.

4.3 Linear equalization

Here again $N_1 = N_2$ (no canceler) and $\tilde{W}(t)$ is chosen to minimize (6). The expression for the optimum MSE in this case has been shown to be[9]

$$(MSE)_0 = 2\sigma^2(L) < \frac{1}{1 + \rho H^2} >. \tag{18}$$

This formula is directly used in (16) to calculate the upperbound on error rate:

$$P_e < 2 \exp\left[-\frac{1}{2\sigma^2(L)} \left(< \frac{1}{1 + \rho H^2} > \right)^{-1}\right].$$

4.4 Inverse equalization

In this case $N_1 = N_2 = 0$ (no cancellation) we choose $\tilde{W}(t)$ to be the inverse of the channel frequency characteristics,

$$W(\omega) = H^{-1}(\omega), |\omega| \leq \frac{\pi}{T}$$

$$= 0, |\omega| > \frac{\pi}{T}.$$

Here the channel is clearly perfectly equalized so that intersymbol interference is completely eliminated; the penalty is increased output noise power. For this simple scheme, it is possible to express the error rate exactly but for reasons of uniformity we use the upperbound

$$P_e \leq 2 \exp - \left\{ \frac{\rho}{2\sigma^2(L)} \frac{1}{< \frac{1}{H^2} >} \right\}. \tag{19}$$

4.5 Decision feedback

For this equalization system $N_1 = 0$ and $N_2 = \infty$ and again we choose $\tilde{W}(t)$ to minimize (6). In this type of equalizer, the causal or postcursor intersymbol interference is entirely eliminated and an expression for the optimum MSE is known,[10,11] as shown below.

$$(MSE)_0 = \sigma^2(L) \exp\{-\langle \ln[1 + \rho H^2] \rangle\}. \tag{20}$$

This is used in (16) to express an upperbound on error rate.

4.6 The ideal equalizer

In this utopian scheme the precursor and postcursor cancelers become infinite, $N_1 = N_2 = \infty$, so that all the intersymbol interference is eliminated. In this ideal situation we obtain the very best possible result, namely, the matched filter bound, which is a lower bound on P_e. This scheme assumes that it is possible to detect each pulse $\tilde{a}_k \tilde{r}(t - nT)$ optimally by a matched filter without incurring interference from all other pulses. The filter, $\tilde{W}(t)$, in this case is chosen to be matched to the channel characteristic, i.e.,

$$W(\omega) = H^*(\omega), |\omega| \leq \frac{\pi}{T}$$

$$= 0, |\omega| > \frac{\pi}{T},$$

where * denotes the complex conjugate.

For this idealization the upperbound on error rate is simply,

$$P_e \leq 2 \exp\left\{-\frac{\rho}{2\sigma^2(L)} \langle H^2 \rangle\right\}. \tag{21}$$

No other detection scheme can do better. In the next section we use these formulas to calculate the various efficiency indices.

Before concluding this section, we remark that there is one more easy case and one extremely difficult case that might be considered as candidates for making comparisons. Suppose that no filtering, other than out-of-band elimination of noise, were performed at the receiver. What performance can one expect? While we cannot answer this question exactly because channel phase characteristics are unavailable at this writing, we would expect performance to be worse than removing the phase characteristic entirely—a situation we will examine.

The second approach, which is a very difficult one to analyze, involves the use of a finite-state Viterbi decoder. Nevertheless, we will report a bound on the performance of this processor. Specifically, the performance of an infinite canceler (the matched filter) is superior to the maximum likelihood (Viterbi) decoder. As shown later, for the channels considered, the matched filter bound is close in performance to decision feedback. Consequently, we shall see that the performance of maximum likelihood sequence estimation is tightly bracketed because it is superior to decision feedback.

4.7 Information theory bound on communications efficiency index

In this section we discuss a formula for the maximum number of bits per cycle that can be attained for a given $H(\omega)$. If $H(\omega)$ were constant in frequency the formula for the efficiency index in bits per cycle would be simply

$$I = \log_2(1 + \rho|H|^2).$$

It is reasonable to expect that if $H(\omega)$ is frequency-dependent, the maximum efficiency index would be

$$I = \frac{1}{\Omega}\int \log_2(1 + \rho|H(\omega)|^2)d\omega, \qquad (22)$$

where the integral is over a frequency band of size $\Omega = 2\pi\mathscr{W}$. Indeed this is the case. To outline a derivation, we note first that A. Kolmogorov has generalized Shannon's notion of capacity to provide a very fundamental definition that gives a useful starting point for developing capacity formulas in nonstandard situations such as the one at hand.[12] M. S. Pinsker[13] was able to derive from the Kolmogorov approach a formula for the amount of information in a stationary Gaussian process about another stationary Gaussian process related to it. Specifically, if $S_x(\omega)$ and $S_y(\omega)$ are the power spectral densities of the processes and $S_{xy}(\omega)$ is the cross-spectral density, the formula for the amount of information is

$$-\int \ln\left(1 - \frac{|S_{xy}(\omega)|^2}{S_x(\omega)S_y(\omega)}\right) d\omega.$$

If we require the transmitter output to be Gaussian, then since the additive noise is Gaussian, Pinsker's formula can be applied to the case where x is the transmitted process and y is the received process to obtain (22). Requiring the transmitter output to be Gaussian is really not a limitation since, when the additive noise is Gaussian, one can prove that capacity is attainable with a Gaussian transmitter output using the methods discussed in Refs. 14 through 16. Thus, (22) gives the efficiency index formula.

References 16 and 17 also provide approaches to establishing (22).

4.8 Information theory limit on index when transmitter is optimized

So far we have treated ρ, eq. (3), as a constant. In this section we set the stage for exploring the advisability of optimizing the output power spectral density to maximize the efficiency index. Since we are now allowing in-band shaping of the transmitter filter frequency characteristic, we will consider ρ as a function of ω and write $\bar{\rho}$ to denote the previously considered case where ρ is constant over the band.

Although the analysis in this section is focused on the information theory limit on the efficiency index, the decision feedback index involves the identical functional form and so our analysis will be applicable to decision feedback as well.

We will compare the previously discussed index

$$I(\text{flat}) = \frac{1}{\Omega}\int \log_2(1 + \bar{\rho}|H(\omega)|^2)d\omega$$

with

$$I(\text{opt}) = \frac{1}{\Omega}\int \log_2(1 + \rho_0(\omega)|H(\omega)|^2)d\omega,$$

where $\rho_0(\omega)$ is the function maximizing I under a constraint on $\int \rho(\omega)\, d\omega$, the received signal-to-noise ratio in the absence of fading. This constraint is equivalent to a constraint on the transmitter output power, since in the absence of fading the channel has a flat loss characteristic.

This optimization problem is known[17] and yields easily to the calculus of variations. The solution is called "water pouring." The name stems from the graphical interpretation that if $\bar{\rho}\Omega$ is the constraint on $\int \rho(\omega)\, d\omega$, the optimum $\rho(\omega)$, which we denote by $\rho_0(\omega)$, is obtained by forming a vessel with base $|H(\omega)|^{-2}$ and vertical sides at the band edges. One pours "water," that is, area, of amount $\bar{\rho}\Omega$ into the vessel and $\rho_0(\omega)$ is given by the depth of the water at ω. It is clear that this construction obeys the constraint. Generally, if $\bar{\rho}$ is sufficiently small, the poured water will not touch both of the vertical sides of the vessel. In such situations, it would be advantageous to limit the transmitted power to a frequency band less than Nyquist. In our case, however, the unfaded signal-to-noise ratio is so great that, for the simulated channels, the water level always meets both vertical sides. In other words, the transmitted power always occupies the full Nyquist band. Thus, $\rho_0(\omega) = A - |H(\omega)|^{-2}$, where A is chosen so that $\int \rho_0(\omega)d\omega = \bar{\rho}\Omega$, that is, $A\Omega - \int |H(\omega)|^{-2}d\omega = \bar{\rho}\Omega$ or

$$\rho_0(\omega) = \bar{\rho} + \frac{1}{\Omega}\int |H(v)|^{-2}dv - |H(\omega)|^{-2}.$$

Therefore,

$$\frac{I(\text{flat})}{I(\text{opt})} = \frac{\int \log_2(1 + \bar{\rho}|H(\omega)|^2)d\omega}{\int \log_2\left[\frac{|H(\omega)|^2}{\Omega}\int |H(v)|^{-2}dv + \bar{\rho}|H(\omega)|^2\right]d\omega}.$$

We expand the logarithm for $\bar{\rho}$ large to find an asymptotic representation. We get, after a cumbersome derivation,

$$\frac{I(\text{flat})}{I(\text{opt})} = 1 - \frac{1}{2\bar{\rho}^2\log_2\bar{\rho}}\left(1 + \frac{\int \log_2|H|^2 d\omega}{\Omega \log_2\bar{\rho}}\right)^{-1}$$

$$\cdot\left[\frac{\int \frac{1}{|H|^4}d\omega}{\Omega} - \left(\frac{\int \frac{1}{|H|^2}d\omega}{\Omega}\right)^2\right] + O\left(\frac{1}{\bar{\rho}^3\log_2\bar{\rho}}\right). \quad (23)$$

The last multiplier in the perturbation expression represents the variance associated with a random sampling of a specific $|H(\omega)|^{-2}$. This multiplier is zero if $|H(\omega)|^{-2}$ is a constant. We note, for later use, that for $\bar{\rho} = 10^{6.3}$, (i.e., a 63-dB s/n in the absence of fading) we have

$$\left(\frac{1}{2\bar{\rho}^2\log_2\bar{\rho}}\right) < 10^{-14}.$$

4.9 Communication efficiency index

The relationships in eqs. (17) through (21) are unifying expressions for the error rate for the five equalization cases. From Section 1.0 we have $I = 2\log_2 L$ and $\sigma^2(L) = [(L^2 - 1)/3]$. These two equations in conjunction with the P_e formulas enable us to determine I as a function of the P_e objective, ρ, the channel, and the equalization scheme. Specifically,

$$I = \log_2\left[\frac{G(H, \rho)}{\ln(P_e/2)} + 1\right],$$

where $G(H, \rho)$ is a function that depends on the communication method. With the channel response considered to be a random function, I is a random variable, and we can determine its probability distribution function for each communication scheme. The quantities ρ and P_e are parameters of the distribution.

V. MODEL FOR THE FADING CHANNEL

Now we describe the mathematical model for frequency-selective fading owing to multipath reception. This mathematical model for the random functions $|H(\omega)|^2$ is due to W. D. Rummler[1,2] and is based on measurements of a 26.4-mile hop between Palmetto and Atlanta, Georgia. The measurements of frequency-selective fades were made on a 25.3-MHz channel situated in the 6-GHz band during the heavy fading month of June (1977).

Rummler's model uses a two-ray representation of the signal, which was quite adequate for fitting the experimental records. (The model is not necessarily intended to depict the underlying physical mechanism for a fade. The true mechanism could involve a much more complex ray combination.) Also, it is not possible to deduce the phase characteristic associated with any particular amplitude characteristic. It has been experimentally determined that this kind of a channel cannot always be viewed as minimum phase.[18]

In the model, the $|H(\omega)|^2$ functions are 68-degree sections of scaled, displaced cosine waves. Specifically, conditional on a fade occurring

$$|H(\omega)|^2 = a^2|1 + b^2 - 2b\cos(\omega\tau + \theta)|,$$

where:

(*i*) $b = 1/10^{B/20} > 0$ with B an exponential random variable with mean 3.8.

(*ii*) The parameter, a, is a log normal random variable with dependence on the parameter, b. Specifically, $a = 1/10^{A/20}$, where A is normal with a mean of $24.6(B^4 + 500)/(B^4 + 800)$dB and a standard deviation of 5 dB.

(*iii*) The phase, θ, is independent of a and b and has a constant density on each section $|\theta| > \pi/2$ and $|\theta| < \pi/2$ with $P\{|\theta| < \pi/2\} = 5 \cdot P\{|\theta| > \pi/2\}$.

(*iv*) The scale factor τ is a constant = 6.31 nanoseconds.

In the model, the channel is in the faded state for 8060 seconds in a normal heavy fading month. Thus the channel can be viewed as being in one of two states where:

P {unfaded state} = 0.99689
P {faded state (Rummler model operative)} = 0.00311.

In what follows we employ this model to estimate the outage distributions for various communication methods. The model should be regarded as a working assumption valuable in gaining initial insight into the potential of the communication techniques we consider. However, we emphasize that more measurements may be required to refine Rummler's model to accommodate different geographical situations and wider bandwidths than 25 MHz, and to sharpen the accuracy of the representation of the more severe fades that are of major concern in what follows.

VI. OUTAGE OBJECTIVES AND SOME PROSPECTIVE INDEX VALUES

From the proposed performance objectives for the digital transmission network,[19] we have that the round trip system availability objective is 99.98 percent. So the probability of outage is 0.0002 round trip or 0.0001 one way. The 0.0001 breaks down as 0.00005 for fading, 0.000025 for equipment failure, and 0.000025 for maintenance and plant errors. Thus, for a 4000-mile system composed of 156 hops, each with a nominal length of 25.6 miles, we get a per hop outage probability of 3.2×10^{-7} for fading. This corresponds to about 10 seconds of outage per year. If we assume the year is composed of three heavy fading months and nine months with no fading, we obtain that the probability of outage in a heavy fading month is 1.28×10^{-6}. For a 250-mile short-haul system, the outage objective is 16 times less stringent on a hop, namely, 2.05×10^{-5}.

For the purpose of discussion we shall later consider the possibility of accommodating two DS-3 digital signals in a 20-, 30-, and 40-MHz channel. Each DS-3 signal corresponds to 672 64 kb/s voice circuits, so that two DS-3 signals correspond to about 90 Mb/s. Thus, for 20-, 30-, and 40-MHz channels we require 4.5, 3, and 2.25 bits per cycle, respectively. We will use 10^{-4} as the probability of bit-error threshold for registering outage. The sensitivity to this threshold will also be analyzed.

VII. COMPUTER PROGRAM

A comprehensive FORTRAN program was written to compute and display outage distributions. The program is composed of three main segments.

The first segment simulates the power transfer characteristic for the channel in the faded state. It uses a PORT routine to generate random numbers uniformly distributed on [0, 1] and functions of these are evaluated to produce the random variables with the three densities underlying Rummler's model. The variables A and B are appropriately correlated. The random channel characteristics are then computed and a file containing them is produced. The file contains 25,000 characteristics.

The second segment calculates the efficiency index for each channel and then computes the probability distribution function of the indexes. Various options and parameters can be chosen in exercising this stage. These include:

(*i*) Method of communication (i.e., type of linear equalization, decision feedback equalization, MLSE, and the information theoretic optimum processing)

(*ii*) Transmitter spectrum, i.e., optimization of the transmitter spectral density for a given average power constraint or flat power spectral density

(*iii*) Probability of bit-error objective

(*iv*) Unfaded signal-to-noise ratio at the receiver input

(*v*) Channel bandwith.

Option (*ii*) is only available for the decision feedback and the information theoretic optimum communication schemes. Extending the option to the other schemes seemed inadvisable because of the closeness of the results, as will be seen later.

The number 25,000 was found through computational experience to stabilize the density tail in the range of interest and yet not be wasteful of computer resources. Since the number 25,000 is very close to the number of experimental records of fade characteristics, we could have worked from original experimental data. We elected to work with the Rummler model since it is weighted to track the worse fades, which are our interest here, and since the model is widely accepted.

The final segment of the computer program provides labeled plots of the outage distribution functions. It uses the graphic package DISSPLA.

VIII. PRINCIPAL RESULTS

8.1 Preparatory remarks

For the purpose of presenting our principal results we will need the following notation for the outage distribution functions:

F_{PH}: phase distortion removed

F_{LIN}: optimum linear equalization

F_{DF}: postcursive intersymbol interference (ISI) removed

F_{MF}: all ISI removed (matched filter bound)

F_{IT}: information theory limit (Shannon).

The efficiency index distributions were computed for 30-MHz channels. Strictly speaking, the notion of an index in bits per cycle is imprecise in that $F(I)$ (the probability distribution of bits per cycle) would change if calculated at 20 MHz or at 40 MHz. However, by calculation we established that, for the purposes of the discussion that follows, treating $F(I)$ as invariant over the 20- to 40-MHz range of bandwidths is an adequate approximation.

In the actual development of systems of the kind we have idealized, much more detailed performance analysis is required than that reported here. One important aspect we have not considered is the effect of excess bandwidth associated with practical filter designs with rolloff factors other than zero. To get a preliminary idea of the effect of rolloff of an amount α on the communication efficiency indexes, we would simply scale the distributions abscissas by an amount $1/(1 + \alpha)$.

Equivalently, we would inflate the desired number of bits per cycle by $1 + \alpha$ before going to the curves. Thus, in considering the accommodation of two DS-3 signals, with $\alpha = 1/3$, we would inflate the 4.5-, 3-, and 2.25-bit per cycle values corresponding to 20-, 30-, and 40-MHz channels by 1.333 to obtain 6, 4, and 3 bits per cycle. We would then consult the derived curves at these values to obtain the outage probabilities. For the purpose of discussion in the section that follows, we use these three inflated bits per cycle values along with the long-haul and short-haul objectives of 1.3×10^{-6} and 2.1×10^{-5}, respectively, given in Section VI. Subsequently, an alternative means of accounting for α will be given.

8.2 The graphs

The most striking features of the outage distribution functions $F(I; P_e, \rho)$ are exhibited in Fig. 2. The beneficial effects of adaptive equalization are apparent. The three equalization schemes yield roughly similar results; however, as one looks more closely at the extreme outage tail, F_{LIN}, F_{DF}, and F_{MF} begin to depart from each other. F_{IT} is displaced over two bits per cycle to the right of F_{MF}, while F_{PH} is very substantially to the left of F_{LIN}. For the 40-MHz band (3 bits/cycle), both outage objectives are met with linear equalization. For the 30-MHz band (4 bits/cycle) and the long-haul objective, linear and decision feedback equalization are not adequate and some coding or use of maximum likelihood sequence estimation are possible solutions. However, it may be practical to overcome the shortfall by some other means, such as improving the amplifier noise figure. For the 20-MHz channel (6 bits/cycle) the long-haul objective is not met. Also, this efficiency is so close to the information theory limit that any attempt to achieve it by coding may be ill-advised because of complexity. On the other hand, with some moderate coding the short-haul objective for 20-MHz channels should be attainable. For the other two bands, short-haul objectives are roundly met.

The plot for the equalizer that inverts the channel is not shown, as it is not perceptibly different from that for the optimum linear equalizer. This is expected since the optimum linear filter is essentially inverting the channel at the high signal-to-noise ratios we are considering.

Figures 3, 4, and 5 show the sensitivity of $F(I, P_e, \rho)$ to P_e. The

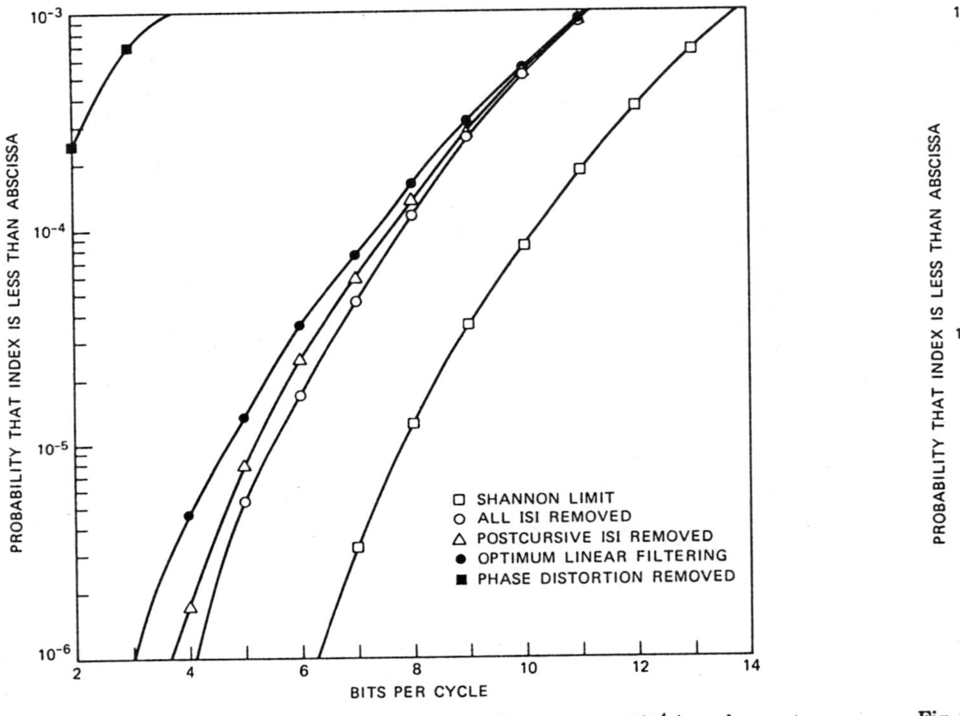

Fig. 2—Comparison of index distribution tails at 63-dB s/n, p.e. $< 10^{-4}$ (p.e. does not apply to the Shannon limit curve).

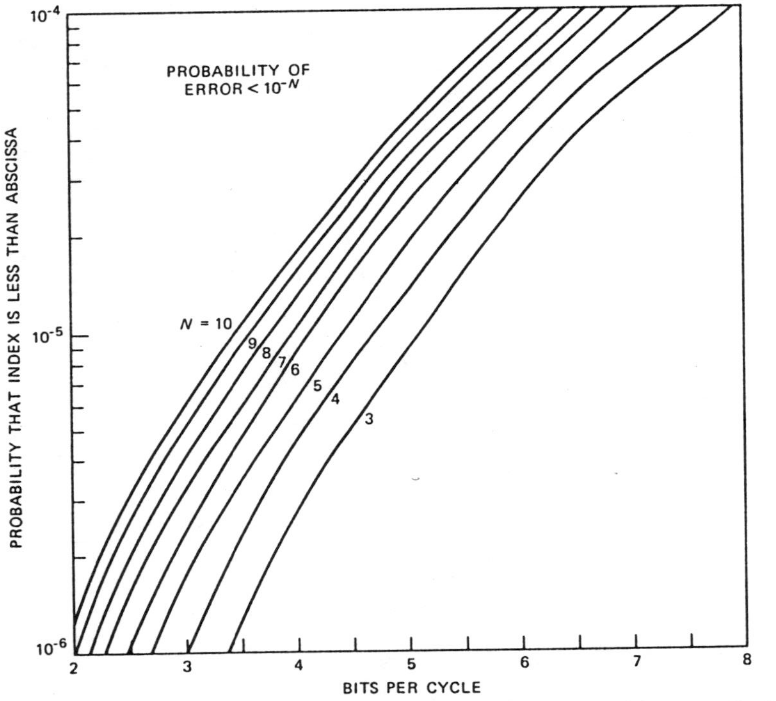

Fig. 3—Index distribution tail sensitivity to probability of error for linear equalization at 63-dB s/n.

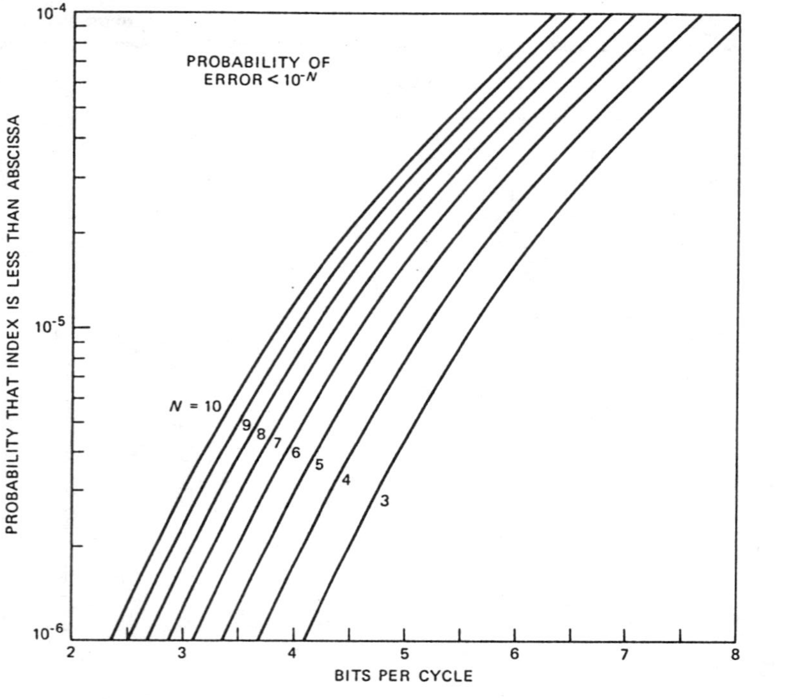

Fig. 4—Index distribution tail sensitivity to probability of error for decision feedback at 63-dB s/n.

sensitivity is especially small at the low error rates needed for data (as opposed to voice) transmission. An asymptotic analysis shows that, for large ρ, the curves translate to the left in accordance with a $\log_2 N$ shift, where 10^{-N} is the P_e objective. This insensitivity is an illustration of the well-known result[20] that once a pulse code modulation (PCM) operating point is achieved it takes a very small improvement to make the error rate an order of magnitude smaller. In fact, if at some operating data rate the probability of error turns out to be $10^{-5} - 10^{-6}$, one should be able to design an error-correcting code of small redundancy and moderate complexity that could improve the error rate by several orders of magnitude.

Figures 6 through 9 illustrate the sensitivity to signal-to-noise ratio. The translation in all cases is roughly 1/3-bit/cycle/dB. Note the curves for the Shannon limit have an ordinate range of 4 to 10 bits per cycle, while the others range from 2 to 8 bits per cycle. No sensitivity for F_{PH} is given since, unlike all the other distributions, there is negligible improvement as ρ increases. This is because, as ρ increases, the effect of intersymbol interference (ISI) remains and nothing is being done to mitigate it. In the other four cases, ISI tends to be

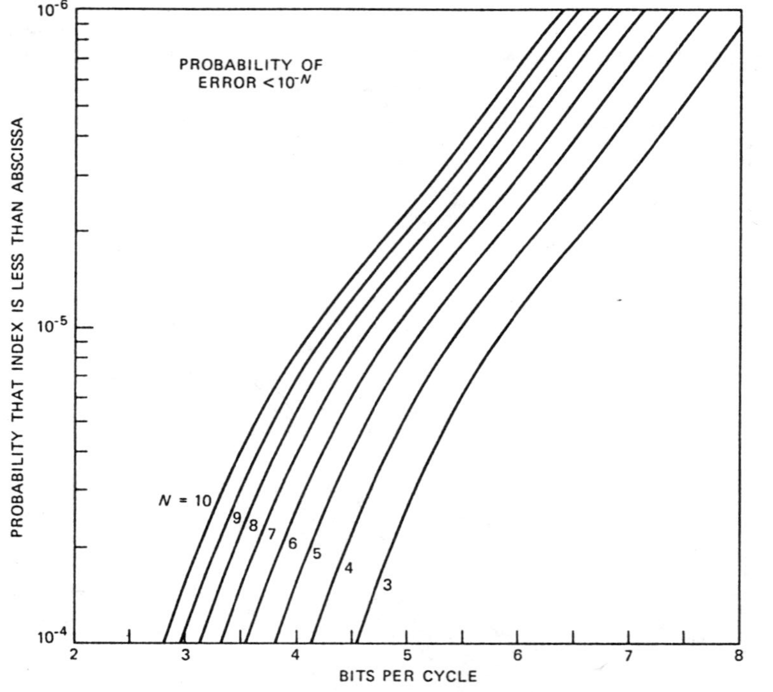

Fig. 5—Index distribution tail sensitivity to probability of error for the case of all ISI removed.

eliminated as ρ increases (the channels can be perfectly equalized without an inordinate amount of noise enhancement).

In Section 8.1 we mentioned the $(1 + \alpha)^{-1}$ scaling as a method of accounting for rolloff. This assumed $\mathscr{W} = 1/T$ so that $(1 + \alpha)\mathscr{W}$ is the actual bandwidth. Suppose instead that the real bandwidth is fixed at $\mathscr{W}$ but the data rate is slowed by an amount $(1 + \alpha)$, leaving the average transmitter power and N_o constant. Then the true s/n is increased by $10 \log_{10}(1 + \alpha)$. From this alternative perspective the suggested $(1 + \alpha)^{-1}$ scaling would be supplemented by a shift to the right of the probability distribution function tail by approximately $(10/3)\log_{10}(1 + \alpha)$ bits per cycle. Whether in estimating the effect of rolloff one takes the perspective that the symbol rate or the bandwidth is fixed is a matter of convenience.

Next, we consider optimization of the transmitter power spectral density. There is a practical question as to whether such an optimization could be achieved since the fade characteristic, which is first determined at the receiver, would need to be relayed back to the transmitter in time to be useful. However, the question is academic since we demonstrated that, even if an optimized transmitter could be

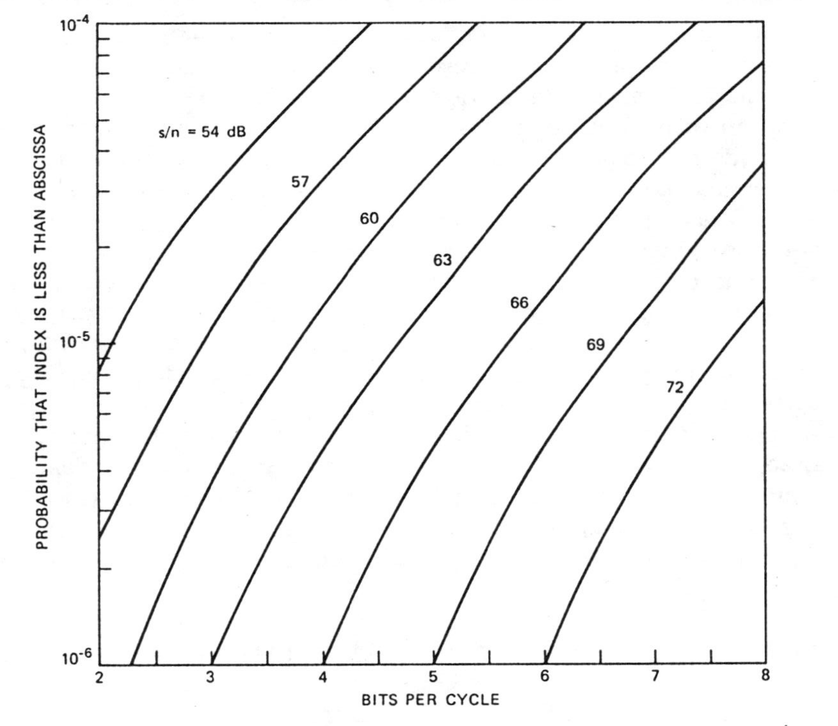

Fig. 6—Index distribution tail sensitivity to s/n for linear equalization, p.e. < 10^{-4}.

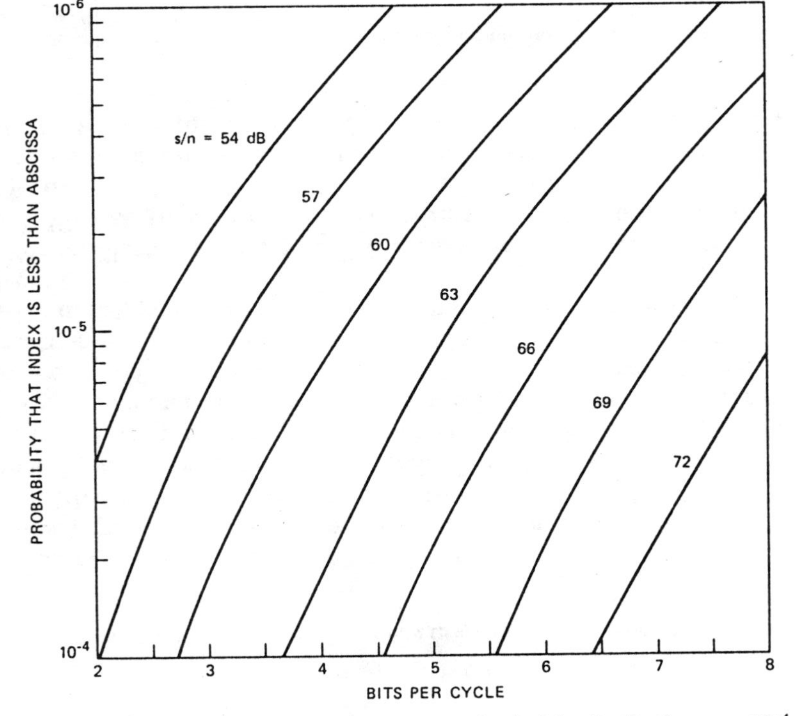

Fig. 7—Index distribution tail sensitivity to s/n for decision feedback, p.e. < 10^{-4}.

adapted in real time, the performance benefit would be negligible.

To understand why it is not worthwhile to optimize the transmitted power spectral density, we first consider the information theory limit that was anlayzed in Section 4.7. Our detailed numerical work has shown that a plot of the tail of the index distribution for the information theory limit under the assumption of an optimized transmitter would be imperceptibly different from the F_{DF} tail plotted in Fig. 2. This closeness of the two distributions follows from the fact that, for the severest fades in our data base of 25,000 channels, $|H|^{-2}$ is of the order of 10^{-6} and the terms involving $|H|^{-4}$ ($\sim 10^{-12}$) in the perturbation expression, eq. (23), are not enough to overcome the 10^{-14} multiplier.

Since the decision feedback index has the same form as (22), we can also conclude that the distribution tail corresponding to decision feedback would not be significantly altered if the optimum transmitter were used.

The decision feedback index and information theory limit on the index give imperceptible benefits when the power spectral density is optimized; therefore, it seems extremely unlikely that there is any worthwhile benefit associated with optimizing the transmitter in the other cases.

IX. INITIAL ESTIMATE OF THE EFFECT OF FREQUENCY DIVERSITY

In implementations, digital radio systems are often protected with frequency diversity. In such systems impairments such as fading and equipment outages prompt the switching of communication traffic to a protection channel situated at a different frequency. The notation *mxn* means that *m* protection channels back up *n* working channels. So long as a protection channel is not occupied by an impaired channel, or is not itself impaired, it is available for temporary use in any of the *n* working systems. Some illustrations are 2×10 and 1×11 at 4 GHz, while at 6 GHz 2×6 and 1×7 are examples.

For FM systems the factor expressing the improvement in outage associated with frequency diversity is given by the expression $100/DG \cdot f_0 H^2$ in eq. (24) [corresponding to (34) and (35) in Ref. 21]. The parameter f_0 represents the carrier frequency in gigahertz and D denotes the path distance measured in miles. The parameter G depends

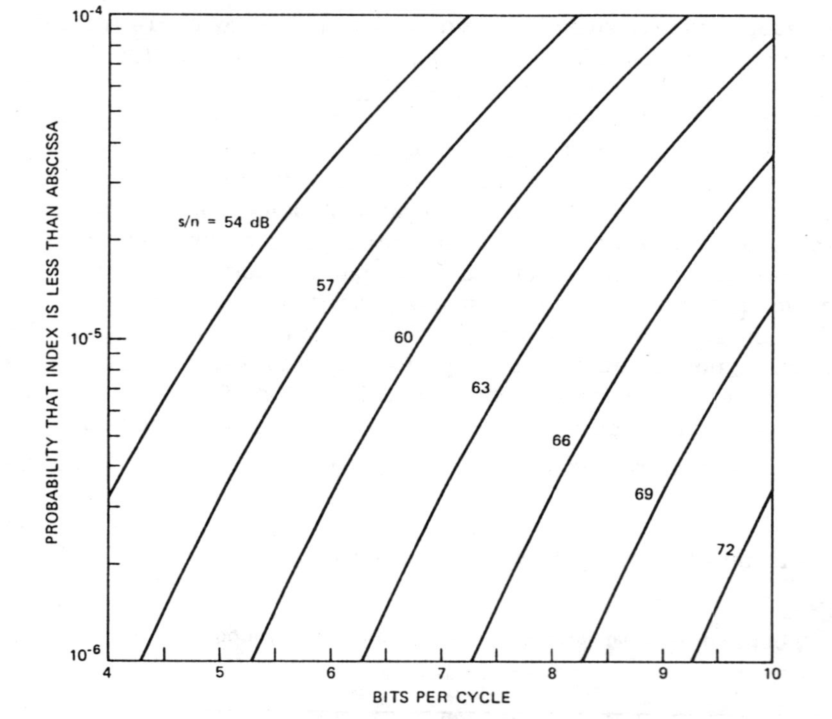

Fig. 8—Index distribution tail sensitivity to s/n for the case of all ISI removed, p.e. $< 10^{-4}$.

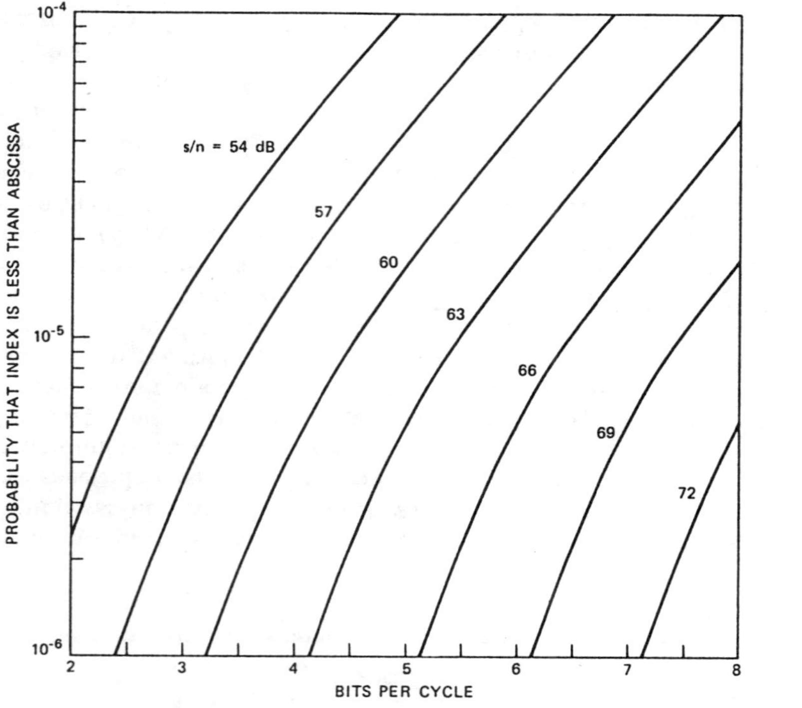

Fig. 9—Distribution tail sensitivity to s/n for the Shannon limit.

on the details of the frequency protection. G incorporates combinatorial effects corresponding to using m channels to back up n as well as empirical expressions involving the individual frequencies of channels involved. The term H is commonly expressed in decibels as $-20 \log H$ and is called fade margin. The fade margin is the smallest loss relative to the unfaded received signal at which the system fails. The notation H for the voltage level agrees with the previous use of H in this paper so long as the channel has a flat characteristic.

As pointed out in Ref. 22 the notation of a flat fade margin is considered meaningless in digital radio systems since the frequency-selective aspect of the fade characteristic appears necessary to describe performance of a channel. As of this writing we are not aware of any method in the extant literature for extending our results to include the effect of frequency diversity. However, for the special case of optimum linear equalization there is a way to introduce an equivalent flat fade margin so as to enable the use of (24) in making a (preliminary) estimate of the diversity effect. The estimation method was discovered in the course of generalizing the computer program to compute index distributions for arbitrary bandwidths. The generalization was needed to investigate—and, as it turned out, to substantiate—the bandwidth insensitivity of the distribution in the 20- to 40-MHz range. The generalized program was also exercised for bandwidths an order of magnitude smaller and it was observed that the F_{LIN} distribution changed imperceptibly.* So the F_{LIN} tail in the range of interest here can be correctly obtained by using the univariate samples $|H(0)|^2$. Treating (18) as an equality and solving for $|H(0)|^2$ and then substituting in (24) gives an estimate of the improvement owing to frequency diversity.

For an illustration refer to Fig. 2 for which s/n = 63 dB and $Pe < 10^{-4}$. We see that with linear equalization, at the long-haul outage objective of 1.3×10^{-6}, 3.2 bits per cycle can be supported and the corresponding number for the short-haul objective is 5.3 bits per cycle. Using (24) and the G values from Ref. 21 we show in Table I the

* This is not true of the individual values of I and no claim is made for invariance of F_{DF}, F_{MF}, or F_{IT}.

Table I—Estimates of improved bits/cycle indices using frequency diversity

	Channel at 4 GHz		Channel at 6 GHz	
Short Haul	(1 × 11)	7.1	(1 × 7)	6.5
	(2 × 10)	7.8	(2 × 6)	7.1
Long Haul	(1 × 11)	5.3	(1 × 7)	4.6
	(2 × 10)	5.8	(2 × 6)	5.3

following estimates of improved indices when frequency diversity is employed.

X. ACKNOWLEDGMENT

Discussions with N. Amitay, A. Gieger, L. J. Greenstein, V. K. Prabhu, B. G. King, G. Vannucci, A. Vigants, C. B. Woodworth, W. D. Rummler and Y. S. Yeh were valuable in the course of the work reported here.

XI. POSTSCRIPT

Application of multilevel QAM in the radio channels might be inhibited by the amplitude (AM-AM) and (AM-PM) nonlinearities present in RF power amplifiers. A method for solving this problem without sacrificing amplifier power efficiency will be described in a forthcoming paper.[23]

REFERENCES

1. W. D. Rummler, "A New Selective Fading Model: Application to Propagation Data," B.S.T.J., *58,* No. 5 (May–June 1979), pp. 1032–71.
2. W. D. Rummler, "More on the Multipath Fading Channel Model," IEEE Trans. Commun., *COM-29,* No. 3 (March 1981), pp. 346–52.
3. A. Gersho and V. B. Lawrence, unpublished work.
4. G. Ungerbock, "Channel Coding with Multilevel/Phase Signals," IEEE Trans. Inform. Theory, *IT-28,* No. 1 (January 1982), pp. 55–67.
5. B. R. Saltzberg, "Intersymbol Interference Error Bounds with Application to Ideal Bandlimited Signaling," IEEE Trans. Inform. Theory, *IT-14,* No. 4 (July 1968), pp. 563–8.
6. M. S. Mueller and J. Salz, "A Unified Theory of Data Aided Equalization," B.S.T.J., *60,* No. 9 (November 1981), pp. 2023–38.
7. R. D. Gitlin and S. B. Weinstein, "Fractionally Spaced Equalization: An Improved Digital Transversal Equalizer," B.S.T.J., *60,* No. 2 (February 1981), pp. 275–96.
8. R. D. Gitlin and J. E. Mazo, "Comparison of Some Cost Functions for Automatic Equalization," IEEE Trans. Commun., *COM-21,* No. 3 (March 1973), pp. 233–7.
9. T. Berger and D. W. Tufts, "Optimal Pulse Amplitude Modulation, Part I, Transmitter-Receiver Design and Bounds from Information Theory," IEEE Trans. Inform. Theory, *IT-13,* No. 2 (April 1967), pp. 196–208.
10. J. Salz, "Optimum Mean-Square Decision Feedback Equalization," B.S.T.J., *52,* No. 8 (October 1973), pp. 1341–73.
11. D. D. Falconer and G. J. Foschini, "Theory of Minimum Mean-Square-Error QAM Systems Employing Decision Feedback Equalization," B.S.T.J., *52,* No. 10 (December 1973), pp. 1821–48.
12. S. P. Lloyd, "On a Measure of Stochastic Dependence," Theory of Probability and Its Application, No. 7, 1962, pp. 312–22.
13. M. S. Pinsker, "A Quantity of Information of a Gaussian Random Stationary Process, Contained in a Second Process Connected with it in a Stationary Manner," Doklady Akad. Nank S.S.S.R., New Series, *99,* 1954, pp. 213–16.
14. R. K. Mueller and G. J. Foschini, "The Capacity of Linear Channels with Additive Gaussian Noise," B.S.T.J. *49,* No. 1 (January 1970), pp. 81–94.
15. J. L. Holsinger, "Digital Communication over Fixed Time-Continuous Channels with Memory-with Special Application to Telephone Channels," Lincoln Laboratory, Technical Report 366, Lexington, MA, October 1964.
16. J. B. Thomas, "Statistical Communication Theory," New York: John Wiley and Sons, 1969, Chapter 8.
17. R. M. Fano, "Transmission of Information," New York: M.I.T. Press and John Wiley and Sons, Inc., 1961.
18. B. G. King, unpublished work.
19. M. A. Rezny and J. S. Wu, unpublished work.
20. B. M. Oliver, J. R. Pierce, and C. E. Shannon, "The Philosophy of PCM," Proc. IEEE (November 1948), pp. 1324–31.
21. A. Vigants and M. V. Pursley, "Transmission Unavailability of Frequency—Protected Microwave FM Radio Systems Caused by Multipath Fading," B.S.T.J., *58,* No. 8 (October 1979), pp. 1779–96.
22. A. Gieger and W. T. Barnett, "Effects of Multipath Propagation on Digital Radio," IEEE Trans. Commun., *COM*-29, No. 9 (September 1981), pp. 1345–52.
23. A. A. M. Saleh and J. Salz, "Adaptive Linearization of Power Amplifier Nonlinearity in Digital Radio Systems," B.S.T.J., *62,* No. 4 (April 1983).

Part VI
Future Trends

Future Trends in Microwave Digital Radio

Editor's Note: We lead off this part with three short papers, introduced by Heiichi Yamamoto, that forecast the future of microwave digital radio in Asia, North America and Europe. The same grouping, with the same set of authors, was used to close the Special Series on Microwave Digital Radio in the February 1987 issue of the IEEE COMMUNICATIONS MAGAZINE. The versions presented here have been updated by the authors and edited for inclusion in this book.

Foreword

HEIICHI YAMAMOTO

THIS sixth and final part of the book deals with future trends in the field of microwave digital radio. The tutorial articles that led off the first five parts dealt with the critical technologies involved and clearly presented the major technical considerations. Because of the progress that has been made worldwide since the early 1970's, microwave digital radio, along with fiber optics and satellites, is now an essential component in realizing digital telecommunications networks. The importance of this transmission technology will continue to increase in the future. More effort than ever will be made in this field by many digital radio designers.

Three short papers follow, giving the opinions of six specialists from Asia, North America, and Europe on future trends in the field. The authors are all renowned by virtue of remarkable contributions to this field. Without a doubt, they are well qualified to foretell future trends, as well as to report on the present status of digital radio in their respective parts of the world. Some of the authors, in their discussions, refer to spectacular innovations, such as super multilevel modulation and co-frequency repeating (that is, transmission and reception on the same frequency), which will greatly stimulate digital communications engineers. Other authors take note of the connectivity of digital radio with optical fiber systems, which will play a very important future role in nationwide digital networks.

Through these three discussions, readers will see that the technical trends in microwave digital radio are directed mainly at increasing spectrum utilization efficiency. This means that even higher-level modulations are to be pursued in the future. All the authors indicate that quadrature amplitude modulation (QAM) will be used to achieve signal constellations up to the level of 256 points, and possibly 1024 points, and that new-generation systems using these large constellations will be achieved in the near future. To realize such systems, a variety of technical problems must be overcome. In particular, adaptive signal processing, such as cross-polarization interference cancellation (XPIC) and new antenna designs, must be researched. In addition, from the viewpoint of devices, CMOS-VLSI, MMICs, SAW filters, and GaAs FETs will be key technologies in obtaining economical repeater configurations.

Finally, I would like to add my personal opinion. Microwave digital radio used as a signal transmission medium is characterized by its excellent transmission performance under normal conditions. This implies that, when we establish countermeasures for multipath fading and interference and overcome outages due to these impairments, radio will offer an extraordinarily reliable transmission medium. It is very important that microwave system designers continue their efforts to increase the reliability of transmission, as well as increasing transmission capacity.

Future Trends in Microwave Radio: A View from Asia

KENJI KOHIYAMA AND OSAMU KURITA

THE PRESENT DIGITAL RADIO NETWORK IN JAPAN

PRESENTLY, microwave radio-relay routes are used for about half of all toll telephone circuits and almost all television program circuits in Japan. The network of microwave routes covers the entire country, as shown in Fig. 1. Analog microwave systems employing frequency modulation (FM) have been placed on most of these routes. In particular, FDM/FM systems in the 4-, 5-, and 6-GHz bands have played an important role in large capacity long-haul transmission.

NTT is now concentrating efforts on network digitalization in Japan and the introduction of digital radio-relay systems. The highest percentage of digital radio routes at present are those in toll areas. As of the end of 1986, only about 20 percent of all long-haul transmission routes used digital radio. However, the radio-relay network shown in Fig. 1 will be almost completely digitalized by the end of the year 2000 [1], [2]. The digital radio systems used in Japan today are summarized in Table I. Among these, the 2-, 11-, 15-, and 20-GHz bands are used for short-haul transmission, and the 4-, 5- and 20-GHz bands are used for long-haul transmission.

The early digital radio systems employed 4-PSK modulation. After much research, systems using 16-QAM were developed in 1982, offering twice the capacity of 4-PSK. By combining 16-QAM with dual polarization, it was possible to achieve a spectral efficiency of 5 bit/s/Hz [3], [4], [5].

A long-span digital radio system was developed in 1984 to digitalize the overwater transmission lines between the mainland and the isolated Okinawa islands. This system provides a capacity of 200 Mb/s per 40 MHz channel using 16-QAM modulation and dual-polarization, and it achieves repeater spacing of about 100 km on overwater paths [6]. The main new techniques for this system are 1) multicarrier transmission, that is, several digital carriers per radio channel, to enhance resistance to multipath fading [7]; and 2) cross-polar cancellation, that is, an adaptive countermeasure to the deterioration of cross-polarization discrimination in multipath fading.

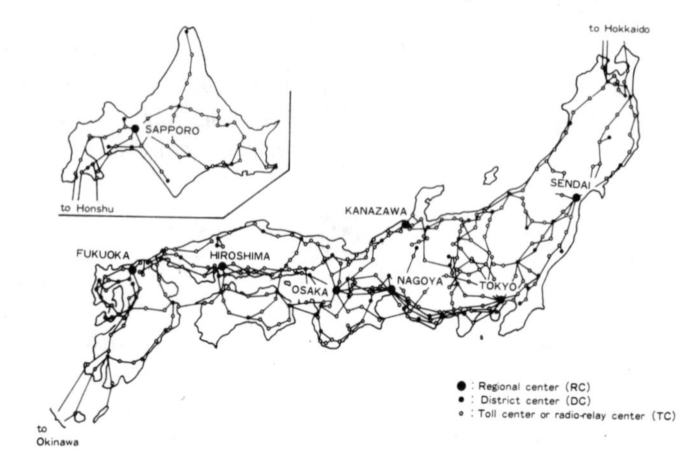

Fig. 1. Long-haul microwave routes in Japan.

TECHNOLOGIES FOR NEAR-FUTURE SYSTEMS

Spectrum utilization efficiency, in bits/s/Hz, is one of the evaluation criteria for digital radio systems performance. Fig. 2 shows this quantity for single-polarization multilevel QAM systems, plotted against receiver input carrier-to-noise ratio under ideal conditions (no multipath fading, ideal coherent detection). Also shown is the theoretical upper bound on spectrum efficiency given by Shannon's limit.

A new 400 Mb/s system using 256-QAM and multicarrier transmission is now under development in Japan [8]. This system achieves a spectrum efficiency of 10 bits/s/Hz by employing dual polarization. The first field transmission test, using a single-carrier repeater, was conducted on a 54.6 km land span in August 1985 [9]. A second field transmission test, using four-carrier repeaters, was conducted on a 33.5 km overwater span in the summer of 1986. The purpose was to estimate the performance of multicarrier repeaters using common amplification, and to assess countermeasures for waveform distortion and interference, such as space diversity, transversal equalization, and cross-polar interference cancellation.

In the following subsections, we discuss the sophisticated technologies needed to deploy high-capacity radios like the 256-QAM system under development [10], [11], [12]. The topics covered are (i) multilevel modems; (ii) waveform distortion countermeasures; (iii) interference countermeasures; (iv) diversity; (v) power amplifiers; and (vi) antennas.

Multilevel Modems

Performance requirements for multilevel modems increase with the number of modulation levels. For example, assuming a bit error ratio (BER) of 10^{-4}, the receiver input carrier-to-noise ratio required for 256-QAM is 11.9 dB greater than that required for 16-QAM. As another example, consider rms phase jitter and the associated power penalty (the increase in carrier power needed to maintain BER at 10^{-4}): Assuming a power penalty of 0.5 dB, the allowed phase jitter is 12 dB smaller for 256-QAM than for 16-QAM [10]. These numbers underscore the importance of achieving accuracy in the modulators, cosine-rolloff spectrum shaping, carrier recovery and multilevel decision circuits. Monolithic IC multipliers for the modulation function have been designed for this purpose. Also, ROM predistortion to correct for modulator nonlinear-

TABLE I
MAIN DIGITAL MICROWAVE SYSTEMS IN JAPAN

Frequency Band (GHz)	Capacity per RF Channel (Mb/s)	Channel Spacing (MHz)	Modulation Technique	Typical Repeater Spacing (km)	Transmitting Power (dBm)
2.11~2.29	3	1.25	4PSK	25	21
	32	7	16QAM	50	23
3.6~4.2 4.4~5.0	200	40	16QAM	50	26
10.7~11.7	100	40	4PSK	25	30
	12.6	5	4PSK	15	26
14.4~15.23	100	40	4PSK	8	23
17.7~21.2	400	160	4PSK	6	26

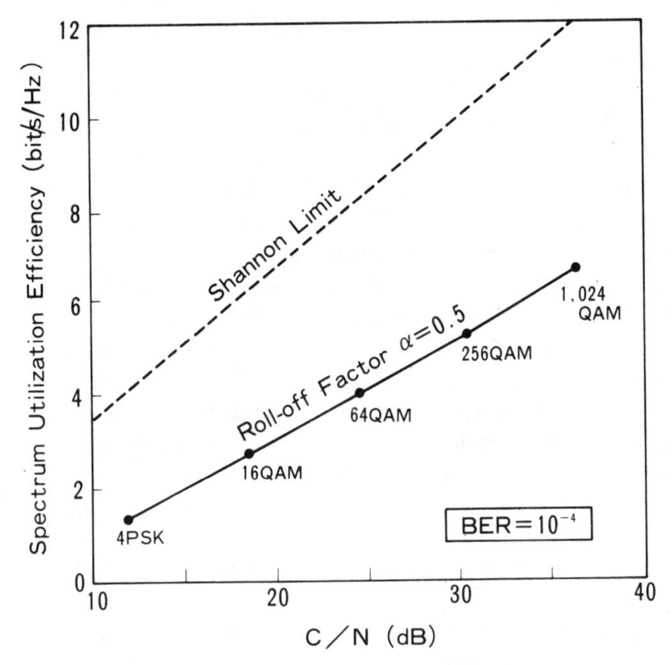

Fig. 2. Spectrum utilization efficiency.

ity, and automatic threshold control to minimize dc drift of the demodulated signal, are effective techniques.

Another valuable technique in multilevel modems is forward error correction (FEC) coding, to minimize residual bit errors under normal conditions [13]. The type and rate of the code selected should consider the tradeoff between power efficiency and spectrum efficiency. For the 400-Mb/s system under development, power penalties of 1 dB (at BER $= 10^{-4}$) and 5 dB (at BER $= 10^{-9}$) have been measured on a link using a particular FEC code, and the residual bit error ratio under normal conditions was reduced to below 10^{-10} [14].

Wavefore Distortion Countermeasures

Multipath induced waveform distortion increases as the modulating level increases. For example, at BER $= 10^{-4}$, the allowable linear amplitude dispersion over the Nyquist bandwidth is only 1 dB for a 256-QAM system, which is 4 dB smaller than that for 16-QAM. Therefore, a variety of countermeasures must be studied and developed.

The most effective countermeasure is time-domain transversal equalization. At present, three types of 7-tap transversal equalizers (IF, baseband analog, and all-digital) are under consideration. Another effective technique for high-capacity systems in multipath environments is multicarrier transmission. The current designs feature four separate digital carriers, each with one-fourth the bandwidth that a signal carrier at the same total bit rate would have. Since vulnerability to multipath grows sharply with signal bandwidth, these carriers are more robust to such distortion. The cost is greater demand on device technology. Considering the rapid development of this technology, however, applications of this technique should extend far into the future.

Interference Countermeasures

In multilevel modulation systems, the required carrier-to-noise ratio increases sharply with the number of modulation levels. For example, the required carrier-to-interference ratio for dual-polarization systems is 14 dB greater for 256-QAM than for 16-QAM. A cross-polarization interference canceler (XPIC) is thus very important for systems employing dual polarization. A new XPIC technique using a transversal equalizer can cancel interference even when the frequency characteristics of the cross-polar interference are different from those of the original signal. A plan to reduce the size and cost of this XPIC circuit via full digitalization is now being investigated [15].

Other forms of co-channel interference, including that from FM systems, are also important. We cite further countermeasures later in the discussion of antennas.

Diversity

Space diversity (SD) has been widely used in a variety of microwave systems because it minimizes received power degradation and waveform distortion. It also provides large improvements when combined with transversal equalization [16].

A new SD combining method is being studied wherein a sensing path is isolated from the output signal path, and an algorithm is used which enables more stable and robust operation in the presence of in-band amplitude dispersion. It is now being examined through field transmission testing [17].

Power Amplifiers

In the most rudimentary digital radio system (for example, single-carrier with 4-PSK modulation), the power amplifier is not a topic of major importance since transmitting power is relatively low and the envelope is nearly constant. However, in multicarrier transmission of multilevel modulations, the high power amplifier and the use of nonlinear predistortion are very important. Fig. 3 shows the increase in required back-off for a particular transmitting power amplifier. The benefit of using predistortion in the 4-carrier case is evident.

A novel new predistortion linearizer, using a diode and a circulator in the distortion generator to obtain high signal component isolation, is being investigated. This predistortion linearizer is suitable for MIC configuration and compactness [18].

Antennas

To minimize cross-polar interference and FM interference from existing microwave routes, low sidelobes and low cross-polarized antenna radiation patterns are required. An offset antenna has been developed in Japan that features low sidelobe characteristics, achieved by shaping the curvatures of reflectors and by suppressing diffraction from the antenna structures. Low cross-polarized characteristics are achieved by properly setting the orientations and separations of the three reflectors used. Other features of this configuration are smaller height and lighter weight, as compared to the horn reflector antenna.

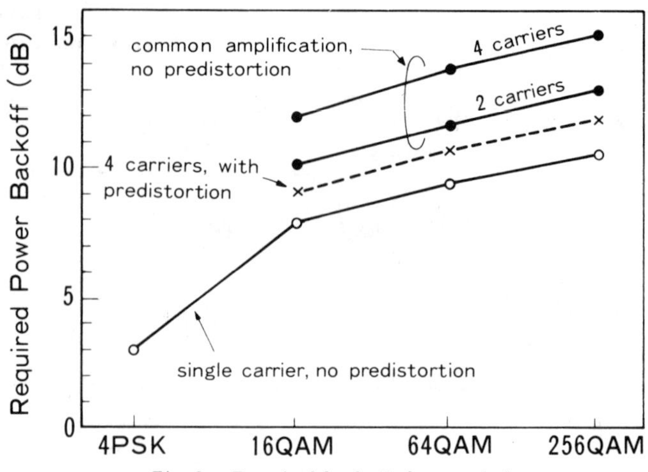

Fig. 3. Required backoff characteristics.

TECHNOLOGIES FOR FUTURE SYSTEMS

Since the frequency resource is limited, it is vital to develop even higher-level modulations to increase digital radio transmission capacity. We have recently arrived at the generation of 256-QAM, but we will encounter great difficulties if we pursue the next logical scheme for doubling capacity, such as 256 × 256-QAM (256 levels in each quadrature rail of the transmitted signal).

Has radio transmission effectiveness thus reached its upper limit? The answer is, "No!" Digital radio has a great future, for we can expect to increase transmission capacity many times over by a variety of measures. Technologies for future digital radio, including technical problems and estimated time frames of introduction, are presented in Table II. The successful development of all these technologies will result in an enormous increase of transmission capacity.

However, increasing the reliability of digital radio transmission is just as important as increasing the transmission capacity. The more digital integrated services will be used, the more this aspect will be stressed. Fortunately, the technologies tabulated in Table II can improve the reliability whether the transmission capacity is within the present level or exceeds it. Blocking time values on the order of 1/10 ~ 1/100 of the present CCIR objective can be obtained in the future.

Several comments on these future technologies are described next.

Multilevel and "Super Multilevel" Modulation—At present, 256-QAM systems are almost in hand. However, the trend to even higher numbers of levels will not cease. "Super multilevel" modulations, such as 256×256-QAM, can be developed by the year 2020. Technical problems will include waveform distortion countermeasures and the elimination of residual bit errors. Forward error correction (FEC) and combined modulation/coding may be key technologies.

Dual-Polarization—Co-channel dual polarization can double route capacity. For this purpose, XPIC techniques are being studied intensively for future application, and these studies will be completed in the 1980's. Considering the rapid development of Digital Signal Processing (DSP), high-precision correlation detection techniques, such as matched filtering, will certainly be achieved in the near future.

One-Frequency Repeating—If we can succeed in improving interference canceling techniques, we can double route capacity by using co-frequency transmission and reception on the same link. High-performance antennas and transmitting power control techniques will also contribute to this approach. Since interference in this case is from the same system, canceling it may not be so difficult. One-frequency repeating will probably be developed in the 1990's.

High-Density Branching—Interference canceling techniques will enable high-density branching, thereby increasing transmission capacity at network nodes. In environments where a variety of radio systems such as satellite and mobile radio co-exist, such interference canceling will be essential, possibly determining the fate of future microwave digital radio.

Higher Frequency Bands—The use of higher frequency bands, such as sub-millimeter wave and millimeter wave, could increase system capacity greatly. Radio frequency circuit technology, such as MMIC's (monolithic microwave IC's) using GaAs FET's, will be very important. Countermeasures for rain attenuation must also be studied intensely.

TABLE II
PROMISING FUTURE TECHNOLOGIES IN DIGITAL RADIO

Technique	Increases In—	Problems	Countermeasures and Key Technologies	Realization Time Frame*
Super Multi-Level Modulation	Radio Channel Capacity	Waveform Distortion Residual Errors Fading	Transversal Equalizers, Matched Filters, Space Diversity, Multi-Carrier, Coded Modulation, Forward Error Correction, Microwave Digital Processing	256QAM: 1980s 256×256 QAM: 2020s
Dual-Polarization	Radio Channel Capacity (2 fold)	XPD	High XPD Antennas, Transversal-Type Cancellers	1980s
One-Frequency Repeating	Route Capacity (2 fold)	Transmit/Receive Interference	Transversal-Type Cancellers, Transmitting Power Control, Network Control	1990s
High Density Branching	Nodal Capacity (3-4 fold)	Branch Route Interference	Interference Cancellers, High Directivity Antennas, Network Control	2000s
Higher Frequency Bands	Route & Nodal Capacity (2 fold)	Rain Attenuation RF Circuit	MIC's, Monolithic MIC's, GaAs Amplifiers, Linearizers, Analog-Digital Hybrid Repeaters	Sub-Millimeter: 1990s Millimeter: 2010s
Route Diversity	Nodal Capacity (2 fold)	Switching Control Fading Rain Attenuation	Hitless Switching, Network Control, Satellites	2000s

*Time frames beyond the year 2000 are estimates or goals.

It will be necessary to clarify long-span transmission characteristics of the higher frequency bands. Considering the sizes of rain cells, the idea that rain attenuation is proportional to distance on such spans may be found to be overly pessimistic.

Route Diversity—Route diversity will also be widely utilized in the future. To realize route diversity, network control technologies, including the use of satellite communications, will be vigorously investigated. It may extend hop length, reduce system cost, and open up the new higher frequency bands. As the use of radio communications increases, route diversity will increase, and its realization will become easier. It may provide the strongest reinforcement to future radio communications networks.

CONCLUSION

Some of the techniques discussed here are already making steady growth; therefore, prospects for increases in digital radio transmission capacity are promising. Further, the construction of radio transmission networks which integrate terrestrial radio, satellite, and mobile communications seems likely in the future.

REFERENCES

[1] S. Matumoto, *et al,* "Digitalization of the existing radio-relay network using 4/5 GHz 200 Mb/s system," in *ICC '82,* pp. 2B.4.1–4.5, June 1982.

[2] S. Katayama, *et al,* "Digital radio-relay systems in NTT," in *ICC '84,* pp. 978–983, May 1984.

[3] T. Murase, *et al,* "200 Mb/s 16 QAM digital radio system with new countermeasure techniques for multipath fading," in *ICC '81,* pp. 46.1.1–5.1, June 1981.

[4] I. Horikawa, *et al,* "Design and performance of a 200 Mb/s 16 QAM digital radio system," *IEEE Trans. Commun.,* vol. COM-27, no. 12, pp. 1953–1958, Dec. 1979.

[5] H. Yamamoto, "Advanced 16 QAM techniques for digital microwave radio," *IEEE Commun. Mag.,* vol. 19, no. 3, pp. 36–45, May 1981.

[6] M. Araki, *et al,* "100 km overwater span digital radio system," in *ICC '85,* pp. 460–465, June 1985.

[7] T. Yoshida, *et al,* "System design and new techniques for an overwater 100 km span digital radio," in *ICC '83,* pp. 664–670, June 1983.

[8] Y. Saito, *et al,* "Feasibility considerations of high-level QAM multicarrier system," in *ICC '84,* pp. 665–671, May 1984.

[9] Y. Saito, et al, "400 Mb/s 256 QAM digital microwave radio system performance," in *ICC '86,* pp. 451–456, June 1986.

[10] Y. Saito, *et al,* "256 QAM modem for high capacity digital radio system," *IEEE Trans. Commun.,* vol. COM-34, no. 8, pp. 799–805, Aug. 1986.

[11] Y. Yoshida, *et al,* "6 GHz 140 MBPS digital radio repeater with 256 QAM modulation," in *ICC '86,* pp. 1482–1486, June 1986.

[12] Y. Daido, *et al,* "256 QAM modem for high capacity digital radio systems," in *GLOBECOM '84,* pp. 547-551, Nov. 1984.

[13] K. Nakamura, *et al,* "A class of error correcting codes for DPSK channels," in *ICC '79,* pp. 45.4.1–4.5, June 1979.

[14] Y. Nakamura, *et al,* "256 QAM modem for multicarrier 400 Mb/s digital radio," *IEEE J. Select. Areas Commun.,* vol. SAC-5, no. 3, pp. 329-335, Apr. 1987.

[15] H. Matue, *et al,* "Digitalized cross polarization interference canceler for multilevel digital radio," *IEEE J. Select. Areas Commun.,* vol. SAC-5, no. 3, pp. 493–501, Apr. 1987.

[16] S. Komaki, *et al,* "Performance of 16 QAM digital radio system using new space diversity," in *ICC '80,* pp. 52.2.1–52.2.6, June 1980.

[17] H. Ichikawa, *et al,* "256 QAM multicarrier 400 Mb/s microwave radio system field tests," in *ICC '87,* pp. 1803–1808, June 1987.

[18] N. Imai, *et al,* "High power amplifier linearization for a multicarrier digital microwave system," in *GLOBECOM '86,* pp. 547–553, Dec. 1986.

Future Trends in Microwave Digital Radio: A View from North America

MARTIN H. MEYERS AND VASANT K. PRABHU

Introduction and Historical Background

THE terrestrial microwave radio network, specifically at frequencies 4 and 6 GHz, has become ubiquitous in North America (see Fig. 1) and is now the primary "workhorse" of telecommunications. For example, the total number of microwave radio stations at 4 and 6 GHz used in the USA by AT&T now exceeds 3,000. The percentage of long-haul traffic carried on radio is in excess of 70 percent, and investment in this valuable embedded resource by AT&T alone exceeds 2.5 billion dollars [1].

Microwave radio was first developed as an analog network, using FM initially as the primary form of modulation, and progressing to the introduction of SSB in the 1970's. The migration since 1970 of this network to digital technology has been driven largely by the introduction of digital switching, increased use of the public switched network for the data transmission, and increased demand for high-quality transmission. The advances in digitizing the radio network in Canada and the USA, both for short- and long-haul routes, has been dramatic [2]–[4].

The first five tutorials in this book [5]–[9] have dealt with various aspects of state-of-the-art digital radio. In this discussion on future trends in North America, we extrapolate from the past, speculate on the future, and try to conjecture how digital radio may fare during the next several decades. Since the radio network has proven to be one of the most flexible high-quality transmission media, we anticipate that a modernized and digitized radio network, along with the optical fiber network gradually being introduced in North America, will form the transmission medium of digital communications in years to come.

Fig. 1. AT&T 4-GHz radio network in the USA (1986).

Technology Trends for Increasing Capacity

Ultimate Digital Radio Capacity

In order to provide a reference for comparison, let us compute the ideal spectral efficiency that can only be approached by using Shannon-type signaling. Using Shannon's formula [10] for a channel limited by white Gaussian noise, we can say that the spectral efficiency η is bounded by

$$\eta = \frac{\text{Bit Rate}}{\text{Channel Bandwidth}} \leq \log_2 (1 + S/N) \text{ bits/s/Hz} \quad (1)$$

where S/N is the signal-to-noise ratio at the receiver. A typical achievable S/N, allowing for fading statistics, is on the order of 30 dB, in which case $\eta < 10$ bits/s/Hz.

For conventional (uncoded) systems, assuming that the channel bandwidth must be approximately 4/3 the signaling rate for satisfactory performance, the spectral efficiency can be expressed as

$$\eta = 0.75 \log_2 M \text{ bits/s/Hz} \quad (2)$$

where M is the number of points in the signal constellation ($M = 4$ for QPSK and $\eta = 1.5$ bits/s/Hz; $M = 64$ for 64-QAM and $\eta = 4.5$ bits/s/Hz).

We compare in Fig. 2 the ideal efficiency of (1) to the spectral efficiencies, (2), that can be achieved using M-ary QAM and a maximum allowable bit error ratio of 10^{-6}. As the number of modulation levels (M) increases, the spectral efficiency increases at the cost of required S/N, but we move no closer to the Shannon limit. This is because the limit can only be approached by using complex and costly coding with possibly large delays.

Co-Channel Dually Polarized Transmissions

We conjecture that, with spectrum efficiency of primary concern, the use of dually polarized (dual-pol) transmissions to double radio channel capacity will increase in North America. The cross-polarization (cross-pol) discrimination that can be achieved in the present radio network is of the order of 30 to 35 dB under normal conditions. Since dual-pol operation is highly susceptible to cross-pol interference, with susceptibility increasing monotonically with M, the use of cross-pol cancellers will be required for high-level modulation systems. At present, such cancellers are being realized at lower modulation levels, with increasing effort being spent on their use at higher levels. Improvements in antenna design, especially in antenna cross-pol discrimination characteristics, should aid the overall process.

Migration to Higher-Order Modulation Systems

Higher-order modulation systems have better spectral efficiency but are more susceptible to the effects of interference,

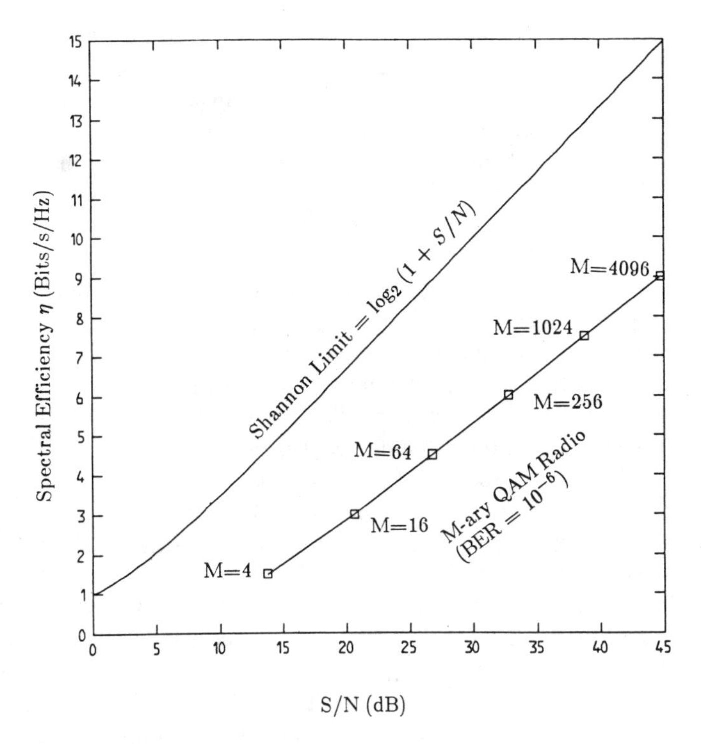

Fig. 2. Spectral efficiencies.

nonlinearities, noise, phase jitter, propagation anomalies, etc. We thus expect future radio systems to use a larger number of levels, such as 256-QAM, 1024-QAM, etc., in conjunction with sophisticated countermeasures to reduce the effects of impairments.

Improvements in manufacturing technologies, such as the design of filters and other components with smaller tolerances, will result in less sensitivity to degradations. In particular, as DSP (Digital Signal Processor) speeds increase, all-digital modems will become feasible and allow highly accurate and reproducible implementations of digital radio receivers. Advances in VLSI (Very Large Scale Integrated) circuitry, coupled with advances in MMIC's (Millimeter Wave Integrated Circuits) and DSP's, may lead to highly integrated radio transmitters and receivers with advanced signal processing and monitoring capabilities.

Error-Correction Coding

It has been observed that increased numbers of modulation levels and sophisticated coding are complementary techniques that may be used to increase the spectral efficiency of digital radio and to shape the spectrum. Error-correction schemes used in present digital radio systems are designed to provide "essentially error-free" transmission during unfaded conditions, and not to provide correction capability under anomalous conditions. More sophisticated techniques, encompassing combined modulation and coding, as well as Viterbi-type decoders in the receivers, promise high spectral efficiency and coding gains under both normal and abnormal propagation conditions.

We expect that more complex coding schemes (such as those suggested by Ungerboeck [11]–[12] and used in present-day voice-band modems) will likely be considered, both to reduce the sensitivity of higher-level modulation schemes, and to shape the spectrum. Since the radio environment in North America is increasingly becoming interference-limited, we also conjecture that future techniques will be specifically designed for immunity to interference.

Diversity Techniques

The first line of defense against the effects of multipath fading has been, and should remain, the proper engineering of routes, particularly hop selection and the choice of transmitter power, antenna, and tower parameters. Frequency diversity techniques are often used as the second line of defense. Recent results have shown that frequency diversity holds great potential for digital radio systems, even more so than for analog systems. This is based upon the observation that, while broadband fading produces power loss over a wide range of channels, a dispersive notch is likely to occur in only one channel at a given time. Also, "essentially error-free" multipath protection may be realizable by the synergistic combination of errorless frequency diversity switching and error correction coding, where the switching is initiated on the basis of the precorrection error rate [13].

With higher-order modulation systems requiring more protection from deep fading, space diversity reception will still be required on some hops. Typically, this involves mounting two antennas with a large vertical separation.

A new form of diversity, referred to as horizontal diversity [14], angle diversity [15]–[19], or aperture diversity [20], has been shown to be more effective than conventional space diversity in combating dispersive fading. Where two antennas are already available on a tower they may be placed side by side to achieve the effect of angle diversity. For example, Balaban, *et al* report in [18] that significant improvements have been achieved by using two separate antennas in an angle-diversity configuration. Alternatively, Lin, *et al* report in [19] that, by combining two different beams in a single antenna, diversity improvements have been achieved without the need for a separate antenna. In addition, successful modeling principles for both configurations, and their relationship to digital radio propagation characteristics, have also been established [18], [19]. Both techniques result in reductions in channel distortion and fade depth. In short, this new form of diversity can be physically realized in many different ways and, as has been experimentally shown, it is highly suited to digital radio applications, where dispersive fading dominates.

Research on diversity combined with error-correction coding may also result in new algorithms and lead to improved performance.

Adaptive Equalization

In addition to diversity, adaptive equalization can be used to counteract the effects of dispersive fading. Significant gains in this field have been obtained through the use of dynamic transversal equalization, an option virtually becoming a standard in the installation of present radio systems. It has been found that, for a nondiversity equalized receiver, outage time can be increased by hysteresis between the receiver's in-lock and resynchronization behavior. More sophisticated equalizers which work under more severe fading conditions will be used in future radio systems.

Recovery Techniques

The degree of random phase variations (jitter) in the recovered carrier and timing wave leads to degradations in performance, and these degradations increase monotonically with the number of levels (M). Current recovery techniques have kept these degradations to tolerable levels. As modulation complexity increases, jitter reduction techniques will need to be further refined.

Interference Countermeasures

Typically, a digital radio operates with a fixed transmitter power and has an SNR on the order of 65 dB under normal propagation conditions. A significant source of impairment in the existing analog systems is the digital radio interference. Accordingly, a trend has developed recently to incorporate dynamic Automatic Power Control (APC) [21]–[23]. This technique permits a digital radio link to operate at maximum power only when the received signal fades below a specified level, and usually not more than 0.01 percent of the time. Since the radio operates at a low power level (typically, 10 to 15 dB below maximum) most of the time, the effect of the APC is to reduce the average value of interference into all systems.

Another approach is to cancel interference from sources with co-located receivers. Such a technique is currently being used to cancel FM interference into digital radio systems [24]. We expect that such techniques may become valuable for all-digital routes as spectrum utilization increases.

Predistortion-Type Techniques

Predistortion-type linearizing schemes are now common, so that transmitter power amplifiers are able to operate at their higher power levels with less distortion and degradation. Since higher-order modulation schemes are more susceptible to nonlinearities in the radio equipment, we foresee increased use of these and more advanced techniques.

Digital Radio at Higher Frequency Bands

The increasing demand for digital transmission in metropolitan areas in North America may lead to the deployment of digital radio at frequency bands above 15 GHz (for example at 18 and 23 GHz), both to provide connectivity to the switched network, and to provide wide bandwidths and increased capacities. Since the propagation characteristics at higher bands may ultimately be controlled by rain attenuation, new design criteria may be required to channelize these bands and optimize their use.

Other Considerations

Connectivity with Optical Fiber

With optical fiber technology being gradually introduced in North America for the transmission of voice, data, video, etc., connectivity of digital radio to this medium will play a very important role. These two technologies will complement each other, as well as compete, because of their diverse characteristics and traffic carrying capacities. In metropolitan areas, where interference may be the major source of impairment, optical fiber technology will probably dominate over digital radio. In rugged terrain, where the installation of fiber may pose serious problems, and in feeder routes, digital radio may be the medium of choice.

Monitoring and Maintenance

Performance monitoring, coupled with automatic capability to replace or switch out failed circuits, can lead to increased use of radio in remote locations. Further, automatic and remote configuration of transmission routes offers the potential for reductions in cost as well as increased flexibility, especially if coordinated with the fiber network.

Maintenance procedures used for the radio network are undergoing extensive revisions to reduce costs, increase reliability, and provide better performance. Further revisions in these procedures and equipment are expected so that diagnostics and maintenance can be combined, and so that some of these tasks can be done remotely and less frequently.

Future Directions

The signal constellations used in present-day digital radio are either QAM or PSK (or minor variations of them). Since future digital radio performance is likely to be limited by interference and multipath fading, signal constellations which are more resistant to these impairments are likely to be used. Research in the general area of modulation and coding is warranted.

Research in antenna design for achieving diversity, improved cross-pol discrimination, and reduced sidelobe interference is needed. Also, new methods of combining equalization with frequency, space, and other diversities to suit propagation conditions, to minimize costs, and to achieve desired reliability are needed. Finally, advances in propagation prediction and their use in reducing the effects of propagation irregularities will lead to more efficient and cost-effective uses of digital radio, in North America and elsewhere.

Acknowledgments

Discussions held with W. T. Barnett, C. P. Bates, L. J. Greenstein, J. J. Kenny, W. D. Rummler, and R. P. Slade of AT&T Bell Laboratories, S. C. Ladd of AT&T Network Systems, G. D. Lystad and J. L. Robinson of AT&T Network Operations Group, and M. Shafi of Telecom Corporation of New Zealand are gratefully acknowledged.

References

[1] M. J. Pagones and V. K. Prabhu, "Effect of interference from geostationary satellites on the terrestrial radio network," in *Conf. Rec., GLOBECOM 1985*, pp. 47.7.1–47.7.5, Dec. 1985.

[2] I. Godier, "DRS-8 digital radio for long-haul transmission," in *Conf. Rec., ICC 1977*, pp. 5.4.102–5.4.105, June 1977.

[3] R. A. Roadhouse, T. G. Fellows, and J. L. Spencer, "The trans-Canada digital radio network," in *Conf. Rec., ICC 1977*, pp. 5.1.91–5.1.95, June 1977.

[4] C. P. Bates, W. G. Robinson, III, and M. A. Skinner, "DR-6-135 system design and applications," in *Conf. Rec., GLOBECOM 1984*, pp. 16.7.1–16.7.8, Nov. 1984.

[5] D. P. Taylor and P. R. Hartmann, "Telecommunications by microwave digital radio," Part I of this book. [An earlier version was published in *IEEE Commun. Mag.*, vol. 24, no. 8, Aug. 1986, pp. 11–16.]

[6] T. Noguchi, Y. Daido, and J. A. Nossek, "Modulation techniques for microwave digital radio," Part II of this book. [An earlier version was published in *IEEE Commun. Mag.*, vol. 24, no. 11, Sept. 1986, pp. 21–30.]

[7] W. D. Rummler, R. P. Coutts, and M. Liniger, "Multipath fading channel models for microwave digital radio," Part III of this book. [An earlier version was published in *IEEE Commun. Mag.*, vol. 24, no. 11, Nov. 1986, pp. 30–42.]

[8] J. K. Chamberlain, F. M. Clayton, H. Sari, and P. Vandamme, "Receiver techniques for microwave digital radio," Part IV of this book. [An earlier version was published in *IEEE Commun. Mag.*, vol. 24, no. 11, Nov. 1986, pp. 43–54.]

[9] L. J. Greenstein and M. Shafi, "Outage calculation methods for microwave digital radio," Part V of this book. [An earlier version was published in *IEEE Commun. Mag.*, vol. 25, no. 2, Feb. 1987, pp. 30–39.]

[10] R. W. Lucky, J. Salz, and E. J. Weldon, *Principles of Data Communication.* New York, NY: McGraw-Hill, 1967.

[11] G. Ungerboeck, "Trellis-coded modulation with redundant signal sets: Part I: Introduction," *IEEE Commun. Mag.*, vol. 25, no. 2, pp. 5–11, Feb. 1987.

[12] ——, "Trellis-coded modulation with redundant signal sets: Part II: State of the art," *IEEE Commun. Mag.*, vol. 25, no. 2, pp. 12–21, Feb. 1987.

[13] C. P. Bates, G. L. Frazer, G. D. Martin, and W. C. Trested, "Effectiveness of error correction and errorless frequency diversity switching in a multipath environment," *Conf. Rec., ICC 1987*, Seattle, WA, Paper 23.4, pp. 826–829, June 1987.

[14] M. F. Gardina and S. H. Lin, "Measured performance of horizontal space diversity on a microwave radio path," *Conf. Rec., GLOBECOM 1985*, pp. 36.6.1–36.6.4, Dec. 1985.

[15] A. Malaga and S. A. Parl, "Experimental comparison of angle and space diversity for line-of-sight microwave links," *Conf. Rec., MILCOM 1985*, Boston, MA, Paper 19.5.

[16] R. W. Hubbard, "Angle diversity reception for LOS digital microwave radio," *Conf. Rec., MILCOM 1985*, Boston, MA, Paper 19.6.

[17] K. P. Dombek, "Reduction of multipath interference by adaptive beam orientation," *Conf. Rec., European Conf. Radio Relay Syst.*, Nov. 1986, pp. 400–406, Nov. 1986.

[18] P. Balaban, E. A. Sweedyk, and G. S. Axeling, "Angle diversity with two antennas: Model and experimental results," in *Conf. Rec., ICC 1987*, Seattle, WA, Paper 23.7, pp. 846–852, June 1987.

[19] E. H. Lin, A. J. Giger, and G. D. Alley, "Angle diversity on line-of-sight microwave paths using dual-beam dish antennas," in *Conf. Rec., ICC 1987*, Seattle, WA, Paper 23.5, pp. 831–841, June 1987.

[20] P. M. Dekan, J. H. Berg, and M. Evans, "Aperture diversity using similar antennas," in *Conf. Rec., ICC 1987*, Seattle, WA, Paper 23.6, pp. 842–845, June 1987.

[21] P. Dupuis, M. Joindot, A. Leclert, and M. Rooryck, "Fade margin of high capacity digital radio system," in *Conf. Rec., ICC 1979*, Boston, MA, Paper 48.6, pp. 48.6.1–48.6.5, June 1979.

[22] G. Bonnerot, L. Bourgeade, C. Lerouge, and M. Daout, "Performance of the STN65-140 digital radio," in *Conf. Rec., ICC 1984*, Amsterdam, The Netherlands, pp. 1358–1370, June 1984.

[23] T. R. Cooper, P. D. Lindsay, S. A. Harvey, J. E. Knecht, and S. A. Lee, "90 Mb/s transmission on the TD network," in *Conf. Rec., GLOBECOM 1986*, Houston, TX, Paper 51.1, pp. 1836–1842, Dec. 1986.

[24] D. J. Ramirez and D. J. Wisler, "Modifying existing FM analog radio relay systems to carry high speed broadband digital signals," in *Conf. Rec., GLOBECOM 1986*, Houston, TX, Paper 15.5, pp. 534–542, Dec. 1986.

Future Trends in Microwave Digital Radio: A View from Europe

GREG HART AND JAN A. STEINKAMP

Introduction

MICROWAVE radio relay transmission presently accounts for between 20 and 50 percent of overall traffic carried within each of the European transmission networks, and digital radio now provides about 20 percent of this capacity. The first digital radio systems in Europe were brought into service in the early 1960's, in the 13-GHz band, and use has been growing steadily since that time. Digital radio is still considered by most CEPT administrations to be of strategic necessity, alongside optical fiber transmission, and continues to have a healthy future. This is particularly so in the climate of telecommunications liberalization which is now emerging within Europe. An insight into some of the commercial and political trends in the use of microwave radio can be found in [1].

Early European digital radio systems were optimized for mainly national requirements, and several different, sometimes incompatible, designs have emerged in the evolutionary process. Basic system designs now are converging towards a few generic standards, partly as a result of greater commonality of operational requirements, and partly because of system harmonization activities within CEPT [2].

Radio is also used to service regional and local network sectors within Europe, and this is predicted to be an expanding future requirement. New non-PTT markets are being established, and higher frequency bands are being opened up, providing greater stimulus to develop new systems for these applications.

This "View from Europe" gives an outline of the trends in system applications foreseen by CEPT network operators, and the trends in system technology foreseen by some of the radio manufacturing interests within Europe [3].

Early Evolutionary Trends

Trends in digital radio technology have been influenced by three distinct evolutionary stages. Many of the early systems used low-level modulation techniques, such as QPSK or 8-PSK, which are not so demanding in terms of the transmission channel's technical characteristics (linearity, amplitude and phase response, signal-to-noise conditions, and so on). Existing analog transceiver technology could often be re-used, together with IF digital modems. Thus, in the initial stage, requirements could be satisfied without the introduction of substantially new technologies [4].

The transition from these first-generation designs to second-generation designs was brought about by the operational need for higher-order modulation techniques, such as 16-QAM and 64-QAM. Here, the analog transceiver technology was inadequate, since a high degree of linearity is required in the RF transmitting amplifiers to account for the amplitude component of the modulation. The technology therefore changed from essentially limiting RF amplifiers to high-linearity GaAs FETs, or TWTs, with power backoff or predistortion to produce the required linearity [5].

The sensitivity of high-level modulation methods to phase and amplitude distortions during dispersive multipath fading also necessitated the introduction of new technology. Adaptive equalizers based upon a transversal filter structure have turned out to provide nearly optimal solutions to this problem because of their high performance, stability, and rapid acquisition time [6]. Also, the extensive high-speed signal processing in these modems requires a large number of transistor functions; this has opened the way to large scale integration of partial modem functions, for example, using ECL gate arrays with up to 2500 gate functions, and delay times of 350 ps.

Third-generation systems using 256-QAM [5], and possibly 1024-QAM, are now in the early development stage. The trends in these systems are discussed later.

Trends in Trunk Network Applications

Many of the frequency bands available for trunk applications have already been heavily utilized for analog systems, but as these near the end of useful service life they are gradually being replaced with digital systems. Eventually, all the trunk radio bands below 12 GHz are likely to be primarily digital, but there will be a transition period when compatibility with analog systems is an essential requirement.

The trunk radio network infrastructures of the European countries differ markedly (maximum hop lengths, nodal angles, and so forth), due to geographical factors and to cost considerations imposed at the time these networks evolved. Modern system designs must therefore cope with the variety of constraints imposed by these networks, including the problematical hops, and the steadily increasing interference environment.

The amount of spectrum sharing and re-use is growing, and may reach the point where it will significantly influence future evolution in the trunk radio bands. The net interference levels and coordination criteria resulting from sharing with satellite and other fixed services could, in the future, dictate the performance standards obtainable from trunk radio using high-level modulation techniques.

One further technical constraint on system evolution is downward compatibility. Only in exceptional cases can bands be cleared to allow optimization for a new system technology.

Newly introduced equipment must therefore successfully cohabit with existing systems in terms of channel plans and interference levels. Thus, the basic 30- and 40-MHz channel spacings, as adopted in CCIR recommendations, are likely to continue.

The predominant CEPT requirement is to fit integer multiples of the 140-Mbit/s hierarchical level into these 30- and 40-MHz channel spacings, although there are also some 2 × 34-Mbit/s applications. Fitting these bit rates in can be quite tricky, as illustrated in Table I.

The next evolution of the generic 16-QAM 140-Mbit/s design, already in extensive operational service, is likely to be the use of dual-polar operation on 40-MHz channel spacings. This approach offers the prospect of doubling spectrum efficiency by retrospective system upgrades, including reducing the rolloff factor of the spectrum shaping filters to improve adjacent channel protection, and improving transmitter linearity. There is some technical risk with dual-polar operation that has still to be quantified, although some European administrations are already committed to it, and to developments in adaptive cross-polar interference cancelers (AXPICs).

For channel spacings around 30 MHz, the obvious short term trend is towards 64-QAM, since this gives good compatibility with existing analog utilization, particularly in the lower 6-GHz band, which is the most popular CEPT requirement for 64-QAM at present. In the longer term, there may also be prospects of dual-polar operation with 64-QAM, given good antenna performance and AXPICs.

Trunk radio systems are normally required to comply with the high-grade performance requirements given in CCIR Recommendation 634 and, as the complexity of systems increases, it becomes more difficult to satisfy these requirements. Space diversity, transversal equalizers, and decision feedback equalizers are already in selective use as fading countermeasures in 16-QAM systems but will need to be used more extensively with 64-QAM systems, thereby adding to the cost. The performance of these adaptive techniques also needs to be improved, in keeping with the latest understanding of propagation, and how it affects the behavior of more complex systems. Here there is a vital need for system developers to receive good feedback from network operators in terms of how systems respond in field conditions. This had already led to some improvements in aspects such as dynamic signature performance and hysteresis.

Third-Generation Digital Radio Systems

The development of third-generation systems is being driven by the network requirement to provide greater transmission capacity by improving bandwidth utilization, and by the need to increase economy in order to maintain the competitive position of radio systems in relation to monomode optical fiber systems.

Spectrum efficiency improvements over second-generation systems are possible by use of modulation schemes of still higher orders, such as 256-QAM or even 1024-QAM, and also by dual-polar operation [8]. Thus, the transmission of 2 × 2 × 140 Mbit/s in a channel spacing of 40 MHz is a possibility, achieving an efficiency of 12 bit/s/Hz.

Table I
Minimum Modulation Methods for CEPT Digital Hierarchy*

Channel spacing	Bit rate 2 x 34 Mbit/s	1 x 140 Mbit/s	2 x 140 Mbit/s	4 x 140 Mbit/s
40 MHz	QPSK	16-QAM	256-QAM	256-QAM with dual-polar operation
30 MHz	8-PSK	64-QAM	1024-QAM	?
20 MHz	16-QAM	256-QAM	256-QAM with dual-polar operation	?

*Taken from CCIR Report 384, assuming 1 percent bit rate overhead, with Nyquist filtering.

The technical requirements of these systems are in many respects much more severe than those of the second generation. The linearity requirement for the signal path becomes comparable to that of SSB-AM systems; furthermore, requirements concerning signal-to-noise ratios and carrier stability are extremely stringent. A still greater problem is the higher sensitivity to propagation anomalies. Apart from dispersive fading, depolarization effects (particularly cross-polarization interference) have to be counteracted wherever dual-polar operation is adopted [9].

To overcome these problems, adaptive signal processing methods will be applied in the future to a much greater extent. This is true not only for functional units such as amplifiers and equalizers, but also between different system parts, and even between systems, in order to optimize transmission channel performance. Adaptive methods will be used not only for transmitter signal path linearization [10] and frequency- and time-domain equalization, but also for cross-polarization interference cancellation, for minimum-distortion combining of main and diversity signals, and for adaptive control of antenna patterns. Furthermore, forward error correction (FEC) coding will generally be applied to improve the BER performance, especially the residual BER under free space conditions [11].

To realize these additional functions for the optimization of transmission characteristics, additional signal processing and circuit linearization measures will be mandatory. The cost of this extra processing, however, should be more than offset by economies now possible in production and testing techniques.

A major part of the necessary savings can be achieved by the application of MMICs in RF circuits, and surface mounted device (SMD) technology in IF and baseband circuits, enabling the automatic positioning of printed circuit board elements. Wider use of numerically controlled machines for the production of RF subassemblies, and use of bus-controlled automatic test set-ups for testing of components and systems, will also enable further economies to be achieved [12].

The large-scale integration of more and more complex system functions in VLSI circuits will have the most dramatic impact on cost reduction and on increased reliability. In the future, we will see the use of fast, custom-specific VLSI-ICs in advanced CMOS technology, providing more than 50,000 transistor functions per circuit. This will facilitate the merging of a number of functional units which, in earlier generations of radio systems, were realized with ECL gate array technology. Thus, a major portion of the additional adaptive functions discussed above can be expected to be incorporated into VLSI-ICs, together with other system functions, thereby holding down production costs.

Trends in Regional and Local Network Applications

The 2- , 7- , 15- , and 13-GHz bands are already well utilized for junction network applications in Europe, using equipment with capacities up to 34 Mbit/s, and based on the Harmonization Recommendations prepared within CEPT. Now, there is an emerging demand for higher capacities in these bands, and the exploitation of higher bands, to provide for growth in traffic requirements.

The evolution is likely to be from 2 × 8 Mbit/s to 34 Mbit/s in the 15-GHz band, and from 34 Mbit/s to 140 Mbit/s in the 13-GHz band. High-level modulation methods (64-QAM, or 16-QAM with reduced rolloff factors) will be necessary to retain the existing 28-MHz channel spacings.

The CEPT requirements for local network radio are for such applications as ISDN, higher-capacity data links, PABX interconnects, videophone, and videoconferencing services. The higher frequency bands, at 18 GHz, 23 GHz, 28 GHz, 38 GHz and above, are most appropriate for these applications, since considerable savings accrue where the spectrum management criteria need not be too stringent. Higher frequencies also permit small and compact designs of equipment and antennas, which can be more readily accommodated on rooftop locations. Simple, cost-effective design will be the main goal of future development.

Conclusions

The importance of radio systems in future telecommunications networks will increase beyond their present status provided that three conditions are satisfied: firstly, if their economy in relation to other transmission media is maintained; secondly, if they can provide for growth in transmission capacity; and thirdly, if they can deliver acceptable standards of performance and reliability.

By the introduction of a third generation of digital radio systems, based on high-level modulation methods with dual-polar operation, and by the exploitation of higher frequency bands, much of this capacity can be expected to be made available. Technological progress in the fields of CMOS-VLSI circuits, MMICs, and SAW filters, as well as automated machining and testing methods, will help to maintain digital radio as an economically competitive medium into the future; and the continuing development and refinement of fading countermeasures will help to provide the necessary standards of transmission performance.

New applications for digital radio, particularly in local networks, also show good promise, and requirements are likely to expand in the future as further technological development permits the cost of these systems to reduce to a level comparable to line systems.

Acknowledgment

The authors would like to thank colleagues within Europe, and in particular from within CEPT subgroup TR4, for their assistance and contributions towards this paper.

References

[1] G. Hart, B. Humphries, and M. Relph, "The role of radio relay systems in the United Kingdom," in *Proc. European Conf. Radio Relay Syst.*, Nov. 1987.

[2] CEPT Recommendation TR4/1, "Digital radio relay systems with input bit rate up to 2 × 8448 kbit/s in the 15 GHz frequency band;" Recommendation TR4/2, "Digital radio relay systems with input bit rate of 34 Mbit/s in the 13 GHz frequency band;" Recommendation TR4/3, "Digital radio relay systems with input bit rate up to 2 × 140 Mbit/s in the 18 GHz frequency band."

[3] H. Panschar and J. Steinkamp, "State-of-the-art and future trends in digital radio relay systems," *Siemens Telecom Rpt. (Special Ed. on Radiocomm.)*, vol. 10, Aug. 1987.

[4] R. B. Rennaker, "Digital microwave, an overview," *Telephone, Engineer, and Management*, pp. 47–52, July 1, 1981.

[5] H.-J. Thaler, "Characterization and performance comparison in high linearity RF power amplifiers for 16-QAM digital radio systems," in *Conf. Rec., GLOBECOM '83*, pp. 870–874, Nov./Dec. 1983.

[6] G. Sebald, B. Lankl, and J. A. Nossek, "Advanced adaptive equalisation of multilevel/QAM digital radio systems," in *Conf. Rec., ICC '86*, pp. 1472–1476, June 1986.

[7] Y. Takeda, N. Iizuka, Y. Daido, S. Taneka, and H. Nakamura, "Performance of 256-QAM modem for digital radio relay systems," in *Conf. Rec., GLOBECOM '85*, pp. 1455–1459, Dec. 1985.

[8] H. Panschar, "High capacity digital radio systems: State of the art and future trends," in *Conf. Prof. IREE*, pp. 464-467, 1985.

[9] M. Kavehrad and J. Salz, "Cross polarization cancellation and equalization in digital channels over dually polarized multipath channels," *AT&T Tech. J.*, vol. 64, no. 10, pp. 2211–2260, Dec. 1985.

[10] J. Grabowski and R. C. Davis, "An experimental M-QAM modem using amplifier linearization and baseband equalization techniques in national telesystems," in *Conf. Rec., National Telesystems Conf.*, Paper E.3.2, Nov. 1982.

[11] M. Salerno, P. Vicini, and F. Cagliari, "Design concepts, system architecture and technology for a new 64-QAM digital radio family," in *Conf. Proc. IREE*, pp. 472–475, 1985.

[12] F. Ivanek, "Microwave communications technology," presented at 4th World Telecommunications Forum Tech. Symp., Oct. 29–Nov. 1, 1983, Geneva, Switzerland.

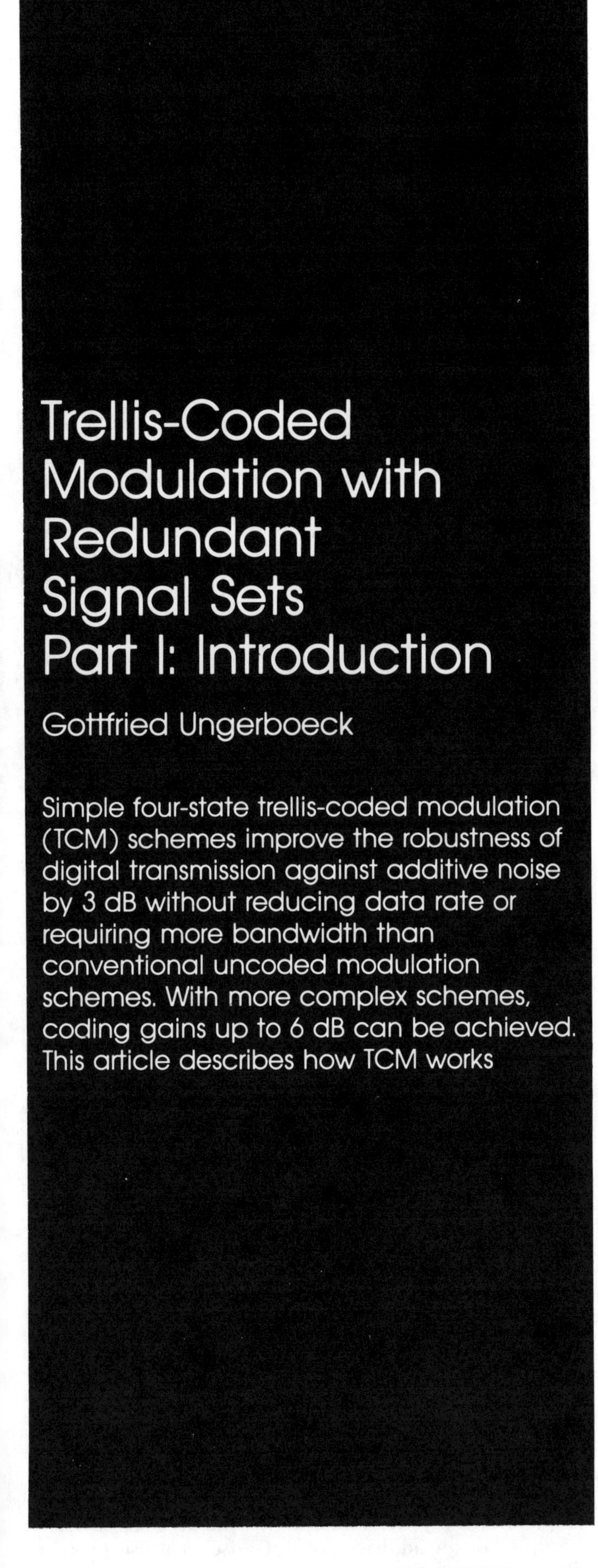

Trellis-Coded Modulation with Redundant Signal Sets Part I: Introduction

Gottfried Ungerboeck

Simple four-state trellis-coded modulation (TCM) schemes improve the robustness of digital transmission against additive noise by 3 dB without reducing data rate or requiring more bandwidth than conventional uncoded modulation schemes. With more complex schemes, coding gains up to 6 dB can be achieved. This article describes how TCM works

Trellis-Coded Modulation (TCM) has evolved over the past decade as a combined coding and modulation technique for digital transmission over band-limited channels. Its main attraction comes from the fact that it allows the achievement of significant coding gains over conventional uncoded multilevel modulation without compromising bandwidth efficiency. The first TCM schemes were proposed in 1976 [1]. Following a more detailed publication [2] in 1982, an explosion of research and actual implementations of TCM took place, to the point where today there is a good understanding of the theory and capabilities of TCM methods. In Part 1 of this two-part article, an introduction into TCM is given. The reasons for the development of TCM are reviewed, and examples of simple TCM schemes are discussed. Part II [15] provides further insight into code design and performance, and addresses recent advances in TCM.

TCM schemes employ redundant nonbinary modulation in combination with a finite-state encoder which governs the selection of modulation signals to generate coded signal sequences. In the receiver, the noisy signals are decoded by a soft-decision maximum-likelihood sequence decoder. Simple four-state TCM schemes can improve the robustness of digital transmission against additive noise by 3 dB, compared to conventional uncoded modulation. With more complex TCM schemes, the coding gain can reach 6 dB or more. These gains are obtained without bandwidth expansion or reduction of the effective information rate as required by traditional error-correction schemes. Shannon's information theory predicted the existence of coded modulation schemes with these characteristics more than three decades ago. The development of effective TCM techniques and today's signal-processing technology now allow these gains to be obtained in practice.

Signal waveforms representing information sequences are most impervious to noise-induced detection errors if they are very different from each other. Mathematically, this translates into the requirement that signal sequences should have large distance in Euclidean signal space. The essential new concept of TCM that led to the aforementioned gains was to use signal-set expansion to provide redundancy for coding, and to design coding and signal-mapping functions jointly so as to maximize directly the "free distance" (minimum Euclidean distance) between coded signal sequences. This allowed the construction of modulation codes whose free distance significantly exceeded the minimum distance between uncoded modulation signals, at the same information rate, bandwidth, and signal power. The term "trellis" is used because these schemes can be described by a state-transition (trellis) diagram similar to the trellis diagrams of binary convolutional codes. The difference is that in TCM schemes, the trellis branches are labeled with redundant nonbinary modulation signals rather than with binary code symbols.

The basic principles of TCM were published in 1982 [2]. Further descriptions followed in 1984 [3-6], and coincided with a rapid transition of TCM from the research stage to practical use. In 1984, a TCM scheme with a coding gain of 4 dB was adopted by the International Telegraph and Telephone Consultative Commit-

Reprinted from *IEEE Comm. Mag.*, vol. 25, no. 2, pp. 5-11, Feb. 1987.

tee (CCITT) for use in new high-speed voiceband modems [5,7,8]. Prior to TCM, uncoded transmission at 9.6 kbit/s over voiceband channels was often considered as a practical limit for data modems. Since 1984, data modems have appeared on the market which employ TCM along with other improvements in equalization, synchronization, and so forth, to transmit data reliably over voiceband channels at rates of 14.4 kbit/s and higher. Similar advances are being achieved in transmission over other bandwidth-constrained channels. The common use of TCM techniques in such applications, as satellite [9-11], terrestrial microwave, and mobile communications, in order to increase throughput rate or to permit satisfactory operation at lower signal-to-noise ratios, can be safely predicted for the near future.

Classical Error-Correction Coding

In classical digital communication systems, the functions of modulation and error-correction coding are separated. Modulators and demodulators convert an analog waveform channel into a discrete channel, whereas encoders and decoders correct errors that occur on the discrete channel.

In conventional multilevel (amplitude and/or phase) modulation systems, during each modulation interval the modulator maps m binary symbols (bits) into one of $M = 2^m$ possible transmit signals, and the demodulator recovers the m bits by making an independent M-ary nearest-neighbor decision on each signal received. Figure 1 depicts constellations of real- or complex-valued modulation amplitudes, henceforth called signal sets, which are commonly employed for one- or two-dimensional M-ary linear modulation. Two-dimensional carrier modulation requires a bandwidth of 1/T Hz around the carrier frequency to transmit signals at a modulation rate of 1/T signals/sec (baud) without intersymbol interference. Hence, two-dimensional 2^m-ary modulation systems can achieve a spectral efficiency of about m bit/sec/Hz. (The same spectral efficiency is obtained with one-dimensional $2^{m/2}$-ary baseband modulation.)

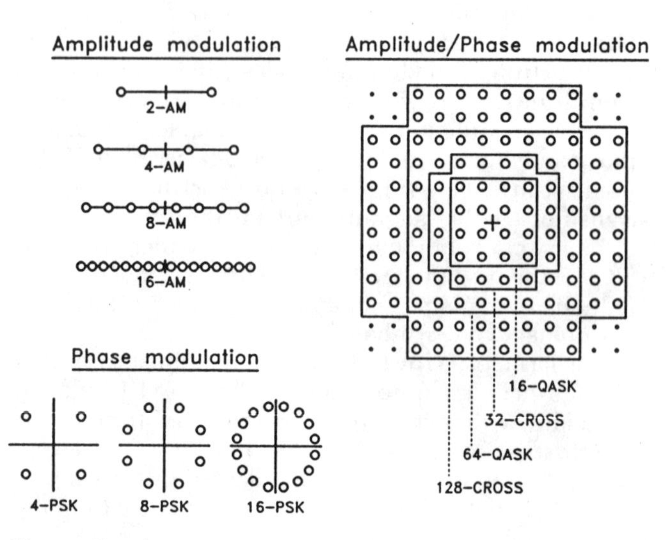

Fig. 1. Signal sets for one-dimensional amplitude modulation, and two-dimensional phase and amplitude/phase modulation.

Conventional encoders and decoders for error correction operate on binary, or more generally Q-ary, code symbols transmitted over a discrete channel. With a code of rate $k/n < 1$, $n - k$ redundant check symbols are appended to every k information symbols. Since the decoder receives only discrete code symbols, Hamming distance (the number of symbols in which two code sequences or blocks differ, regardless of how these symbols differ) is the appropriate measure of distance for decoding and hence for code design. A minimum Hamming distance d^H_{min}, also called "free Hamming distance" in the case of convolutional codes, guarantees that the decoder can correct at least $[(d^H_{min} - 1)/2]$ code-symbol errors. If low signal-to-noise ratios or nonstationary signal disturbance limit the performance of the modulation system, the ability to correct errors can justify the rate loss caused by sending redundant check symbols. Similarly, long delays in error-recovery procedures can be a good reason for trading transmission rate for forward error-correction capability.

Generally, there exist two possibilities to compensate for the rate loss: increasing the modulation rate if the channel permits bandwidth expansion, or enlarging the signal set of the modulation system if the channel is band-limited. The latter necessarily leads to the use of nonbinary modulation ($M > 2$). However, when modulation and error-correction coding are performed in the classical independent manner, disappointing results are obtained.

As an illustration, consider four-phase modulation (4-PSK) without coding, and eight-phase modulation (8-PSK) used with a binary error-correction code of rate 2/3. Both systems transmit two information bits per modulation interval (2 bit/sec/Hz). If the 4-PSK system operates at an error rate of 10^{-5}, at the same signal-to-noise ratio the "raw" error rate at the 8-PSK demodulator exceeds 10^{-2} because of the smaller spacing between the 8-PSK signals. Patterns of at least three bit errors must be corrected to reduce the error rate to that of the uncoded 4-PSK system. A rate-2/3 binary convolutional code with constraint length $\nu = 6$ has the required value of $d^H_{min} = 7$ [12]. For decoding, a fairly complex 64-state binary Viterbi decoder is needed. However, after all this effort, error performance only breaks even with that of uncoded 4-PSK.

Two problems contribute to this unsatisfactory situation.

Soft-Decision Decoding and Motivation for New Code Design

One problem in the coded 8-PSK system just described arises from the independent "hard" signal decisions made prior to decoding which cause an irreversible loss of information in the receiver. The remedy for this problem is soft-decision decoding, which means that the decoder operates directly on unquantized "soft" output samples of the channel. Let the samples be $r_n = a_n + w_n$ (real- or complex-valued, for one- or two-dimensional modulation, respectively), where the a_n are the discrete signals sent by the modulator, and the w_n represent samples of an additive white Gaussian noise process. The decision rule of the optimum sequence decoder is to determine, among the set C of all coded signal sequences which a cascaded encoder and modulator can produce, the sequence $\{\hat{a}_n\}$ with minimum squared Euclidean distance (sum of squared errors) from $\{r_n\}$, that is, the sequence $\{\hat{a}_n\}$ which satisfies

$$|r_n - \hat{a}_n|^2 = \underset{\{\hat{a}_n\} \epsilon C}{Min} \sum |r_n - a_n|^2.$$

The Viterbi algorithm, originally proposed in 1967 [13] as an "asymptotically optimum" decoding technique for convolutional codes, can be used to determine the coded signal sequence $\{\hat{a}_n\}$ closest to the received unquantized signal sequence $\{r_n\}$ [12,14], provided that the generation of coded signal sequences $\{a_n\} \epsilon C$ follows the rules of a finite-state machine. However, the notion of "error-correction" is then no longer appropriate, since there are no hard-demodulator decisions to be corrected. The decoder determines the most likely coded signal sequence directly from the unquantized channel outputs.

The most probable errors made by the optimum soft-decision decoder occur between signals or signal sequences $\{a_n\}$ and $\{b_n\}$, one transmitted and the other decoded, that are closest together in terms of squared Euclidean distance. The minimum squared such distance is called the squared "free distance:"

$$d_{free}^2 = \underset{\{a_n\} \neq \{b_n\}}{Min} \sum |a_n - b_n|^2; \quad \{a_n\}, \{b_n\} \epsilon C.$$

When optimum sequence decisions are made directly in terms of Euclidean distance, a second problem becomes apparent. Mapping of code symbols of a code optimized for Hamming distance into nonbinary modulation signals does not guarantee that a good Euclidean distance structure is obtained. In fact, generally one cannot even find a monotonic relationship between Hamming and Euclidean distances, no matter how code symbols are mapped.

For a long time, this has been the main reason for the lack of good codes for multilevel modulation. Squared Euclidean and Hamming distances are equivalent only in the case of binary modulation or four-phase modulation, which merely corresponds to two orthogonal binary modulations of a carrier. In contrast to coded multilevel systems, binary modulation systems with codes optimized for Hamming distance and soft-decision decoding have been well established since the late 1960s for power-efficient transmission at spectral efficiencies of less than 2 bit/sec/Hz.

The motivation of this author for developing TCM initially came from work on multilevel systems that employ the Viterbi algorithm to improve signal detection in the presence of intersymbol interference. This work provided him with ample evidence of the importance of Euclidean distance between signal sequences. Since improvements over the established technique of adaptive equalization to eliminate intersymbol interference and then making independent signal decisions in most cases did not turn out to be very significant, he turned his attention to using coding to improve performance. In this connection, it was clear to him that codes should be designed for maximum free Euclidean distance rather than Hamming distance, and that the redundancy necessary for coding would have to come from expanding the signal set to avoid bandwidth expansion.

To understand the potential improvements to be expected by this approach, he computed the channel capacity of channels with additive Gaussian noise for the case of discrete multilevel modulation at the channel input and unquantized signal observation at the channel output. The results of these calculations [2] allowed making two observations: firstly, that in principle coding gains of about 7-8 dB over conventional uncoded multilevel modulation should be achievable, and secondly, that most of the achievable coding gain could be obtained by expanding the signal sets used for uncoded modulation only by the factor of two. The author then concentrated his efforts on finding trellis-based signaling schemes that use signal sets of size 2^{m+1} for transmission of m bits per modulation interval. This direction turned out to be succesful and today's TCM schemes still follow this approach.

The next two sections illustrate with two examples how TCM schemes work. Whenever distances are discussed, Euclidean distances are meant.

Four-State Trellis Code for 8-PSK Modulation

The coded 8-PSK scheme described in this section was the first TCM scheme found by the author in 1975 with a significant coding gain over uncoded modulation. It was designed in a heuristic manner, like other simple TCM systems shortly thereafter. Figure 2 depicts signal sets and state-transition (trellis) diagrams for a) uncoded 4-PSK modulation and b) coded 8-PSK modulation with four trellis states. A trivial one-state trellis diagram is shown in Fig. 2a only to illustrate uncoded 4-PSK from the viewpoint of TCM. Every connected path through a trellis in Fig. 2 represents an allowed signal sequence. In

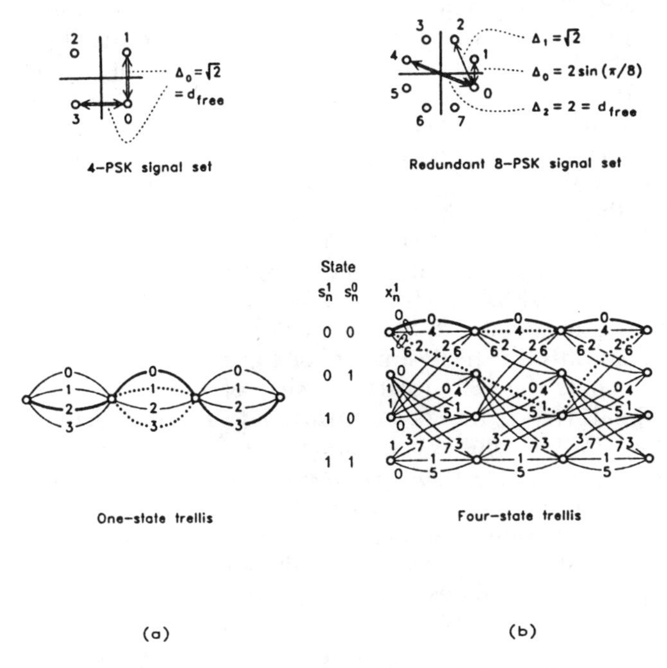

Fig. 2. (a) Uncoded four-phase modulation (4-PSK), (b) Four-state trellis-coded eight-phase modulation (8-PSK).

both systems, starting from any state, four transitions can occur, as required to encode two information bits per modulation interval (2 bit/sec/Hz). For the following discussion, the specific encoding of information bits into signals is not important.

The four "parallel" transitions in the one-state trellis diagram of Fig. 2a for uncoded 4-PSK do not restrict the sequences of 4-PSK signals that can be transmitted, that is, there is no sequence coding. Hence, the optimum decoder can make independent nearest-signal decisions for each noisy 4-PSK signal received. The smallest distance between the 4-PSK signals is $\sqrt{2}$, denoted as Δ_0. We call it the "free distance" of uncoded 4-PSK modulation to use common terminology with sequence-coded systems. Each 4-PSK signal has two nearest-neighbor signals at this distance.

In the four-state trellis of Fig. 2b for the coded 8-PSK scheme, the transitions occur in pairs of two parallel transitions. (A four-state code with four distinct transitions from each state to all successor states was also considered; however, the trellis as shown with parallel transitions permitted the achievement of a larger free distance.) Fig. 2b shows the numbering of the 8-PSK signals and relevant distances between these signals: $\Delta_0 = 2\sin(\pi/8)$, $\Delta_1 = \sqrt{2}$, and $\Delta_2 = 2$. The 8-PSK signals are assigned to the transitions in the four-state trellis in accordance with the following rules:

a) Parallel transitions are associated with signals with maximum distance Δ_2(8-PSK) = 2 between them, the signals in the subsets (0,4), (1,5), (2,6), or (3,7).
b) Four transitions originating from or merging in one state are labeled with signals with at least distance Δ_1(8-PSK) = $\sqrt{2}$ between them, that is, the signals in the subsets (0,4,2,6) or (1,5,3,7).
c) All 8-PSK signals are used in the trellis diagram with equal frequency.

Any two signal paths in the trellis of Fig. 2(b) that diverge in one state and remerge in another after more than one transition have at least squared distance $\Delta_1^2 + \Delta_0^2 + \Delta_1^2 = \Delta_2^2 + \Delta_0^2$ between them. For example, the paths with signals 0-0-0 and 2-1-2 have this distance. The distance between such paths is greater than the distance between the signals assigned to parallel transitions, Δ_2(8-PSK) = 2, which thus is found as the free distance in the four-state 8-PSK code: $d_{free} = 2$. Expressed in decibels, this amounts to an improvement of 3 dB over the minimum distance $\sqrt{2}$ between the signals of uncoded 4-PSK modulation. For any state transition along any coded 8-PSK sequence transmitted, there exists only one nearest-neighbor signal at free distance, which is the 180° rotated version of the transmitted signal. Hence, the code is invariant to a signal rotation by 180°, but to no other rotations (cf., Part II). Figure 3 illustrates one possible realization of an encoder-modulator for the four-state coded 8-PSK scheme.

Soft-decision decoding is accomplished in two steps: In the first step, called "subset decoding", within each subset of signals assigned to parallel transitions, the signal closest to the received channel output is determined. These signals are stored together with their squared distances from the channel output. In the second step, the Viterbi algorithm is used to find the signal path through the code trellis with the minimum sum of squared distances from the sequence of noisy channel outputs received. Only the signals already chosen by subset decoding are considered.

Tutorial descriptions of the Viterbi algorithm can be found in several textbooks, for example, [12]. The essential points are summarized here as follows: assume that the optimum signal paths from the infinite past to all trellis states at time n are known; the algorithm extends these paths iteratively from the states at time n to the states at time n + 1 by choosing one best path to each new state as a "survivor" and "forgetting" all other paths that cannot be extended as the best paths to the new states; looking backwards in time, the "surviving" paths tend to merge into the same "history path" at some time n − d; with a sufficient decoding delay D (so that the randomly changing value of d is highly likely to be smaller than D), the information associated with a transition on the common history path at time n − D can be selected for output.

Let the received signals be disturbed by uncorrelated Gaussian noise samples with variance σ^2 in each signal dimension. The probability that at any given time the decoder makes a wrong decision among the signals associated with parallel transitions, or starts to make a sequence of wrong decisions along some path diverging for more than one transition from the correct path, is called the error-event probability. At high signal-to-noise ratios, this probability is generally well approximated by

$$Pr(e) \simeq N_{free} \cdot Q[d_{free}/(2\sigma)],$$

where Q(.) represents the Gaussian error integral

$$Q(x) = \frac{1}{\sqrt{2\pi}} \int_x^{\infty} exp(-y^2/2)dy,$$

and N_{free} denotes the (average) number of nearest-neighbor signal sequences with distance d_{free} that diverge at any state from a transmitted signal sequence, and remerge with it after one or more transitions. The above approximate formula expresses the fact that at high

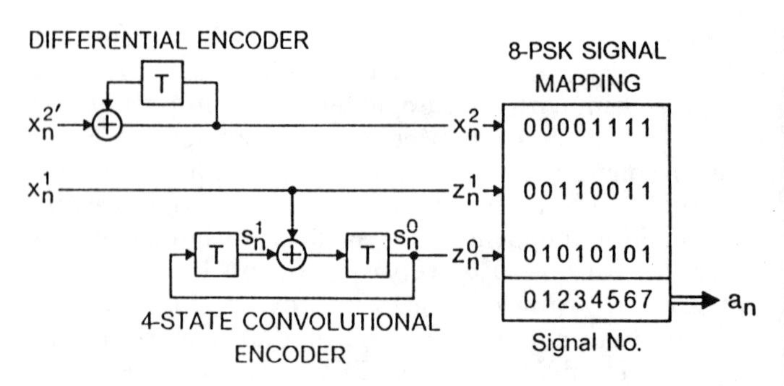

Fig. 3. Illustrates an encoder for the four-state 8-PSK code.

signal-to-noise ratios the probability of error events associated with a distance larger then d_{free} becomes negligible.

For uncoded 4-PSK, we have $d_{free} = \sqrt{2}$ and $N_{free} = 2$, and for four-state coded 8-PSK we found $d_{free} = 2$ and $N_{free} = 1$. Since in both systems free distance is found between parallel transitions, single signal-decision errors are the dominating error events. In the special case of these simple systems, the numbers of nearest neighbors do not depend on which particular signal sequence is transmitted.

Figure 4 shows the error-event probability of the two systems as a function of signal-to-noise ratio. For uncoded 4-PSK, the error-event probability is extremely well approximated by the last two equations above. For four-state coded 8-PSK, these equations provide a lower bound that is asymptotically achieved at high signal-to-noise ratios. Simulation results are included in Fig. 4 for the coded 8-PSK system to illustrate the effect of error events with distance larger than free distance, whose probability of occurrence is not negligible at low signal-to-noise ratios.

Figure 5 illustrates a noisy four-state coded 8-PSK signal as observed at complex baseband before sampling in the receiver of an experimental 64 kbit/s satellite modem [9]. At a signal-to-noise ratio of $E_s/N_0 = 12.6$ dB (E_s: signal energy, N_0: one-sided spectral noise density), the signal is decoded essentially error-free. At the same signal-to-noise ratio, the error rate with uncoded 4-PSK modulation would be around 10^{-5}.

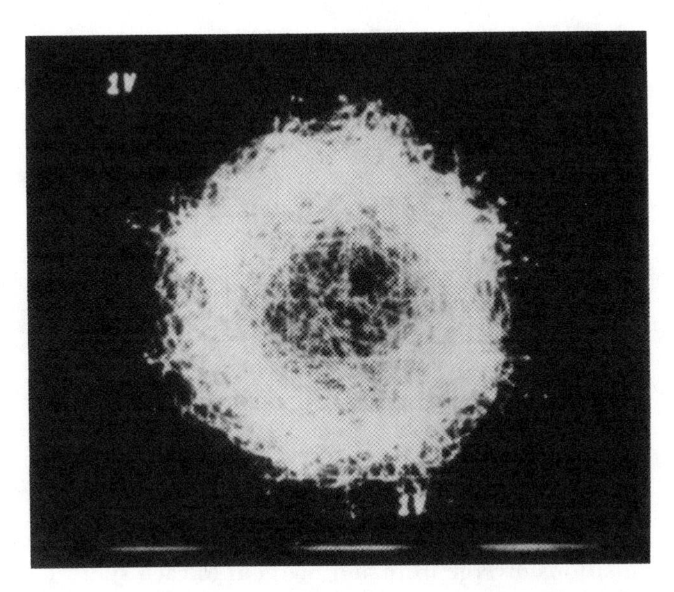

Fig. 5. Noisy four-state coded 8-PSK signal at complex baseband with a signal-to-noise ratio of $E_sN_0 = 12.6$ dB.

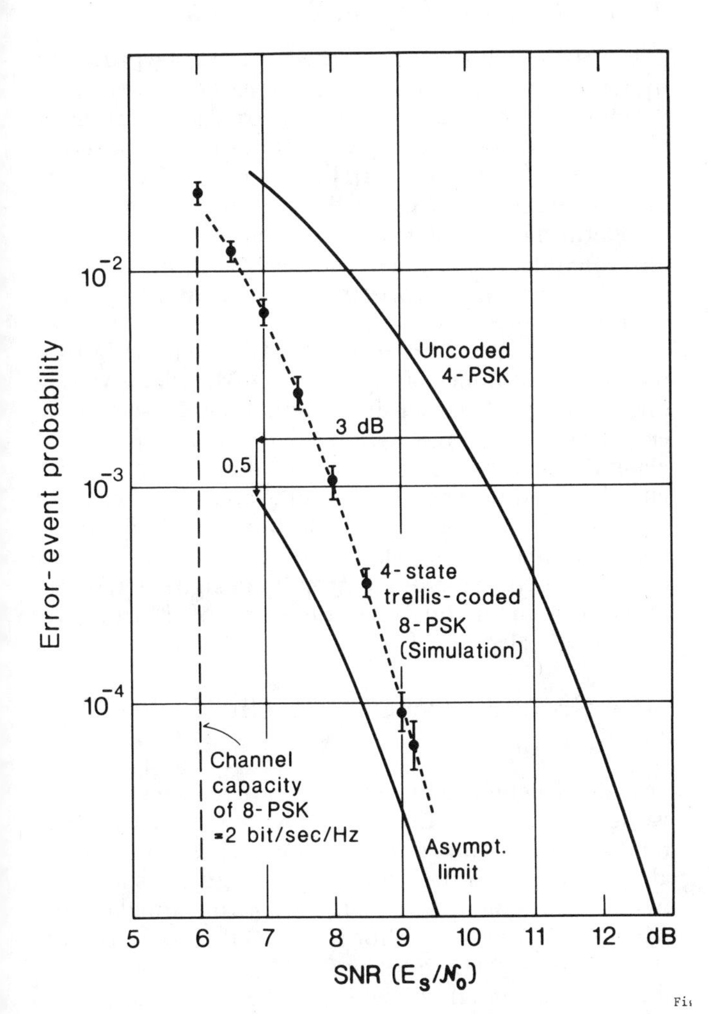

Fig. 4. Error-event probability versus signal-to-noise ratio for uncoded 4-PSK and four-state coded 8-PSK.

In TCM schemes with more trellis states and other signal sets, d_{free} is not necessarily found between parallel transitions, and N_{free} will generally be an average number larger than one, as will be shown by the second example.

Eight-State Trellis Code for Amplitude/Phase Modulation

The eight-state trellis code discussed in this section was designed for two-dimensional signal sets whose signals are located on a quadratic grid, also known as a lattice of type "Z_2". The code can be used with all of the signal sets depicted in Fig. 1 for amplitude/phase modulation. To transmit m information bits per modulation interval, a signal set with 2^{m+1} signals is needed. Hence, for m = 3 the 16-QASK signal set is used, for m = 4 the 32-CROSS signal set, and soforth. For any m, a coding gain of approximately 4 dB is achieved over uncoded modulation.

Figure 6 illustrates a "set partitioning" of the 16-QASK and 32-CROSS signal sets into eight subsets. The partitioning of larger signal sets is done in the same way. The signal set chosen is denoted by A0, and its subsets by D0, D1, . . . D7. If the smallest distance among the signals in A0 is Δ_0, then among the signals in the union of the subsets D0,D4,D2,D6 or D1,D5,D3,D7 the minimum distance is $\sqrt{2}\,\Delta_0$, in the union of the subsets D0,D4; D2,D6; D1,D5; or D3,D7 it is $\sqrt{4}\,\Delta_0$, and within the individual subsets it is $\sqrt{8}\,\Delta_0$. (A conceptually similar partitioning of the 8-PSK signal set into smaller signal sets with increasing intra-set distances was implied in the example of coded 8-PSK. The fundamental importance

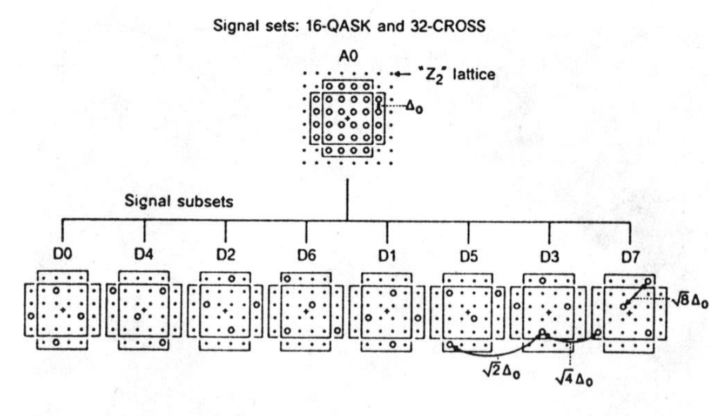

Fig. 6. Set partitioning of the 16-QASK and 32-CROSS signal sets.

of this partitioning for TCM codes will be explained in Part II.)

In the eight-state trellis depicted in Fig. 7, four transitions diverge from and merge into each state. To each transition, one of the subsets D0, . . . D7 is assigned. If A0 contains 2^{m+1} signals, each of its subsets will comprise 2^{m-2} signals. This means that the transitions shown in Fig. 7 in fact represent 2^{m-2} parallel transitions in the same sense as there were two parallel transitions in the coded 8-PSK scheme. Hence, 2^m signals can be sent from each state, as required to encode m bits per modulation interval.

The assignment of signal subsets to transitions satisfies the same three rules as discussed for coded 8-PSK, appropriately adapted to the present situation. The four transitions from or to the same state are always assigned either the subsets D0,D4,D2,D6 or D1,D5,D3,D7. This guarantees a squared signal distance of at least $2\Delta_0^2$ when sequences diverge and when they remerge. If paths remerge after two transitions, the squared signal distance is at least $4\Delta_0^2$ between the diverging transitions, and hence the total squared distance between such paths will be at least $6\Delta_0^2$. If paths remerge after three or more transitions, at least one intermediate transition contributes an additional squared signal distance Δ_0^2, so the squared distance between sequences is at least $5\Delta_0^2$.

Hence, the free distance of this code is $\sqrt{5}\ \Delta_0$. This is smaller than the minimum signal distance within in the subsets D0, . . . D7, which is $\sqrt{8}\ \Delta_0$. For one particular code sequence D0-D0-D3-D6, Fig. 6 illustrates four error paths at distance $\sqrt{5}\ \Delta_0$ from that code sequence; all starting at the same state and remerging after three or four transitions. It can be shown that for any code sequence and from any state along this sequence, there are four such paths, two of length three and two of length four. The most likely error events will correspond to these error paths, and will result in bursts of decision errors of length three or four.

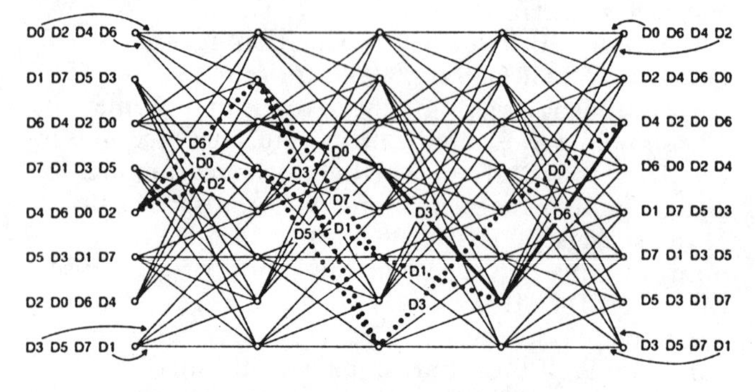

Fig. 7. Eight-state trellis code for amplitude/phase modulation with "Z_2"-type signal sets; $d_{free} = \sqrt{5}\ \Delta_0$.

The coding gains asymptotically achieved at high signal-to-noise ratios are calculated in decibels by

$$G_{c/u} = 10\ log_{10}\ [(d^2_{free,c}/d^2_{free,u})/E_{s,c}/E_{s,u})],$$

where $d^2_{free,c}$ and $d^2_{free,u}$ are the squared free distances, and $E_{s,c}$ and $E_{s,u}$ denote the average signal energies of the coded and uncoded schemes, respectively. When the signal sets have the same minimum signal spacing Δ_0, $d^2_{free,c}/d^2_{free,u} = 5$, and $E_{s,c}/E_{s,u} \simeq 2$ for all relevant values of m. Hence, the coding gain is 10 $\log_{10}(5/2) \simeq 4$ dB.

The number of nearest neighbors depends on the sequence of signals transmitted, that is N_{free} represents an average number. This is easy to see for uncoded modulation, where signals in the center of a signal set have more nearest neighbors than the outer ones. For uncoded 16-QASK, N_{free} equals 3. For eight-state coded 16-QASK, N_{free} is around 3.75. In the limit of large "Z_2"-type signal sets, these values increase toward 4 and 16 for uncoded and eight-state coded systems, respectively.

Trellis Codes of Higher Complexity

Heuristic code design and checking of code properties by hand, as was done during the early phases of the development of TCM schemes, becomes infeasible for codes with many trellis states. Optimum codes must then be found by computer search, using knowledge of the general structure of TCM codes and an efficient method to determine free distance. The search technique should also include rules to reject codes with improper or equivalent distance properties without having to evaluate free distance.

In Part II, the principles of TCM code design are outlined, and tables of optimum TCM codes given for one-, two-, and higher-dimensional signal sets. TCM encoder/modulators are shown to exhibit the following general structure: (a) of the m bits to be transmitted per encoder/modulator operation, $\tilde{m} \leq m$ bits are expanded into $\tilde{m}$ + 1 coded bits by a binary rate-$\tilde{m}/(\tilde{m}+1)$ convolutional encoder; (b) the $\tilde{m}$ + 1 coded bits select one of $2^{\tilde{m}+1}$ subsets of a redundant 2^{m+1}-ary signal set; (c) the remaining m−$\tilde{m}$ bits determine one of $2^{m-\tilde{m}}$ signals within the selected subset.

New Ground Covered by Trellis-Coded Modulation

TCM schemes achieve significant coding gains at values of spectral efficiency for which efficient coded-modulation schemes were not previously known, that is, above and including 2 bit/sec/Hz. Figure 8 shows the free distances obtained by binary convolutional coding with 4-PSK modulation for spectral efficiencies smaller than 2 bit/sec/Hz, and by TCM schemes with two-dimensional signal sets for spectral efficiencies equal to or larger than 2 bit/sec/Hz. The free distances of uncoded modulation at the respective spectral effi-

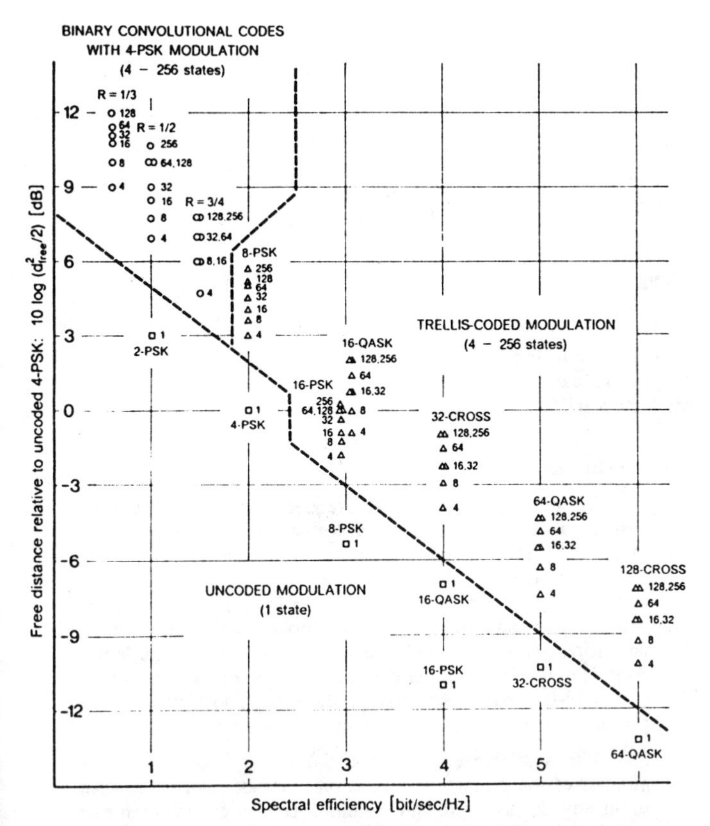

Fig. 8. Free distance of binary convolutional codes with 4-PSK modulation, and TCM with a variety of two-dimensional modulation schemes, for spectral efficiencies from 2/3 to 6 bit/sec/Hz.

ciencies are also depicted. The average signal energy of all signal sets is normalized to unity. Free distances are expressed in decibels relative to the value $d^2_{free} = 2$ of uncoded 4-PSK modulation. The binary convolutional codes of rates 1/3, 1/2, and 3/4 with optimum Hamming distances are taken from textbooks, such as, [12]. The TCM codes and their properties are found in the code tables presented in Part II (largely reproduced from [2]).

All coded systems achieve significant distance gains with as few as 4, 8, and 16 code states. Roughly speaking, it is possible to gain 3 dB with 4 states, 4 dB with 8 states, nearly 5 dB with 16 states, and up to 6 dB with 128 or more states. The gains obtained with two-state codes usually are very modest. With higher numbers of states, the incremental gains become smaller. Doubling the number of states does not always yield a code with larger free distance. Generally, limited distance growth and increasing numbers of nearest neighbors, and neighbors with next-larger distances, are the two mechanisms that prevent real coding gains from exceeding the ultimate limit set by channel capacity. This limit can be characterized by the signal-to-noise ratio at which the channel capacity of a modulation system with a 2^{m+1}-ary signal set equals m bit/sec/Hz [2] (see also Fig. 4).

Conclusion

Trellis-coded modulation was invented as a method to improve the noise immunity of digital transmission systems without bandwidth expansion or reduction of data rate. TCM extended the principles of convolutional coding to nonbinary modulation with signal sets of arbitrary size. It allows the achievement of coding gains of 3-6 dB at spectral efficiencies equal to or larger than 2 bit/sec/Hz. These are the values at which one wants to operate on many band-limited channels. Thus, a gap in the theory and practice of channel coding has been closed.

References

[1] G. Ungerboeck and I. Csajka, "On improving data-link performance by increasing the channel alphabet and introducing sequence coding," 1976 Int. Symp. Inform. Theory, Ronneby, Sweden, June 1976.

[2] G. Ungerboeck, "Channel coding with multilevel/phase signals," *IEEE Trans. Information Theory,* vol. IT-28, pp. 55-67, Jan. 1982.

[3] G. D. Forney, Jr., R. G. Gallager, G. R. Lang, F. M. Longstaff, and S. U. Qureshi, "Efficient modulation for band-limited channels," *IEEE Trans. Selected Areas in Comm.,* vol. SAC-2, pp. 632-647, Sept. 1984.

[4] L. F. Wei, "Rotationally invariant convolutional channel coding with expanded signal space—Part I: 180 degrees," *IEEE Trans. Selected Areas in Comm.,* vol. SAC-2, pp. 659-672, Sept. 1984.

[5] L. F. Wei, "Rotationally invariant convolutional channel coding with expanded signal space—Part II: nonlinear codes," *IEEE Trans. Selected Areas in Comm.,* vol. SAC-2, pp. 672-686, Sept. 1984.

[6] A. R. Calderbank and J. E. Mazo, "A new description of trellis codes," *IEEE Trans. Information Theory,* vol. IT-30, pp. 784-791, Nov. 1984.

[7] CCITT Study Group XVII, "Recommendation V.32 for a family of 2-wire, duplex modems operating on the general switched telephone network and on leased telephone-type circuits," Document AP VIII-43-E, May 1984.

[8] CCITT Study Group XVII, "Draft recommendation V.33 for 14400 bits per second modem standardized for use on point-to-point 4-wire leased telephone-type circuits," Circular No. 12, COM XVII/YS, Geneva, May 17, 1985.

[9] G. Ungerboeck, J. Hagenauer, and T. Abdel Nabi, "Coded 8-PSK experimental modem for the INTELSAT SCPC system," Proc. 7th Int. Conf. on Digital Satellite Communications (ICDS-7), pp. 299-304, Munich, May 12-16, 1986.

[10] R. J. F. Fang, "A coded 8-PSK system for 140-Mbit/s information rate transmission over 80-MHz nonlinear transponders," Proc. 7th Int. Conf. on Digital Satellite Communications (ICDS-7), pp. 305-313, Munich, May 12-16, 1986.

[11] T. Fujino, Y. Moritani, M. Miyake, K. Murakami, Y. Sakato, and H. Shiino, "A 120 Mbit/s 8PSK modem with soft-Viterbi decoding," Proc. 7th Int. Conf. on Digital Satellite Communications (ICDS-7), pp. 315-321, Munich, May 12-16, 1986.

[12] G. C. Clark and J. B. Cain, *Error-Correction Coding for Digital Communications,* Plenum Press, New York and London, 1981.

[13] A. J. Viterbi, "Error bounds for convolutional codes and an asymptotically optimum decoding algorithm," *IEEE Trans. Information Theory,* vol. IT-13, pp. 260-269, April 1967.

[14] G. D. Forney, Jr., "The Viterbi algorithm," *Proc. of the IEEE,* vol. 61, pp. 268-278, March 1973.

[15] G. Ungerboeck, "Trellis-coded modulation with redundant signal sets, Part II: State of the art," *IEEE Communications Magazine,* vol. 25, no. 2, Feb. 1987.

MORE ON FREQUENCY DIVERSITY FOR DIGITAL RADIO

T. C. Lee and S. H. Lin

Bell Communications Research, Inc.
331 Newman Springs Road
Red Bank, New Jersey 07701

ABSTRACT

The characteristics of frequency diversity improvement factor (FDIF) for digital radio based on a new, broadband three-ray model are presented. The calculated results confirm the recent experimental finding that the FDIF for digital radio are higher than those predicted by the existing analog FM radio model by one order of magnitude. The dependence of FDIF on in-band-power difference, mean delays among multiple rays, frequency separation, number of working channels and channel bandwidth are shown. The use of this new model will provide substantial cost savings to digital radio users by avoiding unnecessary space diversity protection or by allowing longer radio hop lengths.

1. INTRODUCTION

The recent experiments in Georgia and in Wyoming [1, 2, 3] show that the measured frequency diversity improvement factors (FDIF) for 6-GHz 90-Mb/s digital radio are greater than those predicted by the analog FM radio model by one order of magnitude. The analog radio model does not properly account for the sensitivity of digital radio to multipath dispersive fading and is too pessimistic for digital radio applications. A new broad band model of multipath dispersive fading has, therefore, been developed recently for calculation for FDIF for digital radio [3].

In this paper, we present the results of FDIF for digital radio on microwave line-of-sight paths based on new broad band, three-ray model.

2. BROAD BAND MODELING OF MULTIPATH DISPERSION

2.1 Digital Radio Outages Caused By Multipath Dispersion

In the recent few years, the adaptive equalization technology, including amplitude and transversal equalizers, advanced so rapidly that the multipath disperson caused digital radio outages can occur only if (i) a notch of the multipath dispersion falls inside the channel occupied by the digital signal, and (ii) the resultant In-Band-Power-Difference (IBPD) exceeds a threshold which depends on the modulation and equalizer technologies. In other words, the slope and the minor higher order curvatures in the channel transfer function produced by a notch falling outside of a digital radio channel are so well equalized by the advanced adaptive equalizers that no outage occurs.

The IBPD at a given instance of time is the peak-to-peak power difference in dB of the dispersive amplitude transfer function within a radio channel. For example, for typical 16-QAM system with an adaptive amplitude equalizer, the IBPD threshold for outage is about 9 dB when the notch is inside the channel.

Therefore, for a 1-by-1 frequency diversity protected system, a dispersion caused outage can occur only if two or more channels are hit by notches simultaneously and the resultant IBPDs also exceed the outage threshold simultaneously. The analysis of FDIF for digital radio, thus, requires a broad band model of multipath dispersion covering an entire common carrier band. For example, for the North American 6-GHz common carrier band, the model must cover the 250-MHz band occupied by the seven working channels plus the protection channel transmitting in the same direction.

The available experimental data [4-7] indicate that the number of the propagation paths on a line-of-sight microwave radio hop during multipath condition can be greater than two and that the probability distributions of the relative delays are approximately exponential.

In modeling multipath dispersion within a radio channel (e.g., 30 MHz), many authors [8-13] have used various forms of two-ray model to study the effects of multipath dispersion on digital radio performance. In Reference [3], we demonstrated that the two-ray model, although is adequate for narrow band modeling, is inapplicable for broadband modeling to investigate the characteristics of FDIF for digital radio. A three-ray model is, therefore, necessary.

2.2 Three-Ray Model For Frequency Diversity Analysis

The three-ray model consists of a main ray with unity amplitude and zero reference delay and two additional rays with randomly varying amplitudes and delays. The amplitudes of the second and the third rays are assumed to be exponentially distributed. The random variations of these amplitudes and delays are assumed to be independent. The FDIF of the three-ray model is calculated by using a computer Monte Carlo simulation process. Essentially, the simulation process computes the probability of two radio channels being hit simultaneously by two notches with respect to the probability of a notch in a channel. Furthermore, the IBPDs in the two affected channels must exceed a given threshold simultaneously for the outage to occur.

For the 26-mile path in Georgia, the mean delay of the second ray is assumed to be 0.4 nanosecond (ns) according to References [8] and [10]. The FDIF predicted by the three-way model is a strong function of the remaining model parameter, namely: the mean delay of the third ray. The simulation results indicate that a mean delay of 2 ns for the third ray yields an FDIF in close agreement with the measured data from the Georgia 26-mile path.

A similar process yields the mean delays of 1.5 ns and 4.5 ns for the second and the third rays for the 64-mile path in Wyoming.

Reprinted from *IEEE 4th Global Telecomm. Conf.*, pp. 1108–1112, Dec. 1985.

3. RESULTS AND DISCUSSION

3.1 Effect of In-Band-Power-Difference

With the parameters of the three-ray model determined in Section 2.2, we calculate the 1-by-1 FDIFs for the Georgia 26-mile path and for the Wyoming 64-mile path as function of the IBPD at the outage threshold as shown in Figure 1.It indicates that FDIF is an increasing function of the IBPD at the outage threshold as expected. This means using a more powerful adaptive equalizer to increase the IBPD outage threshold will increase the FDIF.

In the following results, we assume that the IBPD at the outage threshold is 9 dB for illustration purpose.

3.2 Effect of Mean Delay

Figure 1 also demonstrates the strong dependence of FDIF on the mean delays among the three rays. It implies that a more dispersive path, which has longer mean delays among the interfering rays, will have a smaller FDIF.

3.3 Effect of Frequency Separation

The dependence of 1-by-1 FDIF on frequency separation given in Reference [3] is reproduced on Figure 2 in this paper for completeness. It is seen that as the frequency separation decreases, the FDIF for digital radio increases whereas the FDIF for analog FM radio decreases. Thus, a smaller frequency separation provides a larger FDIF for digital radio.

An intuitive explanation for the increasing digital radio FDIF with decreasing frequency separation is that the frequency separation between notches is inversely proportional to the delays among the rays. The longer delay, corresponding to smaller frequency separation between notches, has a smaller occurrence probability which leads to the smaller probability of simultaneous outages on two channels with small frequency separation.

3.4 Effect of Number of Working Channels

Figure 3 shows that FDIF decreases as the number of working channels increases. Also shown on Figure 3 as dashed lines are the upper and the lower envelopes of the variation range of FDIF predicted by the analog FM radio model. The calculated 1-by-N FDIF depends on the frequency positions and the frequency separations among the N working channels and the protection channel. The FDIF for digital radio is consistently and substantially higher than those for analog radio for the entire range of N from 1 to 7 of the 6-GHz common carrier band.

3.5 Effect of Channel Bandwidth

Most of the above results are for 6-GHz radio with 30-MHz channel bandwidth. Figure 4 shows that the calculated FDIF for 11-GHz digital radio with 40-MHz channel bandwidth is less than those for 6-GHz digital radio with 30-MHz bandwidth by a factor of 1.5. In other words, the FDIF increases as the radio channel bandwidth decreases.

4. CONCLUSION

The results of these frequency diversity experiments and modeling lead to the following conclusions:

(a) frequency diversity can provide a large FDIF for digital radio,

(b) frequency diversity can provide substantial cost savings for digital radio routes by allowing longer hop length or by avoiding unnecessary space diversity, unnecessary addition of new radio repeater stations in a long hop, and unnecessary replacement of long radio hops by more expensive cable transmission facility,

(c) the FDIF calculations based on analog FM radio model is too conservative (i.e., pessimistic) for digital radio application,

(d) to maximize the FDIF for digital radio, the frequency separation between the working channel and the protection channel should be minimized, and

(e) to realize the large FDIF, for digital radio system must be equipped with a hitless frequency diversity switch with a fast switch initiator driven by a fast bit-error-ratio estimator.

REFERENCES

[1] P. L. Dirner and S. H. Lin, "Measured Frequency Diversity Improvement For Digital Radio," IEEE Trans. Comm. Vol. COM-33, No. 1, January 1985, pp. 106-109.

[2] P. E. Greenfield, "Digital Radio Performance On A Long, Highly Dispersive Fading Path," 1984 IEEE International Conference On Communications, Amsterdam, Conference Record pp. 1451-1454.

[3] T. C. Lee and S. H. Lin, "A Model of Frequency Diversity Improvement for Digital Radio," 1985 International Symposium on Antennas and Propagation, Kyoto, Japan, August 20-22, 1985, Summaries of Symposium Papers.

[4] A. B. Crawford and W. C. Jakes, "Selective Fading of Microwaves," Bell System Technical Journal, Vol. 31, No. 1, January 1952, pp. 68-90.

[5] R. L. Kaylor, "A Statistical Study of Selective Fading of Super-High Frequency Radio Signals," Bell System Technical Journal, September 1953, pp. 1187-1202.

[6] O. Sasaki and T. Akiyama, "Multipath Delay Characteristics on Line-of-Sight Microwave Radio System," IEEE Trans. Comm. Vol. COM-27, No. 12, December, 1979, pp. 1876-1886.

[7] J. Sandberg, "Extraction of Multipath Parameters from Swept Measurements on a Line-of-Sight Path," IEEE Trans. Antennas and Propagation, Vol. AP-28, No. 6, November 1980, pp. 743-750.

[8] W. C. Jakes, "An Approximate Method to Estimate An Upper Bound On the Effect of Multipath Delay Distortion on Digital Transmission," IEEE Trans. Comm., Vol. COM-27, No. 1, January 1879, pp. 76-81.

[9] W. D. Rummler, "More on the Multipath Fading Channel Model," IEEE Trans. Comm., Vol. COM-29, No. 3, March 1981, pp. 346-352.

[10] M. H. Myers, "Multipath Fading Characteristics of Broadband Radio Channel," 1984 GLOBECOM, Atlanta, Conference Record, pp. 45.1.1-45.1.6.

[11] M. H. Meyers, "Multipath Fading Outage Estimates Incorporating Path and Equipment characteristics," 1984 GLOBECOM, Atlanta, Conference Record, pp. 45.2.1 to 45.2.5.

[12] J. C. Campbell, A. L. Martin and R. P. Couts, "140 Mbits/Digital Radio Field Experiment - Further Results," 1984 IEEE International Conference On Communications, Amsterdam, Conference Record.

[13] A. L. Martin, R. P. Couts and J. C. Campbell, "Return Of A 16-QAM 140 Mbit/s Digital Radio Field Experiment," 1983 IEEE International Conference On Communications, Boston, Conference Record pp. F2.2.1 - F2.2.8.

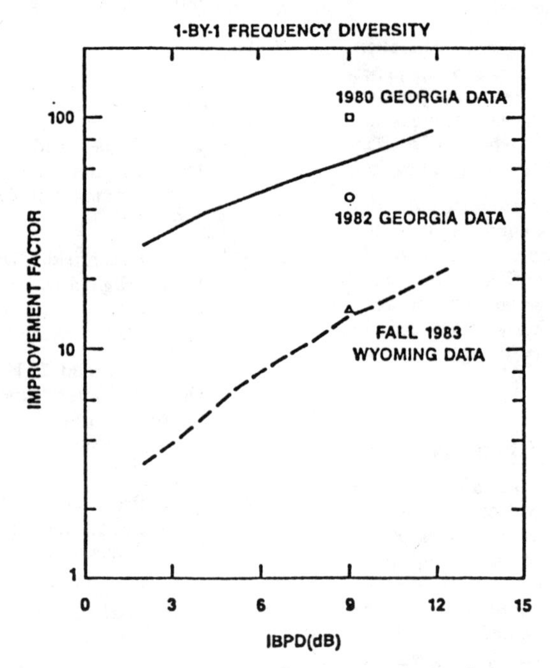

Figure 1 Dependence of frequency diversity improvement factor on IBPD at outage threshold.

1-By-1 Frequency Diversity

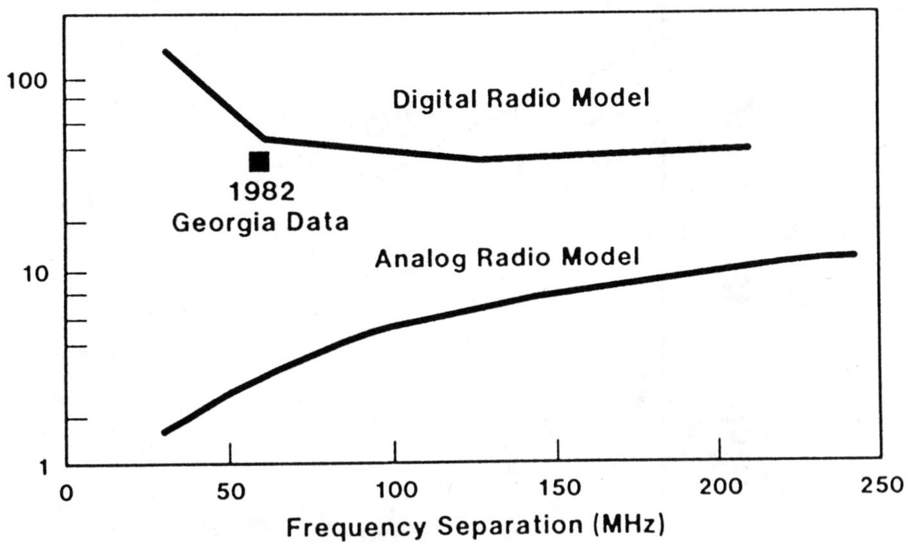

Figure 2 Dependence of 1-by-1 frequency diversity improvement factor on frequency separation.

1-BY-N FREQUENCY DIVERSITY

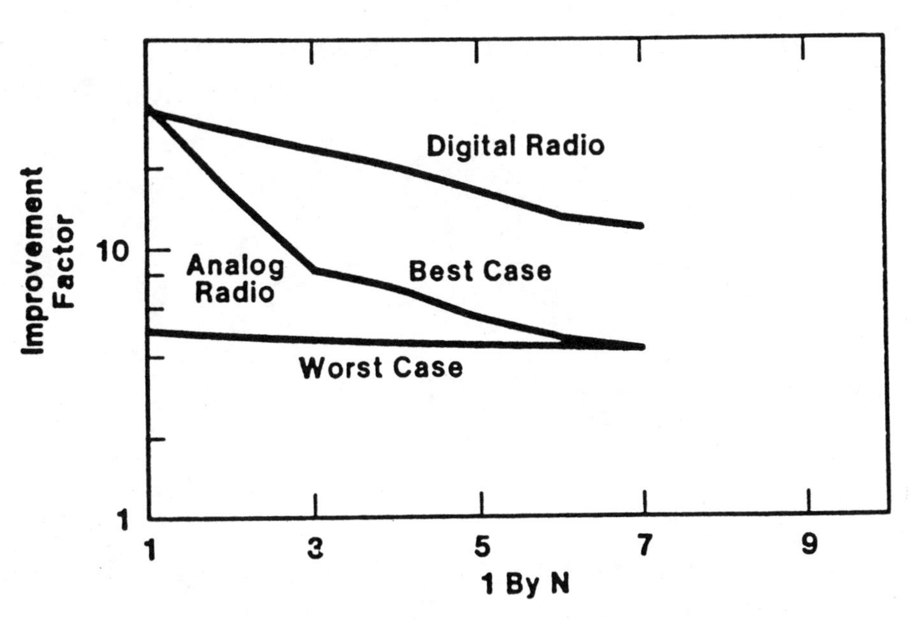

Figure 3 Effect of number of working channels on 1-by-N frequency diversity improvements factor.

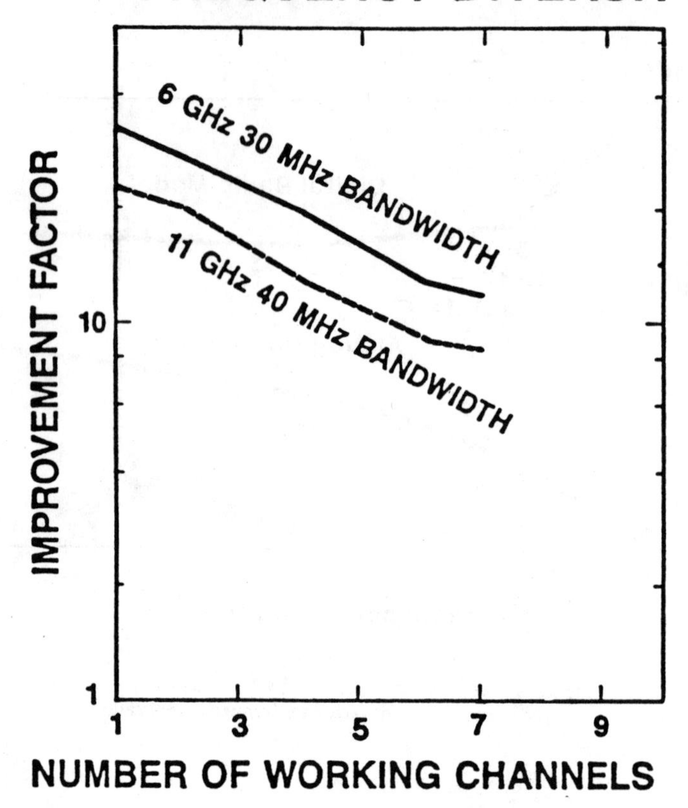

Figure 4 Effect of channel bandwidth on 1-by-N frequency diversity improvement factor.

ANGLE DIVERSITY ON LINE-OF-SIGHT MICROWAVE PATHS USING DUAL-BEAM DISH ANTENNAS

E. H. Lin, A. J. Giger, and G. D. Alley
AT&T Bell Laboratories, Andover, MA

ABSTRACT

The paper presents both a theory and experimental results for angle diversity on line-of-sight terrestrial microwave paths. Using a dual-beam parabolic dish antenna very large reductions in channel distortion and fade depth are predicted, as well as measured, during periods of multipath fading. This makes angle diversity particularly attractive for digital radio transmission. If properly designed, angle diversity performance can be far superior to space diversity. In addition the use of a single antenna, instead of the two required for space diversity, has substantial economic advantages as well.

1. INTRODUCTION

Angle diversity has been extensively used in troposcatter radio links. In such links, two or more slightly tilted receiving antenna beams pick up scattered energy from different volumes in the troposphere that are illuminated by the same transmitting beam[1]. The antenna beams are typically separated in elevation by their 3-dB beamwidths and are formed by using offset primary feeds in a common parabolic dish. The essentially uncorrelated signals are then processed in a maximum ratio combiner. Angle diversity has the advantage over space diversity in that it can work with a single antenna instead of two antennas.

In contrast to troposcatter systems, angle diversity has not been used in line-of-sight terrestrial microwave links except in some recent experiments[2,3]. This almost complete disinterest in angle diversity may have been due to the fact that volume scattering, which is the source of diversity in troposcatter applications, would not be an important mechanism in line-of-sight links. We will show in the paper that substantial benefits can be obtained with angle diversity if fading is caused by two rays interfering with each other. Field measurements using three tones[4] have shown that up to 99% of all fading events on line-of-sight microwave hops can be represented by a two-ray modeling function. It suggests that most deep fades are due to destructive interference between two rays of about equal amplitude. This nulling of the resultant amplitude can be easily undone by a small change in the amplitude, phase or both of the individual rays. Such unbalancing guarantees a substantial reduction of dispersive fade depths and distortion in the radio channel. Therefore, when a deep dispersive fade is observed on the first beam of an angle diversity receiving antenna, switching over to the second beam will change the amplitude (and possibly the phase) relationship between the two rays and this will reduce with almost absolute certainty the fade depth and distortion experienced by the channel. Instead of switching between the two antenna patterns, maximum power or other combining methods can be employed as well. Similar results are also obtained by initially forming the sum and difference antenna patterns as is done in monopulse radar, followed by diversity switching or combining.

A number of experiments, somewhat related to angle diversity have been described in the past. They did not detract from the ubiquitous use of space diversity, the preferred tool for multipath protection. For instance, in 1967 Harkless and Lenzing[5] described an experiment with *mode diversity* using the pyramidal horn reflector antenna. The regular antenna pattern was used together with a TE_{02} mode pattern which has a null on boresight. Selecting the larger of the two signals through simulated switch diversity, a relatively modest improvement in the cumulative fade statistics was observed. *Polarization diversity* was found by Cronin[6] to be an unexpected but desirable by-product of a dual polarized digital radio system. Two radio channels were operated cochannel on V and H polarizations, each exhibiting the same cumulative fading statistics but having very little simultaneous fading. This behavior could be expressed by a diversity improvement factor of about ten. The improvement was explained by the slight differences in the antenna patterns for the two polarizations, preventing destructive interference (deep fades) from occurring simultaneously. A very similar effect was also discovered by Gardina and Lin[7] in 1985 during an experiment to compare feeder ripples in horn-reflector and dish antennas. The two antennas were mounted at the same height on the tower but displaced by 12 feet. Their boresight axes were aligned and pointing towards the transmitter but the two antenna patterns were slightly different. For the same reason as given in Cronin's experiment a diversity improvement was found for this case of *horizontal space diversity* or *pattern diversity*. The possibility of pattern diversity in line of sight links had been pointed out earlier in 1983 by one of the authors in an unpublished memorandum based on work done with adaptive arrays[8]. Another ICC '87 paper further elaborates on pattern and angle diversity[9] using two antennas. Limited experiments with *beam-scan diversity* were made at Bell Labs in the late 1970s using a mechanically steered dish antenna. A recent paper by Dombek[10] indicates that electronic beam scanning could be an effective form of diversity.

Reprinted from *IEEE Int. Conf. Comm.*, vol. 2, pp. 831–841, June 1987.

The paper first discusses the propagation model used in the following analysis. A theory of angle diversity is then presented stressing general principles as well as using specific antenna patterns. Finally, field measurements of angle diversity are presented.

2. THE PROPAGATION MODEL

In order to analyze angle diversity we must have a propagation model that provides a good description of fading in line of sight terrestrial links. In a separate publication[4] we have shown that a simple two-ray transfer function can fit 99% of all observed fade events. This does not mean that 99% of the observed fades physically consist of two rays but there is a good chance that a high percentage actually do. For these reasons we have chosen to use the following model function

$$F(\omega) = 1 - re^{-j\Delta\omega\tau} \qquad (1)$$

which consists of a direct ray of unity amplitude and a second ray of amplitude r and delay τ, leading to a notch frequency $\omega_0 = 2\pi f_0$. The equation further uses the frequency offset $\Delta\omega = \omega - \omega_0$. If more than two rays were involved the analysis would be more complex but the results would not be substantially different.

Figure 1 shows the radio hop between Salton and Brawley, CA that was used in our angle diversity experiment. The transmitting antenna in Salton is 725 feet higher than the dual-beam receiving antenna in Brawley resulting in a negative boresight angle of 0.21°. In order to account for different atmospheric refractive conditions the ground profile is shown three times, first for the standard atmosphere (K=4/3), second for subrefractive conditions (K=0.46), and third for superrefractive conditions (K=∞). In addition to the direct ray, a second ray can reach the receiver by either ground reflection or reflection from an atmospheric inversion layer. The Fresnel ellipses around the direct ray give the points of equal delay for the second ray, or $\tau = n/2f_0$ with n indicating the number of half wavelengths. For instance, the ray reflected from P_1 on the extended flat ground surface will be delayed by ten half wavelengths or 0.81 ns for

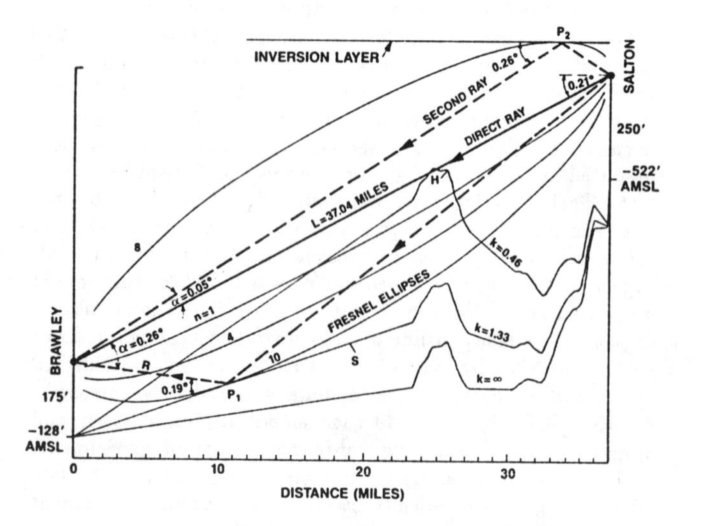

Fig. 1. Test hop Salton-Brawley, CA.

$f_0 = 6175$ MHz. The flat surface and the small grazing angle of 0.19° will produce a phase reversal and a large amplitude for the reflected ray, leading to relatively severe fading conditions. Figure 2 shows a long term distribution of delays τ for this particular hop estimated with the three tone technique described in Reference 4. Note the pronounced peak at 0.8 ns. As atmospheric conditions change the

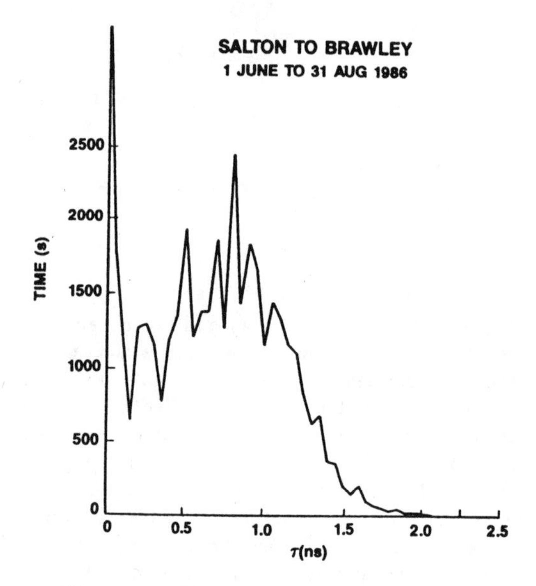

Fig. 2. Measured delay distribution.

K-factors will change and with it the delay caused by the inclined surface adjacent to Brawley. Even numbered Fresnel zones will tend to cause deep fades whereas odd zones are responsible for signal enhancements. In practice the situation is not that simple because the reflection phase can be different from 180°. As K becomes smaller (K=0.46) the large reflecting plain near Brawley will be shadowed by the hill H, eliminating in essence all ground reflections. As a result of these atmospheric changes plots of the τ distribution vary greatly from day to day and from the shape shown in Figure 2. Contributing to this fluctuation are reflections from inversion layers in the atmosphere. They can cause a rapid drop ΔN in the radio refractivity of the atmosphere. This causes a ray coming from below the layer to be totally reflected up to a grazing angle of

$$\alpha_{max} = \sqrt{2\,\Delta N \cdot 10^{-6}}. \qquad (2)$$

In a typical case where $\Delta N = 20$, we find $\alpha_{max} = 0.36$degree. Such layers, therefore, are not able to reflect signals at angles larger than α_{max} where the layers become totally transparent. Figure 1 shows a horizontal inversion layer with total reflection at point P_2. Since P_2 is on the n=8 Fresnel ellipse, a delay of 0.65 ns is obtained.

A determination of the angles of arrival α of reflected rays is complicated by the complexity of the path geometries. For instance, for the very similar delays of Figure 1, the ground reflection arrives at the receiving antenna 0.26° off boresight, whereas the ray from the inversion layer arrives at 0.05°. These angles can be calculated from the formula

$$\alpha = 2 \arcsin \sqrt{\frac{c\tau}{2L}(\frac{L}{R} - 1)} \tag{3}$$

where c = speed of light, L = hop length, and R is the distance from the receiving antenna to the reflection point. The formula is important because it gives an idea of the angles at which the second rays arrive at the receiving antenna, a parameter of great importance in the analysis of angle diversity. A further complication results if the direct ray is not unity but either smaller or larger than the free space value and has an amplitude r_1. This leads to a true delay τ' which is different from the τ shown in Figure 2, or approximately $\tau' = \tau/r_1$ (Reference 4).

3. ANGLE DIVERSITY THEORY

3.1 General Antenna Patterns and the Concept of Discrimination Ratio

We now demonstrate how angle diversity can achieve improvement in reducing multipath fade depth and dispersion. We will introduce a factor called *discrimination ratio* to account for the difference in gains of the two antennas in the directions of two incident rays. We then show that by making this ratio large, one can obtain a small fade depth as well as low dispersion from at least one of the antennas. The technique employed here, to a large extent follows the one used in[8] for evaluating adaptive array performance.

Assuming that there are two rays incident upon an antenna with two different radiation patterns V_1 and V_2 as shown in Figure 3, the received signals are

$$H_1 = a_1 - b_1 r e^{-j\Delta\omega\tau} = a_1(1 - r\frac{b_1}{a_1}e^{-j\Delta\omega\tau}) \tag{4}$$

$$H_2 = a_2 - b_2 r e^{-j\Delta\omega\tau} = a_2(1 - r\frac{b_2}{a_2}e^{-j\Delta\omega\tau}), \tag{5}$$

where a_1, a_2, b_1, and b_2 are functions of incident elevation angle α and H_1 and H_2 are functions of both α and $\Delta\omega$. The two ray model given in equation 1 was used to represent the

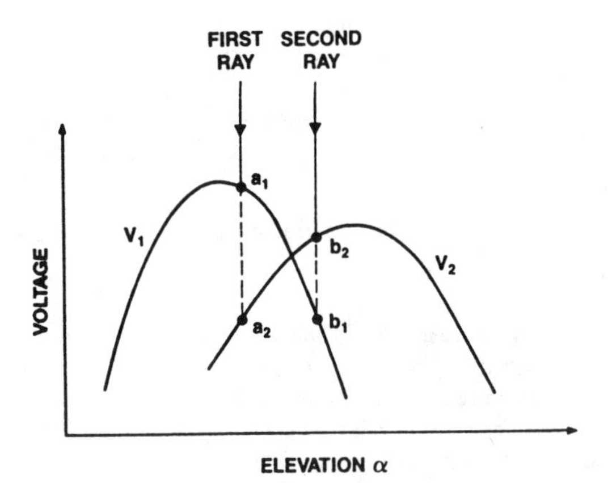

Fig. 3. Voltage patterns of angle diversity antenna and incoming rays.

incoming signals. We have also assumed that there is no continuous phase change with α in the antenna patterns except for a sudden change from 0° to 180° going from main lobe to sidelobe. We also should be aware that in the more general case where two physically separate antennas are used, and/or there is a continuous phase change with incident angle in the antenna patterns, the term $e^{-j\Delta\omega\tau}$ in H_1 and H_2 would have to be replaced by the more general terms $e^{-j(\omega\tau_1+\phi_1)}$ and $e^{-j(\omega\tau_2+\phi_2)}$.

In order to prevent simultaneous deep fades on both antenna patterns it is very important that the following condition be obeyed:

$$\frac{b_1}{a_1} \neq \frac{b_2}{a_2} \tag{6}$$

The relationship $b_1/a_1 = b_2/a_2$ would be the one necessary to balance a Wheatstone bridge made up of resistors a_1, b_1 in one branch and a_2, b_2 in the other branch, fed by the common "voltage" $re^{-j\Delta\omega\tau}$. In the important practical case where $a_1 = a_2$ a balanced bridge would mean that $H_1 = H_2$, which is a highly undesirable condition resulting in no diversity improvement. Unbalancing the bridge according to equation 6 is, therefore, absolutely necessary. It is also a well-known fact that it is easy to unbalance a bridge and thus generate with absolute certainty different received signals H_1 and H_2. Equation 6 leads us to introduce the concept of discrimination ratio

$$d = \left|\frac{a_1/b_1}{a_2/b_2}\right|, \tag{7}$$

or in decibel notation $D = |20 \log d|$. This ratio is only related to the antenna patterns and the incident angles and is independent of the incoming signal levels. It does, however, characterize the antenna with regards to its ability to combat multipath fading. A value for d close to unity, or equivalently a value of 0 dB for D, indicates that the two antenna patterns are identical which has to be avoided by all means. It is, therefore, very desirable to have a large D which can only be obtained by making a_1/b_1 very different from a_2/b_2, implying that the two antenna patterns are very different. In certain applications one purposely generates a null on boresight for the second beam, or $a_2 = 0$ which results in d and D being infinite. Introducing $r' = rb_1/a_1$ equations 4 and 5 become:

$$H_1 = a_1(1 - r'e^{-j\Delta\omega\tau}) \quad \text{and} \tag{8}$$

$$H_2 = a_2(1 - r'de^{-j\Delta\omega\tau}) \tag{9}$$

These equations are plotted in Figure 4 for the case where $a_1 = a_2 = 1$ and $r' = 1$. The resulting notch depth (at $\Delta\omega = 0$) for H_1 is infinite and for H_2 it becomes:

$$F_{d\infty} = -20 \log |1 - d|. \tag{10}$$

when $d < 1$. This result means that an infinitely deep fade on pattern V_1 is reduced to a much lower fade depth $F_{d\infty}$ on pattern V_2. Switch diversity based on maximum power will

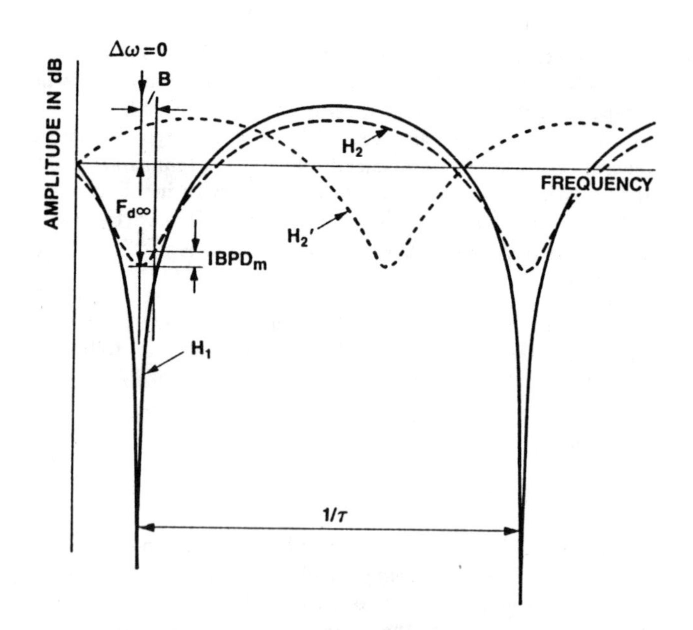

Fig. 4. Transfer functions H_1 and H_2

achieve this selection automatically. Figure 5 shows a plot of $F_{d\infty}$ versus the discrimination ratio D. Substantial reductions in fade depth are achieved for relatively small D. Further extension of the example used in Figure 4 reveals that the maximum possible fade depth F_{dm} is somewhat deeper than $F_{d\infty}$ and is the same for both beams. This occurs at some value $r' \neq 1$ such that

$$F_{dm} = -20 \log \frac{|1-d|}{1+d}. \tag{11}$$

This worst possible fade condition is also shown in Figure 5. If D can be kept larger than 2 dB, for instance, then the better of the two notch depths will never exceed 19 dB.

Equations 10 and 11 also hold if the exponents in equations 8 and 9 are not identical, which can be due to phase differences in the antenna patterns or the use of two

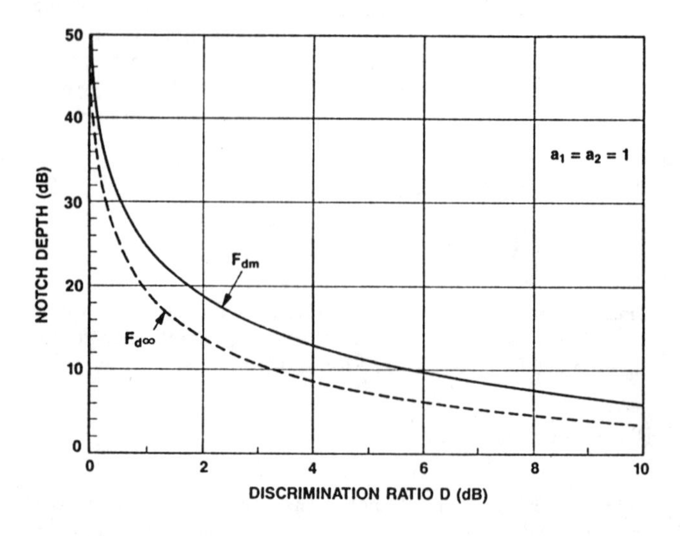

Fig. 5. Maximum notch depths $F_{d\infty}$ and F_{dm}.

physically separated antennas with beams V_1 and V_2. The dashed curve H_2' in Figure 4 is due to such a difference in exponents. We see that this further reduces the fade depth encountered on the second beam after diversity switching. The value F_{dm} is therefore a worst-worst case. During tests with a single dual beam antenna we have actually observed the H_2' displacement phenomenon, which is caused by phase changes across the antenna main lobes. If using two antennas for angle diversity, one tilted upward and one downward, a small additional vertical offset of a few feet can also achieve the same benefit.

The discrimination ratio is also helpful in characterizing the maximum dispersion in a radio channel. We express the dispersion in a band B in terms of the In Band Power Difference (IBPD) which is the maximum amplitude difference in decibels over the band B. We consider IBPD a useful, although somewhat qualitative indicator of digital radio degradation. Switching from signal H_1 to signal H_2 typically achieves a dramatic reduction in IBPD. A reduction to 1 dB is not uncommon. This is contrasted by a much lesser reduction in fade depth to perhaps 5 or 15 dB. In the following mathematical analysis we have assumed switch diversity that selects the antenna beam with the lowest IBPD. The analysis shows that the maximum possible IBPD is bounded for each D and τ as the notch position and r' are varied. This worst case $IBPD_m$ occurs when the notch frequency coincides with the edge of the band B, as shown in Figure 4, although r' in general is not equal to unity. The results are shown in Figure 6, plotted as a function of τ for three values of D when B = 22.5 MHz. We see that $IBPD_m$

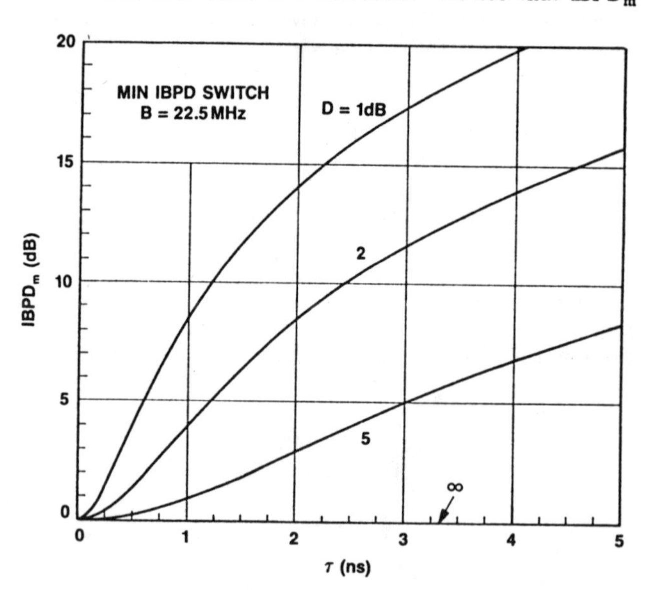

Fig. 6. Maximum In-Band Power Difference $IBPD_m$.

monotonically decreases as D increases and τ decreases. This is to be expected from an inspection of Figure 4 where dispersion is reduced for larger D because notch depth is reduced, and for smaller τ because the frequency period of the fade shape is increased. As in the case of fade depth, differences in the exponents of equations 8 and 9 also have a beneficial effect on IBPD, making $IBPD_m$ truly a worst-worst case.

3.2 Analysis of the Dual Beam Dish Antenna

In this section we analyze angle diversity by considering specific antenna patterns and combining techniques. Figure 7a shows the use of sinx/x patterns for beams V_1 and V_2. These patterns are associated with a uniformly illuminated rectangular aperture where the 3-dB beamwidth is $\alpha_0 = 0.88\lambda/a$ and a is the diameter of the aperture. The relationship for a typical circular dish antenna would be $\alpha_0 = 1.3\lambda/a$. In Figure 7 we have made the beam separation α_s equal to α_0, which makes the two beams cross 3 dB below their individual peaks. Techniques developed for radar monopulse applications could be used to form these beams in a parabolic dish antenna. Also useful in angle diversity work are the sum and difference patterns V_s and V_d, derived from V_1 and V_2 by addition and subtraction and shown in Figure 7b. Since the first ray is normally incident upon the antenna around boresight, the patterns of Figure 7b will yield a very small value of a_2, including $a_2 = 0$. This is very desirable because, from equation 7, the discrimination ratio D will become very large. Figure 8a shows the connection to the antenna when the individual beams are used directly and Figure 8b indicates the processing necessary to form the sum and difference beams.

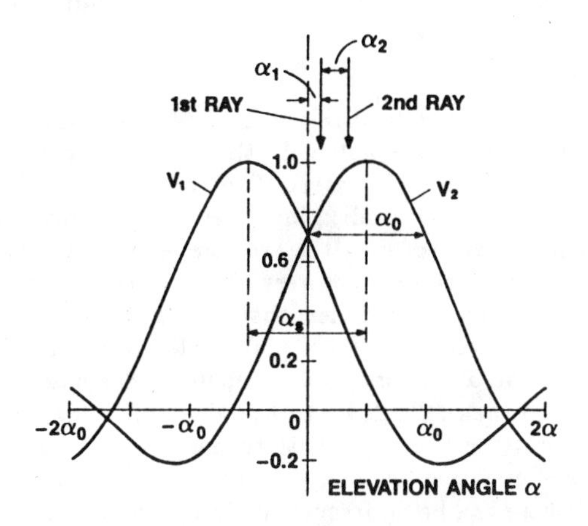

a) INDIVIDUAL RADIATION PATTERNS

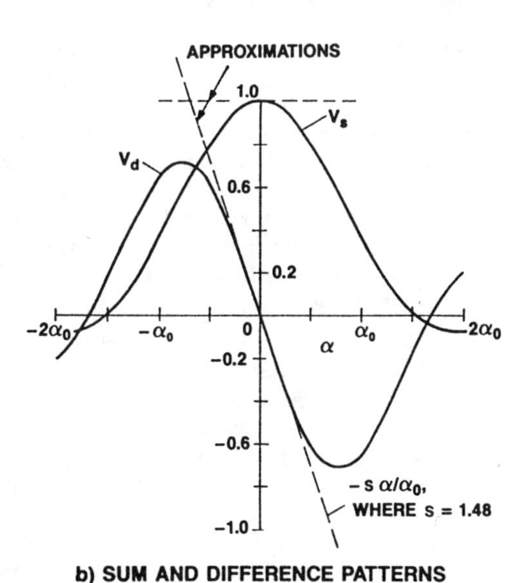

b) SUM AND DIFFERENCE PATTERNS

Fig. 7. Voltage patterns of angle diversity dish antenna.

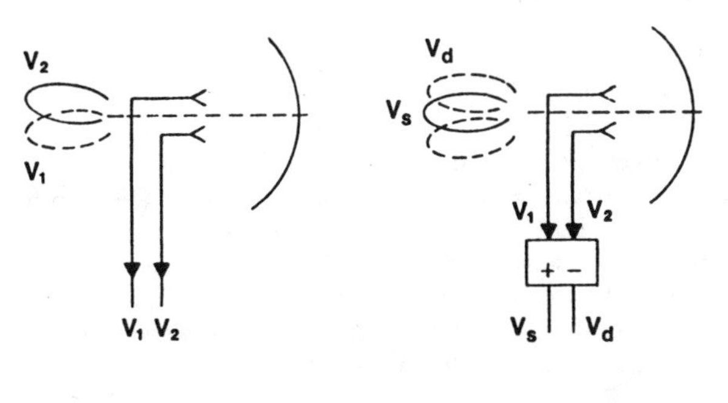

Fig. 8. Dual beam dish antennas.

3.2.1 Signal Power Analysis

Let us first look at the use of sum and difference beams because they represent a technique that maximizes the discrimination ratio. We have simplified the V_s and V_d patterns as indicated by the dashed lines in Figure 7b. This leads to particularly simple expressions that help gain a good qualitative insight into the mechanics of angle diversity. The following signals are then obtained from equations 4 and 5 at a frequency that coincides with the fade notch ($\Delta\omega = 0$) and for $\alpha_1 = 0$:

$$H_1 = 1 - r \tag{12}$$

$$H_2 = -rb_2 = rs\frac{\alpha_2}{\alpha_0} \tag{13}$$

We now introduce the following dB-notations: $F_n = -20 \log |1-r|$ and $F_d = -20 \log |H|$. F_n is the notch depth of the signal received through the sum pattern approximation of Figure 7b and F_d is the fade depth after the diversity switch has selected the larger of the two signals, either H_1 or H_2. With this equations 12 and 13 are transformed into

$$F_{d1} = F_n \quad \text{and} \tag{14}$$

$$F_{d2} = -20 \log (1 \pm 10^{-\frac{F_n}{20}}) - 20 \log s \left| \frac{\alpha_2}{\alpha_0} \right| \tag{15}$$

The minus sign in the second equation is for minimum phase fades (MPF, r<1) and the positive sign for non-minimum phase fades (NMPF, r>1). Figure 9 is a plot of the last equations with s = 1.48, where s is the slope of the difference pattern. We clearly notice the two branches corresponding to the sum and difference patterns. The diversity switch with its associated control circuit automatically chooses the branch that gives the higher output power, which means the smaller fade depth F_d. The digital radio systems manufactured by AT&T Technologies are equipped with a soft IF switch that follows this principle. The asymptotic value of the horizontal branch is from equation 15:

$$F_{d\infty} = -20 \log s \left| \frac{\alpha_2}{\alpha_0} \right|, \quad (16)$$

Figure 9 reveals the rather surprising result that for a given angle α_2 and for deep enough notch depths, the fade depth F_d after diversity switching is essentially constant. A low value of this constant is advantageous and this can be achieved by making α_0 small, which means a large antenna diameter, and making the slope s of the difference pattern large. The benefits of the fade depth truncation are especially great where most needed, namely for highly dispersive fades that have large τ and, according to equation 3, large α_2.

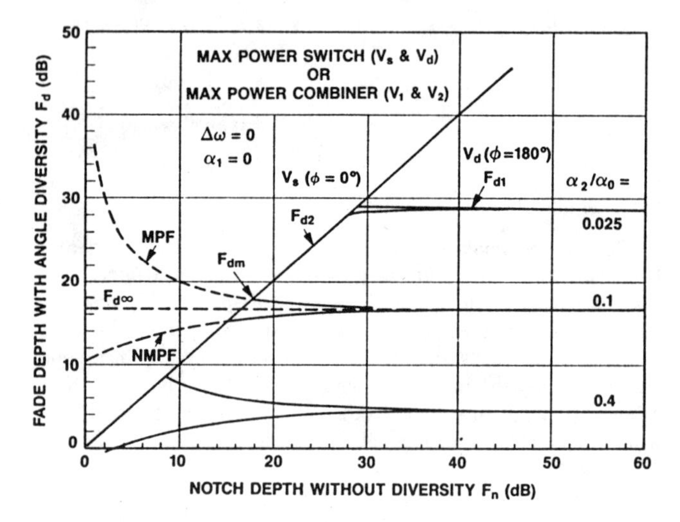

Fig. 9. Angle diversity input-output power characteristic.

The results just derived for maximum power switch diversity using patterns V_s and V_d can also be used without change for a continuous combining scheme using patterns V_1 and V_2. In this case a phase shifter ϕ is inserted in one of the antenna leads of Figure 8a and a feedback control circuit adjusts the phase shifter continuously to obtain maximum power combining. It could be shown that the phase shifter will be driven to either one of two positions, $\phi = 0°$ or $180°$ if $\Delta\omega = 0$. This means that the combined patterns in this special case are identical to either V_s or V_d. If $\Delta\omega \neq 0$ then ϕ can be any value and the resulting curves would fall below those shown in Figure 9. Individual radio channels in the TD and AR6A analog radio systems of the AT&T Communications network are equipped with RF combiners of this type.

Performance curves very similar to Figure 9 can also be obtained if signals V_1 and V_2 are used with a maximum power switch or signals V_s and V_d with a continuous phase shifter combiner.

3.2.2 Dispersion Analysis

In order to gain a good qualitative insight into the dispersive aspects of angle diversity we concentrate first on the two systems covered in Figure 9. Their IBPD characteristics are shown in Figure 10 based upon a uniform

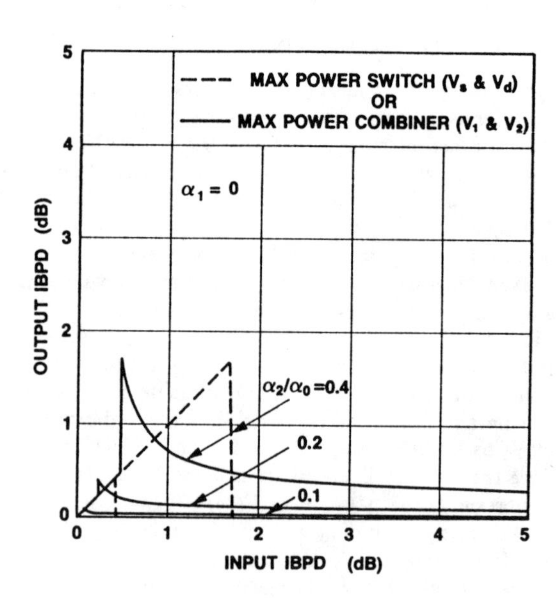

Fig. 10. Angle diversity input-output IBPD characteristic.

spectrum with B = 22.5 MHz whose center is aligned with the fade notch. When the notch depth is small, the maximum power switch (the 45° dashed line) operates on the sum branch of Figure 9, indicating that no improvement in dispersion is achieved. But the problem is minor because the shallow fades involved have very little frequency shape, as reflected by values of IBPD less than 1.7 dB. When the notch depth increases reception takes place in the difference branch of Figure 9 and all distortion is eliminated. The dashed curve in Figure 10 then falls down to the abscissa. This is due to the fact that the difference pattern has a null on boresight, therefore eliminating one of the two rays. With only one ray received there can be no frequency dispersion, a fact that can be verified in Figure 6 for $D = \infty$. Therefore, for a practical range of angles of arrival, e.g. $\alpha_2/\alpha_0 \leq 0.4$, IBPD can be assumed to be bounded by 1.7 dB. This is in contrast to the signal power fade depth $F_{d\infty}$ which will go to infinity as α_2 approaches zero. The solid curves in Figure 10 show the case of a maximum power combiner working on signals V_1 and V_2. Measurements made in the field show typically the behavior indicated in Figure 10 with most data points falling near the origin and long narrow tails extending along the abscissa and ordinate. The case shown in Figure 10 does not give the maximum possible IBPD which would occur when the fade notch is at the bandedge, yet the values shown in the Figure are close to this maximum.

In the following we like to present a general analysis which demonstrates that there are upper bounds for IBPD as α_1 and α_2 are varied. We will also find that these bounds have a minimum at a specific beam separation α_s. The analysis is based on the angle diversity system of Figure 8a, assuming that the signals from V_1 and V_2 are switched on the basis of the optimum (smallest) IBPD. The results for $IBPD_m$ obtained previously and plotted in Figure 6 are used as inputs to the analysis. This requires that the discrimination ratio D is determined as a function of α for the patterns given in Figure 7a, followed by a computation of delay τ from equation 3, with R=L/2 which gives the maximum possible delay for a hop of length L. Results of computations are shown in

Figures 11 through 13 for the following three cases:

Fig. 11: $IBPD_m$ vs α_2 for $\alpha_1 = 0$

Fig. 12: $IBPD'_m$ vs α_1
where $IBPD' = IBPD_m$ (worst α_2)

Fig. 13: $IBPD''_m$ vs α_s
where $IBPD''_m = IBPD'_m$ (worst α_1)

The worst α_2 and α_1 are values that maximize IBPD. The computations involved an exhaustive search on the computer.

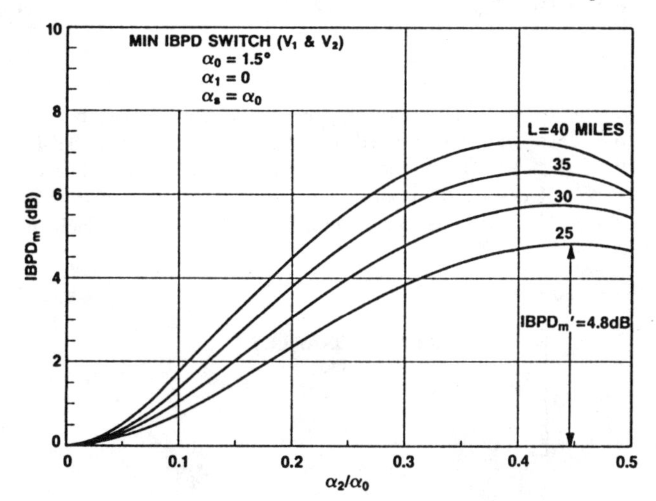

Fig. 11. Maximum In-Band Power Difference $IBPD_m$ for radiation patterns of Fig. 7a.

From Figure 11 we see again that $IBPD_m$ is bounded as α_2 is changed. The low values of $IBPD_m$ at small α_2 are due to the fact that the two rays incident on the antenna are close together. This results in a small τ (equation 3) and, in turn, a small dispersion (Figure 4). For the higher angles α_2 the delay is larger but the discrimination ratio D becomes larger

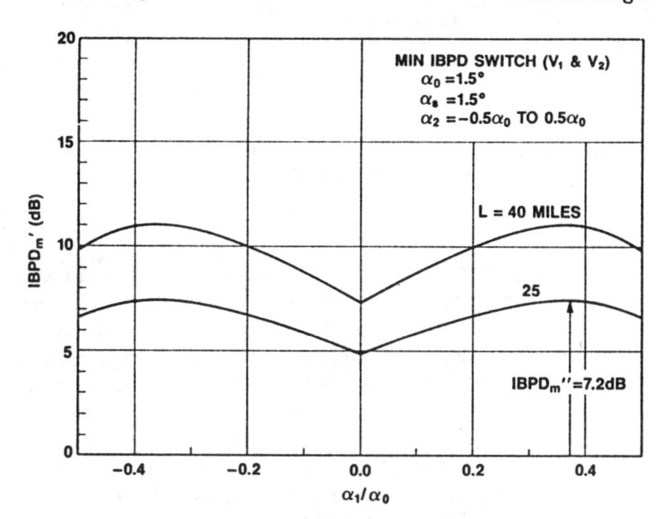

Fig. 12. Maximum In-Band Power Difference $IBPD'_m$ for Fig. 7a patterns.

as well, keeping $IBPD_m$ within bounds (Figure 6). For instance, the highest possible IBPD is only 4.8 dB for a 25 mile hop and an antenna with $\alpha_0 = \alpha_s = 1.5°$. Figure 12 tells us that moving the first ray angle α_1 away from boresight increases dispersion only slightly. Figure 13, finally, indicates that proper selection of the beam separation α_s is very important. For a given hop length angle diversity performance is determined by beamwidth α_0 and beam separation α_s. Large beam separations will have to be avoided because they reduce the gain at boresight and with it the thermal fade margin. As pointed out before in connection with Figures 10 and 11 the results of Figures 12 and 13 further demonstrate the boundedness of IBPD. This makes angle diversity very different from conventional space diversity which does not exhibit this bound. Angle diversity is therefore particularly attractive for digital radios which are strongly affected by dispersion.

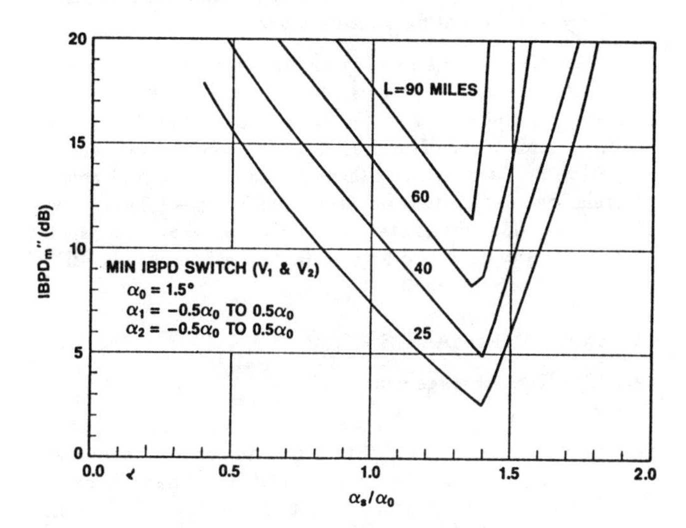

Fig. 13. Maximum In-Band Power Difference $IBPD''_m$ for Fig. 7a patterns.

The results shown in Figures 11, 12, and 13 are based on a diversity switch algorithm that selects the best IBPD. The algorithms that we used in section 3.2.1 resulted in maximum signal power (minimum fade depth), however. Analysis has shown, that using the latter algorithms, the worst IBPD is only slightly higher than the $IBPD_m$ shown in Figures 11, 12, and 13. Since power fading is the limiting element in an angle diversity system it is wise to use diversity algorithms that maximize signal power.

3.2.3 Correlation Analysis

We can increase our understanding of the digital radio performance improvement obtained by angle diversity by using the concept of correlation between input and output signals. We have considerable freedom in selecting the signals, or mathematical quantities, that we want to correlate. When it comes to comparing the statistics of dispersion before and after diversity we have found it meaningful to use as one of the variables (x) the higher of the two IBPDs (a positive dB number) and as the other variable (y) the negative absolute difference of the two IBPDs (a negative dB number). Or more formally:

$$x = \max(\text{IBPD}_1, \text{IBPD}_2) \tag{17}$$

$$y = -|\text{IBPD}_1 - \text{IBPD}_2| \tag{18}$$

An ideal angle diversity system would always be able to reduce an IBPD on one of the beams, e.g. $x = 10$ dB, to zero on the other beam, making $y = -10$ dB in the example. We can also interpret y as a "compensating signal" that reduces the value of x. In the ideal diversity case we would have $y = -x$. Now we introduce the standard definition of correlation coefficient

$$\rho = \frac{E\left[(x-x_0)(y-y_0)\right]}{\left[E(x-x_0)^2 E(y-y_0)^2\right]^{1/2}}, \tag{19}$$

where x_0 and y_0 are averages. For the ideal diversity case, we find $\rho = -1$, indicating anticorrelation.

We have used the two ray model described in section 2 to calculate the correlation coefficient for the transformed IBPD variables x and y and made use of the relationship between delay τ and angle of arrival α of the second ray given in equation 3. As a result of these simulations we find that the correlation coefficient is between -0.99 and -1.0 over a wide range of model parameters. We can see that this is quite plausible by referring to Figure 10. where the output IBPD is either zero or very small

4. EXPERIMENTAL RESULTS

4.1 The Test Arrangement

Early in June, 1986 we started to simultaneously monitor the performance of four different antenna arrangements using an unmodulated AR6A single sideband channel on the microwave radio path between Salton and Brawley, California (Figure 1). As in [4] we used the three radio pilots on AR6A to provide an estimate of IBPD, single tone and other statistics. A frequency of 6375 MHz and vertical polarization were used in the tests. In the first arrangement we monitored the performance of the unprotected pyramidal horn antenna mounted at the top of the tower and used as the main receiving antenna at Brawley. The second arrangement monitored the existing space diversity system at Brawley. The diversity antenna is a 10 foot diameter conical horn antenna mounted 42 feet below the main antenna. The outputs of the main antenna and the space diversity antenna were combined using the standard AR6A RF maximum power combiner. The radio pilots of this combined signal were then monitored. The third arrangement monitored the performance of a single 8-foot parabolic angle diversity antenna with two beams displaced in elevation ($\alpha_0 = 1.5°$, $\alpha_s = 1.7°$). The two beams V_1 and V_2 of Figure 8a, were used as inputs to the standard AR6A RF maximum power combiner mentioned in section 3.2.1, and the radio pilots at the output of the combiner were monitored. The fourth configuration also monitored the angle diversity antenna but using the sum and difference beams, V_s and V_d of Figure 8b, which were formed from the two individual beams by an RF magic Tee mounted at the rear of the antenna. The use of waveguide directional couplers mounted at the antenna allowed configurations three and four to share the same aperture. The angle diversity antenna was mounted eight feet below the main pyramidal horn antenna.

In January, 1987 we also began to monitor the performance of AT&T Technology's DR6-30-135, 64 QAM high capacity digital radio at a frequency of 6123 MHz, vertically polarized.

4.2 Scatter Diagrams and Correlation Coefficients

We have plotted some test results, taken during the first week of January, 1987, in the form of scatter diagrams. Single tone levels H_1 and H_2 from the two antenna beams V_1 and V_2 were first studied. These levels are expressed in terms of fade depths F_1 and F_2 and transformed into variables x and y as follows:

$$x = \max(F_1, F_2) \tag{20}$$

$$y = -|F_1 - F_2| \tag{21}$$

Since we were interested especially in deep dispersive fades the data was conditioned on $x > 10$ dB and max (IBPD_1, IBPD_2) > 10 dB. A total of 12,000 sample pairs F_1 and F_2 were collected at a rate of eight samples per second, but only every 20th sample was plotted in Figure 14. Each dot,

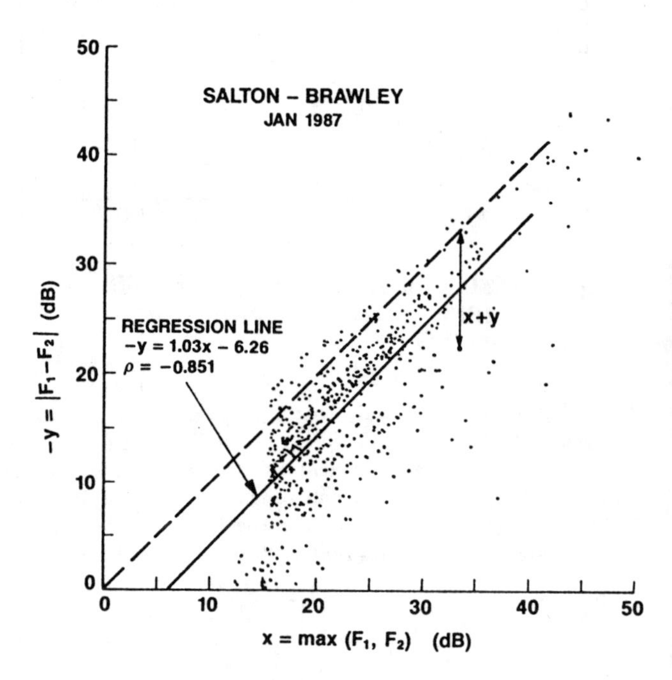

Fig. 14. Scatter diagram for transformed single tone fade depths F_1 and F_2.

therefore, represents 2.5 s of data. The correlation coefficient, slope and intercept point of the linear regression line, however, were calculated for the entire data set. The mean dispersive fade depth at the output of the mathematical diversity switch, using the regression line, will be equal to $x+y = 6.26 - 0.03x$. This mean is close to 6.26 dB regardless of the actual fade depth x at the input. The correlation coefficient of -0.851 is reflected in a relatively large spread of data around the mean. For instance, given a fixed value of $x = 30$ dB we find from Figure 14 a spread in $-y$ of about 14 to 31 dB, meaning that the fade depths at the diversity switch output, x+y, varied from 16 dB to -1 dB (an upfade).

We have also plotted measurements of IBPD during the same January, 1987 period in form of the transformed variables x and y given in equations 17 and 18 and conditioned on $x > 10$ dB. The 12,000 sample pairs of $IBPD_1$ and $IBPD_2$ from beams V_1 and V_2, were again compressed by a factor of twenty before being plotted in Figure 15. As we

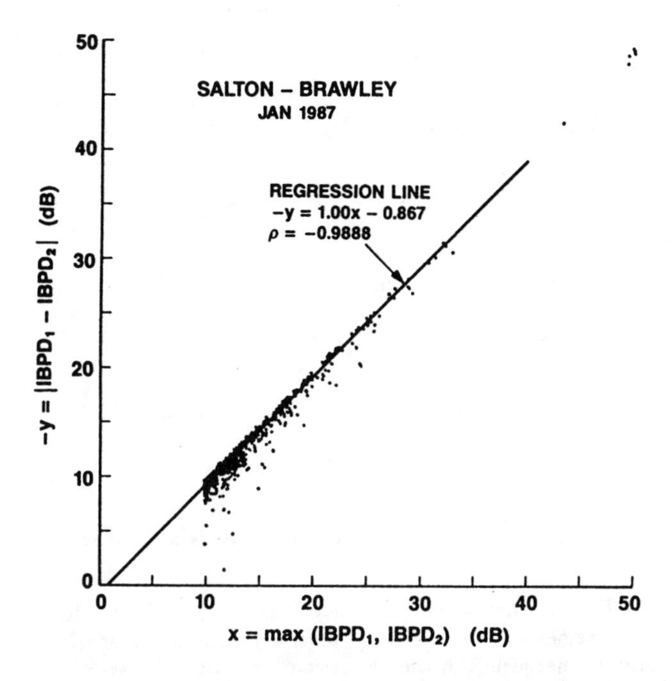

Fig. 15. Scatter diagram for transformed In-Band Power Differences $IBPD_1$ and $IBPD_2$.

had already predicted by computer simulation a high degree of anticorrelation was found in the variables x and y as reflected in a measured correlation coefficient of -0.9888. The mean value of IBPD at the output of the mathematical diversity switch, obtained from the regression line is equal to $x+y = 0.867$ dB. Figure 15 shows that the higher the IBPD at the input, the closer to 0.867 dB the IBPD at the output will be. We had mentioned a dramatic reduction in IBPD in connection with Figure 4. The situation is not as good at lower values of input IBPD. For instance, for $x = 12$ dB we find a range in $-y$ of 1 dB to 12 dB, resulting in an output IBPD of $x+y = 0$ to 11 dB. The upper value compares with the y limit of 9.5 dB shown in Figure 13.

4.3 Test Results in Form of Cumulative Statistics

Figure 16 shows the cumulative statistics of one of the three AR6A radio pilots as received through four different system configurations. Curves 1 and 2 are for reception through the low beam V_1 and the high beam V_2, respectively. Curve 3 shows the pilot tone level at the output of the AR6A maximum power combiner, using inputs from V_1 and V_2. A mathematical switch, selecting the higher of the two signals V_1 and V_2, produces curve 4. Curve 1 demonstrates that during the 19-day measurement period multipath fading caused by ground reflections exceeded the fading activity on the high beam which is mainly due to atmospheric reflections. Both beams exhibit almost pure Rayleigh fading. The simulated maximum power switch generates a diversity signal

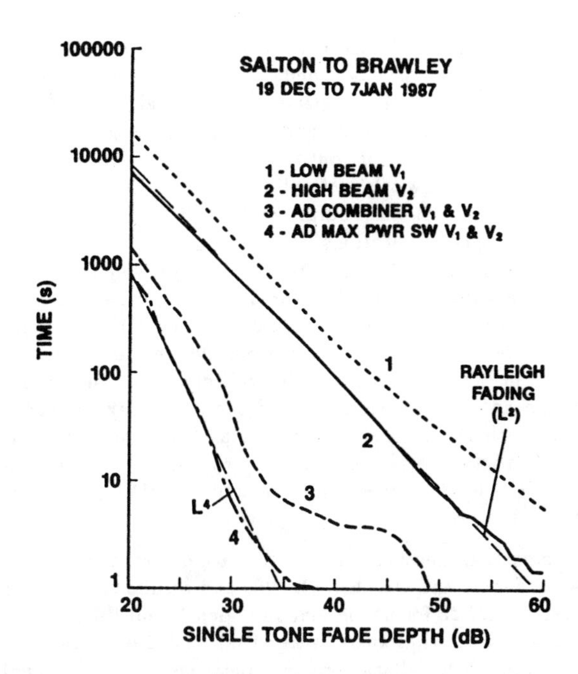

Fig. 16. Time that single-tone fade depth exceeds abscissa.

that follows the familiar L^4-law (L being used instead of H), typical of other diversity systems. Curve 3 was taken at the output of the RF combiner and theoretically should have been slightly below curve 4. The fact that it is not is due to a speed limitation in the combiner control circuit which has been designed for analog radio applications on typical radio hops. (The soft-switch used in the digital radio tests has no such speed limitation.) We have found that fading events on the Salton to Brawley hop are rather extreme in two respects. First, they occur for long periods of time every day all year round and secondly, they are considerably faster than observed on other radio hops.

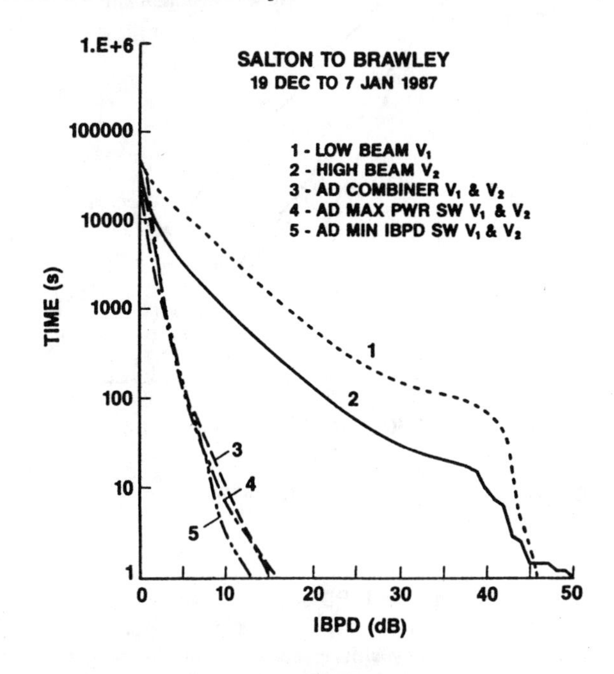

Fig. 17. Time that IBPD exceeds abscissa.

Figure 17 shows the IBPD for the same test period used in Figure 16. A fifth curve is added, representing the output of a mathematical switch that selects the terminal V_1 or V_2 with the lower IBPD. This technique, which is expected to reduce dispersion the most, is only marginally better than the maximum power switch method shown in curve 4. The RF combiner, curve 3, does surprisingly well when taking into account its speed limitations. The maximum $IBPD''_m$ expected for our system and obtained from Figure 13 would be about 9 dB. This value is reached at the 5 s level in Figure 17 but is exceeded at the 1 s level by about 3 dB (curve 5). The higher IBPD values may be due to periods where three rays are involved in the fading process, one reflected from the ground, one from an atmospheric layer and the direct ray. It can be shown in this case that the advantages of angle diversity would be slightly weakened but they remain still substantial. Because the IBPD curves are truncated at $IBPD''_m$, no L^4-law results.

Comparison measurements with space diversity were made between June 1 and October 10, 1986 and results are shown in Figures 18 and 19 for single tone fade depth and IBPD. From Figure 18 we see that there is essentially no difference in the single tone fade characteristics between the two angle diversity systems and space diversity. The three diversity curves fall less steep than the L^4-law, part of which we attribute to the speed limitations in the RF combiners that were used in all three cases. The long term equivalence of angle diversity and space diversity with respect to single tone statistics is considered coincidental. Angle diversity could be better or worse than space diversity depending on size and beam geometry of the angle diversity antenna.

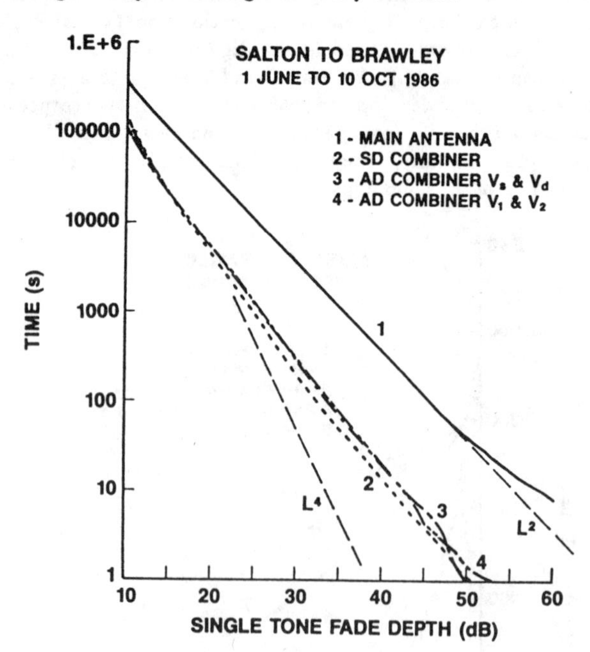

Fig. 18. Time that single-tone fade depth exceeds abscissa.

Figure 19 gives the IBPD statistics which show angle diversity to be superior to space diversity. Space diversity depends on a frequency shifted curve H_1 in Figure 4 for its diversity signal. This is less effective than switching from H_1

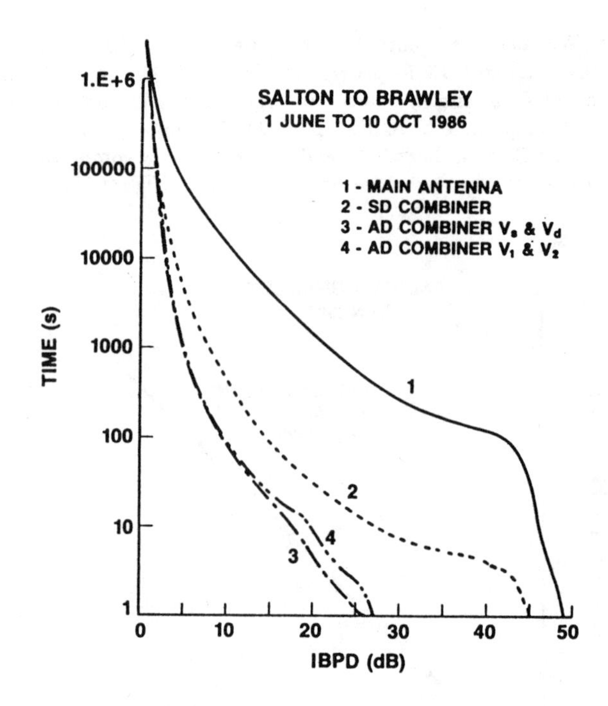

Fig. 19. Time that IBPD exceeds abscissa.

to H_2 in the angle diversity case. The large IBPD improvement factor for angle diversity seen in Figure 19 is also a consequence of the dispersion limitation characteristics discussed in section 3.2.2.

4.4 Results from Digital Radio Tests

Early in January, 1987 we began to monitor the performance of AT&T Technology's DR6, 64 QAM, 135 Mbit/s, digital radio on the Salton to Brawley path. In this experiment two DR6 digital radio receivers were equipped with soft IF diversity switches, IF slope equalizers and baseband adaptive transversal equalizers. (The error correction option available in DR 6 was not provided in the experiment. It would be able to give a substantial reduction of isolated errors.) The experiment was configured to compare the performance of the two angle diversity arrangements, V_1 & V_2 and V_s & V_d, with the performance of the unprotected pyramidal horn antenna which was connected to a third identically equipped DR6 bay. Differences in waveguide losses due to the special experimental setup, and differences in antenna gains and receiver noise figures, resulted in a reduction of thermal fade margin of 9 dB for the angle diversity systems relative to the pyramidal horn antenna system. Despite this handicap, excellent results were obtained for errored seconds (ES), severely errored seconds (SES) and very severe bursts (VSB) over a period of 42 days in January and February of 1987. They are summarized in Table 1.

Errored seconds are seconds that have at least one error and severely errored seconds are defined as seconds in which the BER exceeds 10^{-3}. Very severe bursts are events which last between 2.5 and 10 s and have a BER exceeding 10^{-2}. All measurements were made at the DS3 rate of 45 Mbit/s. The diversity improvement factors for ES, SES and VSB are very large for both angle diversity systems and the differences, seemingly favoring V_s and V_d, may be due to statistical

TABLE 1. Results of the Digital Radio Experiment between Salton and Brawley, CA

	Angle Diversity Antenna		Unprotected Antenna
	V_1 & V_2	V_s & V_d	
ES	161	137	23275
SES	38	23	14709
VSB	2	2	914

fluctuations. The important SES improvement factor is 387 for V_1 & V_2 and 639 for V_s & V_d. This compares with much smaller improvement factors obtained with space diversity.

5. CONCLUSIONS

Angle diversity using dual beam antennas and various combining and switching techniques has been analyzed and measured. As a result of these investigations we conclude that angle diversity is a very effective tool to combat dispersive fading in line-of-sight terrestrial microwave links. If properly designed, angle diversity exhibits substantial improvements over conventional space diversity. In addition it is very cost effective because it does not require a second antenna as space diversity does and doesn't need the frequency space of frequency diversity.

6. ACKNOWLEDGEMENTS

The assistance of AT&T-Communications and D. L. Jacobs in implementing the angle diversity experiments and the contributions of W. C. Peng and C. H. Bianchi to the analysis of the data collected from Brawley are gratefully acknowledged.

REFERENCES

[1] A. B. Crawford, D. C. Hogg, and W. H. Kummer, "Studies in Tropospheric Propagation Beyond the Horizon," BSTJ, September 1959, pp. 1067-1178.

[2] A. Malaga and S. A. Parl, "Experimental Comparison of Angle and Space Diversity for Line-of-Sight Microwave Links," Milcom 1985, paper 19.5.

[3] R. W. Hubbard, "Angle Diversity Reception for LOS Digital Microwave Radio," Milcom 1985, paper 19.6.

[4] E. H. Lin and A. J. Giger, "Radio Channel Characterization by Three Tones," IEEE Journal on Selected Areas in Communications, April 1986.

[5] E. T. Harkless and H. F. Lenzing, "Excitation of Higher Order Antenna Modes by Multipath Propagation," IEEE Transactions on Communication Technology, August 1967, pp. 597-603.

[6] P. M. Cronin, "Dual-Polarized Digital Radio Operation in a Fading Environment," ICC '80, paper 52.5.

[7] M. F. Gardina and S. H. Lin, "Measured Performance of Horizontal Space Diversity on a Microwave Radio Path," Globecom '85.

[8] E. H. Lin, "Spatial Correlations in Adaptive Arrays", IEEE Transaction on Antennas and Propagation, March 1982, pp. 212-223.

[9] P. Balaban, E. A. Sweedyk and G. S. Axeling, "Angle Diversity with Two Antennas: Model and Experimental Results."

[10] K. P. Dombek, "Reduction of Multipath Interference by Adaptive Beam Orientation" European Conference on Radio Relay Systems, Nov. 1986, pp. 400-406.

Advanced Time- and Frequency-Domain Adaptive Equalization in Multilevel QAM Digital Radio Systems

GEORG SEBALD, BERTHOLD LANKL, MEMBER, IEEE, AND JOSEF A. NOSSEK, SENIOR MEMBER, IEEE

Abstract—Digital radio systems employing multilevel QAM are at least optionally equipped with adaptive time- and/or frequency-domain equalizers. Their purpose is to reduce the vulnerability of these systems to linear distortion caused by multipath propagation.

Linear transversal filters are prominent candidates for the realization of time-domain equalizers, especially for high-capacity applications. They are well known for their good performance and their relatively easy implementation at a high data rate.

On the other hand, decision feedback equalizers are known to be very capable of eliminating linear distortion, especially of the so-called minimum-phase type. But realization problems are likely to occur in a high-speed application.

A solution is proposed which merges the relative advantages of both the linear transversal and the decision feedback approaches. The goal of a frequency-domain equalizer, which is the restoration of the shape of the power density spectrum of the received signal without any recovered carrier and timing signals, can also be achieved with the aid of a transversal filter.

The performance obtained with the joint utilization of the novel time- and frequency-domain equalizers is described.

INTRODUCTION

THE vulnerability of multilevel QAM digital radio systems with respect to linear distortion caused by multipath propagation can greatly be reduced by the use of adaptive equalizers. The choice of the equalizer structure and of its adaptation algorithm is heavily influenced by the baud rate of the system and by the technology available for the implementation. An equally important factor is the very special nature of the radio channel, particularly the time-variant transfer function.

The equalizer concepts can be classified whether time- or frequency-domain criteria are used for their control. Time-domain equalizers are mainly, but not necessarily, realized at baseband, while frequency-domain equalizers are mainly realized at IF.

Nevertheless, since QAM systems are supposed to be linear, baseband and IF implementations can be made equivalent from the theoretical point of view. Therefore, the realization aspects are determining the particular way to go.

First, the time-domain approach is covered, focusing on baseband implementation, and second, the frequency-domain case is dealt with, discussing the corresponding IF implementation.

It is well known that transversal equalizers controlled by the zero-forcing algorithm offer an attractive solution in the case of high data rate, i.e., 140 Mbit/s radio systems [1], [4], [8], [9]. Since a transversal equalizer is a nonrecursive system, pipelining can be applied to relax speed problems associated especially with the A/D converters used for both the data decision as well as the equalizer control. To ensure correct decisions of these A/D converters and to avoid code error problems due to metastable states of some of their comparators [2], latency time can be introduced without affecting the equalization performance. On the other hand, the very able decision feedback approach [3] is inherently exposed to this speed problem because in a recursive system pipelining cannot be applied to the same extent.

The present contribution focuses on a solution which is based on a transversal equalizer [1] which is backed up by a "far-off" recursive coefficient. This way the speed problems are bypassed since several baud periods are available to carry out the necessary critical operations. Nevertheless, the cancellation of residual lagging echoes is carried out similarly to a conventional decision feedback equalizer.

The superior acquisition behavior of the transversal equalizer [1], which mainly relies on the precision of the realized integrators in the correlation unit, is carried on in the recursive case [6].

The purpose of a frequency-domain equalizer is to restore the desired shape of the power density spectrum of the received signal without the need for already-decided data. Depending on the type of simultaneously employed time-domain equalizer, it seems to be desirable to have a frequency-domain equalizer of the minimum-phase (MP), the linear-phase (LP) or the nonminimum-phase (NMP) type. A solution, basically offering these possibilities, is described.

The performance obtained with the joint utilization of the new time-domain baseband equalizer and a very simple, but still powerful, version of the transversal frequency-domain equalizer is investigated. For the sake of brevity, detailed simulated and measured results are given for 16 QAM only. The enhanced performance achieved is obviously not restricted to this case.

Manuscript received October 12, 1986; revised December 19, 1986.
The authors are with the Transmission Systems Division, Radio Relay Design, Siemens AG, Munich, West Germany.
IEEE Log Number 8613339.

Reprinted from *IEEE J. Selected Areas Comm.*, vol. SAC-5, no. 3, pp. 448–456, Apr. 1987.

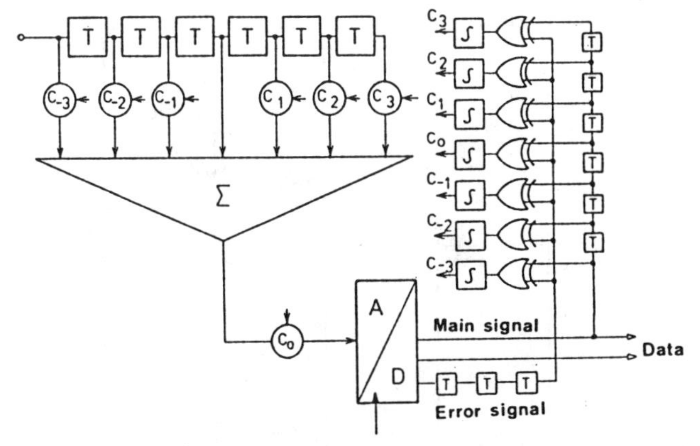

Fig. 1. Single-rail linear adaptive transversal equalizer controlled by the zero-forcing algorithm.

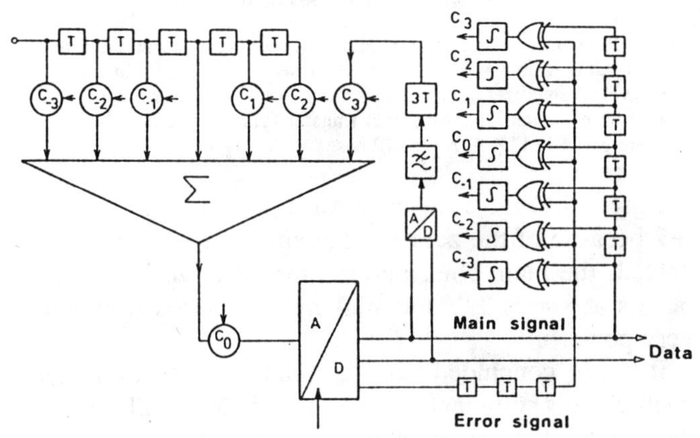

Fig. 2. Modified equalizer with a single "far-off" recursive tap (decision feedbacked).

Time-Domain Baseband Equalizer

Concept

The basic idea is explained first with the aid of a single-rail baseband equalizer for a four-level baseband signal, as given in Figs. 1 and 2.

Fig. 1 shows a seven-tap single-rail linear transversal equalizer together with the A / D converter and a simple realization of the zero-forcing algorithm based on 1 bit-quantized representations of the error and the main signal. In this special case the multiplication needed in the correlation process turns out to be a simple "EXOR," which is very desirable in a high-speed system.

Since there is no feedback loop within the signal path, latency time can be introduced to ease the job of the flash A / D converter. This is important since the A / D converters need time to make correct decisions and to avoid code error problems due to metastable states of some of their comparators [2]. This becomes an even more important problem in an adaptively equalized system because the better equalized the system is, the more likely the situation occurs that the flash A / D converter is exposed to an input signal just on the threshold of one of its comparators. The nonrecursive nature of a transversal equalizer offers a rather simple solution to this problem because latency time can be introduced without affecting the equalization performance.

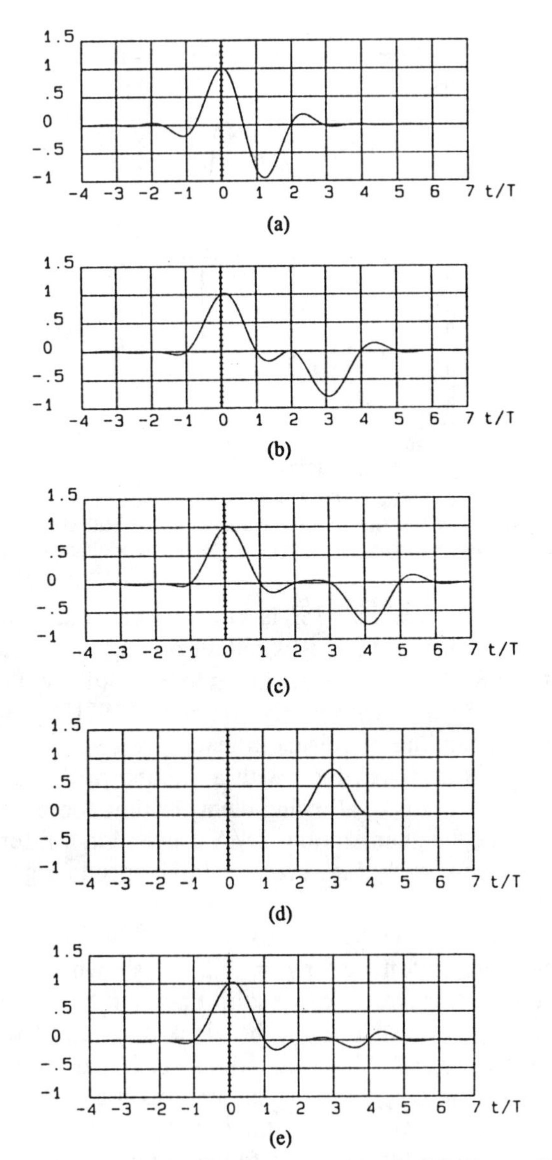

Fig. 3. Simulated pulse shapes at various points of a full-response Nyquist system suffering from two-path propagation (τ = 6.3 ns, B = 35 dB MP $f_n = f_c$). (a) Input pulse shape, unequalized. (b) Output pulse of a five-tap symmetric transversal equalizer controlled by the zero-forcing algorithm. (c) Same as (b), but for a seven-tap equalizer. (d) Feedback signal for compensation of the residual ISI at $t = 3T$ of the five-tap equalizer of (b). (e) Output pulse shape of the modified equalizer, given in Fig. 2.

Now it is worthwhile to consider the pulse shapes in Fig. 3, which have been simulated for a two-path propagation situation [5], creating a center notch with 35 dB depth. Such a center notch in a QAM system does not cause cross-rail distortion (crosstalk between quadrature channels), and therefore it is sufficient to deal with the single-rail baseband model. While in Fig. 3(a) the unequalized pulse is depicted with its large intersymbol interference (ISI) at the first subsequent timing instant, it is obvious from Fig. 3(b) and (c) that over the span of the equalizer length the zeros in the pulse response are restored. But at the first timing instant outside the equalizer span, the residual ISI is still very large, and with increasing filter length it decreases only marginally, not substantially.

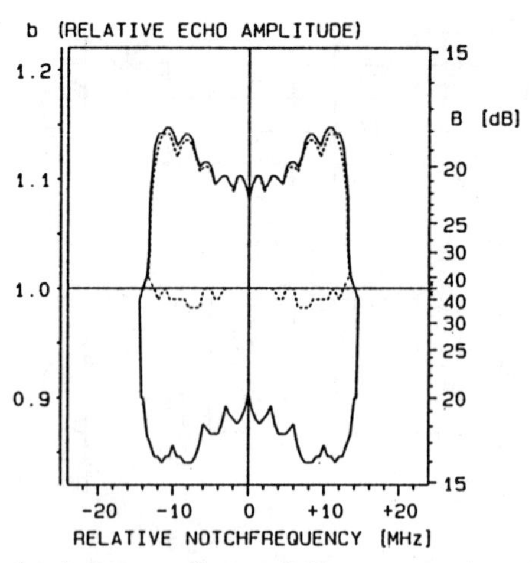

Fig. 4. Simulated signatures $D_{\text{peak}} = 0.33$ = constant (τ = 6.3 ns) obtained with a complex transversal equalizer. (——) With six linear taps (three pre- and two postcursors). (- - - -) With one additional recursive tap.

Because the residual ISI is concentrated onto almost only one timing instant, it is not surprising that a single quantized feedback pulse, as shown in Fig. 3(d), will yield an equalized signal with nearly no residual ISI [Fig. 3(e)]. The corresponding circuit modification is given in Fig. 2 showing the feedback loop with a "roundtrip" delay of three symbol periods, allowing plenty of time for the A/D conversion, the quantization (D/A conversion), filtering, and coefficient weighing, even in a high-speed (e.g., 140 Mbit/s) system.

The simulated signatures for a 16-QAM 140 Mbit/s modem with rolloff factor $\rho = 0.5$ are shown in Fig. 4 for the pure transversal case (solid line) and for the modified approach (dashed line), the block diagram of which is shown in Fig. 7. The simulations are based on a constant value of D_{peak}, which amounts to the normalized maximum linear distortion at the sampling instants for the equalized pulse shape. For the four-level baseband signal, the greatest permissible value for D_{peak} is $\frac{1}{3}$ for a perfect system, resulting in a just-closed eye pattern.

The improvement obtained by the modification, which has qualitatively been described with the aid of Fig. 3, can now quantitatively be seen in Fig. 5 where the residual linear distortion is plotted versus the notch-offset frequency, while the notch depth is in accordance with Fig. 4. The D_{peak} evaluation is carried out over a sufficient number of adjacent symbols.

$$D_{\text{peak}} = \frac{1}{|i(0)|}\left\{ |q(0)| + \sum_{\substack{n=-\infty \\ n\neq 0}}^{+\infty} \left[|i(n)| + |q(n)| \right] \right\} \tag{1}$$

with $i(t)$ and $q(t)$ being the response of the equalized in-phase and quadrature channel with respect to a single excitation of the in-phase channel.

Moreover, the individual contributions

$$x_n = \frac{1}{|i(0)|}\left[|i(nT)| + |q(nT)| \right]; \quad n = -5, -4, +3, +4, +5 \tag{2}$$

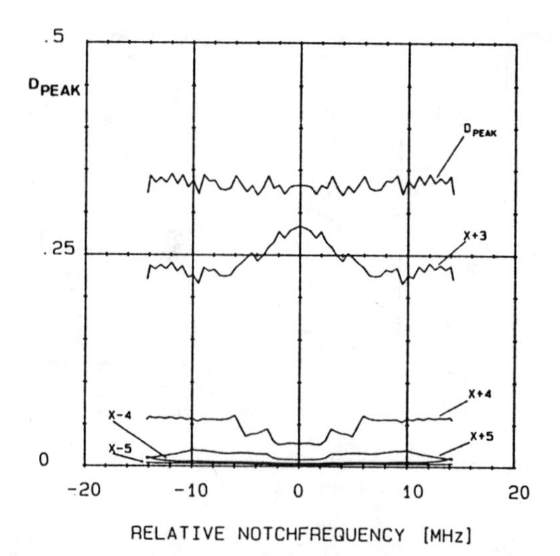

Fig. 5. Peak linear distortion and individual contributions at different timing instants from $t = -5T$ to $t = +5T$ at the output of a linear six-tap transversal equalizer controlled by the zero-forcing algorithm (three pre- and two postcursors) for channel transfer functions defined by the signature given in Fig. 4 in the MP case.

are given in Fig. 5, $x_n = 0$ for $n = -3, -2, -1, +1, +2$ because of the zero-forcing six-tap equalizer. As expected, the main contribution along the whole signature occurs at $t = +3T$, thus making the proposed approach very powerful.

It can be concluded that one single recursive "far-off" coefficient is sufficient to take care of almost all the linear distortion which is left over by a pure transversal equalizer controlled by the zero-forcing algorithm, thus yielding a greatly improved MP signature.

An important assumption in obtaining the just-described improvement has been the utilization of the zero-forcing algorithm. The minimum mean-square error (MMSE) algorithm is not as powerful as the zero-forcing algorithm in this respect. The reason for this result can at least be qualitatively explained starting with Fig. 3(a). From here it can be seen that mainly the first lagging timing instant (in the case of an MP channel) is contaminated by ISI. All other zero crossings in the response of the channel to a Nyquist pulse excitation are more or less undisturbed. The amount of ISI at the first lagging timing instant increases with increasing channel notch depth. In a transversal filter this ISI is eliminated by adding a compensating signal, which is a delayed and weighed replica of the unequalized input signal. Thereby, the ISI at the first lagging instant is eliminated, but major ISI at the next timing instant is introduced. The very same argument applies for the operation of the next coefficient, and so on. Therefore, it can be concluded that by enforcing as many zeros as possible in the time response with a transversal filter of a given length, a major residual ISI concentrated at the first timing instant exceeding the span of the equalizer is established. There are two main reasons for this effect: first, the presence of only one strong echo, and second, the utilization of the zero-forcing algorithm. This very result is already depicted in Fig. 5, showing that X_{+3}, the ISI three baud periods after the main lobe of the time

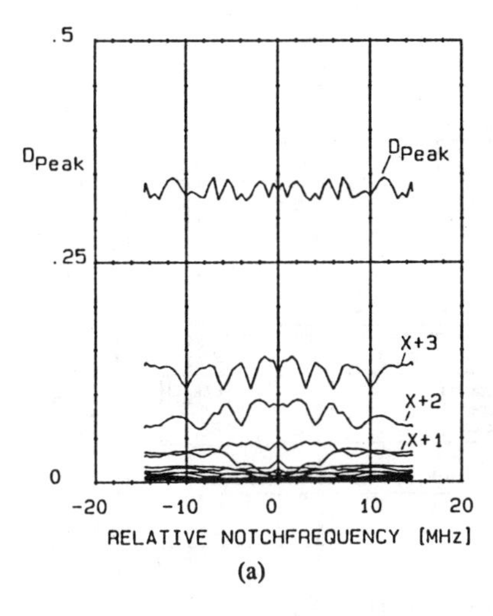

(a)

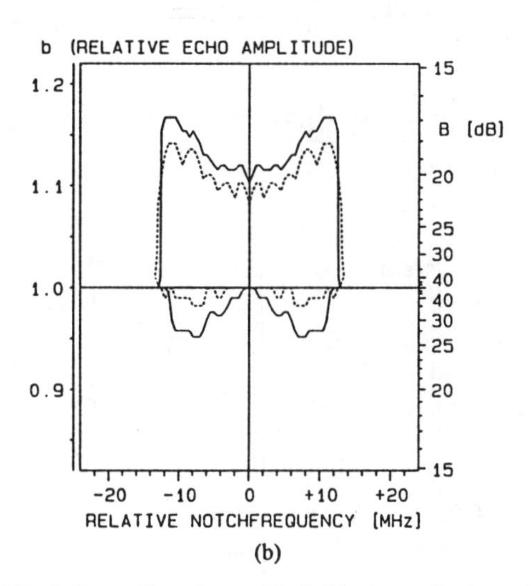

(b)

Fig. 6. (a) Peak linear distortion and individual contributions at different timing instants from $t = -5T$ to $t = +5T$ at the output of a linear six-tap transversal equalizer controlled by the MMSE algorithm (three pre- and two postcursors) for channel transfer functions defined by the signature (given in Fig. 4). (b) Simulated signatures for $D_{peak} = 0.33$ = constant ($\tau = 6.3$ ns) obtained with the modified equalizer (Figs. 2 and 7). (——) Controlled by the MMSE algorithm. (- - - -) Controlled by the zero-forcing algorithm.

response, just the first outside the span of a six-tap transversal equalizer with three pre- and two postcursors, is by far the largest contribution to the overall worst case ISI.

Since the MMSE algorithm is not enforcing zeros and, therefore, is not concentrating the ISI outside of the equalizer span, the potential of the introduction of a single recursive coefficient is considerably reduced. This is quantitatively shown in Fig. 6(a) where the individual contributions of various timing instants to the overall worst case ISI ($= D_{peak}$) are given. This is also proved by the comparison of signatures [Fig. 6(b)] for the same equalizer network (with one "far-off" recursive coefficient) controlled by the MMSE, and the zero forcing algorithm, respectively.

Thus, the control of the equalizer by the zero-forcing algorithm has, in the case of a high-capacity radio system, not only the advantage of a simpler implementation, but also of an enhanced signature.

Lock-In Performance

In the case of very severe selective fading, although a powerful equalizer is used, not only might a BER of 10^{-3} be exceeded, but also, some of the control loops of the adaptive equalizer, the carrier- and the timing-recovery will lock out. It is, of course, desirable that only a slight reduction of the channel distortion is necessary for the whole system to recover. That allows all the control loops to lock in in the situations where the resulting BER still exceeds 10^{-3}. Experiments have shown that in the system described the equalizer is locking in already in the case of noncoherent demodulation, i.e., if the carrier recovery has not locked in again, thus providing an already equalized signal for the generation of the control voltage of the demodulating VCO (see Fig. 7).

Fig. 8 shows a complex representation of the equalized pulse, verifying that the enforced zeros in the pulse shape are independent of the carrier phase angle, and because a fixed, real center tap is used, the setting of the pre- and postcursors of the equalizer is unchanged. This qualitatively explains the convergence of the equalizer control even under noncoherent demodulation, greatly aiding the carrier recovery loop.

Fig. 9 summarizes this ability of the seven-tap linear transversal equalizer by comparing the BER = 10^{-3} signature to the signatures for the lock in of the equalizer as well as the carrier recovery loop. Both lock-in signatures are well within the BER = 10^{-3} signature. Therefore, the outage probability is not affected by the acquisition behavior of the system.

But it is also obvious that the overall capability of an already-locked-in equalizer is increased by reducing the residual linear distortion and consequently pushing the BER below 10^{-3}. That is exactly what the modified transversal equalizer with its "far-off" recursive coefficient does. It seems important to note that this modification gains its full advantage only in conjunction with the superior acquisition performance. The corresponding signatures are given in Fig. 10 where the expected improvement in the MP case is obtained, while the result in the NMP case remains pretty much unchanged.

An important requirement to realize that acquisition performance in a practical implementation is actually the precision of the correlation unit, consisting of modulo-2 multipliers (EXOR gates) and analog integrators. The reference voltage of the latter must be controlled tightly in order not to deteriorate the theoretically excellent acquisition behavior.

Fig. 11 shows measured values of the relative integrator offset for different two-path situations with varying notch depth. For each channel situation the maximum value of the relative integrator offset of the most critical coefficient is plotted versus the notch depth. These values are obtained for lock in of the equalizer during noncoherent demodulation. It can be concluded that a relative integrator offset above 10 percent would prevent lock in even with a perfect distortion-free channel. It should be noted that a more powerful equalizer asks for more ac-

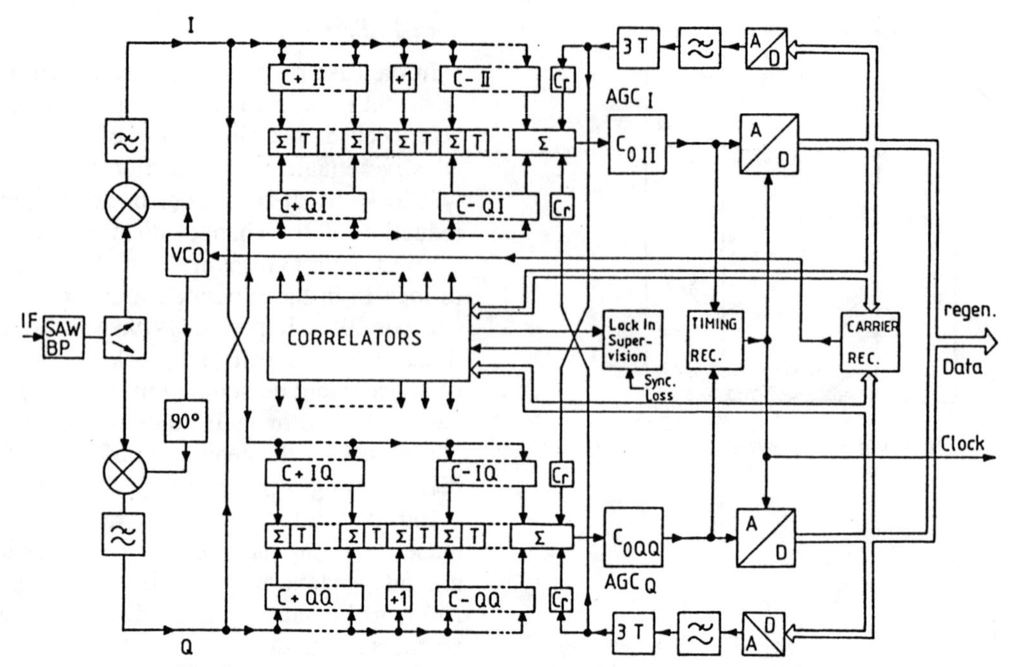

Fig. 7. Block diagram of the demodulator with the modified equalizer with a single recursive tap.

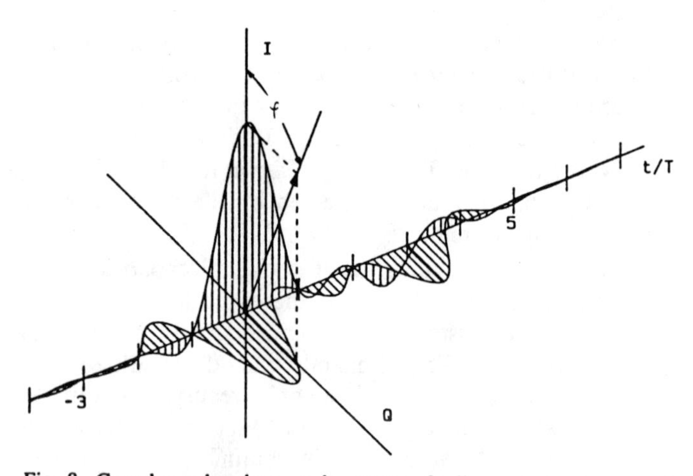

Fig. 8. Complex pulse shape at the output of a linear six-tap transversal equalizer controlled by the zero-forcing algorithm.

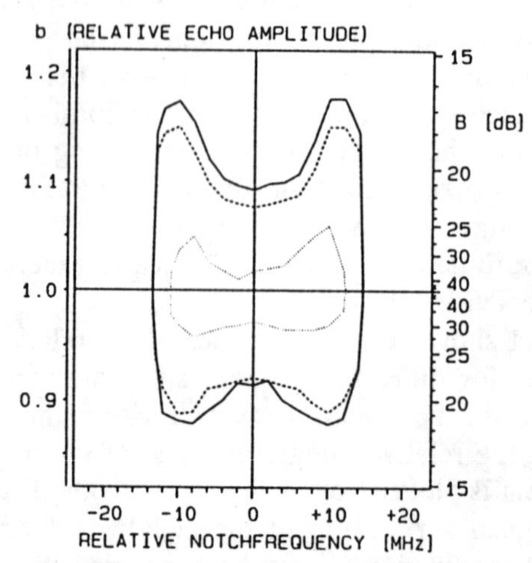

Fig. 9. Measured signatures (16-QAM, 140 Mbit/s, symmetric seven-tap linear transversal equalizer, τ = 6.3 ns). (——) For BER = 10^{-3}. (- - - -) For lock in of the carrier recovery loop. (· · · ·) For lock in of the transversal equalizer.

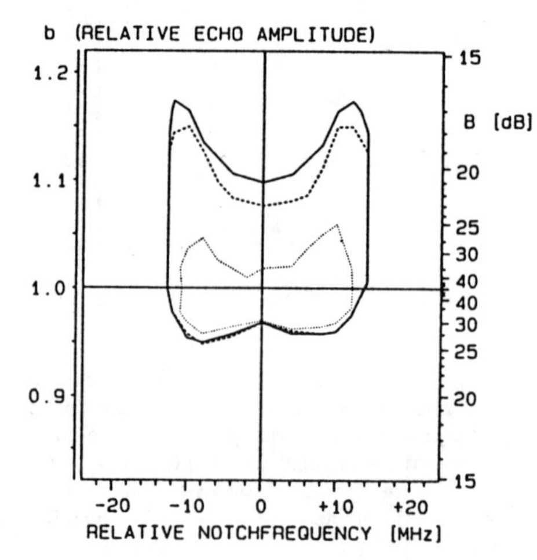

Fig. 10. Measured signatures (16-QAM, 140 Mbit/s, seven-tap modified equalizer, τ = 6.3 ns). (——) For BER = 10^{-3}. (- - - -) For lock in of the carrier recovery loop. (· · · ·) For lock in of the equalizer.

curate integrators. For the proposed solution with one recursive tap, the relative integrator offset has to be below 1–2 percent.

Simulations of the lock-in performance of a three-tap linear transversal equalizer with 1 percent [Fig. 12(a)] and 5 percent [Fig. 12(b)] relative integrator offset, respectively, are in agreement with the aforementioned experimental results.

Realization

A special auto-zero technique has been implemented to control the reference voltage of the integrator with sufficient precision, resulting in a worst case relative offset below 1 percent (Fig. 13). The basic idea is to derive automatically the reference voltage from a so-called "re-

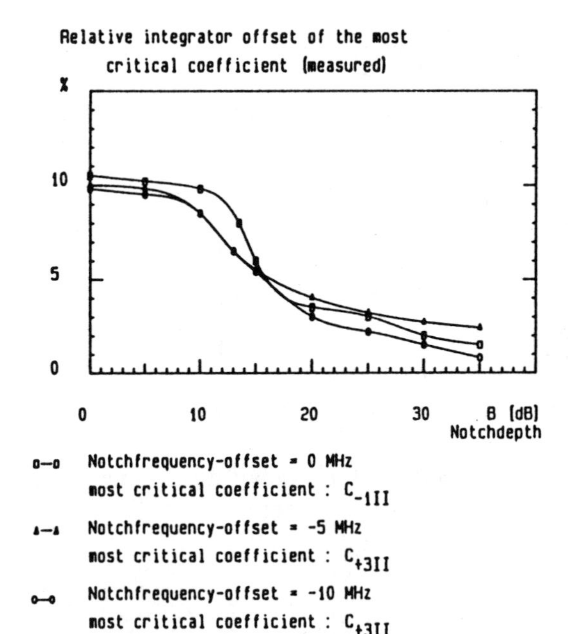

Fig. 11. Maximum permissible integrator offset for lock in of transversal equalizer during noncoherent demodulation.

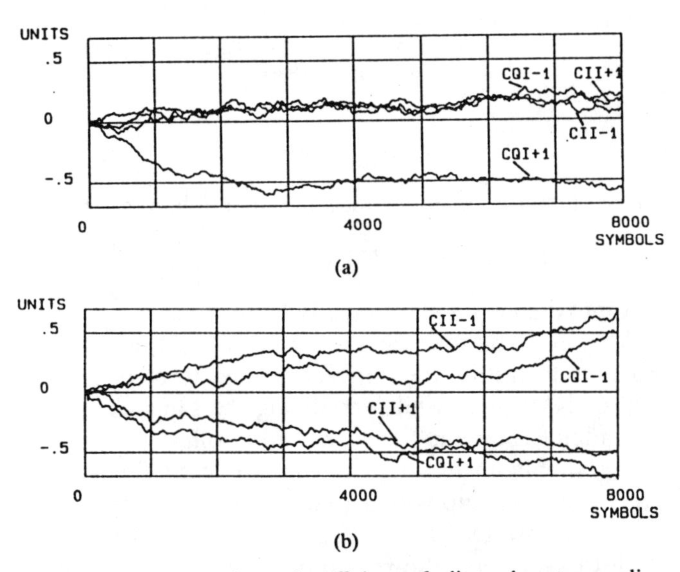

Fig. 12. Simulated transients of coefficients of a linear three-tap equalizer with noncoherent demodulation under the two path conditions: τ = 6.3 ns, B = 12.5 dB MP, f_n = 10 MHz + f_c. (a) 1 percent relative integrator offset (convergence). (b) 5 percent relative integrator offset (divergence).

fresh sequence'' having equally likely ''zeros'' and ''ones.'' This technique takes into account the influence of supply voltages, tolerances of logic ''low'' and ''high'' voltages, and deviations in rise and fall time. The actual realization is based on the potentials of custom-integrated LSI chips and of hybrid integrated circuits, minimizing the physical size of the equipment. Fig. 7 shows the complete realization of the proposed modified equalizer as used in multilevel QAM systems. As already described in

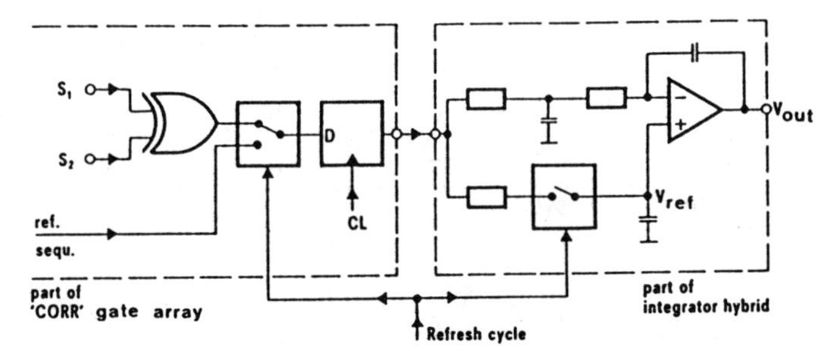

Fig. 13. Correlator with auto-zero technique.

[1], the parallel-in/serial-out structure has been chosen for reasons of modularity. The quantized feedback signals are shaped with low-pass filters to be ISI free. The loop delay time is distributed over all components contained in the recursive loop. Since there are three baud periods available, there is plenty of time assigned to the A/D conversion process, avoiding ''dribble errors'' caused by metastable states of the comparators [2].

Therefore, the modified approach is as good as a linear transversal filter as far as ''background'' error performance is concerned. This is a considerable advantage compared to conventional decision feedback solutions, at least in the case of very high data rates.

Frequency-Domain IF Equalizer

Concept

It is the purpose of a frequency-domain equalizer, which is normally, but not necessarily, implemented at IF, to correct the power density spectrum of the received signal, which is analyzed with the aid of, for example, a BP filter bank. As long as there is no major redundant information in the transmitted signal, it is not possible to gain any information on the phase or group delay distortion of the channel; only the attenuation distortion can be recognized and equalized properly.

Depending on the time-domain equalizer used, different approaches for the frequency-domain equalizer, as far as its phase behavior is concerned, seem to be desirable to come up with a well-balanced signature for MP and NMP channel situations:

1) no time-domain equalizer → frequency-domain equalizer network of MP type (i.e., the phase and the magnitude response are linked to each other via the Hilbert transform) because such a system can withstand only rather shallow notches, which are more likely of the MP type;
2) linear, symmetric transveral time-domain equalizer → frequency-domain equalizer network of LP type to leave the phase distortion of the channel unchanged resulting in a symmetric (MP/NMP) signature;
3) time-domain equalizer with recursive coefficient → frequency-domain equalizer network of NMP type to improve especially the NMP signature of the time-domain equalizer.

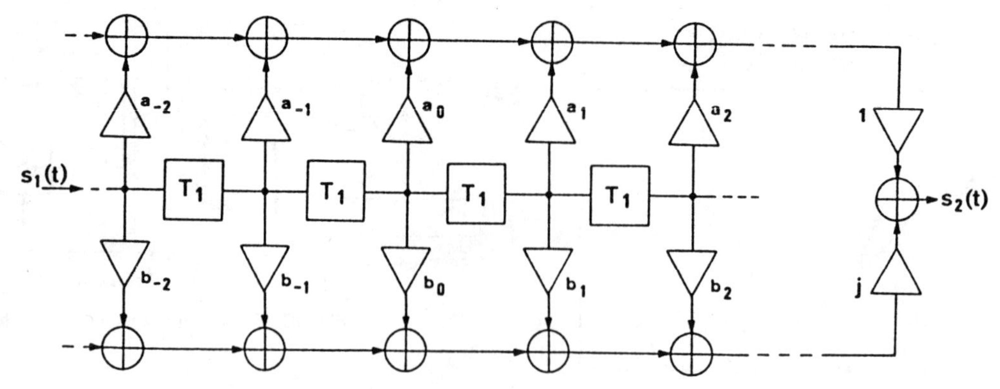

Fig. 14. Complex transversal filter.

It is well known that a complex transversal filter can arbitrarily closely approximate any desired realizable phase and magnitude response. Depending on special relations between its pre- and postcursors, it can be programmed to belong to one of the aforementioned classes.

Fig. 14 shows such a transversal filter in the serial-in/parallel-out configuration, the transfer function of which

$$H(\omega) = e^{-j\omega NT_1} \sum_{n=-N}^{+N} (a_n + jb_n)\, e^{-j\omega nT_1} \tag{3}$$

is of

a) MP if

$$a_{-n} = 0,\ b_{-n} = 0, \qquad n = 1, \cdots, N$$

and

$$a_0^2 + b_0^2 \geq \sum_{n=1}^{N} (a_n^2 + b_n^2) \tag{4}$$

b) LP if

$$a_{-n} = a_n,\ b_{-n} = -b_n, \qquad n = 0, 1, \cdots, N \tag{5}$$

c) NMP if

$$a_n = 0,\ b_n = 0, \qquad n = 1, \cdots, N$$

and

$$a_0^2 + b_0^2 \geq \sum_{n=1}^{N} (a_{-n}^2 + b_{-n}^2). \tag{6}$$

If the coefficients in case a) (MP) and case c) (NMP) are mirror images of each other, such that

$$a_{n\text{MP}} = a_{-n\text{NMP}}$$

and

$$-b_{n\text{MP}} = b_{-n\text{NMP}}, \qquad n = 0, 1, \cdots, N \tag{7}$$

holds, the magnitudes of the respective transfer function are identical:

$$\left|H_{\text{MP}}(\omega)\right| = \left|H_{\text{NMP}}(\omega)\right| = \left\{\left[a_0 + \sum_{n=1}^{N} (a_n \cos n\omega T_1 + b_n \sin n\omega T_1)\right]^2 + \left[b_0 + \sum_{n=1}^{N} (b_n \cos n\omega T_1 - a_n \sin n\omega T_1)\right]^2\right\}^{1/2} \tag{8}$$

and the respective group delay distortions are equal in magnitude but opposite in sign:

$$\Delta\tau_{G\text{MP}}(\omega) = -\Delta\tau_{G\text{NMP}}(\omega) = \frac{d}{d\omega}\left[\arctan \frac{b_0 + \sum_{n=1}^{N} (b_n \cos n\omega T_1 - a_n \sin n\omega T_1)}{a_0 + \sum_{n=1}^{N} (a_n \cos n\omega T_1 + b_n \sin n\omega T_1)}\right]. \tag{9}$$

If in addition to (7) all coefficients are small compared to the reference tap a_0 and

$$a_{n\text{MP}} = 2a_{n\text{LP}}$$

and

$$b_{n\text{MP}} = 2b_{n\text{LP}} \tag{10}$$

holds, then the transfer function magnitude of the LP case

$$\left|H_{\text{LP}}(\omega)\right| \approx \left|H_{\text{MP}}(\omega)\right| = \left|H_{\text{NMP}}(\omega)\right| \tag{11}$$

is approximately equal to the two other cases.

Therefore, by following the rules (4), (5), and (6) it is possible to restrict the transversal filter to be of the MP, LP, or NMP type while equalizing the same channel attenuation. Thus, the transversal filter offers the potential to be simply matched to the jointly used time-domain equalizer to end up with a well-balanced signature.

In any case the choice of the delay time T_1 should be such to avoid periodic repetition of the transfer function within the spectrum to be equalized.

Realization

In the following the NMP structure is considered because it is the best tradeoff together with the baseband equalizer with its "far-off" recursive tap.

In addition, the discussion is restricted to the realization of a simple slope equalizer, as it is widely used in radio systems. This results in a very simple first-order real transversal filter, which can be realized at IF, e.g., with the aid of SAW elements [7] or quite conventional delay elements.

The transfer function of such a first-order NMP trans-

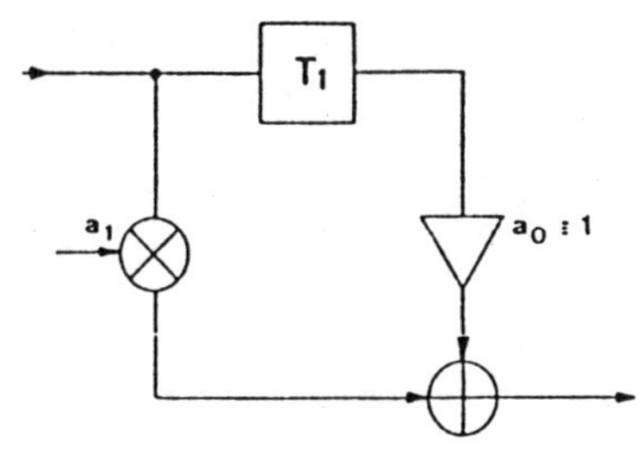

Fig. 15. Transversal IF slope equalizer of NMP type with a single real tap.

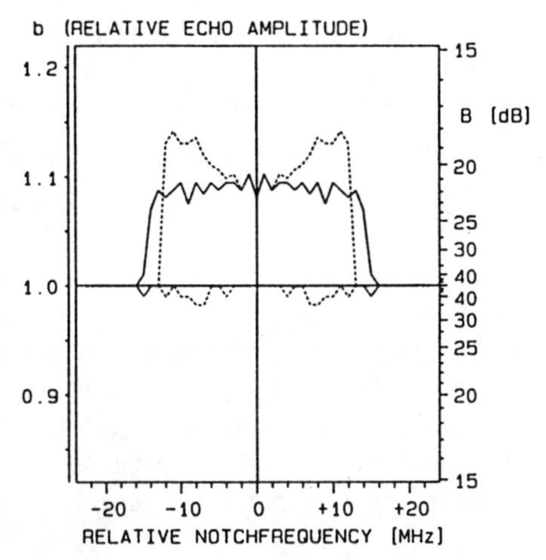

Fig. 16. Simulated signatures for $D_{peak} = 0.333$ = constant ($\tau = 6.3$ ns) obtained with a modified seven-tap time-domain equalizer only (----) and with an additional IF slope equalizer, as given in Fig. 15 (——).

versal filter is

$$H_{\mathrm{NMP}}(\omega) = e^{-j\omega T_1} \left[a_0 + a_1 \left(\cos \omega T_1 + j \sin \omega T_1\right)\right] \tag{12}$$

with a magnitude and group delay response of

$$\left|H_{\mathrm{NMP}}(\omega)\right| = \sqrt{a_0^2 + a_1^2 + 2a_0 a_1 \cos \omega T_1} \tag{13}$$

$$\Delta\tau_{G_{\mathrm{NMP}}}(\omega) = -\frac{a_1 T_1 \left(a_0 \cos \omega T_1 + a_1\right)}{a_0^2 + a_1^2 + 2a_0 a_1 \cos \omega T_1}, \tag{14}$$

which shows the positive peak of the group delay as expected in NMP situations.

One choice of T_1 resulting in a slope controlled by a_1 is

$$T_1 = \tfrac{5}{4} \cdot f_{\mathrm{IF}}. \tag{15}$$

Fig. 15 shows the block diagram of this simple equalizer, which is controlled by sensing the power spectrum at f_{IF} $\pm f_{\mathrm{Nyquist}}$ with two BP filters (implemented either at IF or baseband). Fig. 16 shows the signatures of the system equipped with the time-domain equalizer only (dashed line) and with both the time- and the frequency-domain equalizer of the NMP type (solid line). The improvement exceeds the values obtained with an LP or an MP IF equalizer. Nevertheless, it is very easy to reconfigure the structure to end up with an IF equalizer, either of LP or of MP type.

Conclusions

A rather simple modification of a linear transversal equalizer in a QAM digital radio system is proposed which combines the advantages of nonrecursive and quantized recursive equalization. In addition, an NMP frequency-domain equalizer is described which complements the time-domain equalizer by especially taking care of NMP channel situations.

Acknowledgment

The authors would like to thank their colleagues, who have made significant contributions during the course of this work.

References

[1] W. Grafinger, J. A. Nossek, and G. Sebald, "Design and realization of a high-speed multilevel QAM digital radio modem with time-domain equalization," in *Proc. Int. Conf. Commun.*, 1985, pp. 31.5.1–31.5.6.

[2] B. Zojer, R. Petschacher, and W. Luschnig, "A 6 bit/200 MHz full Nyquist A/D converter," *IEEE J. Solid-State Circuits*, vol. SC-20, pp. 780–786, June 1985.

[3] A. Leclert and P. Vandamme, "Decision feedback equalization of dispersive radio channels," *IEEE Trans. Commun.*, vol. COM-33, pp. 676–684, July 1985.

[4] T. Noguchi *et al.*, "6 GHz 135 Mbps digital radio system with 64 QAM modulation," in *Proc. Int. Conf. Commun.*, 1983, pp. F2.4.1–F2.4.5.

[5] W. Rummler, "A multipath channel model for line-of-sight digital radio systems," in *Proc. Int. Conf. Commun.*, 1978, pp. 47.5.1–47.5.4.

[6] G. Sebald, B. Lankl, and J. A. Nossek, "Advanced adaptive equalization of multilevel-QAM digital radio systems," in *Proc. Int. Conf. Commun.*, 1986, pp. 46.5.1–46.5.5.

[7] J. A. Nossek and H.-J. Thaler, "Spektrumformung in Digital-Richtfunksystemen," presented at the NTG-Diskussionssitzung 1985, München, West Germany.

[8] P. R. Hartmann *et al.*, "135 Mbps-6 GHz transmission system using 64 QAM modulation," in *Proc. Int. Conf. Commun.*, 1983, pp. F2.6.1–F2.6.7.

[9] C. P. Bates *et al.*, "Impact of technology on high capacity digital radio systems," in *Proc. Int. Conf. Commun.*, 1983, pp. F2.3.1–2.3.5.

Digitalized Cross-Polarization Interference Canceller for Multilevel Digital Radio

HIDEAKI MATSUE, HIROYUKI OHTSUKA, AND TAKEHIRO MURASE, MEMBER, IEEE

Abstract—**This paper proposes a newly developed, highly precise cross-polarization interference canceller (XPIC), utilizing digitalized transversal filters. Performance of the digitalized XPIC is solely determined by quantization and tap numbers and is not affected by any circuit imperfection. Moreover, it can be universally applied to different bit rates and various modulation schemes. Performance of the XPIC is analytically estimated for various modulation schemes and various interference conditions. Then, the digitalized XPIC is realized and applied to a 12.5 MB 256 QAM digital radio system. Experimental results have closely conformed with theoretical estimations and it is confirmed that the XPIC is a powerful countermeasure against multipath fading for a future digital radio.**

I. Introduction

IN recent digital microwave radio communication systems, multilevel modulations and dual-polarization techniques have been applied to improve frequency utilization efficiency. Frequency utilization efficiency of 5 bits/s/Hz has already been achieved using a 16 QAM technique [1]. Moreover, a 256 QAM modulation system [2] is being developed to obtain 10 bit/s/Hz efficiency. However, this advanced system is greatly influenced by the propagation conditions. In order to combat multipath fading, it is necessary for realization of the 256 QAM system to develop both higher performance equalizers [3], [4] and cross-polarization interference cancellers (XPIC's) [5]–[7]. The digital signal processing (DSP) technique is one of the most effective ways to attain higher performance, precise cancellation, and equalization. However, the digitalized XPIC, which applies the DSP technique, has not yet been developed, much less in a high-speed region, such as 12.5 MHz.

In this paper, a more precise XPIC, applicable to multilevel QAM system, using digitalized transversal filters has been proposed. The digitalized XPIC has the following advanced features. 1) The performance is exclusively determined by quantization and tap numbers. 2) The XPIC can avoid circuit imperfection degradation. 3) It can be universally applied to different bit rates and various modulation schemes.

This paper shows the analytical method of the digitalized XPIC performance for various modulations and interference conditions. Additionally, the digitalized XPIC configuration for the 256 QAM system is discussed, and experimental performance is described. The experimental results conform closely with the theoretical ones.

Manuscript received August 3, 1986; revised December 19, 1986.
The authors are with the Radio Transmission Section, NTT Radio Communication Networks Laboratories, Kanagawa, 238-03 Japan.
IEEE Log Number 8613241.

II. XPIC Operational Principles

The block diagram of a cross-polarization interference cancellation system is shown in Fig. 1. A parallel data transmission system sharing the same carrier frequency band by dual-polarization techniques is assumed. Both polarized channels transmit two independent data signals (vertical and horizontal) with the same modulation scheme at the same carrier frequency and transmission rate.

Quadrature amplitude modulation (QAM) signals are considered as transmitting signals, because this signal has the best modulation with respect to spectrum utilization efficiency [1].

The complex equivalent baseband system for V-pol signal transmits amplitudes a_k and b_k at time kT in the inphase channel and the quadrature channel, respectively, where, $a_k, b_k = (\pm 1, \pm 3, \pm 5 \cdots)$, and the symbol $r(t)$ represents a single-pulse response of the overall system. Similarly, c_k and d_k are transmitting amplitudes of the H-pol signal. Vertical and horizontal transmitting signals are represented as

$$S_V(t) = \sum_k (a_k + jb_k) \cdot r(t - kT) \cdot e^{j\omega_V t} \qquad (1)$$

$$S_H(t) = \sum_k (c_k + jd_k) \cdot r(t - kT) \cdot e^{j\omega_H t} \qquad (2)$$

where

w_V = vertical carrier angular frequency,
w_H = horizontal carrier angular frequency.

Multipath fading is chiefly characterized by a two-ray model [8]. In the propagation path, vertical (V) and horizontal (H) signals undergo two-ray fading. Cross-polarization interference (XPI), which also undergoes two-ray fading, reciprocally occurs between the V-pol and H-pol signals [9], [10]. It is assumed that XPI cancellation is from the H-pol signal to the V-pol signal. In this paper, to explain the XPIC operation, it may be assumed that V-pol (main) signal does not undergo a two-ray fading, however, H-pol signal and the XPI from H-pol to V-pol undergo a two-ray fading. Furthermore, the XPI from V-pol to H-pol is assumed to be negligible. The received

Reprinted from *IEEE J. Selected Areas Comm.*, vol. SAC-5, no. 3, pp. 493–501, Apr. 1987.

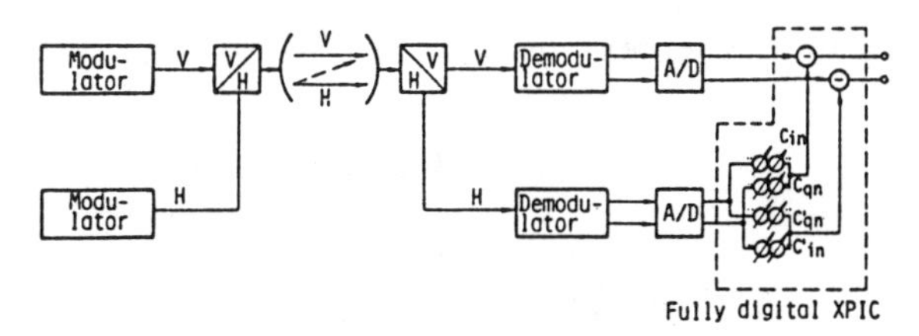

Fig. 1. Fundamental operation of fully digital cross-polarization interference canceller (XPIC).

V-pol and H-pol signals are represented as

$$R_V(t) = \sum_k (a_k + jb_k) \cdot r(t - kT) \cdot e^{j\omega_V t} + a\Big[\sum_k (c_k + jd_k) \cdot r(t - kT + \Delta T) \cdot e^{j(\omega_H t + \Delta\theta)} - \rho_I \sum_k (c_k + jd_k) \cdot r(t - kT - \tau_I + \Delta T) \cdot e^{j(\omega_H t + \Delta\theta + \theta_{IO})}\Big] \quad (3)$$

$$R_H(t) = \sum_k (c_k + jd_k) \cdot r(t - kT + \Delta T') \cdot e^{j(\omega_H t + \Delta\theta')} - \rho_H \sum_k (c_k + jd_k) \cdot r(t - kT - \tau_H + \Delta T') \cdot e^{j(\omega_H t + \Delta\theta' + \theta_{HO})} \quad (4)$$

where

α = amplitude ratio of XPI versus V-pol signal,
ΔT = time difference between V-signal and XPI,
$\Delta T'$ = time difference between V-signal and H-signal,
ρ_I = relative amplitude of two-ray fading for XPI,
ρ_H = relative amplitude of two-ray fading for H-signal,
τ_I = delay difference of two-ray fading for XPI,
τ_H = delay difference of two-ray fading for H-signal,
θ_{IO} = relative notch phase from band center of two-ray fading for XPI,
θ_{HO} = relative notch phase from band center of two-ray fading for H-pol signal,
$\Delta\theta$ = phase difference between V-signal and XPI,
$\Delta\theta'$ = phase difference between V-signal and H-signal.

Received V-pol and H-pol signals are demodulated by each quadrature coherent detector. The two pairs of detected signals are sampled and quantized by the A/D converters at respective sampling times. The two pairs of sampled and quantized signals are represented as follows (Appendix I).

$$I_V(mT) = a_0 + \hat{\alpha}\Big[\sum_k c_k \cdot X_I(mT - kT + \Delta T) + \sum_k d_k \cdot Y_I(mT - kT + \Delta T)\Big] \quad (5)$$

$$Q_V(mT) = b_0 + \hat{\alpha}\Big[\sum_k c_k \cdot Y_I(mT - kT + \Delta T) - \sum_k d_k \cdot X_I(mT - kT + \Delta T)\Big] \quad (6)$$

$$I_H(mT + t_0) = \sum_k c_k \cdot X_H(mT - kT + \Delta T' + t_0) + \sum_k d_k \cdot Y_H(mT - kT + \Delta T' + t_0) \quad (7)$$

$$Q_H(mT + t_0) = \sum_k c_k \cdot Y_H(mT - kT + \Delta T' + t_0) - \sum_k d_k \cdot X_H(mT - kT + \Delta T' + t_0) \quad (8)$$

where

X_I = quantized in-phase impulse response for the XPI,
Y_I = quantized quadrature impulse response for the XPI,
X_H = quantized in-phase impulse response for the H-signal,
Y_H = quantized quadrature impulse response for the H-signal,
to = optimum initial sampling time for H-signal,
$\hat{\alpha}$ = quantized value for α.

A pair of sampled and quantized data signals for XPI of I_H and Q_H are introduced into the baseband XPIC. The XPIC is constructed using four, fully digital transversal filters, each having $2N + 1$ taps $(-N, \cdots, 0, \cdots, N)$. C_{in} and $C_{in'}$ $(n = -N, \cdots, N)$ are assumed to be the in-phase tap coefficient. C_{qn} and C'_{qn} $(n = -N, \cdots, N)$ are quadrature tap coefficient components. In this paper, tap weighting control algorithm is assumed to be zero-forcing method.

Since in-phase and quadrature component equations are symmetrical, the following equations can be obtained (Appendix II).

$$C_{in} = C'_{in} \quad (9)$$

$$C_{qn} = -C'_{qn}. \quad (10)$$

The following equations must be satisfied, since a pair of filtered output data signals are transformed to the same XPI component value at the sample timing point.

$$\sum_k c_k \cdot X_R(mT + \Delta T' + t_0 - kT) + \sum_k d_k \cdot Y_R(mT + \Delta T' + t_0 - kT) = \hat{\alpha}\Big[\sum_k c_k \cdot X_I(mT + \Delta T - kT) + \sum_k d_k \cdot Y_I(mT + \Delta T - kT)\Big] \quad (11)$$

$$\sum_k c_k \cdot Y_R(mT + \Delta T' + t_0 - kT) - \sum_k d_k \cdot X_R(mT + \Delta T' + t_0 - kT)$$
$$= \hat{\alpha}\left[\sum_k c_k \cdot Y_I(mT + \Delta T - kT) - \sum_k d_k \cdot X_I(mT + \Delta T - kT)\right] \quad (12)$$

where

X_R = in-phase transversal filter, output impulse response,
Y_R = quadrature transversal filter, output impulse response.

$X_R(t)$ and $Y_R(t)$ are represented as

$$X_R(t) = \sum_{n=-N}^{N} \hat{C}_{in} \cdot X_H(t - nT) + \sum_{n=-N}^{N} \hat{C}_{qn} \cdot Y_H(t - nT) \quad (13)$$
$$Y_R(t) = \sum_{n=-N}^{N} \hat{C}_{in} \cdot Y_H(t - nT) - \sum_{n=-N}^{N} \hat{C}_{qn} \cdot X_H(t - nT) \quad (14)$$

where

$\hat{C}_{in}$ = quantized value of C_{in},
$\hat{C}_{qn}$ = quantized value of C_{qn}.

For (11) and (12) in any c_k and d_k, the following equations must be satisfied.

$$\sum_{n=-N}^{N} \hat{C}_{in} \cdot X_H(mT + \Delta T' + t_0 - nT) + \sum_{n=-N}^{N} \hat{C}_{qn} \cdot Y_H(mT + \Delta T' + t_0 - nT) = \hat{\alpha} \cdot X_I(mT + \Delta T) \quad (15)$$

and

$$\sum_{n=-N}^{N} \hat{C}_{in} \cdot Y_H(mT + \Delta T' + t_0 - nT) - \sum_{n=-N}^{N} \hat{C}_{qn} \cdot X_H(mT + \Delta T' + t_0 - nT) = \hat{\alpha} \cdot Y_I(mT + \Delta T). \quad (16)$$

Solving (15) and (16), $\hat{C}_{in}$ and $\hat{C}_{qn}$ are obtained. A quantization model of the input signal and tap coefficient is shown in Appendix III.

X_R and Y_R filtered output data are used for interference signal subtraction, so that V-pol signal XPI is eliminated. The residual peak intersymbol interference at the sampling time is represented as

$$D_R = \left[\sum_{m=-\infty}^{\infty} \left|\hat{\alpha} \cdot X_I(mT + \Delta T) - X_R(mT + \Delta T' + t_0)\right| + \sum_{m=-\infty}^{\infty} \left|\hat{\alpha} \cdot Y_I(mT + \Delta T) - Y_R(mT + \Delta T' + t_0)\right|\right] \times (M - 1) \quad (17)$$

where

M = multilevel numbers of the baseband signal.

In contrast, the peak intersymbol interference for the XPI at the sampling time without XPIC is represented as

$$D_I = \left[\sum_{m=-\infty}^{\infty} \left|\hat{\alpha} \cdot X_I(mT + \Delta T)\right| + \sum_{m=-\infty}^{\infty} \left|\hat{\alpha} \cdot Y_I(mT + \Delta T)\right|\right] \times (M - 1). \quad (18)$$

Desired versus undesired signal power ratio (D/U) is represented as

$$D/U|_i = 10 \log \left[\frac{\int_{-\infty}^{\infty} r^2(t)\, dt}{\int_{-\infty}^{\infty} x_I^2(t)\, dt + \int_{-\infty}^{\infty} y_I^2(t)\, dt}\right] (\text{dB}). \quad (19)$$

The equivalent XPIC D/U improvement is represented as

$$D/U|_{imp} = 10 \log \cdot \left(\frac{D_I}{D_R}\right)^2 (\text{dB}). \quad (20)$$

An example of the XPIC basic operation is given in Fig. 2. Graph (a) shows X-pol (H-signal) impulse response under two-ray fading (amplitude ratio $\rho_H = 0.9$, delay difference $\tau_H/T = 0.1$, normalized notch frequency from band center: $\Delta f_{Hdip} = 0.2$), Graph (b) demonstrates X-pol interference impulse response under two-ray fading ($\rho_I = 0.99$, $\tau_I/T = 0.1$, $\Delta f_{Idip} = 0.6$, $\Delta T = T/8$). XPIC tap-weight distribution is given in (c), and (d) shows the residual interference impulse response. The latter indicates residual interference to be zero at the sampling times within the tap number range. In this case, XPIC input signal quantization (Q_{in}) and tap coefficient quantization (Q_{tap}), respectively, are 10 bits, and 7 tap numbers are used.

III. XPIC Performance Estimation for Various Modulations and Interference Conditions

In this section, XPIC performance estimations for various modulations and various interference conditions are described.

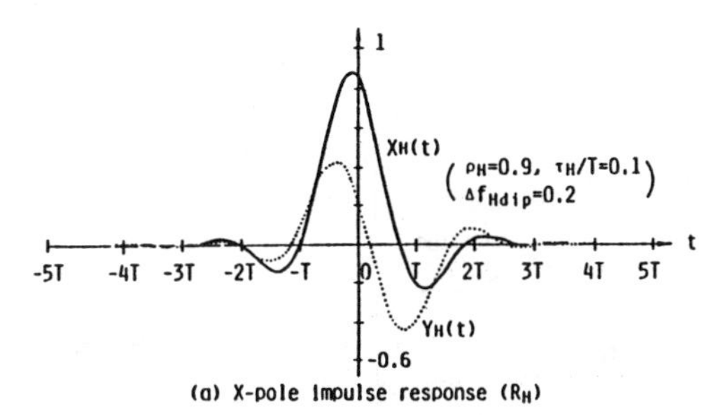

(a) X-pole impulse response (R_H)

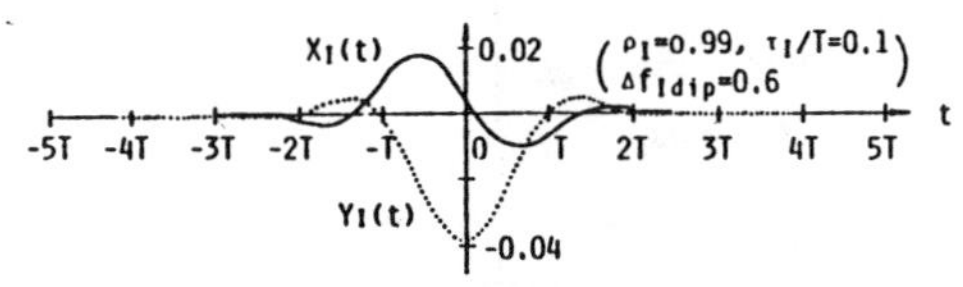

(b) X-pole interference impulse response (XPI)

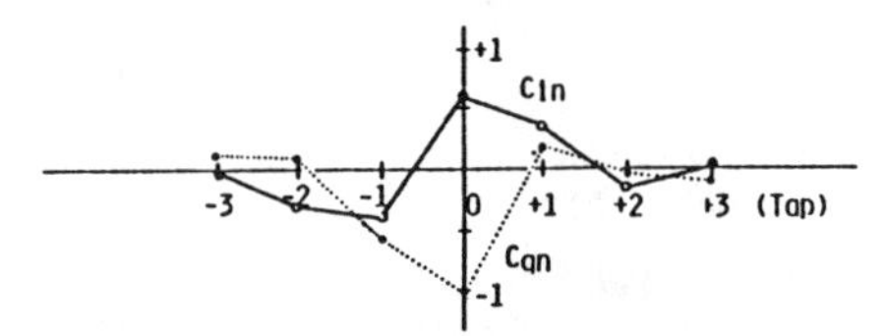

(c) XPIC tap-weighting distribution

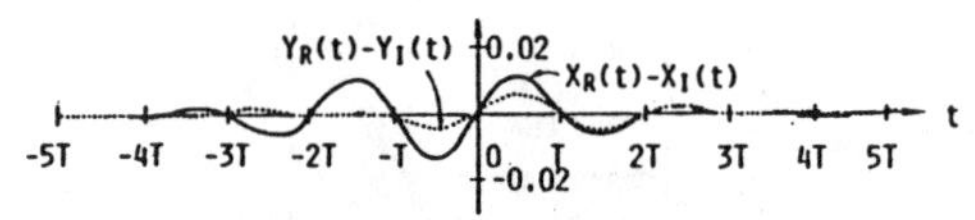

(d) Residual interference impulse response

(XPIC: Qin=10 bit, Qtap=10 bit, Tap number=7)

Fig. 2. XPIC basic operation.

Attention should be given to XPIC performance concerning the following two conditions.

1) D/U improvement when interference is large.

In condition 1, D/U improvement is defined as the D/U difference between using or not using the XPIC at a C/N of 40 dB.

2) Interference observation capability when interference is small. In condition 2, the input D/U observation capability is defined as the D/U value at the cross point of the XPIC use, nonuse curves.

Bit error rate (BER) characteristics are expressed as follows [2].

$$Pe \doteqdot \frac{1}{8} \operatorname{erfc}\left(\frac{\delta}{\sqrt{2}\,\sigma}\right) \tag{21}$$

where

2δ = minimum distance between signal points,

σ^2 = white Gaussian noise power, and

$$\operatorname{erfc}(x) = \frac{2}{\sqrt{\pi}} \int_x^{\infty} e^{-t^2}\, dt.$$

The 256 QAM average CNR can be obtained by

$$C/N = 85\delta^2/\sigma^2. \tag{22}$$

The 64 QAM average CNR is given by

$$C/N = 21\delta^2/\sigma^2. \tag{23}$$

TABLE I
TYPICAL INTERFERENCE CONDITIONS AND REQUIRED XPIC FREQUENCY CHARACTERISTICS

	X-pol signal (R_H)	X-pol Interference (XPI)	Required XPIC frequency characteristics
Type I	$\rho_H=0$; flat	$\rho_I=0$, $\Delta T=T/8$; flat	flat
Type II	$\rho_H=0$; flat	$\rho_I=0.99$(40dB), $\tau_I/T=0.1$, $\Delta f_{Idip}=0$, $\Delta T=T/8$	
Type III	$\rho_H=0.9$(20dB), $\tau_H/T=0.1$, $\Delta f_{Hdip}=0.2$	$\rho_I=0$, $\Delta T=T/8$; flat	

The 16 QAM average CNR is determined according to

$$C/N = 5\delta^2/\sigma^2. \tag{24}$$

BER characteristics, given XPI, are as follows.

$$Pe \doteqdot \frac{1}{8} \operatorname{erfc}\left(\frac{\delta}{\sqrt{2}\,\sigma}(1 - D)\right) \tag{25}$$

where

D = peak intersymbol interference caused by XPI.

D_I must be substituted for D when the XPIC is not used, and D_R must be substituted for D when the XPIC is used.

Three typical interference conditions, Types I, II, and III are considered and presented in Table I. For Type I, when the X-pol and XPI signals are flat, flat XPIC frequency characteristics are required. In Type II, when the X-pol signal is flat and the XPI signal experiences two-ray fading, the required XPIC frequency characteristics are given by this symbol: 𝕋. Type III, when the X-pol signal undergoes two-ray fading and the XPI signal is flat, has required XPIC frequency characteristics as given by ⫫. Using (25), with BER at a constant 10^{-4} value, the relationship between C/N versus D/U for various modulations in Type III is estimated, and the results are plotted in Fig. 3. XPIC, having Q_{in} of 8 bits, Q_{tap} of 10 bits and tap number of 7, is used. Fig. 3 shows D/U improvements of 18 dB for 16 QAM, 16.5 dB for 64 QAM, and 14.5 dB for 256 QAM. Input D/U observation capabilities are about 32 dB for 16 QAM, 37.5 dB for 64 QAM, and 44 dB for 256 QAM.

Using (25) with BER at a constant 10^{-4} value, the relationship between C/N versus D/U for various interference conditions in a 256 QAM system is estimated, and the results are plotted in Fig. 4. Fig. 5 shows the estimated relationship between D/U improvement and modulation for three interference conditions employing an XPIC. In these figures, the XPIC has a Q_{in} of 8 bits, Q_{tap} of 10 bits, and a tap number of 7.

Since the XPIC frequency characteristics ⫫ are more difficult to obtain than 𝕋, Type III represents the most severe interference condition. Moreover, the more multilevel numbers increase, the less D/U improvement occurs. Fig. 6 shows the estimated results of input D/U observation capability versus modulation for various in-

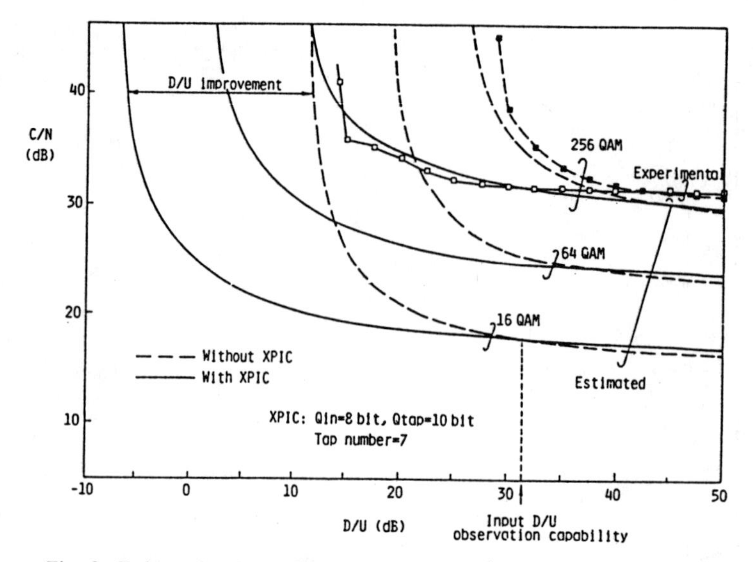

Fig. 3. Estimated and experimental results of C/N versus D/U at BER of 10^{-4} in various modulation systems. (Interference condition: Type III)

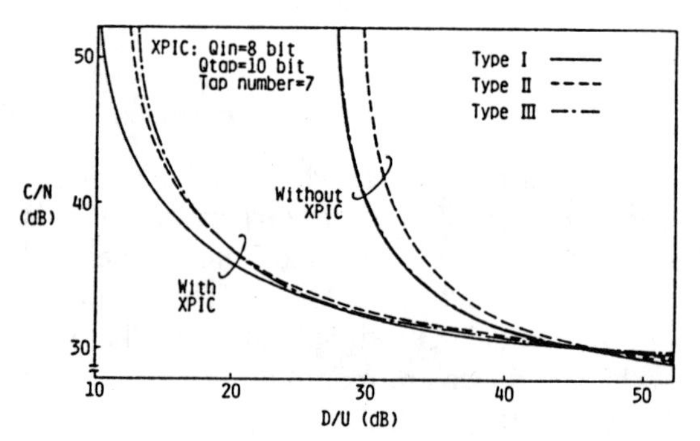

Fig. 4. Estimated results C/N versus D/U at BER 10^{-4} for 256 QAM system. (Parameter: interference conditions)

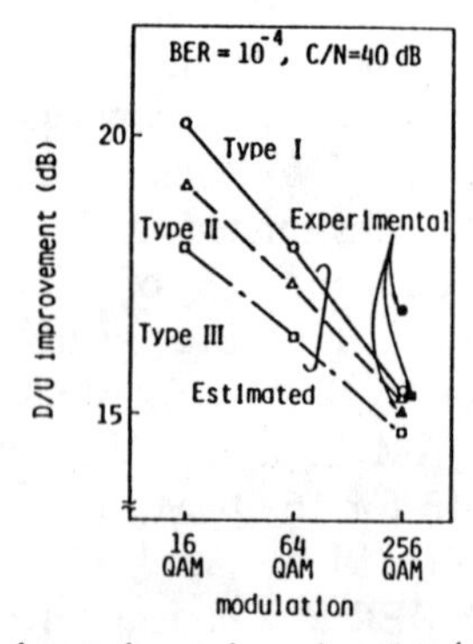

Fig. 5. Estimated and experimental results of D/U improvement versus modulation for various interference conditions. (XPIC: $Q_{\text{in}} = 8$ bit, $Q_{\text{tap}} = 10$ bit, tap number = 7)

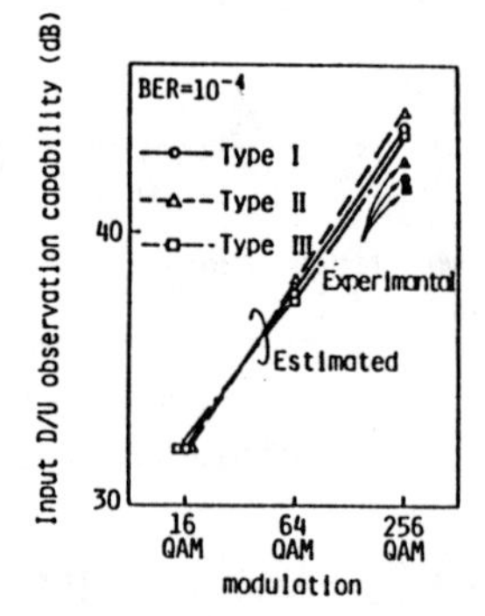

Fig. 6. Estimated and experimental results of input D/U observation capability versus modulation for various interference conditions. (XPIC: $Q_{\text{in}} = 8$ bit, $Q_{\text{tap}} = 10$ bit, tap number = 7)

terference conditions using an XPIC which has a Q_{in} of 8 bits, Q_{tap} of 10 bits, and a tap number of 7. This indicates that differences among the three interference conditions do not exist. However, the more multilevel numbers increase, the more input D/U observation capabilities increase.

IV. Digitalized XPIC Configuration

A digitalized XPIC configuration having seven taps, is shown in Fig. 7. The XPIC can operate at very high frequencies, such as 12.5 MHz, using CMOS digital multipliers and digital full adders with a unit delay time of T [11]. Sampled and quantized in-phase and quadrature signals for X-pol of I_H and Q_H are fed into four transversal filters. Input signal I_H is fed through transversal filters numbered 1 and 3, just as Q_H is fed through filters numbered 2 and 4. The outputs of filters numbered 1 and 2 are added, as are the outputs of filters numbered 3 and 4. The resulting pair of filtered output data signals are transformed to the same XPI component value at the sampling time by controlling tap weights. The output data pair are then used for subtraction from a main-pol signal data pair, so that crosstalk in the main-pol signal is removed.

XPIC controller configuration is shown in Fig. 8. A pair of main-pol error signal bits, e_{IV} and e_{QV}, and a pair of X-pol signal bits, I_H and Q_H, are introduced into the controller. The controller is composed of four blocks. These block outputs control the tap weights for the four transversal filters, and output signal variations are expressed as follows.

$$\Delta C_{\text{in}} = \beta \cdot \sum_{k=1}^{2L} \cdot e_{IVn-k} \cdot I_{Hk} \tag{26}$$

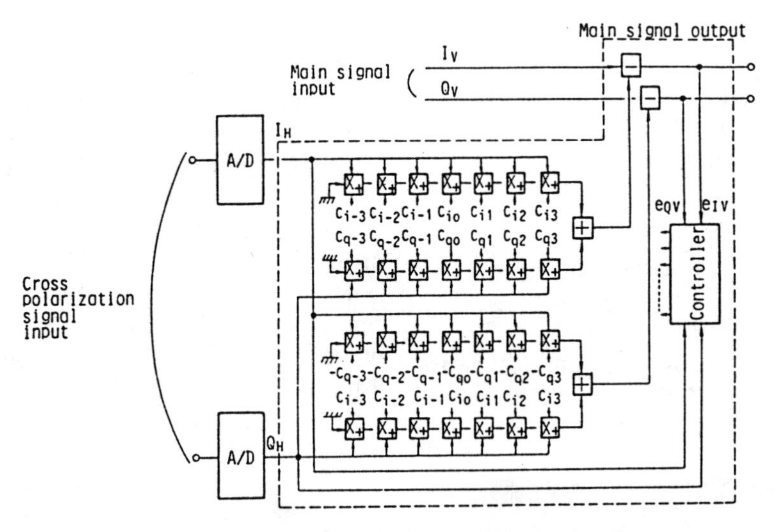

Fig. 7. Configuration of fully digital cross-polarization interference canceller.

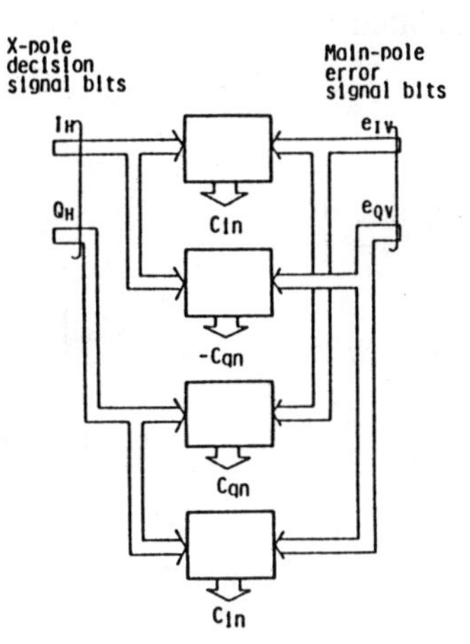

Fig. 8. Configuration of XPIC controller.

$$\Delta C_{qn} = \beta \cdot \sum_{k=1}^{2^L} \cdot e_{IVn-k} \cdot Q_{Hk} \quad (27)$$

$$\Delta C'_{\text{in}} = \beta \cdot \sum_{k=1}^{2^L} \cdot e_{QVn-k} \cdot Q_{Hk} = \Delta C_{\text{in}} \quad (28)$$

$$\Delta C'_{qn} = \beta \cdot \sum_{k=1}^{2^L} \cdot e_{QVn-k} \cdot I_{Hk} = -\Delta C_{qn} \quad (29)$$

where, β is a constant, 2^L is the integration pulse number.

The main specifications of the digitalized XPIC are shown in Table II.

V. Experimental XPIC Performance for 256 QAM System

A. Experimental Setup

The experimental setup of the cross-polarization interference canceller is shown in Fig. 9. A pair of 12.5 MB, 50 percent Nyquist-cosine rolloff spectrum shaped, 256 QAM signals are used. Moreover, carrier frequencies and transmission rates are commonly used along with two-ray fading simulators. To monitor the eye pattern with and without the XPIC, two D/A converters are employed.

TABLE II
Specifications of Digitalized XPIC

Item	Content
Tap number	Less than 7 taps
Input signal quantization (Qin)	Less than 10 bits
Tap coefficient quantization (Qtap)	Less than 10 bits
Operation speed	Less than 15 MHz
Control algorithm	Mean Square Error method
Power source	+5 V only
Operation region	Baseband

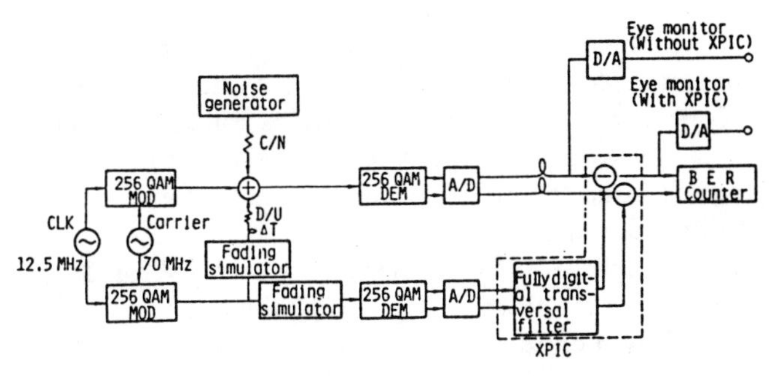

Fig. 9. Experimental setup of cross polarization interference canceller.

B. Experimental Performance for Various Interference Conditions

In this experiment, the three typical interference conditions shown in Table I are investigated. Fig. 3 shows the experimental results of C/N versus D/U at a BER of 10^{-4} in Type III for the 256 QAM system in addition to the estimated results. The performance factors mentioned above, D/U improvement and input D/U observation capability, must be considered.

According to this figure, D/U improvement of 16 dB and input D/U observation capability of 41.5 dB can be achieved. Fig. 10 shows the experimental results of C/N versus D/U at a BER of 10^{-4} for the three interference conditions in the 256 QAM systems. These results conform closely with estimated results shown in Fig. 4.

Experimental results of D/U improvement for various

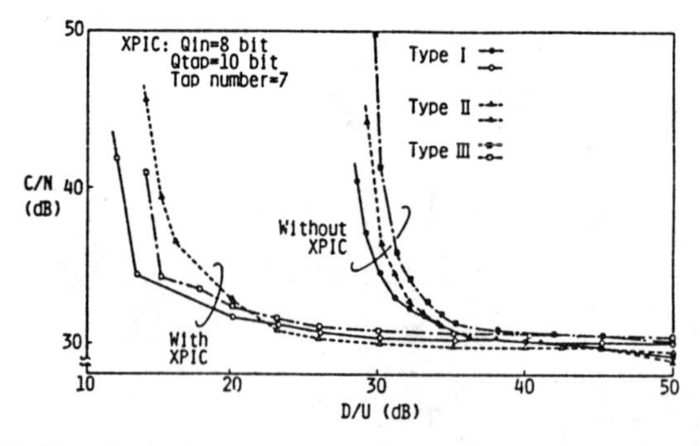

Fig. 10. Experimental results of C/N versus D/U at BER 10^{-4} for 256 QAM system. (Parameter: interference conditions)

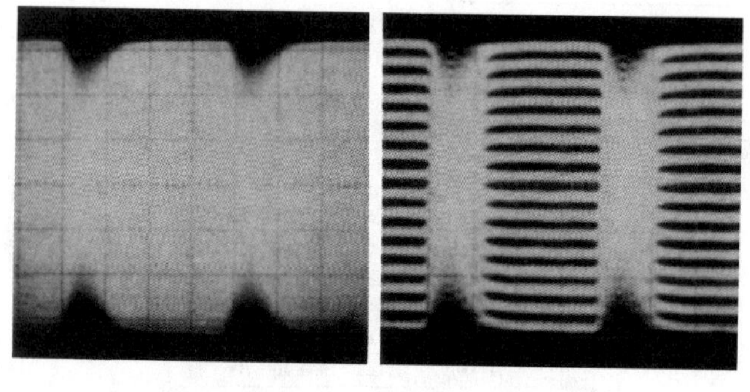

(a) Without XPIC (b) With XPIC

(H: 20ns/div)

Fig. 11. An example of XPIC performance. (Input D/U = 20 dB, C/N = 32.4 dB. (XPIC: Q_{in} = 8 bit, Q_{tap} = 10 bit, tap number = 7)

interference conditions is shown in Fig. 5. Experimental results, in addition to the estimated ones, of the input D/U observation capability in various interference conditions is shown in Fig. 6. Consequently, it can be confirmed that the theoretical results agree closely with the experimental ones.

An example of XPIC experimental results for a 256 QAM system is shown in Fig. 11. These eye patterns are observed by monitoring the D/A converters.

VI. Conclusion

The precise XPIC using digital signal processing (DSP) techniques has been proposed. It can be applied to multilevel QAM systems and different bit rates modulation schemes. This newly developed, digitalized XPIC, constituted by CMOS LSI's, can operate at such very high speed as 12.5 MHz. It was also confirmed that theoretical estimations agree closely with experimental results.

The main results obtained in this paper are summarized as follows.

1) Two XPIC performance estimation parameters are proposed: D/U improvement when cross polarization interference (XPI) is large, and input D/U observation capability when XPI is small. XPIC performance requires both D/U improvement and input D/U observation capability.

2) It was theoretically and experimentally determined that D/U improvement is dependent on interference conditions, with Type III being the most severe between the studied cases. However, input D/U observation capability is independent of interference conditions.

3) It was theoretically clarified that the more multilevel modulation numbers increase, the less D/U improvement occurs. In contrast, the more multilevel numbers increase, the more input D/U observation capability increases.

4) 15 dB D/U improvement and 42 dB input D/U observation capability can be obtained by using the XPIC in which Q_{in} is 8 bits, Q_{tap} is 10 bits and the tap number is 7 for a 12.5 MB, 256 QAM system.

5) Easy adjustment, low power consumption, compactness, and low cost for the XPIC can be achieved.

Appendix I

The main-pol (V) signal and cross-pol (H) signal are demodulated by each quadrature coherent detector. Two pairs of demodulated signals are represented as follows.

$$i_V(t) = \sum_k a_k \cdot r(t - kT) \cdot \cos\theta_V + \sum_k b_k \cdot r(t - kT) \cdot \sin\theta_V + \alpha\left[\sum_k c_k \cdot x_I(t - kT + \Delta T) + \sum_k d_k \cdot y_I(t - kT + \Delta T)\right] \tag{A-1}$$

$$q_V(t) = \sum_k a_k \cdot r(t - kT) \cdot \sin\theta_V + \sum_k b_k \cdot r(t - kT) \cdot \cos\theta_V + \alpha\left[\sum_k c_k \cdot y_I(t - kT + \Delta T) - \sum_k d_k \cdot x_I(t - kT + \Delta T)\right] \tag{A-2}$$

$$i_H(t) = \sum_k c_k \cdot x_H(t - kT + \Delta T') + \sum_k d_k \cdot y_H(t - kT + \Delta T') \tag{A-3}$$

$$q_H(t) = \sum_k c_k \cdot y_H(t - kT + \Delta T') - \sum_k d_k \cdot x_H(t - kT + \Delta T') \tag{A-4}$$

where

$x_I(t)$ = in-phase impulse response for the XPI under two-ray fading,

$y_I(t)$ = quadrature impulse response for the XPI under two-ray fading,

$x_H(t)$ = in-phase impulse response for the H-pol signal under two-ray fading,

$y_H(t)$ = quadrature impulse response for the H-pol signal under two-ray fading.

These impulse responses are represented as

$$x_I(t) = r(t) \cdot \cos(\Delta\theta - \theta_V) - \rho_I \cdot r(t - \tau_I) \cdot \cos(\theta_{IO} + \Delta\theta - \theta_V) \tag{A-5}$$

$$y_I(t) = r(t) \cdot \sin(\Delta\theta - \theta_V) + \rho_I \cdot r(t - \tau_I) \cdot \sin(\theta_{IO} + \Delta\theta - \theta_V) \quad \text{(A-6)}$$

$$x_H(t) = r(t) \cdot \cos(\Delta\theta' - \theta_H) - \rho_H \cdot r(t - \tau_H) \cdot \cos(\theta_{HO} + \Delta\theta' - \theta_H) \quad \text{(A-7)}$$

$$y_H(t) = r(t) \cdot \sin(\Delta\theta' - \theta_H) + \rho_H \cdot r(t - \tau_H) \cdot \sin(\theta_{HO} + \Delta\theta' - \theta_H) \quad \text{(A-8)}$$

where

- ρ_I = relative amplitude ratio of two-ray fading for XPI,
- ρ_H = relative amplitude ratio of two-ray fading for H-pol signal,
- τ_I = delay difference of two-ray fading for XPI,
- τ_H = delay difference of two-ray fading for H-pol signal,
- θ_{IO} = relative notch phase from band center of two-ray fading for XPI,
- θ_{HO} = relative notch phase from band center of two-ray fading for H-pol signal,
- θ_V = reference carrier phase for V-pol signal,
- θ_H = reference carrier phase for H-pol signal.

After sampling and quantizing (A.1), (A.2), (A.3), and (A.4), they respectively become (5), (6), (7), and (8). (Appendix III)

Appendix II

Demodulated in-phase and quadrature signals for the H-pol are fed to the four transversal filters, of which the tap coefficients are $\hat{C}_{\text{in}}$, $\hat{C}_{qn}$, $\hat{C}'_{\text{in}}$, and $\hat{C}'_{qn}$. Since a pair of filtered output data signals are transformed to the same XPI component value at the sampling time, the following equations are obtained.

$$\sum_n \hat{C}_{\text{in}} \cdot I_H + \sum_n \hat{C}_{qn} Q_H = \hat{\alpha} \left\{ \sum_k c_k \cdot X_I + \sum_k d_k Y_I \right\} \quad \text{(A-9)}$$

$$\sum_n \hat{C}'_{\text{in}} \cdot Q_H + \sum_n \hat{C}'_{qn} I_H = \hat{\alpha} \left\{ \sum_k c_k \cdot Y_I - \sum_k d_k X_I \right\}. \quad \text{(A-10)}$$

If (7) and (8) are substituted into (A.9) and (A.10), the following equations are obtained.

$$\sum_k c_k \left\{ \sum_n \hat{C}_{\text{in}} X_H + \sum_n \hat{C}_{qn} \cdot Y_H \right\} + \sum_k d_k \left\{ \sum_n \hat{C}_{\text{in}} \cdot Y_H - \sum_n \hat{C}_{qn} \cdot X_H \right\} = \hat{\alpha} \left\{ \sum_k c_k \cdot X_I + \sum_k d_k \cdot Y_I \right\} \quad \text{(A-11)}$$

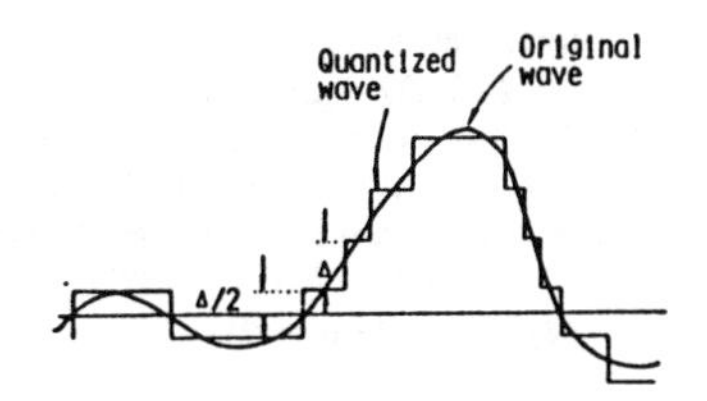

Fig. 12. Quantization model.

$$\sum_k c_k \left\{ \sum_n \hat{C}'_{\text{in}} Y_H + \sum_n \hat{C}'_{qn} \cdot X_H \right\} + \sum_k d_k \left\{ -\sum_n \hat{C}_{\text{in}} \cdot X_H + \sum_n \hat{C}'_{qn} \cdot Y_H \right\} = \hat{\alpha} \left\{ \sum_k c_k \cdot Y_I - \sum_k d_k \cdot X_I \right\}. \quad \text{(A-12)}$$

Comparing the two equations above, the following equations must be satisfied.

$$\hat{C}_{\text{in}} = \hat{C}'_{\text{in}} \quad (9)$$

$$\hat{C}_{qn} = -\hat{C}'_{qn}. \quad (10)$$

To satisfy (A.11) or (A.12) for any c_k and d_k, (15) and (16) can be obtained.

Appendix III

The quantization model is shown in Fig. 12. Δ indicates a quantization unit.

A. XPIC Input Signal Quantization (Q_{in})

The quantization unit of an input signal is represented as follows.

$$\Delta_{\text{in}} = \frac{2 \times M}{2^{Q_{\text{in}}}} \quad \text{(A-13)}$$

where

M = multilevel numbers in the baseband signal, for example, M is 16 in the case of the 256 QAM system.

B. XPIC Tap Coefficient Quantization (Q_{tap})

The quantization unit of a tap coefficient is represented as follows.

$$\Delta_{\text{tap}} = \frac{2}{2^{Q_{\text{tap}}}}. \quad \text{(A-14)}$$

Acknowledgment

The authors wish to thank Dr. M. Shinji, Dr. M. Kuramoto, Dr. K. Kohiyama, and Dr. O. Kurita for their guidance and helpful comments.

References

[1] H. Yamamoto and K. Morita, "4/5/6 L-D1 digital microwave radio system," *Rev. Elec. Commun. Lab.*, vol. 30, no. 5, p. 836, 1982.

[2] Y. Saito, S. Komaki, and M. Murotani, "Feasibility considerations of high-level QAM multi-carrier system," in *Proc. ICC'84*, 1984, p. 22.8.1.

[3] M. Araki, K. Tajima, and H. Matsue, "Correction techniques for multipath fading in 4/5/6 L-D1 digital microwave radio system," *Rev. Elec. Commun. Lab.*, vol. 30, no. 5, p. 873, 1982.

[4] G. L. Fenderson, M. H. Meyers, and M. A. Skinner, "Recent advances in multipath propagation countermeasures for high-capacity digital radio systems," in *Proc. ICC'85*, 1985, p. 39.2.1.

[5] M. Borgne, "An adaptive cross-polarization canceler for digital radio systems," in *Proc. ICC'85*, 1985, p. 39.6.1.

[6] J. Namiki and S. Takahara, "Adaptive receiver for cross-polarization digital transmission," in *Proc. ICC'81*, 1981, p. 46.3.1.

[7] M. Araki and A. Hashimoto, "Performance of cross polarization canceler having transversal filter," *Trans. IECE Japan.*, vol. J69-B, no. 1, 1986.

[8] O. Sasaki and T. Akiyama, "Multipath delay characteristics on line-of-sight microwave radio system," *IEEE Trans. Commun.*, vol. COM-27, p. 1876, 1979.

[9] K. T. Wu, "Measured statistics on multipath dispersion of cross polarization interference," in *Int. Commun. Conf. Rec.*, pp. 194–197, May 1984.

[10] M. Liniger, "Sweep measurements of multipath effects on cross-polarized RF-channels including space diversity," in *Proc. GLOBECOM'84 Conf. Rec.*, Nov. 1984, pp. 45.7.1–45.7.5.

[11] S. Iwase, T. Yamazaki, Y. Hashimoto, E. Yura, I. Kumata, and H. Yamasaki, "Multiplier-adder LSI for digital video," IECEJ Tech. Rep., IE83-58, Sept. 1983.

Tutorial Authors' Affiliations

J. K. Chamberlain is a Principal Research Associate in the Telecommunications Laboratory at GEC Hirst Research Centre in Wembley, Middlesex, ENGLAND.

F. M. Clayton is Chief Mathematician in the Telecommunications Laboratory at GEC Hirst Research Centre in Wembley, Middlesex, ENGLAND.

Reginald P. Coutts is Section Head, Radio and Satellite Networks, at the Telecom Australia Research Laboratories near Melbourne, AUSTRALIA.

Yoshimasa Daido is a Senior Engineer in the Radio and Satellite Communications Systems Laboratory at Fujitsu Laboratories Ltd. in Kawasaki, JAPAN.

Larry J. Greenstein is a Department Head in the Communications Systems Research Laboratory at AT&T Bell Laboratories in Holmdel, New Jersey, USA.

Greg Hart is Section Head, Microwave Radio Systems and Structures, in the Network Systems Engineering Division at British Telecom in London, ENGLAND.

Paul R. Hartmann is Director of Transmission Systems Technology in the Telecommunications Group at Rockwell International in Dallas, Texas, USA.

Kenji Kohiyama is General Manager of the Development Planning Department at the NTT Network Systems Development Center in Tokyo, JAPAN.

Osamu Kurita is a Project Manager in the NTT Network Systems Development Center in Yokosuka, JAPAN.

Markus Liniger is a Lecturer in the Department of Electrical Engineering at the Engineering College of the State of Berne in Biel, SWITZERLAND.

Martin H. Meyers is a Distinguished Member of Technical Staff in the Radio Systems and Terminals Laboratory at AT&T Bell Laboratories in North Andover, Massachusetts, USA.

Toshitake Noguchi is a Manager in the Microwave and Satellite Communications Division at NEC Corporation in Yokohama, JAPAN.

Josef A. Nossek is a Deputy Director in the Transmission Systems Design Department at Siemens AG in Munich, FEDERAL REPUBLIC OF GERMANY.

Vasant K. Prabhu is a Distinguished Member of Technical Staff in the Transmission Technology and Engineering Center at AT&T Bell Laboratories in Holmdel, New Jersey, USA.

William D. Rummler is a Distinguished Member of Technical Staff in the Transmission Technology and Engineering Center at AT&T Bell Laboratories in Holmdel, New Jersey, USA.

Hikmet Sari is Head of the Signal Acquisition and Processing Division at Laboratories d'Electronique et de Physique Apliquée (LEP) in Limeil-Brevannes, FRANCE.

Mansoor Shafi is a Research Engineer in the Research and Development Section at Telecom Corporation of New Zealand in Wellington, NEW ZEALAND.

Jan A. Steinkamp is Head of Radio Relay Equipment and Satellite Earth Station Development at Siemens AG in Munich, FEDERAL REPUBLIC OF GERMANY.

Desmond P. Taylor is Professor of Electrical and Computer Engineering at McMaster University in Hamilton, Ontario, CANADA.

Patrick S. Vandamme is Head of the Digital Communications Group within the Radio Transmissions Department at the Centre National d'Etudes des Télécommunications (CNET) in Lannion, FRANCE.

Heiichi Yamamoto is Director of the NTT Communications Satellite Laboratory at NTT Radio Communications Systems Laboratories in Yokosuka, JAPAN.

Author Index

Subject Index

I

J

L

M

O

P

Q

R

S

T

V

Editors' Biographies

Larry J. Greenstein (M'67–SM'80–F'87) received the B.S., M.S., and Ph.D. degrees in electrical engineering from Illinois Institute of Technology in 1958, 1961, and 1967, respectively.

From 1958 to 1970, he worked at IIT Research Institute, primarily in the areas of radio frequency interference and anticlutter airborne radar. Since 1970, he has been with AT&T Bell Laboratories in Holmdel, NJ, where he is currently Head of the Local Communications Research Department. His major technical activities there have been in digital encoding and filtering, mobile telephony, satellite communications, microwave digital radio, and optical communications. His work in microwave digital radio included the development of multipath fading models and methods of performance calculations, and research on adaptive equalization, space diversity combining, and cross-polarization cancellation. In 1984, he shared the IEEE Communications Society's Leonard G. Abraham Prize Paper Award for published work in mobile telephony.

Dr. Greenstein is a member of Eta Kappa Nu, Tau Beta Pi, and Sigma Xi. He has been an Associate Editor of the IEEE TRANSACTIONS ON COMMUNICATIONS and a Guest Editor of the IEEE JOURNAL ON SELECTED AREAS IN COMMUNICATIONS. He is currently a Senior Technical Editor of IEEE COMMUNICATIONS MAGAZINE, and a member of the Editorial Board of the IEEE PRESS.

Mansoor Shafi (S'69–A'70–M'82–SM'87) was born in Multan, Pakistan on May 5, 1950. He received the BSc degree from the University of Engineering and Technology, Lahore, Pakistan, in 1970, and the Ph.D. degree from the University of Auckland, Auckland, New Zealand, in 1979, both in electrical engineering.

He was awarded a Gold Medal for first standing in the pre-engineering examinations of 1966 from Government College, Multan, and won a four-year Government scholarship to study engineering at Lahore University. From 1971 to 1974, he worked as a Lecturer in the University of Engineering and Technology, Lahore, and conducted research on power systems and generalized machines. From 1975 to 1979, he worked as Junior Lecturer in the University of Auckland and was engaged in research in systems identification and signal processing. Since 1979, he has been with the New Zealand Post Office (now called Telecom Corporation of New Zealand Ltd.) in the Telecommunications Section, Wellington. His research interests are in transmission systems.

During 1982, he was employed as a Senior Postdoctoral Fellow at McMaster University, Hamilton, Ontario, Canada, where he worked on equalization techniques for digital radio. During the period from July 1985 to September 1986, he worked as a research engineer in the Communications Research Lab at McMaster University and worked on the modeling of radio channels. He has published several papers in the area of digital radio systems. Dr. Shafi serves as a New Zealand delegate to the meetings of the CCIR Study Group IX in Geneva.